P9-DMX-595

Fundamentals of
Differential Equations

FIFTH EDITION

Fundamentals of Differential Equations

FIFTH EDITION

R. Kent Nagle

Edward B. Saff
University of South Florida

Arthur David Snider
University of South Florida

▲▲ **ADDISON-WESLEY**

An imprint of Addison Wesley Longman, Inc.

Reading, Massachusetts • Menlo Park, California • New York • Harlow, England
Don Mills, Ontario • Sydney • Mexico City • Madrid • Amsterdam

Sponsoring Editor Laurie Rosatone
Project Manager Rachel S. Reeve
Associate Project Editor Ellen Keohane
Senior Designer Barbara T. Atkinson
Associate Production Supervisor Kim Ellwood
Project Coordination Elm Street Publishing Services, Inc.
Compositor Nesbitt Graphics, Inc.
Art Academy Artworks
Senior Technology Specialist Joe Vetere
Prepress Buyer Caroline Fell
Print Buyer Evelyn Beaton
Marketing Manager Carter Fenton
Associate Marketing Manager Michael Boezi
Cover Illustration George V. Kelvin, from a computer graphic by Alar Toomre, MIT

Library of Congress Cataloging-in-Publication Data

Nagle, R. Kent.
 Fundamentals of differential equations. — 5th ed. / R. Kent
 Nagle, Edward B. Saff, Arthur David Snider.
 p. cm.
 Includes index.
 ISBN 0-201-33868-8
 1. Differential equations. I. Saff, E. B., 1944– .
 II. Snider, Arthur David, 1940– . III. Title.
 QA371.N24 2000
 515'.35—dc21 99-17732
 CIP

Reprinted with corrections.
Copyright © 2000 by Addison Wesley Longman. All rights reserved. No part of this publication may be reproduced, stored in a retrieval system, or transmitted, in any form or by any means, electronic, mechanical, photocopying, recording, or otherwise, without the prior written permission of the publisher. Printed in the United States of America.

6 7 8 9 10-QUF-0302

Dedicated to R. Kent Nagle

He has left his imprint not only on these pages but upon all who knew him. He was that rare mathematician who could effectively communicate at all levels, imparting his love for the subject with the same ease to undergraduates, graduates, precollege students, public school teachers, and his colleagues at the University of South Florida.

Kent was at peace in life—a peace that emanated from the depth of his understanding of the human condition and the strength of his beliefs in the institutions of family, religion, and education. He was a research mathematician, an accomplished author, a Sunday school teacher, and a devoted husband and father.

Kent was also my dear friend and my jogging partner who has left me behind still struggling to keep pace with his high ideals.

E. B. Saff

Preface

OUR GOAL

Fundamentals of Differential Equations is designed to serve the needs of a one-semester course in basic theory as well as applications of differential equations. We have striven to create a flexible text that affords the instructor substantial latitude in designing a syllabus (sample syllabi are provided in this preface), in course emphasis (theory, applications and techniques, numerical methods), and in using commercially available computer software and/or the software utility programs specifically designed for this text.

INNOVATIONS FOR THIS EDITION

In response to requests of users and reviewers, and recognizing recent developments in teaching and learning, we offer the following:

New Sections on Qualitative Features New sections describe how qualitative information can be gleaned about the solutions to intractible autonomous equations by observing their direction fields and critical points on the phase line (for single equations) or the phase plane (for two-dimensional systems). With the assistance of suitable software, this approach provides a refreshing, almost recreational alternative to the traditional analytic methodology as we discuss applications in nonlinear mechanics, ecosystems, and epidemiology.

New Section on Vibrations Motivation for Chapter 4 on linear differential equations is provided in a new introductory section describing the mass-spring oscillator. We exploit the reader's familiarity with common vibratory motions to anticipate the exposition of the theoretical and analytical aspects of linear equations. Not only does this model provide an anchor for the discourse on constant coefficient equations, but a liberal interpretation of its features enables us to predict the qualitative behavior of variable-coefficient and nonlinear equations as well.

Many New Figures In keeping with the increased emphasis on qualitative methods, about 40 new figures have been added.

New Group Projects Thanks again to the feedback we received from instructors using the earlier editions, five new group projects have been introduced at the end of the chapters. These projects include:

Chapter 2
- Two Snowplows

Chapter 4
- An Alternative to the Method of Undetermined Coefficients
- Apollo Reentry

Chapter 5
- Hamiltonian Systems

Chapter 7
- Determining System Parameters

Expanded Review of Linear Algebraic Equations and Matrices The chapter on matrix methods for linear systems (Chapter 9) has been expanded with two (optional) introductory sections reviewing the theory of linear algebraic systems and matrix algebra.

New Treatment of Taylor Polynomials In Chapter 8 the reader is gracefully initiated into the theory of power series solutions with an exposition of Taylor polynomial approximants to solutions, deferring the sophisticated issues of convergence to later sections.

Revised Exercises For instructors' convenience, some of the basic "skills-building" exercises have been modified (to render previous answers obsolete), and a selection of more challenging exercises has been added.

Website Support System As a useful companion to the text, students and instructors are provided access to a specially designed Website that not only includes helpful links keyed to the sections of the text but also provides a utility for solving many of the numerical problems. This utility, which requires only a standard browser with Java and JavaScript activated, is student-friendly, requires little learning time, and will entice students to conduct numerical experiments. It includes facility for direction field plots, Euler's methods, Runge-Kutta methods (single equations and systems), phase plane diagrams, eigenvalues and eigenvectors, and standard computational algorithms such as Newton's method for solving equations, Simpson's rule for numerical integration, and plot/table construction for one or more functions. Also, the user will find recommendations for syllabi, an index of applications, study tips, and suggestions for incorporating *Interactive Differential Equations* modules (described in the "Supplements" section below). Point your Web browser to

http://www.awlonline.com/nagle

to access the Website.

PREREQUISITES

While some universities make linear algebra a prerequisite for differential equations, many schools (especially engineering) only require calculus. With this in mind, we have designed the text so that only Chapter 6 (Theory of Higher-Order Linear Differential Equations) and Chapter 9 (Matrix Methods for Linear Systems) require more than high school linear algebra. Moreover, Chapter 9 contains review sections on matrices and vectors as well as specific references for the deeper results used from the theory of linear algebra. We have also written Chapter 5 so as to give an introduction to systems of differential equations—including methods of solving, phase plane analysis, applications, numerical procedures, and Poincaré maps—that does not require a background in linear algebra.

SAMPLE SYLLABI

As a rough guide in designing a syllabus related to this text, we provide three sample syllabi that can be used for a 15-week course that meets three hours per week: the first emphasizes methods, computations, and applications; the second, theory and methods; and the third, methods and partial differential equations. Chapters 1, 2, and 4 provide the core for any course. The rest of the chapters are, for the most part, independent of each other.

Week	Methods, Computations, and Applications Sections	Theory and Methods (linear algebra prerequisite) Sections	Methods and Partial Differential Equations Sections
1	1.1, 1.2, 1.3	1.1, 1.2, 1.3	1.1, 1.2, 1.3
2	1.4, 1.5, 2.2	2.2, 2.3	1.5, 2.2
3	2.3, 2.4	2.4, 4.2	2.3, 2.4
4	3.2, 3.4, 3.5	4.3, 4.4	3.2, 3.4
5	3.6, 4.1, 4.2	4.5, 4.6, 4.7	4.2, 4.3
6	4.3, 4.5	4.8, 4.9	4.5, 4.6, 4.7
7	4.6, 4.7, 4.8	5.2, 5.3, 6.1	4.8, 4.9
8	4.9, 4.10, 4.11	6.2, 6.3	5.2, 5.3
9	4.12, 5.1, 5.2	7.2, 7.3	7.2, 7.3
10	5.3, 5.4, 5.6	7.4, 7.5	7.4, 7.5
11	7.2, 7.3	7.6, 8.1, 8.2	7.6, 8.1, 8.2
12	7.4, 7.5	8.3, 8.5	8.3, 8.5, 8.6
13	7.6, 7.7	8.6, 9.1, 9.2, 9.3	10.2, 10.3
14	8.1, 8.2, 8.3	9.4, 9.5	10.4, 10.5
15	8.5, 8.6	9.6, 9.7	10.6, 10.7

RETAINED FEATURES

Flexible Organization Most of the material is modular in nature to allow for various course configurations and emphasis (theory, applications and techniques, and concepts).

Optional Use of Computer Software The availability of computer packages such as MATHCAD®, MATHEMATICA®, MATLAB®, and MAPLE® provides an opportunity for the student to conduct numerical experiments and tackle realistic applications that give additional insights into the subject. Consequently, we have inserted several exercises and projects throughout the text that are designed for the student to employ available software in phase plane analysis, eigenvalue computations, and the numerical solutions of various equations.

Choice of Applications Because of syllabus constraints, some courses will have little or no time for sections (such as those in Chapters 3 and 5) that exclusively deal with applications. Therefore, we have made the sections in these chapters almost completely independent of each other. To afford the instructor even greater flexibility, we have built in a variety of applications in the exercises for the theoretical sections. In addition, we have included many projects that deal with such applications.

Group Projects At the end of each chapter are group projects relating to the material covered in the chapter. A project might involve a more challenging application, delve deeper into the theory, or introduce more advanced topics in differential equations. Although these projects can be tackled by an individual student, classroom testing has shown that working in groups lends a valuable added dimension to the learning experience. Indeed, it simulates the interactions that take place in the professional arena.

Technical Writing Exercises Communication skills are, of course, an essential aspect of professional activities. Yet few texts provide opportunities for the reader to develop these skills. Thus we have added at the end of most chapters a set of clearly marked technical writing exercises that invite students to make documented responses to questions dealing with the concepts in the chapter. In so doing, students are encouraged to make comparisons between various methods and to present examples that support their analysis.

Historical Footnotes Throughout the text historical footnotes are set off by colored daggers(†). These footnotes typically provide the name of the person who developed the technique, the date, and the context of the original research.

Motivating Problem Most chapters begin with a discussion of a problem from physics or engineering that motivates the topic presented and illustrates the methodology.

Chapter Summary and Review Problems All of the main chapters contain a set of review problems along with a synopsis of the major concepts presented.

Computer Graphics Most of the figures in the text were generated via computer. Computer graphics not only ensure greater accuracy in the illustrations, they demonstrate the use of numerical experimentation in studying the behavior of solutions.

Proofs While more pragmatic students may balk at proofs, most instructors regard these justifications as an essential ingredient in a textbook on differential equations. As with any text at this level, certain details in the proofs must be omitted. When this occurs, we flag the instance and refer readers either to a problem in the exercises or to another text. For convenience, the end of a proof is marked by the symbol ■ End.

Linear Theory We have developed the theory of linear differential equations in a gradual manner. In Chapter 4 (Linear Second Order Equations) we present the basic theory for linear second order equations and discuss various techniques for solving these equations. Higher-order equations are briefly mentioned in this chapter. A more detailed discussion of linear higher-order differential equations is given in Chapter 6 (Theory of Higher-Order Linear Differential Equations). For a beginning course emphasizing methods of solution, the presentation in Chapter 4 is sufficient and Chapter 6 can be skipped.

Numerical Algorithms Several numerical methods for approximating solutions to differential equations are presented along with program outlines that are easily implemented on a computer. These methods are introduced early in the text so that teachers and/or students can use them for numerical experimentation and for tackling complicated applications. Most algorithms discussed are implemented on the Website for this text.

Exercises An abundance of exercises is graduated in difficulty from straightforward, routine problems to more challenging ones. Deeper theoretical questions, along with applications, usually occur toward the end of the exercise sets. Throughout the text we have included problems and projects that require the use of a calculator or computer. These exercises are denoted by the symbol 🖥. The Website software specifically designed for use with this text greatly facilitates the solutions to these numerical problems.

Optional Sections These sections can be omitted without affecting the logical development of the material. They are marked with an asterisk in the table of contents. As mentioned earlier, the sections in Chapters 3 and 5 are almost completely independent of each other.

Laplace Transforms We provide a detailed chapter on Laplace transforms (Chapter 7) since this is a recurring topic for engineers. Our treatment emphasizes discontinuous forcing terms and includes a section on the Dirac delta function.

Power Series Power series solutions is a topic that occasionally causes student anxiety. Possibly, this is due to inadequate preparation in calculus where the more subtle subject of convergent series is (frequently) covered at a rapid pace. Our solution has been to provide a thorough treatment of power series solutions that also includes a review of their properties as well as a discussion of real analytic functions. Unlike many texts, ours provides an extensive section on the *method of Frobenius* (Section 8.6) as well as a section on finding a second linearly independent solution.

 While we have given considerable space to power series solutions, we have also taken great care to accommodate the instructor who only wishes to give a basic introduction to the topic. *An introduction to solving differential equations using power series and the method of Frobenius can be accomplished by covering the materials in Sections 8.1, 8.2, 8.3, and 8.6.*

Partial Differential Equations An introduction to this subject is provided in Chapter 10, which covers the method of separation of variables, Fourier series, the heat equation, the wave equation, and Laplace's equation. Examples in two and three dimensions are included.

SUPPLEMENTS

Student's Solutions Manual By V. Maymeskul and E. B. Saff. Containing complete, worked-out solutions to selected exercises, this provides students with an excellent study tool. ISBN 0-201-33869-6

Instructor's Resource Guide Contains short answers to all exercises and additional group projects. ISBN 0-201-66975-7

Interactive Differential Equations CD-ROM By Beverly West (Cornell University), Steven Strogatz (Cornell University), Jean Marie McDill (California Polytechnic State University–San Luis Obispo), John Cantwell (St. Louis University), and Hubert Hohn (Massachusetts College of Art). A revised version of a popular software directly tied to the text. Focuses on helping students visualize concepts. Applications are drawn from engineering, physics, chemistry, and biology. Runs on Windows or Macintosh and is bundled *free* with every book.

Companion Website Provides additional resources for both instructors and students, including instructional utility software, helpful links keyed to sections of the text, suggestions for incorporating *Interactive Differential Equations* modules, suggested syllabi, index of applications, and study tips for students. Access:

http://www.awlonline.com/nagle

MAPLE-Based Instructor's Guide By Kenneth Pothoven (University of South Florida). A collection of MAPLE® worksheets and projects to aid instructors in integrating MAPLE into their courses. Also available via the companion Website. ISBN 0-201-61840-0

ACKNOWLEDGMENTS

The staging of this text involved considerable behind-the-scenes activity. A special note of gratitude is extended to Alar Toomre (Massachusetts Institute of Technology), who not only provided us with two new Group Projects along with a variety of challenging new problems for the exercises, but also generated the computer graphic appearing on the front cover. We want to thank Frank Glaser (California State Polytechnic University, Pomona) for many of the historical footnotes. We are also indebted to Herbert E. Rauch (Lockheed Research Laboratory) for help with Section 3.3 on heating and cooling of buildings, Project B in Chapter 3 on aquaculture, and other application problems. Our appreciation goes to Richard H. Elderkin (Pomona College), Jerrold Marsden (University of California, Berkeley), T. G. Proctor (Clemson University), and Philip W. Schaefer (University of Tennessee), who read and reread the manuscript for the original text, making numerous suggestions that greatly

improved the book. We are also indebted to the many people who reviewed the manuscript for this new edition.

REVIEWERS

Tuncay Aktosun, North Dakota State University

Dean R. Brown, Youngstown State University

Yunmei Chen, University of Florida

Elias Deeba, University of Houston–Downtown

Tyrone Duncan, University of Kansas

Kanat Durgun, The Citadel

Alejandro Engel, Rochester Institute of Technology

David J. Kaup, Clarkson University

Subrata Mukherjee, Cornell University

Chuanxi Qian, Mississippi State University

Philip W. Schaefer, University of Tennessee–Knoxville

L. F. Shampine, Southern Methodist University

David J. Uherka, University of North Dakota

In addition, we would like to offer our sincere thanks to all of the instructors who shared their own group projects with us for inclusion in this edition:

Richard Bernatz, Luther College

Randall K. Campbell-Wright, University of Tampa

Alfred Clark, Jr., University of Rochester

Leopoldo P. Franca, University of Colorado–Denver

T. L. Pearson, Acadia University

David Stapleton, University of Central Oklahoma

Alar Toomre, Massachusetts Institute of Technology

The newly developed companion Website for the text was specially designed by Link-Systems International, Inc. Its unique browser-driven utility package, which was developed by Emil Moskona, not only enhances the usefulness of the text but also provides a convenient tool for calculus/differential equation problem solving in general.

Finally, we want to thank the staff at Addison-Wesley for their dedicated assistance in bringing this edition to fruition. Special kudos go to mathematics editor Laurie Rosatone and her assistants Rachel Reeve and Ellen Keohane for their help, encouragement, and sensitivity, as well to Paul Lorczak, who carefully proofread the text and checked the formulas and exercises. The production phase was carefully guided by Kim Ellwood and Phyllis Crittenden, who admirably put up with our frequent outbursts of anxiety.

E. B. SAFF, A. D. SNIDER
TAMPA, FLORIDA

Contents

*Denotes optional sections that can be omitted without compromising the logical flow.

CHAPTER 3 MATHEMATICAL MODELS AND NUMERICAL METHODS INVOLVING FIRST ORDER EQUATIONS 93

CHAPTER 4 LINEAR SECOND ORDER EQUATIONS 156

CHAPTER 5 INTRODUCTION TO SYSTEMS AND PHASE
PLANE ANALYSIS 261

CHAPTER 10 PARTIAL DIFFERENTIAL EQUATIONS 600

Introduction

1.1 BACKGROUND

In the sciences and engineering, mathematical models are developed to aid in the understanding of physical phenomena. These models often yield an equation that contains some derivatives of an unknown function. Such an equation is called a **differential equation.** Two examples of models developed in calculus are the free fall of a body and the decay of a radioactive substance.

In the case of free fall, an object is released from a certain height above the ground and falls under the force of gravity.[†] Newton's second law, which states that an object's mass times its acceleration equals the total force acting on it, can be applied to the falling object. This leads to the equation (see Figure 1.1)

$$m\frac{d^2h}{dt^2} = -mg \ ,$$

where m is the mass of the object, h is the height above the ground, d^2h/dt^2 is its acceleration, g is the (constant) gravitational acceleration, and $-mg$ is the force due to gravity. This is a differential equation containing the second derivative of the unknown height h as a function of time.

Figure 1.1 Apple in free fall

[†]We are assuming here that gravity is the *only* force acting on the object and that this force is constant. More general models would take into account other forces, such as air resistance.

Fortunately, the above equation is easy to solve for h. All we have to do is divide by m and integrate twice with respect to t. That is,

$$\frac{d^2h}{dt^2} = -g \ ,$$

so

$$\frac{dh}{dt} = -gt + c_1$$

and

$$h = h(t) = \frac{-gt^2}{2} + c_1 t + c_2 \ .$$

We shall see that the constants of integration, c_1 and c_2, are determined if we know the *initial* height and the *initial* velocity of the object. We then have a formula for the height of the object at time t.

In the case of radioactive decay (Figure 1.2), we begin from the premise that the rate of decay is proportional to the amount of radioactive substance present. This leads to the equation

$$\frac{dA}{dt} = -kA \ , \qquad k > 0 \ ,$$

where $A(>0)$ is the unknown amount of radioactive substance present at time t and k is the proportionality constant. To solve this differential equation, we rewrite it in the form

$$\frac{1}{A} dA = -k \, dt$$

and integrate to obtain

$$\int \frac{1}{A} dA = \int -k \, dt$$

$$\ln A + C_1 = -kt + C_2 \ .$$

Solving for A yields

$$A = A(t) = e^{\ln A} = e^{-kt} e^{C_2 - C_1} = C e^{-kt} \ ,$$

Figure 1.2 Radioactive decay

where C is the combination of integration constants $e^{C_2 - C_1}$. The value of C, as we shall see later, is determined if the *initial* amount of radioactive substance is given. We then have a formula for the amount of radioactive substance at any future time t.

Even though the above examples were easily solved by methods learned in calculus, they do give us some insight into the study of differential equations in general. First, notice that the solution of a differential equation is a *function,* like $h(t)$ or $A(t)$, not merely a number. Second, integration is an important tool in solving differential equations (not surprisingly!). Third, we cannot expect to get a unique solution to a differential equation, since there will be arbitrary "constants of integration." The second derivative d^2h/dt^2 in the free fall equation gave rise to two constants, c_1 and c_2, and the first derivative in the decay equation gave rise, ultimately, to one constant, C.

Whenever a mathematical model involves the **rate of change** of one variable with respect to another, a differential equation is apt to appear. Unfortunately, in contrast to the examples for free fall and radioactive decay, the differential equation may be very complicated and difficult to analyze.

Differential equations arise in a variety of subject areas, including not only the physical sciences but also such diverse fields as economics, medicine, psychology, and operations research. We now list a few specific examples.

1. A classic application of differential equations is found in the study of an electric circuit consisting of a resistor, an inductor, and a capacitor driven by an electromotive force (see Figure 1.3). Here an application of Kirchhoff's laws[†] leads to the equation

(1) $$L\frac{d^2q}{dt^2} + R\frac{dq}{dt} + \frac{1}{C}q = E(t) \; ,$$

 where L is the inductance, R is the resistance, C is the capacitance, $E(t)$ is the electromotive force, $q(t)$ is the charge on the capacitor, and t is the time.

Figure 1.3 Schematic for a series RLC circuit

2. In the study of the gravitational equilibrium of a star, an application of Newton's law of gravity and of the Stefan-Boltzmann law for gases leads to the equilibrium equation

(2) $$\frac{1}{r^2}\frac{d}{dr}\left(\frac{r^2}{\rho}\frac{dP}{dr}\right) = -4\pi\rho G \; ,$$

 where P is the sum of the gas kinetic pressure and the radiation pressure, r is the dis-

[†]We will discuss the applications of Kirchhoff's laws in Section 5.5.

tance from the center of the star, ρ is the density of matter, and G is the gravitational constant.

3. In psychology, one model of the learning of a task involves the equation

(3)
$$\frac{dy/dt}{y^{3/2}(1-y)^{3/2}} = \frac{2p}{\sqrt{n}} \ .$$

Here the variable y represents the learner's skill level as a function of time t. The constants p and n depend on the individual learner and the nature of the task.

4. In the study of vibrating strings and the propagation of waves, we find the *partial* differential equation

(4)
$$\frac{\partial^2 u}{\partial t^2} - c^2 \frac{\partial^2 u}{\partial x^2} = 0 \ , ^{\dagger}$$

where t represents time, x the location along the string, c the wave speed, and u the displacement of the string, *which is a function of time and location.*

To begin our study of differential equations we need some common terminology. If an equation involves the derivative of one variable with respect to another, then the former is called a **dependent variable** and the latter an **independent variable.** Thus in the equation

(5)
$$\frac{d^2 x}{dt^2} + a\frac{dx}{dt} + kx = 0 \ ,$$

t is the independent variable and x is the dependent variable. We refer to a and k as **coefficients** in equation (5). In the equation

(6)
$$\frac{\partial u}{\partial x} - \frac{\partial u}{\partial y} = x - 2y \ ,$$

x and y are independent variables and u is a dependent variable.

A differential equation involving only ordinary derivatives with respect to a single independent variable is called an **ordinary differential equation.** A differential equation involving partial derivatives with respect to more than one independent variable is a **partial differential equation.** Equation (5) is an ordinary differential equation, and equation (6) is a partial differential equation.

The **order** of a differential equation is the order of the highest-order derivatives present in the equation. Equation (5) is a second order equation because $d^2 x/dt^2$ is the highest-order derivative present. Equation (6) is a first order equation because only first order partial derivatives occur.

It will be useful to classify ordinary differential equations as being either linear or nonlinear. Remember that lines (in two dimensions) and planes (in three dimensions) are especially easy to visualize, when compared to nonlinear objects such as cubic curves or quadric surfaces. For example, all the points on a line can be found if we know just two of them. Correspondingly, *linear* differential equations are more amenable to solution than nonlinear ones. Now the

†*Historical Footnote:* This partial differential equation was first discovered by Jean le Rond d'Alembert (1717–1783) in 1747.

equations for lines $ax + by = c$ and planes $ax + by + cz = d$ have the feature that the variables appear in *additive combinations of their first powers only.* By analogy, a **linear differential equation** is one in which the dependent variable y and its derivatives appear in additive combinations of their first powers.

More precisely, a differential equation is **linear** if it has the format

$$(7) \qquad a_n(x)\frac{d^n y}{dx^n} + a_{n-1}(x)\frac{d^{n-1}y}{dx^{n-1}} + \cdots + a_1(x)\frac{dy}{dx} + a_0(x)y = F(x) \, ,$$

where $a_n(x), a_{n-1}(x), \ldots, a_0(x)$ and $F(x)$ depend only on the independent variable x. The additive combinations are permitted to have multipliers (coefficients) that depend on x; no restrictions are made on the nature of this x-dependence. If an ordinary differential equation is not linear, then we call it **nonlinear.** For example,

$$\frac{d^2 y}{dx^2} + y^3 = 0$$

is a nonlinear second order ordinary differential equation because of the y^3 term, whereas

$$\frac{d^2 y}{dx^2} + y = x^3$$

is linear (despite the x^3 term). The equation

$$\frac{d^2 y}{dx^2} - y\frac{dy}{dx} = \cos x$$

is nonlinear because of the $y\, dy/dx$ term.

Although the majority of equations one is likely to encounter in practice fall into the *nonlinear* category, knowing how to deal with the simpler linear equations is an important first step (just as tangent lines help our understanding of complicated curves by providing local approximations).

EXERCISES 1.1

In Problems 1 through 12, a differential equation is given along with the field or problem area in which it arises. Classify each as an ordinary differential equation (ODE) or a partial differential equation (PDE), give the order, and indicate the independent and dependent variables. If the equation is an ordinary differential equation, indicate whether the equation is linear or nonlinear.

1. $3\dfrac{d^2 x}{dt^2} + 4\dfrac{dx}{dt} + 9x = 2\cos 3t$

(mechanical vibrations, electrical circuits, seismology).

2. $\dfrac{d^2 y}{dx^2} - 2x\dfrac{dy}{dx} + 2y = 0$

(Hermite's equation, quantum mechanical harmonic oscillator).

3. $\dfrac{dy}{dx} = \dfrac{y(2 - 3x)}{x(1 - 3y)}$

(competition between two species, ecology).

4. $\dfrac{\partial^2 u}{\partial x^2} + \dfrac{\partial^2 u}{\partial y^2} = 0$

(Laplace's equation, potential theory, electricity, heat, aerodynamics).

5. $\dfrac{dp}{dt} = kp(P - p)$, where k and P are constants

(logistic curve, epidemiology, economics).

6. $\dfrac{dx}{dt} = (4 - x)(1 - x)$

(chemical reaction rates).

7. $y\left[1 + \left(\dfrac{dy}{dx}\right)^2\right] = C$, where C is a constant

(brachistochrone problem,[†] calculus of variations).

8. $\sqrt{1 - y}\,\dfrac{d^2y}{dx^2} + 2x\dfrac{dy}{dx} = 0$

(Kidder's equation, flow of gases through a porous medium).

9. $x\dfrac{d^2y}{dx^2} + \dfrac{dy}{dx} + xy = 0$

(aerodynamics, stress analysis).

10. $8\dfrac{d^4y}{dx^4} = x(1 - x)$

(deflection of beams).

11. $\dfrac{\partial N}{\partial t} = \dfrac{\partial^2 N}{\partial r^2} + \dfrac{1}{r}\dfrac{\partial N}{\partial r} + kN$, where k is a constant

(nuclear fission).

12. $\dfrac{d^2y}{dx^2} - 0.1(1 - y^2)\dfrac{dy}{dx} + 9y = 0$

(van der Pol's equation, triode vacuum tube).

In Problems 13 through 16, write a differential equation that fits the physical description.

13. The rate of change of the population p of bacteria at time t is proportional to the population at time t.

14. The velocity at time t of a particle moving along a straight line is proportional to the fourth power of its position x.

15. The rate of change in the temperature T of coffee at time t is proportional to the difference between the temperature M of the air at time t and the temperature of the coffee at time t.

16. The rate of change of the mass A of salt at time t is proportional to the square of the mass of salt present at time t.

17. Drag Race. Two drivers, Alison and Kevin, are participating in a drag race. Beginning from a standing start, they each proceed with a constant acceleration. Kevin covers the last $1/4$ of the distance in 3 seconds, whereas Alison covers the last $1/3$ of the distance in 4 seconds. Who wins and by how much time?

1.2 SOLUTIONS AND INITIAL VALUE PROBLEMS

An nth order ordinary differential equation is an equality involving the independent variable x, the dependent variable y, and the first n derivatives of y. Examples are

$$x^2\dfrac{d^2y}{dx^2} + x\dfrac{dy}{dx} + y = x^3 \quad \text{(second order)}$$

$$\sqrt{1 - \left(\dfrac{d^2y}{dx^2}\right)^2} - y = 0 \quad \text{(second order)}$$

$$\dfrac{d^4y}{dx^4} = xy \quad \text{(fourth order)}.$$

[†]*Historical Footnote:* In 1630 Galileo formulated the brachistochrone problem ($\beta\rho\acute{\alpha}\chi\iota\sigma\tau o\varsigma$ = shortest, $\chi\rho\acute{o}\nu o\varsigma$ = time), that is, to determine a path down which a particle will fall from one given point to another in the shortest time. It was reproposed by John Bernoulli in 1696 and solved by him the following year.

Thus a general form for an nth order equation would be

(1) $$F\left(x, y, \frac{dy}{dx}, \ldots, \frac{d^n y}{dx^n}\right) = 0 \ ,$$

where F is a function of the independent variable x, the dependent variable y, and the derivatives of y up to order n; that is, $x, y, \ldots, d^n y/dx^n$. We assume that the equation holds for all x in an interval I (which may or may not include its endpoints: $a \le x \le b$, $a < x \le b$, etc.). In many cases, we can isolate the highest-order term $d^n y/dx^n$, and write (1) as

(2) $$\frac{d^n y}{dx^n} = f\left(x, y, \frac{dy}{dx}, \ldots, \frac{d^{n-1}y}{dx^{n-1}}\right) \ ,$$

which is often preferable to (1) for theoretical and computational purposes.

EXPLICIT SOLUTION

Definition 1. A function $\phi(x)$ that when substituted for y in equation (1) [or (2)] satisfies the equation for all x in the interval I is called an **explicit solution** to the equation on I.

EXAMPLE 1 Show that $\phi(x) = x^2 - x^{-1}$ is an explicit solution to

(3) $$\frac{d^2 y}{dx^2} - \frac{2}{x^2}y = 0 \ .$$

SOLUTION The functions $\phi(x) = x^2 - x^{-1}$, $\phi'(x) = 2x + x^{-2}$, and $\phi''(x) = 2 - 2x^{-3}$ are defined for all $x \ne 0$. Substitution of $\phi(x)$ for y in equation (3) gives

$$\left(2 - 2x^{-3}\right) - \frac{2}{x^2}\left(x^2 - x^{-1}\right) = \left(2 - 2x^{-3}\right) - \left(2 - 2x^{-3}\right) = 0 \ .$$

Since this is valid for any $x \ne 0$, the function $\phi(x) = x^2 - x^{-1}$ is an explicit solution to (3) on $(-\infty, 0)$ and also on $(0, \infty)$. ■ End

EXAMPLE 2 Show that for *any* choice of the constants c_1 and c_2, the function

$$\phi(x) = c_1 e^{-x} + c_2 e^{2x}$$

is an explicit solution to

(4) $$y'' - y' - 2y = 0 \ .$$

SOLUTION We compute $\phi'(x) = -c_1 e^{-x} + 2c_2 e^{2x}$ and $\phi''(x) = c_1 e^{-x} + 4c_2 e^{2x}$. Substitution of ϕ, ϕ', and ϕ'' for y, y', and y'' in equation (4) yields

$$\left(c_1 e^{-x} + 4c_2 e^{2x}\right) - \left(-c_1 e^{-x} + 2c_2 e^{2x}\right) - 2\left(c_1 e^{-x} + c_2 e^{2x}\right)$$
$$= \left(c_1 + c_1 - 2c_1\right)e^{-x} + \left(4c_2 - 2c_2 - 2c_2\right)e^{2x} = 0 \ .$$

Since equality holds for all x in $(-\infty, \infty)$, then $\phi(x) = c_1 e^{-x} + c_2 e^{2x}$ is an explicit solution to (4) on the interval $(-\infty, \infty)$ for any choice of the constants c_1 and c_2. ■ End

As we shall see in Chapter 2, the methods for solving differential equations do not always yield an explicit solution for the equation. We may have to settle for a solution that is defined implicitly. Consider the following example.

EXAMPLE 3 Show that the relation

(5) $y^2 - x^3 + 8 = 0$

implicitly defines a solution to

(6) $\dfrac{dy}{dx} = \dfrac{3x^2}{2y}$

on the interval $(2, \infty)$.

SOLUTION When we solve (5) for y, we obtain $y = \pm\sqrt{x^3 - 8}$. Let's try $\phi(x) = \sqrt{x^3 - 8}$ to see if it is an explicit solution. Since $d\phi/dx = 3x^2/\left(2\sqrt{x^3 - 8}\right)$, both ϕ and $d\phi/dx$ are defined on $(2, \infty)$. Substituting them into (6) yields

$$\frac{3x^2}{2\sqrt{x^3 - 8}} = \frac{3x^2}{2\left(\sqrt{x^3 - 8}\right)} \ ,$$

which is valid for all x in $(2, \infty)$. Hence relation (5) is an implicit solution to (6) on $(2, \infty)$. (The reader can check that $\psi(x) = -\sqrt{x^3 - 8}$ is also an explicit solution to (6).) ■ End

IMPLICIT SOLUTION

Definition 2. A relation $G(x, y) = 0$ is said to be an **implicit solution** to equation (1) on the interval I if it defines one or more explicit solutions on I.

EXAMPLE 4 Show that

(7) $x + y + e^{xy} = 0$

is an implicit solution to

(8) $\left(1 + xe^{xy}\right)\dfrac{dy}{dx} + 1 + ye^{xy} = 0 \ .$

SOLUTION First, we observe that we are unable to solve (7) directly for y in terms of x alone. However, for (7) to hold, we realize that any change in x requires a change in y, so we expect the relation (7) to define implicitly at least one function $y(x)$. This is difficult to show directly but can be rigorously verified using the **implicit function theorem**[†] of advanced calculus, which guarantees that such a function $y(x)$ exists that is also differentiable (see Problem 30).

[†]See *Vector Calculus,* Fourth Edition, by J. E. Marsden and A. J. Tromba (W. H. Freeman and Company, San Francisco, 1996).

Once we know that y is a differentiable function of x, we can use the technique of implicit differentiation. Indeed, from (7) we obtain on differentiating with respect to x,

$$1 + \frac{dy}{dx} + e^{xy}\left(y + x\frac{dy}{dx}\right) = 0$$

or

$$\left(1 + xe^{xy}\right)\frac{dy}{dx} + 1 + ye^{xy} = 0 \ ,$$

which is identical to the differential equation (8). Thus relation (7) is an implicit solution on some interval guaranteed by the implicit function theorem. ■ End

EXAMPLE 5 Verify that $4x^2 - y^2 = C$, where C is an arbitrary constant, gives a one-parameter family of implicit solutions to

(9) $$y\frac{dy}{dx} - 4x = 0$$

and graph several of these solution curves.

SOLUTION When we implicitly differentiate the equation $4x^2 - y^2 = C$ with respect to x, we find

$$8x - 2y\frac{dy}{dx} = 0 \ ,$$

which is equivalent to (9). In Figure 1.4 we have sketched the implicit solutions for $C = 0, \pm 1,$ ± 4. The curves are hyperbolas with common asymptotes $y = \pm 2x$. Notice that the implicit solution curves (with C arbitrary) fill the entire plane and are nonintersecting for $C \neq 0$. For $C = 0$, the implicit solution gives rise to the two explicit solutions $y = 2x$ and $y = -2x$, both of which pass through the origin. ■ End

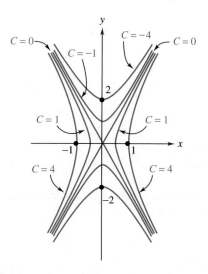

Figure 1.4 Implicit solutions $4x^2 - y^2 = C$

For brevity, we hereafter use the term *solution* to mean either an explicit or an implicit solution.

In the beginning of Section 1.1, we saw that the solution of the *second* order free fall equation invoked two arbitrary constants of integration c_1, c_2:

$$h(t) = \frac{-gt^2}{2} + c_1 t + c_2 ,$$

while the solution of the *first* order radioactive decay equation contained a single constant C:

$$A(t) = Ce^{-kt} .$$

It is clear that integration of the simple *fourth* order equation

$$\frac{d^4 y}{dx^4} = 0$$

brings in four undetermined constants:

$$y(x) = c_1 x^3 + c_2 x^2 + c_3 x + c_4 .$$

It will be shown later in the text that, in general, the methods for solving nth order differential equations evoke n arbitrary constants. In most cases, we will be able to evaluate these constants if we know n initial values $y(x_0), y'(x_0), \ldots, y^{(n-1)}(x_0)$.

INITIAL VALUE PROBLEM

Definition 3. By an **initial value problem** for an nth order differential equation

$$F\left(x, y, \frac{dy}{dx}, \ldots, \frac{d^n y}{dx^n}\right) = 0 ,$$

we mean: Find a solution to the differential equation on an interval I that satisfies at x_0 the n initial conditions

$$y(x_0) = y_0 ,$$

$$\frac{dy}{dx}(x_0) = y_1 ,$$

$$\vdots$$

$$\frac{d^{n-1} y}{dx^{n-1}}(x_0) = y_{n-1} ,$$

where $x_0 \in I$ and $y_0, y_1, \ldots, y_{n-1}$ are given constants.

In the case of a first order equation, the initial conditions reduce to the single requirement

$$y(x_0) = y_0 ,$$

and in the case of a second order equation, the initial conditions have the form

$$y(x_0) = y_0 , \qquad \frac{dy}{dx}(x_0) = y_1 .$$

The terminology *initial conditions* comes from mechanics, where the independent variable x represents *time* and is customarily symbolized as t. Then if t_0 is the starting time, $y(t_0) = y_0$ represents the initial location of an object and $dy(t_0)/dt$ gives its initial velocity.

EXAMPLE 6 Show that $\phi(x) = \sin x - \cos x$ is a solution to the initial value problem

(10) $\dfrac{d^2y}{dx^2} + y = 0 ; \qquad y(0) = -1 , \qquad \dfrac{dy}{dx}(0) = 1 .$

SOLUTION Observe that $\phi(x) = \sin x - \cos x$, $d\phi/dx = \cos x + \sin x$, and $d^2\phi/dx^2 = -\sin x + \cos x$ are all defined on $(-\infty, \infty)$. Substituting into the differential equation gives

$$(-\sin x + \cos x) + (\sin x - \cos x) = 0 ,$$

which holds for all $x \in (-\infty, \infty)$. Hence $\phi(x)$ is a solution to the differential equation in (10) on $(-\infty, \infty)$. When we check the initial conditions, we find

$$\phi(0) = \sin 0 - \cos 0 = -1 ,$$

$$\frac{d\phi}{dx}(0) = \cos 0 + \sin 0 = 1 ,$$

which meets the requirements of (10). Therefore $\phi(x)$ is a solution to the given initial value problem. ■ End

EXAMPLE 7 As shown in Example 2, the function $\phi(x) = c_1 e^{-x} + c_2 e^{2x}$ is a solution to

$$\frac{d^2y}{dx^2} - \frac{dy}{dx} - 2y = 0$$

for any choice of the constants c_1 and c_2. Determine c_1 and c_2 so that the initial conditions

$$y(0) = 2 \quad \text{and} \quad \frac{dy}{dx}(0) = -3$$

are satisfied.

SOLUTION To determine the constants c_1 and c_2, we first compute $d\phi/dx$ to get $d\phi/dx = -c_1 e^{-x} + 2c_2 e^{2x}$. Substituting in our initial conditions gives the following system of equations:

$$\begin{cases} \phi(0) \;\; = c_1 e^0 + c_2 e^0 \;\; = 2 , \\ \dfrac{d\phi}{dx}(0) = -c_1 e^0 + 2c_2 e^0 = -3 , \end{cases} \qquad \text{or} \qquad \begin{cases} c_1 + c_2 \;\; = 2 , \\ -c_1 + 2c_2 = -3 . \end{cases}$$

Adding the last two equations yields $3c_2 = -1$, so $c_2 = -1/3$. Since $c_1 + c_2 = 2$, we find $c_1 = 7/3$. Hence the solution to the initial value problem is $\phi(x) = (7/3)e^{-x} - (1/3)e^{2x}$. ■ End

We now state an existence and uniqueness theorem for first order initial value problems. We presume the differential equation has been cast into the format

$$\frac{dy}{dx} = f(x, y) \ .$$

Of course the right-hand side, $f(x, y)$, must be well defined at the starting value x_0 for x and at the stipulated initial value $y_0 = y(x_0)$ for y. The hypotheses of the theorem, moreover, require *continuity* of both f and $\partial f/\partial y$ for x in some interval $a < x < b$ containing x_0, and for y in some interval $c < y < d$ containing y_0. Notice that the set of points in the xy-plane that satisfy $a < x < b$ and $c < y < d$ constitutes a *rectangle*. Figure 1.5 depicts this "rectangle of continuity" with the initial point (x_0, y_0) in its interior and a sketch of a portion of the solution curve contained therein.

EXISTENCE AND UNIQUENESS OF SOLUTION

Theorem 1. Given the initial value problem

$$\frac{dy}{dx} = f(x, y) \ , \qquad y(x_0) = y_0 \ ,$$

assume that f and $\partial f/\partial y$ are continuous functions in a rectangle

$$R = \{(x, y)\colon a < x < b, c < y < d\}$$

that contains the point (x_0, y_0). Then the initial value problem has a unique solution $\phi(x)$ in some interval $x_0 - h < x < x_0 + h$, where h is a positive number.

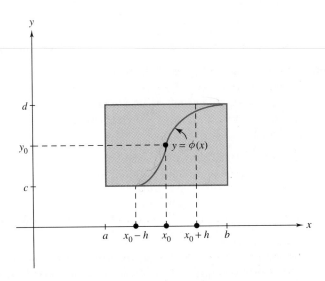

Figure 1.5 Layout for the existence-uniqueness theorem

The preceding theorem tells us two things. First, when an equation satisfies the hypotheses of Theorem 1, we are assured that a solution to the initial value problem exists. Naturally it is desirable to know whether the equation we are trying to solve actually has a solution before we spend too much time trying to solve it. Second, when the hypotheses are satisfied, there is a **unique** solution to the initial value problem. This uniqueness tells us that if we can find a solution, then it is the *only* solution for the initial value problem. Graphically, the theorem says that there is only one solution curve that passes through the point (x_0, y_0). In fact, for this first order equation, two solutions cannot cross anywhere in the rectangle. Notice that the existence and uniqueness of the solution holds only in *some* neighborhood $(x_0 - h, x_0 + h)$. Unfortunately, the theorem does not tell us the span $(2h)$ of this neighborhood (merely that it is not zero). Problem 18 elaborates on this feature.

Problem 19 gives an example of an equation with no solution. Problem 29 displays an initial value problem for which the solution is not unique. Of course, the hypotheses of Theorem 1 are not met for these cases.

When initial value problems are used to model physical phenomena, many practitioners tacitly presume the conclusions of Theorem 1 to be valid. Indeed, for the initial value problem to be a reasonable model, we certainly expect it to have a solution, since physically "something does happen." Moreover, the solution should be unique in those cases when repetition of the experiment under identical conditions yields the same results.[†]

The proof of Theorem 1 involves converting the initial value problem into an integral equation and then using Picard's method to generate a sequence of successive approximations that converge to the solution. The conversion to an integral equation and Picard's method are discussed in Project B at the end of this chapter. A detailed discussion and proof of the theorem are given in Chapter 13.[††]

EXAMPLE 8 For the initial value problem

(11) $\dfrac{dy}{dx} = x^2 - xy^3 , \qquad y(1) = 6 ,$

does Theorem 1 imply the existence of a unique solution?

SOLUTION Here $f(x, y) = x^2 - xy^3$ and $\partial f/\partial y = -3xy^2$. Both of these functions are continuous in any rectangle containing the point $(1, 6)$, so the hypotheses of Theorem 1 are satisfied. It then follows from Theorem 1 that the initial value problem (11) has a unique solution in an interval about $x = 1$ of the form $(1 - h, 1 + h)$, where h is some positive number. ∎End

EXAMPLE 9 For the initial value problem

(12) $\dfrac{dy}{dx} = 3y^{2/3} , \qquad y(2) = 0 ,$

does Theorem 1 imply the existence of a unique solution?

[†]At least this is the case when we are considering a deterministic model, as opposed to a probabilistic model.

[††]All references to Chapters 11–13 refer to the expanded text *Fundamentals of Differential Equations and Boundary Value Problems,* 3rd edition.

(12) $\dfrac{dy}{dx} = 3y^{2/3}$, $y(2) = 0$

SOLUTION Here $f(x, y) = 3y^{2/3}$ and $\partial f/\partial y = 2y^{-1/3}$. Unfortunately $\partial f/\partial y$ is not continuous or even defined when $y = 0$. Consequently there is no rectangle containing $(2, 0)$ in which both f and $\partial f/\partial y$ are continuous. Because the hypotheses of Theorem 1 do not hold, we cannot use Theorem 1 to determine whether the initial value problem does or does not have a unique solution. It turns out that this initial value problem does *not* have a unique solution. We refer the reader to Problem 29 for the details. ■ End

In Example 9, suppose the initial condition is changed to $y(2) = 1$. Then, since f and $\partial f/\partial y$ are continuous in any rectangle that contains the point $(2, 1)$ but does not intersect the x-axis, say $R = \{(x, y): 0 < x < 10, 0 < y < 5\}$, it follows from Theorem 1 that this *new* initial value problem has a unique solution in some interval about $x = 2$.

EXERCISES 1.2

1. (a) Show that $\phi(x) = 2x^3$ is an explicit solution to
$$x\frac{dy}{dx} = 3y$$
on the interval $(-\infty, \infty)$.
 (b) Show that $\phi(x) = e^x - x$ is an explicit solution to
$$\frac{dy}{dx} + y^2 = e^{2x} + (1 - 2x)e^x + x^2 - 1$$
on the interval $(-\infty, \infty)$.
 (c) Show that $\phi(x) = x^2 - x^{-1}$ is an explicit solution to $x^2 d^2y/dx^2 = 2y$ on the interval $(0, \infty)$.

2. (a) Show that $y^2 + x - 3 = 0$ is an implicit solution to $dy/dx = -1/(2y)$ on the interval $(-\infty, 3)$.
 (b) Show that $xy^3 - xy^3 \sin x = 1$ is an implicit solution to
$$\frac{dy}{dx} = \frac{(x \cos x + \sin x - 1)y}{3(x - x \sin x)}$$
on the interval $(0, \pi/2)$.

In Problems 3 through 8, determine whether the given function is a solution to the given differential equation.

3. $y = \sin x + x^2$, $\dfrac{d^2y}{dx^2} + y = x^2 + 2$.

4. $x = 3 \cos t - 5 \sin t$, $x'' + x = 0$.

5. $x = \cos 2t$, $\dfrac{dx}{dt} + tx = \sin 2t$.

6. $\theta = 2e^{3t} - e^{2t}$, $\dfrac{d^2\theta}{dt^2} - \theta\dfrac{d\theta}{dt} + 3\theta = -2e^{2t}$.

7. $y = e^{2x} - 3e^{-x}$, $\dfrac{d^2y}{dx^2} - \dfrac{dy}{dx} - 2y = 0$.

8. $y = 3 \sin 2x + e^{-x}$, $y'' + 4y = 5e^{-x}$.

In Problems 9 through 13, determine whether the given relation is an implicit solution to the given differential equation. Assume that the relationship does define y implicitly as a function of x and use implicit differentiation.

9. $x^2 + y^2 = 6$, $\dfrac{dy}{dx} = \dfrac{x}{y}$.

10. $y - \ln y = x^2 + 1$, $\dfrac{dy}{dx} = \dfrac{2xy}{y - 1}$.

11. $e^{xy} + y = x - 1$, $\dfrac{dy}{dx} = \dfrac{e^{-xy} - y}{e^{-xy} + x}$.

12. $x^2 - \sin(x + y) = 1$, $\dfrac{dy}{dx} = 2x \sec(x + y) - 1$.

13. $\sin y + xy - x^3 = 2$,
$$y'' = \frac{6xy' + (y')^3 \sin y - 2(y')^2}{3x^2 - y}$$.

14. Show that $\phi(x) = c_1 \sin x + c_2 \cos x$ is a solution to $d^2y/dx^2 + y = 0$ for any choice of the constants c_1 and c_2. Thus $c_1 \sin x + c_2 \cos x$ is a two-parameter family of solutions to the differential equation.

15. Show that $\phi(x) = Ce^{3x} + 1$ is a solution to $dy/dx - 3y = -3$ for any choice of the constant C. Thus $Ce^{3x} + 1$ is a one-parameter family of solu-

tions to the differential equation. Graph several of the solution curves using the same coordinate axes.

16. Verify that $x^2 + cy^2 = 1$, where c is an arbitrary nonzero constant, is a one-parameter family of implicit solutions to

$$\frac{dy}{dx} = \frac{xy}{x^2 - 1}$$

and graph several of the solution curves using the same coordinate axes.

17. Verify that $\phi(x) = 2/(1 - ce^x)$, where c is an arbitrary constant, is a one-parameter family of solutions to

$$\frac{dy}{dx} = \frac{y(y - 2)}{2} .$$

Graph the solution curves corresponding to $c = 0$, $\pm 1, \pm 2$ using the same coordinate axes.

18. Let $c > 0$. Show that the function $\phi(x) = (c^2 - x^2)^{-1}$ is a solution to the initial value problem $dy/dx = 2xy^2$, $y(0) = 1/c^2$, on the interval $-c < x < c$. Note that this solution becomes unbounded as x approaches $\pm c$. Thus the solution exists on the interval $(-h, h)$ with $h = c$, but not for larger h. This illustrates that in Theorem 1 the existence interval can be quite small (if c is small) or quite large (if c is large). Notice also that there is no clue from the equation $dy/dx = 2xy^2$ itself, or from the initial value, that the solution will "blow up" at $x = \pm c$.

19. Show that the equation $(dy/dx)^2 + y^2 + 3 = 0$ has no (real-valued) solution.

20. Determine for which values of m the function $\phi(x) = e^{mx}$ is a solution to the given equation.

(a) $\dfrac{d^2y}{dx^2} + 6\dfrac{dy}{dx} + 5y = 0$.

(b) $\dfrac{d^3y}{dx^3} + 3\dfrac{d^2y}{dx^2} + 2\dfrac{dy}{dx} = 0$.

21. Determine for which values of m the function $\phi(x) = x^m$ is a solution to the given equation.

(a) $3x^2\dfrac{d^2y}{dx^2} + 11x\dfrac{dy}{dx} - 3y = 0$.

(b) $x^2\dfrac{d^2y}{dx^2} - x\dfrac{dy}{dx} - 5y = 0$.

22. Verify that the function $\phi(x) = c_1e^x + c_2e^{-2x}$ is a solution to

$$\frac{d^2y}{dx^2} + \frac{dy}{dx} - 2y = 0$$

for any choice of the constants c_1 and c_2. Determine c_1 and c_2 so that each of the following initial conditions is satisfied.

(a) $y(0) = 2$, $y'(0) = 1$.
(b) $y(1) = 1$, $y'(1) = 0$.

In Problems 23 through 28, determine whether Theorem 1 implies that the given initial value problem has a unique solution.

23. $\dfrac{dy}{dx} = x^3 - y^3$, $\qquad\qquad y(0) = 6$.

24. $\dfrac{dy}{d\theta} - \theta y = \sin^2\theta$, $\qquad\quad y(\pi) = 5$.

25. $x\dfrac{dx}{dt} + 4t = 0$, $\qquad\qquad x(2) = -\pi$.

26. $\dfrac{dy}{dx} = 3x - \sqrt[3]{y - 1}$, $\qquad y(2) = 1$.

27. $y\dfrac{dy}{dx} = x$, $\qquad\qquad\qquad y(1) = 0$.

28. $\dfrac{dx}{dt} + \cos x = \sin t$, $\qquad x(\pi) = 0$.

29. For the initial value problem (12) of Example 9, show that $\phi_1(x) \equiv 0$ and $\phi_2(x) = (x - 2)^3$ are solutions. Hence the initial value problem does not have a *unique* solution.

30. Implicit Function Theorem. Let $G(x, y)$ have continuous first partial derivatives in the rectangle $R = \{(x, y): a < x < b, c < y < d\}$ containing the point (x_0, y_0). If $G(x_0, y_0) = 0$ and the partial derivative $G_y(x_0, y_0) \neq 0$, then there exists a differentiable function $y = \phi(x)$, defined in some interval $I = (x_0 - h, x_0 + h)$, that satisfies $G(x, \phi(x)) = 0$ for all $x \in I$.

The implicit function theorem gives conditions under which the relationship $G(x, y) = 0$ defines y implicitly as a function of x. Use the implicit function theorem to show that the relationship $x + y + e^{xy} = 0$, given in Example 4, defines y implicitly as a function of x near the point $(0, -1)$.

31. Consider the equation of Example 5,

(13) $y\dfrac{dy}{dx} - 4x = 0$.

(a) Does Theorem 1 imply the existence of a unique solution to (13) that satisfies $y(x_0) = 0$?

(b) Show that when $x_0 \neq 0$, equation (13) can't possibly have a solution in a neighborhood of $x = x_0$ that satisfies $y(x_0) = 0$.

(c) Show that there are two distinct solutions to (13) satisfying $y(0) = 0$ (see Figure 1.4 on page 9).

1.3 DIRECTION FIELDS

The existence and uniqueness theorem discussed in the preceding section certainly has great value, but it stops short of telling us anything about the *nature* of the solution to a differential equation. For practical reasons, we may need to know the value of the solution at a certain point, or the intervals where the solution is increasing, or the points where the solution attains a maximum value. Certainly knowing an explicit representation (a formula) for the solution would be a considerable help in answering these questions. However, for many of the differential equations that the reader is likely to encounter in "real-world" applications, it will be impossible to find such a formula. Moreover, even if we are lucky enough to obtain an implicit solution, it may be difficult to use this relationship to determine an explicit form. Thus we must rely on other methods to analyze or approximate the solution.

One technique that is useful in visualizing (graphing) the solutions to a first order differential equation is to sketch the direction field for the equation. To describe this method, we need to make a general observation. Namely, a first order equation

$$\frac{dy}{dx} = f(x, y)$$

specifies a slope at each point in the xy-plane where f is defined. In other words, it gives the direction that a solution to the equation must have at each point. Consider, for example, the equation

(1) $\dfrac{dy}{dx} = x^2 - y$.

The graph of a solution to (1) that passes through the point $(-2, 1)$ must have slope $(-2)^2 - 1 = 3$ at that point, and a solution through $(-1, 1)$ has zero slope at that point.

A plot of short line segments drawn at various points in the xy-plane showing the slope of the solution curve there is called a **direction field** for the differential equation. Because the direction field gives the "flow of solutions," it facilitates the drawing of any particular solution (such as the solution to an initial value problem). In Figure 1.6(a) we have sketched the direction field for the equation (1) and in Figure 1.6(b) drawn several solution curves in color.

Some other interesting direction field patterns are displayed in Figure 1.7. Depicted in Figure 1.7(a) is the pattern for the radioactive decay equation $dy/dx = -2y$ (recall that in Section 1.1, we analyzed this equation in the form $dA/dt = -kA$). From the flow patterns, we can see that all solutions tend asymptotically to the positive x-axis as x gets larger. In other words, any material decaying according to this law eventually dwindles to practically nothing. This is consistent with the solution formula we derived earlier,

$$A = Ce^{-kt} , \quad \text{or} \quad y = Ce^{-2x} .$$

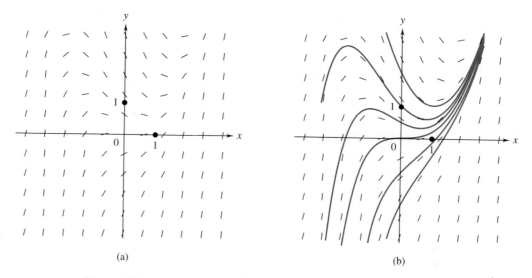

(a)

(b)

Figure 1.6 (a) Direction field for $dy/dx = x^2 - y$ (b) Solutions to $dy/dx = x^2 - y$

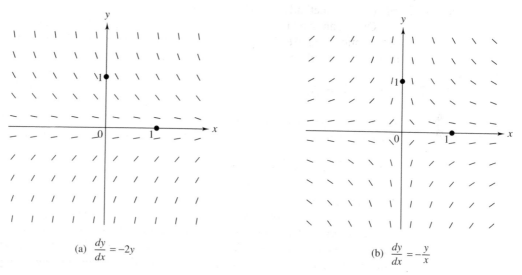

(a) $\dfrac{dy}{dx} = -2y$

(b) $\dfrac{dy}{dx} = -\dfrac{y}{x}$

Figure 1.7 (a) Direction field for $dy/dx = -2y$ (b) Direction field for $dy/dx = -y/x$

From the direction field in Figure 1.7(b), we can anticipate that all solutions to $dy/dx = -y/x$ also approach the x-axis as x approaches infinity (plus or minus infinity, in fact). But more interesting is the observation that no solution can make it across the y-axis; $|y(x)|$ "blows up" as x goes to zero from either direction. Exception: On close examination, it appears the function $y(x) \equiv 0$ might just make it through this barrier. As a matter of fact, in Problem 19 the reader is invited to show that the solutions to this differential equation are given by $y = C/x$, with C an arbitrary constant. So they do diverge at $x = 0$, unless $C = 0$.

Let's interpret the existence-uniqueness theorem of Section 1.2 for these direction fields. For Figure 1.7(a), where $dy/dx = f(x, y) = -2y$, we select a starting point x_0 and an initial value $y(x_0) = y_0$, as in Figure 1.8(a). Because the right-hand side $f(x, y) = -2y$ is continuously differentiable for all x and y, we can enclose any initial point (x_0, y_0) in a "rectangle of continuity." We conclude that the equation has one and only one solution curve passing through (x_0, y_0), as depicted in the figure.

For the equation

$$\frac{dy}{dx} = f(x, y) = -\frac{y}{x} \, ,$$

the right-hand side $f(x, y) = -y/x$ does not meet the continuity conditions when $x = 0$ (i.e., for points on the y-axis). However, for any *nonzero* starting value x_0 and *any* initial value $y(x_0) = y_0$, we *can* enclose (x_0, y_0) in a rectangle of continuity that excludes the y-axis, as in Figure 1.8(b). Thus we can be assured of one and only one solution curve passing through such a point.

The direction field for the equation

$$\frac{dy}{dx} = 3y^{2/3}$$

is intriguing because Example 9 of the preceding section showed that the hypotheses of Theorem 1 do *not* hold in any rectangle enclosing the initial point $x_0 = 2$, $y_0 = 0$. Indeed, Problem 29 of that section demonstrated the violation of uniqueness by exhibiting *two* solutions, $y(x) \equiv 0$ and $y(x) = (x - 2)^3$, passing through $(2, 0)$. Figure 1.9(a) displays this direction field, and Figure 1.9(b) demonstrates how both solution curves can successfully "negotiate" this flow pattern.

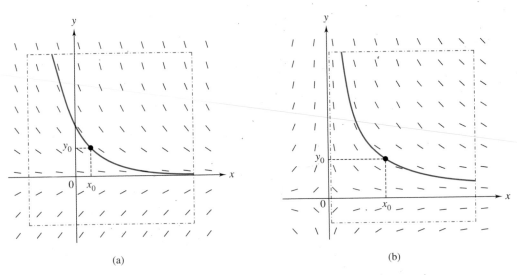

(a) (b)

Figure 1.8 (a) A solution for $dy/dx = -2y$ (b) A solution for $dy/dx = -y/x$

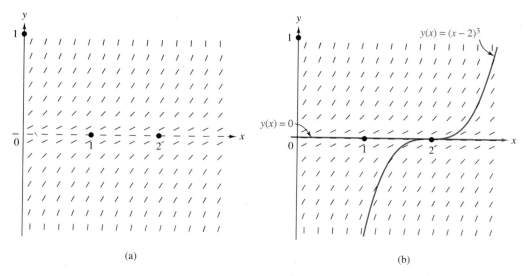

Figure 1.9 (a) Direction field for $dy/dx = 3y^{2/3}$ (b) Solutions for $dy/dx = 3y^{2/3}$

Clearly a sketch of the direction field of a first order differential equation can be helpful in visualizing the solutions. However, such a sketch is not sufficient to enable us to trace, unambiguously, the solution curve passing through a given initial point (x_0, y_0). If we tried to trace one of the solution curves in Figure 1.6(b) on page 17, for example, we could easily "slip" over to an adjacent curve. For nonunique situations like that in Figure 1.9(b), as one negotiates the flow along the curve $y = (x - 2)^3$ and reaches the inflection point, one cannot decide whether to turn or to (literally) go off on the tangent $(y = 0)$.

EXAMPLE 1 The logistic equation for the population p (in thousands) at time t of a certain species is given by

$$(2) \qquad \frac{dp}{dt} = p(2 - p) \ .$$

(Of course p is nonnegative. The interpretation of the terms in the logistic equation is discussed in Section 3.2.) From the direction field sketched in Figure 1.10 on page 20, answer the following.

 (a) If the initial population is 3000 (that is, $p(0) = 3$), what can you say about the limiting population $\lim_{t \to +\infty} p(t)$?

 (b) Can a population of 1000 ever decline to 500?

 (c) Can a population of 1000 ever increase to 3000?

SOLUTION **(a)** The direction field indicates that all solution curves (other than $p(t) \equiv 0$) will approach the horizontal line $p = 2$ as $t \to +\infty$, that is, this line is an asymptote for all positive solutions. Thus $\lim_{t \to +\infty} p(t) = 2$.

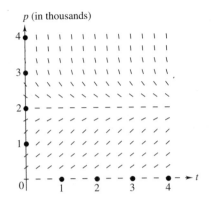

Figure 1.10 Direction field for logistic equation

(b) The direction field further shows that populations greater than 2000 will steadily decrease, while those less than 2000 will increase. In particular, a population of 1000 can never decline to 500.

(c) As mentioned in part (b), a population of 1000 will increase with time. But the direction field indicates it can never reach 2000 or any larger value, i.e., the solution curve cannot cross the line $p = 2$. Indeed, the constant function $p(t) \equiv 2$ is a solution to equation (2), and the uniqueness part of Theorem 1, page 12, precludes intersections of solution curves. ■ End

Notice that the direction field in Figure 1.10 has the nice feature that the slopes do not depend on t; that is, the pattern of slopes is the same along each vertical line. The same is true for Figures 1.8(a) and 1.9. This is the key property of so-called **autonomous equations** $y' = f(y)$, where the right-hand side is a function of the dependent variable only. In the next section we shall investigate such equations in more detail.

Hand sketching the direction field for a differential equation is often tedious. Fortunately several software programs have been developed to obviate this task. When hand sketching is necessary, however, the **method of isoclines** can be helpful in reducing the work.

The Method of Isoclines

An isocline for the differential equation

$$y' = f(x, y)$$

is a set of points in the xy-plane where all the solutions have the same slope dy/dx; thus it is a level curve for the function $f(x, y)$. For example, if

(3) $y' = f(x, y) = x + y$,

the isoclines are simply the curves (straight lines) $x + y = c$ or $y = -x + c$. Here c is an arbitrary constant. But c can be interpreted as the numerical value of the slope dy/dx of every solution curve as it crosses the isocline. (Note that c is not the slope of the isocline itself; the latter is, obviously, -1.) Figure 1.11(a) depicts the isoclines for equation (3).

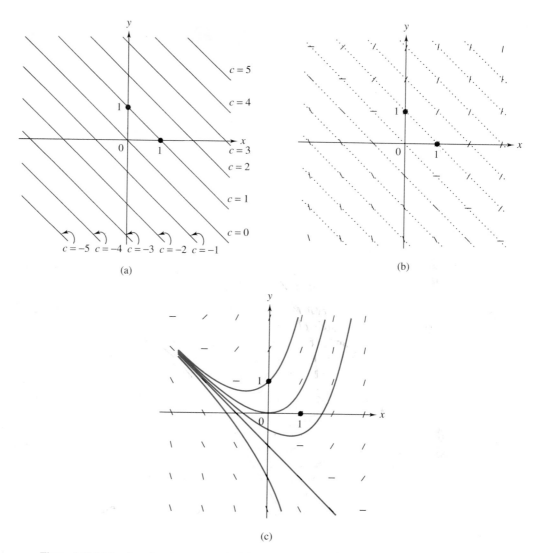

Figure 1.11 (a) Isoclines for $y' = x + y$ (b) Direction field for $y' = x + y$ (c) Solutions to $y' = x + y$

To implement the method of isoclines for sketching direction fields, we draw hash marks with slope c along the isocline $f(x, y) = c$ for a few selected values of c. If we then erase the underlying isocline curves, the hash marks constitute a part of the direction field for the differential equation. Figure 1.11(b) depicts this process for the isoclines shown in Figure 1.11(a), and Figure 1.11(c) displays some solution curves.

Remark. The isoclines themselves are not always straight lines. For equation (1) at the beginning of this section (page 16), they are parabolas $x^2 - y = c$. When the isocline curves are complicated, this method is not practical.

EXERCISES 1.3

1. The direction field for $dy/dx = 4x/y$ is shown in Figure 1.12.

 (a) Verify that the straight lines $y = \pm 2x$ are solution curves, provided $x \neq 0$.

 (b) Sketch the solution curve with initial condition $y(0) = 2$.

 (c) Sketch the solution curve with initial condition $y(2) = 1$.

 (d) What can you say about the behavior of the above solutions as $x \to +\infty$? How about $x \to -\infty$?

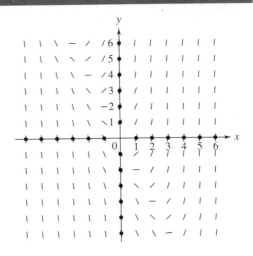

Figure 1.13 Direction field for $dy/dx = 2x + y$

cous medium is given by the equation

$$\frac{dv}{dt} = 1 - \frac{v}{8}.$$

From the direction field shown in Figure 1.14, sketch the solutions with the initial conditions $v(0) = 5, 8,$ and 15. Why is the value $v = 8$ called the "terminal velocity"?

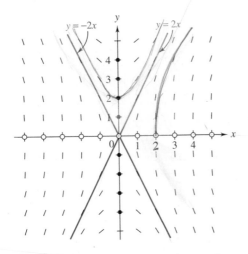

Figure 1.12 Direction field for $dy/dx = 4x/y$

2. The direction field for $dy/dx = 2x + y$ is shown in Figure 1.13.

 (a) Sketch the solution curve that passes through $(0, -2)$. From this sketch, write the equation for the solution.

 (b) Sketch the solution curve that passes through $(-1, 3)$.

 (c) What can you say about the solution in part (b) as $x \to +\infty$? How about $x \to -\infty$?

3. A model for the velocity v at time t of a certain object falling under the influence of gravity in a vis-

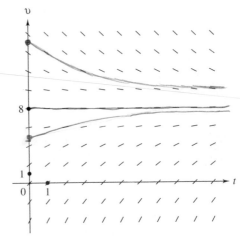

Figure 1.14 Direction field for $\dfrac{dv}{dt} = 1 - \dfrac{v}{8}$

4. If the viscous force in Problem 3 is nonlinear, a possible model would be provided by the differential equation

$$\frac{dv}{dt} = 1 - \frac{v^3}{8} .$$

Redraw the direction field in Figure 1.14 to incorporate this v^3 dependence. Sketch the solutions with initial conditions $v(0) = 0, 1, 2, 3$. What is the terminal velocity in this case?

5. The logistic equation for the population (in thousands) of a certain species is given by

$$\frac{dp}{dt} = 3p - 2p^2 .$$

(a) Sketch the direction field by using either a computer software package or the method of isoclines.

(b) If the initial population is 2000 (that is, $p(0) = 2$), what can you say about the limiting population $\lim_{t \to +\infty} p(t)$?

(c) If $p(0) = 0.5$, what is $\lim_{t \to +\infty} p(t)$?

(d) Can a population of 3000 ever decline to 500?

6. Consider the differential equation

$$\frac{dy}{dx} = x + \sin y .$$

(a) A solution curve passes through the point $(1, \pi/2)$. What is its slope at this point?

(b) Argue that every solution curve is increasing for $x > 1$.

(c) Show that the second derivative of every solution satisfies

$$\frac{d^2y}{dx^2} = 1 + x \cos y + \frac{1}{2} \sin 2y .$$

(d) A solution curve passes through $(0, 0)$. Prove that this curve has a relative minimum at $(0, 0)$.

7. Consider the differential equation

$$\frac{dp}{dt} = p(p - 1)(2 - p)$$

for the population p (in thousands) of a certain species at time t.

(a) Sketch the direction field by using either a computer software package or the method of isoclines.

(b) If the initial population is 3000 (that is, $p(0) = 3$), what can you say about the limiting population $\lim_{t \to +\infty} p(t)$?

(c) If $p(0) = 1.5$, what is $\lim_{t \to +\infty} p(t)$?

(d) If $p(0) = 0.5$, what is $\lim_{t \to +\infty} p(t)$?

(e) Can a population of 900 ever increase to 1100?

8. The motion of a set of particles moving along the x-axis is governed by the differential equation

$$\frac{dx}{dt} = t^3 - x^3 ,$$

where $x(t)$ denotes the position at time t of the particle.

(a) If a particle resides at $x = 1$ when $t = 2$, what is its velocity at this time?

(b) Show that the acceleration of a particle is given by

$$\frac{d^2x}{dt^2} = 3t^2 - 3t^3x^2 + 3x^5 .$$

(c) If a particle is located at $x = 2$ when $t = 2.5$, can it reach the location $x = 1$ at any later time? $\left[Hint: t^3 - x^3 = (t - x)(t^2 + xt + x^2). \right]$

9. Let $\phi(x)$ denote the solution to the initial value problem

$$\frac{dy}{dx} = x - y , \qquad y(0) = 1 .$$

(a) Show that $\phi''(x) = 1 - \phi'(x) = 1 - x + \phi(x)$.

(b) Argue that the graph of ϕ is decreasing for x near zero and that as x increases from zero, $\phi(x)$ decreases until it crosses the line $y = x$, where its derivative is zero.

(c) Let $x*$ be the abscissa of the point where the solution curve $y = \phi(x)$ crosses the line $y = x$. Consider the sign of $\phi''(x*)$ and argue that ϕ has a relative minimum at $x*$.

(d) What can you say about the graph of $y = \phi(x)$ for $x > x*$?

(e) Verify that $y = x - 1$ is a solution to $dy/dx = x - y$ and explain why the graph of $\phi(x)$ always stays above the line $y = x - 1$.

(f) Sketch the direction field for $dy/dx = x - y$ by using the method of isoclines or a computer software package.

(g) Sketch the solution $y = \phi(x)$ using the direction field in part (f).

10. Use a computer software package to sketch the direction field for the following differential equations.

Sketch some of the solution curves.

(a) $dy/dx = \sin x$.
(b) $dy/dx = \sin y$.
(c) $dy/dx = \sin x \sin y$.
(d) $dy/dx = x^2 + 2y^2$.
(e) $dy/dx = x^2 - 2y^2$.

In Problems 11 through 16, draw the isoclines with their direction markers and sketch several solution curves, including the curve satisfying the given initial conditions.

11. $dy/dx = 2x$, $y(0) = -1$.
12. $dy/dx = y$, $y(0) = 1$.
13. $dy/dx = -x/y$, $y(0) = 4$.
14. $dy/dx = x/y$, $y(0) = -1$.
15. $dy/dx = 2x^2 - y$, $y(0) = 0$.
16. $dy/dx = x + 2y$, $y(0) = 1$.

17. From a sketch of the direction field, what can one say about the behavior as x approaches $+\infty$ of a solution to

$$\frac{dy}{dx} = 3 - y + \frac{1}{x} \ ?$$

18. From a sketch of the direction field, what can one say about the behavior as x approaches $+\infty$ of a solution to

$$\frac{dy}{dx} = -y \ ?$$

19. By rewriting the differential equation $dy/dx = -y/x$ in the form

$$\frac{1}{y}dy = \frac{-1}{x}dx \ ,$$

integrate both sides to obtain the solution $y = C/x$ for an arbitrary constant C.

1.4 THE PHASE LINE

Sketching the direction field of a differential equation $dy/dt = f(t, y)$ is particularly easy when the slope function f is independent of t:

(1) $$\frac{dy}{dt} = f(y) \ .$$

Such a direction field is exhibited in Figure 1.15(a), which highlights the essential feature: the

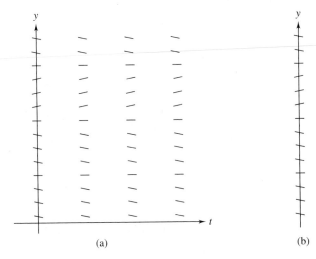

(a) (b)

Figure 1.15 Direction field for an autonomous equation

slopes are all identical along horizontal lines. In fact, a sketch of the slope field along the y-axis tells the whole story (Figure 1.15(b)).

Equations of the form (1), for which the independent variable t does not appear explicitly, are called **autonomous.** If t is interpreted as time, the nomenclature reflects the fact that such equations are self-governing, in the sense that the derivative y' is steered by a function f determined solely by the current state y, and not by any external controller watching a clock.

Figure 1.16 exhibits another feature of autonomous differential equations; their solutions can be shifted in time. Mathematically this is expressed by noting that if $y(t)$ solves (1), so does $y(t - t_0)$ (for any constant t_0).

The solution curves shown in Figure 1.16 demonstrate some qualitative features of solutions to first order autonomous differential equations. The constant solutions $y \equiv y_1$, $y \equiv y_2$, $y \equiv y_3$ are called **equilibrium points** for the equation (by analogy with mechanical systems which are "in equilibrium" when they cease to move). Equilibrium points are easily identified by their horizontal direction fields—that is, points y where the slope f is zero: $f(y_1) = f(y_2) = f(y_3) = 0$ in the figure.

Note how all solutions $y(t)$ that get sufficiently near the particular equilibrium point y_1 are compelled to approach y_1 as $t \to +\infty$. In other words, if the equilibrium solution $y(t) \equiv y_1$ is somehow "perturbed," it will asymptotically return to the value y_1. This behavior is quite distinct from that for the equilibrium $y(t) \equiv y_2$, where any perturbation drives the solution away from y_2, or for $y(t) \equiv y_3$, where negative perturbations are repelled while positive ones are lured back to y_3. We say y_1 is a *stable* equilibrium, and that y_2 and y_3 are *unstable.* Stable equilibrium points like y_1 are also called *sinks,* whereas unstable equilibrium points that repel *all* neighboring solutions, like y_2, are called *sources.* Equilibria like y_3, that are neither sinks nor sources, are called *nodes.* We shall defer the mathematically precise definition of stability to the later chapters, because some subtle distinctions are involved.

Notice how the nature of each equilibrium is related to the mathematical features of the function $f(y)$. Of course equilibria occur at the zeros of f, as illustrated in Figure 1.17 on page 26. For an equilibrium sink like y_1, solutions below the equilibrium are forced upward; that is, if $y(t) < y_1$, $y' = f(y) > 0$. Similarly, solutions above y_1 are forced down. In general,

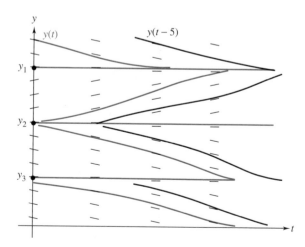

Figure 1.16 Time-shifted solutions of an autonomous equation with equilibrium points y_1, y_2, and y_3

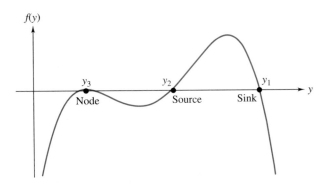

Figure 1.17 Equilibrium points are zeros of $f(y)$

An equilibrium point y^ is stable, or a sink, if $f(y) > 0$ for $y < y^*$ and $f(y) < 0$ for $y > y^*$ (for all y sufficiently near y^*).*

The opposite is true for source equilibria, which repel neighboring solutions: $f(y) < 0$ for $y < y^*$ and $f(y) > 0$ for $y > y^*$. The graph of $f(y)$ in Figure 1.17 illustrates the configurations for the equilibria y_1, y_2, and y_3 in Figure 1.16.

Clearly for a *qualitative* description of the equilibrium points, we only need information about the *sign* of $f(y)$. This is depicted by the **phase line** for the differential equation in Figure 1.18. The large dots on the phase line indicate the equilibrium points ($f(y) = 0$), and the arrows in between show whether $f(y)$ is positive (arrows point right) or negative (arrows point left). The phase line is easy to plot, and from it we can get a lot of qualitative information about the solutions. Note how the appropriateness of the "sink, source" nomenclature is suggested by the phase line plot.

Figure 1.18 Phase line for an autonomous equation

EXAMPLE 1 Sketch the phase line for the equation

$$y' = (y - 1)(y - 2)(y - 3)$$

and state the qualitative nature of its equilibria.

SOLUTION Clearly, $f(y) = (y - 1)(y - 2)(y - 3)$ equals zero at $y = 1, 2$, and 3. We indicate these points in Figure 1.19. For $y > 3$, all factors of $f(y)$ are positive, and f changes sign as y passes through each of the equilibria. This simple observation gives us the complete phase line plot, as shown. From the criteria above we conclude that $y = 3$ is a source equilibrium, $y = 2$ is a sink, and $y = 1$ is, again, a source. The solution graphs (generated by software) for this equation are illustrated in Figure 1.20. ∎ End

EXAMPLE 2 Use the phase line for the equation

$$y' = -(y - 1)^{5/3}(y - 2)^2(y - 3)$$

to predict the asymptotic behavior as $t \to +\infty$ of the solution satisfying $y(0) = 2.1$.

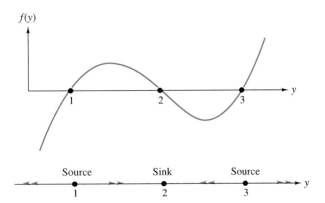

Figure 1.19 Equilibrium points and phase line for Example 1

Figure 1.20 Solution graphs for Example 1

SOLUTION The equilibria are again $y = 1, 2$, and 3. For $y > 3$, f is negative, and it changes sign at $y = 3$. Thus $y = 3$ is a sink. Because of the square in $(y - 2)^2$, the function $f(y)$ is positive above *and* below $y = 2$, so this equilibrium is a node, repelling nearby solutions with $y(t) > 2$ and attracting those with $y(t) < 2$. The factor $(y - 1)^{5/3}$ changes sign at $y = 1$, so this equilibrium is a source. The phase line plot is displayed in Figure 1.21, and from it we predict that the solution starting from $y = 2.1$ tends to $y = 3$ as $t \rightarrow +\infty$. ■ End

Figure 1.21 Phase line and equilibrium points for Example 2

EXAMPLE 3 Sketch the phase line for the equation

$$y' = y \sin y$$

and state the qualitative nature of its equilibria.

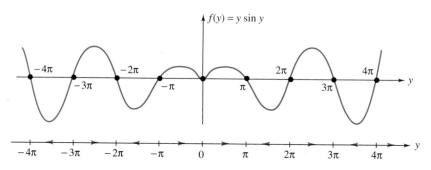

Figure 1.22 Equilibrium points and phase line for Example 3

SOLUTION The graph of $f(y) = y \sin y$ and its phase line plot are shown in Figure 1.22. Clearly there is an equilibrium at every integer multiple of π. The sinks are the positive, odd multiples of π and the negative, even multiples of π. The sources are the positive even multiples and the negative odd multiples. The equilibrium $y = 0$ is a node. Some solutions are displayed in Figure 1.24(b) on page 29. ∎ End

The *node* in Example 3 warrants further examination. Notice, in Figure 1.22, that the equilibrium point of $y' = f(y) = y \sin y$ at $y = 0$ is very tenuous; if we increase $f(y)$ everywhere by as little as 0.1, its zero at $y = 0$ "disappears," and there is no equilibrium: look at Figure 1.23(a). On the other hand, if we *decrease* $f(y)$ everywhere by 0.1, it has *two* zeros, and from Figure 1.23(b) we envision a source and a sink near $y = 0$. This phenomenon is an example of *bifurcation*. The reader should track what happens to the equilibrium solutions in Figure 1.24.

Notice also from Figures 1.23 and 1.24 that the other equilibria of $f(y) = y \sin y$ are not especially interesting—the sinks remain sinks, and the sources remain sources. Clearly what is special about the equilibrium at $y = 0$ is the fact that it is a local minimum of $f(y)$, and thus can be made either to disappear or to "split" by a minor perturbation of f. Such a bifurcation can also occur at the local maximum at the node $y = y_3$ in Figure 1.17 (page 26), if its function f were perturbed.

Our final example will demonstrate how tracking bifurcation phenomena can provide some very specific information about a system governed by a differential equation, without ever solving the differential equation itself! It utilizes the *logistic model,* which will be discussed in detail in Section 3.2. The key feature of the logistic model is that its phase line has the structure of Figure 1.25—a stable equilibrium at some positive value y_1, and an unstable equilibrium at $y = 0$. The population of a particular species often exhibits such characteristics; a population of, say, zero guinea pigs is certainly an equilibrium, but introduction of a few adults into the environment results in a steadily increasing population[†]—until the competition for food and space dictates another, stable, population equilibrium.

[†]"Buy a couple and wait": Cole Porter, *Let's Fall in Love* (1928).

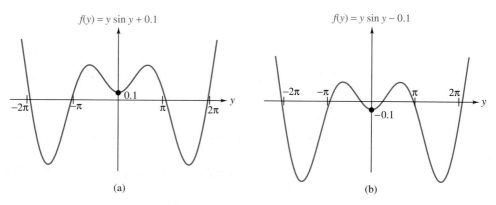

(a)

(b)

Figure 1.23 Perturbed equilibrium points

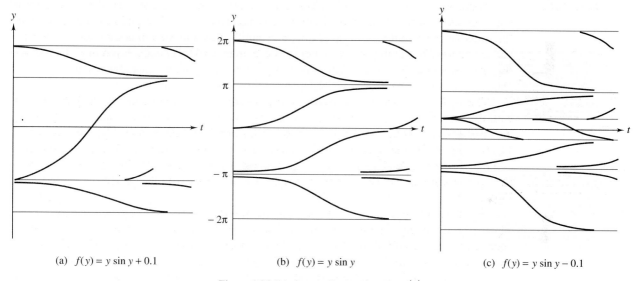

(a) $f(y) = y \sin y + 0.1$

(b) $f(y) = y \sin y$

(c) $f(y) = y \sin y - 0.1$

Figure 1.24 Solution graphs for perturbed $f(y)$

The simplest function yielding a phase line like that depicted in Figure 1.25 has the parabolic form $f(y) = -Ay(y - y_1)$ with $A > 0$. The logistic model, then, is governed by the differential equation

$$y' = -Ay(y - y_1) .$$

Figure 1.25 Phase line for logistic model

EXAMPLE 4 A model for the alligator population y at time t on the grounds of the Kennedy Space Center is given by the logistic equation

(2) $y' = -y(y - 1500)/3200$,

where t is measured in years. If hunting were allowed on the grounds, how many alligator shootings per year would eradicate the population?

SOLUTION If hunters kill s alligators per year, equation (2) will have to be modified to read

(3) $y' = -y(y - 1500)/3200 - s$

to account for the thinning of the population. Let's classify the equilibria for equation (3). The zeros of the quadratic function $f(y) = -y(y-1500)/3200 - s$ occur at $y = 750 \pm \sqrt{562{,}500 - 3200s}$. If we plot a succession of phase lines for increasing values of s (Figure 1.26), we obtain a *bifurcation diagram* for this problem; note how the pair of equilibria (a sink and a source) for low values of s coalesce when $s \approx 175$, then disappear for higher values. Specifically, this coalescence occurs when the *discriminant* $562{,}500 - 3200s$ is zero, i.e., when $s = 562{,}500/3200 \approx 175.78$. (Note that this choice for s forces f to have both a zero and a relative maximum at $y = 750$.)

So if hunters kill over 176 alligators per year, they will (eventually) eliminate the stable alligator population. Yet if they kill only 175 per year, the population stabilizes at around 750. The moral: *Herd management can be a very delicate issue.* ■End

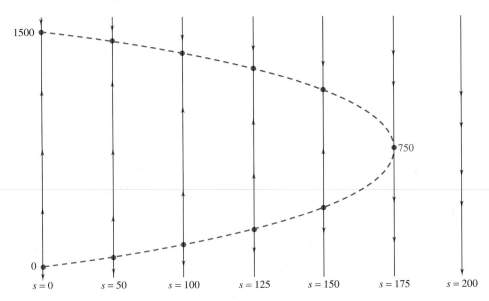

Figure 1.26 Bifurcation diagram for Example 4

In closing, we emphasize that although we only considered qualitative features of the differential equation in Example 4—and never tried to solve it—we were able to deduce very specific values for the parameter s. We shall see other benefits of qualitative analysis in subsequent sections of this book.

EXERCISES 1.4

For each of the autonomous differential equations in Problems 1 through 10, (a) sketch the phase line, (b) identify the equilibrium solutions, and (c) classify the equilibrium points as sinks, sources, or nodes.

1. $y' = y(y - 1)$.
2. $y' = y^2 - 1$.
3. $y' + y^2 - 1 = 0$.
4. $y' = y^2 - 7y + 10$.
5. $y' = y^2 - 2y + 1$.
6. $y' = \cos y$.
7. $y' - y \cos y = 0$.
8. $y' = \sin^2 y$.
9. $y' = y(y^2 - 2y - 8)$.
10. $y' = y(y^2 - 2y + 1)$.

For the differential equations in Problems 11 through 17, use the phase line to predict the asymptotic $(t \to +\infty)$ behavior of the solution satisfying the specified initial condition.

11. $y' = y^2 - 7y + 10$, $y(0) = 3$.
12. $y' = y^2 - 2y + 1$, $y(0) = 0$.
13. $y' = \cos y$, $y(0) = 0$.
14. $y' = y \cos y$, $y(0) = 0$.
15. $y' = \sin^2 y$, $y(0) = \pi/2$.
16. $y' = y(y^2 - 2y - 8)$, $y(0) = -10$.
17. $y' = y(y^2 - 2y + 1)$, $y(0) = 0.5$.

18. Suppose $y' = f(y)$ has an equilibrium point at y_0. Characterize, as far as possible, the nature of the equilibrium point in terms of the signs of $f'(y_0)$ and $f''(y_0)$.

19. If the population y of rats on a farm at time t (in weeks) satisfies
$$y' = -y(y - 100)/50 ,$$
how many rats per week should be killed to eradicate the population?

20. What are the limitations of equation (3) as a realistic population model? (Consider, for example, what the phase line in Figure 1.26 predicts for an initial population of $y(0) = 100$ alligators, if $s = 150$ alligators per year.) Can you think of any ways to repair the model?

For the equations in Problems 21 through 25, sketch a bifurcation diagram like that in Figure 1.26, interpreting s in the equation as a parameter varying from -2 to 2.

21. $y' = -y^2 + 2y - s$.
22. $y' = y^2 - 2y + 1 + s$.
23. $y' = (y^2 + 1)/2 + sy$.
24. $y' = \sin^2 y + s$.
25. $y' = y^3 - sy$.

1.5 THE APPROXIMATION METHOD OF EULER

Euler's method (or the tangent line method) is a procedure for constructing approximate solutions to an initial value problem for a first order differential equation

(1) $y' = f(x, y)$, $y(x_0) = y_0$.

It could be described as a "mechanical" or "computerized" implementation of the informal procedure for hand sketching the solution curve from a picture of the direction field. As such, it remains subject to the failing that it may skip across solution curves. However, under fairly general conditions, iterations of the procedure do converge to true solutions.

The method is illustrated in Figure 1.27 on page 32. Starting at the initial point (x_0, y_0), we follow the straight line with slope $f(x_0, y_0)$, the tangent line, for some distance to the point (x_1, y_1). Then we reset the slope to the value $f(x_1, y_1)$ and follow this line to (x_2, y_2). In this way,

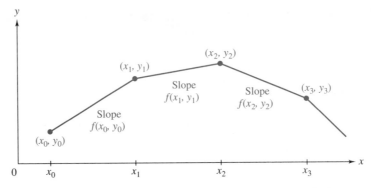

Figure 1.27 Polygonal line approximation given by Euler's method

we construct polygonal (broken line) approximations to the solution. As we take smaller spacings between points (and thus employ more points), we may expect to converge to the true solution.

To be more precise, assume the initial value problem (1) has a unique solution $\phi(x)$ in some interval centered at x_0. Let h be a fixed positive number (called the *step size*) and consider the equally spaced points[†]

$$x_n := x_0 + nh , \qquad n = 0, 1, 2, \ldots .$$

The construction of values y_n that approximate the solution values $\phi(x_n)$ proceeds as follows. At the point (x_0, y_0), the slope of the solution to (1) is given by $dy/dx = f(x_0, y_0)$. Hence the tangent line to the solution curve at the initial point (x_0, y_0) is

$$y = y_0 + (x - x_0)f(x_0, y_0) .$$

Using this tangent line to approximate $\phi(x)$, we find that for the point $x_1 = x_0 + h$

$$\phi(x_1) \approx y_1 := y_0 + hf(x_0, y_0) .$$

Next, starting at the point (x_1, y_1), we construct the line with slope given by the direction field at the point (x_1, y_1); that is, with slope equal to $f(x_1, y_1)$. If we follow this line[††] $\big($namely, $y = y_1 + (x - x_1)f(x_1, y_1)\big)$ in stepping from x_1 to $x_2 = x_1 + h$, we arrive at the approximation

$$\phi(x_2) \approx y_2 := y_1 + hf(x_1, y_1) .$$

Repeating the process (as illustrated in Figure 1.27), we get

$$\phi(x_3) \approx y_3 := y_2 + hf(x_2, y_2) ,$$
$$\phi(x_4) \approx y_4 := y_3 + hf(x_3, y_3) , \qquad \text{etc.}$$

This simple procedure is **Euler's method** and can be summarized by the recursive formulas

(2) $\qquad x_{n+1} = x_n + h ,$

(3) $\qquad y_{n+1} = y_n + hf(x_n, y_n) , \qquad n = 0, 1, 2, \ldots .$

[†]The symbol $:=$ means "is defined to be."

[††]Because y_1 is an *approximation* to $\phi(x_1)$, we cannot assert that this line is tangent to the solution curve $y = \phi(x)$.

EXAMPLE 1 Use Euler's method with step size $h = 0.1$ to approximate the solution to the initial value problem

(4) $y' = x\sqrt{y}$, $y(1) = 4$

at the points $x = 1.1, 1.2, 1.3, 1.4$, and 1.5.

SOLUTION Here $x_0 = 1$, $y_0 = 4$, $h = 0.1$, and $f(x, y) = x\sqrt{y}$. Thus the recursive formula (3) for y_n is

$$y_{n+1} = y_n + hf(x_n, y_n) = y_n + (0.1)x_n\sqrt{y_n} .$$

Substituting $n = 0$, we get

$$x_1 = x_0 + 0.1 = 1 + 0.1 = 1.1 ,$$
$$y_1 = y_0 + (0.1)x_0\sqrt{y_0} = 4 + (0.1)(1)\sqrt{4} = 4.2 .$$

Putting $n = 1$ yields

$$x_2 = x_1 + 0.1 = 1.1 + 0.1 = 1.2 ,$$
$$y_2 = y_1 + (0.1)x_1\sqrt{y_1} = 4.2 + (0.1)(1.1)\sqrt{4.2} = 4.42543 .$$

Continuing in this manner, we obtain the results listed in Table 1.1. For comparison, we have included the exact value (to five decimal places) of the solution $\phi(x) = (x^2 + 7)^2/16$ to (4), which can be obtained using separation of variables (see Section 2.2). As one might expect, the approximation deteriorates as x moves farther away from 1. ■End

TABLE 1.1 COMPUTATIONS FOR $y' = x\sqrt{y}$, $y(1) = 4$			
n	x_n	Euler's Method	Exact Value
0	1	4	4
1	1.1	4.2	4.21276
2	1.2	4.42543	4.45210
3	1.3	4.67787	4.71976
4	1.4	4.95904	5.01760
5	1.5	5.27081	5.34766

Given the initial value problem (1) and a specific point x, how can Euler's method be used to approximate $\phi(x)$? Starting at x_0, we can take one giant step that lands on x, or we can take several smaller steps to arrive at x. If we wish to take N steps, then we set $h = (x - x_0)/N$ so that the step size h and the number of steps N are related in a specific way. For example, if $x_0 = 1.5$ and we wish to approximate $\phi(2)$ using 10 steps, then we would take $h = (2 - 1.5)/10 = 0.05$. It is expected that the more steps we take, the better will be the approximation. (But keep in mind that more steps mean more computations and hence greater accumulated roundoff error.)

EXAMPLE 2 Use Euler's method to find approximations to the solution of the initial value problem

(5) $y' = y$, $y(0) = 1$

at $x = 1$, taking 1, 2, 4, 8, and 16 steps.

 Remark. Observe that the solution to (5) is just $\phi(x) = e^x$, so Euler's method will generate algebraic approximations to the transcendental number $e = 2.71828. \ldots$.

SOLUTION Here $f(x, y) = y$, $x_0 = 0$, and $y_0 = 1$. The recursive formula for Euler's method is

$$y_{n+1} = y_n + hy_n = (1 + h)y_n \ .$$

To obtain approximations at $x = 1$ with N steps, we take the step size $h = 1/N$. For $N = 1$, we have

$$\phi(1) \approx y_1 = (1 + 1)(1) = 2 \ .$$

For $N = 2$, $\phi(x_2) = \phi(1) \approx y_2$. In this case, we get

$$y_1 = (1 + 0.5)(1) = 1.5 \ ,$$
$$\phi(1) \approx y_2 = (1 + 0.5)(1.5) = 2.25 \ .$$

For $N = 4$, $\phi(x_4) = \phi(1) \approx y_4$, where

$$y_1 = (1 + 0.25)(1) = 1.25 \ ,$$
$$y_2 = (1 + 0.25)(1.25) = 1.5625 \ ,$$
$$y_3 = (1 + 0.25)(1.5625) = 1.95313 \ ,$$
$$\phi(1) \approx y_4 = (1 + 0.25)(1.95313) = 2.44141 \ .$$

(In the above computations, we have rounded to five decimal places.) Similarly, taking $N = 8$ and 16, we obtain even better estimates for $\phi(1)$. These approximations are shown in Table 1.2. For comparison, Figure 1.28 displays the polygonal line approximations to e^x using Euler's method with $h = 1/4$ ($N = 4$) and $h = 1/8$ ($N = 8$). Notice that the smaller step size yields the better approximation. ∎ End

TABLE 1.2 EULER'S METHOD FOR $y' = y$, $y(0) = 1$		
N	h	Approximation for $\phi(1) = e$
1	1.0	2.0
2	0.5	2.25
4	0.25	2.44141
8	0.125	2.56578
16	0.0625	2.63793

 How good (or bad) is Euler's method? In judging a numerical scheme, we must begin with two fundamental questions. Does the method converge? And, if so, what is the rate of

convergence? These important issues are discussed in Section 3.5, where improvements in Euler's method are introduced (see also Problems 12 and 13).

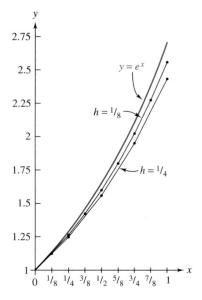

Figure 1.28 Approximations of e^x using Euler's method with $h = 1/4$ and $1/8$

EXERCISES 1.5

In many of the problems below, it will be helpful to have a calculator or computer available. The reader may also find it convenient to write a program for solving initial value problems using Euler's method. (Remember, all trigonometric calculations are done in radians.)

In Problems 1 through 4, use Euler's method to approximate the solution to the given initial value problem at the points $x = 0.1, 0.2, 0.3, 0.4,$ and 0.5 using steps of size 0.1 $(h = 0.1)$.

1. $dy/dx = x/y$, $y(0) = -1$.
2. $dy/dx = -x/y$, $y(0) = 4$.
3. $dy/dx = y(2 - y)$, $y(0) = 3$.
4. $dy/dx = x + y$, $y(0) = 1$.
5. Use Euler's method with step size $h = 0.2$ to approximate the solution to the initial value problem
$$y' = \frac{1}{x}(y^2 + y) , \qquad y(1) = 1$$
at the points $x = 1.2, 1.4, 1.6,$ and 1.8.

6. Use Euler's method with step size $h = 0.1$ to approximate the solution to the initial value problem
$$y' = x - y^2 , \qquad y(1) = 0$$
at the points $x = 1.1, 1.2, 1.3, 1.4,$ and 1.5.

7. Use Euler's method to find approximations to the solution of the initial value problem
$$\frac{dx}{dt} = 1 + t \sin(tx) , \qquad x(0) = 0$$
at $t = 1$, taking 1, 2, 4, and 8 steps.

8. Use Euler's method to find approximations to the solution of the initial value problem
$$y' = 1 - \sin y , \qquad y(0) = 0$$
at $x = \pi$, taking 1, 2, 4, and 8 steps.

9. Use Euler's method with $h = 0.1$ to approximate the solution to the initial value problem
$$y' = \frac{1}{x^2} - \frac{y}{x} - y^2 , \qquad y(1) = -1$$
on the interval $1 \le x \le 2$. Compare these approxi-

mations with the actual solution $y = -1/x$ (verify!) by graphing the polygonal line approximation and the actual solution on the same coordinate system.

10. Use Euler's method with $h = 0.1$ to approximate the solution to the initial value problem

$$y' = x - y , \qquad y(0) = 0$$

on the interval $0 \le x \le 1$. Compare these approximations with the actual solution $y = e^{-x} + x - 1$ (verify!) by graphing the polygonal line approximation and the actual solution on the same coordinate system.

11. Use Euler's method with 20 steps to approximate the solution to the initial value problem

$$\frac{dx}{dt} = 1 + x^2 , \qquad x(0) = 0$$

at $t = 1$. Compare the approximation with the actual solution $x = \tan t$ (verify!) evaluated at $t = 1$.

12. In Example 2, we approximated the transcendental number e by using Euler's method to solve the initial value problem

$$y' = y , \qquad y(0) = 1 .$$

Show that the Euler approximation y_n obtained by using the step size $1/n$ is given by the formula

$$y_n = \left(1 + \frac{1}{n}\right)^n , \qquad n = 1, 2, \dots .$$

Recall from calculus that

$$\lim_{n \to \infty} \left(1 + \frac{1}{n}\right)^n = e ,$$

and hence Euler's method converges (theoretically) to the correct value.

13. Prove that the "rate of convergence" for Euler's method in Problem 12 is comparable to $1/n$ by showing that

$$\lim_{n \to \infty} \frac{e - y_n}{1/n} = \frac{e}{2} .$$

[*Hint:* Use L'Hôpital's rule and the Maclaurin expansion for $\ln(1 + t)$.]

14. Use Euler's method with the spacings $h = 0.5, 0.1,$ $0.05, 0.01$ to approximate the solution to the initial value problem

$$y' = 2xy^2 , \qquad y(0) = 1$$

on the interval $0 \le x \le 2$. (The explanation for the erratic results lies in Problem 18 of Exercises 1.2.)

Heat Exchange. *There are basically two mechanisms by which a physical body exchanges heat with its environment. The* **contact** *heat transfer across the body's surface is driven by the difference in the body's temperature and that of the environment; this is known as Newton's law of cooling. However, heat transfer also occurs due to* **thermal radiation,** *which according to Stefan's law of radiation is governed by the difference of the fourth powers of these temperatures. In most cases, one of these modes dominates the other. Problems 15 and 16 invite the reader to simulate each mode numerically for a given set of initial conditions.*

15. Newton's Law of Cooling. Newton's law of cooling states that the rate of change in the temperature $T(t)$ of a body is proportional to the difference between the temperature of the medium $M(t)$ and the temperature of the body. That is,

$$\frac{dT}{dt} = K\big[M(t) - T(t)\big] ,$$

where K is a constant. Let $K = 1$ (min)$^{-1}$ and the temperature of the medium be constant, $M(t) \equiv 70°$. If the body is initially at $100°$, use Euler's method with $h = 0.1$ to approximate the temperature of the body after

(a) 1 minute.

(b) 2 minutes.

16. Stefan's Law of Radiation. Stefan's law of radiation states that the rate of change in temperature of a body at $T(t)$ degrees in a medium at $M(t)$ degrees is proportional to $M^4 - T^4$. That is,

$$\frac{dT}{dt} = K(M(t)^4 - T(t)^4) ,$$

where K is a constant. Let $K = (40)^{-4}$ and assume the medium temperature is constant, $M(t) \equiv 70°$. If $T(0) = 100°$, use Euler's method with $h = 0.1$ to approximate $T(1)$ and $T(2)$.

CHAPTER SUMMARY

In this chapter, we introduced some basic terminology for differential equations. The **order** of a differential equation is the order of the highest derivative present. The subject of this text is **ordinary** differential equations, which involve derivatives with respect to a single independent variable. Such equations are classified as **linear** or **nonlinear.**

An **explicit solution** of a differential equation is a function of the independent variable that satisfies the equation on some interval. An **implicit solution** is a relation between the dependent and independent variables that implicitly defines a function that is an explicit solution. A differential equation typically has infinitely many solutions. In contrast, there are theorems that ensure a unique solution exists for certain **initial value problems** in which one must find a solution to the differential equation that also satisfies given initial conditions. For an nth order equation, these conditions refer to the values of the solution and its first $n - 1$ derivatives at some point.

Even if one is not successful in finding explicit solutions to a differential equation, several techniques can be used to help analyze the solutions. One such method for first order equations views the differential equation $dy/dx = f(x, y)$ as specifying directions (slopes) at points on the plane. The conglomerate of such slopes is the **direction field** for the equation. Knowing the "flow of solutions" is helpful in sketching the solution to an initial value problem. Furthermore, carrying out this method algebraically leads to numerical approximations to the desired solution. This numerical process is called **Euler's method.** If the differential equation is **autonomous** (i.e., does not depend explicitly on the independent variable), its **phase line** encapsulates information about the nature of its **equilibria** (constant solutions).

TECHNICAL WRITING EXERCISES

1. Select four fields (for example, astronomy, geology, biology, and economics) and for each field discuss a situation in which differential equations are used to solve a problem. Select examples that are not covered in Section 1.1.

2. Compare the different types of solutions discussed in this chapter—explicit, implicit, graphical, and numerical. What are advantages and disadvantages of each?

A. Taylor Series Method

Euler's method is based on the fact that the tangent line gives a good local approximation for the function. But why restrict ourselves to linear approximants when higher-degree polynomial approximants are available? For example, we can use the **Taylor polynomial** of degree n about $x = x_0$, which is defined by

$$P_n(x) := y(x_0) + y'(x_0)(x - x_0) + \frac{y''(x_0)}{2!}(x - x_0)^2 + \cdots + \frac{y^{(n)}(x_0)}{n!}(x - x_0)^n .$$

This polynomial is the nth partial sum of the Taylor series representation

$$\sum_{k=0}^{\infty} \frac{y^{(k)}(x_0)}{k!}(x - x_0)^k .$$

To determine the Taylor series for the solution $\phi(x)$ to the initial value problem

$$dy/dx = f(x, y) , \qquad y(x_0) = y_0 ,$$

we need only determine the values of the derivatives of ϕ (assuming they exist) at x_0; that is, $\phi(x_0), \phi'(x_0), \ldots$. The initial condition gives the first value $\phi(x_0) = y_0$. Using the equation $y' = f(x, y)$, we find $\phi'(x_0) = f(x_0, y_0)$. To determine $\phi''(x_0)$, we differentiate the equation $y' = f(x, y)$ implicitly with respect to x to obtain

$$y'' = \frac{\partial f}{\partial x} + \frac{\partial f}{\partial y}\frac{dy}{dx} = \frac{\partial f}{\partial x} + \frac{\partial f}{\partial y}f$$

and thereby we can compute $\phi''(x_0)$.

(a) Compute the Taylor polynomials of degree 4 for the solutions to the given initial value problems. Use these Taylor polynomials to approximate the solution at $x = 1$.

(i) $\dfrac{dy}{dx} = x - 3y$; $y(0) = 1$. 　　(ii) $\dfrac{dy}{dx} = y(3 - y)$; $y(0) = 4$.

(b) Compare the use of Euler's method with that of the Taylor series to approximate the solution $\phi(x)$ to the initial value problem

$$\frac{dy}{dx} + y = \cos x + \sin x , \qquad y(0) = 1 .$$

Do this by completing Table 1.3. Give the approximations for $\phi(1)$ and $\phi(3)$ to the nearest thousandth. Verify that $\phi(x) = \sin x + e^{-x}$ and use this formula together with a calculator or tables to find the exact values of $\phi(x)$ to the nearest thousandth. Finally, decide which of the first four methods in Table 1.3 will yield the closest approximation to $\phi(10)$ and give the reasons for your choice. (Remember that the computation of trigonometric functions must be done in the radian mode.)

(c) Compute the Taylor polynomial of degree 6 for the solution to the equation

$$\frac{d^2y}{dx^2} = -xy$$

with the initial conditions $y(0) = 1$, $y'(0) = 0$. Do you see how, in general, the Taylor series method for an nth order differential equation will employ each of the n initial conditions mentioned in Definition 3, Section 1.2?

TABLE 1.3

Method	Approximation of $\phi(1)$	Approximation of $\phi(3)$
Euler's method using steps of size 0.1		
Euler's method using steps of size 0.01		
Taylor polynomial of degree 2		
Taylor polynomial of degree 5		
Exact value of $\phi(x)$ to nearest thousandth		

B. *Picard's Method*

The initial value problem

(1) $y'(x) = f(x, y)$, $y(x_0) = y_0$

can be rewritten as an **integral** equation. This is obtained by integrating both sides of (1) with respect to x from $x = x_0$ to $x = x_1$:

(2) $\displaystyle\int_{x_0}^{x_1} y'(x)dx = y(x_1) - y(x_0) = \int_{x_0}^{x_1} f\big(x, y(x)\big)dx$.

Substituting $y(x_0) = y_0$ and solving for $y(x_1)$ gives

(3) $\displaystyle y(x_1) = y_0 + \int_{x_0}^{x_1} f\big(x, y(x)\big)dx$.

If we use t instead of x as the variable of integration, we can let $x = x_1$ be the upper limit of integration. Equation (3) then becomes

(4) $\displaystyle y(x) = y_0 + \int_{x_0}^{x} f\big(t, y(t)\big)dt$.

Equation (4) can be used to generate successive approximations of a solution to (1). Let the function $\phi_0(x)$ be an initial guess or approximation of a solution to (1). Then a new approximation function is given by

$$\phi_1(x) := y_0 + \int_{x_0}^{x} f\big(t, \phi_0(t)\big)dt \ ,$$

where we have replaced $y(t)$ by the approximation $\phi_0(t)$ in the argument of f. In a similar fashion, we can use $\phi_1(x)$ to generate a new approximation $\phi_2(x)$, and so on. In general, we obtain the $(n + 1)$st approximation from the relation

(5) $\phi_{n+1}(x) := y_0 + \displaystyle\int_{x_0}^{x} f\left(t, \phi_n(t)\right)dt$.

This procedure is called **Picard's method.**[†] Under certain assumptions on f and $\phi_0(x)$, the sequence $\{\phi_n(x)\}$ is known to converge to a solution to (1). These assumptions and the proof of convergence are given in Chapter 13.[††]

Without further information about the solution to (1), it is common practice to take $\phi_0(x) \equiv y_0$.

(a) Use Picard's method with $\phi_0(x) \equiv 1$ to obtain the next four successive approximations of the solution to

(6) $y'(x) = y(x)$, $y(0) = 1$.

Show that these approximations are just the partial sums of the Maclaurin series for the actual solution e^x.

(b) Use Picard's method with $\phi_0(x) \equiv 0$ to obtain the next three successive approximations of the solution to the nonlinear problem

(7) $y'(x) = x - [y(x)]^2$, $y(0) = 0$.

Graph these approximations for $0 \le x \le 1$.

(c) In Problem 29 in Exercises 1.2, we showed that the initial value problem

(8) $y'(x) = 3[y(x)]^{2/3}$, $y(2) = 0$

does not have a unique solution. Show that Picard's method beginning with $\phi_0(x) \equiv 0$ converges to the solution $y(x) \equiv 0$, whereas Picard's method beginning with $\phi_0(x) = x - 2$ converges to the second solution $y(x) = (x - 2)^3$. [*Hint:* For the guess $\phi_0(x) = x - 2$, show that $\phi_n(x)$ has the form $c_n(x - 2)^{r_n}$, where $c_n \to 1$ and $r_n \to 3$ as $n \to \infty$.]

C. *Magnetic Dipole*

A bar magnet is often modeled as a magnetic dipole with one end labeled the north pole N and the opposite end labeled the south pole S. The magnetic field for the magnetic dipole is symmetric with respect to rotation about the axis passing lengthwise through the center of the bar. Hence we can study the magnetic field by restricting ourselves to a plane containing this axis of symmetry with the origin at the center of the bar (see Figure 1.29).

[†]*Historical Footnote:* This approximation method is a by-product of the famous Picard-Lindelöf existence theorem formulated at the end of the nineteenth century.

[††]All references to Chapters 11–13 refer to the expanded text *Fundamentals of Differential Equations and Boundary Value Problems*, 3rd edition.

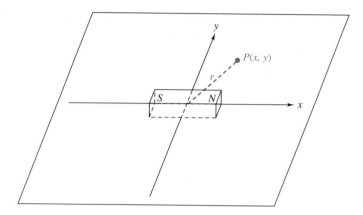

Figure 1.29 Magnetic dipole

For a point P that is located a distance r from the origin, where r is much greater than the length of the magnet, the **magnetic field lines** satisfy the differential equation

(9)
$$\frac{dy}{dx} = \frac{2xy}{3x^2 - y^2}$$

and the **equipotential lines** satisfy the equation

(10)
$$\frac{dy}{dx} = \frac{y^2 - 3x^2}{2xy}.$$

(a) Show that the two families of curves are perpendicular where they intersect. [*Hint:* Consider the slopes of the tangent lines of the two curves at a point of intersection.]

(b) Sketch the direction field for equation (9) for $-5 \le x \le 5$, $-5 \le y \le 5$. You can use a software package to generate the direction field or use the method of isoclines discussed in Section 1.3. The direction field should remind you of the experiment where iron filings are sprinkled on a sheet of paper that is held above a bar magnet. The iron filings correspond to the hash marks.

(c) Use the direction field found in part (b) to help sketch the magnetic field lines that are solutions to (9).

(d) Apply the statement of part (a) to the curves in part (c) to sketch the equipotential lines that are solutions to (10). The magnetic field lines and the equipotential lines are examples of *orthogonal trajectories*. (See Problem 32 in Exercises 2.4, page 71.)

First Order

Differential Equations

INTRODUCTION: MOTION OF A FALLING BODY

> **An object falls through the air toward Earth. Assuming that the only forces acting on the object are gravity and air resistance, determine the velocity of the object as a function of time.**

Newton's second law states that force is equal to mass times acceleration. We can express this by the equation

$$m\frac{dv}{dt} = F \ ,$$

where F represents the total force on the object, m is the mass of the object, and dv/dt is the acceleration, expressed as the derivative of velocity with respect to time.

Near Earth's surface, the force due to gravity is just the weight of the object. This force can be expressed by mg, where g is the acceleration due to gravity. There is no general law that precisely models the air resistance acting on the object, since this force seems to depend on the velocity of the object, the density of the air, and the shape of the object, among other things. However, in some instances air resistance can be reasonably represented by $-bv$, where b is a positive constant depending on the density of the air and the shape of the object. We use the negative sign because air resistance is a force that opposes the motion. The forces acting on the object are depicted in Figure 2.1. (Note that we have generalized the free fall model in Section 1.1 by including air resistance.)

Applying Newton's law, we obtain the first order differential equation

$$(1) \qquad m\frac{dv}{dt} = mg - bv \ .$$

Figure 2.1 Forces on falling object

To solve this equation, we exploit a technique called **separation of variables,** which was used to analyze the radioactive decay model in Section 1.1 and will be developed in full detail in Section 2.2. Treating dv and dt as differentials, we rewrite equation (1) so as to isolate the variables v and t on opposite sides of the equation:

$$\frac{dv}{mg - bv} = \frac{dt}{m} .$$

(Hence the nomenclature "separation of variables.")

Next we integrate the separated equation

$$\int \frac{dv}{mg - bv} = \int \frac{dt}{m}$$

and derive

$$-\frac{1}{b} \ln |mg - bv| = \frac{t}{m} + c .$$

Therefore

$$|mg - bv| = e^{-bc} e^{-bt/m}$$

or

$$mg - bv = Ae^{-bt/m} ,$$

where the new constant A has magnitude e^{-bc} and the same sign (\pm) as $(mg - bv)$. Solving for v, we obtain

(2) $$v = \frac{mg}{b} - \frac{A}{b} e^{-bt/m} ,$$

which is called a **general solution** to the differential equation because, as we shall see in Section 2.3, every solution to (1) can be expressed in the form given in (2).

In a specific case, we would be given the values of m, g, and b. To determine the constant A in the general solution, we can use the initial velocity of the object v_0. That is, we solve the **initial value problem**

$$m\frac{dv}{dt} = mg - bv \ , \qquad v(0) = v_0 \ .$$

Substituting $v = v_0$ and $t = 0$ into the general solution to the differential equation, we can solve for A. With this value for A, the solution to the initial value problem is

$$(3) \qquad v = \frac{mg}{b} + \left(v_0 - \frac{mg}{b}\right)e^{-bt/m} \ .$$

The preceding formula gives the velocity of the object falling through the air as a function of time if the initial velocity of the object is v_0. In Figure 2.2, we have sketched the graph of $v(t)$ for various values of v_0. It appears from Figure 2.2 that the velocity $v(t)$ approaches mg/b *regardless of the initial velocity v_0.* (This is easy to see from formula (3) by letting $t \to +\infty$.) The constant mg/b is referred to as the **limiting** or **terminal velocity** of the object.

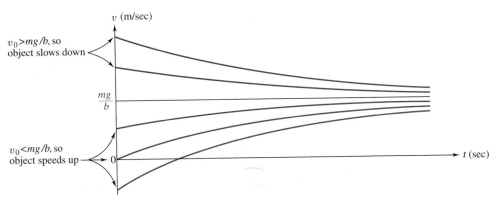

Figure 2.2 Graph of $v(t)$ for six different initial velocities v_0. ($g = 9.8$ m/sec^2, $m/b = 5$ sec)

From this model for a falling body, we can make certain observations. Because $e^{-bt/m}$ rapidly tends to zero, the velocity is approximately the weight, mg, divided by the coefficient of air resistance, b. Thus in the presence of air resistance, the heavier the object, the faster it will fall, assuming shapes and sizes are the same. Also, when air resistance is lessened (b is made smaller), the object will fall faster. These observations certainly agree with our experience.

Many other physical problems,[†] when formulated mathematically, lead to first order differential equations or initial value problems. Several of these are discussed in Chapter 3. In this chapter, we shall learn how to recognize and obtain solutions for some special types of first order

[†]The physical problem just discussed has other mathematical models. For example, one could take into account the variations in the gravitational field of Earth and the more general equations for air resistance.

equations. We begin by studying separable equations, then linear equations, and then exact equations. The methods for solving these are the most basic. In the last two sections, we illustrate how devices such as integrating factors, substitutions, and transformations can be used to transform certain equations into either separable, exact, or linear equations that we can solve. Through our discussion of these special types of equations, the reader will gain insight into the behavior of solutions to more general equations and the possible difficulties in finding these solutions.

2.2 SEPARABLE EQUATIONS

A simple class of first order differential equations that can be solved using integration is the class of **separable equations.** These are equations

(1) $$\frac{dy}{dx} = f(x, y) \ ,$$

which can be rewritten so as to isolate the variables x and y (together with their differentials dx and dy) on opposite sides of the equation, as in

$$h(y)dy = g(x)dx \ .$$

So the original right-hand side $f(x, y)$ must have the factored form

$$f(x, y) = g(x) \cdot \frac{1}{h(y)} \ .$$

More formally, we write $p(y) = 1/h(y)$ and present the following definition.

SEPARABLE EQUATION

Definition 1. If the right-hand side of the equation

$$\frac{dy}{dx} = f(x, y)$$

can be expressed as a function $g(x)$ that depends only on x times a function $p(y)$ that depends only on y, then the differential equation is called **separable.**[†]

In other words, a first order equation is separable if it can be written in the form

(2) $$\frac{dy}{dx} = g(x)p(y) \ .$$

[†]*Historical Footnote:* A procedure for solving separable equations was discovered implicitly by Gottfried Leibniz in 1691. The explicit technique called separation of variables was formalized by John Bernoulli in 1694.

For example, the equation

$$\frac{dy}{dx} = \frac{2x + xy}{y^2 + 1}$$

is separable, since (if one is sufficiently alert to detect the factorization)

$$\frac{2x + xy}{y^2 + 1} = x\frac{2 + y}{y^2 + 1} = g(x)p(y) \ .$$

However, the equation

$$\frac{dy}{dx} = 1 + xy$$

admits no such factorization of the right-hand side and so is not separable.

Informally speaking, one solves separable equations by performing the separation and then integrating each side.

METHOD FOR SOLVING SEPARABLE EQUATIONS

To solve the equation

$$\frac{dy}{dx} = g(x)p(y)$$

multiply by dx and by $h(y) := 1/p(y)$ to obtain

$$h(y)\,dy = g(x)\,dx \ .$$

Then integrate both sides:

$$\int h(y)\,dy = \int g(x)\,dx \ ,$$

(3) $H(y) = G(x) + C \ .$

The last equation gives an implicit solution to the differential equation.

We will look at the mathematical justification of this "streamlined" procedure shortly, but first we study some examples.

EXAMPLE 1 Solve

$$\frac{dy}{dx} = \frac{x - 5}{y^2} \ .$$

SOLUTION Following the streamlined approach, we separate the variables and rewrite the equation in the form

$$y^2 \, dy = (x - 5)dx \ .$$

Integrating, we have

$$\int y^2 \, dy = \int (x - 5)dx$$

$$\frac{y^3}{3} = \frac{x^2}{2} - 5x + C \ ,$$

and solving this last equation for y gives

$$y = \left(\frac{3x^2}{2} - 15x + 3C \right)^{1/3} .$$

Since C is a constant of integration that can be any real number, $3C$ can also be any real number. Replacing $3C$ by the single symbol K, we then have

$$y = \left(\frac{3x^2}{2} - 15x + K \right)^{1/3} .$$

If we wish to abide by the custom of letting C represent an arbitrary constant, we can go one step further and use C instead of K in the final answer. This solution family is graphed in Figure 2.3. ■ End

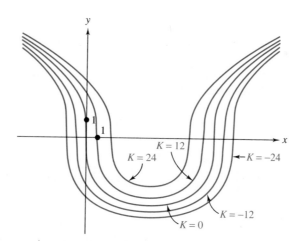

Figure 2.3 Family of solutions for Example 1

As Example 1 attests, separable equations are among the easiest to solve. However, the procedure does require a facility for computing integrals. Many of the procedures to be discussed in the text also require a familiarity with the techniques of integration. For the reader's convenience, a brief table of integrals can be found on the inside front cover.

EXAMPLE 2 Solve the initial value problem

(4) $\dfrac{dy}{dx} = \dfrac{y-1}{x+3}$, $y(-1) = 0$.

SOLUTION Separating the variables and integrating gives

$$\frac{dy}{y-1} = \frac{dx}{x+3} ,$$

$$\int \frac{dy}{y-1} = \int \frac{dx}{x+3} ,$$

(5) $\ln |y-1| = \ln |x+3| + C$.

At this point, we can either solve for y explicitly (retaining the constant C) or use the initial condition to determine C and then solve explicitly for y. Let's try the first approach.

Exponentiating equation (5), we have

$$e^{\ln |y-1|} = e^{\ln |x+3| + C} = e^{C} e^{\ln |x+3|} ,$$

(6) $|y-1| = e^{C} |x+3| = C_1 |x+3|$,

where $C_1 := e^{C}$.[†] Now, depending on the values of y, we have $|y-1| = \pm(y-1)$; and similarly, $|x+3| = \pm(x+3)$. Thus (6) can be written as

$$y - 1 = \pm C_1(x+3) \quad \text{or} \quad y = 1 \pm C_1(x+3) ,$$

where the choice of sign depends (as we said) on the values of x and y. Because C_1 is a *positive* constant (recall that $C_1 = e^{C} > 0$), we can replace $\pm C_1$ by K, where K now represents an *arbitrary* nonzero constant. We then obtain

(7) $y = 1 + K(x+3)$.

Finally, we determine K such that the initial condition $y(-1) = 0$ is satisfied. Putting $x = -1$ and $y = 0$ in equation (7) gives

$$0 = 1 + K(-1+3) = 1 + 2K ,$$

and so $K = -1/2$. Thus the solution to the initial value problem is

(8) $y = 1 - \dfrac{1}{2}(x+3) = -\dfrac{1}{2}(x+1)$.

[†]Recall that the symbol $:=$ means "is defined to be."

Alternative Approach. The second approach is to first set $x = -1$ and $y = 0$ in equation (5) and solve for C. In this case, we obtain

$$\ln|0 - 1| = \ln|-1 + 3| + C ,$$
$$0 = \ln 1 = \ln 2 + C ,$$

and so $C = -\ln 2$. Thus, from (5), the solution y is given implicitly by

$$\ln(1 - y) = \ln(x + 3) - \ln 2 .$$

Here we have replaced $|y - 1|$ by $1 - y$ and $|x + 3|$ by $x + 3$, since we are interested in x and y near the initial values $x = -1$, $y = 0$ (for such values, $y - 1 < 0$ and $x + 3 > 0$). Solving for y, we find

$$\ln(1 - y) = \ln(x + 3) - \ln 2 = \ln\left(\frac{x + 3}{2}\right) ,$$

$$1 - y = \frac{x + 3}{2} ,$$

$$y = 1 - \frac{1}{2}(x + 3) = -\frac{1}{2}(x + 1) ,$$

which agrees with the solution (8) found by the first method. ■ End

EXAMPLE 3 Solve

$$(9) \qquad \frac{dy}{dx} = \frac{6x^5 - 2x + 1}{\cos y + e^y} .$$

SOLUTION Separating variables and integrating, we find

$$(\cos y + e^y)dy = (6x^5 - 2x + 1)dx ,$$
$$\int (\cos y + e^y)dy = \int (6x^5 - 2x + 1)dx ,$$
$$\sin y + e^y = x^6 - x^2 + x + C .$$

At this point, we reach an impasse. We would like to solve for y explicitly, but we cannot. This is often the case in solving nonlinear first order equations. Consequently, when we say "solve the equation," we must on occasion be content if only an implicit form of the solution has been found. ■ End

The separation of variables technique, as well as several other techniques discussed in this book, entails rewriting a differential equation by performing certain algebraic operations on it. "Rewriting $dy/dx = g(x)p(y)$ as $h(y)dy = g(x)dx$" amounts to dividing both sides by $p(y)$. The reader may recall from his or her algebra days that doing this can be treacherous. For example, the equation $x(x - 2) = 4(x - 2)$ has two solutions: $x = 2$ and $x = 4$. But if we "rewrite" the equation as $x = 4$ by dividing both sides by $(x - 2)$, we lose track of the root $x = 2$. Thus we should record the zeros of $(x - 2)$ itself before dividing by this factor.

By the same token we must take note of the zeros of $p(y)$ in the separable equation $dy/dx = g(x)p(y)$ prior to dividing. After all, if (say) $g(x)p(y) = (x-2)^2(y-13)$, then observe that the constant function $y(x) \equiv 13$ solves the differential equation $dy/dx = g(x)p(y)$:

$$\frac{dy}{dx} = \frac{d(13)}{dx} = 0 \ ,$$

$$g(x)p(y) = (x-2)^2(13-13) = 0 \ .$$

Indeed, in solving the equation of Example 2,

$$\frac{dy}{dx} = \frac{y-1}{x+3} \ ,$$

we obtained $y = 1 + K(x+3)$ as the set of solutions, where K was a *nonzero* constant (since K replaced $\pm e^C$). But notice that the constant function $y \equiv 1$ (which in this case corresponds to $K = 0$) is also a solution to the differential equation. The reason we lost this solution can be traced back to a division by $y-1$ in the separation process. (See Problem 31 for an example of where a solution is lost and cannot be retrieved by setting the constant $K = 0$.)

Formal Justification of Method

We close this section by reviewing the separation of variables procedure in a more rigorous framework. The original differential equation (2) is rewritten in the form

(10) $\qquad h(y)\dfrac{dy}{dx} = g(x) \ ,$

where $h(y) := 1/p(y)$. Letting $H(y)$ and $G(x)$ denote antiderivatives (indefinite integrals) of $h(y)$ and $g(x)$, respectively—that is,

$$H'(y) = h(y) \ , \qquad G'(x) = g(x) \ ,$$

we recast equation (10) as

$$H'(y)\frac{dy}{dx} = G'(x) \ .$$

By the chain rule for differentiation, the left-hand side is the derivative of the composite function $H(y(x))$:

$$\frac{d}{dx}H(y(x)) = H'(y(x))\frac{dy}{dx} \ .$$

Thus if $y(x)$ is a solution to equation (2), then $H(y(x))$ and $G(x)$ are two functions of x that have the same derivative. Therefore they differ by a constant:

(11) $\qquad H(y(x)) = G(x) + C \ .$

Equation (11) agrees with equation (3), which was derived informally, and thus we have verified that the latter can be used to construct implicit solutions.

EXERCISES 2.2

In Problems 1 through 6, determine whether the given differential equation is separable.

1. $\dfrac{dy}{dx} = 2y^3 + y + 4$. **2.** $\dfrac{dy}{dx} = \sin(x + y)$.

3. $\dfrac{dy}{dx} = \dfrac{ye^{x+y}}{x^2 + 2}$. **4.** $\dfrac{ds}{dt} = t\ln(s^{2t}) + 8t^2$.

5. $s^2 + \dfrac{ds}{dt} = \dfrac{s + 1}{st}$.

6. $(xy^2 + 3y^2)\,dy - 2x\,dx = 0$.

In Problems 7 through 16, solve the equation.

7. $\dfrac{dy}{dx} = \dfrac{1 - x^2}{y^2}$. **8.** $\dfrac{dy}{dx} = \dfrac{1}{xy^3}$.

9. $\dfrac{dy}{dx} = y(2 + \sin x)$. **10.** $\dfrac{dx}{dt} = 3xt^2$.

11. $\dfrac{dy}{dx} = \dfrac{\sec^2 y}{1 + x^2}$. **12.** $x\dfrac{dv}{dx} = \dfrac{1 - 4v^2}{3v}$.

13. $\dfrac{dx}{dt} + x^2 = x$. **14.** $\dfrac{dy}{dx} = 3x^2(1 + y^2)$.

15. $y^{-1}\,dy + ye^{\cos x}\sin x\,dx = 0$.

16. $(x + xy^2)\,dx + e^{x^2}y\,dy = 0$.

In Problems 17 through 26, solve the initial value problem.

17. $y' = x^3(1 - y)$, $y(0) = 3$.

18. $\dfrac{dy}{dx} = (1 + y^2)\tan x$, $y(0) = \sqrt{3}$.

19. $\dfrac{dy}{d\theta} = y\sin\theta$, $y(\pi) = -3$.

20. $\dfrac{dy}{dx} = \dfrac{3x^2 + 4x + 2}{2y + 1}$, $y(0) = -1$.

21. $\dfrac{dy}{dx} = 2\sqrt{y + 1}\cos x$, $y(\pi) = 0$.

22. $x^2\,dx + 2y\,dy = 0$, $y(0) = 2$.

23. $\dfrac{dy}{dx} = 2x\cos^2 y$, $y(0) = \pi/4$.

24. $\dfrac{dy}{dx} = 8x^3 e^{-2y}$, $y(1) = 0$.

25. $\dfrac{dy}{dx} = x^2(1 + y)$, $y(0) = 3$.

26. $\sqrt{y}\,dx + (1 + x)\,dy = 0$, $y(0) = 1$.

27. Solutions Not Expressible in Terms of Elementary Functions. As discussed in calculus, certain indefinite integrals (antiderivatives) such as $\int e^{x^2}\,dx$ cannot be expressed in finite terms using elementary functions. When such an integral is encountered while solving a differential equation, it is often helpful to use definite integration (integrals with variable upper limit). For example, consider the initial value problem

$$\frac{dy}{dx} = e^{x^2}y^2 , \qquad y(2) = 1 .$$

The differential equation separates if we divide by y^2 and multiply by dx. We integrate the separated equation from $x = 2$ to $x = x_1$ and find

$$\int_{x=2}^{x=x_1} e^{x^2}\,dx = \int_{x=2}^{x=x_1} \frac{dy}{y^2}$$

$$= -\frac{1}{y}\bigg|_{x=2}^{x=x_1}$$

$$= -\frac{1}{y(x_1)} + \frac{1}{y(2)} .$$

If we let t be the variable of integration and replace x_1 by x and $y(2)$ by 1, then we can express the solution to the initial value problem by

$$y(x) = \left\{ 1 - \int_2^x e^{t^2}\,dt \right\}^{-1} .$$

Use definite integration to find an explicit solution to the initial value problems in (a)–(c).

(a) $dy/dx = e^{x^2}$, $y(0) = 0$.
(b) $dy/dx = e^{x^2}y^{-2}$, $y(0) = 1$.
(c) $dy/dx = \sqrt{1 + \sin x}\,(1 + y^2)$, $y(0) = 1$.
(d) Use a numerical integration algorithm (such as Simpson's rule, described in Appendix B) to approximate the solution to part (b) at $x = 0.5$ to three decimal places.

28. Sketch the solution to the initial value problem

$$\frac{dy}{dt} = 2y - 2yt , \qquad y(0) = 3$$

and determine its maximum value.

29. Uniqueness Questions. In Chapter 1, we indicated that in applications most *initial value problems* will have a unique solution. In fact, the existence of unique solutions was so important that we stated an existence and uniqueness theorem, Theorem 1, page 12. The method for separable equations can give us a solution, but it may not give us all the solutions (also see Problem 31). To illustrate this, consider the equation $dy/dx = y^{1/3}$.

(a) Use the method of separation of variables to show that

$$y = \left(\frac{2x}{3} + C\right)^{3/2}$$

is a solution.

(b) Show that the initial value problem $dy/dx = y^{1/3}$ with $y(0) = 0$ is satisfied for $C = 0$ by $y = (2x/3)^{3/2}$ for $x \geq 0$.

(c) Now show that the constant function $y \equiv 0$ also satisfies the initial value problem given in part (b). Hence this initial value problem does not have a unique solution.

(d) Finally, show that the conditions of Theorem 1 on page 12 are not satisfied.

(The solution $y \equiv 0$ was lost because of the division by zero in the separation process.)

30. Formal Solutions. The method of separation of variables usually produces an implicit solution to an equation, which in certain cases can be solved to obtain an explicit solution. This problem illustrates that the method actually gives expressions that "formally" satisfy the equation, but some care is required when choosing the arbitrary constant if one is, in fact, going to obtain an implicit solution.

(a) Solve the equation $x + y \, dy/dx = 0$ using separation of variables to get an expression of the form $x^2 + y^2 = C$.

(b) Show that when $C = -1$, there are no real values for x and y that satisfy $x^2 + y^2 = C$.

(c) Although C was an arbitrary constant of integration, show that it is only for $C > 0$ that we obtain an implicit relationship between x and y.

31. As stated in this section, the separation of equation (2) on page 45 requires division by $p(y)$, and this may disguise the fact that the roots of the equation $p(y) = 0$ are actually constant solutions to the differential equation.

(a) To explore this further, separate the equation

$$\frac{dy}{dx} = (x - 3)(y + 1)^{2/3}$$

to derive the solution,

$$y = -1 + (x^2/6 - x + C)^3 .$$

(b) Show that $y \equiv -1$ satisfies the original equation $dy/dx = (x - 3)(y + 1)^{2/3}$.

(c) Show that there is no choice of the constant C that will make the solution in part (a) yield the solution $y \equiv -1$. Thus we lost the solution $y \equiv -1$ when we divided by $(y + 1)^{2/3}$.

32. Interval of Definition. By looking at an initial value problem $dy/dx = f(x, y)$ with $y(x_0) = y_0$, it is not always possible to determine the domain of the solution $y(x)$ or the interval over which the function $y(x)$ satisfies the differential equation.

(a) Solve the equation $dy/dx = xy^3$.

(b) Give explicitly the solutions to the initial value problem with $y(0) = 1$; $y(0) = 1/2$; $y(0) = 2$.

(c) Determine the domains of the solutions in part (b).

(d) As found in part (c), the domains of the solutions depend on the initial conditions. For the initial value problem $dy/dx = xy^3$ with $y(0) = a$, $a > 0$, show that as a approaches zero from the right, the domain approaches the whole real line $(-\infty, \infty)$, and as a approaches $+\infty$, the domain shrinks to a single point.

(e) Sketch the solutions to the initial value problem $dy/dx = xy^3$ with $y(0) = a$ for $a = \pm1/2, \pm1$, and ±2.

33. Mixing. Suppose a brine containing 3 kg of salt per liter (L) runs into a tank initially filled with 400 L of water containing 20 kg of salt. If the brine enters at 10 L/min, the mixture is kept uniform by stirring, and the mixture flows out at the same rate, find the mass of salt in the tank after 10 min (see Figure 2.4). [*Hint:* Let A denote the number of kilograms of salt in the tank at t minutes after the process begins, and use the fact that

rate of increase in A = rate of input − rate of exit.

A further discussion of mixing problems is given in Section 3.2.]

10 L/min
3 kg/L

A(t)

400 L

A(0) = 20 kg

10 L/min

Figure 2.4 Schematic representation of a mixing problem

34. Newton's Law of Cooling. According to Newton's law of cooling, if an object at temperature T is immersed in a medium having the constant temperature M, then the rate of change of T is proportional to the difference of temperature M − T. This gives the differential equation

$$dT/dt = k(M - T) .$$

(a) Solve the differential equation for T.
(b) A thermometer reading 100° is placed in a medium having a constant temperature of 70°. After 6 min, the thermometer reads 80°. What is the reading after 20 min?

(Further applications of Newton's law of cooling appear in Section 3.3.)

35. Blood plasma is stored at 40°. Before the plasma can be used, it must be at 90°. When the plasma is placed in an oven at 120°, it takes 45 min for the plasma to warm to 90°. Assume Newton's law of cooling (Problem 34) applies. How long will it take for the plasma to warm to 90° if the oven temperature is set at **(a)** 100°, **(b)** 140°, and **(c)** 80°?

36. A pot of boiling water at 100°C is removed from a stove at time $t = 0$ and left to cool in the kitchen. After 5 min, the water temperature has decreased to 80°C, and another 5 min later it has dropped to 65°C. Assuming Newton's law of cooling (Problem 34) applies, determine the (constant) temperature of the kitchen.

37. Compound Interest. If $P(t)$ is the amount of dollars in a savings bank account that pays a yearly interest rate of r% *compounded continuously,* then

$$\frac{dP}{dt} = \frac{r}{100}P , \qquad t \text{ in years.}$$

Assume the interest is 5% annually, $P(0) = \$1000$, and no monies are withdrawn.

(a) How much will be in the account after 2 yr?
(b) When will the account reach $4000?
(c) If $1000 is added to the account every 12 months, how much will be in the account after $3\frac{1}{2}$ yr?

38. Free Fall. In Section 2.1, we discussed a model for an object falling toward Earth. Assuming only air resistance and gravity are acting on the object, we found that the velocity v must satisfy the equation

$$m\frac{dv}{dt} = mg - bv ,$$

where m is the mass, g is the acceleration due to gravity, and $b > 0$ is a constant (see Figure 2.1). If $m = 100$ kg, $g = 9.8$ m/sec², $b = 5$ kg/sec, and $v(0) = 10$ m/sec, solve for $v(t)$. What is the limiting (i.e., terminal) velocity of the object?

39. Grand Prix Race. Driver A had been leading archrival B for a while by a *steady* 3 miles. Only 2 miles from the finish, driver A ran out of gas and decelerated thereafter at a rate proportional to the square of his remaining speed. One mile later, driver A's speed was exactly halved. If driver B's speed remained constant, who won the race?

2.3 LINEAR EQUATIONS

A type of first order differential equation that occurs frequently in applications is the linear equation. Recall from Section 1.1 that a **linear first order equation** is an equation that can be expressed in the form

(1) $$a_1(x)\frac{dy}{dx} + a_0(x)y = b(x) \ ,$$

where $a_1(x)$, $a_0(x)$, and $b(x)$ depend only on the independent variable x, not on y.

For example, the equation

$$x^2\sin x - (\cos x)y = (\sin x)\frac{dy}{dx}$$

is linear, because it can be rewritten in the form

$$(\sin x)\frac{dy}{dx} + (\cos x)y = x^2\sin x \ .$$

However, the equation

$$y\frac{dy}{dx} + (\sin x)y^3 = e^x + 1$$

is not linear; it cannot be put in the form of equation (1) due to the presence of the y^3 and $y\,dy/dx$ terms.

There are two situations for which the solution of a linear differential equation is quite immediate. The first arises if the coefficient $a_0(x)$ is identically zero, for then equation (1) reduces to

(2) $$a_1(x)\frac{dy}{dx} = b(x) \ ,$$

which is equivalent to

$$y(x) = \int \frac{b(x)}{a_1(x)}dx + C$$

(as long as $a_1(x)$ is not zero).

The second is less trivial. Note that if $a_0(x)$ happens to equal the derivative of $a_1(x)$, that is, $a_0(x) = a_1'(x)$, then the two terms on the left-hand side of equation (1) simply comprise the derivative of the product $a_1(x)y$:

$$a_1(x)y' + a_0(x)y = a_1(x)y' + a_1'(x)y = \frac{d}{dx}\big[a_1(x)y\big] \ .$$

Therefore equation (1) becomes

(3) $$\frac{d}{dx}\big[a_1(x)y\big] = b(x)$$

and the solution is again elementary:

$$a_1(x)y = \int b(x)\,dx + C\ ,$$

$$y(x) = \frac{1}{a_1(x)}\left\{\int b(x)\,dx + C\right\}\ .$$

One can seldom rewrite a linear differential equation so that it reduces to a form as simple as (2). However, the form (3) can be achieved through multiplication of the original equation (1) by a well-chosen function $\mu(x)$. Such a function $\mu(x)$ is then called an "integrating factor" for equation (1). The easiest way to see this is first to divide the original equation (1) by $a_1(x)$ and put it into **standard form**

(4) $$\frac{dy}{dx} + P(x)y = Q(x)\ .$$

Next we wish to determine $\mu(x)$ so that the left-hand side of the multiplied equation

(5) $$\mu(x)\frac{dy}{dx} + \mu(x)P(x)y = \mu(x)Q(x)$$

is just the derivative of the product $\mu(x)y$:

$$\mu(x)\frac{dy}{dx} + \mu(x)P(x)y = \frac{d}{dx}\left[\mu(x)y\right] = \mu(x)\frac{dy}{dx} + \mu'(x)y\ .$$

Clearly this requires that μ satisfy

(6) $$\mu' = \mu P\ .$$

To find such a function, we recognize that equation (6) is a separable differential equation which we can write as $(1/\mu)\,d\mu = P(x)\,dx$. Integrating both sides gives

(7) $$\mu(x) = e^{\int P(x)\,dx}\ .$$

With this choice[†] for $\mu(x)$, equation (5) becomes

$$\frac{d}{dx}\left[\mu(x)y\right] = \mu(x)Q(x)\ ,$$

which has the solution

(8) $$y(x) = \frac{1}{\mu(x)}\left\{\int \mu(x)Q(x)\,dx + C\right\}\ .$$

[†]Any choice of the integration constant in $\int P(x)\,dx$ will produce a suitable $\mu(x)$.

Here C is an arbitrary constant, so (8) gives a one-parameter family of solutions to (4). This form is known as the **general solution** to (4).

We can summarize the method for solving linear equations as follows.

METHOD FOR SOLVING LINEAR EQUATIONS

(a) Write the equation in the standard form

$$\frac{dy}{dx} + P(x)y = Q(x) .$$

(b) Calculate the integrating factor $\mu(x)$ by the formula

$$\mu(x) = \exp\left(\int P(x)dx\right) .$$

(c) Multiply the equation in standard form by $\mu(x)$ and, recalling that the left-hand side is just $\frac{d}{dx}(\mu(x)y)$, obtain

$$\underbrace{\mu(x)\frac{dy}{dx} + P(x)\mu(x)y}_{\frac{d}{dx}(\mu(x)y)} = \mu(x)Q(x) ,$$

$$\frac{d}{dx}(\mu(x)y) = \mu(x)Q(x) .$$

(d) Integrate the last equation and solve for y by dividing by $\mu(x)$ to obtain (8).

EXAMPLE 1 Find the general solution to

(9) $$\frac{1}{x}\frac{dy}{dx} - \frac{2y}{x^2} = x\cos x , \qquad x > 0 .$$

SOLUTION To put this linear equation in standard form, we multiply by x to obtain

(10) $$\frac{dy}{dx} - \frac{2}{x}y = x^2\cos x .$$

Here $P(x) = -2/x$, so

$$\int P(x)dx = \int \frac{-2}{x}dx = -2\ln|x| .$$

Thus an integrating factor is

$$\mu(x) = e^{-2\ln|x|} = e^{\ln(x^{-2})} = x^{-2} .$$

Multiplying equation (10) by $\mu(x)$ yields

$$\underbrace{x^{-2}\frac{dy}{dx} - 2x^{-3}y}_{} = \cos x ,$$

$$\frac{d}{dx}(x^{-2}y) = \cos x .$$

We now integrate both sides and solve for y to find

$$x^{-2}y = \int \cos x \, dx = \sin x + C .$$

(11) $y = x^2 \sin x + Cx^2 .$

It is easily checked that this solution is valid for all $x > 0$. In Figure 2.5 we have sketched solutions for various values of the constant C in (11). ■End

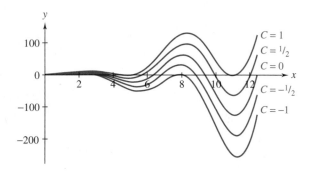

Figure 2.5 Graph of $y = x^2 \sin x + Cx^2$ for five values of the constant C

In the next example, we encounter a linear equation that arises in the study of the radioactive decay of an isotope.

EXAMPLE 2 A rock contains two radioactive isotopes, RA_1 and RA_2, that belong to the same radioactive series; that is, RA_1 decays into RA_2, which then decays into stable atoms. Assume that the rate at which RA_1 decays into RA_2 is $50e^{-10t}$ kg/sec. Because the rate of decay of RA_2 is proportional to the mass $y(t)$ of RA_2 present, the rate of change in RA_2 is

$$\frac{dy}{dt} = \text{rate of creation} - \text{rate of decay} ,$$

(12) $\dfrac{dy}{dt} = 50e^{-10t} - ky ,$

where $k > 0$ is the decay constant. If $k = 2/\text{sec}$ and initially $y(0) = 40$ kg, find the mass $y(t)$ of RA_2 for $t \geq 0$.

SOLUTION Equation (12) is linear, so we begin by writing it in standard form

(13) $\dfrac{dy}{dt} + 2y = 50e^{-10t} ,$ $y(0) = 40 ,$

where we have substituted $k = 2$ and displayed the initial condition. We now see that $P(t) = 2,$

so $\int P(t)dt = \int 2\, dt = 2t$. Thus an integrating factor is $\mu(t) = e^{2t}$. Multiplying equation (13) by $\mu(t)$ yields

$$\underbrace{e^{2t}\frac{dy}{dt} + 2e^{2t}y}_{} = 50e^{-10t+2t} = 50e^{-8t} ,$$

$$\frac{d}{dt}(e^{2t}y) \quad = 50e^{-8t} .$$

Integrating both sides and solving for y, we find

$$e^{2t}y = -\frac{25}{4}e^{-8t} + C ,$$

$$y = -\frac{25}{4}e^{-10t} + Ce^{-2t} . ,$$

Substituting $t = 0$ and $y(0) = 40$ gives

$$40 = -\frac{25}{4}e^0 + Ce^0 = -\frac{25}{4} + C ,$$

so $C = 40 + 25/4 = 185/4$. Thus the mass $y(t)$ of RA_2 at time t is given by

(14) $$y(t) = \left(\frac{185}{4}\right)e^{-2t} - \left(\frac{25}{4}\right)e^{-10t} , \qquad t \ge 0 . \quad \blacksquare \text{End}$$

EXAMPLE 3 For the initial value problem

$$y' + y = \sqrt{1 + \cos^2 x}, \qquad y(1) = 4 ,$$

find the value of $y(2)$.

SOLUTION The integrating factor for the differential equation is, from equation (7),

$$\mu(x) = e^{\int 1\, dx} = e^x .$$

The general solution form (8) thus reads

$$y(x) = e^{-x}\left[\int e^x \sqrt{1 + \cos^2 x}\, dx + C \right] .$$

However, this indefinite integral cannot be expressed in finite terms with elementary functions (recall a similar situation in Problem 27 of Exercises 2.2). Because we *can* use numerical algorithms such as Simpson's rule (Appendix B) to perform *definite* integration, we revert to the form (5), which in this case reads

$$\frac{d}{dx}\left[e^x y\right] = e^x \sqrt{1 + \cos^2 x} ,$$

and take the definite integral from the initial value $x = 1$ to the desired value $x = 2$:

$$e^x y \Big|_{x=1}^{x=2} = e^2 y(2) - e^1 y(1) = \int_{x=1}^{x=2} e^x \sqrt{1 + \cos^2 x}\, dx .$$

Inserting the given value of $y(1)$ and solving, we express

$$y(2) = e^{-2+1}(4) + e^{-2} \int_1^2 e^x \sqrt{1 + \cos^2 x}\; dx \;.$$

Using Simpson's rule, we find that the definite integral is approximately 4.841, so

$$y(2) \approx 4e^{-1} + 4.841e^{-2} \approx 2.127 \;. \quad \blacksquare \text{End}$$

In Example 3, we had no difficulty expressing the integral for the integrating factor $\mu(x) = e^{\int 1\, dx} = e^x$. Clearly, situations will arise where this integral, too, cannot be expressed with elementary functions. In such cases, we must again resort to a numerical procedure such as Euler's method (Section 1.5) or to a "nested loop" implementation of Simpson's rule. The reader is invited to explore such a possibility in Problem 27.

Because we have established explicit formulas for the solutions to *linear* first order differential equations, we get as a dividend a direct proof of the following theorem.

EXISTENCE AND UNIQUENESS OF SOLUTION

Theorem 1. Suppose $P(x)$ and $Q(x)$ are continuous on an interval (a, b) that contains the point x_0. Then for any choice of initial value y_0, there exists a unique solution $y(x)$ on (a, b) to the initial value problem

$$(15) \qquad \frac{dy}{dx} + P(x)y = Q(x) \;, \qquad y(x_0) = y_0 \;.$$

In fact, the solution is given by (8) for a suitable value of C.

The essentials of the proof of Theorem 1 are contained in the deliberations leading to equation (8); Problem 34 provides the details. This theorem differs from Theorem 1 on page 12 in that for the *linear* initial value problem (15), we have the existence and uniqueness of the solution on the *whole* interval (a, b), rather than on some smaller unspecified interval about x_0.

The theory of linear differential equations is an important branch of mathematics not only because these equations occur in applications, but also because of the elegant structure associated with them. For example, first order linear equations always have a general solution given by equation (8). Some further properties of first order linear equations are described in Problems 28 and 36. Higher-order linear equations are treated in Chapters 4, 6, and 8.

EXERCISES 2.3

In Problems 1 through 6, determine whether the given equation is separable, linear, neither, or both.

1. $x^2\dfrac{dy}{dx} + \cos x = y$.

2. $\dfrac{dx}{dt} + xt = e^x$.

3. $x\dfrac{dx}{dt} + t^2 x = \sin t$.

4. $3t = e^t\dfrac{dy}{dt} + y \ln t$.

5. $(t^2 + 1)\dfrac{dy}{dt} = yt - y$.

6. $3r = \dfrac{dr}{d\theta} - \theta^3$.

In Problems 7 through 16, obtain the general solution to the equation.

7. $\dfrac{dy}{dx} - y = e^{3x}$.

8. $\dfrac{dy}{dx} = \dfrac{y}{x} + 2x + 1$.

9. $\dfrac{dr}{d\theta} + r \tan \theta = \sec \theta$.

10. $x\dfrac{dy}{dx} + 2y = x^{-3}$.

11. $(t + y + 1)dt - dy = 0$.

12. $\dfrac{dy}{dx} = x^2 e^{-4x} - 4y$.

13. $y\dfrac{dx}{dy} + 2x = 5y^3$.

14. $x\dfrac{dy}{dx} + 3y + 2x^2 = x^3 + 4x$.

15. $(x^2 + 1)\dfrac{dy}{dx} + xy = x$.

16. $(x^2 + 1)\dfrac{dy}{dx} = x^2 + 2x - 1 - 4xy$.

In Problems 17 through 22, solve the initial value problem.

17. $\dfrac{dy}{dx} - \dfrac{y}{x} = xe^x$, $y(1) = e - 1$.

18. $\dfrac{dy}{dx} + 4y - e^{-x} = 0$, $y(0) = \dfrac{4}{3}$.

19. $t^3\dfrac{dx}{dt} + 3t^2x = t$, $x(2) = 0$.

20. $\dfrac{dy}{dx} + \dfrac{3y}{x} + 2 = 3x$, $y(1) = 1$.

21. $\cos x \dfrac{dy}{dx} + y \sin x = 2x \cos^2 x$,

$y\left(\dfrac{\pi}{4}\right) = \dfrac{-15\sqrt{2}\pi^2}{32}$.

22. $\sin x \dfrac{dy}{dx} + y \cos x = x \sin x$, $y\left(\dfrac{\pi}{2}\right) = 2$.

23. **Radioactive Decay.** In Example 2, assume the rate at which RA_1 decays into RA_2 is $40e^{-20t}$ kg/sec and the decay constant for RA_2 is $k = 5$/sec. Find the mass $y(t)$ of RA_2 for $t \geq 0$ if initially $y(0) = 10$ kg.

24. In Example 2, the decay constant for isotope RA_1 was 10/sec, which expresses itself in the exponent of the rate term $50e^{-10t}$ kg/sec. When the decay constant for RA_2 is $k = 2$/sec, we see that in formula

(14) for y the term $(185/4)e^{-2t}$ eventually dominates (has greater magnitude for t large).

(a) Redo Example 2 taking $k = 20$/sec. Now which term in the solution eventually dominates?

(b) Redo Example 2 taking $k = 10$/sec.

25. (a) Using definite integration, show that the solution to the initial value problem

$$\dfrac{dy}{dx} + 2xy = 1 , \qquad y(2) = 1 ,$$

can be expressed as

$$y(x) = e^{-x^2}\left\{e^4 + \int_2^x e^{t^2}\, dt\right\} .$$

(b) Use numerical integration (such as Simpson's rule, Appendix B) to approximate the solution at $x = 3$.

26. Use numerical integration (such as Simpson's rule, Appendix B) to approximate the solution, at $x = 1$, to the initial value problem

$$\dfrac{dy}{dx} + \dfrac{\sin 2x}{2(1 + \sin^2 x)}y = 1 , \qquad y(0) = 0 .$$

Ensure your appoximation is accurate to three decimal places.

27. Consider the initial value problem

$$\dfrac{dy}{dx} + \sqrt{1 + \sin^2 x}\, y = x , \qquad y(0) = 2 .$$

(a) Using definite integration, show that the integrating factor for the differential equation can be written

$$\mu(x) = \exp\left(\int_0^x \sqrt{1 + \sin^2 t}\, dt\right)$$

and that the solution to the initial value problem is

$$y(x) = \dfrac{1}{\mu(x)}\int_0^x \mu(s)\, s\, ds + \dfrac{2}{\mu(x)} .$$

(b) Obtain an approximation to the solution at $x = 1$ by using numerical integration (such as Simpson's rule, Appendix B) in a nested loop to estimate values of $\mu(x)$ and, thereby, the value of

$$\int_0^1 \mu(s)\, s\, ds .$$

[*Hint:* First, use Simpson's rule to approximate $\mu(x)$ at $x = 0.1, 0.2, \ldots , 1$. Then use these values and apply Simpson's rule again to approximate $\int_0^1 \mu(s)s \, ds$.]

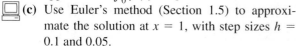 (c) Use Euler's method (Section 1.5) to approximate the solution at $x = 1$, with step sizes $h = 0.1$ and 0.05.

(A direct comparison of the merits of the two numerical schemes in (b) and (c) is very complicated, since it should take into account the number of functional evaluations in each algorithm as well as the inherent accuracies.)

28. Constant Multiples of Solutions.

(a) Show that $y = e^{-x}$ is a solution of the linear equation

$$(16) \qquad \frac{dy}{dx} + y = 0 \ ,$$

and $y = x^{-1}$ is a solution of the nonlinear equation

$$(17) \qquad \frac{dy}{dx} + y^2 = 0 \ .$$

(b) Show that for any constant C, Ce^{-x} is a solution of equation (16), while Cx^{-1} is a solution of equation (17) only when $C = 0$ or 1.

(c) Show that for any linear equation of the form

$$\frac{dy}{dx} + P(x)y = 0 \ ,$$

if $\hat{y}(x)$ is a solution, then for any constant C, the function $C\hat{y}(x)$ is also a solution.

29. Use your ingenuity to solve the equation

$$\frac{dy}{dx} = \frac{1}{e^{4y} + 2x} \ .$$

30. Bernoulli Equations. The equation

$$(18) \qquad \frac{dy}{dx} + 2y = xy^{-2}$$

is an example of a Bernoulli equation. (Further discussion of Bernoulli equations is in Section 2.6.)

(a) Show that the substitution $v = y^3$ reduces equation (18) to the equation

$$(19) \qquad \frac{dv}{dx} + 6v = 3x \ .$$

(b) Solve equation (19) for v. Then make the substitution $v = y^3$ to obtain the solution to equation (18).

31. Discontinuous Coefficients. As we will see in Chapter 3, occasions arise when the coefficient $P(x)$ in a linear equation fails to be continuous because of jump discontinuities. Fortunately, we may still obtain a "reasonable" solution. For example, consider the initial value problem

$$\frac{dy}{dx} + P(x)y = x \ , \qquad y(0) = 1 \ ,$$

where

$$P(x) := \begin{cases} 1 \ , & 0 \le x \le 2 \ , \\ 3 \ , & x > 2 \ . \end{cases}$$

(a) Find the general solution for $0 \le x \le 2$.

(b) Choose the constant in the solution of part (a) so that the initial condition is satisfied.

(c) Find the general solution for $x > 2$.

(d) Now choose the constant in the general solution from part (c) so that the solution from part (b) and the solution from part (c) agree at $x = 2$. By patching the two solutions together, we are able to obtain a continuous function that satisfies the differential equation except at $x = 2$, where its derivative is undefined.

(e) Sketch the graph of the solution from $x = 0$ to $x = 5$.

32. Discontinuous Forcing Terms. There are occasions when the forcing term $Q(x)$ in a linear equation fails to be continuous because of jump discontinuities. Fortunately, we may still obtain a reasonable solution imitating the procedure discussed in Problem 31. Use this procedure to find the continuous solution to the initial value problem.

$$\frac{dy}{dx} + 2y = Q(x) \ , \qquad y(0) = 0 \ ,$$

where

$$Q(x) := \begin{cases} 2 \ , & 0 \le x \le 3 \ , \\ -2 \ , & x > 3 \ . \end{cases}$$

Sketch the graph of the solution from $x = 0$ to $x = 7$.

33. **Singular Points.** Those values of x for which $P(x)$ in equation (4) is not defined are called **singular points** of the equation. For example, $x = 0$ is a singular point of the equation $xy' + 2y = 3x$, since when the equation is written in the standard form, $y' + (2/x)y = 3$, we see that $P(x) = 2/x$ is not defined at $x = 0$. On an interval containing a singular point, the questions of the existence and uniqueness of a solution are left unanswered, since Theorem 1 does not apply. To show the possible behavior of solutions near a singular point, consider the following equations.

(a) Show that $xy' + 2y = 3x$ has only one solution defined at $x = 0$. Then show that the initial value problem for this equation with initial condition $y(0) = y_0$ has a unique solution when $y_0 = 0$ and no solution when $y_0 \neq 0$.

(b) Show that $xy' - 2y = 3x$ has an infinite number of solutions defined at $x = 0$. Then show that the initial value problem for this equation with initial condition $y(0) = 0$ has an infinite number of solutions.

34. **Existence and Uniqueness.** Under the assumptions of Theorem 1, we will prove that equation (8) gives a solution to equation (4) on (a, b). We can then choose the constant C in equation (8) so that the initial value problem (15) is solved.

(a) Show that since $P(x)$ is continuous on (a, b), then $\mu(x)$ defined in (7) is a positive, continuous function satisfying $d\mu/dx = P(x)\mu(x)$ on (a, b).

(b) Since

$$\frac{d}{dx} \int \mu(x)Q(x)\, dx = \mu(x)Q(x)\,,$$

verify that y given in equation (8) satisfies equation (4) by differentiating both sides of equation (8).

(c) Show that when we let $\int \mu(x)Q(x)\, dx$ be the antiderivative whose value at x_0 is 0 (i.e., $\int_{x_0}^{x} \mu(t)Q(t)\, dt$) and choose C to be $y_0\mu(x_0)$, the initial condition $y(x_0) = y_0$ is satisfied.

(d) Start with the assumption that $y(x)$ is a solution to the initial value problem (15) and argue that the discussion leading to equation (8) implies that $y(x)$ must obey equation (8). Then argue that the initial condition in (15) determines the constant C uniquely.

35. **Mixing.** Suppose a brine containing 2 kg of salt per liter runs into a tank initially filled with 500 L of water containing 50 kg of salt. The brine enters the tank at a rate of 5 L/min. The mixture, kept uniform by stirring, is flowing out at the rate of 5 L/min (see Figure 2.6).

5 L/min
2 kg/L

$A(t)$

500 L

$A(0) = 50$ kg

5 L/min

Figure 2.6 Mixing problem with equal flow rates

(a) Find the concentration, in kilograms per liter, of salt in the tank after 10 min. [*Hint:* Let A denote the number of kilograms of salt in the tank at t minutes after the process begins and use the fact that

rate of increase in A = rate of input − rate of exit.

A further discussion of mixing problems is given in Section 3.2.]

(b) After 10 min, a leak develops in the tank and an additional liter per minute of mixture flows out of the tank (see Figure 2.7). What will be the concentration, in kilograms per liter, of salt in the tank 20 min after the leak develops? [*Hint:* Use the method discussed in Problems 31 and 32.]

5 L/min
2 kg/L

$A(t)$

? L

$A(10) = ?$ kg

5 L/min

1 L/min

Figure 2.7 Mixing problem with unequal flow rates

36. **Variation of Parameters.** Here is another procedure for solving linear equations that is particularly useful for higher-order linear equations. This method

is called **variation of parameters.** It is based on the idea that just by knowing the *form* of the solution, we can substitute into the given equation and solve for any unknowns. Here we illustrate the method for first order equations (see Sections 4.9 and 6.4 for the generalization to higher-order equations).

(a) Show that the general solution to

(20) $\quad \dfrac{dy}{dx} + P(x)y = Q(x)$

has the form

$\quad y(x) = Cy_h(x) + y_p(x)$,

where y_h is a solution to equation (20) when $Q(x) \equiv 0$, C is a constant, and $y_p(x) = v(x)y_h(x)$ for a suitable function $v(x)$. [*Hint:* Show that we can take $y_h = \mu^{-1}(x)$ and then use equation (8).]

Knowing this form, we can determine the unknown function y_h by solving a separable equation. Then direct substitution into the original equation will give a simple equation that can be solved for v.

Use this procedure to find the general solution to

(21) $\quad \dfrac{dy}{dx} + \dfrac{3}{x}y = x^2$, $\quad x > 0$,

by completing the following steps:

(b) Find a nontrivial solution y_h to the separable equation

(22) $\quad \dfrac{dy}{dx} + \dfrac{3}{x}y = 0$, $\quad x > 0$.

(c) Assuming (21) has a solution of the form $y_p(x) = v(x)y_h(x)$, substitute this into equation (21), and simplify to obtain $v'(x) = x^2/y_h(x)$.

(d) Now integrate to get $v(x)$.

(e) Verify that $y(x) = Cy_h(x) + v(x)y_h(x)$ is a general solution to (21).

37. Secretion of Hormones. The secretion of hormones into the blood is often a periodic activity. If a hormone is secreted on a 24-hr cycle, then the rate of change of the level of the hormone in the blood may be represented by the initial value problem

$$\frac{dx}{dt} = \alpha - \beta \cos \frac{\pi t}{12} - kx \ , \qquad x(0) = x_0 \ ,$$

where $x(t)$ is the amount of the hormone in the blood at time t, α is the average secretion rate, β is the amount of daily variation in the secretion, and k is a positive constant reflecting the rate at which the body removes the hormone from the blood. If $\alpha = \beta = 1$, $k = 2$, and $x_0 = 10$, solve for $x(t)$.

38. Use the separation of variables technique to derive the solution (7) to the differential equation (6).

39. The temperature T (in units of 100°F) of a university classroom on a cold winter day varies with time t (in hours) as

$$\frac{dT}{dt} = \begin{cases} 1 - T \ , & \text{if heating unit is ON.} \\ -T \ , & \text{if heating unit is OFF.} \end{cases}$$

Suppose that $T = 0$ at 9:00 A.M., the heating unit is ON from 9–10 A.M., OFF from 10–11 A.M., ON again from 11 A.M.–noon, and so on for the rest of the day. How warm will the classroom be at noon? At 5:00 P.M.?

2.4 EXACT EQUATIONS

Suppose the mathematical function $F(x, y)$ represents some physical quantity, such as temperature, in a region of the xy-plane. Then the level curves of F, where $F(x, y) = constant$, could be interpreted as isotherms on a weather map, as depicted in Figure 2.8 on page 64.

How does one calculate the slope of the tangent to a level curve? It's easy. Because $F(x, y) = constant$ on the level curve, the total differential of F along the curve is zero:

(1) $\quad dF(x, y) := \dfrac{\partial F}{\partial x}dx + \dfrac{\partial F}{\partial y}dy = 0$.

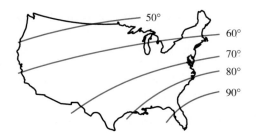

Figure 2.8 Level curves of $F(x, y)$

Thus the equation for its slope $f(x, y)$ at the point (x, y) is given by

(2) $$\frac{dy}{dx} = f(x, y) = -\frac{\partial F/\partial x}{\partial F/\partial y} .$$

As equation (2) has the form of a differential equation, we should be able to reverse this logic and come up with a very easy technique for solving some differential equations. After all, any first order differential equation $dy/dx = f(x, y)$ can be rewritten in the (differential) form

(3) $$M(x, y)\,dx + N(x, y)\,dy = 0$$

(in a variety of ways). Now, if the left-hand side of equation (3) can be identified as a total differential:

$$M(x, y)\,dx + N(x, y)\,dy = \frac{\partial F}{\partial x}\,dx + \frac{\partial F}{\partial y}\,dy = dF(x, y) ,$$

then its solutions are given (implicitly) by the level curves

$$F(x, y) = C$$

for an arbitrary constant C.

EXAMPLE 1 Solve the differential equation

$$\frac{dy}{dx} = -\frac{2xy^2 + 1}{2x^2 y} .$$

SOLUTION Some of the choices of differential forms corresponding to this equation are

$$(2xy^2 + 1)\,dx + 2x^2 y\,dy = 0 ,$$

$$\frac{2xy^2 + 1}{2x^2 y}\,dx + dy = 0 ,$$

$$dx + \frac{2x^2 y}{2xy^2 + 1}\,dy = 0 , \quad \text{etc.}$$

However, the first form is best for our purposes because it is a total differential of the function $F(x, y) = x^2y^2 + x$:

$$(2xy^2 + 1)dx + 2x^2y\,dy = d[x^2y^2 + x]$$

$$= \frac{\partial}{\partial x}(x^2y^2 + x)dx + \frac{\partial}{\partial y}(x^2y^2 + x)dy .$$

Thus the solutions are given implicitly by the formula $x^2y^2 + x = C$. See Figure 2.9. ∎End

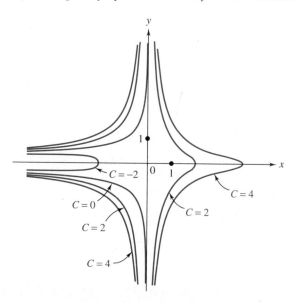

Figure 2.9 Solutions of Example 1

Next we introduce some terminology.

EXACT DIFFERENTIAL FORM

Definition 2. The differential form $M(x, y)dx + N(x, y)dy$ is said to be **exact** in a rectangle R if there is a function $F(x, y)$ such that

(4) $\dfrac{\partial F}{\partial x}(x, y) = M(x, y)$ and $\dfrac{\partial F}{\partial y}(x, y) = N(x, y)$

for all (x, y) in R. That is, the total differential of $F(x, y)$ satisfies

$$dF(x, y) = M(x, y)dx + N(x, y)dy .$$

If $M(x, y)dx + N(x, y)dy$ is an exact differential form, then the equation

$$M(x, y)dx + N(x, y)dy = 0$$

is called an **exact equation**.

As you might suspect, in applications a differential equation is rarely given to us in exact differential form. However, the solution procedure is so quick and simple for such equations that we devote this section to it. From Example 1, we see that what is needed is (i) a test to determine if a differential form $M(x, y)dx + N(x, y)dy$ is exact and, if so, (ii) a procedure for finding the function $F(x, y)$ itself.

The test for exactness arises from the following observation. If

$$M(x, y)dx + N(x, y)dy = \frac{\partial F}{\partial x}dx + \frac{\partial F}{\partial y}dy \ ,$$

then the calculus theorem concerning the equality of continuous mixed partial derivatives

$$\frac{\partial}{\partial y}\frac{\partial F}{\partial x} = \frac{\partial}{\partial x}\frac{\partial F}{\partial y}$$

would dictate a "compatibility condition" on the functions M and N:

$$\frac{\partial}{\partial y}M(x, y) = \frac{\partial}{\partial x}N(x, y) \ .$$

In fact, the following Theorem 2 states that the compatibility condition is also *sufficient* for the differential form to be exact.

TEST FOR EXACTNESS

Theorem 2. Suppose the first partial derivatives of $M(x, y)$ and $N(x, y)$ are continuous in a rectangle R. Then

$$M(x, y)dx + N(x, y)dy = 0$$

is an exact equation in R if and only if the compatibility condition

(5) $$\frac{\partial M}{\partial y}(x, y) = \frac{\partial N}{\partial x}(x, y)$$

holds for all (x, y) in R.[†]

Before we address the proof of Theorem 2, note that in Example 1 the differential form that led to the total differential was

$$(2xy^2 + 1)dx + (2x^2y)dy = 0 \ .$$

[†]*Historical Footnote:* This theorem was proven by Leonhard Euler in 1734.

The compatibility conditions are easily confirmed:

$$\frac{\partial M}{\partial y} = \frac{\partial}{\partial y}(2xy^2 + 1) = 4xy \ ,$$

$$\frac{\partial N}{\partial x} = \frac{\partial}{\partial x}(2x^2y) = 4xy \ .$$

Also clear is the fact that the other differential forms considered,

$$\frac{2xy^2 + 1}{2x^2y}dx + dy = 0 \ , \qquad dx + \frac{2x^2y}{2xy^2 + 1}dy = 0 \ ,$$

do *not* meet the compatibility conditions.

Proof of Theorem 2. There are two parts to the theorem: Exactness implies compatibility, and compatibility implies exactness. First, we have seen that if the differential equation is exact, then the two members of equation (5) are simply the mixed second partials of a function $F(x, y)$. As such, their equality is assured by the theorem of calculus that states that mixed second partials are equal if they are continuous. Because the hypothesis of Theorem 2 guarantees the latter condition, equation (5) is validated.

Rather than proceed directly with the proof of the second part of the theorem, let's derive a formula for a function $F(x, y)$ that satisfies $\partial F/\partial x = M$ and $\partial F/\partial y = N$. Integrating the first equation with respect to x yields

$$(6) \qquad F(x, y) = \int M(x, y)dx + g(y) \ .$$

Notice that instead of using C to represent the constant of integration, we have written $g(y)$. This is because y is held fixed while integrating with respect to x, and so our "constant" may well depend on y. To determine $g(y)$, we differentiate both sides of (6) with respect to y to obtain

$$(7) \qquad \frac{\partial F}{\partial y}(x, y) = \frac{\partial}{\partial y}\int M(x, y)dx + \frac{\partial}{\partial y}g(y) \ .$$

As g is a function of y alone, we can write $\partial g/\partial y = g'(y)$, and solving (7) for $g'(y)$ gives

$$g'(y) = \frac{\partial F}{\partial y}(x, y) - \frac{\partial}{\partial y}\int M(x, y)dx \ .$$

Since $\partial F/\partial y = N$, this last equation becomes

$$(8) \qquad g'(y) = N(x, y) - \frac{\partial}{\partial y}\int M(x, y)dx \ .$$

Notice that although the right-hand side of (8) indicates a possible dependence on x, the appearances of this variable must cancel because the left-hand side, $g'(y)$, depends only on y. By integrating (8), we can determine $g(y)$ up to a numerical constant, and therefore we can determine the function $F(x, y)$ up to a numerical constant from the functions $M(x, y)$ and $N(x, y)$.

To finish the proof of Theorem 2, we need to show that the condition (5) implies that $M\,dx + N\,dy = 0$ is an exact equation. This we do by actually exhibiting a function $F(x, y)$ that satisfies $\partial F/\partial x = M$ and $\partial F/\partial y = N$. Fortunately we needn't look too far for such a function.

The discussion in the first part of the proof suggests (6) as a candidate, where $g'(y)$ is given by (8). Namely, we *define* $F(x, y)$ by

(9) $$F(x, y) := \int_{x_0}^{x} M(t, y)\,dt + g(y) ,$$

where (x_0, y_0) is a fixed point in the rectangle R and $g(y)$ is determined, up to a numerical constant, by the equation

(10) $$g'(y) := N(x, y) - \frac{\partial}{\partial y} \int_{x_0}^{x} M(t, y)\,dt .$$

Before proceeding, we must address an extremely important question concerning the definition of $F(x, y)$. That is, how can we be sure (in this portion of the proof) that $g'(y)$, as given in equation (10), is really a function of just y alone? To show that the right-hand side of (10) is independent of x (that is, that the appearances of the variable x cancel), all we need to do is show that its partial derivative with respect to x is zero. This is where condition (5) is utilized. We leave to the reader this computation and the verification that $F(x, y)$ satisfies conditions (4) (see Problems 35 and 36). ∎ End

The construction in the proof of Theorem 2 actually provides an explicit procedure for solving exact equations. Let's recap and look at some examples.

METHOD FOR SOLVING EXACT EQUATIONS

(a) If $M\,dx + N\,dy = 0$ is exact, then $\partial F/\partial x = M$. Integrate this last equation with respect to x to get

(11) $$F(x, y) = \int M(x, y)\,dx + g(y) .$$

(b) To determine $g(y)$, take the partial derivative with respect to y of both sides of equation (11) and substitute N for $\partial F/\partial y$. We can now solve for $g'(y)$,

(c) Integrate $g'(y)$ to obtain $g(y)$ up to a numerical constant. Substituting $g(y)$ into equation (11) gives $F(x, y)$.

(d) The solution to $M\,dx + N\,dy = 0$ is given implicitly by

$$F(x, y) = C .$$

(Alternatively, starting with $\partial F/\partial y = N$, the implicit solution can be found by first integrating with respect to y; see Example 3.)

EXAMPLE 2 Solve

(12) $$(2xy - \sec^2 x)\,dx + (x^2 + 2y)\,dy = 0 .$$

SOLUTION Here $M(x, y) = 2xy - \sec^2 x$ and $N(x, y) = x^2 + 2y$. Because

$$\frac{\partial M}{\partial y} = 2x = \frac{\partial N}{\partial x} ,$$

equation (12) is exact. To find $F(x, y)$. we begin by integrating M with respect to x:

(13) $\quad F(x, y) = \displaystyle\int \left(2xy - \sec^2 x\right) dx + g(y)$

$\qquad\qquad = x^2 y - \tan x + g(y)$.

Next we take the partial derivative of (13) with respect to y and substitute $x^2 + 2y$ for N:

$$\frac{\partial F}{\partial y}(x, y) = N(x, y) \ ,$$

$$x^2 + g'(y) = x^2 + 2y \ .$$

Thus $g'(y) = 2y$, and since the choice of the constant of integration is not important, we can take $g(y) = y^2$. Hence, from (13), we have $F(x, y) = x^2 y - \tan x + y^2$, and the solution to equation (12) is given implicitly by $x^2 y - \tan x + y^2 = C$. ■ End

Remark. The procedure for solving exact equations requires several steps. As a check on our work, we observe that when we solve for $g'(y)$, we must obtain a function that is independent of x. If this is not the case, then we have erred either in our computation of $F(x, y)$ or in computing $\partial M/\partial y$ or $\partial N/\partial x$.

In the construction of $F(x, y)$, we can first integrate $N(x, y)$ with respect to y to get

(14) $\quad F(x, y) = \displaystyle\int N(x, y) dy + h(x)$

and then proceed to find $h(x)$. We illustrate this alternative method in the next example.

EXAMPLE 3 Solve

(15) $\quad \left(1 + e^x y + x e^x y\right) dx + \left(x e^x + 2\right) dy = 0$.

SOLUTION Here $M = 1 + e^x y + x e^x y$ and $N = x e^x + 2$. Because

$$\frac{\partial M}{\partial y} = e^x + x e^x = \frac{\partial N}{\partial x} \ ,$$

equation (15) is exact. If we now integrate $N(x, y)$ with respect to y, we obtain

$$F(x, y) = \int \left(x e^x + 2\right) dy + h(x) = x e^x y + 2y + h(x) \ .$$

When we take the partial derivative with respect to x and substitute for M, we get

$$\frac{\partial F}{\partial x}(x, y) = M(x, y)$$

$$e^x y + x e^x y + h'(x) = 1 + e^x y + x e^x y \ .$$

Thus $h'(x) = 1$, so we take $h(x) = x$. Hence $F(x, y) = x e^x y + 2y + x$, and the solution to equation (15) is given implicitly by $x e^x y + 2y + x = C$. In this case, we can solve explicitly for y to obtain $y = (C - x)/(2 + x e^x)$. ■ End

Remark. Since we can use either procedure for finding $F(x, y)$, it may be worthwhile to consider each of the integrals $\int M(x, y)dx$ and $\int N(x, y)dy$. If one is easier to evaluate than the other, this would be sufficient reason for us to use one method over the other. (The skeptical reader should try solving equation (15) by first integrating $M(x, y)$.)

EXAMPLE 4 Show that

(16) $(x + 3x^3\sin y)dx + (x^4\cos y)dy = 0$

is *not* exact but that multiplying this equation by the factor x^{-1} yields an exact equation. Use this fact to solve (16).

SOLUTION In equation (16), $M = x + 3x^3\sin y$ and $N = x^4\cos y$. Because

$$\frac{\partial M}{\partial y} = 3x^3\cos y \neq 4x^3\cos y = \frac{\partial N}{\partial x} \ ,$$

equation (16) is not exact. When we multiply (16) by the factor x^{-1}, we obtain

(17) $(1 + 3x^2\sin y)dx + (x^3\cos y)dy = 0 \ .$

For this new equation, $M = 1 + 3x^2\sin y$ and $N = x^3\cos y$. If we test for exactness, we now find that

$$\frac{\partial M}{\partial y} = 3x^2\cos y = \frac{\partial N}{\partial x} \ ,$$

and hence (17) is exact. Upon solving (17), we find that the solution is given implicitly by $x + x^3\sin y = C$. Since equations (16) and (17) differ only by a factor of x, then any solution to one will be a solution for the other whenever $x \neq 0$. Hence the solution to equation (16) is given implicitly by $x + x^3\sin y = C$. ∎End

In Section 2.5, we discuss methods for finding factors that, like x^{-1} in Example 4, change inexact equations into exact equations.

EXERCISES 2.4

In Problems 1 through 8, classify the equation as separable, linear, exact, or none of these. Notice that some equations may have more than one classification.

1. $(x^2y + x^4\cos x)dx - x^3\,dy = 0 \ .$
2. $(x^{10/3} - 2y)dx + x\,dy = 0 \ .$
3. $(ye^{xy} + 2x)dx + (xe^{xy} - 2y)dy = 0 \ .$
4. $\sqrt{-2y - y^2}\,dx + (3 + 2x - x^2)dy = 0 \ .$
5. $y^2\,dx + (2xy + \cos y)dy = 0 \ .$
6. $xy\,dx + dy = 0 \ .$
7. $\theta\,dr + (3r - \theta - 1)d\theta = 0 \ .$
8. $[2x + y\cos(xy)]dx + [x\cos(xy) - 2y]dy = 0 \ .$

In Problems 9 through 20, determine whether the equation is exact. If it is, then solve it.

9. $(2xy + 3)dx + (x^2 - 1)dy = 0 \ .$
10. $(2x + y)dx + (x - 2y)dy = 0 \ .$

11. $(\cos x \cos y + 2x)dx - (\sin x \sin y + 2y)dy = 0$.

12. $(e^x\sin y - 3x^2)dx + (e^x\cos y + y^{-2/3}/3)dy = 0$.

13. $(t/y)dy + (1 + \ln y)dt = 0$.

14. $e^t(y - t)dt + (1 + e^t)dy = 0$.

15. $\cos\theta\, dr - (r\sin\theta - e^\theta)d\theta = 0$.

16. $(ye^{xy} - 1/y)dx + (xe^{xy} + x/y^2)dy = 0$.

17. $(1/y)dx - (3y - x/y^2)dy = 0$.

18. $[2x + y^2 - \cos(x + y)]dx$
$\quad + [2xy - \cos(x + y) - e^y]dy = 0$.

19. $\left(2x + \dfrac{y}{1 + x^2y^2}\right)dx + \left(\dfrac{x}{1 + x^2y^2} - 2y\right)dy = 0$.

20. $\left[\dfrac{2}{\sqrt{1 - x^2}} + y\cos(xy)\right]dx$
$\quad + [x\cos(xy) - y^{-1/3}]dy = 0$.

In Problems 21 through 26, solve the initial value problem.

21. $(1/x + 2y^2x)dx + (2yx^2 - \cos y)dy = 0$,
$\quad y(1) = \pi$.

22. $(ye^{xy} - 1/y)dx + (xe^{xy} + x/y^2)dy = 0$,
$\quad y(1) = 1$.

23. $(e^ty + te^ty)dt + (te^t + 2)dy = 0$, $y(0) = -1$.

24. $(e^tx + 1)dt + (e^t - 1)dx = 0$, $x(1) = 1$.

25. $(y^2\sin x)dx + (1/x - y/x)dy = 0$, $y(\pi) = 1$.

26. $(\tan y - 2)dx + (x\sec^2y + 1/y)dy = 0$,
$\quad y(0) = 1$.

27. For each of the following equations, find the most general function $M(x, y)$ so that the equation is exact:

(a) $M(x, y)dx + (\sec^2y - x/y)dy = 0$.

(b) $M(x, y)dx + (\sin x \cos y - xy - e^{-y})dy = 0$.

28. For each of the following equations, find the most general function $N(x, y)$ so that the equation is exact:

(a) $[y\cos(xy) + e^x]dx + N(x, y)dy = 0$.

(b) $[ye^{xy} - 4x^3y + 2]dx + N(x, y)dy = 0$.

29. Consider the equation

$$(y^2 + 2xy)dx - x^2\, dy = 0$$.

(a) Show that this equation is not exact.

(b) Show that multiplying both sides of the equation by y^{-2} yields a new equation that is exact.

(c) Use the solution of the resulting exact equation to solve the original equation.

(d) Were any solutions lost in the process?

30. Consider the equation

$$(5x^2y + 6x^3y^2 + 4xy^2)dx$$
$$+ (2x^3 + 3x^4y + 3x^2y)dy = 0$$.

(a) Show that the equation is not exact.

(b) Multiply the equation by x^ny^m and determine values for n and m that make the resulting equation exact.

(c) Use the solution of the resulting exact equation to solve the original equation.

31. Argue that in the proof of Theorem 2, the function g can be taken as

$$g(y) = \int_{y_0}^{y} N(x, t)dt - \int_{y_0}^{y}\left[\frac{\partial}{\partial t}\int_{x_0}^{x} M(s, t)ds\right]dt ,$$

which can be expressed as

$$g(y) = \int_{y_0}^{y} N(x, t)dt - \int_{x_0}^{x} M(s, y)ds$$
$$+ \int_{x_0}^{x} M(s, y_0)ds .$$

This leads ultimately to the representation

(18) $$F(x, y) = \int_{y_0}^{y} N(x, t)dt + \int_{x_0}^{x} M(s, y_0)ds .$$

Evaluate this formula directly with $x_0 = 0$, $y_0 = 0$ to rework

(a) Example 1.

(b) Example 2.

(c) Example 3.

32. Orthogonal Trajectories. A geometric problem occurring often in engineering is that of finding a family of curves (orthogonal trajectories) that intersects a given family of curves orthogonally at each point. For example, we may be given the lines of force of an electric field and want to find the equation for the equipotential curves. Consider the fam-

ily of curves described by $F(x, y) = k$, where k is a parameter. Recall from the discussion of equation (2) that for each curve in the family, the slope is given by

$$\frac{dy}{dx} = -\frac{\partial F}{\partial x} \bigg/ \frac{\partial F}{\partial y} .$$

(a) Recall that the slope of a curve that is orthogonal (perpendicular) to a given curve is just the negative reciprocal of the slope of the given curve. Using this fact, show that the curves orthogonal to the family $F(x, y) = k$ satisfy the differential equation

$$\frac{\partial F}{\partial y}(x, y)dx - \frac{\partial F}{\partial x}(x, y)dy = 0 .$$

(b) Using the preceding differential equation, show that the orthogonal trajectories to the family of circles $x^2 + y^2 = k$ are just straight lines through the origin (see Figure 2.10).

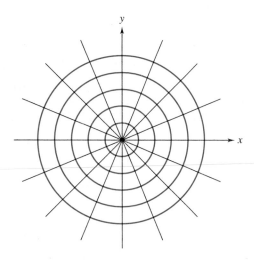

Figure 2.10 Orthogonal trajectories for concentric circles are lines through the center

(c) Show that the orthogonal trajectories to the family of hyperbolas $xy = k$ are the hyperbolas $x^2 - y^2 = k$ (see Figure 2.11).

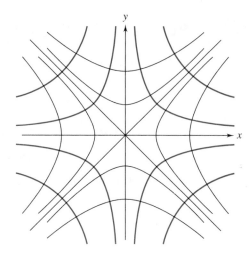

Figure 2.11 Families of orthogonal hyperbolas

33. Use the method in Problem 32 to find the orthogonal trajectories for each of the given families of curves.

 (a) $2x^2 + y^2 = k$. **(b)** $y = kx^4$.
 (c) $y = e^{kx}$. **(d)** $y^2 = kx$.

 [*Hint:* First express the family in the form $F(x, y) = k$.]

34. Use the method described in Problem 32 to show that the orthogonal trajectories to the family of curves $x^2 + y^2 = kx$, k a parameter, satisfy

$$(2yx^{-1})dx + (y^2x^{-2} - 1)dy = 0 .$$

Find the orthogonal trajectories by solving the above equation. Sketch the family of curves, along with their orthogonal trajectories. [*Hint:* Try multiplying the equation by $x^m y^n$ as in Problem 30.]

35. Using condition (5), show that the right-hand side of (10) is independent of x by showing that its partial derivative with respect to x is zero. [*Hint:* Since the partial derivatives of M are continuous, Leibniz's theorem allows you to interchange the operations of integration and differentiation.]

36. Verify that $F(x, y)$ as defined by (9) and (10) satisfies conditions (4).

2.5 SPECIAL INTEGRATING FACTORS

If we take the standard form for the linear differential equation of Section 2.3,

$$\frac{dy}{dx} + P(x)y = Q(x) ,$$

and rewrite it in differential form by multiplying through by dx, we obtain

$$\left[P(x)y - Q(x) \right] dx + dy = 0 .$$

This form is certainly not exact, but it becomes exact upon multiplication by the integrating factor $\mu(x) = e^{\int P(x)dx}$. We have

$$\left[\mu(x)P(x)y - \mu(x)Q(x) \right] dx + \mu(x)dy = 0$$

as the form, and the compatibility condition is precisely the identity $\mu(x)P(x) = \mu'(x)$ (see Problem 18).

This leads us to generalize the notion of an integrating factor.

INTEGRATING FACTOR

Definition 3. If the equation

(1) $M(x, y)dx + N(x, y)dy = 0$

is not exact, but the equation

(2) $\mu(x, y)M(x, y)dx + \mu(x, y)N(x, y)dy = 0 ,$

which results from multiplying equation (1) by the function $\mu(x, y)$, *is* exact, then $\mu(x, y)$ is called an **integrating factor**[†] of the equation (1).

EXAMPLE 1 Show that $\mu(x, y) = xy^2$ is an integrating factor for

(3) $(2y - 6x)dx + (3x - 4x^2y^{-1})dy = 0 .$

Use this integrating factor to solve the equation.

SOLUTION We leave it for the reader to show that (3) is not exact. Multiplying (3) by $\mu(x, y) = xy^2$, we obtain

(4) $(2xy^3 - 6x^2y^2)dx + (3x^2y^2 - 4x^3y)dy = 0 .$

For this equation we have $M = 2xy^3 - 6x^2y^2$ and $N = 3x^2y^2 - 4x^3y$. Because

$$\frac{\partial M}{\partial y}(x, y) = 6xy^2 - 12x^2y = \frac{\partial N}{\partial x}(x, y) ,$$

equation (4) is exact. Hence $\mu(x, y) = xy^2$ is indeed an integrating factor of equation (3).

[†]*Historical Footnote:* A general theory of integrating factors was developed by Alexis Clairaut in 1739. Leonhard Euler also studied classes of equations that could be solved using a specific integrating factor.

Let's now solve equation (4) using the procedure of Section 2.4. To find $F(x, y)$, we begin by integrating M with respect to x:

$$F(x, y) = \int (2xy^3 - 6x^2y^2)dx + g(y) = x^2y^3 - 2x^3y^2 + g(y) \ .$$

When we take the partial derivative with respect to y and substitute for N, we find

$$\frac{\partial F}{\partial y}(x, y) = N(x, y)$$

$$3x^2y^2 - 4x^3y + g'(y) = 3x^2y^2 - 4x^3y \ .$$

Thus $g'(y) = 0$, so we can take $g(y) \equiv 0$. Hence $F(x, y) = x^2y^3 - 2x^3y^2$, and the solution to equation (4) is given implicitly by

$$x^2y^3 - 2x^3y^2 = C \ .$$

While equations (3) and (4) have essentially the same solutions, *it is possible to lose or gain solutions when multiplying by $\mu(x, y)$.* In this case, $y \equiv 0$ is a solution of equation (4) but not of equation (3). The extraneous solution arises because when we multiply (3) by $\mu = xy^2$ to obtain (4), we are actually multiplying both sides of (3) by zero if $y \equiv 0$. This gives us $y \equiv 0$ as a solution to (4), while it is not a solution to (3). ■ End

How do we find an integrating factor? If $\mu(x, y)$ is an integrating factor of (1) with continuous first partial derivatives, then testing (2) for exactness, we must have

$$\frac{\partial}{\partial y}\left[\mu(x, y)M(x, y)\right] = \frac{\partial}{\partial x}\left[\mu(x, y)N(x, y)\right] \ .$$

By use of the product rule, this reduces to the equation

$$(5) \qquad M\frac{\partial \mu}{\partial y} - N\frac{\partial \mu}{\partial x} = \left(\frac{\partial N}{\partial x} - \frac{\partial M}{\partial y}\right)\mu \ .$$

But solving the partial differential equation (5) for μ is usually more difficult than solving the original equation (1). There are, however, two important exceptions.

Let's assume equation (1) has an integrating factor that depends only on x; that is, $\mu = \mu(x)$. In this case, equation (5) reduces to the separable equation

$$(6) \qquad \frac{d\mu}{dx} = \left\{\frac{\partial M/\partial y - \partial N/\partial x}{N}\right\}\mu \ ,$$

where $(\partial M/\partial y - \partial N/\partial x)/N$ is (presumably) just a function of x. In a similar fashion, if equation (1) has an integrating factor that depends only on y, then equation (5) reduces to the separable equation

$$(7) \qquad \frac{d\mu}{dy} = \left\{\frac{\partial N/\partial x - \partial M/\partial y}{M}\right\}\mu \ ,$$

where $(\partial N/\partial x - \partial M/\partial y)/M$ is just a function of y.

We can reverse the above argument. In particular, if $(\partial M/\partial y - \partial N/\partial x)/N$ is a function that depends only on x, then we can solve the separable equation (6) to obtain the integrating factor

$$\mu(x) = \exp\left(\int \left\{\frac{\partial M/\partial y - \partial N/\partial x}{N}\right\}dx\right)$$

for equation (1). We summarize these observations in the following theorem.

SPECIAL INTEGRATING FACTORS

Theorem 3. If $(\partial M/\partial y - \partial N/\partial x)/N$ is continuous and depends only on x, then

$$(8)\qquad \mu(x) = \exp\left(\int \left\{\frac{\partial M/\partial y - \partial N/\partial x}{N}\right\}dx\right)$$

is an integrating factor for equation (1).

If $(\partial N/\partial x - \partial M/\partial y)/M$ is continuous and depends only on y, then

$$(9)\qquad \mu(y) = \exp\left(\int \left\{\frac{\partial N/\partial x - \partial M/\partial y}{M}\right\}dy\right)$$

is an integrating factor for equation (1).

Theorem 3 suggests the following procedure.

METHOD FOR FINDING SPECIAL INTEGRATING FACTORS

If $M\,dx + N\,dy = 0$ is neither separable nor linear, compute $\partial M/\partial y$ and $\partial N/\partial x$. If $\partial M/\partial y = \partial N/\partial x$, then the equation is exact. If it is not exact, consider

$$(10)\qquad \frac{\partial M/\partial y - \partial N/\partial x}{N}\ .$$

If (10) is a function of just x, then an integrating factor is given by formula (8). If not, consider

$$(11)\qquad \frac{\partial N/\partial x - \partial M/\partial y}{M}\ .$$

If (11) is a function of just y, then an integrating factor is given by formula (9).

EXAMPLE 2 Solve

$$(12)\qquad (2x^2 + y)dx + (x^2y - x)dy = 0\ .$$

SOLUTION A quick inspection shows that equation (12) is neither separable nor linear. We also note that

$$\frac{\partial M}{\partial y} = 1 \neq (2xy - 1) = \frac{\partial N}{\partial x}\ .$$

Because (12) is not exact, we compute

$$\frac{\partial M/\partial y - \partial N/\partial x}{N} = \frac{1 - (2xy - 1)}{x^2 y - x} = \frac{2(1 - xy)}{-x(1 - xy)} = \frac{-2}{x} \ .$$

We obtain a function of only x, so an integrating factor for (12) is given by formula (8). That is,

$$\mu(x) = \exp\left(\int \frac{-2}{x} dx\right) = x^{-2} \ .$$

When we multiply (12) by $\mu = x^{-2}$, we get the exact equation

$$\left(2 + yx^{-2}\right)dx + \left(y - x^{-1}\right)dy = 0 \ .$$

Solving this equation, we ultimately derive the implicit solution

(13) $2x - yx^{-1} + \dfrac{y^2}{2} = C \ .$

Notice that the solution $x \equiv 0$ was lost in multiplying by $\mu = x^{-2}$. Hence (13) and $x \equiv 0$ are solutions to equation (12). ■ End

There are many differential equations that are not covered by Theorem 3 but for which an integrating factor nevertheless exists. The major difficulty, however, is in finding an explicit formula for these integrating factors, which in general will depend on both x and y.

EXERCISES 2.5

In Problems 1 through 6, identify the equation as separable, linear, exact, or having an integrating factor that is a function of either x alone or y alone.

1. $\left(2y^3 + 2y^2\right)dx + \left(3y^2 x + 2xy\right)dy = 0 \ .$
2. $\left(2x + yx^{-1}\right)dx + \left(xy - 1\right)dy = 0 \ .$
3. $\left(y^2 + 2xy\right)dx - x^2\,dy = 0 \ .$
4. $\left(2x + y\right)dx + \left(x - 2y\right)dy = 0 \ .$
5. $\left(2y^2 x - y\right)dx + x\,dy = 0 \ .$
6. $\left(x^2 \sin x + 4y\right)dx + x\,dy = 0 \ .$

In Problems 7 through 12, solve the equation.

7. $\left(3x^2 + y\right)dx + \left(x^2 y - x\right)dy = 0 \ .$
8. $(2xy)dx + \left(y^2 - 3x^2\right)dy = 0 \ .$
9. $\left(2y^2 + 2y + 4x^2\right)dx + \left(2xy + x\right)dy = 0 \ .$
10. $\left(x^4 - x + y\right)dx - x\,dy = 0 \ .$
11. $\left(y^2 + 2xy\right)dx - x^2\,dy = 0 \ .$
12. $\left(2xy^3 + 1\right)dx + \left(3x^2 y^2 - y^{-1}\right)dy = 0 \ .$

In Problems 13 and 14, find an integrating factor of the form $x^n y^m$ and solve the equation.

13. $\left(2y^2 - 6xy\right)dx + \left(3xy - 4x^2\right)dy = 0 \ .$
14. $\left(12 + 5xy\right)dx + \left(6xy^{-1} + 3x^2\right)dy = 0 \ .$

15. Show that if $(\partial N/\partial x - \partial M/\partial y)/(xM - yN)$ depends only on the product xy, that is

$$\frac{\partial N/\partial x - \partial M/\partial y}{xM - yN} = H(xy) \ ,$$

then the equation $M(x, y)dx + N(x, y)dy = 0$ has an integrating factor of the form $\mu(xy)$. Give the general formula for $\mu(xy)$.

16. If $xM(x, y) + yN(x, y) \equiv 0$, find the solution to the equation

$$M(x, y)dx + N(x, y)dy = 0.$$

17. **Fluid Flow.** The streamlines associated with a certain fluid flow are represented by the family of curves $y = x - 1 + ke^{-x}$. The velocity potentials of the

flow are just the orthogonal trajectories of this family.

(a) Use the method described in Problem 32 of Exercises 2.4 to show that the velocity potentials satisfy

$$dx + (x - y)dy = 0 \ .$$

[*Hint:* First express the family $y = x - 1 + ke^{-x}$ in the form $F(x, y) = k$.]

(b) Find the velocity potentials by solving the equation obtained in part (a).

18. Verify that when the linear differential equation $[P(x)y - Q(x)]dx + dy = 0$ is multiplied by $\mu(x) = e^{\int P(x)dx}$, the result is exact.

2.6 SUBSTITUTIONS AND TRANSFORMATIONS

When the equation

$$M(x, y)dx + N(x, y)dy = 0$$

is not a separable, exact, or linear equation, it may still be possible to transform it into one that we know how to solve. This, in fact, was our approach in Section 2.5, where we used an integrating factor to transform our original equation into an exact equation.

In this section, we study four types of equations that can be transformed into either a separable or linear equation by means of a suitable substitution or transformation.

SUBSTITUTION PROCEDURE

(a) Identify the type of equation and determine the appropriate substitution or transformation.
(b) Rewrite the original equation in terms of new variables.
(c) Solve the transformed equation.
(d) Express the solution in terms of the original variables.

Homogeneous Equations

HOMOGENEOUS EQUATION

Definition 4. If the right-hand side of the equation

$$(1) \qquad \frac{dy}{dx} = f(x, y)$$

can be expressed as a function of the ratio y/x alone, then we say the equation is **homogeneous.**

For example, the equation

$$(2) \qquad (x - y)dx + x\,dy = 0$$

For example, the equation
$$(x-y)dx + x\,dy = 0$$

can be written in the form

$$\left(\frac{y}{x}-1\right) = v-1 \quad \text{where } v = \frac{y}{x}$$

$$\frac{dy}{dx} = \frac{y-x}{x} = \left(\frac{y}{x}-1\right).$$

Since we have expressed $(y-x)/x$ as a function of the ratio y/x; that is, $(y-x)/x = G(y/x)$, where $G(v) := v - 1$, then equation (2) is homogeneous.

The equation

(3) $(x - 2y + 1)dx + (x - y)dy = 0$

can be written in the form

$$\frac{dy}{dx} = \frac{x - 2y + 1}{y - x} = \frac{1 - 2(y/x) + (1/x)}{(y/x) - 1}.$$

Here the right-hand side cannot be expressed as a function of y/x alone because of the term $1/x$ in the numerator. Hence equation (3) is not homogeneous.

One test for the homogeneity of equation (1) is to replace x by tx and y by ty. Then (1) is homogeneous if and only if

$$f(tx, ty) = f(x, y)$$

for all $t \neq 0$ (see Problem 43(a)).

To solve a homogeneous equation, we make a rather obvious substitution. Let

$$v = \frac{y}{x}.$$

Our homogeneous equation now has the form

(4) $\dfrac{dy}{dx} = G(v)$,

and all we need is to express dy/dx in terms of x and v. Since $v = y/x$, then $y = vx$. Keeping in mind that both v and y are functions of x, we use the product rule for differentiation to deduce from $y = vx$ that

$$\frac{dy}{dx} = v + x\frac{dv}{dx}.$$

We then substitute the above expression for dy/dx into equation (4) to obtain

(5) $v + x\dfrac{dv}{dx} = G(v)$.

The new equation (5) is separable, and we can obtain its implicit solution from

$$\int \frac{1}{G(v) - v}dv = \int \frac{1}{x}dx .$$

All that remains to do is to express the solution in terms of the original variables x and y.

EXAMPLE 1 Solve

(6) $(xy + y^2 + x^2)dx - x^2\,dy = 0$.

SOLUTION A check will show that equation (6) is not separable, exact, or linear. If we express (6) in the derivative form

(7) $\dfrac{dy}{dx} = \dfrac{xy + y^2 + x^2}{x^2} = \dfrac{y}{x} + \left(\dfrac{y}{x}\right)^2 + 1$,

then we see that the right-hand side of (7) is a function of just y/x. Thus equation (6) is homogeneous.

Now let $v = y/x$ and recall that $dy/dx = v + x(dv/dx)$. With these substitutions, equation (7) becomes

$$v + x\dfrac{dv}{dx} = v + v^2 + 1 \ .$$

The above equation is separable, and, on separating the variables and integrating, we obtain

$$\int \dfrac{1}{v^2 + 1}\,dv = \int \dfrac{1}{x}\,dx \ ,$$

$$\arctan v = \ln|x| + C \ .$$

Hence

$$v = \tan(\ln|x| + C) \ .$$

Finally, we substitute y/x for v and solve for y to get

$$y = x\tan(\ln|x| + C)$$

as an explicit solution to equation (6). Also note that $x \equiv 0$ is a solution. ∎ End

Equations of the Form $dy/dx = G(ax + by)$

When the right-hand side of the equation $dy/dx = f(x, y)$ can be expressed as a function of the combination $ax + by$, where a and b are constants, that is,

$$\dfrac{dy}{dx} = G(ax + by) \ ,$$

then the substitution

$$z = ax + by$$

transforms the equation into a separable one. The method is illustrated in the next example.

EXAMPLE 2 Solve

(8) $\dfrac{dy}{dx} = y - x - 1 + (x - y + 2)^{-1}$.

SOLUTION The right-hand side can be expressed as a function of $x - y$, that is,

$$y - x - 1 + (x - y + 2)^{-1} = -(x - y) - 1 + \left[(x - y) + 2\right]^{-1} ,$$

so let $z = x - y$. To solve for dy/dx, we differentiate $z = x - y$ with respect to x to obtain $dz/dx = 1 - dy/dx$, and so $dy/dx = 1 - dz/dx$. Substituting into (8) yields

$$1 - \frac{dz}{dx} = -z - 1 + (z + 2)^{-1} ,$$

or

$$\frac{dz}{dx} = (z + 2) - (z + 2)^{-1} .$$

Solving this separable equation, we obtain

$$\int \frac{z + 2}{(z + 2)^2 - 1} dz = \int dx ,$$

$$\frac{1}{2} \ln|(z + 2)^2 - 1| = x + C_1 ,$$

from which it follows that

$$(z + 2)^2 = Ce^{2x} + 1 .$$

Finally, replacing z by $x - y$ yields

$$(x - y + 2)^2 = Ce^{2x} + 1$$

as an implicit solution to equation (8). ■ End

Bernoulli Equations

> ### BERNOULLI EQUATION
>
> **Definition 5.** A first order equation that can be written in the form
>
> (9) $\dfrac{dy}{dx} + P(x)y = Q(x)y^n ,$
>
> where $P(x)$ and $Q(x)$ are continuous on an interval (a, b) and n is a real number, is called a **Bernoulli equation.**[†]

Notice that when $n = 0$ or 1, equation (9) is also a linear equation and can be solved by the

[†]*Historical Footnote:* This equation was proposed for solution by James Bernoulli in 1695. It was solved by his brother John Bernoulli. (James and John were two of eight mathematicians in the Bernoulli family.) In 1696 Gottfried Leibniz showed that the Bernoulli equation can be reduced to a linear equation by making the substitution $v = y^{1-n}$.

method discussed in Section 2.3. For other values of n, the substitution

$$v = y^{1-n}$$

transforms the Bernoulli equation into a linear equation, as we now show.
 Dividing equation (9) by y^n yields

(10) $y^{-n}\dfrac{dy}{dx} + P(x)y^{1-n} = Q(x)$.

Taking $v = y^{1-n}$, we find via the chain rule that

$$\frac{dv}{dx} = (1-n)y^{-n}\frac{dy}{dx} \,,$$

and so equation (10) becomes

$$\frac{1}{1-n}\frac{dv}{dx} + P(x)v = Q(x) \,.$$

Because $1/(1-n)$ is just a constant, the last equation is indeed linear.

EXAMPLE 3 Solve

(11) $\dfrac{dy}{dx} - 5y = -\dfrac{5}{2}xy^3$.

SOLUTION This is a Bernoulli equation with $n = 3$, $P(x) = -5$, and $Q(x) = -5x/2$. To transform (11) into a linear equation, we first divide by y^3 to obtain

$$y^{-3}\frac{dy}{dx} - 5y^{-2} = -\frac{5}{2}x \,.$$

Next we make the substitution $v = y^{-2}$. Since $dv/dx = -2y^{-3}\,dy/dx$, the transformed equation is

$$-\frac{1}{2}\frac{dv}{dx} - 5v = -\frac{5}{2}x \,,$$

(12) $\dfrac{dv}{dx} + 10v = 5x$.

Equation (12) is linear, so we can solve it for v using the method discussed in Section 2.3. When we do this, it turns out that

$$v = \frac{x}{2} - \frac{1}{20} + Ce^{-10x} \,.$$

Substituting $v = y^{-2}$ gives the solution

$$y^{-2} = \frac{x}{2} - \frac{1}{20} + Ce^{-10x} \,.$$

Not included in the last equation is the solution $y \equiv 0$ that was lost in the process of dividing (11) by y^3. ∎ End

Equations with Linear Coefficients

We have used various substitutions for y to transform the original equation into a new equation that we could solve. In some cases, we must transform *both* x and y into new variables, say u and v. This is the situation for **equations with linear coefficients,** that is, equations of the form

$$(13) \qquad (a_1 x + b_1 y + c_1)dx + (a_2 x + b_2 y + c_2)dy = 0 \ ,$$

where the a_i's, b_i's and c_i's are constants. We leave it as an exercise to show that when $a_1 b_2 = a_2 b_1$, equation (13) can be put in the form $dy/dx = G(ax + by)$, which we solved via the substitution $z = ax + by$.

Before considering the general case when $a_1 b_2 \neq a_2 b_1$, let's first look at the special situation when $c_1 = c_2 = 0$. Equation (13) then becomes

$$(a_1 x + b_1 y)dx + (a_2 x + b_2 y)dy = 0 \ ,$$

which can be rewritten in the form

$$\frac{dy}{dx} = -\frac{a_1 x + b_1 y}{a_2 x + b_2 y} = -\frac{a_1 + b_1(y/x)}{a_2 + b_2(y/x)} \ .$$

This equation is homogeneous, so we can solve it using the method discussed earlier in this section.

The above discussion suggests the following procedure for solving (13). If $a_1 b_2 \neq a_2 b_1$, then we seek a translation of axes of the form

$$x = u + h \qquad \text{and} \qquad y = v + k \ ,$$

where h and k are constants, that will change $a_1 x + b_1 y + c_1$ into $a_1 u + b_1 v$ and change $a_2 x + b_2 y + c_2$ into $a_2 u + b_2 v$. Some elementary algebra shows that such a transformation exists if the system of equations

$$(14) \qquad \begin{aligned} a_1 h + b_1 k + c_1 &= 0 \ , \\ a_2 h + b_2 k + c_2 &= 0 \end{aligned}$$

has a solution. This is assured by the assumption $a_1 b_2 \neq a_2 b_1$, which is geometrically equivalent to assuming that the two lines described by the system (14) intersect. Now if (h, k) satisfies (14), then the substitutions $x = u + h$ and $y = v + k$ transform equation (13) into the homogeneous equation

$$(15) \qquad \frac{dv}{du} = -\frac{a_1 u + b_1 v}{a_2 u + b_2 v} = -\frac{a_1 + b_1(v/u)}{a_2 + b_2(v/u)} \ ,$$

which we know how to solve.

EXAMPLE 4 Solve

$$(16) \qquad (-3x + y + 6)dx + (x + y + 2)dy = 0 \ .$$

SOLUTION Since $a_1 b_2 = (-3)(1) \neq (1)(1) = a_2 b_1$, we will use the translation of axes $x = u + h$,

$y = v + k$, where h and k satisfy the system

$$-3h + k + 6 = 0 \, ,$$
$$h + k + 2 = 0 \, .$$

Solving the above system for h and k gives $h = 1, k = -3$. Hence we let $x = u + 1$ and $y = v - 3$. As $dy = dv$ and $dx = du$, substituting in equation (16) for x and y yields

$$(-3u + v)du + (u + v)dv = 0$$

or

$$\frac{dv}{du} = \frac{3 - (v/u)}{1 + (v/u)} \, .$$

The above equation is homogeneous, so we let $z = v/u$. Then $dv/du = z + u(dz/du)$, and, substituting for v/u, we obtain

$$z + u\frac{dz}{du} = \frac{3 - z}{1 + z} \, .$$

Separating variables gives

$$\int \frac{z + 1}{z^2 + 2z - 3} dz = -\int \frac{1}{u} du \, ,$$

$$\frac{1}{2} \ln |z^2 + 2z - 3| = -\ln |u| + C_1 \, ,$$

from which it follows that

$$z^2 + 2z - 3 = Cu^{-2} \, .$$

When we substitute back in for z, u, and v, we find

$$(v/u)^2 + 2(v/u) - 3 = Cu^{-2} \, ,$$
$$v^2 + 2uv - 3u^2 = C \, ,$$
$$(y + 3)^2 + 2(x - 1)(y + 3) - 3(x - 1)^2 = C \, .$$

This last equation gives an implicit solution to (16). ■ End

EXERCISES 2.6

In Problems 1 through 8, identify the equation as homogeneous, Bernoulli, linear coefficients, or of the form $y' = G(ax + by)$.

1. $(y - 4x - 1)^2 \, dx - dy = 0$.

2. $2tx \, dx + (t^2 - x^2)dt = 0$.

3. $(t + x + 2)dx + (3t - x - 6)dt = 0$.

4. $dy/dx + y/x = x^3y^2$.

5. $(ye^{-2x} + y^3)dx - e^{-2x} \, dy = 0$.

6. $\theta \, dy - y \, d\theta = \sqrt{\theta y} \, d\theta$.

7. $(y^3 - \theta y^2)d\theta + 2\theta^2 y \, dy = 0$.

8. $\cos(x + y)dy = \sin(x + y)dx$.

Use the method discussed under "Homogeneous Equations" to solve Problems 9 through 16.

9. $(3x^2 - y^2)dx + (xy - x^3y^{-1})dy = 0$.

10. $(xy + y^2)dx - x^2\,dy = 0$.

11. $(y^2 - xy)dx + x^2\,dy = 0$.

12. $(x^2 + y^2)dx + 2xy\,dy = 0$.

13. $\dfrac{dx}{dt} = \dfrac{x^2 + t\sqrt{t^2 + x^2}}{tx}$.

14. $\dfrac{dy}{d\theta} = \dfrac{\theta\sec(y/\theta) + y}{\theta}$.

15. $\dfrac{dy}{dx} = \dfrac{x^2 - y^2}{3xy}$. **16.** $\dfrac{dy}{dx} = \dfrac{y(\ln y - \ln x + 1)}{x}$.

Use the method discussed under "Equations of the Form $dy/dx = G(ax + by)$" to solve Problems 17 through 20.

17. $dy/dx = \sqrt{x + y} - 1$. **18.** $dy/dx = (x + y + 2)^2$.

19. $dy/dx = (x - y + 5)^2$. **20.** $dy/dx = \sin(x - y)$.

Use the method discussed under "Bernoulli Equations" to solve Problems 21 through 28.

21. $\dfrac{dy}{dx} + \dfrac{y}{x} = x^2y^2$.

22. $\dfrac{dy}{dx} - y = e^{2x}y^3$.

23. $\dfrac{dy}{dx} = \dfrac{2y}{x} - x^2y^2$.

24. $\dfrac{dy}{dx} + \dfrac{y}{x - 2} = 5(x - 2)y^{1/2}$.

25. $\dfrac{dx}{dt} + tx^3 + \dfrac{x}{t} = 0$. **26.** $\dfrac{dy}{dx} + y = e^x y^{-2}$.

27. $\dfrac{dr}{d\theta} = \dfrac{r^2 + 2r\theta}{\theta^2}$. **28.** $\dfrac{dy}{dx} + y^3x + y = 0$.

Use the method discussed under "Equations with Linear Coefficients" to solve Problems 29 through 32.

29. $(-3x + y - 1)dx + (x + y + 3)dy = 0$.

30. $(x + y - 1)dx + (y - x - 5)dy = 0$.

31. $(2x - y)dx + (4x + y - 3)dy = 0$.

32. $(2x + y + 4)dx + (x - 2y - 2)dy = 0$.

In Problems 33 through 40, solve the equation given in:

33. Problem 1. **34.** Problem 2.

35. Problem 3. **36.** Problem 4.

37. Problem 5. **38.** Problem 6.

39. Problem 7. **40.** Problem 8.

41. Use the substitution $v = x - y + 2$ to solve equation (8).

42. Use the substitution $y = vx^2$ to solve
$$\frac{dy}{dx} = \frac{2y}{x} + \cos(y/x^2) \; .$$

43. (a) Show that the equation $dy/dx = f(x, y)$ is homogeneous if and only if $f(tx, ty) = f(x, y)$. [*Hint:* Let $t = 1/x$.]

 (b) A function $H(x, y)$ is called **homogeneous of order n** if $H(tx, ty) = t^n H(x, y)$. Show that the equation
$$M(x, y)dx + N(x, y)dy = 0$$
is homogeneous if $M(x, y)$ and $N(x, y)$ are both homogeneous of the same order.

44. Show that equation (13) reduces to an equation of the form
$$\frac{dy}{dx} = G(ax + by) \; ,$$
when $a_1b_2 = a_2b_1$. [*Hint:* If $a_1b_2 = a_2b_1$, then $a_2/a_1 = b_2/b_1 = k$, so that $a_2 = ka_1$ and $b_2 = kb_1$.]

45. Coupled Equations. In analyzing coupled equations of the form
$$\frac{dy}{dt} = ax + by \; ,$$
$$\frac{dx}{dt} = \alpha x + \beta y \; ,$$
where a, b, α, and β are constants, we may wish to determine the relationship between x and y rather than the individual solutions $x(t), y(t)$. For this purpose, divide the first equation by the second to obtain

(17) $\quad \dfrac{dy}{dx} = \dfrac{ax + by}{\alpha x + \beta y} \; .$

This new equation is homogeneous, so we can solve it via the substitution $v = y/x$. We refer to the solu-

tions of (17) as **integral curves.** Determine the integral curves for the system

$$\frac{dy}{dt} = -4x - y ,$$

$$\frac{dx}{dt} = 2x - y .$$

46. Riccati Equation. An equation of the form

(18) $$\frac{dy}{dx} = P(x)y^2 + Q(x)y + R(x)$$

is called a generalized Riccati equation.[†]

(a) If one solution, say $u(x)$, of (18) is known, show that the substitution $y = u + 1/v$ reduces (18) to a linear equation in v.

(b) Given that $u(x) = x$ is a solution to

$$\frac{dy}{dx} = x^3(y - x)^2 + \frac{y}{x} ,$$

use the result of part (a) to find all the other solutions to this equation. (The particular solution $u(x) = x$ can be found by inspection or by using a Taylor series method; see Section 8.1.)

47. Guerrilla Combat Model.[††] In a simplified model of a pair of guerrilla forces in combat, the following system arises:

$$\frac{dy}{dt} = -axy ,$$

$$\frac{dx}{dt} = -bxy ,$$

where $x(t)$ and $y(t)$ are the strengths of opposing forces at time t and a and b are positive constants. The terms $-axy$ and $-bxy$ represent the *combat loss rate* for the troops y and x, respectively. This model assumes no reinforcements.

(a) Show that x and y satisfy the equation

$$\frac{dy}{dx} = \frac{a}{b} .$$

(b) Let $y(0) = y_0$ and $x(0) = x_0$. Solving the equation in part (a), derive the **linear combat law**

$$by - ax = c ,$$

where $c = by_0 - ax_0$.

(c) Use the linear combat law to show that the y troops win if $c > 0$ and lose if $c < 0$.

48. Mixed Combat Model. In a simplified model of a conflict between a guerilla force and a conventional force, the following system arises:

$$\frac{dy}{dt} = -ax ,$$

$$\frac{dx}{dt} = -bxy ,$$

where $x(t)$ is the strength of the guerrilla force at time t, $y(t)$ is the strength of the conventional force, and a and b are positive constants. The terms $-ax$ and $-bxy$ represent the *combat loss rate* for troops y and x, respectively. This model assumes no reinforcements.

(a) Show that x and y satisfy the equation

$$\frac{dy}{dx} = \frac{a}{by} .$$

(b) Let $y(0) = y_0$ and $x(0) = x_0$. Solving the equation in part (a), derive the **parabolic combat law**

$$by^2 - 2ax = c ,$$

where $c = by_0^2 - 2ax_0$.

(c) Use the parabolic combat law to show that the y troops win if $c > 0$ and lose if $c < 0$.

[†]*Historical Footnote:* Count Jacopo Riccati studied a particular case of this equation in 1724 during his investigation of curves whose radii of curvature depend only on the variable y and not the variable x.

[††]For a discussion of combat models, see *Differential Equation Models,* M. Braun, C. S. Coleman, and D. A. Drew, eds. (Springer-Verlag, New York, 1983), Chapter 8.

CHAPTER SUMMARY

In this chapter, we have discussed various types of first order differential equations. The most important were the separable, linear, and exact equations. Their principal features and method of solution are outlined below.

Separable Equations: $dy/dx = g(x)p(y)$. Separate the variables and integrate.

Linear Equations: $dy/dx + P(x)y = Q(x)$. The integrating factor $\mu = \exp\left(\int P(x)dx\right)$ reduces the equation to $d(\mu y)/dx = \mu Q$, so that $\mu y = \int \mu Q \, dx + C$.

Exact Equations: $dF(x, y) = 0$. Solutions are given implicitly by $F(x, y) = C$. If $\partial M/\partial y = \partial N/\partial x$, then $M\,dx + N\,dy = 0$ is exact and F is given by

$$F = \int M \, dx + g(y), \quad \text{where} \quad g'(y) = N - \frac{\partial}{\partial y}\int M \, dx$$

or

$$F = \int N \, dy + h(x), \quad \text{where} \quad h'(x) = M - \frac{\partial}{\partial x}\int N \, dy.$$

When an equation is not separable, linear, or exact, it may be possible to find an integrating factor or perform a substitution that will enable us to solve the equation.

Special Integrating Factors: $\mu M\,dx + \mu N\,dy = 0$ is exact. If $(\partial M/\partial y - \partial N/\partial x)/N$ depends only on x, then

$$\mu(x) = \exp\left(\int \left\{\frac{\partial M/\partial y - \partial N/\partial x}{N}\right\} dx\right)$$

is an integrating factor. If $(\partial N/\partial x - \partial M/\partial y)/M$ depends only on y, then

$$\mu(y) = \exp\left(\int \left\{\frac{\partial N/\partial x - \partial M/\partial y}{M}\right\} dy\right)$$

is an integrating factor.

Homogeneous Equations: $dy/dx = G(y/x)$. Let $v = y/x$. Then $dy/dx = v + x(dv/dx)$, and the transformed equation in the variables v and x is separable.

Equations of the Form: $dy/dx = G(ax + by)$. Let $z = ax + by$. Then $dz/dx = a + b(dy/dx)$, and the transformed equation in the variables z and x is separable.

Bernoulli Equations: $dy/dx + P(x)y = Q(x)y^n$. For $n \neq 0$ or 1, let $v = y^{1-n}$. Then $dv/dx = (1 - n)y^{-n}(dy/dx)$, and the transformed equation in the variables v and x is linear.

Linear Coefficients: $(a_1x + b_1y + c_1)dx + (a_2x + b_2y + c_2)dy = 0.$ For $a_1b_2 \neq a_2b_1$, let $x = u + h$ and $y = v + k$, where h and k satisfy

$$a_1h + b_1k + c_1 = 0 \ ,$$
$$a_2h + b_2k + c_2 = 0 \ .$$

Then the transformed equation in the variables u and v is homogeneous.

REVIEW PROBLEMS

In Problems 1 through 30, solve the equation.

1. $\dfrac{dy}{dx} = \dfrac{e^{x+y}}{y - 1}$.

2. $\dfrac{dy}{dx} - 4y = 32x^2$.

3. $(x^2 - 2y^{-3})dy + (2xy - 3x^2)dx = 0$.

4. $\dfrac{dy}{dx} + \dfrac{3y}{x} = x^2 - 4x + 3$.

5. $\left[\sin(xy) + xy\cos(xy)\right]dx + \left[1 + x^2\cos(xy)\right]dy = 0$.

6. $2xy^3\,dx - (1 - x^2)dy = 0$.

7. $t^3y^2\,dt + t^4y^{-6}\,dy = 0$.

8. $\dfrac{dy}{dx} + \dfrac{2y}{x} = 2x^2y^2$.

9. $(x^2 + y^2)dx + 3xy\,dy = 0$.

10. $\left[1 + (1 + x^2 + 2xy + y^2)^{-1}\right]dx$
$\quad + \left[y^{-1/2} + (1 + x^2 + 2xy + y^2)^{-1}\right]dy = 0$.

11. $\dfrac{dx}{dt} = 1 + \cos^2(t - x)$.

12. $(y^3 + 4e^xy)dx + (2e^x + 3y^2)dy = 0$.

13. $\dfrac{dy}{dx} - \dfrac{y}{x} = x^2\sin 2x$.

14. $\dfrac{dx}{dt} - \dfrac{x}{t - 1} = t^2 + 2$.

15. $\dfrac{dy}{dx} = 2 - \sqrt{2x - y + 3}$.

16. $\dfrac{dy}{dx} + y\tan x + \sin x = 0$.

17. $\dfrac{dy}{d\theta} + 2y = y^2$.

18. $\dfrac{dy}{dx} = (2x + y - 1)^2$.

19. $(x^2 - 3y^2)dx + 2xy\,dy = 0$.

20. $\dfrac{dy}{d\theta} + \dfrac{y}{\theta} = -4\theta y^{-2}$.

21. $(y - 2x - 1)dx + (x + y - 4)dy = 0$.

22. $(2x - 2y - 8)dx + (x - 3y - 6)dy = 0$.

23. $(y - x)dx + (x + y)dy = 0$.

24. $\left(\sqrt{y/x} + \cos x\right)dx + \left(\sqrt{x/y} + \sin y\right)dy = 0$.

25. $y(x - y - 2)dx + x(y - x + 4)dy = 0$.

26. $\dfrac{dy}{dx} + xy = 0$.

27. $(3x - y - 5)dx + (x - y + 1)dy = 0$.

28. $\dfrac{dy}{dx} = \dfrac{x - y - 1}{x + y + 5}$.

29. $(4xy^3 - 9y^2 + 4xy^2)dx$
$\quad + (3x^2y^2 - 6xy + 2x^2y)dy = 0$.

30. $\dfrac{dy}{dx} = (x + y + 1)^2 - (x + y - 1)^2$.

In Problems 31 through 40, solve the initial value problem.

31. $(x^3 - y)dx + x\,dy = 0$, $y(1) = 3$.

32. $\dfrac{dy}{dx} = \left(\dfrac{x}{y} + \dfrac{y}{x}\right)$, $y(1) = -4$.

33. $(t + x + 3)dt + dx = 0$, $x(0) = 1$.

34. $\dfrac{dy}{dx} - \dfrac{2y}{x} = x^2\cos x$, $y(\pi) = 2$.

35. $(2y^2 + 4x^2)dx - xy\,dy = 0$, $\qquad y(1) = -2$.

36. $[2\cos(2x + y) - x^2]dx$
$\quad + [\cos(2x + y) + e^y]dy = 0$, $\qquad y(1) = 0$.

37. $(2x - y)dx + (x + y - 3)dy = 0$, $\qquad y(0) = 2$.

38. $\sqrt{y}\,dx + (x^2 + 4)dy = 0$, $\qquad y(0) = 4$.

39. $\dfrac{dy}{dx} - \dfrac{2y}{x} = x^{-1}y^{-1}$, $\qquad y(1) = 3$.

40. $\dfrac{dy}{dx} - 4y = 2xy^2$, $\qquad y(0) = -4$.

TECHNICAL WRITING EXERCISES

1. An instructor at Ivey U. asserted: "All you need to know about first order differential equations is how to solve those that are exact." Give arguments that support and arguments that refute the instructor's claim.

2. What properties do solutions to linear equations have that are not shared by solutions to either separable or exact equations? Give some specific examples to support your conclusions.

3. Consider the differential equation

$$\frac{dy}{dx} = ay + be^{-x} , \qquad y(0) = c ,$$

where a, b, and c are constants. Describe what happens to the asymptotic behavior as $x \rightarrow +\infty$ of the solution when the constants a, b, and c are varied. Illustrate with figures and/or graphs.

GROUP PROJECTS FOR CHAPTER 2

A. The Snowplow Problem

To apply the techniques discussed in this chapter to real-world problems, it is necessary to "translate" these problems into questions that can be answered mathematically. The process of reformulating a real-world problem as a mathematical one often requires making certain simplifying assumptions. To illustrate this, consider the following snowplow problem:

> One morning it began to snow very hard and continued snowing steadily throughout the day. A snowplow set out at 8:00 A.M. to clear a road, clearing 2 mi by 11:00 A.M. and an additional mile by 1:00 P.M. *At what time did it start snowing?*

To solve this problem, you can make two physical assumptions concerning the rate at which it is snowing and the rate at which the snowplow can clear the road. Because it is snowing steadily, it is reasonable to assume it is snowing at a constant rate. From the data given (and from our experience), the deeper the snow, the slower the snowplow moves. With this in mind, assume the rate (in mi/hr) at which a snowplow can clear a road is inversely proportional to the depth of the snow.

B. Two Snowplows

Alar Toomre, Massachusetts Institute of Technology

One day it began to snow exactly at noon at a heavy and *steady* rate. A snowplow left its garage at 1:00 P.M., and another one followed in its tracks at 2:00 P.M. (see Figure 2.12).

(a) *At what time did the second snowplow crash into the first?* To answer this question, assume as in Project A that the rate (in mi/hr) at which a snowplow can clear the road is inversely proportional to the depth of the snow (and hence to the time elapsed since the road was clear of snow). [*Hint:* Begin by writing differential equations for $x(t)$ and $y(t)$, the distances traveled by the first and second snowplows, respectively, at t hours past noon. To solve the differential equation involving y, let t rather than y be the dependent variable!]

(b) *Could the crash have been avoided by dispatching the second snowplow at 3:00 P.M. instead?*

Figure 2.12 Method of successive snowplows

C. *Asymptotic Behavior of Solutions to Linear Equations*

To illustrate how the asymptotic behavior of the forcing term $Q(x)$ affects the solution to a linear equation, consider the equation

(1) $\dfrac{dy}{dx} + ay = Q(x)$,

where the constant a is positive and $Q(x)$ is continuous on $[0, \infty)$.

(a) Show that the general solution to equation (1) can be written in the form

(2) $y(x) = y(x_0)e^{-a(x-x_0)} + e^{-ax} \displaystyle\int_{x_0}^{x} e^{at}Q(t)\,dt$,

where x_0 is a nonnegative constant.

(b) If $|Q(x)| \le k$ for $x \ge x_0$, where k and x_0 are nonnegative constants, show that

$|y(x)| \le |y(x_0)|e^{-a(x-x_0)} + \dfrac{k}{a}\big[1 - e^{-a(x-x_0)}\big]$ for $x \ge x_0$.

(c) Let $z(x)$ satisfy the same equation as (1) but with forcing function $\tilde{Q}(x)$. That is,

$\dfrac{dz}{dx} + az = \tilde{Q}(x)$,

where $\tilde{Q}(x)$ is continuous on $[0, \infty)$. Show that if

$|\tilde{Q}(x) - Q(x)| \le K$ for $x \ge x_0$,

then

$|z(x) - y(x)| \le |z(x_0) - y(x_0)|e^{-a(x-x_0)} + \dfrac{K}{a}\big[1 - e^{-a(x-x_0)}\big]$ for $x \ge x_0$.

(d) Now show that if $Q(x) \to \beta$ as $x \to \infty$, then any solution $y(x)$ of equation (1) satisfies $y(x) \to \beta/a$ as $x \to \infty$. [*Hint:* Take $\tilde{Q}(x) \equiv \beta$ and $z(x) \equiv \beta/a$ in part (c).]

(e) As an application of part (d), suppose that a brine solution containing $q(t)$ kilograms of salt per liter at time t runs into a tank of water at a fixed rate and that the mixture, kept uniform by stirring, flows out at the same rate. Given that $q(t) \to \beta$ as $t \to \infty$, use the result of part (d) to determine the limiting concentration of the salt in the tank as $t \to \infty$ (see Exercises 2.3, Problem 35).

D. *Torricelli's Law of Fluid Flow*

Randall K. Campbell-Wright, University of Tampa

How long does it take for water to drain through a hole in the bottom of a tank? Consider the tank pictured in Figure 2.13, which drains through a small, round hole. **Torricelli's law**[†] states that

[†]*Historical Footnote:* Evangelista Torricelli (1608–1647) invented the barometer and worked on computing the value of the acceleration of gravity as well as observing this principle of fluid flow.

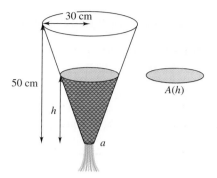

Figure 2.13 Conical tank

when the surface of the water is at a height h, the water drains with the velocity it would have if it fell freely from a height h (ignoring various forms of friction).

(a) Show that the standard gravity differential equation

$$\frac{d^2h}{dt^2} = -g$$

leads to the conclusion that an object that falls from a height h will land with a velocity of $-\sqrt{2gh}$.

(b) Let $A(h)$ be the cross-sectional area of the water in the tank at height h and a the area of the drain hole. The rate at which water is flowing out of the tank at time t can be expressed as the cross-sectional area at height h times the rate at which the height of the water is changing. Alternatively, the rate at which water flows out of the hole can be expressed as the area of the hole times the velocity of the draining water. Set these two equal to each other and insert Torricelli's law to derive the differential equation

(3) $A(h)\dfrac{dh}{dt} = -a\sqrt{2gh}$.

(c) The conical tank of Figure 2.13 has a radius of 30 cm when it is filled to an initial depth of 50 cm. A small round hole at the bottom has a diameter of 1 cm. Determine $A(h)$ and a and then solve the differential equation in (3), thus deriving a formula relating time and the height of the water in this tank.

(d) Use your solution to (c) to predict how long it will take for the tank to drain entirely.

(e) Which would drain faster, the tank pictured or an upside down conical tank of the same dimensions draining through a hole of the same size? How long would it take to drain the upside down tank?

(f) Find a water tank and time how long it takes to drain. (You may be able to borrow a "separatory funnel" from your chemistry department or use a large water cooler.) The tank should be large enough to take several minutes to drain, and the drain hole should be large enough to allow water to flow freely. The top of the tank should be open (so that the water will not "glug"). Repeat steps (c) and (d) for your tank and compare the prediction of Torricelli's law to your experimental results.

E. *Clairaut Equations and Singular Solutions*

An equation of the form

(4) $y = x\dfrac{dy}{dx} + f(dy/dx)$,

where the continuously differentiable function $f(t)$ is evaluated at $t = dy/dx$, is called a **Clairaut equation.**[†] Interest in these equations is due to the fact that (4) has a one-parameter family of solutions that consist of *straight lines*. Further, the **envelope** of this family—that is, the curve whose tangent lines are given by the family—is also a solution to (4) and is called the **singular solution.**

To solve a Clairaut equation:

(a) Differentiate equation (4) with respect to x and simplify to show that

(5) $\left[x + f'(dy/dx)\right]\dfrac{d^2y}{dx^2} = 0$, where $f'(t) = \dfrac{d}{dt} f(t)$.

(b) From (5), conclude that $dy/dx = c$ or $f'(dy/dx) = -x$. Assume $dy/dx = c$ and substitute back into (4) to obtain the family of *straight-line solutions*

$y = cx + f(c)$.

(c) Show that another solution to (4) is given parametrically by

$x = -f'(p)$,

$y = f(p) - pf'(p)$,

where the parameter $p = dy/dx$. This solution is the *singular solution.*

(d) Use the above method to find the family of straight-line solutions and the singular solution to the equation

$y = x\left(\dfrac{dy}{dx}\right) + 2\left(\dfrac{dy}{dx}\right)^2$.

Here $f(t) = 2t^2$. Sketch several of the straight-line solutions along with the singular solution on the same coordinate system. Observe that the straight-line solutions are all tangent to the singular solution.

(e) Repeat part (d) for the equation

$x\left(\dfrac{dy}{dx}\right)^3 - y\left(\dfrac{dy}{dx}\right)^2 + 2 = 0$.

[†]*Historical Footnote:* These equations were studied by Alexis Clairaut in 1734.

CHAPTER 3

Mathematical Models and Numerical Methods Involving First Order Equations

3.1 MATHEMATICAL MODELING

Adopting the Babylonian practices of careful measurement and detailed observations, the ancient Greeks sought to comprehend nature by logical analysis. Aristotle's convincing arguments that the world was not flat, but spherical, led the intellectuals of that day to ponder the question: What is the circumference of Earth? And it was astonishing that Eratosthenes managed to obtain a fairly accurate answer to this problem without having to set foot beyond the ancient city of Alexandria. His method involved certain assumptions and simplifications: Earth is a perfect sphere, the sun's rays travel parallel paths, the city of Syene was 5000 stadia due south of Alexandria, and so on. With these idealizations, Eratosthenes created a mathematical context in which the principles of geometry could be applied.[†]

Today, as scientists seek to further our understanding of nature and as engineers seek, on a more pragmatic level, to find answers to technical problems, the technique of representing our "real world" in mathematical terms has become an invaluable tool. This process of mimicking reality by using the language of mathematics is known as **mathematical modeling.**

Formulating problems in mathematical terms has several benefits. First, it requires that we clearly state our premises. Real-world problems are often complex, involving several different and possibly interrelated processes. Before mathematical treatment can proceed, one must determine which variables are significant and which can be ignored. Often, for the relevant variables, relationships are postulated in the form of laws, formulas, theories, etc. These assumptions constitute the **idealizations** of the model.

Mathematics contains a wealth of theorems and techniques for making logical deductions and manipulating equations. Hence it provides a context in which analysis can take place free

[†]For further reading, see, for example, *The Mapmakers,* by John Noble Wilford (Vintage Books, New York, 1982), Chapter 2.

of any preconceived notions of the outcome. It is also of great practical importance that mathematics provides a format for obtaining numerical answers via a computer.

The process of building an effective mathematical model takes skill, imagination, and objective evaluation. Certainly an exposure to several existing models that illustrate various aspects of modeling can lead to a better feel for the process. Several excellent books and articles are devoted exclusively to the subject.[†] In this chapter, we concentrate on examples of models that involve first order differential equations. In studying these and in building your own models, the following broad outline of the process may be helpful.

Formulate the Problem

Here you must pose the problem in such a way that it can be "answered" mathematically. This requires an understanding of the problem area as well as the mathematics. At this stage, you may need to confer with experts in that area and read the relevant literature.

Develop the Model

There are two things to be done here. First you must decide which variables are important and which are not. The former are then classified as independent variables or dependent variables. The unimportant variables are those that have very little or no effect on the process. (For example, in studying the motion of a falling body, its color is usually of little interest.) The independent variables are those whose effect is significant and that will serve as input for the model.[††] For the falling body, its shape, mass, initial position, initial velocity, and time from release are possible independent variables. The dependent variables are those that are affected by the independent variables and that are important to solving the problem. Again, for a falling body, its velocity, location, and time of impact are all possible dependent variables.

Second, you must determine or specify the relationships (for example, a differential equation) that exist among the relevant variables. This requires a good background in the area and insight into the problem. You may begin with a crude model and then, based upon testing, refine the model as needed. For example, you might begin by ignoring any friction acting on the falling body. Then, if necessary to obtain a more acceptable answer, try to take into account any frictional forces that may affect the motion.

Test the Model

Before attempting to "verify" a model by comparing its output with experimental data, the following questions should be considered:

> Are the assumptions reasonable?
>
> Are the equations dimensionally consistent? (For example, we don't want to add units of force to units of velocity.)

[†]See, for example, *A First Course in Mathematical Modeling,* 2nd Edition, by F. R. Giordano and M. D. Weir (Brooks/Cole Publishing Company, Pacific Grove, California, 1997) or *Concepts of Mathematical Modeling,* by W. J. Meyer (McGraw-Hill Book Company, New York, 1984).

[††]In the mathematical formulation of the model, certain of the independent variables may be called **parameters.**

Is the model internally consistent in the sense that equations do not contradict one another?

Do the relevant equations have solutions?

Are the solutions unique?

How difficult is it to obtain the solutions?

Do the solutions provide an answer for the problem being studied?

When possible, try to validate the model by comparing its predictions with any experimental data. Begin with rather simple predictions that involve little computation or analysis. Then, as the model is refined, check to see that the accuracy of the model's predictions is acceptable to you. In some cases, validation is impossible or socially, politically, economically, or morally unreasonable. For example, how does one validate a model that predicts when our sun will die out?

Each time the model is used to predict the outcome of a process and hence solve a problem, it provides a test of the model that may lead to further refinements or simplifications. In many cases, a model is simplified to give a quicker or less expensive answer—provided, of course, that sufficient accuracy is maintained.

One should always keep in mind that a model is *not* reality but only a representation of reality. The more refined models *may* provide an understanding of the underlying processes of nature. For this reason, applied mathematicians strive for better, more refined models. Still, the real test of a model is its ability to find an acceptable answer for the posed problem.

In this chapter, we discuss various models that involve differential equations. In Section 3.2, Compartmental Analysis, and Section 3.4, Newtonian Mechanics, we consider two general models that result from interpreting the derivative as a rate of change of one variable with respect to another. Section 3.3 describes a model for the heating and cooling of a building. Sections 3.5 and 3.6 contain a brief introduction to some numerical techniques for solving first order initial value problems. This will enable us to study more realistic models that cannot be solved using the techniques of Chapter 2. Finally, in Section 3.7 we discuss some refined techniques that are used in professionally prepared computer codes for solving initial value problems.

3.2 COMPARTMENTAL ANALYSIS

Many complicated processes can be broken down into distinct stages and the entire system modeled by describing the interactions between the various stages. Such systems are called **compartmental** and are graphically depicted by **block diagrams.** In this section, we study the basic unit of these systems, a single compartment, and analyze some simple processes that can be handled by such a model.

The basic one-compartment system consists of a function $x(t)$ that represents the amount of a substance in the compartment at time t, an input rate at which the substance enters the compartment, and an output rate at which the substance leaves the compartment (see Figure 3.1 on page 96).

Figure 3.1 Schematic representation of a one-compartment system

Because the derivative of x with respect to t can be interpreted as the rate of change in the amount of the substance in the compartment with respect to time, the one-compartment system suggests

(1) $\dfrac{dx}{dt} = \text{input rate} - \text{output rate}$

as a mathematical model for the process.

Mixing Problems

A problem for which the one-compartment system provides a useful representation is the mixing of fluids in a tank. Let $x(t)$ represent the amount of a substance in a tank (compartment) at time t. To use the compartmental analysis model, we must be able to determine the rates at which this substance enters and leaves the tank. In mixing problems, one is often given the rate at which a fluid containing the substance flows into the tank, along with the concentration of the substance in that fluid. Hence multiplying the flow rate (volume/time) by the concentration (amount/volume) yields the input rate (amount/time).

The output rate of the substance is usually more difficult to determine. If we are given the exit rate of the mixture of fluids in the tank, then how do we determine the concentration of the substance in the mixture? One simplifying assumption that we might make is that the concentration is kept uniform in the mixture. Then we can compute the concentration of the substance in the mixture by dividing the amount $x(t)$ by the volume of the mixture in the tank at time t. Multiplying this concentration by the exit rate of the mixture then gives the desired output rate of the substance. This model is used in the next two examples.

EXAMPLE 1 Consider a large tank holding 1000 L of pure water into which a brine solution of salt begins to flow at a constant rate of 6 L/min. The solution inside the tank is kept well stirred and is flowing out of the tank at a rate of 6 L/min. If the concentration of salt in the brine entering the tank is 1 kg/L, determine when the concentration of salt in the tank will reach 1/2 kg/L (see Figure 3.2).

Figure 3.2 Mixing problem with equal flow rates

SOLUTION We can view the tank as a compartment containing salt. If we let $x(t)$ denote the mass of salt in the tank at time t, we can determine the concentration of salt in the tank by dividing $x(t)$ by the volume of fluid in the tank at time t. We use the mathematical model described by equation (1) to solve for $x(t)$.

First we must determine the rate at which salt enters the tank. We are given that brine flows into the tank at a rate of 6 L/min. Since the concentration is 1 kg/L, we conclude that the input rate of salt into the tank is

(2) $(6 \text{ L/min})(1 \text{ kg/L}) = 6 \text{ kg/min}.$

We must now determine the output rate of salt from the tank. The brine solution in the tank is kept well stirred, so let's assume the concentration of salt in the tank is uniform. That is, the concentration of salt in any part of the tank at time t is just $x(t)$ divided by the volume of fluid in the tank. Because the tank initially contains 1000 L and the rate of flow into the tank is the same as the rate of flow out, the volume is a constant 1000 L. Hence the output rate of salt is

(3) $(6 \text{ L/min})\left(\dfrac{x(t)}{1000} \text{ kg/L}\right) = \dfrac{3x(t)}{500} \text{ kg/min}.$

The tank initially contained pure water, so we set $x(0) = 0$. Substituting the rates in (2) and (3) into equation (1) then gives the initial value problem

(4) $\dfrac{dx}{dt} = 6 - \dfrac{3x}{500}, \qquad x(0) = 0,$

as a mathematical model for the mixing problem.

The equation in (4) is separable (and linear) and easy to solve. Using the initial condition $x(0) = 0$ to evaluate the arbitrary constant, we obtain

(5) $x(t) = 1000\left(1 - e^{-3t/500}\right).$

Thus the concentration of salt in the tank at time t is

$$\frac{x(t)}{1000} = 1 - e^{-3t/500} \text{ kg/L}.$$

To determine when the concentration of salt is 1/2 kg/L, we set the right-hand side equal to 1/2 and solve for t. This gives

$$1 - e^{-3t/500} = \frac{1}{2} \quad \text{or} \quad e^{-3t/500} = \frac{1}{2},$$

and hence

$$t = \frac{500 \ln 2}{3} \approx 115.52 \text{ min}.$$

Consequently the concentration of salt in the tank will be 1/2 kg/L after 115.52 min. ∎ End

From equation (5), we observe that the mass of salt in the tank steadily increases and has the limiting value

$$\lim_{t \to \infty} x(t) = \lim_{t \to \infty} 1000(1 - e^{-3t/500}) = 1000 \text{ kg.}$$

Thus the limiting concentration of salt in the tank is 1 kg/L, which is the same as the concentration of salt in the brine flowing into the tank. This certainly agrees with our expectations! It might be interesting to see what would happen to the concentration if the flow rate into the tank is greater than the flow rate out.

EXAMPLE 2 For the mixing problem described in Example 1, assume now that the brine leaves the tank at a rate of 5 L/min instead of 6 L/min, with all else being the same (see Figure 3.3). Determine the concentration of salt in the tank as a function of time.

SOLUTION The difference between the rate of flow into the tank and the rate of flow out is $6 - 5 = 1$ L/min, so the volume of fluid in the tank after t minutes is $(1000 + t)$ L. Hence the rate at which salt leaves the tank is

$$(5 \text{ L/min}) \left[\frac{x(t)}{1000 + t} \text{ kg/L} \right] = \frac{5x(t)}{1000 + t} \text{ kg/min.}$$

Using this in place of (3) for the output rate gives the initial value problem

$$\textbf{(6)} \qquad \frac{dx}{dt} = 6 - \frac{5x}{1000 + t} \ , \qquad x(0) = 0 \ ,$$

as a mathematical model for the mixing problem.

The differential equation in (6) is linear, so we can use the procedure outlined on page 56 to solve for $x(t)$. The integrating factor is $\mu(t) = (1000 + t)^5$. Thus

$$\frac{d}{dt}\left[(1000 + t)^5 x\right] = 6(1000 + t)^5 \ ,$$

$$(1000 + t)^5 x = (1000 + t)^6 + c \ ,$$

$$x(t) = (1000 + t) + c(1000 + t)^{-5} \ .$$

Using the initial condition $x(0) = 0$, we find $c = -(1000)^6$, and thus the solution to (6) is

$$x(t) = (1000 + t) - (1000)^6(1000 + t)^{-5} \ .$$

6 L/min
1 kg/L

$x(t)$

? L

$x(0) = 0$ kg 5 L/min

Figure 3.3 Mixing problem with unequal flow rates

Hence the concentration of salt in the tank at time t is

(7) $\qquad \dfrac{x(t)}{1000 + t} = 1 - (1000)^6(1000 + t)^{-6} \text{ kg/L.}$ ∎ End

As in Example 1, the concentration given by (7) approaches 1 kg/L as $t \to \infty$. However, in Example 2 the volume of fluid in the tank becomes unbounded, and when the tank begins to overflow, the model in (6) is no longer appropriate.

For mixing problems, the one-compartment model is extremely useful. The only assumption made that is the least bit questionable is that the well-stirred mixture in the tank has a uniform concentration. Even with stirring, the salt entering the tank at the top has a delayed effect on the concentration of the brine leaving at the bottom. It is more realistic to assume that the concentration of brine leaving the tank at time t is equal to the average concentration at an earlier instant, say at time $t - t_0$, where t_0 is a positive constant. The original problem of Example 1 would now be described by

$$ x'(t) = 6 - \frac{3}{500}x(t - t_0) \, , \qquad x(t) = 0 \quad \text{for} \quad t \in \left[-t_0, 0\right] \, . $$

Such equations are called **delay differential equations** due to the "delay" appearing in the argument of $x(t - t_0)$. These equations are discussed further in the projects at the end of this chapter (see Project A).

Population Models

How does one predict the growth of a population? If we are interested in a single population, we can think of the species as being contained in a compartment (a petri dish, an island, a country, etc.) and study the growth process as a one-compartment system.

Let $p(t)$ be the population at time t. While the population is always an integer, it is usually large enough so that very little error is introduced in assuming that $p(t)$ is a continuous function. We now need to determine the growth (input) rate and the death (output) rate for the population.

Let's consider a population of bacteria that reproduce by simple cell division. In our model, we assume that the growth rate is proportional to the population present. This assumption is consistent with observations of bacterial growth. As long as there are sufficient space and ample food supply for the bacteria, we can also assume that the death rate is zero. (Remember that in cell division, the parent cell does not die, but becomes two new cells.) Hence a mathematical model for a population of bacteria is

(8) $\qquad \dfrac{dp}{dt} = k_1 p \, , \qquad p(0) = p_0 \, ,$

where $k_1 > 0$ is the proportionality constant for the growth rate and p_0 is the population at time $t = 0$. For human populations, the assumption that the death rate is zero is certainly wrong! However, if we assume that people die only of natural causes, we might expect the death rate also to be proportional to the size of the population. That is, we revise (8) to read

(9) $\qquad \dfrac{dp}{dt} = k_1 p - k_2 p = (k_1 - k_2)p = kp \, ,$

where $k := k_1 - k_2$ and k_2 is the proportionality constant for the death rate. Let's assume $k_1 > k_2$ so that $k > 0$. This gives the mathematical model

$$(10) \qquad \frac{dp}{dt} = kp , \qquad p(0) = p_0 ,$$

which is called the **Malthusian,**[†] or **exponential, law** of population growth. This equation is separable, and solving the initial value problem for $p(t)$ gives

$$(11) \qquad p(t) = p_0 e^{kt} .$$

To test the Malthusian model, let's apply it to the demographic history of the United States.

EXAMPLE 3 In 1790 the population of the United States was 3.93 million, and in 1890 it was 62.95 million. Using the Malthusian model, estimate the U.S. population as a function of time.

SOLUTION If we set $t = 0$ to be the year 1790, then by formula (11) we have

$$(12) \qquad p(t) = (3.93)e^{kt} ,$$

where $p(t)$ is the population in millions. One way to obtain a value for k would be to make the model fit the data for some specific year, such as 1890 ($t = 100$ years).[††] We have

$$p(100) = 62.95 = (3.93)e^{100k} .$$

Solving for k yields

$$k = \frac{\ln(62.95) - \ln(3.93)}{100} \approx 0.027737 .$$

Substituting this value in equation (12), we find

$$(13) \qquad p(t) = (3.93)e^{(0.027737)t} . \quad \blacksquare \text{End}$$

In Table 3.1 we list the U.S. population as given by the U.S. Bureau of the Census, and the population predicted by the Malthusian model using equation (13). From Table 3.1, we see that the predictions based on the Malthusian model are in reasonable agreement with the census data until about 1900. After 1900, the predicted population is too large, and so the Malthusian model is unacceptable.

We remark that a Malthusian model can be generated using the census data for any two different years. We selected 1790 and 1890 for purposes of comparison with the logistic model that we now describe.

[†]*Historical Footnote:* Thomas R. Malthus (1766–1834) was a British economist who studied population models.

[††]The choice of the year 1890 is purely arbitrary, of course; a more democratic (and better) way of extracting parameters from data is described after Example 4.

		Malthusian	Logistic	$\dfrac{1}{p}\dfrac{dp}{dt}$	Logistic
Year	U.S. Census	(Example 3)	(Example 4)		(Least Squares)
1790	3.93	3.93	3.93		4.01
1800	5.31	5.19	5.30	0.0312	5.34
1810	7.24	6.84	7.13	0.0299	7.10
1820	9.64	9.03	9.58	0.0292	9.42
1830	12.87	11.92	12.82	0.0289	12.47
1840	17.07	15.73	17.07	0.0302	16.44
1850	23.19	20.76	22.61	0.0310	21.58
1860	31.44	27.39	29.71	0.0264	28.16
1870	39.82	36.15	38.66	0.0235	36.47
1880	50.16	47.70	49.71	0.0231	46.80
1890	62.95	62.95	62.97	0.0205	59.37
1900	75.99	83.07	78.40	0.0191	74.28
1910	91.97	109.63	95.68	0.0162	91.43
1920	105.71	144.67	114.25	0.0146	110.52
1930	122.78	190.91	133.33	0.0106	130.94
1940	131.67	251.94	152.05	0.0108	151.93
1950	151.33	332.47	169.61	0.0157	172.61
1960	179.32	438.75	185.40	0.0145	192.16
1970	203.21	579.00	199.07	0.0116	209.93
1980	226.50	764.08	210.51	0.0102	225.52
1990	249.63	1008.32	219.82		238.77
2000	?	1330.63	227.24		249.74

TABLE 3.1 A COMPARISON OF THE MALTHUSIAN AND LOGISTIC MODELS WITH U.S. CENSUS DATA (POPULATION IS GIVEN IN MILLIONS)

The Malthusian model considered only death by natural causes. What about premature deaths due to malnutrition, inadequate medical supplies, communicable diseases, violent crimes, etc.? These factors involve a competition within the population, so we might assume there is another component of the death rate that is proportional to the number of two-party interactions. There are $p(p-1)/2$ such possible interactions for a population of size p. Thus if we combine the birth rate (8) with the death rate and rearrange constants, we get the **logistic model**

$$\frac{dp}{dt} = k_1 p - k_3 \frac{p(p-1)}{2}$$

or

$$(14) \qquad \frac{dp}{dt} = -Ap(p - p_1), \qquad p(0) = p_0,$$

where $A = k_3/2$ and $p_1 = (2k_1/k_3) + 1$.

Equation (14) has two equilibrium solutions: $p(t) \equiv p_1$ (a *sink,* in the terminology of Section 1.4) and $p(t) \equiv 0$ (mathematically a *source,* but the negative solutions are physically meaningless). The nonequilibrium solutions can be found by separating variables and using a partial fractions expansion:

$$\int \frac{dp}{p(p - p_1)} = -A \int dt$$

$$\frac{-1}{p_1} \int \frac{dp}{p} + \frac{1}{p_1} \int \frac{dp}{p - p_1} = -A \int dt$$

$$\frac{-1}{p_1} \ln |p| + \frac{1}{p_1} \ln |p - p_1| = -At + c_1$$

$$\ln \left| \frac{p}{p - p_1} \right| = Ap_1 t + c_2$$

$$\frac{p}{p - p_1} = c_3 e^{Ap_1 t} .$$

If $p(t) = p_0$ at $t = 0$, then $c_3 = p_0/(p_0 - p_1)$ and, solving for $p(t)$, we find

$$(15) \qquad p(t) = \frac{p_0 p_1}{p_0 + (p_1 - p_0)e^{-Ap_1 t}} .$$

The function $p(t)$ given in (15) is called the **logistic function,** and graphs of logistic curves are displayed in Figure 3.4.[†]

Let's test the logistic model on the population growth of the United States.

EXAMPLE 4　Taking the 1790 population of 3.93 million as the initial population and given the 1840 and 1890 populations of 17.07 and 62.95 million, respectively, use the logistic model to estimate the population at time t.

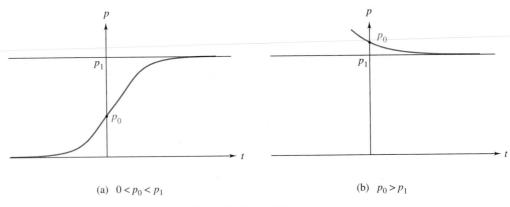

(a)　$0 < p_0 < p_1$　　　　　　　　　(b)　$p_0 > p_1$

Figure 3.4 The logistic curves

[†]*Historical Footnote:* The logistic model for population growth was first developed by P. F. Verhulst around 1840.

SOLUTION With $t = 0$ corresponding to the year 1790, we know that $p_0 = 3.93$. We must now determine the parameters A, p_1 in equation (15). For this purpose, we use the given facts that $p(50) = 17.07$ and $p(100) = 62.95$; that is,

(16) $$17.07 = \frac{3.93\,p_1}{3.93 + (p_1 - 3.93)e^{-50Ap_1}},$$

(17) $$62.95 = \frac{3.93\,p_1}{3.93 + (p_1 - 3.93)e^{-100Ap_1}}.$$

Equations (16) and (17) are two nonlinear equations in the two unknowns A, p_1. To solve such a system, we would usually resort to a numerical approximation scheme such as Newton's method. However, for the case at hand, it is possible to find the solutions directly because the data are given at times t_a and t_b with $t_b = 2t_a$ (see Problem 12). Carrying out the algebra described in Problem 12, we ultimately find that

(18) $p_1 = 250.9992$ and $A = 0.0001214$.

Thus the logistic model for the given data is

(19) $$p(t) = \frac{986.43}{3.93 + (247.07)e^{-(0.030471)t}}.\quad \blacksquare\text{End}$$

The population data predicted by (19) are displayed in column 4 of Table 3.1. As you can see, these predictions are in better agreement with the census data than the Malthusian model is. And, of course, the agreement is perfect in the years 1790, 1840, and 1890. However, the choice of these particular years for estimating the parameters p_0, A, and p_1 is quite arbitrary, and we would expect that a more robust model would use *all* of the data, in some way, for the estimation. One way to implement this idea is the following.

Observe from equation (14) that the logistic model predicts a linear relationship between $(dp/dt)/p$ and p:

$$\frac{1}{p}\frac{dp}{dt} = Ap_1 - Ap,$$

with Ap_1 as the intercept and $-A$ as the slope. In column five of Table 3.1, we list values of $(dp/dt)/p$, which are estimated from centered differences according to

(20) $$\frac{1}{p(t)}\frac{dp}{dt}(t) \approx \frac{1}{p(t)}\frac{p(t+10) - p(t-10)}{20}$$

(see Problem 16). In Figure 3.5 on page 104, these estimated values of $(dp/dt)/p$ are plotted against p in what is called a *scatter diagram*. The linear relationship predicted by the logistic model suggests that we approximate the plot by a straight line. A standard technique for doing this is the so-called *least-squares linear fit,* which is discussed in Appendix D. This yields the straight line

$$\frac{1}{p}\frac{dp}{dt} = 0.0291108 - 0.0001007p,$$

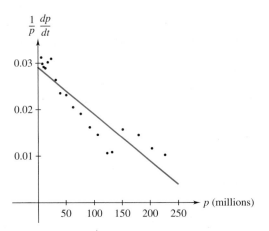

Figure 3.5 Scatter data and straight line fit

which is also depicted in Figure 3.5. Now with $A = 0.0001007$ and $p_1 = (0.0291108/A) = 289.1$, we can solve equation (15) for p_0:

(21) $$p_0 = \frac{p(t)p_1e^{-Ap_1t}}{p_1 - p(t)\left[1 - e^{-Ap_1t}\right]} \ .$$

By averaging the right-hand side of (21) over all the data, we obtain the estimate $p_0 \approx 4.009$. Finally, the insertion of these estimates for the parameters in equation (15) leads to the predictions listed in column six of Table 3.1.

Note that this model yields $p_1 = 289.1$ million as the limit on the future population of the United States.

EXERCISES 3.2

1. A brine solution of salt flows at a constant rate of 8 L/min into a large tank that initially held 100 L of brine solution in which was dissolved 5 kg of salt. The solution inside the tank is kept well stirred and flows out of the tank at the same rate. If the concentration of salt in the brine entering the tank is 0.5 kg/L, determine the mass of salt in the tank after t minutes. When will the concentration of salt in the tank reach 0.2 kg/L?

2. A brine solution of salt flows at a constant rate of 6 L/min into a large tank that initially held 50 L of brine solution in which was dissolved 5 kg of salt. The solution inside the tank is kept well stirred and flows out of the tank at the same rate. If the concentration of salt in the brine entering the tank is 0.5 kg/L, determine the mass of salt in the tank after t minutes. When will the concentration of salt in the tank reach 0.3 kg/L?

3. A nitric acid solution flows at a constant rate of 6 L/min into a large tank that initially held 200 L of a 0.5% nitric acid solution. The solution inside the tank is kept well stirred and flows out of the tank at a rate of 8 L/min. If the solution entering the tank is 20% nitric acid, determine the volume of nitric acid in the tank after t minutes. When will the percentage of nitric acid in the tank reach 10%?

4. A brine solution of salt flows at a constant rate of 4 L/min into a large tank that initially held 100 L of pure water. The solution inside the tank is kept well stirred and flows out of the tank at a rate of 3 L/min. If the concentration of salt in the brine entering the tank is 0.2 kg/L, determine the mass of salt in the tank after t minutes. When will the concentration of salt in the tank reach 0.1 kg/L?

5. A swimming pool whose volume is 10,000 gal contains water that is 0.01% chlorine. Starting at $t = 0$, city water containing 0.001% chlorine is pumped into the pool at a rate of 5 gal/min. The pool water flows out at the same rate. What is the percentage of chlorine in the pool after 1 hr? When will the pool water be 0.002% chlorine?

6. The air in a small room 12 ft by 8 ft by 8 ft is 3% carbon monoxide. Starting at $t = 0$, fresh air containing no carbon monoxide is blown into the room at a rate of 100 ft³/min. If air in the room flows out through a vent at the same rate, when will the air in the room be 0.01% carbon monoxide?

7. Beginning at time $t = 0$, fresh water is pumped at the rate of 3 gal/min into a 60-gal tank initially filled with brine. The resulting less-and-less salty mixture overflows at the same rate into a second 60-gal tank that initially contained only pure water, and from there it eventually spills onto the ground. Assuming perfect mixing in both tanks, when will the water in the second tank taste saltiest? And exactly how salty will it then be, compared with the original brine?

8. Blood carries a drug into an organ at a rate of 3 cm³/sec and leaves at the same rate. The organ has a liquid volume of 125 cm³. If the concentration of the drug in the blood entering the organ is 0.2 g/cm³, what is the concentration of the drug in the organ at time t if there was no trace of the drug initially? When will the concentration of the drug in the organ reach 0.1 g/cm³?

9. In 1980 the Department of Natural Resources released 1000 splake (a crossbreed of fish) into a lake. In 1987 the population of splake in the lake was estimated to be 3000. Using the Malthusian law for population growth, estimate the population of splake in the lake in the year 2004.

10. Use a sketch of the phase line (see Section 1.4) to argue that any solution to the mixing problem model

$$\frac{dx}{dt} = a - bx \; ; \qquad a, b > 0 \; ,$$

approaches the equilibrium solution $x(t) \equiv a/b$ as t approaches $+\infty$; that is, a/b is a sink.

11. Use a sketch of the phase line (see Section 1.4) to argue that any solution to the logistic model

$$\frac{dp}{dt} = (a - bp)p \; ; \qquad p(t_0) = p_0 \; ,$$

where a, b, and p_0 are positive constants, approaches the equilibrium solution $p(t) \equiv a/b$ as t approaches $+\infty$.

12. For the logistic curve (15), assume $p_a := p(t_a)$ and $p_b := p(t_b)$ are given with $t_b = 2t_a \; (t_a > 0)$. Show that

$$p_1 = \left[\frac{p_a p_b - 2p_0 p_b + p_0 p_a}{p_a^2 - p_0 p_b} \right] p_a \; ,$$

$$A = \frac{1}{p_1 t_a} \ln \left[\frac{p_b(p_a - p_0)}{p_0(p_b - p_a)} \right] \; .$$

[*Hint:* Equate the expressions (21) for p_0 at times t_a and t_b. Set $\chi = \exp(-Ap_1 t_a)$ and $\chi^2 = \exp(-Ap_1 t_b)$ and solve for χ. Insert into one of the earlier expressions and solve for p_1.]

13. In Problem 9, suppose we have the additional information that the population of splake in 1994 was estimated to be 5000. Use a logistic model to estimate the population of splake in the year 2004. What is the predicted limiting population? [*Hint:* Use the formulas in Problem 12.]

14. In 1970 the population of alligators on the Kennedy Space Center grounds was estimated to be 300. In 1980 the population had grown to an estimated 1500. Using the Malthusian law for population growth, estimate the alligator population on the Kennedy Space Center grounds in the year 2004.

15. In Problem 14, suppose we have the additional information that the population of alligators on the grounds of the Kennedy Space Center in 1975 was estimated to be 1200. Use a logistic model to estimate the population of alligators in the year 2004.

What is the predicted limiting population? [*Hint:* Use the formulas in Problem 12.]

16. Show that for a differentiable function $p(t)$, we have

$$\lim_{h \to 0} \frac{p(t + h) - p(t - h)}{2h} = p'(t) \, ,$$

which is the basis of the centered difference approximation used in (20).

17. (a) For the U.S. census data, use the forward difference approximation to the derivative, that is,

$$\frac{1}{p(t)} \frac{dp}{dt}(t) \approx \frac{1}{p(t)} \frac{p(t + 10) - p(t)}{10} \, ,$$

to recompute column 5 of Table 3.1.

(b) Using the data from part (a), determine the constants A, p_1 in the least-squares fit

$$\frac{1}{p} \frac{dp}{dt} \approx Ap_1 - Ap$$

(see Appendix D).

(c) With the values for A and p_1 found in part (b), determine p_0 by averaging formula (21) over the data.

(d) Substitute A, p_1, and p_0 as determined above into the logistic formula (15) and calculate the populations predicted for each of the years listed in Table 3.1.

(e) Compare this model with that of the centered difference-based model in column 6 of Table 3.1.

18. A population model used in actuarial predictions is based on the **Gompertz equation**

$$\frac{dP}{dt} = P(a - b \ln P) \, ,$$

where a and b are constants.

(a) Solve the Gompertz equation for $P(t)$.

(b) If $P(0) = P_0 > 0$, give a formula for $P(t)$ in terms of a, b, P_0, and t.

(c) Describe the behavior of $P(t)$ as $t \to +\infty$. [*Hint:* Consider cases for $b > 0$ and $b < 0$.]

19. The initial mass of a certain species of fish is 7 million tons. The mass of fish, if left alone, would increase at a rate proportional to the mass, with a proportionality constant of 2/yr. However, commercial fishing removes fish mass at a constant rate of 15 million tons per year. When will all the fish be gone? If the fishing rate is changed so that the mass of fish remains constant, what should that rate be?

20. From theoretical considerations, it is known that light from a certain star should reach Earth with intensity I_0. However, the path taken by the light from the star to Earth passes through a dust cloud, with absorption coefficient 0.1/light year. The light reaching Earth has intensity $1/2 \, I_0$. How thick is the dust cloud? (The rate of change of light intensity with respect to thickness is proportional to the intensity. One light year is the distance traveled by light during 1 yr.)

21. A snowball melts in such a way that the rate of change in its volume is proportional to its surface area. If the snowball was initially 4 in. in diameter and after 30 min its diameter is 3 in., when will its diameter be 2 in.? Mathematically speaking, when will the snowball disappear?

22. Suppose the snowball in Problem 21 melts so that the rate of change in its *diameter* is proportional to its surface area. Using the same given data, determine when its diameter will be 2 in. Mathematically speaking, when will the snowball disappear?

In Problems 23 through 27, assume the rate of decay of a radioactive substance is proportional to the amount of the substance present. The half-life of a radioactive substance is the time it takes for one-half of the substance to disintegrate.

23. If initially there are 50 g of a radioactive substance and after 3 days there are only 10 g remaining, what percentage of the original amount remains after 4 days?

24. If initially there are 300 g of a radioactive substance and after 5 years there are 200 g remaining, how much time must elapse before only 10 g remain?

25. Carbon dating is often used to determine the age of a fossil. For example, a humanoid skull was found in a cave in South Africa along with the remains of a campfire. Archaeologists believe the age of the skull to be the same age as the campfire. It is determined that only 2% of the original amount of carbon-14 remains in the burnt wood of the campfire. Estimate the age of the skull if the half-life of carbon-14 is about 5600 years.

26. To see how sensitive the technique of carbon dating of Problem 25 is,

 (a) Redo Problem 25 assuming the half-life of carbon-14 is 5550 years.

 (b) Redo Problem 25 assuming 3% of the original mass remains.

 (c) If each of the figures in parts (a) and (b) represents a 1% error in measuring the two parameters of half-life and percent of mass remaining, to which parameter is the model more sensitive?

27. The only undiscovered isotopes of the two unknown elements hohum and inertium (symbols Hh and It) are radioactive. Hohum decays into inertium with a decay constant of 2/yr, and inertium decays into the nonradioactive isotope of bunkum (symbol Bu) with a decay constant of 1/yr. An initial mass of 1 kg of hohum is put into a nonradiaoctive container, with no other source of hohum, inertium, or bunkum. How much of each of the three elements is in the container after t years? (The **decay constant** is the constant of proportionality in the statement that the rate of loss of mass of the element at any time is proportional to the mass of the element at that time.)

3.3 HEATING AND COOLING OF BUILDINGS

Our goal is to formulate a mathematical model that describes the 24-hr temperature profile inside a building as a function of the outside temperature, the heat generated inside the building, and the furnace heating or air conditioner cooling. From this model, we would like to answer the following three questions:

 (a) How long does it take to change the building temperature substantially?

 (b) How does the building temperature vary during spring and fall when there is no furnace heating or air conditioning?

 (c) How does the building temperature vary in summer when there is air conditioning or in the winter when there is furnace heating?

A natural approach to modeling the temperature inside a building is to use compartmental analysis. Let $T(t)$ represent the temperature inside the building at time t and view the building as a single compartment. Then the rate of change in the temperature is determined by all the factors that generate or dissipate heat.

We will consider three main factors that affect the temperature inside the building. First is the heat produced by people, lights, and machines inside the building. This causes a rate of increase in temperature that we will denote by $H(t)$. Second is the heating (or cooling) supplied by the furnace (or air conditioner). This rate of increase (or decrease) in temperature will be represented by $U(t)$. In general, the additional heating rate $H(t)$ and the furnace (or air conditioner) rate $U(t)$ are described in terms of energy per unit time (such as British thermal units per hour). However, by multiplying by the heat capacity of the building (in units of degrees temperature change per unit heat energy), we can express the two quantities $H(t)$ and $U(t)$ in terms of temperature per unit time.

The third factor is the effect of the outside temperature $M(t)$ on the temperature inside the building. Experimental evidence has shown that this factor can be modeled using **Newton's law of cooling.** This law states that there is a rate of change in the temperature $T(t)$ that is proportional to the difference between the outside temperature $M(t)$ and the inside temperature

$T(t)$. That is, the rate of change in the building temperature due to $M(t)$ is

$$K[M(t) - T(t)] \;.$$

The positive constant K depends on the physical properties of the building, such as the number of doors and windows and the type of insulation, but K does not depend on M, T, or t. Hence, when the outside temperature is greater than the inside temperature, $M(t) - T(t) > 0$ and there is an increase in the building temperature due to $M(t)$. On the other hand, when the outside temperature is less than the inside temperature, then $M(t) - T(t) < 0$ and the building temperature decreases.

Summarizing, we find

(1) $$\frac{dT}{dt} = K[M(t) - T(t)] + H(t) + U(t) \;,$$

where the additional heating rate $H(t)$ is always nonnegative and $U(t)$ is positive for furnace heating and negative for air conditioner cooling. A more detailed model of the temperature dynamics of the building could involve more variables to represent different temperatures in different rooms or zones. Such an approach would use compartmental analysis, with the rooms as different compartments (see Problems 35–37, Exercises 5.3).

Because equation (1) is linear, it can be solved using the method discussed in Section 2.3. Rewriting (1) in the standard form

(2) $$\frac{dT}{dt}(t) + P(t)T(t) = Q(t) \;,$$

where

$$P(t) := K \;,$$

(3) $$Q(t) := KM(t) + H(t) + U(t) \;,$$

we find that the integrating factor is

$$\mu(t) = \exp\left(\int K \, dt\right) = e^{Kt} \;.$$

To solve (2), multiply each side by e^{Kt} and integrate:

$$e^{Kt}\frac{dT}{dt}(t) + Ke^{Kt}T(t) = e^{Kt}Q(t) \;,$$

$$e^{Kt}T(t) = \int e^{Kt}Q(t)dt + C \;.$$

Solving for $T(t)$ gives

(4) $$T(t) = e^{-Kt}\int e^{Kt}Q(t)dt + Ce^{-Kt}$$

$$= e^{-Kt}\left\{\int e^{Kt}[KM(t) + H(t) + U(t)]dt + C\right\} \;.$$

EXAMPLE 1 Suppose that at the end of the day (at time t_0) when people leave the building, the outside temperature stays constant at M_0, the additional heating rate H inside the building is zero, and the furnace/air conditioner rate U is zero. Determine $T(t)$, given the initial condition $T(t_0) = T_0$.

SOLUTION With $M = M_0$, $H = 0$, and $U = 0$, equation (4) becomes

$$T(t) = e^{-Kt}\left\{\int e^{Kt}KM_0\,dt + C\right\} = e^{-Kt}\left[M_0 e^{Kt} + C\right]$$

$$= M_0 + Ce^{-Kt} \ .$$

Setting $t = t_0$ and using the initial value T_0 of the temperature, we find that the constant C is $(T_0 - M_0)\exp(Kt_0)$. Hence,

(5) $T(t) = M_0 + (T_0 - M_0)e^{-K(t - t_0)}$. ∎ End

When $M_0 < T_0$, the solution in (5) decreases exponentially from the initial temperature T_0 to the final temperature M_0. To determine a measure of the time it takes for the temperature to change "substantially," consider the simple linear equation $dA/dt = -\alpha A$, whose solutions have the form $A(t) = A(0)e^{-\alpha t}$. Now, as $t \to +\infty$, the function $A(t)$ either decays exponentially $(\alpha > 0)$ or grows exponentially $(\alpha < 0)$. In either case, the time it takes for $A(t)$ to change from $A(0)$ to $A(0)/e \left(\approx 0.368\,A(0)\right)$ is just $1/\alpha$ because

$$A\left(\frac{1}{\alpha}\right) = A(0)e^{-\alpha(1/\alpha)} = \frac{A(0)}{e} \ .$$

The quantity $1/|\alpha|$, which is independent of $A(0)$, is called the **time constant** for the equation. For linear equations of the more general form $dA/dt = -\alpha A + g(t)$, we again refer to $1/|\alpha|$ as the time constant.

Returning to Example 1, we see that the temperature $T(t)$ satisfies the equations

$$\frac{dT}{dt}(t) = -KT(t) + KM_0 \ , \qquad \frac{d(T - M_0)}{dt}(t) = -K[T(t) - M_0] \ ,$$

for M_0 a constant. In either case, the time constant is just $1/K$, which represents the time it takes for the temperature difference $T - M_0$ to change from $T_0 - M_0$ to $(T_0 - M_0)/e$. We also call $1/K$ the **time constant for the building** (without heating or air conditioning). A typical value for the time constant of a building is 2 to 4 hr, but the time constant can be much shorter if windows are open or if there is a fan circulating air. Or it can be much longer if the building is well insulated.

In the context of Example 1, we can use the notion of time constant to answer our initial question (a): The building temperature changes exponentially with a time constant of $1/K$. An answer to question (b) about the temperature inside the building during spring and fall is given in the next example.

EXAMPLE 2 Find the building temperature $T(t)$ if the additional heating rate $H(t)$ is equal to the constant H_0, there is no heating or cooling $\left(U(t) \equiv 0\right)$, and the outside temperature M varies as a sine wave over a 24-hr period, with its minimum at $t = 0$ (midnight) and its maximum at $t = 12$ (noon). That is,

$$M(t) = M_0 - B\cos\omega t \ ,$$

where B is a positive constant, M_0 is the average outside temperature, and $\omega = 2\pi/24 = \pi/12$ radians/hr. (This could be the situation during the spring or fall when there is neither furnace heating nor air conditioning.)

SOLUTION The function $Q(t)$ in (3) is now

$$Q(t) = K(M_0 - B \cos \omega t) + H_0 .$$

Setting $B_0 := M_0 + H_0/K$, we can rewrite Q as

(6) $$Q(t) = K(B_0 - B \cos \omega t) ,$$

where KB_0 represents the daily average value of $Q(t)$; that is,

$$KB_0 = \frac{1}{24} \int_0^{24} Q(t)dt .$$

When the forcing function $Q(t)$ in (6) is substituted into the expression for the temperature in equation (4), the result (after using integration by parts) is

$$T(t) = e^{-Kt} \left\{ \int e^{Kt}(KB_0 - KB \cos \omega t)dt + C \right\}$$

(7) $$T(t) = B_0 - BF(t) + Ce^{-Kt} ,$$

where

$$F(t) := \frac{\cos \omega t + (\omega/K)\sin \omega t}{1 + (\omega/K)^2} .$$

The constant C is chosen so that at midnight $(t = 0)$, the value of the temperature T is equal to some initial temperature T_0. Thus

$$C = T_0 - B_0 + BF(0) = T_0 - B_0 + \frac{B}{1 + (\omega/K)^2} .$$ ■ End

Notice that the third term in solution (7) involving the constant C tends to zero exponentially. The constant term B_0 in (7) is equal to $M_0 + H_0/K$ and represents the daily average temperature inside the building (neglecting the exponential term). When there is no additional heating rate inside the building $(H_0 = 0)$, this average temperature is equal to the average outside temperature M_0. The term $BF(t)$ in (7) represents the sinusoidal variation of temperature inside the building responding to the outside temperature variation. Since $F(t)$ can be written in the form

(8) $$F(t) = \left[1 + (\omega/K)^2\right]^{-1/2} \cos(\omega t - \phi) ,$$

where $\tan \phi = \omega/K$ (see Problem 16), the sinusoidal variation inside the building lags behind the outside variation by ϕ/ω hours. Further, the magnitude of the variation inside the building is slightly less, by a factor of $\left[1 + (\omega/K)^2\right]^{-1/2}$, than the outside variation. The angular frequency of variation ω is $2\pi/24$ radians/hr (which is about 1/4). Typical values for the dimensionless ratio ω/K lie between 1/2 and 1. For this range, the lag between inside and outside temperature is approximately 1.8 to 3 hr and the magnitude of the inside variation is between 89% and 71%

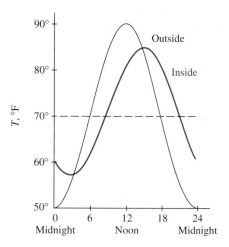

Figure 3.6 Temperature variation inside and outside an unheated building

of the variation outside. Figure 3.6 shows the 24-hr sinusoidal variation of the outside temperature for a typical moderate day as well as the temperature variations inside the building for a dimensionless ratio ω/K of unity, which corresponds to a time constant $1/K$ of approximately 4 hr. In sketching the latter curve, we have assumed that the exponential term has died out.

EXAMPLE 3 Suppose that in the building of Example 2, a simple thermostat is installed that is used to compare the actual temperature inside the building with a desired temperature T_D. If the actual temperature is below the desired temperature, the furnace supplies heating; otherwise it is turned off. If the actual temperature is above the desired temperature, the air conditioner supplies cooling; otherwise it is off. (In practice, there is some dead zone around the desired temperature in which the temperature difference is not sufficient to activate the thermostat, but that is to be ignored here.) Assuming the amount of heating or cooling supplied is proportional to the difference in temperature—that is,

$$U(t) = K_U\left[T_D - T(t)\right] ,$$

where K_U is the (positive) proportionality constant—find $T(t)$.

SOLUTION If the proportional control $U(t)$ is substituted directly into the differential equation (1) for the building temperature, we get

(9) $$\frac{dT(t)}{dt} = K\left[M(t) - T(t)\right] + H(t) + K_U\left[T_D - T(t)\right] .$$

A comparison of equation (9) with the first order linear differential equation (2) shows that for this example, the quantity P is equal to $K + K_U$, while the quantity $Q(t)$ representing the forcing function includes the desired temperature T_D. That is,

$$P = K + K_U ,$$
$$Q(t) = KM(t) + H(t) + K_U T_D .$$

When the additional heating rate is a constant H_0 and the outside temperature M varies as a sine wave over a 24-hr period in the same way as it did in Example 2, the forcing function is

$$Q(t) = K(M_0 - B \cos \omega t) + H_0 + K_U T_D .$$

The function $Q(t)$ has a constant term and a cosine term just as in equation (6). This equivalence becomes more apparent after substituting

(10) $Q(t) = K_1(B_2 - B_1 \cos \omega t) ,$

where

$$\omega := \frac{2\pi}{24} = \frac{\pi}{12} , \qquad\qquad K_1 := K + K_U ,$$

(11) $B_2 := \dfrac{K_U T_D + K M_0 + H_0}{K_1} , \qquad B_1 := \dfrac{BK}{K_1} .$

The expressions for the constant P and the forcing function $Q(t)$ of equation (10) are the same as the expressions in Example 2, except that the constants K, B_0, and B are replaced, respectively, by the constants K_1, B_2, and B_1. Hence the solution to the differential equation (9) will be the same as the temperature solution in Example 2, except that the three constant terms are changed. Thus

(12) $T(t) = B_2 - B_1 F_1(t) + C \exp(-K_1 t) ,$

where

$$F_1(t) := \frac{\cos \omega t + (\omega/K_1)\sin \omega t}{1 + (\omega/K_1)^2} .$$

The constant C is chosen so that at time $t = 0$ the value of the temperature T equals T_0. Thus

$$C = T_0 - B_2 + B_1 F_1(0) . \quad \blacksquare \text{End}$$

In the above example, the time constant for equation (9) is $1/P = 1/K_1$, where $K_1 = K + K_U$. Here $1/K_1$ is referred to as the **time constant for the building with heating and air conditioning.** For a typical heating and cooling system, K_U is somewhat less than 2; for a typical building, the constant K is between $1/2$ and $1/4$. Hence the sum gives a value for K_1 of about 2, and the time constant for the building with heating and air conditioning is about $1/2$ hr.

When the heating or cooling is turned on, it takes about 30 min for the exponential term in (12) to die off. If we neglect this exponential term, the average temperature inside the building is B_2. Since K_1 is much larger than K and H_0 is small, it follows from (11) that B_2 is roughly T_D, the desired temperature. In other words, after a certain period of time, the temperature inside the building is roughly T_D, with a small sinusoidal variation. (The outside average M_0 and inside heating rate H_0 have only a small effect.) Thus, to save energy, the heating or cooling system may be left off during the night. When it is turned on in the morning, it will take roughly 30 min for the inside of the building to attain the desired temperature. These observations provide an answer to question (c), regarding the temperature inside the building during summer and winter, that was posed at the beginning of this section.

The assumption made in Example 3 that the amount of heating or cooling is $U(t) = K_U\big[T_D - T(t)\big]$ may not always be suitable. We have used it here and in the exercises to illustrate the use of the time constant. More adventuresome readers may want to experiment with other models for $U(t)$, especially if they have available the numerical techniques discussed in Sections 3.5 and 3.6. Project G, page 155, addresses temperature regulation with fixed-rate controllers.

EXERCISES 3.3

1. A cup of hot coffee initially at 95°C cools to 80°C in 5 min while sitting in a room of temperature 21°C. Using just Newton's law of cooling, determine when the temperature of the coffee will be a nice 50°C.

2. A cold beer initially at 35°F warms up to 40°F in 3 min while sitting in a room of temperature 70°F. How warm will the beer be if left out for 20 min?

3. A white wine at room temperature 70°F is chilled in ice (32°F). If it takes 15 min for the wine to chill to 60°F, how long will it take for the wine to reach 56°F?

4. A red wine is brought up from the wine cellar, which is a cool 10°C, and left to breathe in a room of temperature 23°C. If it takes 10 min for the wine to reach 15°C, when will the temperature of the wine reach 18°C?

5. It was noon on a cold December day in Tampa: 16°C. Detective Taylor arrived at the crime scene to find the sergeant leaning over the body. The sergeant said there were several suspects. If they knew the exact time of death, then they could narrow the list. Detective Taylor took out a thermometer and measured the temperature of the body: 34.5°C. He then left for lunch. Upon returning at 1:00 P.M., he found the body temperature to be 33.7°C. When did the murder occur? [*Hint:* Normal body temperature is 37°C.]

6. On a mild Saturday morning while people are working inside, the furnace keeps the temperature inside the building at 21°C. At noon the furnace is turned off and the people go home. The temperature outside is a constant 12°C the rest of the afternoon. If the time constant for the building is 3 hr, when will the temperature inside the building reach 16°C? If some windows are left open and the time constant drops to 2 hr, when will the temperature inside reach 16°C?

7. On a hot Saturday morning while people are working inside, the air conditioner keeps the temperature inside the building at 24°C. At noon the air conditioner is turned off and the people go home. The temperature outside is a constant 35°C the rest of the afternoon. If the time constant for the building is 4 hr, what will be the temperature inside the building at 2:00 P.M.? At 6:00 P.M.? When will the temperature inside the building reach 27°C?

8. A garage with no heating or cooling has a time constant of 2 hr. If the outside temperature varies as a sine wave with a minimum of 50°F at 2:00 A.M. and a maximum of 80°F at 2:00 P.M., determine the times at which the building reaches its lowest temperature and its highest temperature, assuming the exponential term has died off.

9. A warehouse is being built that will have neither heating nor cooling. Depending on the amount of insulation, the time constant for the building may range from 1 to 5 hr. To illustrate the effect insulation will have on the temperature inside the warehouse, assume the outside temperature varies as a sine wave, with a minimum of 16°C at 2:00 A.M. and a maximum of 32°C at 2:00 P.M. Assuming the exponential term (which involves the initial temperature T_0) has died off, what is the lowest temperature inside the building if the time constant is 1 hr? If the time constant is 5 hr? What is the highest temperature inside the building if the time constant is 1 hr? If it is 5 hr?

10. Early Monday morning, the temperature in the lecture hall has fallen to 40°F, the same as the temperature outside. At 7:00 A.M., the janitor turns on the furnace with the thermostat set at 70°F. The time constant for the building is $1/K = 2$ hr and that for

the building along with its heating system is $1/K_1 = 1/2$ hr. Assuming the outside temperature remains constant, what will be the temperature inside the lecture hall at 8:00 A.M.? When will the temperature inside the hall reach 65°F?

11. During the summer the temperature inside a van reaches 55°C while that outside is a constant 35°C. When the driver gets into the van, she turns on the air conditioner with the thermostat set at 16°C. If the time constant for the van is $1/K = 2$ hr and that for the van with its air conditioning system is $1/K_1 = 1/3$ hr, when will the temperature inside the van reach 27°C?

12. Two friends sit down to talk and enjoy a cup of coffee. When the coffee is served, the impatient friend immediately adds a teaspoon of cream to his coffee. The relaxed friend waits 5 min before adding a teaspoon of cream (which has been kept at a constant temperature). The two now begin to drink their coffee. *Who has the hotter coffee?* Assume the cream is cooler than the air and use Newton's law of cooling.

13. A solar hot water heating system consists of a hot water tank and a solar panel. The tank is well insulated and has a time constant of 64 hr. The solar panel generates 2000 Btu/hr during the day, and the tank has a heat capacity of 2°F per thousand Btu. If

the water in the tank is initially 110°F and the room temperature outside the tank is 80°F, what will be the temperature in the tank after 12 hr of sunlight?

14. In Problem 13, if a larger tank with a heat capacity of 1°F per thousand Btu and a time constant of 72 hr is used instead (with all other factors being the same), what would be the temperature in the tank after 12 hr?

15. **Stefan's law** of radiation states that the rate of change of temperature of a body at T degrees Kelvin in a medium at M degrees Kelvin is proportional to $M^4 - T^4$. That is,

$$\frac{dT}{dt} = k(M^4 - T^4) \ ,$$

where k is a positive constant. Solve this equation using separation of variables. Explain why Newton's law and Stefan's law are nearly the same when T is close to M and M is constant. [*Hint:* Factor $M^4 - T^4$.]

16. Show that $C_1 \cos \omega t + C_2 \sin \omega t$ can be written in the form $A \cos(\omega t - \phi)$, where $A = \sqrt{C_1^2 + C_2^2}$ and $\tan \phi = C_2/C_1$. [*Hint:* Use a standard trigonometric identity with $C_1 = A \cos \phi$, $C_2 = A \sin \phi$.] Use this fact to verify the alternate representation (8) of $F(t)$ discussed in Example 2.

3.4 NEWTONIAN MECHANICS

Mechanics is the study of the motion of objects and the effect of forces acting on those objects. It is the foundation of several branches of physics and engineering. **Newtonian,** or **classical, mechanics** deals with the motion of **ordinary** objects, that is, objects that are large compared to an atom and slow moving compared with the speed of light. A model for Newtonian mechanics can be based on **Newton's laws of motion:**[†]

1. When a body is subject to no resultant external force, it moves with a constant velocity.

2. When a body is subject to one or more external forces, the time rate of change of the body's momentum is equal to the vector sum of the external forces acting on it.

3. When one body interacts with a second body, the force of the first body on the

[†]For a discussion of Newton's laws of motion, see *University Physics,* 9th Edition, by F. W. Sears, M. W. Zemansky, and H. D. Young (Addison-Wesley Publishing Co., Reading, Mass., 1996).

second is equal in magnitude, but opposite in direction, to the force of the second body on the first.

Experimental results for more than two centuries verify that these laws are extremely useful for studying the motion of ordinary objects in an **inertial reference frame,** that is, a reference frame in which an undisturbed body moves with a constant velocity. It is Newton's second law, which applies only to inertial reference frames, that enables us to formulate the equations of motion for a moving body. We can express Newton's second law by

$$\frac{dp}{dt} = F(t, x, v) \ ,$$

where $F(t, x, v)$ is the resultant force on the body at time t, location x, and velocity v, and $p(t)$ is the momentum of the body at time t. The momentum is the product of the mass of the body and its velocity, that is,

$$p(t) = mv(t) \ ,$$

so we can express Newton's second law as

(1) $$m\frac{dv}{dt} = ma = F(t, x, v) \ ,$$

where $a = dv/dt$ is the acceleration of the body at time t.

Typically one substitutes $v = dx/dt$ for the velocity in (1) and obtains a second order differential equation in the dependent variable x. However, in the present section we shall focus on situations where the force F does not depend on x. This enables us to regard (1) as a *first* order equation

(2) $$m\frac{dv}{dt} = F(t, v)$$

in $v(t)$.

To apply Newton's laws of motion to a problem in mechanics, the following general procedure may be useful.

PROCEDURE FOR NEWTONIAN MODELS

(a) Determine *all* relevant forces acting on the object being studied. It is helpful to draw a simple diagram of the object that depicts these forces.
(b) Choose an appropriate axis or coordinate system in which to represent the motion of the object and the forces acting on it. Keep in mind that this coordinate system must be an inertial reference frame.
(c) Apply Newton's second law as expressed in equation (2) to determine the equations of motion for the object.

In this section, we express Newton's second law in either of two systems of units: the U.S. Customary System or the meter-kilogram-second (MKS) system. The various units in these

systems are summarized in Table 3.2, along with approximate values for the gravitational acceleration (on the surface of Earth).

TABLE 3.2 MECHANICAL UNITS IN THE U.S. CUSTOMARY AND MKS SYSTEMS		
Unit	U.S. Customary System	MKS System
Distance	feet (ft)	meters (m)
Mass	slugs	kilograms (kg)
Time	seconds (sec)	seconds (sec)
Force	pounds (lb)	newtons (N)
g (Earth)	32 ft/sec^2	9.81 m/sec^2

EXAMPLE 1 An object of mass m is given an initial downward velocity v_0 and allowed to fall under the influence of gravity. Assuming the gravitational force is constant and the force due to air resistance is proportional to the velocity of the object, determine the equation of motion for this object.

SOLUTION There are two forces acting on the object: a constant force due to the downward pull of gravity and a force due to air resistance that is proportional to the velocity of the object and acts in opposition to the motion of the object. Hence the motion of the object will take place along a vertical axis. On this axis, we choose the origin to be the point where the object was initially dropped and let $x(t)$ denote the distance the object has fallen in time t (see Figure 3.7).

The forces acting on the object along this axis can be expressed as follows. The force due to gravity is

$$F_1 = mg \ ,$$

where g is the acceleration due to gravity at the Earth's surface (see Table 3.2). The force due to air resistance is

$$F_2 = -bv(t) \ ,$$

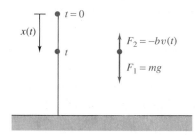

Figure 3.7 Forces on a falling object

where $b(>0)$ is the proportionality constant[†] and the negative sign is present because air resistance acts in opposition to the motion of the object. Hence the net force acting on the object (see Figure 3.7) is

(3) $F = F_1 + F_2 = mg - bv(t)$.

We now apply Newton's second law by substituting (3) into (2) to obtain

$$m\frac{dv}{dt} = mg - bv .$$

Since the initial velocity of the object is v_0, a model for the velocity of the falling body is expressed by the initial value problem

(4) $m\dfrac{dv}{dt} = mg - bv , \qquad v(0) = v_0$,

where g and b are positive constants.

The model (4) is the same as the one we obtained in Section 2.1. Using separation of variables or the method of Section 2.3 for linear equations, we get

(5) $v(t) = \dfrac{mg}{b} + \left(v_0 - \dfrac{mg}{b}\right)e^{-bt/m}$.

Since we have taken $x = 0$ when $t = 0$, we can determine the equation of motion of the object by integrating $v = dx/dt$ with respect to t. Thus from (5) we obtain

$$x(t) = \int v(t)\,dt = \frac{mg}{b}t - \frac{m}{b}\left(v_0 - \frac{mg}{b}\right)e^{-bt/m} + c ,$$

and setting $x = 0$ when $t = 0$, we find

$$0 = -\frac{m}{b}\left(v_0 - \frac{mg}{b}\right) + c ,$$

$$c = \frac{m}{b}\left(v_0 - \frac{mg}{b}\right) .$$

Hence the equation of motion is

(6) $x(t) = \dfrac{mg}{b}t + \dfrac{m}{b}\left(v_0 - \dfrac{mg}{b}\right)(1 - e^{-bt/m})$. ■ End

In Figure 3.8 on page 118, we have sketched the graphs of the velocity and the position as functions of t. Observe that the velocity $v(t)$ approaches the horizontal asymptote $v = mg/b$ as

[†]The units of b are lb-sec/ft in the U.S. system, and N-sec/m in the MKS system.

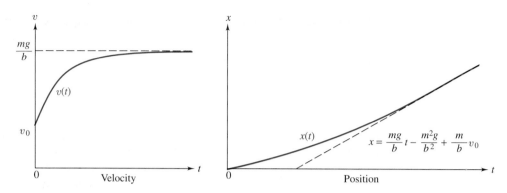

Figure 3.8 Graphs of the position and velocity of a falling object when $v_0 < mg/b$

$t \to +\infty$ and that the position $x(t)$ asymptotically approaches the line

$$x = \frac{mg}{b}t - \frac{m^2g}{b^2} + \frac{m}{b}v_0$$

as $t \to +\infty$. The value mg/b of the horizontal asymptote for $v(t)$ is called the **limiting,** or **terminal, velocity** of the object. The reader should verify that $v \equiv mg/b$ is a stable equilibrium solution (a "sink") for the autonomous equation (4) (see Section 1.4).

Notice that the terminal velocity depends on the mass but not the initial velocity of the object; the velocity of every free-falling body approaches the limiting value mg/b. Heavier objects *do,* in the presence of friction, ultimately fall faster than lighter ones, but for a given object the terminal velocity is the same whether it initially is tossed upward or downward, or simply dropped from rest.

Now that we have obtained the equation of motion for a falling object with air resistance proportional to v, we can answer a variety of questions.

EXAMPLE 2 An object of mass 3 kg is released from rest 500 m above the ground and allowed to fall under the influence of gravity. Assume the gravitational force is constant, with $g = 9.81$ m/sec^2, and the force due to air resistance is proportional to the velocity of the object[†] with proportionality constant $b = 3$ N-sec/m. Determine when the object will strike the ground.

SOLUTION We can use the model discussed in Example 1 with $v_0 = 0$, $m = 3$, $b = 3$, and $g = 9.81$. From (6), the equation of motion in this case is

$$x(t) = \frac{(3)(9.81)}{3}t - \frac{(3)^2(9.81)}{(3)^2}\left(1 - e^{-3t/3}\right) = (9.81)t - (9.81)\left(1 - e^{-t}\right) \ .$$

Because the object is released 500 m above the ground, we can determine when the object

[†]The effects of more sophisticated air resistance models (such as a quadratic drag law) are analyzed numerically in Exercises 3.5, Problem 20.

strikes the ground by setting $x(t) = 500$ and solving for t. Thus we put

$$500 = (9.81)t - 9.81 + (9.81)e^{-t}$$

or

$$t + e^{-t} = \frac{509.81}{9.81} = 51.97 \ ,$$

where we have rounded the computations to two decimal places. Unfortunately this equation cannot be solved explicitly for t. We might try to approximate t using Newton's approximation method (see Appendix A), but in this case, it is not necessary. Since e^{-t} will be very small for t near 51.97 $(e^{-51.97} \approx 10^{-22})$, we simply ignore the term e^{-t} and obtain as our approximation $t = 51.97$ sec. ∎End

EXAMPLE 3 A parachutist whose mass is 75 kg drops from a helicopter hovering 4000 m above the ground and falls toward the earth under the influence of gravity. Assume the gravitational force is constant. Assume also that the force due to air resistance is proportional to the velocity of the parachutist, with the proportionality constant $b_1 = 15$ N-sec/m when the chute is closed and with constant $b_2 = 105$ N-sec/m when the chute is open. If the chute does not open until 1 min after the parachutist leaves the helicopter, after how many seconds will she hit the ground?

SOLUTION We are interested only in when the parachutist will hit the ground, not where. Thus we consider only the vertical component of her descent. For this, we need to use two equations: one to describe the motion before the chute opens and the other to apply after it opens. Before the chute opens, the model is the same as in Example 1 with $v_0 = 0$, $m = 75$ kg, $b = b_1 = 15$ N-sec/m, and $g = 9.81$ m/sec^2. If we let $x_1(t)$ be the distance the parachutist has fallen in t seconds and let $v_1 = dx_1/dt$, then substituting into equations (5) and (6), we have

$$v_1(t) = \frac{(75)(9.81)}{15}\left(1 - e^{-(15/75)t}\right)$$
$$= (49.05)(1 - e^{-0.2t}) \ ,$$

and

$$x_1(t) = \frac{(75)(9.81)}{15}t - \frac{(75)^2(9.81)}{(15)^2}\left(1 - e^{-(15/75)t}\right)$$
$$= 49.05t - 245.25(1 - e^{-0.2t}) \ .$$

Hence, after 1 min, when $t = 60$, the parachutist is falling at the rate

$$v_1(60) = (49.05)\left(1 - e^{-0.2(60)}\right) = 49.05 \text{ m/sec} \ ,$$

and has fallen

$$x_1(60) = (49.05)(60) - (245.25)\left(1 - e^{-0.2(60)}\right) = 2697.75 \text{ m} \ .$$

(In these and other computations for this problem, we round our answers to two decimal places.)

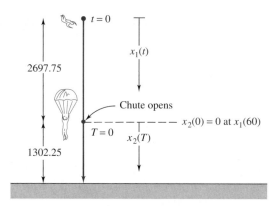

Figure 3.9 The fall of the parachutist

Now when the chute opens, the parachutist is $4000 - 2697.75$ or 1302.25 m above the ground and traveling at a velocity of 49.05 m/sec. To determine the equation of motion after the chute opens, let $x_2(T)$ denote the position of the parachutist T seconds after the chute opens (so that $T = t - 60$), taking $x_2(0) = 0$ at $x_1(60)$ (see Figure 3.9). Further assume the initial velocity of the parachutist after the chute opens is the same as the final velocity before it opens, that is, $x_2'(0) = x_1'(60) = 49.05$ m/sec. Because the forces acting on the parachutist are the same as those acting on the object in Example 1, we can again use equations (5) and (6). With $v_0 = 49.05$, $m = 75$, $b = b_2 = 105$, and $g = 9.81$, we find from (6) that

$$x_2(T) = \frac{(75)(9.81)}{105}T + \frac{75}{105}\left[49.05 - \frac{(75)(9.81)}{105}\right]\left(1 - e^{-(105/75)T}\right)$$

$$= 7.01T + 30.03\left(1 - e^{-1.4T}\right) \ .$$

To determine when the parachutist will hit the ground, we set $x_2(T) = 1302.25$, the height the parachutist was above the ground when her parachute opened. This gives

$$7.01T + 30.03 - 30.03e^{-1.4T} = 1302.25$$

(7) $$T - 4.28e^{-1.4T} - 181.49 = 0 \ .$$

Again we cannot solve (7) explicitly for T. However, observe that $e^{-1.4T}$ is very small for T near 181.49, so we ignore the exponential term and obtain $T = 181.49$. Hence the parachutist will strike the ground 181.49 sec after the parachute opens, or 241.49 sec after dropping from the helicopter. ■ End

In the computation for T in equation (7), we found that the exponential $e^{-1.4T}$ was negligible. Consequently, ignoring the corresponding exponential term in equation (5), we see that the parachutist's velocity at impact is

$$\frac{mg}{b_2} = \frac{(75)(9.81)}{105} = 7.01 \text{ m/sec} \ ,$$

which is the limiting velocity for her fall with the chute open.

EXERCISES 3.4

Unless otherwise stated, in the following problems we assume the gravitational force is constant with $g = 9.81$ m/sec^2 in the MKS system and $g = 32$ ft/sec^2 in the U.S. Customary System.

1. An object of mass 5 kg is released from rest 1000 m above the ground and allowed to fall under the influence of gravity. Assuming the force due to air resistance is proportional to the velocity of the object with proportionality constant $b = 50$ N-sec/m, determine the equation of motion of the object. When will the object strike the ground?

2. A 400-lb object is released from rest 500 ft above the ground and allowed to fall under the influence of gravity. Assuming the force in pounds due to air resistance is $-10v$, where v is the velocity of the object in ft/sec, determine the equation of motion of the object. When will the object hit the ground?

3. If the object in Problem 1 has a mass of 500 kg instead of 5 kg, when will it strike the ground? [*Hint:* Here the exponential term is too large to ignore. Use Newton's method to approximate the time t when the object strikes the ground (see Appendix A).]

4. If the object in Problem 2 is released from rest 30 ft above the ground instead of 500 ft, when will it strike the ground? [*Hint:* Use Newton's method to solve for t.]

5. An object of mass 5 kg is given an initial downward velocity of 50 m/sec and then allowed to fall under the influence of gravity. Assume the force in newtons due to air resistance is $-10v$, where v is the velocity of the object in m/sec. Determine the equation of motion of the object. If the object is initially 500 m above the ground, determine when the object will strike the ground.

6. An object of mass 8 kg is given an upward initial velocity of 20 m/sec and then allowed to fall under the influence of gravity. Assume that the force in newtons due to air resistance is $-16v$, where v is the velocity of the object in m/sec. Determine the equation of motion of the object. If the object is initially 100 m above the ground, determine when the object will strike the ground.

7. A parachutist whose mass is 75 kg drops from a helicopter hovering 2000 m above the ground and falls toward the ground under the influence of gravity. Assume the force due to air resistance is proportional to the velocity of the parachutist, with the proportionality constant $b_1 = 30$ N-sec/m when the chute is closed and $b_2 = 90$ N-sec/m when the chute is open. If the chute does not open until the velocity of the parachutist reaches 20 m/sec, after how many seconds will she reach the ground?

8. A parachutist whose mass is 100 kg drops from a helicopter hovering 3000 m above the ground and falls under the influence of gravity. Assume the force due to air resistance is proportional to the velocity of the parachutist, with the proportionality constant $b_3 = 20$ N-sec/m when the chute is closed and $b_4 = 100$ N-sec/m when the chute is open. If the chute does not open until 30 sec after the parachutist leaves the helicopter, after how many seconds will he hit the ground? If the chute does not open until 1 min after he leaves the helicopter, after how many seconds will he hit the ground?

9. An object of mass 100 kg is released from rest from a boat into the water and allowed to sink. While gravity is pulling the object down, a buoyancy force of $1/40$ times the weight of the object is pushing the object up (weight $= mg$). If we assume water resistance exerts a force on the object that is proportional to the velocity of the object, with proportionality constant 10 N-sec/m, find the equation of motion of the object. After how many seconds will the velocity of the object be 70 m/sec?

10. An object of mass 2 kg is released from rest from a platform 30 m above the water and allowed to fall under the influence of gravity. After the object strikes the water, it begins to sink with gravity pulling down and a buoyancy force pushing up. Assume the force of gravity is constant, no change in momentum occurs on impact with the water, the buoyancy force is $1/2$ the weight (weight $= mg$), and the force due to air resistance or water resistance is proportional to the velocity, with proportionality

constant $b_1 = 10$ N-sec/m in the air and $b_2 = 100$ N-sec/m in the water. Find the equation of motion of the object. What is the velocity of the object 1 min after it is released?

11. In Example 1, we solved for the velocity of the object as a function of time (equation (5)). In some cases, it is useful to have an expression, independent of t, that relates v and x. Find this relation for the motion in Example 1. [*Hint:* Letting $v(t) = V(x(t))$, then $dv/dt = (dV/dx)V$.]

12. A shell of mass 2 kg is shot upward with an initial velocity of 200 m/sec. The magnitude of the force on the shell due to air resistance is $|v|/20$. When will the shell reach its maximum height above the ground? What is the maximum height?

13. When the velocity v of an object is very large, the magnitude of the force due to air resistance is proportional to v^2 with the force acting in opposition to the motion of the object. A shell of mass 3 kg is shot upward from the ground with an initial velocity of 500 m/sec. If the magnitude of the force due to air resistance is $(0.1)v^2$, when will the shell reach its maximum height above the ground? What is the maximum height?

14. An object of mass m is released from rest and falls under the influence of gravity. If the magnitude of the force due to air resistance is bv^n, where b and n are positive constants, find the limiting velocity of the object (assuming this limit exists). [*Hint:* Argue that the existence of a (finite) limiting velocity implies that $dv/dt \to 0$ as $t \to +\infty$.]

15. A rotating flywheel is being turned by a motor that exerts a constant torque T (see Figure 3.10). A re-

tarding torque due to friction is proportional to the angular velocity ω. If the moment of inertia of the flywheel is I and its initial angular velocity is ω_0, find the equation for the angular velocity ω as a function of time. [*Hint:* Use Newton's second law for rotational motion, that is, moment of inertia \times angular acceleration = sum of the torques.]

16. Find the equation for the angular velocity ω in Problem 15, assuming the retarding torque is proportional to $\sqrt{\omega}$.

17. In Problem 16, let $I = 50$ kg-m^2 and the retarding torque be $5\sqrt{\omega}$ N-m. If the motor is turned off with the angular velocity at 225 rad/sec, determine how long it will take for the flywheel to come to rest.

18. When an object slides on a surface, it encounters a resistance force called **friction.** This force has a magnitude of μN, where μ is the **coefficient of kinetic friction** and N is the magnitude of the normal force that the surface applies to the object. Suppose that an object of mass 30 kg is released from the top of an inclined plane that is inclined 30° to the horizontal (see Figure 3.11). Assume the gravitational force is constant, air resistance is negligible, and the coefficient of kinetic friction $\mu = 0.2$. Determine the equation of motion for the object as it slides down the plane. If the top surface of the plane is 5 m long, what is the velocity of the object when it reaches the bottom?

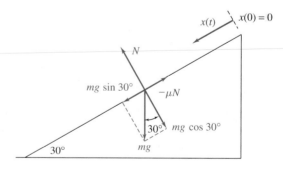

Figure 3.11 Forces on an object on an inclined plane

19. An object of mass 60 kg starts from rest at the top of a 45° inclined plane. Assume that the coefficient of kinetic friction is 0.05 (see Problem 18). If the force due to air resistance is proportional to the

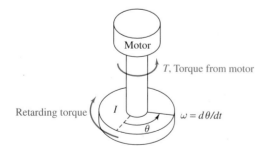

Figure 3.10 Motor-driven flywheel

velocity of the object, say $-3v$, find the equation of motion of the object. How long will it take the object to reach the bottom of the inclined plane if the incline is 10 m long?

20. An object at rest on an inclined plane will not slide until the component of the gravitational force down the incline is sufficient to overcome the force due to **static friction.** Static friction is governed by an experimental law somewhat like that of kinetic friction (Problem 18); it has a magnitude of *at most* μN, where μ is the coefficient of static friction and N is, again, the magnitude of the normal force exerted by the surface on the object. If the plane is inclined at an angle α, determine the critical value α_0 for which the object will slide if $\alpha > \alpha_0$ but will not move for $\alpha < \alpha_0$.

21. A sailboat has been running (on a straight course) under a light wind at 1 m/sec. Suddenly the wind picks up, blowing hard enough to apply a constant force of 600 N to the sailboat. The only other force acting on the boat is water resistance that is proportional to the velocity of the boat. If the proportionality constant for water resistance is $b = 100$ N-sec/m and the mass of the sailboat is 50 kg, find the equation of motion of the sailboat. What is the limiting velocity of the sailboat under this wind?

22. In Problem 21, it is observed that when the velocity of the sailboat reaches 5 m/sec, the boat begins to rise out of the water and "plane." When this happens, the proportionality constant for the water resistance drops to $b_0 = 60$ N-sec/m. Now find the equation of motion of the sailboat. What is the limiting velocity of the sailboat under this wind as it is planing?

23. Sailboats A and B each have a mass of 60 kg and cross the starting line at the same time on the first leg of a race. Each has an initial velocity of 2 m/sec. The wind applies a constant force of 650 N to each boat, and the force due to water resistance is proportional to the velocity of the boat. For sailboat A, the proportionality constants are $b_1 = 80$ N-sec/m before planing when the velocity is less than 5 m/sec, and $b_2 = 60$ N-sec/m when the velocity is above 5 m/sec. For sailboat B, the proportionality constants are $b_3 = 100$ N-sec/m before planing when the

velocity is less than 6 m/sec and $b_4 = 50$ N-sec/m when the velocity is above 6 m/sec. If the first leg of the race is 500 m long, which sailboat will be leading at the end of the first leg?

24. Rocket Flight. A model rocket having initial mass m_0 kg is launched vertically from the ground. The rocket expels gas at a constant rate of α kg/sec and at a constant velocity of β m/sec relative to the rocket. Assume the magnitude of the gravitational force is proportional to the mass with proportionality constant g. Because the mass is not constant, Newton's second law leads to the equation

$$(m_0 - \alpha t)\frac{dv}{dt} - \alpha\beta = -g(m_0 - \alpha t) \ ,$$

where $v = dx/dt$ is the velocity of the rocket, x is its height above the ground, and $m_0 - \alpha t$ is the mass of the rocket at t seconds after launch. If the initial velocity is zero, solve the above equation to determine the velocity of the rocket and its height above ground for $0 \le t < m_0/\alpha$.

25. Escape Velocity. According to Newton's law of gravitation, the attractive force between two objects varies inversely as the square of the distances between them. That is, $F_g = GM_1M_2/r^2$, where M_1 and M_2 are the masses of the objects, r is the distance between them (center to center), F_g is the attractive force, and G is the constant of proportionality. Consider a projectile of constant mass m being fired vertically from Earth (see Figure 3.12). Let t represent time and v the velocity of the projectile.

(a) Show that the motion of the projectile, under Earth's gravitational force, is governed by the equation

$$\frac{dv}{dt} = -\frac{gR^2}{r^2} \ ,$$

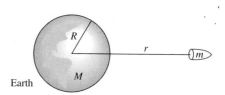

Figure 3.12 Projectile escaping from Earth

where r is the distance between the projectile and the center of Earth, R is the radius of Earth, M is the mass of Earth, and $g = GM/R^2$.

(b) Use the fact that $dr/dt = v$ to obtain

$$v\frac{dv}{dr} = -\frac{gR^2}{r^2}\ .$$

(c) If the projectile leaves Earth's surface with velocity v_0, show that

$$v^2 = \frac{2gR^2}{r} + v_0^2 - 2gR\ .$$

(d) Use the result of part (c) to show that the velocity of the projectile remains positive if and only if $v_0^2 - 2gR > 0$. The velocity $v_e = \sqrt{2gR}$ is called the **escape velocity** of Earth.

(e) If $g = 9.81$ m/sec^2 and $R = 6370$ km for Earth, what is Earth's escape velocity?

(f) If the acceleration due to gravity for the moon is $g_m = g/6$ and the radius of the moon is $R_m = 1738$ km, what is the escape velocity of the moon?

3.5 IMPROVED EULER'S METHOD

While the analytical techniques presented in Chapter 2 were useful for the variety of mathematical models presented earlier in this chapter, *the majority of the differential equations encountered in applications cannot be solved either implicitly or explicitly.* This is especially true of higher-order equations and systems of equations, which we study in later chapters. In this section and the next, we discuss methods for obtaining a numerical *approximation* of the solution to an initial value problem for a first order differential equation. Our goal is to develop algorithms that you can use with a calculator or computer. These algorithms also extend naturally to higher-order equations (see Section 5.6). We describe the rationale behind each method but leave the more detailed discussion to texts on numerical analysis.[†]

Consider the initial value problem

(1) $y' = f(x, y)\ ,$ $y(x_0) = y_0\ .$

To guarantee that (1) has a unique solution, we assume f and $\partial f/\partial y$ are continuous in a rectangle $R := \{(x, y): a < x < b, c < y < d\}$ containing (x_0, y_0). It follows from Theorem 1 in Chapter 1 that the initial value problem (1) has a unique solution $\phi(x)$ in some interval $x_0 - \delta < x < x_0 + \delta$, where δ is a positive number. Because δ is not known *a priori*, there is no assurance that the solution will exist at a particular point $x(\neq x_0)$, even if x is in the interval (a, b). However, if $\partial f/\partial y$ is continuous and *bounded*[††] on the vertical strip

$$S := \{(x, y): a < x < b, -\infty < y < \infty\}\ ,$$

then it turns out that (1) has a unique solution on the whole interval (a, b). In describing numerical methods, we assume this last condition is satisfied and that f possesses as many continuous partial derivatives as needed.

[†] See, for example, *Numerical Solution of Ordinary Differential Equations* by L. Shampine (Chapman & Hall, New York, 1994) or *Numerical Analysis*, 6th Edition, by R. L. Burden and J. D. Faires (Prindle, Weber & Schmidt, Boston, 1997).

[††] A function $g(x, y)$ is bounded on S if there exists a number M such that $|g(x, y)| \leq M$ for all (x, y) in S.

In Section 1.5 we used the concept of direction fields to motivate a scheme for approximating the solution to the initial value problem (1). This scheme, called **Euler's method,** is one of the most basic, so it is worthwhile to discuss its advantages, disadvantages, and possible improvements. We begin with a derivation of Euler's method that is somewhat different from that presented in Section 1.5.

Let $h > 0$ be fixed (h is called the **step size**) and consider the equally spaced points

(2) $$x_n := x_0 + nh , \qquad n = 0, 1, 2, \ldots .$$

Our goal is to obtain an approximation to the solution $\phi(x)$ of the initial value problem (1) at those points x_n that lie in the interval (a, b). Namely, we will describe a method that generates values y_0, y_1, y_2, \ldots that approximate $\phi(x)$ at the respective points x_0, x_1, x_2, \ldots ; that is,

$$y_n \approx \phi(x_n) , \qquad n = 0, 1, 2, \ldots .$$

Of course, the first "approximant" y_0 is exact, since $y_0 = \phi(x_0)$ is given. Thus we must describe how to compute $y_1, y_2, \ldots .$

For Euler's method, we begin by integrating both sides of equation (1) from x_n to x_{n+1} to obtain

$$\phi(x_{n+1}) - \phi(x_n) = \int_{x_n}^{x_{n+1}} \phi'(t)\,dt = \int_{x_n}^{x_{n+1}} f\big(t, \phi(t)\big)\,dt ,$$

where we have substituted $\phi(x)$ for y. Solving for $\phi(x_{n+1})$, we have

(3) $$\phi(x_{n+1}) = \phi(x_n) + \int_{x_n}^{x_{n+1}} f\big(t, \phi(t)\big)\,dt .$$

Without knowing $\phi(t)$, we cannot integrate $f\big(t, \phi(t)\big)$. Hence, we must approximate the integral in (3). Assuming we have already found $y_n \approx \phi(x_n)$, the simplest approach is to approximate the area under the function $f\big(t, \phi(t)\big)$ by the rectangle with base $[x_n, x_{n+1}]$ and height $f\big(x_n, \phi(x_n)\big)$ (see Figure 3.13). This gives

$$\phi(x_{n+1}) \approx \phi(x_n) + (x_{n+1} - x_n)f\big(x_n, \phi(x_n)\big) .$$

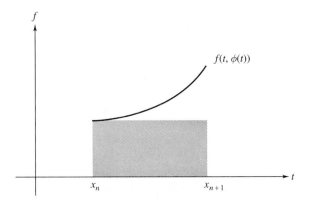

Figure 3.13 Approximation by a rectangle

Substituting h for $x_{n+1} - x_n$ and the approximation y_n for $\phi(x_n)$, we arrive at the numerical scheme

(4) $y_{n+1} = y_n + hf(x_n, y_n)$, $n = 0, 1, 2, \ldots,$

which is Euler's method.

Starting with the given value y_0, we use (4) to compute $y_1 = y_0 + hf(x_0, y_0)$ and then use y_1 to compute $y_2 = y_1 + hf(x_1, y_1)$, and so on. Several examples of Euler's method can be found in Section 1.5.

As discussed in Section 1.5, if we wish to use Euler's method to approximate the solution to the initial value problem (1) at a particular value of x, say $x = c$, then we must first determine a suitable step size h so that $x_0 + Nh = c$ for some integer N. For example, we can take $N = 1$ and $h = c - x_0$ in order to arrive at the approximation after just one step:

$$\phi(c) = \phi(x_0 + h) \approx y_1 \ .$$

If, instead, we wish to take 10 steps in Euler's method, we choose $h = (c - x_0)/10$ and ultimately obtain

$$\phi(c) = \phi(x_0 + 10h) = \phi(x_{10}) \approx y_{10} \ .$$

In general, depending on the size of h, we will get different approximations to $\phi(c)$. It is reasonable to expect that as h gets smaller (or, equivalently, as N gets larger), the Euler approximations approach the exact value $\phi(c)$. On the other hand, as h gets smaller, the number (and cost) of computations increases and hence so do machine errors that arise from round-off. Thus it is important to analyze how the error in the approximation scheme varies with h.

If Euler's method is used to approximate the solution $\phi(x) = e^x$ to the problem

(5) $y' = y$, $y(0) = 1$,

at $x = 1$, then we obtain approximations to the constant $e = \phi(1)$. It turns out that these approximations take a particularly simple form that enables us to compare the error in the approximation with the step size h. Indeed, setting $f(x, y) = y$ in (4) yields

$$y_{n+1} = y_n + hy_n = (1 + h)y_n \ , n = 0, 1, 2, \ldots.$$

Since $y_0 = 1$, we get

$$y_1 = (1 + h)y_0 = 1 + h$$
$$y_2 = (1 + h)y_1 = (1 + h)(1 + h) = (1 + h)^2$$
$$y_3 = (1 + h)y_2 = (1 + h)(1 + h)^2 = (1 + h)^3 \ ,$$

and, in general,

(6) $y_n = (1 + h)^n$, $n = 0, 1, 2, \ldots.$

For the problem in (5) we have $x_0 = 0$, so to obtain approximations at $x = 1$, we must set $nh = 1$. That is, h must be the reciprocal of an integer $(h = 1/n)$. Replacing n by $1/h$ in (6), we see that Euler's method gives the (familiar) approximation $(1 + h)^{1/h}$ to the constant e. In

Table 3.3, we list this approximation for $h = 1, 10^{-1}, 10^{-2}, 10^{-3}$, and 10^{-4}, along with the corresponding errors

$$e - (1 + h)^{1/h} .$$

From the second and third columns in Table 3.3, we see that the approximation gains roughly one decimal place in accuracy as h decreases by a factor of 10; that is, the error is roughly proportional to h. This observation is further confirmed by the entries in the last column of Table 3.3. In fact, using methods of calculus (see Exercises 1.5, Problem 13), it can be shown that

(7) $$\lim_{h \to 0} \frac{\text{error}}{h} = \lim_{h \to 0} \frac{e - (1 + h)^{1/h}}{h} = \frac{e}{2} \approx 1.35914 .$$

TABLE 3.3 EULER'S APPROXIMATIONS TO e = 2.71828. . .

h	Euler's Approximation $(1 + h)^{1/h}$	Error $e - (1 + h)^{1/h}$	Error $\dfrac{}{h}$
1	2.00000	0.71828	0.71828
10^{-1}	2.59374	0.12454	1.24539
10^{-2}	2.70481	0.01347	1.34680
10^{-3}	2.71692	0.00136	1.35790
10^{-4}	2.71815	0.00014	1.35902

The general situation is similar: When Euler's method is used to approximate the solution to the initial value problem (1), the error in the approximation is at worst a constant times the step size h. Moreover, in view of (7), this is the best one can say.

Numerical analysts have a convenient notation for describing the convergence behavior of a numerical scheme. For fixed x, we denote by $y(x; h)$ the approximation to the solution $\phi(x)$ of (1) obtained via the scheme when using a step size of h. We say that the numerical scheme **converges** at x if

$$\lim_{h \to 0} y(x; h) = \phi(x) .$$

In other words, as the step size h decreases to zero, the approximations for a convergent scheme approach the exact value $\phi(x)$. The rate at which $y(x; h)$ tends to $\phi(x)$ is often expressed in terms of a suitable power of h. If the error $\phi(x) - y(x; h)$ tends to zero like a constant times h^p, we write

$$\phi(x) - y(x; h) = O(h^p)$$

and say that the **method is of order p**. Of course, the higher the power p, the faster is the rate of convergence as $h \to 0$.

As seen from our earlier discussion, the rate of convergence of Euler's method is $O(h)$; that is, *Euler's method is of order $p = 1$*. In fact, the limit in (7) shows that for equation (5), the error is roughly $1.36h$ for small h. This means that to have an error less than 0.01 requires

$h < 0.01/1.36$, or $n = 1/h > 136$ computation steps. Thus Euler's method converges too slowly to be of practical use.

How can we improve Euler's method? To answer this, let's return to the derivation expressed in formulas (3) and (4) and analyze the "errors" that were introduced to get the approximation. The crucial step in the process was to approximate the integral

$$\int_{x_n}^{x_{n+1}} f\big(t, \phi(t)\big) dt$$

by using a rectangle (recall Figure 3.13). This step gives rise to what is called the *local truncation error* in the method. From calculus, we know that a better (more accurate) approach to approximating the integral is to use a trapezoid; that is, to apply the trapezoidal rule (see Figure 3.14). This gives

$$\int_{x_n}^{x_{n+1}} f\big(t, \phi(t)\big) dt \approx \frac{h}{2}\left[f\big(x_n, \phi(x_n)\big) + f\big(x_{n+1}, \phi(x_{n+1})\big)\right],$$

which leads to the numerical scheme

(8) $y_{n+1} = y_n + \dfrac{h}{2}\left[f(x_n, y_n) + f(x_{n+1}, y_{n+1})\right],$ $n = 0, 1, 2, \ldots.$

We call equation (8) the **trapezoid scheme.** It is an example of an **implicit method.** That is, unlike Euler's method, equation (8) gives only an implicit formula for y_{n+1}, since y_{n+1} appears as an argument of f. Assuming we have already computed y_n, some root-finding technique such as Newton's method (see Appendix A) might be needed to compute y_{n+1}. Despite the inconvenience of working with an implicit method, the trapezoid scheme has two advantages over Euler's method. First, it is a method of order $p = 2$; that is, it converges at a rate that is proportional to h^2 and hence is faster than Euler's method. Second, as described in Project E, the trapezoid scheme has the desirable feature of being **stable.**

Can we somehow modify the trapezoid scheme in order to obtain an explicit method? One idea is first to get an estimate, say, y_{n+1}^*, of the value y_{n+1} using Euler's method and then use formula (8) with y_{n+1} replaced by y_{n+1}^* on the right-hand side. This two-step process is an example of a **predictor-corrector method.** That is, we predict y_{n+1} using Euler's method and

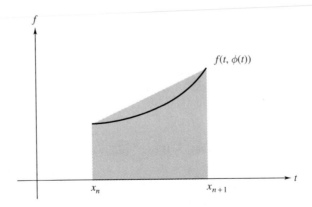

Figure 3.14 Approximation by a trapezoid

then use that value in (8) to obtain a "more correct" approximation. Setting $y_{n+1} = y_n + hf(x_n, y_n)$ in the right-hand side of (8), we obtain

(9) $$y_{n+1} = y_n + \frac{h}{2}[f(x_n, y_n) + f(x_n + h, y_n + hf(x_n, y_n))] , \qquad n = 0, 1, \ldots,$$

where $x_{n+1} = x_n + h$. This explicit scheme is known as the **improved Euler's method.**

EXAMPLE 1 Compute the improved Euler's method approximation to the solution $\phi(x) = e^x$ of

$$y' = y , \qquad y(0) = 1$$

at $x = 1$ using step sizes of $h = 1, 10^{-1}, 10^{-2}, 10^{-3}$, and 10^{-4}.

SOLUTION The starting values are $x_0 = 0$ and $y_0 = 1$. Since $f(x, y) = y$, formula (9) becomes

$$y_{n+1} = y_n + \frac{h}{2}[y_n + (y_n + hy_n)] = y_n + hy_n + \frac{h^2}{2}y_n ,$$

that is,

(10) $$y_{n+1} = \left(1 + h + \frac{h^2}{2}\right)y_n .$$

Since $y_0 = 1$, we see inductively that

$$y_n = \left(1 + h + \frac{h^2}{2}\right)^n , \qquad n = 0, 1, 2, \ldots .$$

To obtain approximations at $x = 1$, we must have $1 = x_0 + nh = nh$, and so $n = 1/h$. Hence the improved Euler's approximations to $e = \phi(1)$ are just

$$\left(1 + h + \frac{h^2}{2}\right)^{1/h} .$$

In Table 3.4 on page 130, we have computed this approximation for the specified values of h, along with the corresponding errors

$$e - \left(1 + h + \frac{h^2}{2}\right)^{1/h} .$$

Comparing the entries of this table with those of Table 3.3, we observe that the improved Euler's method converges much more rapidly than the original Euler's method. In fact, from the first few entries in the second and third columns of Table 3.4, it appears that the approximation gains two decimal places in accuracy each time h is decreased by a factor of 10. In other words, the error is roughly proportional to h^2 (see the last column of the table and also Problem 4). The entries in the last two rows of the table must be regarded with caution. Indeed, when $h = 10^{-3}$ or $h = 10^{-4}$, the true error is so small that our calculator rounded it to zero, to five decimal places. The entries in color in the last column may be inaccurate due to the loss of significant figures in the calculator arithmetic. ∎ End

	TABLE 3.4 IMPROVED EULER'S APPROXIMATION TO $e = 2.71828...$		
h	Approximation $\left(1 + h + \dfrac{h^2}{2}\right)^{1/h}$	Error	Error $\dfrac{\text{Error}}{h^2}$
1	2.50000	0.21828	0.21828
10^{-1}	2.71408	0.00420	0.42010
10^{-2}	2.71824	0.00004	0.44966
10^{-3}	2.71828	0.00000	0.45271
10^{-4}	2.71828	0.00000	0.45313

As the preceding example suggests, the improved Euler's method converges at the rate $O(h^2)$ and, indeed, it can be proved that in general *this method is of order $p = 2$.*

A step-by-step outline for a subroutine that implements the improved Euler's method over a given interval $[x_0, c]$ is described below. For programming purposes, it is usually more convenient to input the number of steps N in the interval rather than the step size h itself. For an interval starting at $x = x_0$ and ending at $x = c$, the relation between h and N is

(11) $Nh = c - x_0$.

(Note that the subroutine includes an option for printing x and y.) Of course, the implementation of this algorithm with N steps on the interval $[x_0, c]$ only produces approximations to the actual solution at $N + 1$ equally spaced points. If we wish to use these points to help graph an

IMPROVED EULER'S METHOD SUBROUTINE

Purpose To approximate the solution $\phi(x)$ to the initial value problem

$$y' = f(x, y) , \qquad y(x_0) = y_0 ,$$

for $x_0 \leq x \leq c$.

INPUT x_0, y_0, c, N (number of steps), PRNTR ($=1$ to print table)

Step 1 Set step size $h = (c - x_0)/N$, $x = x_0$, $y = y_0$

Step 2 For $i = 1$ to N, do Steps 3–5

Step 3 Set

$$F = f(x, y)$$
$$G = f(x + h, y + hF)$$

Step 4 Set

$$x = x + h$$
$$y = y + h(F + G)/2$$

Step 5 If PRNTR $= 1$, print x, y

approximate solution over the whole interval $[x_0, c]$, then we must somehow "fill in" the gaps between these points. A crude method is to simply join the points by straight line segments producing a polygonal line approximation to $\phi(x)$. More sophisticated techniques for prescribing the intermediate points are used in professional codes.

Now we want to devise a program that will compute $\phi(c)$ *to a desired accuracy.* As we have seen, the accuracy of the approximation depends on the step size h. Our strategy, then, will be to estimate $\phi(c)$ for a given step size and then halve the step size and recompute the estimate; halve again, and so on. When two consecutive estimates of $\phi(c)$ differ by less than some prescribed tolerance ε, we take the final estimate as our approximation to $\phi(c)$. Admittedly this does not guarantee that $\phi(c)$ is known to within ε, but it provides a reasonable stopping procedure in practice.[†] The following procedure also contains a safeguard to stop if the desired tolerance is not reached after M halvings of h.

IMPROVED EULER'S METHOD WITH TOLERANCE

Purpose To approximate the solution to the initial value problem

$$y' = f(x, y) , \qquad y(x_0) = y_0 ,$$

at $x = c$, with tolerance ε

INPUT x_0, y_0, c, ε ,
 M (maximum number of halvings of step size)

Step 1 Set $z = y_0$, PTNTR $= 0$
Step 2 For $m = 0$ to M, do Steps 3–7[††]
Step 3 Set $N = 2^m$
Step 4 Call IMPROVED EULER'S METHOD SUBROUTINE
Step 5 Print h, y
Step 6 If $|y - z| < \varepsilon$, go to Step 10
Step 7 Set $z = y$
Step 8 Print "$\phi(c)$ is approximately"; y; "but may not be within the tolerance"; ε
Step 9 Go to Step 11
Step 10 Print "$\phi(c)$ is approximately"; y; "with tolerance"; ε
Step 11 STOP
OUTPUT Approximations of the solution to the initial value problem at $x = c$ using 2^m steps

If one desires a stopping procedure that simulates the **relative error**

$$\left| \frac{\text{approximation} - \text{true value}}{\text{true value}} \right| ,$$

[†]Professional codes monitor accuracy much more carefully and vary step size in an adaptive fashion for this purpose (see Section 3.7).

[††]To save time, one can start with $m = K < M$ rather than with $m = 0$.

then replace Step 6 by

Step 6' If $\left| \dfrac{z - y}{y} \right| < \varepsilon$, go to Step 10.

EXAMPLE 2 Use the improved Euler's method with tolerance to approximate the solution to the initial value problem

(12) $y' = x + 2y$, $y(0) = 0.25$,

at $x = 2$. For a tolerance of $\varepsilon = 0.001$, use a stopping procedure based on the absolute error.

SOLUTION The starting values are $x_0 = 0$, $y_0 = 0.25$. Because we are computing the approximations for $c = 2$, the initial value for h is

$$h = (2 - 0)2^{-0} = 2 .$$

For equation (12), we have $f(x, y) = x + 2y$, so the numbers F and G in the subroutine are

$$F = x + 2y$$
$$G = (x + h) + 2(y + hF) = x + 2y + h(1 + 2x + 4y) ,$$

and we find

$$x = x + h$$
$$y = y + \frac{h}{2}(F + G) = y + \frac{h}{2}(2x + 4y) + \frac{h^2}{2}(1 + 2x + 4y) .$$

Thus, with $x_0 = 0$, $y_0 = 0.25$, and $h = 2$, we get for the first approximation

$$y = 0.25 + (0 + 1) + 2(1 + 1) = 5.25 .$$

To describe the further outputs of the algorithm, we use the notation $y(2; h)$ for the approximation obtained with step size h. Thus $y(2; 2) = 5.25$, and we find from the algorithm

$$
\begin{aligned}
y(2; 1) &= 11.25000 & y(2; 2^{-5}) &= 25.98132 \\
y(2; 2^{-1}) &= 18.28125 & y(2; 2^{-6}) &= 26.03172 \\
y(2; 2^{-2}) &= 23.06067 & y(2; 2^{-7}) &= 26.04468 \\
y(2; 2^{-3}) &= 25.12012 & y(2; 2^{-8}) &= 26.04797 \\
y(2; 2^{-4}) &= 25.79127 & y(2; 2^{-9}) &= 26.04880 .
\end{aligned}
$$

Since $|y(2; 2^{-9}) - y(2; 2^{-8})| = 0.00083$, which is less than $\varepsilon = 0.001$, we stop.

The exact solution of (12) is $\phi(x) = \frac{1}{2}\left(e^{2x} - x - \frac{1}{2}\right)$, so we have determined that

$$\phi(2) = \frac{1}{2}\left(e^4 - \frac{5}{2}\right) \approx 26.04880 .$$ ∎ End

In the next section, we discuss methods with higher rates of convergences than either Euler's or the improved Euler's methods.

EXERCISES 3.5

 In many of the following problems, it will be helpful to have a calculator or computer available. The reader may use a software package or write a program for solving initial value problems using the improved Euler's method algorithms on pages 130 and 131. (Remember, all trigonometric calculations are done in radians.)

1. Show that when Euler's method is used to approximate the solution of the initial value problem

$$y' = 5y , \qquad y(0) = 1 ,$$

 at $x = 1$, then the approximation with step size h is $(1 + 5h)^{1/h}$.

2. Show that when Euler's method is used to approximate the solution of the initial value problem

$$y' = -\frac{1}{2}y , \qquad y(0) = 3 ,$$

 at $x = 2$, then the approximation with step size h is

$$3\left(1 - \frac{h}{2}\right)^{2/h} .$$

3. Show that when the trapezoid scheme given in formula (8) is used to approximate the solution $\phi(x) = e^x$ of

$$y' = y , \qquad y(0) = 1 ,$$

 at $x = 1$, then we get

$$y_{n+1} = \left(\frac{1 + h/2}{1 - h/2}\right)y_n , \qquad n = 0, 1, 2, \ldots ,$$

 which leads to the approximation

$$\left(\frac{1 + h/2}{1 - h/2}\right)^{1/h}$$

 for the constant e. Compute this approximation for $h = 1, 10^{-1}, 10^{-2}, 10^{-3}$, and 10^{-4} and compare your results with those in Tables 3.3 and 3.4.

4. In Example 1, the improved Euler's method approximation to e with step size h was shown to be

$$\left(1 + h + \frac{h^2}{2}\right)^{1/h} .$$

 First prove that the error $:= e - (1 + h + h^2/2)^{1/h}$ approaches zero as $h \to 0$. Then use L'Hôpital's rule

to show that

$$\lim_{h \to 0} \frac{\text{error}}{h^2} = \frac{e}{6} \approx 0.45305 .$$

Compare this constant with the entries in the last column of Table 3.4.

5. Show that when the improved Euler's method is used to approximate the solution of the initial value problem

$$y' = 4y , \qquad y(0) = \frac{1}{3} ,$$

 at $x = 1/2$, then the approximation with step size h is

$$\frac{1}{3}(1 + 4h + 8h^2)^{1/(2h)} .$$

6. Since the integral $y(x) := \int_0^x f(t)\,dt$ with variable upper limit satisfies (for continuous f) the initial value problem

$$y' = f(x) , \qquad y(0) = 0 ,$$

 any numerical scheme that is used to approximate the solution at $x = 1$ will give an approximation to the definite integral

$$\int_0^1 f(t)\,dt .$$

 Derive a formula for this approximation of the integral using

 (a) Euler's method.
 (b) the trapezoid scheme.
 (c) the improved Euler's method.

7. Use the improved Euler's method subroutine with step size $h = 0.1$ to approximate the solution to the initial value problem

$$y' = x - y^2 , \qquad y(1) = 0 ,$$

 at the points $x = 1.1, 1.2, 1.3, 1.4$, and 1.5. (Thus input $N = 5$.) Compare these approximations with those obtained using Euler's method (see Exercises 1.5, Problem 6).

8. Use the improved Euler's method subroutine with step size $h = 0.2$ to approximate the solution to the initial value problem

$$y' = \frac{1}{x}(y^2 + y) , \qquad y(1) = 1 ,$$

at the points $x = 1.2, 1.4, 1.6,$ and $1.8.$ (Thus input $N = 8.$) Compare these approximations with those obtained using Euler's method (see Exercises 1.5, Problem 5).

9. Use the improved Euler's method subroutine with step size $h = 0.2$ to approximate the solution to
$$y' = x + 3\cos(xy) , \qquad y(0) = 0 ,$$
at the points $x = 0, 0.2, 0.4, \ldots, 2.0.$ Use your answers to make a rough sketch of the solution on $[0, 2].$

10. Use the improved Euler's method subroutine with step size $h = 0.1$ to approximate the solution to
$$y' = 4\cos(x + y) , \qquad y(0) = 1 ,$$
at the points $x = 0, 0.1, 0.2, \ldots, 1.0.$ Use your answers to make a rough sketch of the solution on $[0, 1].$

11. Use the improved Euler's method with tolerance to approximate the solution to
$$\frac{dx}{dt} = 1 + t\sin(tx) , \qquad x(0) = 0 ,$$
at $t = 1.$ For a tolerance of $\varepsilon = 0.01,$ use a stopping procedure based on the absolute error.

12. Use the improved Euler's method with tolerance to approximate the solution to
$$y' = 1 - \sin y , \qquad y(0) = 0 ,$$
at $x = \pi.$ For a tolerance of $\varepsilon = 0.01,$ use a stopping procedure based on the absolute error.

13. Use the improved Euler's method with tolerance to approximate the solution to
$$y' = 1 - y + y^3 , \qquad y(0) = 0 ,$$
at $x = 1.$ For a tolerance of $\varepsilon = 0.003,$ use a stopping procedure based on the absolute error.

14. By experimenting with the improved Euler's method subroutine, find the maximum value over the interval $[0, 2]$ of the solution to the initial value problem
$$y' = \sin(x + y) , \qquad y(0) = 2 .$$
Where does this maximum value occur? Give answers to two decimal places.

15. The solution to the initial value problem
$$\frac{dy}{dx} = (x + y + 2)^2 , \qquad y(0) = -2$$
crosses the x-axis at a point in the interval $[0, 1.4].$ By experimenting with the improved Euler's method

subroutine, determine this point to two decimal places.

16. The solution to the initial value problem
$$\frac{dy}{dx} + \frac{y}{x} = x^3 y^2 , \qquad y(1) = 3$$
has a vertical asymptote ("blows up") at some point in the interval $[1, 2].$ By experimenting with the improved Euler's method subroutine, determine this point to two decimal places.

17. Use Euler's method (4) with $h = 0.1$ to approximate the solution to the initial value problem
$$y' = -20y , \qquad y(0) = 1 ,$$
on the interval $0 \le x \le 1$ (that is, at $x = 0, 0.1, \ldots, 1.0$). Compare your answers with the actual solution $y = e^{-20x}.$ What went wrong? Next, try the step size $h = 0.025$ and also $h = 0.2.$ What conclusions can you draw concerning the choice of step size?

18. **Local versus Global Error.** In deriving formula (4) for Euler's method, a rectangle was used to approximate the area under a curve (see Figure 3.13). With $g(t) := f(t, \phi(t)),$ this approximation can be written as
$$\int_{x_n}^{x_{n+1}} g(t)\,dt \approx hg(x_n) , \qquad \text{where} \quad h = x_{n+1} - x_n .$$

(a) Show that if g has a continuous derivative that is bounded in absolute value by $B,$ then the rectangle approximation has error $O(h^2),$ that is, for some constant $M,$
$$\left| \int_{x_n}^{x_{n+1}} g(t)\,dt - hg(x_n) \right| \le Mh^2 .$$
This is called the *local truncation error* of the scheme. [*Hint:* Write
$$\int_{x_n}^{x_{n+1}} g(t)\,dt - hg(x_n) = \int_{x_n}^{x_{n+1}} [g(t) - g(x_n)]\,dt .$$
Next, using the mean value theorem, show that $|g(t) - g(x_n)| \le B|t - x_n|.$ Then integrate to obtain the error bound $(B/2)h^2.$]

(b) In applying Euler's method, local truncation errors occur in each step of the process and are propagated throughout the further computations. Show that the *sum* of the local truncation errors

in part (a) that arise after n steps is $O(h)$. This is the *global error*, which is the same as the convergence rate of Euler's method.

19. Logistic Model. In Section 3.2, we discussed the logistic equation

$$\frac{dp}{dt} = Ap_1 p - Ap^2 , \qquad p(0) = p_0 ,$$

and its use in modeling population growth. A more general model might involve the equation

(13) $\qquad \frac{dp}{dt} = Ap_1 p - Ap^r , \qquad p(0) = p_0 ,$

where $r > 1$. To see the effect of changing the parameter r in (13), take $p_1 = 3$, $A = 1$, and $p_0 = 1$. Then use the improved Euler's method subroutine with $h = 0.25$ to approximate the solution to (13) on the interval $0 \le t \le 5$ for $r = 1.5, 2$, and 3. What is the limiting population in each case? For $r > 1$, determine a general formula for the limiting population.

20. Falling Body. In Example 1 of Section 3.4, we modeled the velocity of a falling body by the initial value problem

$$m\frac{dv}{dt} = mg - bv , \qquad v(0) = v_0 ,$$

under the assumption that the force due to air resistance is $-bv$. However, in certain cases the force due to air resistance behaves more like $-bv^r$, where $r \ (>1)$ is some constant. This leads to the model

(14) $\qquad m\frac{dv}{dt} = mg - bv^r , \qquad v(0) = v_0 .$

To study the effect of changing the parameter r in (14), take $m = 1$, $g = 9.81$, $b = 2$, and $v_0 = 0$. Then use the improved Euler's method subroutine with $h = 0.2$ to approximate the solution to (14) on the interval $0 \le t \le 5$ for $r = 1.0, 1.5$, and 2.0. What is the relationship between these solutions and the constant solution $v(t) \equiv (9.81/2)^{1/r}$?

21. Building Temperature. In Section 3.3, we modeled the temperature inside a building by the initial value problem

(15) $\qquad \frac{dT}{dt} = K[M(t) - T(t)] + H(t) + U(t) ,$

$$T(t_0) = T_0 ,$$

where M is the temperature outside the building, T is the temperature inside the building, H is the additional heating rate, U is the furnace heating or air conditioner cooling rate, K is a positive constant, and T_0 is the initial temperature at time t_0. In a typical model, $t_0 = 0$ (midnight), $T_0 = 65°F$, $H(t) = 0.1$, $U(t) = 1.5[70 - T(t)]$, and

$$M(t) = 75 - 20 \cos(\pi t/12) .$$

The constant K is usually between $1/4$ and $1/2$, depending on such things as insulation. To study the effect of insulating this building, consider the typical building described above and use the improved Euler's method subroutine with $h = 2/3$ to approximate the solution to (15) on the interval $0 \le t \le 24$ (1 day) for $K = 0.2, 0.4$, and 0.6.

3.6 HIGHER-ORDER NUMERICAL METHODS: TAYLOR AND RUNGE-KUTTA

In Sections 1.5 and 3.5, we discussed a simple numerical procedure, Euler's method, for obtaining a numerical approximation of the solution $\phi(x)$ to the initial value problem

(1) $\qquad y' = f(x, y) , \qquad y(x_0) = y_0 .$

Euler's method is easy to implement because it involves only linear approximations to the solution $\phi(x)$. But it suffers from slow convergence, being a method of order 1, that is, the error is $O(h)$. Even the improved Euler's method discussed in Section 3.5 has order of only 2. In this section, we present numerical methods that have faster rates of convergence. These include **Taylor methods,** which are natural extensions of the Euler procedure, and **Runge-Kutta**

methods, which are the more popular schemes for solving initial value problems because they have fast rates of convergence and are easy to program.

As in the previous section, we assume that f and $\partial f/\partial y$ are continuous and bounded on the vertical strip $\{(x, y): a < x < b, -\infty < y < \infty\}$ and that f possesses as many continuous partial derivatives as needed.

To derive the Taylor methods, let $\phi_n(x)$ be the *exact* solution of the related initial value problem

$$(2) \qquad \phi_n' = f(x, \phi_n) , \qquad \phi_n(x_n) = y_n .$$

The Taylor series for $\phi_n(x)$ about the point x_n is

$$\phi_n(x) = \phi_n(x_n) + h\phi_n'(x_n) + \frac{h^2}{2!}\phi_n''(x_n) + \cdots ,$$

where $h = x - x_n$. Since ϕ_n satisfies (2), we can write this series in the form

$$(3) \qquad \phi_n(x) = y_n + hf(x_n, y_n) + \frac{h^2}{2!}\phi_n''(x_n) + \cdots .$$

Observe that the recursive formula for y_{n+1} in Euler's method is obtained by truncating the Taylor series after the linear term. For a better approximation, we will use more terms in the Taylor series. This requires that we express the higher-order derivatives of the solution in terms of the function $f(x, y)$.

If y satisfies $y' = f(x, y)$, we can compute y'' by using the chain rule:

$$(4) \qquad y'' = \frac{\partial f}{\partial x}(x, y) + \frac{\partial f}{\partial y}(x, y)\, y'$$

$$= \frac{\partial f}{\partial x}(x, y) + \frac{\partial f}{\partial y}(x, y)f(x, y)$$

$$=: f_2(x, y) .$$

In a similar fashion, define f_3, f_4, \ldots, that correspond to the expressions $y'''(x), y^{(4)}(x)$, etc. If we truncate the expansion in (3) after the h^p term, then, with the above notation, the recursive formulas for the **Taylor method of order p** are

$$(5) \qquad x_{n+1} = x_n + h ,$$

$$(6) \qquad y_{n+1} = y_n + hf(x_n, y_n) + \frac{h^2}{2!}f_2(x_n, y_n) + \cdots + \frac{h^p}{p!}f_p(x_n, y_n) .$$

As before, $y_n \approx \phi(x_n)$, where $\phi(x)$ is the solution to the initial value problem (1). It can be shown[†] that **the Taylor method of order p has the rate of convergence O(h^p).**

EXAMPLE 1 Determine the recursive formulas for the Taylor method of order 2 for the initial value problem

$$(7) \qquad y' = \sin(xy) , \qquad y(0) = \pi .$$

[†]See *Introduction to Numerical Analysis* by J. Stoer and R. Bulirsch (Springer-Verlag, New York, 1980).

SOLUTION We must compute $f_2(x, y)$ as defined in (4). Since $f(x, y) = \sin(xy)$,

$$\frac{\partial f}{\partial x}(x, y) = y\cos(xy) , \qquad \frac{\partial f}{\partial y}(x, y) = x\cos(xy) .$$

Substituting into (4), we have

$$f_2(x, y) = \frac{\partial f}{\partial x}(x, y) + \frac{\partial f}{\partial y}(x, y)f(x, y)$$
$$= y\cos(xy) + x\cos(xy)\sin(xy)$$
$$= y\cos(xy) + \frac{x}{2}\sin(2xy) ,$$

and the recursive formulas (5) and (6) become

$$x_{n+1} = x_n + h ,$$
$$y_{n+1} = y_n + h\sin(x_n y_n) + \frac{h^2}{2}\left[y_n\cos(x_n y_n) + \frac{x_n}{2}\sin(2x_n y_n)\right] ,$$

where $x_0 = 0$, $y_0 = \pi$ are the starting values. ■ End

The convergence rate, $O(h^p)$, of the pth order Taylor method raises an interesting question: if we could somehow let p go to infinity would we obtain **exact** solutions for the interval $[x_0, x_0 + h]$? This possibility is explored in depth in Chapter 8. Of course a practical difficulty in employing high-order Taylor methods is the tedious computation of the partial derivatives needed to determine f_p (typically these computations grow exponentially with p). One way to circumvent this difficulty is to use one of the **Runge-Kutta methods.**[†]

Observe that the general Taylor method has the form

(8) $y_{n+1} = y_n + hF(x_n, y_n; h) ,$

where the choice of F depends on p. In particular (cf. (6)), for

$$p = 1 , \qquad F = T_1(x, y; h) := f(x, y) ,$$

(9) $p = 2 , \qquad F = T_2(x, y; h) := f(x, y) + \frac{h}{2}\left[\frac{\partial f}{\partial x}(x, y) + \frac{\partial f}{\partial y}(x, y)f(x, y)\right] .$

The idea behind the Runge-Kutta method of order 2 is to choose F in (8) of the form

(10) $F = K_2(x, y; h) := f\big(x + \alpha h, y + \beta hf(x, y)\big) ,$

where the constants α, β are to be selected so that (8) has the rate of convergence $O(h^2)$. The advantage here is that K_2 is computed by two evaluations of the original function $f(x, y)$ and does not involve the derivatives of $f(x, y)$.

To ensure $O(h^2)$ convergence, we compare this new scheme with the Taylor method of order 2 and require

$$T_2(x, y; h) - K_2(x, y; h) = O(h^2) , \quad \text{as} \quad h \to 0 .$$

[†]*Historical Footnote:* These methods were developed by C. Runge in 1895 and W. Kutta in 1901.

That is, we choose α, β so that the Taylor expansions for T_2 and K_2 agree through terms of order h. For (x, y) fixed, when we expand $K_2 = K_2(h)$ as given in (10) about $h = 0$, we find

(11) $\displaystyle K_2(h) = K_2(0) + \frac{dK_2}{dh}(0)h + O(h^2)$

$$= f(x, y) + \left[\alpha \frac{\partial f}{\partial x}(x, y) + \beta \frac{\partial f}{\partial y}(x, y)f(x, y) \right]h + O(h^2) \; ,$$

where the expression in brackets for dK_2/dh, evaluated at $h = 0$, follows from the chain rule. Comparing (11) with (9), we see that for T_2 and K_2 to agree through terms of order h, we must have $\alpha = \beta = 1/2$. Thus

$$K_2(x, y; h) = f\left(x + \frac{h}{2}, y + \frac{h}{2}f(x, y) \right) \; .$$

The Runge-Kutta method we have derived is called the **midpoint method** and it has the recursive formulas

(12) $x_{n+1} = x_n + h \; ,$

(13) $\displaystyle y_{n+1} = y_n + hf\left(x_n + \frac{h}{2}, y_n + \frac{h}{2}f(x_n, y_n) \right) \; .$

By construction, the midpoint method has the same rate of convergence as the Taylor method of order 2; that is, $O(h^2)$. This is the same rate as the improved Euler's method.

In a similar fashion, one can work with the Taylor method of order 4 and, after some elaborate calculations, obtain the **classical fourth order Runge-Kutta method.** The recursive formulas for this method are

(14)
$$x_{n+1} = x_n + h \; ,$$
$$y_{n+1} = y_n + \frac{1}{6}(k_1 + 2k_2 + 2k_3 + k_4) \; ,$$

where

$$k_1 = hf(x_n, y_n) \; ,$$

$$k_2 = hf\left(x_n + \frac{h}{2}, y_n + \frac{k_1}{2} \right) \; ,$$

$$k_3 = hf\left(x_n + \frac{h}{2}, y_n + \frac{k_2}{2} \right) \; ,$$

$$k_4 = hf(x_n + h, y_n + k_3) \; .$$

The classical fourth order Runge-Kutta method is one of the more popular methods because its rate of convergence is $O(h^4)$ and it is easy to program. Typically, it produces very accurate approximations even when the number of iterations is reasonably small. However, as the number of iterations becomes large, other types of errors may creep in.

Program outlines for the fourth order Runge-Kutta method are given below. Just as with the algorithms for the improved Euler's method, the first program (the Runge-Kutta sub-

routine) is useful for approximating the solution over an interval $[x_0, c]$ and takes the number of steps in the interval as input. As in Section 3.5, the number of steps N is related to the step size h and the interval $[x_0, c]$ by

$$Nh = c - x_0 .$$

The subroutine has the option to print out a table of values of x and y. The second algorithm (Runge-Kutta with tolerance) on page 140 is used to approximate, for a given tolerance, the solution at an inputed value $x = c$. This algorithm[†] automatically halves the step sizes successively until the two approximations $y(c; h)$ and $y(c; h/2)$ differ by less than the prescribed tolerance ε. For a stopping procedure based on the relative error, Step 6 of the algorithm should be replaced by

Step 6′ If $\left| \dfrac{y - z}{y} \right| < \varepsilon$, go to Step 10.

CLASSICAL 4TH ORDER RUNGE-KUTTA SUBROUTINE

Purpose To approximate the solution to the initial value problem

$$y' = f(x, y) , \qquad y(x_0) = y_0$$

for $x_0 \leq x \leq c$

INPUT x_0, y_0, c, N (number of steps), PRNTR ($=1$ to print table)

Step 1 Set step size $h = (c - x_0)/N$, $x = x_0$, $y = y_0$

Step 2 For $i = 1$ to N do Steps 3–5

Step 3 Set

$$k_1 = hf(x, y)$$

$$k_2 = hf\left(x + \frac{h}{2}, y + \frac{k_1}{2}\right)$$

$$k_3 = hf\left(x + \frac{h}{2}, y + \frac{k_2}{2}\right)$$

$$k_4 = hf(x + h, y + k_3)$$

Step 4 Set

$$x = x + h$$

$$y = y + \frac{1}{6}(k_1 + 2k_2 + 2k_3 + k_4)$$

Step 5 If PRNTR $= 1$, print x, y

[†]Note that the form of this algorithm is the same as that for the improved Euler's method on page 131 except for Step 4 where the Runge-Kutta subroutine is called. More sophisticated stopping procedures are used in production-grade codes.

CLASSICAL 4TH ORDER RUNGE-KUTTA ALGORITHM WITH TOLERANCE

Purpose To approximate the solution to the initial value problem
$$y' = f(x, y) , \qquad y(x_0) = y_0$$
at $x = c$, with tolerance ε

INPUT $x_0, y_0, c, \varepsilon, M$ (maximum number of iterations)

Step 1 Set $z = y_0$, PRNTR $= 0$

Step 2 For $m = 0$ to M, do Steps 3–7 (or, to save time, start with $m > 0$)

Step 3 Set $N = 2^m$

Step 4 Call FOURTH ORDER RUNGE-KUTTA SUBROUTINE

Step 5 Print h, y

Step 6 If $|z - y| < \varepsilon$, go to Step 10

Step 7 Set $z = y$

Step 8 Print "$\phi(c)$ is approximately"; y; "but may not be within the tolerance"; ε

Step 9 Go to Step 11

Step 10 Print "$\phi(c)$ is approximately"; y; "with tolerance"; ε

Step 11 STOP

OUTPUT Approximations of the solution to the initial value problem at $x = c$, using 2^m steps.

EXAMPLE 2 Use the classical fourth order Runge-Kutta algorithm to approximate the solution $\phi(x)$ of the initial value problem
$$y' = y , \qquad y(0) = 1 ,$$
at $x = 1$ with a tolerance of 0.001.

SOLUTION The inputs are $x_0 = 0$, $y_0 = 1$, $c = 1$, $\varepsilon = 0.001$, and $M = 25$ (say). Since $f(x, y) = y$, the formulas in Step 3 of the subroutine become
$$k_1 = hy , \qquad k_2 = h\left(y + \frac{k_1}{2}\right) , \qquad k_3 = h\left(y + \frac{k_2}{2}\right) , \qquad k_4 = h(y + k_3) .$$

The initial value for N in this algorithm is $N = 1$, so
$$h = (1 - 0)/1 = 1 .$$

Thus, in Step 3 of the subroutine, we compute
$$k_1 = (1)(1) = 1 , \qquad k_2 = (1)(1 + 0.5) = 1.5 ,$$
$$k_3 = (1)(1 + 0.75) = 1.75 , \qquad k_4 = (1)(1 + 1.75) = 2.75 ,$$

and, in Step 4 of the subroutine, we get for the first approximation
$$y = y_0 + \frac{1}{6}(k_1 + 2k_2 + 2k_3 + k_4)$$
$$= 1 + \frac{1}{6}\left[1 + 2(1.5) + 2(1.75) + 2.75\right]$$
$$= 2.70833 ,$$

where we have rounded to five decimal places. Because

$$|z - y| = |y_0 - y| = |1 - 2.70833| = 1.70833 > \varepsilon \ ,$$

we start over and reset $N = 2$, $h = 0.5$.

Doing Steps 3 and 4 for $i = 1$ and 2, we ultimately obtain (for $i = 2$) the approximation

$$y = 2.71735 \ .$$

Since $|z - y| = |2.70833 - 2.71735| = 0.00902 > \varepsilon$, we again start over and reset $N = 4$, $h = 0.25$. This leads to the approximation

$$y = 2.71821 \ ,$$

so that

$$|z - y| = |2.71735 - 2.71821| = 0.00086 \ ,$$

which is less than $\varepsilon = 0.001$. Hence $\phi(1) = e \approx 2.71821$. ∎ End

In Example 2, we were able to obtain a better approximation for $\phi(1) = e$ with $h = 0.25$ than we obtained in Section 3.5 using Euler's method with $h = 0.001$ (see Table 3.3) and roughly the same accuracy as we obtained in Section 3.5 using the improved Euler's method with $h = 0.01$ (see Table 3.4).

EXAMPLE 3 Use the fourth order Runge-Kutta subroutine to approximate the solution $\phi(x)$ of the initial value problem

(15) $y' = y^2 \ , \qquad y(0) = 1 \ ,$

on the interval $0 \le x \le 2$ using $N = 8$ steps (i.e., $h = 0.25$).

SOLUTION Here the starting values are $x_0 = 0$ and $y_0 = 1$. Since $f(x, y) = y^2$, the formulas in Step 3 of the subroutine are

$$k_1 = hy^2 \ , \qquad\qquad k_2 = h\left(y + \frac{k_1}{2}\right)^2 \ ,$$

$$k_3 = h\left(y + \frac{k_2}{2}\right)^2 \ , \qquad k_4 = h(y + k_3)^2 \ .$$

From the output, we find

$$\begin{aligned}
x &= 0.25 & y &= 1.33322 \ , \\
x &= 0.50 & y &= 1.99884 \ , \\
x &= 0.75 & y &= 3.97238 \ , \\
x &= 1.00 & y &= 32.82820 \ , \\
x &= 1.25 & y &= 4.09664 \times 10^{11} \ , \\
x &= 1.50 & y &= \text{overflow} \ .
\end{aligned}$$

What happened? Fortunately the equation in (15) is separable, and, solving for $\phi(x)$, we obtain $\phi(x) = (1 - x)^{-1}$. It is now obvious where the problem lies: The true solution $\phi(x)$ is not defined

at $x = 1$. If we had been more cautious, we would have realized that $\partial f/\partial y = 2y$ is *not* bounded for all y. Hence the existence of a unique solution is not guaranteed for all x between 0 and 2, and in this case, the method does *not* give meaningful approximations for x near (or greater than) 1. ■ End

EXAMPLE 4 Use the fourth order Runge-Kutta algorithm to approximate the solution $\phi(x)$ of the initial value problem

$$y' = x - y^2 , \qquad y(0) = 1 ,$$

at $x = 2$ with a tolerance of 0.0001.

SOLUTION This time we check to see whether $\partial f/\partial y$ is bounded. Here $\partial f/\partial y = -2y$, which is certainly unbounded in any vertical strip. However, let's consider the qualitative behavior of the solution $\phi(x)$. The solution curve starts at $(0, 1)$, where $\phi'(0) = 0 - 1 < 0$, so $\phi(x)$ begins decreasing and continues to decrease until it crosses the curve $y = \sqrt{x}$. After crossing this curve, $\phi(x)$ begins to increase, since $\phi'(x) = x - \phi^2(x) > 0$. As $\phi(x)$ increases, it remains below the curve $y = \sqrt{x}$. This is because if the solution were to get "close" to the curve $y = \sqrt{x}$, then the derivative of $\phi(x)$ would approach zero, so that overtaking the function \sqrt{x} is impossible.

Therefore, although the existence-uniqueness theorem does not guarantee a solution, we are inclined to try the algorithm anyway. The above argument shows that $\phi(x)$ probably exists for $x > 0$, so we feel reasonably sure the fourth order Runge-Kutta method will give a good approximation of the true solution $\phi(x)$. Proceeding with the algorithm, we use the starting values $x_0 = 0$ and $y_0 = 1$. Since $f(x, y) = x - y^2$, the formulas in Step 3 of the subroutine become

$$k_1 = h(x - y^2) , \qquad k_2 = h\left[\left(x + \frac{h}{2}\right) - \left(y + \frac{k_1}{2}\right)^2\right] ,$$

$$k_3 = h\left[\left(x + \frac{h}{2}\right) - \left(y + \frac{k_2}{2}\right)^2\right] , \qquad k_4 = h\left[(x + h) - (y + k_3)^2\right] .$$

In Table 3.5, we give the approximations $y(2; 2^{-m+1})$ for $\phi(2)$ for $m = 0, 1, 2, 3,$ and 4. The algorithm stops at $m = 4$, since

$$|y(2; 0.125) - y(2; 0.25)| = 0.00000 .$$

Hence $\phi(2) \approx 1.25132$ with a tolerance of 0.0001. ■ End

TABLE 3.5 CLASSICAL 4TH ORDER RUNGE-KUTTA APPROXIMATION FOR $\phi(2)$

| m | h | Approximation for $\phi(2)$ | $|y(2; h) - y(2; 2h)|$ |
|---|---|---|---|
| 0 | 2.0 | −8.33333 | |
| 1 | 1.0 | 1.27504 | 9.60837 |
| 2 | 0.5 | 1.25170 | 0.02334 |
| 3 | 0.25 | 1.25132 | 0.00038 |
| 4 | 0.125 | 1.25132 | 0.00000 |

EXERCISES 3.6

As in Exercises 3.5, the reader will find it helpful to have a calculator or computer available. For Problems 1–17, note whether or not $\partial f/\partial y$ is bounded.

1. Determine the recursive formulas for the Taylor method of order 2 for the initial value problem
$$y' = \cos(x + y) , \qquad y(0) = \pi .$$

2. Determine the recursive formulas for the Taylor method of order 2 for the initial value problem
$$y' = xy - y^2 , \qquad y(0) = -1 .$$

3. Determine the recursive formulas for the Taylor method of order 4 for the initial value problem
$$y' = x - y , \qquad y(0) = 0 .$$

4. Determine the recursive formulas for the Taylor method of order 4 for the initial value problem
$$y' = x^2 + y , \qquad y(0) = 0 .$$

5. Use the Taylor methods of orders 2 and 4 with $h = 0.25$ to approximate the solution to the initial value problem
$$y' = x + 1 - y , \qquad y(0) = 1 ,$$
at $x = 1$. Compare these approximations to the actual solution $y = x + e^{-x}$ evaluated at $x = 1$.

6. Use the Taylor methods of orders 2 and 4 with $h = 0.25$ to approximate the solution to the initial value problem
$$y' = 1 - y , \qquad y(0) = 0 ,$$
at $x = 1$. Compare these approximations to the actual solution $y = 1 - e^{-x}$ evaluated at $x = 1$.

7. Use the fourth order Runge-Kutta subroutine with $h = 0.25$ to approximate the solution to the initial value problem
$$y' = 2y - 6 , \qquad y(0) = 1 ,$$
at $x = 1$. (Thus input $N = 4$.) Compare this approximation to the actual solution $y = 3 - 2e^{2x}$ evaluated at $x = 1$.

8. Use the fourth order Runge-Kutta subroutine with $h = 0.25$ to approximate the solution to the initial value problem
$$y' = 1 - y , \qquad y(0) = 0 ,$$
at $x = 1$. Compare this approximation with the one obtained in Problem 6 using the Taylor method of order 4.

9. Use the fourth order Runge-Kutta subroutine with $h = 0.25$ to approximate the solution to the initial value problem
$$y' = x + 1 - y , \qquad y(0) = 1 ,$$
at $x = 1$. Compare this approximation with the one obtained in Problem 5 using the Taylor method of order 4.

10. Use the fourth order Runge-Kutta algorithm to approximate the solution to the initial value problem
$$y' = 1 - xy , \qquad y(1) = 1 ,$$
at $x = 2$. For a tolerance of $\varepsilon = 0.001$, use a stopping procedure based on the absolute error.

11. The solution to the initial value problem
$$y' = \frac{2}{x^4} - y^2 , \qquad y(1) = -0.414$$
crosses the x-axis at a point in the interval $[1, 2]$. By experimenting with the fourth order Runge-Kutta subroutine, determine this point to two decimal places.

12. By experimenting with the fourth order Runge-Kutta subroutine, find the maximum value over the interval $[1, 2]$ of the solution to the initial value problem
$$y' = \frac{1.8}{x^4} - y^2 , \qquad y(1) = -1 .$$
Where does this maximum occur? Give your answers to two decimal places.

13. The solution to the initial value problem
$$\frac{dy}{dx} = y^2 - 2e^x y + e^{2x} + e^x , \qquad y(0) = 3$$
has a vertical asymptote ("blows up") at some point in the interval $[0, 2]$. By experimenting with the fourth order Runge-Kutta subroutine, determine this point to two decimal places.

14. Use the fourth order Runge-Kutta algorithm to approximate the solution to the initial value problem
$$y' = y \cos x , \qquad y(0) = 1 ,$$
at $x = \pi$. For a tolerance of $\varepsilon = 0.01$, use a stopping procedure based on the absolute error.

15. Use the fourth order Runge-Kutta subroutine with $h = 0.1$ to approximate the solution to
$$y' = \cos(5y) - x , \qquad y(0) = 0 ,$$

at the points $x = 0, 0.1, 0.2, \ldots, 3.0$. Use your answers to make a rough sketch of the solution on $[0, 3]$.

16. Use the fourth order Runge-Kutta subroutine with $h = 0.1$ to approximate the solution to
$$y' = 3\cos(y - 5x) , \qquad y(0) = 0 ,$$
at the points $x = 0, 0.1, 0.2, \ldots, 4.0$. Use your answers to make a rough sketch of the solution on $[0, 4]$.

17. The Taylor method of order 2 can be used to approximate the solution to the initial value problem
$$y' = y , \qquad y(0) = 1 ,$$
at $x = 1$. Show that the approximation y_n obtained by using the Taylor method of order 2 with the step size $1/n$ is given by the formula
$$y_n = \left(1 + \frac{1}{n} + \frac{1}{2n^2}\right)^n , \qquad n = 1, 2, \ldots .$$
The solution to the initial value problem is $y = e^x$, so y_n is an approximation to the constant e.

18. If the Taylor method of order p is used in Problem 17, show that
$$y_n = \left(1 + \frac{1}{n} + \frac{1}{2n^2} + \frac{1}{6n^3} + \cdots + \frac{1}{p!n^p}\right)^n ,$$
$$n = 1, 2, \ldots .$$

19. **Fluid Flow.** In the study of the nonisothermal flow of a Newtonian fluid between parallel plates, the equation
$$\frac{d^2y}{dx^2} + x^2 e^y = 0 , \qquad x > 0 ,$$
was encountered. By a series of substitutions, this equation can be transformed into the first order equation
$$\frac{dv}{du} = u\left(\frac{u}{2} + 1\right)v^3 + \left(u + \frac{5}{2}\right)v^2 .$$

Use the fourth order Runge-Kutta algorithm to approximate $v(3)$ if $v(u)$ satisfies $v(2) = 0.1$. For a tolerance of $\varepsilon = 0.0001$, use a stopping procedure based on the relative error.

20. **Chemical Reactions.** The reaction between nitrous oxide and oxygen to form nitrogen dioxide is given by the balanced chemical equation $2NO + O_2 = 2NO_2$. At high temperatures, the dependence of the rate of this reaction on the concentrations of NO, O_2, and NO_2 is complicated. However, at 25°C the rate at which NO_2 is formed obeys the law of mass action and is given by the rate equation
$$\frac{dx}{dt} = k(\alpha - x)^2\left(\beta - \frac{x}{2}\right) ,$$
where $x(t)$ denotes the concentration of NO_2 at time t, k is the rate constant, α is the initial concentration of NO, and β is the initial concentration of O_2. At 25°C, the constant k is 7.13×10^3 (liter)2/(mole)2(second). Let $\alpha = 0.0010$ mole/L, $\beta = 0.0041$ mole/L, and $x(0) = 0$ mole/L. Use the fourth order Runge-Kutta algorithm to approximate $x(10)$. For a tolerance of $\varepsilon = 0.000001$, use a stopping procedure based on the relative error.

21. **Transmission Lines.** In the study of the electric field that is induced by two nearby transmission lines, an equation of the form
$$\frac{dz}{dx} + g(x)z^2 = f(x)$$
arises. Let $f(x) = 5x + 2$ and $g(x) = x^2$. If $z(0) = 1$, use the fourth order Runge-Kutta algorithm to approximate $z(1)$. For a tolerance of $\varepsilon = 0.0001$, use a stopping procedure based on the absolute error.

3.7 PROFESSIONAL CODES FOR SOLVING INITIAL VALUE PROBLEMS

The numerical methods and program outlines described in this text are intended only to provide the reader with a rudimentary flavor of available computational tools for solving initial value problems. There are, in fact, a variety of sophisticated packages to choose from, depending on the type of output desired, the programming language to be used, and the hardware

available. The popular commercial packages MATLAB, MATHEMATICA, MATHCAD, and MAPLE all provide access to special libraries of codes for solving differential equations. For general scientific computation, the "Guide to Available Mathematical Software" (GAMS) accessible at *http://gams.nist.gov* has an extensive list of high-quality numerical programs, many of which are available free.

Before utilizing professional software, however, it is essential to keep in mind that *not all differential equations can be solved numerically.* For example, most packages encounter difficulties when the functions appearing in the differential equation are not smooth; i.e., they have discontinuous derivatives. Furthermore, numerical software will only accept problems over *finite* intervals, so that long-term behavior of solutions is often beyond the computer's reach. Nonetheless, professional packages are a valuable resource when properly applied, and so it is worthwhile to gain a better understanding of their basic ingredients, two of which are multistep methods and variable step size.

The numerical procedures discussed in the two previous sections are examples of **one-step methods** because the approximation y_{n+1} of the true value $y(x_{n+1})$ is obtained using the differential equation and the previous approximation y_n of $y(x_n)$. More generally, we can use **multistep methods.** For such schemes, the differential equation and the previous k approximations

$$y_n, \ y_{n-1}, \ \ldots, \ y_{n-(k-2)}, \ y_{n-(k-1)}$$

are used to obtain the approximation y_{n+1}. (In this case the method is called a **k-step method.**)

A two-step method that is easy to derive is based on Simpson's rule for approximating integrals (see Appendix B). Let y_n be the approximation of the solution $y(x)$ to the initial value problem

(1) $\qquad y' = f(x, y) \ , \qquad y(x_0) = y_0 \ ,$

at $x_n = x_0 + nh$, where $h > 0$ is the step size. If we integrate both sides of the equation in (1) from x_{n-1} to x_{n+1}, then we have

(2) $\qquad y(x_{n+1}) - y(x_{n-1}) = \displaystyle\int_{x_{n-1}}^{x_{n+1}} f\big(x, y(x)\big) \, dx \ .$

We now approximate the integral in (2) using Simpson's rule. This gives

$$y(x_{n+1}) - y(x_{n-1}) \approx \frac{h}{3}\left[f\big(x_{n-1}, y(x_{n-1})\big) + 4f\big(x_n, y(x_n)\big) + f\big(x_{n+1}, y(x_{n+1})\big) \right] \ .$$

Substituting the approximations y_{n-1}, y_n, and y_{n+1} for $y(x_{n-1})$, $y(x_n)$, and $y(x_{n+1})$, respectively, we obtain the numerical scheme

(3) $\qquad y_{n+1} - y_{n-1} = \dfrac{h}{3}\left[f(x_{n-1}, y_{n-1}) + 4f(x_n, y_n) + f(x_{n+1}, y_{n+1}) \right] \ .$

This method is also called **Simpson's rule.** As you can see, it is an implicit, two-step method. Starting with y_0 and y_1, formula (3) determines y_2; knowing y_1 and y_2, we again use (3) to determine y_3; etc. Formula (3) can also be used in a predictor-corrector procedure, where y_{n+1}^*, a predicted value for y_{n+1}, is used in the right-hand side of (3) to obtain a "corrected" estimate y_{n+1}.

Multistep methods were introduced by J. C. Adams prior to the Runge-Kutta methods. In fact, two important classes of multistep methods are the explicit Adams methods called the **Adams-Bashford methods** and the implicit Adams methods called the **Adams-Moulton methods.** (Like the Runge-Kutta methods, there exist Adams methods with different orders of convergence.)

A disadvantage of the multistep methods is that they are more susceptible to instability. (Stability is discussed in Project E.) From a programming viewpoint, the multistep methods also have the disadvantage of not being self-starting; that is, the first k approximations $y_0, y_1, \ldots, y_{k-1}$ are needed in a k-step method in order to determine y_k. Therefore one must use some other method to generate the starting values $y_1, y_2, \ldots, y_{k-1}$ (recall that y_0 is given). This means the Runge-Kutta formulas are easier to code. However, the Runge-Kutta formulas require several evaluations of the function $f(x, y)$ at each step, whereas the multistep formulas require only one evaluation of $f(x, y)$ at each stage. This makes the multistep methods less expensive (and faster) when the function $f(x, y)$ is complicated. For a discussion of multistep methods, we refer the reader to the texts *Numerical Solution of Ordinary Differential Equations* by L. Shampine (Chapman & Hall, 1994), and *Introduction to Numerical Analysis* by J. Stoer and R. Bulirsch (Springer-Verlag, 1980).

In practice, one is confronted with the problem of approximating the solution to within a desired tolerance and obtaining this approximation as inexpensively or as quickly as possible. This is seldom achieved using a *constant* step size h. To illustrate the problem, consider the periodic function $y(x)$ whose graph is given in Figure 3.15. (The solutions to the "Brusselator" equation, which models a chemical reaction that possesses periodic solutions, have graphs similar to the one in Figure 3.15.) If we are trying to approximate $y(x)$, then over most of its period we can use a step size h that is relatively large. However, at the "spike," we must take h to be very small or else we will miss the spike entirely or poorly approximate the values of y on the spike. Therefore it is advantageous to be able to control the step size h.

Much work has been done on writing codes that vary the step size h. These codes can choose h at each step so that the local error is within the given tolerance; shorten the step size, if necessary, to keep the error within tolerance; and decide what the next step size should be. The codes using multistep methods can vary the order of approximation by shifting from one type of multistep method to another. This allows them to begin with a one-step method, which resolves the difficulty of determining the starting values $y_1, y_2, \ldots, y_{k-1}$ in a k-step method.

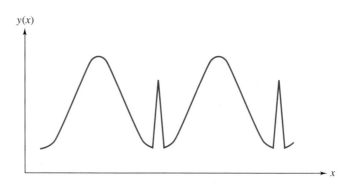

Figure 3.15 Periodic function with spike

Even when using a code with flexible step size and order-of-accuracy strategies, it is important to use whatever information one might have to assist in approximating the solution on those intervals where it is changing rapidly.

Differential equations that involve one or more fast decaying processes in time, with time constants (cf. Section 3.3, page 109) that are short compared to the time span of interest, are called **stiff problems.** For example, consider the problem

$$\frac{dy}{dt} = -1000y + 999e^{-t}, \qquad y(0) = 0,$$

which has the solution

$$y(t) = e^{-t} - e^{-1000t}.$$

To capture the decay of the term e^{-1000t} numerically would require step sizes smaller than, say, $\frac{1}{2000}$, but this would entail over 4000 iterations to reach $t = 2$ (so as to capture e^{-t}). The accumulation of truncation and rounding error would probably be unacceptable. The vast difference between the time constants of e^{-1000t} and e^{-t} makes this a stiff problem. Special computer codes have been written to deal specifically with such situations.[†]

Documented codes for multistep, predictor-corrector, high-order Runge-Kutta, stiff system, and adaptive step size control are obtainable at virtually all scientific computing sites. And as we indicated, versions for the PC and Macintosh are available through the libraries of the popular software packages.

[†]The classic reference for stiff problems is *Numerical Initial Value Problems in Ordinary Differential Equations* by C. W. Gear, Prentice-Hall, 1971.

GROUP PROJECTS FOR CHAPTER 3

A. Delay Differential Equations

In our discussion of mixing problems in Section 3.2, we encountered the initial value problem

(1) $$x'(t) = 6 - \frac{3}{500}x(t - t_0) , \qquad x(t) = 0 \quad \text{for } t \in [-t_0, 0]$$

where t_0 is a positive constant. The equation in (1) is an example of a **delay differential equation.** These equations differ from the usual differential equations by the presence of the shift $(t - t_0)$ in the argument of the unknown function $x(t)$. In general, these equations are more difficult to work with than are regular differential equations, but quite a bit is known about them.[†]

(a) Show that the simple linear delay differential equation

(2) $$x'(t) = ax(t - b) ,$$

where a and b are constants, has a solution of the form $x(t) = Ce^{st}$, for any constant C, provided s satisfies the transcendental equation $s = ae^{-bs}$.

(b) A solution to (2) for $t > 0$ can also be found using the **method of steps.** Assume that $x(t) = f(t)$ for $-b \le t \le 0$. For $0 \le t \le b$, equation (2) becomes

$$x'(t) = ax(t - b) = af(t - b) ,$$

and so

$$x(t) = \int_0^t af(v - b)dv + x(0) .$$

Now that we know $x(t)$ on $[0, b]$, we can repeat this procedure to obtain

$$x(t) = \int_b^t ax(v - b)dv + x(b)$$

for $b \le t \le 2b$. This process can be continued indefinitely.

Use the method of steps to show that the solution to the initial value problem

$$x'(t) = -x(t - 1) , \qquad x(t) = 1 \quad \text{on} \quad [-1, 0]$$

is given by

$$x(t) = \sum_{k=0}^{n} (-1)^k \frac{[t - (k - 1)]^k}{k!} , \qquad \text{for} \quad n - 1 \le t \le n ,$$

where n is a nonnegative integer. (This problem can also be solved using the Laplace transform method of Chapter 7.)

(c) Use the method of steps to compute the solution to the initial value problem given in (1) on the interval $0 \le t \le 15$ for $t_0 = 3$.

[†]See, for example, *Differential-Difference Equations,* by R. Bellman and K. L. Cooke, Academic Press, New York, 1963, or *Ordinary and Delay Differential Equations,* by R. D. Driver, Springer-Verlag, New York, 1977.

B. *Aquaculture*

Aquaculture is the art of cultivating the plants and animals indigenous to water. In the example considered here, it is assumed that a batch of catfish are raised in a pond. We are interested in determining the best time for harvesting the fish so that the cost per pound for raising the fish is minimized.

A differential equation describing the growth of fish may be expressed as

$$(3) \qquad \frac{dW}{dt} = KW^{\alpha} \, ,$$

where $W(t)$ is the weight of the fish at time t, and K and α are empirically determined growth constants. The functional form of this relationship is similar to that of the growth models for other species. Modeling the growth rate or metabolic rate by a term like W^{α} is a common assumption. Biologists often refer to equation (3) as the **allometric equation.** It can be supported by plausibility arguments such as growth rate depending on the surface area of the gut (which varies like $W^{2/3}$) or depending on the volume of the animal (which varies like W).

(a) Solve equation (3) when $\alpha \neq 1$.

(b) The solution obtained in part (a) grows large without bound, but in practice there is some limiting maximum weight W_{MAX} for the fish. This limiting weight may be included in the differential equation describing growth by inserting a dimensionless variable S that can range between 0 and 1 and involves an empirically determined parameter μ. Namely, we now assume

$$(4) \qquad \frac{dW}{dt} = KW^{\alpha}S \, ,$$

where $S := 1 - (W/W_{\mathrm{MAX}})^{\mu}$.
When $\mu = 1 - \alpha$, equation (4) has a closed form solution. Solve equation (4) when $K = 10$, $\alpha = 3/4$, $\mu = 1/4$, $W_{\mathrm{MAX}} = 81$ (ounces), and $W(0) = 1$ (ounce). The constants are given for t measured in months.

(c) The differential equation describing the total cost in dollars $C(t)$ of raising a fish for t months has one constant term K_1 that specifies the cost per month (due to costs such as interest, depreciation, and labor) and a second constant K_2 that multiplies the growth rate (because the amount of food consumed by the fish is approximately proportional to the growth rate). That is,

$$(5) \qquad \frac{dC}{dt} = K_1 + K_2\frac{dW}{dt} \, .$$

Solve equation (5) when $K_1 = 0.4$, $K_2 = 0.1$, $C(0) = 1.1$ (dollars), and $W(t)$ is as determined in part (b).

(d) Sketch the curve obtained in part (b) that represents the weight of the fish as a function of time. Next, sketch the curve obtained in part (c) that represents the total cost of raising the fish as a function of time.

(e) To determine the optimal time for harvesting the fish, sketch the ratio $C(t)/W(t)$. This ratio represents the total cost per ounce as a function of time. When this ratio reaches its minimum—that is, when the total cost per ounce is at its lowest—it is the optimal time to harvest the fish. Determine this optimal time to the nearest month.

C. *Curve of Pursuit*

An interesting geometric model arises when one tries to determine the path of a pursuer chasing its prey. This path is called a **curve of pursuit.** These problems were analyzed using methods of calculus circa 1730 (more than two centuries after Leonardo da Vinci had considered them). The simplest problem is to find the curve along which a vessel moves in pursuing another vessel that flees along a straight line, assuming the speeds of the two vessels are constant.

Let's assume vessel A, traveling at a speed α, is pursuing vessel B, which is traveling at a speed β. In addition, assume vessel A begins (at time $t = 0$) at the origin and pursues vessel B, which begins at the point $(1, 0)$ and travels up the line $x = 1$. After t hours, vessel A is located at the point $P = (x, y)$ and vessel B is located at the point $Q = (1, \beta t)$ (see Figure 3.16). The goal is to describe the locus of points P; that is, to find y as a function of x.

(a) Vessel A is pursuing vessel B, so at time t, vessel A must be heading right at vessel B. That is, the tangent line to the curve of pursuit at P must pass through the point Q (see Figure 3.16). For this to be true, show that

(6)
$$\frac{dy}{dx} = \frac{y - \beta t}{x - 1} .$$

(b) We know the speed at which vessel A is traveling, so we know that the distance it travels in time t is αt. This distance is also the length of the pursuit curve from $(0, 0)$ to (x, y). Using the arclength formula from calculus, show that

(7)
$$\alpha t = \int_0^x \sqrt{1 + [y'(u)]^2}\, du .$$

Solving for t in equations (6) and (7), conclude that

(8)
$$\frac{y - (x - 1)(dy/dx)}{\beta} = \frac{1}{\alpha} \int_0^x \sqrt{1 + [y'(u)]^2}\, du .$$

(c) Differentiating both sides of (8) with respect to x, derive the first order equation

$$(x - 1)\frac{dw}{dx} = -\frac{\beta}{\alpha} \sqrt{1 + w^2} ,$$

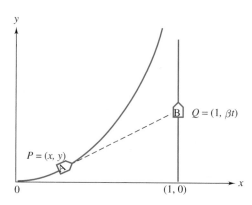

Figure 3.16 The path of vessel A as it pursues vessel B

where $w := dy/dx$.

(d) Using separation of variables and the initial conditions $x = 0$ and $w = dy/dx = 0$ when $t = 0$, show that

(9) $$\frac{dy}{dx} = w = \frac{1}{2}\left[(1 - x)^{-\beta/\alpha} - (1 - x)^{\beta/\alpha}\right] .$$

(e) For $\alpha > \beta$—that is, the pursuing vessel A travels faster than the pursued vessel B— use equation (9) and the initial conditions $x = 0$ and $y = 0$ when $t = 0$, to derive the curve of pursuit

(10) $$y = \frac{1}{2}\left[\frac{(1 - x)^{1 + \beta/\alpha}}{1 + \beta/\alpha} - \frac{(1 - x)^{1 - \beta/\alpha}}{1 - \beta/\alpha}\right] + \frac{\alpha\beta}{\alpha^2 - \beta^2} .$$

(f) Find the location where vessel A intercepts vessel B if $\alpha > \beta$.

(g) Show that if $\alpha = \beta$, then the curve of pursuit is given by

(11) $$y = \frac{1}{2}\left\{\frac{1}{2}\left[(1 - x)^2 - 1\right] - \ln(1 - x)\right\} .$$

Will vessel A ever reach vessel B?

D. *Aircraft Guidance in a Crosswind*

T. L. Pearson, Acadia University, Nova Scotia

An aircraft flying under the guidance of a nondirectional beacon (a fixed radio transmitter, abbreviated NDB) moves so that its longitudinal axis always points toward the beacon (see Figure 3.17). A pilot sets out toward an NDB from a point at which the wind is at right angles to the initial direction of the aircraft; the wind maintains this direction. Assume the wind speed and the speed of the aircraft through the air (its "airspeed") remain constant. (Keep in mind that the latter is different from the aircraft's speed with respect to the ground.)

(a) Locate the flight in the xy-plane, placing the start of the trip at $(2, 0)$ and the destination at $(0, 0)$. Set up the differential equation describing the aircraft's path over the ground. [*Hint:* $dy/dx = (dy/dt)/(dx/dt)$.]

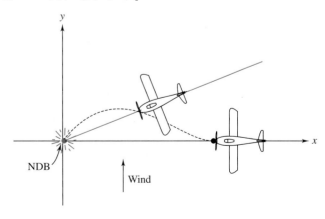

Figure 3.17 Guided aircraft

(b) Make an appropriate substitution and solve this equation. [*Hint:* See Section 2.6.]

(c) Use the fact that $x = 2$ and $y = 0$ at $t = 0$ to determine the appropriate value of the arbitrary constant in the solution set.

(d) Solve to get y explicitly in terms of x. Write your solution in terms of a hyperbolic function.

(e) Let γ be the ratio of windspeed to airspeed. Using a software package, graph the solutions for the cases $\gamma = 0.1, 0.3, 0.5,$ and 0.7 all on the same set of axes. Interpret these graphs.

(f) Discuss the (terrifying!) cases $\gamma = 1$ and $\gamma > 1$.

E. Stability of Numerical Methods

Numerical methods are often tested on simple initial value problems of the form

(12) $y' + \lambda y = 0$, $y(0) = 1$, $(\lambda = \text{constant})$,

which has the solution $\phi(x) = e^{-\lambda x}$. Notice that for each $\lambda > 0$, the solution $\phi(x)$ tends to zero as $x \to +\infty$. Thus a desirable property for any numerical scheme that generates approximations $y_0, y_1, y_2, y_3, \ldots$ to $\phi(x)$ at the points $0, h, 2h, 3h, \ldots$ is that, for $\lambda > 0$,

(13) $y_n \to 0$ as $n \to \infty$.

For single-step linear methods, property (13) is called **absolute stability.**

(a) Show that for $x_n = nh$, Euler's method, when applied to the initial value problem (12), yields the approximations

$$y_n = (1 - \lambda h)^n , \qquad n = 0, 1, 2, \ldots ,$$

and deduce that this method is absolutely stable only when $0 < \lambda h < 2$. (This means that for a given $\lambda > 0$, we must choose the step size h sufficiently small in order for property (13) to hold.) Further show that for $h > 2/\lambda$, the error $y_n - \phi(x_n)$ grows large exponentially!

(b) Show that for $x_n = nh$ the trapezoid scheme of Section 3.5, applied to problem (12), yields the approximations

$$y_n = \left(\frac{1 - \lambda h/2}{1 + \lambda h/2}\right)^n , \qquad n = 0, 1, 2, \ldots ,$$

and deduce that this scheme is absolutely stable for all $\lambda > 0, h > 0$.

(c) Show that the improved Euler's method applied to problem (12) is absolutely stable for $0 < \lambda h < 2$.

Multistep Methods. When multistep numerical methods are used, instability problems may arise that cannot be circumvented by simply choosing a sufficiently small step size h. This is because multistep methods yield "extraneous solutions," which may dominate the calculations. To see what can happen, consider the two-step method

(14) $y_{n+1} = y_{n-1} + 2hf(x_n, y_n)$, $n = 1, 2, \ldots ,$

for the equation $y' = f(x, y)$.

(d) Show that for the initial value problem

(15) $y' + 2y = 0$, $y(0) = 1$,

the recurrence formula (14), with $x_n = nh$, becomes

(16) $\qquad y_{n+1} + 4hy_n - y_{n-1} = 0$.

Equation (16), which is called a **difference equation,** can be solved by using the following approach. We postulate a solution of the form $y_n = r^n$, where r is a constant to be determined.

(e) Show that substituting $y_n = r^n$ in (16) leads to the "characteristic equation"

$$r^2 + 4hr - 1 = 0 ,$$

which has roots

$$r_1 = -2h + \sqrt{1 + 4h^2} \quad \text{and} \quad r_2 = -2h - \sqrt{1 + 4h^2} .$$

By analogy with the theory for second order differential equations, it can be shown that a general solution of (16) is

$$y_n = c_1 r_1^n + c_2 r_2^n ,$$

where c_1 and c_2 are arbitrary constants. Thus the general solution to the difference equation (16) has two independent constants, whereas the differential equation in (15) has only one, namely, $\phi(x) = ce^{-2x}$.

(f) Show that for each $h > 0$,

$$\lim_{n \to \infty} r_1^n = 0 , \quad \text{but} \quad \lim_{n \to \infty} |r_2^n| = \infty .$$

Hence the term r_1^n behaves like the solution $\phi(x_n) = e^{-2x_n}$ as $n \to \infty$. However, the extraneous solution r_2^n grows large without bound.

(g) Applying the scheme of (14) to the initial value problem (15) requires two starting values y_0, y_1. The exact values are $y_0 = 1$, $y_1 = e^{-2h}$. However, regardless of the choice of starting values and the size of h, the term $c_2 r_2^n$ will eventually dominate the full solution to the recurrence equation as x_n increases. Illustrate this instability taking $y_0 = 1$, $y_1 = e^{-2h}$, and using a calculator or computer to compute $y_2, y_3, \ldots, y_{100}$ from the recurrence formula (16) for $h = 0.5$ and $h = 0.05$. (*Note:* Even if initial conditions are chosen so that $c_2 = 0$, round-off error will inevitably "excite" the extraneous dominant solution.)

F. *Period Doubling and Chaos*

In the study of dynamical systems, the phenomena of **period doubling** and **chaos** are observed. These phenomena can be seen when one uses a numerical scheme to approximate the solution to an initial value problem for a nonlinear differential equation such as the following logistic model for population growth:

(17) $\qquad \dfrac{dp}{dt} = 10p(1 - p) , \qquad p(0) = 0.1$.

(see Section 3.2).

(a) Solve the initial value problem (17) and show that $p(t)$ approaches 1 as $t \to +\infty$.

(b) Show that using Euler's method (see Sections 1.5 and 3.5) with step size h to approximate the solution to (17) gives

(18) $\qquad p_{n+1} = (1 + 10h)p_n - (10h)p_n^2 , \qquad p_0 = 0.1$.

(c) For $h = 0.18, 0.23, 0.25$, and 0.3, show that the first 40 iterations of (18) appear to (i) converge to 1 when $h = 0.18$, (ii) jump between 1.18 and 0.69 when $h = 0.23$, (iii) jump between 1.23, 0.54, 1.16, and 0.70 when $h = 0.25$, and (iv) display no discernible pattern when $h = 0.3$.

The transitions from convergence to jumping between two numbers, then four numbers, and so on, are called **period doubling.** The phenomenon displayed when $h = 0.3$ is referred to as **chaos.** This transition from period doubling to chaos as h increases is frequently observed in dynamical systems.

The transition to chaos is nicely illustrated in the bifurcation diagram (see Figure 3.18). This diagram is generated for equation (18) as follows. Beginning at $h = 0.18$, compute the sequence $\{p_n\}$ using (18) and, starting at $n = 201$, plot the next 30 values, that is, $p_{201}, p_{202}, \dots, p_{230}$. Next, increment h by 0.001 to 0.181 and repeat. Continue this process until $h = 0.30$. Notice how the figure splits from one branch to two, then four, and then finally gives way to chaos.

Our concern is with the instabilities of the numerical procedure when h is not chosen small enough. Fortunately, the instability observed for Euler's method—the period doubling and chaos—was immediately recognized because we know that this type of behavior is not expected of a solution to the logistic equation. Consequently, if we had tried Euler's method with $h = 0.23$, 0.25, or 0.3 to solve (17) numerically, we would have realized that h was not chosen small enough.

The situation for the classical fourth order Runge-Kutta method (see Section 3.6) is more troublesome. It may happen that for a certain choice of h period doubling occurs, but it is also

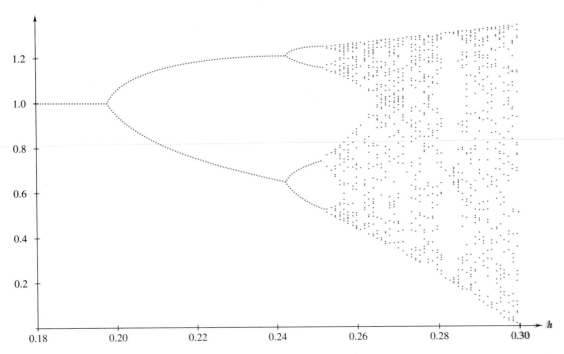

Figure 3.18 Period doubling to chaos

possible that for other choices of h the numerical solution actually converges to a limiting value that is *not* the limiting value for any solution to the logistic equation in (17).

(d) Approximate the solution to (17) by computing the first 60 iterations of the classical fourth order Runge-Kutta method using the step size $h = 0.3$. (Thus, for the subroutine on page 139, input $N = 60$ and $c = 60(0.3) = 18$.) Repeat with $h = 0.325$ and $h = 0.35$. Which values of h (if any) do you feel are giving the "correct" approximation to the solution? Why?

A further discussion of chaos appears in Section 5.7.

G. *Bang-Bang Controls*

In Example 3 of Section 3.3 (page 111), it was assumed that the amount of heating or cooling supplied by a furnace or air conditioner is proportional to the difference between the actual temperature and the desired temperature; recall the equation

$$U(t) = K_U[T_D - T(t)] \ .$$

In many homes, the heating/cooling mechanisms deliver a *constant* rate of heat flow, say,

$$U(t) = \begin{cases} K_1 \ , & \text{if} \quad T(t) > T_D \ , \\ K_2 \ , & \text{if} \quad T(t) < T_D \end{cases}$$

(with $K_1 < 0$).

(a) Modify the differential equation (9) in Example 3 so that it describes the temperature of a home employing this "bang-bang" control law.

(b) Suppose the initial temperature $T(0)$ is greater than T_D. Modify the constants in the solution (12), page 112, so that the formula is valid as long as $T(t) > T_D$.

(c) If the initial temperature $T(0)$ is less than T_D, what values should the constants in (12) take to make the formula valid for $T(t) < T_D$?

(d) How does one piece the solutions in (b) and (c) to obtain a complete time description of the temperature $T(t)$?

CHAPTER 4

Linear Second Order Equations

4.1 INTRODUCTION: THE MASS-SPRING OSCILLATOR

A damped mass-spring oscillator consists of a mass m attached to a spring fixed at one end, as shown in Figure 4.1. Devise a differential equation that governs the motion of this oscillator, taking into account the forces acting on it due to the spring elasticity, friction (damping), and possible external influences.

Figure 4.1 Damped mass-spring oscillator

Newton's second law—force equals mass times acceleration ($F = ma$)—is without a doubt the most commonly encountered differential equation in practice. It is an ordinary differential equation of the *second order* since acceleration is the second derivative of position (y) with respect to time ($a = d^2y/dt^2$).

When the second law is applied to a mass-spring oscillator, the resulting motions are common experiences of everyday life, and we can exploit our familiarity with these vibrations to obtain a qualitative description of the solutions of more general second order equations.

We begin by referring to Figure 4.1, which depicts the mass-spring oscillator. When the spring is unstretched and the inertial mass m is still, the system is at equilibrium; we measure

156

the coordinate y of the mass by its displacement from the equilibrium position. When the mass m is displaced from equilibrium, the spring is stretched or compressed and it exerts a force that resists the displacement. For most springs this force is directly proportional to the displacement y and is thus given by

(1) $F_{spring} = -ky$,

where the positive constant k is known as the *stiffness* and the negative sign reflects the opposing nature of the force. **Hooke's law,** as equation (1) is commonly known, is only valid for sufficiently small displacements; if the spring is compressed so strongly that the coils press against each other, the opposing force obviously becomes much stronger. Similarly, if the coils are completely stretched out, the force also becomes much stronger (until the spring breaks!).

 Practically all mechanical systems also experience friction, and for vibrational motion this force is usually modeled accurately by the equation

(2) $F_{friction} = -b\dfrac{dy}{dt} = -by'$,

where $b\ (\geq 0)$ is the *damping coefficient* and the negative sign has the same significance as in equation (1).

 The other forces on the oscillator are usually regarded as *external* to the system. Although they may be gravitational, electrical, or magnetic, commonly the most important external forces are transmitted to the mass by shaking the base of the spring. We ignore these details for now and apply Newton's second law to derive the differential equation for the mass-spring oscillator:

$$my'' = -ky - by' + F_{external}(t)$$

or

$$my'' + by' + ky = F_{external}(t) \ .$$

 What do mass-spring motions look like? Everyday experience tells us that when there is no friction $(b = 0)$ or external force, the motions are perpetual vibrations like the one depicted in Figure 4.2. These vibrations resemble sinusoidal functions, $A \cos \omega t$ or $B \sin \omega t$, with the *amplitude* depending on the initial displacement and velocity, and the angular frequency ω becoming faster for stiffer springs, but slower for heavier masses.

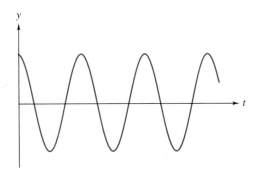

Figure 4.2 Sinusoidal vibration

UNFORCED, UNDAMPED OSCILLATIONS

Equation: $my'' + ky = 0$
Solutions: $y(t) = A\cos \omega t + B\sin \omega t$
Behavior: ω increases with k, decreases with m;
A and B increase with $y(0)$ and $y'(0)$.

(In fact, the reader can easily verify that both $\cos \omega t$ and $\sin \omega t$ satisfy $my'' + ky = 0$ if $\omega = \sqrt{k/m}$.)

When friction is present, these oscillations are damped and the motions resemble Figure 4.3. Such damped vibrations would result if the sinusoid of Figure 4.2 was multiplied by a decaying exponential, leading us to postulate the functional forms $Ae^{-t/\tau}\cos \omega t$ or $Be^{-t/\tau}\sin \omega t$, where $\tau\ (>0)$ is the time constant (cf. Section 3.3, Example 1).[†] Higher friction results in more intense damping—i.e., the amplitude decays faster—and lower angular frequencies ω. In fact, if the friction is strong enough, it may damp out the oscillation altogether, resulting in an "overdamped" motion. See Figure 4.4.

Figure 4.3 Damped oscillations

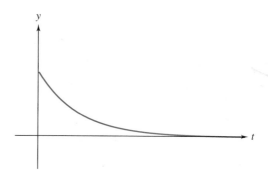

Figure 4.4 Overdamped motion

UNFORCED, UNDERDAMPED OSCILLATIONS

Equation: $my'' + by' + ky = 0$, $(b > 0)$
Solutions: $y(t) = [A\cos \omega t + B\sin \omega t] \times e^{-t/\tau}$, $(\tau > 0)$
Behavior: τ decreases with b;
ω increases with k, decreases with m and b.

The reader should verify that

$$y_1(t) = e^{-3t}\cos 4t \text{ solves } y'' + 6y' + 25y = 0 ,$$
$$y_2(t) = e^{-4t}\cos 3t \text{ solves } y'' + 8y' + 25y = 0 ,$$
$$y_3(t) = e^{-5t} \qquad \text{ solves } y'' + 10y' + 25y = 0 .$$

[†]The nomenclature "time constant" refers, in this context, to the time for the *envelope* $Ae^{-t/\tau}$ of the oscillations to decay by the factor $1/e$.

Thus increasing the damping coefficient ($b = 6, 8, 10$) decreases the time constants ($\tau = 1/3$, $1/4$, $1/5$) and the frequencies ($\omega = 4, 3, 0$) as predicted.

Now if a damped oscillator is *driven* by an external force that is sinusoidal at frequency Ω,

$$F_{external} = C \cos \Omega t + D \sin \Omega t ,$$

our experiences indicate that the initial (transient) response of the oscillator may be erratic, but eventually it will fall into "sync" with the driver and respond at the same frequency.

FORCED, DAMPED OSCILLATIONS

Equation: $\quad my'' + by' + ky = C \cos \Omega t + D \sin \Omega t ,\qquad (b > 0)$

Solutions: $\quad y(t) = (\text{transient}) + \left[A \cos \Omega t + B \sin \Omega t\right]$

Behavior: $\quad y(t) \rightarrow \left[A \cos \Omega t + B \sin \Omega t\right]$ as $t \rightarrow +\infty$.

All of these solution forms will be derived in this chapter. Common examples of systems vibrating in synchronization with their "drivers" are sound system speakers, cyclists bicycling over railroad tracks, electronic amplifier circuits, and ocean tides (driven by the periodic pull of the moon). However, there is more to the story than is revealed above. Systems can be enormously sensitive to the particular frequency Ω at which they are driven. Thus accurately tuned musical notes can shatter fine crystal, wind-induced vibrations at the right frequency can bring down a bridge, and a dripping faucet can cause inordinate headaches. These "resonant" responses (for which the solutions have maximum amplitudes) may be quite destructive, and structural engineers have to be very careful to ensure that their products will not resonate with any of the vibrations likely to occur in the operating environment.

EXAMPLE 1 Find the synchronous response of the mass-spring oscillator with $m = 1$, $b = 1$, $k = 25$ to the force $\sin \Omega t$.

SOLUTION We seek solutions of the differential equation

$$(3) \qquad y'' + y' + 25y = \sin \Omega t$$

of the form $y(t) = A \cos \Omega t + B \sin \Omega t$. Since

$$y' = -\Omega A \sin \Omega t + \Omega B \cos \Omega t ,$$
$$y'' = -\Omega^2 A \cos \Omega t - \Omega^2 B \sin \Omega t ,$$

we can simply insert these forms into equation (3), collect terms, and match coefficients:

$$
\begin{aligned}
\sin \Omega t = {} & y'' + y' + 25y \\
= {} & -\Omega^2 A \cos \Omega t - \Omega^2 B \sin \Omega t \\
& + \left[-\Omega A \sin \Omega t + \Omega B \cos \Omega t\right] \\
& + 25 \left[A \cos \Omega t + B \sin \Omega t\right] \\
= {} & \left[-\Omega^2 B - \Omega A + 25B\right] \sin \Omega t + \left[-\Omega^2 A + \Omega B + 25A\right] \cos \Omega t ,
\end{aligned}
$$

so

$$-\Omega A + (-\Omega^2 + 25)B = 1$$
$$(-\Omega^2 + 25)A + \Omega B = 0 .$$

We find

$$A = \frac{-\Omega}{\Omega^2 + (\Omega^2 - 25)^2} , \quad B = \frac{-\Omega^2 + 25}{\Omega^2 + (\Omega^2 - 25)^2} .$$

Figure 4.5 displays A and B as functions of the driving frequency Ω. A resonance clearly occurs around $\Omega \approx 5$. ∎ End

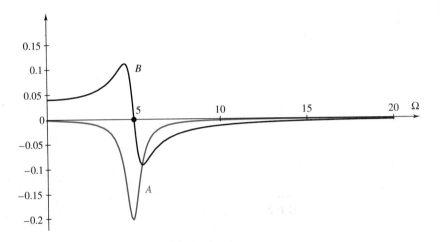

Figure 4.5 Vibration amplitudes around resonance

As we discuss the mathematics behind the solution and the analysis of second order ordinary differential equations, we will frequently be comparing the given equation to the paradigm

(4) $\left[\,\textbf{inertia}\,\right] \times y'' + \left[\,\textbf{damping}\,\right] \times y' + \left[\,\textbf{stiffness}\,\right] \times y = F_{\text{external}}$

for inspiration in interpreting our formulas.

EXERCISES 4.1

In Problems 1 through 3, use the form $A \cos \Omega t + B \sin \Omega t$ to find synchronous solutions to the following forced oscillator equations:

1. $y'' + 2y' + 4y = 3 \sin 5t$.
2. $y'' + 2y' + 4y = 5 \sin 3t$.
3. $y'' + 2y' + 4y = 3 \cos 2t + 4 \sin 2t$.

4. Undamped oscillators that are driven at resonance have unusual (and nonphysical) solutions.

 (a) To investigate this, find the synchronous solution $A \cos \Omega t + B \sin \Omega t$ to the generic forced oscillator equation

 (5) $my'' + by' + ky = \cos \Omega t$.

(b) Sketch graphs of the coefficients A and B, as functions of Ω, for $m = 1$, $b = 0.1$, $k = 25$.

(c) Now set $b = 0$ in your formulas for A and B, and resketch the graphs in part (b) with $m = 1$, $k = 25$. What happens at $\Omega = 5$? Notice that the amplitudes of the synchronous solutions grow without bound as Ω approaches 5.

(d) Show directly, by substituting the form $A \cos \Omega t + B \sin \Omega t$ into equation (5), that when $b = 0$ there are *no* synchronous solutions if $\Omega = \sqrt{k/m}$.

(e) Verify that $(2m\Omega)^{-1} t \sin \Omega t$ solves equation (5) when $b = 0$ and $\Omega = \sqrt{k/m}$. $\Omega = \sqrt{k/m}$. Notice that this *non*synchronous solution grows in time, without bound.

Clearly one cannot neglect damping in analyzing an oscillator forced at resonance, because otherwise the solutions as shown in part (e) are nonphysical. This behavior will be studied later in this chapter.

4.2 LINEAR DIFFERENTIAL OPERATORS

The equation describing the motion of a mass-spring oscillator that we derived in Section 4.1 is an example of a linear differential equation with time t as the independent variable. In this section, we begin our study of general second order linear differential equations, and in so doing return to x as the independent variable.

A **linear** second order equation is an equation that can be written in the form

(1) $$a_2(x)\frac{d^2y}{dx^2} + a_1(x)\frac{dy}{dx} + a_0(x)y = b(x) \ .$$

We will assume that $a_0(x)$, $a_1(x)$, $a_2(x)$, and $b(x)$ are continuous functions of x on an interval I. When a_0, a_1, and a_2 are constants (as was the case for the mass-spring oscillator), we say that the equation has **constant coefficients;** otherwise it has **variable coefficients.**

For now, we are interested only in those linear equations for which $a_2(x)$ is never zero on I. In this case, we can rewrite (1) in the **standard form**

(2) $$\frac{d^2y}{dx^2} + p(x)\frac{dy}{dx} + q(x)y = g(x) \ ,$$

where $p(x) := a_1(x)/a_2(x)$, $q(x) := a_0(x)/a_2(x)$, and $g(x) := b(x)/a_2(x)$ are continuous on I. (Recall the standard form for linear first order equations, Section 2.3, page 55.)

Associated with equation (2) is the equation

(3) $$\frac{d^2y}{dx^2} + p(x)\frac{dy}{dx} + q(x)y = 0 \ ,$$

which is obtained from (2) by replacing $g(x)$ with the zero function. We say that (2) is a **nonhomogeneous** equation and that (3) is the corresponding **homogeneous** equation. Here the meaning of *homogeneous* is not related to the use of the term for first order equations (Section 2.6) but is used in the sense of a homogeneous system of linear equations as studied in linear algebra.

Let's consider the expression on the left-hand side of equation (3), which, for convenience, we now write with prime notation:

(4) $$y''(x) + p(x)y'(x) + q(x)y(x) \ .$$

Given any function y with a continuous second derivative on the interval I, then (4) generates a

new function that we will denote by $L[y]$. This is

(5) $L[y] := y'' + py' + qy$.

What we have done is to associate with each such function y the function $L[y]$.

The term *function,* in its general sense, means a mapping that associates with each element in its domain a unique element in its range. Thus L can be interpreted as a function that is defined *on a set of functions.* Its domain is the collection of functions with continuous second derivatives; its range consists of continuous functions; and the rule of correspondence is given by (5). To avoid confusion, we call such mappings **operators.** Because L involves differentiation, we refer to L as a **differential operator.**

The image of a function y under the operator L is the function $L[y]$. If we want to evaluate this image function at some point x, we write $L[y](x)$.

For example, let $p(x) = x$ and $q(x) = x - 1$. Then

$$L[y](x) = y''(x) + xy'(x) + (x - 1)y(x) .$$

If $y_1(x) = x^3$, we get

$$L[y_1](x) = 6x + x(3x^2) + (x - 1)x^3 = x^4 + 2x^3 + 6x .$$

Thus L maps the function x^3 to the function $x^4 + 2x^3 + 6x$. In particular, at $x = 2$,

$$L[y_1](2) = 2^4 + 2 \cdot 2^3 + 6 \cdot 2 = 44 .$$

If $y_2(x) = \sin 2x$, then we similarly find

$$\begin{aligned} L[y_2](x) &= -4 \sin 2x + x(2 \cos 2x) + (x - 1)\sin 2x \\ &= x \sin 2x - 5 \sin 2x + 2x \cos 2x . \end{aligned}$$

For $x = \pi$, we have $L[y_2](\pi) = 2\pi$.

The differential operator L defined by (5) has two very important properties that are extensions of two familiar facts: "The derivative of the sum is the sum of the derivatives," and "the derivative of a constant times a function is the constant times the derivative of the function."

LINEARITY OF THE DIFFERENTIAL OPERATOR

Lemma 1. Let $L[y] := y'' + py' + qy$. If y, y_1, and y_2 are any twice-differentiable functions on the interval I and if c is any constant, then

(6) $L[y_1 + y_2] = L[y_1] + L[y_2]$,

(7) $L[cy] = cL[y]$.

In (6) and (7) equality is meant in the sense of equal functions on I.

Proof. On I, we have

$$\begin{aligned} L[y_1 + y_2] &= (y_1 + y_2)'' + p(y_1 + y_2)' + q(y_1 + y_2) \\ &= (y_1'' + y_2'') + p(y_1' + y_2') + q(y_1 + y_2) . \end{aligned}$$

On regrouping, this gives

$$L[y_1 + y_2] = (y_1'' + py_1' + qy_1) + (y_2'' + py_2' + qy_2)$$
$$= L[y_1] + L[y_2] ,$$

which verifies property (6). We leave the proof of property (7) for the exercises (see Problem 28). ∎ End

An operator that satisfies properties (6) and (7) for any constant c and any functions y, y_1, and y_2 in its domain is called a **linear operator.** Sometimes we abbreviate (6) and (7) by simply saying "L preserves linear combinations." If (6) or (7) fails to hold, the operator is **nonlinear.**

As we showed in Lemma 1, the operator L defined by (5) is linear. We now give an example of a nonlinear operator.

EXAMPLE 1 Show that the operator T defined by

$$T[y] := yy'' ,$$

where y is any twice-differentiable function, is nonlinear.

SOLUTION To demonstrate that T is a nonlinear operator, it suffices to show that property (7) is *not always* satisfied. Let's try $y_1(x) = x^2$. Because $y_1''(x) \equiv 2$, we have

$$T[cy_1](x) = (cx^2)(2c) = 2c^2x^2 , \qquad cT[y_1](x) = c(x^2)(2) = 2cx^2 .$$

But, in general, $2c^2x^2 \neq 2cx^2$ (take, for example, $c = 3$), so property (7) is violated. Hence T is a nonlinear operator. ∎ End

The linearity of the differential operator L in (5) can be used to prove the following theorem concerning homogeneous equations.

LINEAR COMBINATIONS OF SOLUTIONS

Theorem 1. Let y_1 and y_2 be solutions to the homogeneous equation

(8) $y'' + py' + qy = 0$.

Then any linear combination $c_1y_1 + c_2y_2$ of y_1 and y_2, where c_1 and c_2 are constants, is also a solution to (8).

Proof. If we let $L[y] := y'' + py' + qy$, then $L[y_1] = 0$ and $L[y_2] = 0$, since y_1 and y_2 are solutions to (8). Using the linearity of L as expressed in properties (6) and (7), we have

$$L[c_1y_1 + c_2y_2] = L[c_1y_1] + L[c_2y_2]$$
$$= c_1L[y_1] + c_2L[y_2]$$
$$= 0 + 0 = 0 .$$

Thus $c_1y_1 + c_2y_2$ is a solution to equation (8). ∎ End

Remark. We emphasize that Theorem 1 applies only to solutions of *homogeneous* equations. Consider, for example, two solutions $y_1(x)$ and $y_2(x)$ of $L[y] = \sin x$. The combination $y = y_1 + y_2$ would *not* satisfy $L[y] = \sin x$ (because $L[y] = 2\sin x$). The *operator L* is linear, but the *equation* $L[y] = g$ only accommodates linear combinations of solutions if it is *homogeneous* ($g \equiv 0$).

EXAMPLE 2 Given that $y_1(x) = e^{2x}\cos 3x$ and $y_2(x) = e^{2x}\sin 3x$ are solutions to the homogeneous equation

(9) $y'' - 4y' + 13y = 0$,

find a solution to (9) that satisfies the initial conditions

(10) $y(0) = 5$ and $y'(0) = -2$.

SOLUTION As a consequence of Theorem 1, any linear combination

(11) $y(x) = c_1 e^{2x}\cos 3x + c_2 e^{2x}\sin 3x$,

with c_1 and c_2 arbitrary constants, will be a solution to (9). Thus we try to select c_1 and c_2 so as to satisfy the initial conditions. On differentiating (11), we find

(12) $y'(x) = c_1(2e^{2x}\cos 3x - 3e^{2x}\sin 3x) + c_2(2e^{2x}\sin 3x + 3e^{2x}\cos 3x)$.

Substituting (11) and (12) into the initial conditions (10) yields the system of equations

$$c_1 = 5 , \quad 2c_1 + 3c_2 = -2 ,$$

whose solution is given by $c_1 = 5, c_2 = -4$. Hence, the solution to (9) that satisfies (10) is

$$y(x) = 5e^{2x}\cos 3x - 4e^{2x}\sin 3x. \quad \blacksquare \text{ End}$$

The properties of the operator L defined in (5) were derived from those of the basic differentiation operators with respect to x; that is, $Dy := dy/dx, D^2y := d^2y/dx^2$. In general, let $D^n y := d^n y/dx^n$. Using this notation,[†] we can express L as

$$L[y] = D^2 y + pDy + qy = (D^2 + pD + q)[y] .$$

For example, if

$$L[y] = y'' + 4y' + 3y ,$$

then we can write L in the form

$$L[y] = D^2 y + 4Dy + 3y = (D^2 + 4D + 3)[y] .$$

When p and q are *constants*, we can even treat $D^2 + pD + q$ as a polynomial in D and factor it. This is illustrated in the next example.

EXAMPLE 3 Show that the operator $D^2 + 4D + 3$ is the same as the product $(D + 1)(D + 3)$, where by *product* we mean the composition of operators.

[†]*Historical Footnote:* The symbol D was introduced by the French mathematician F. J. Servois. The notion of a symbolic operation dates from B. Brisson in 1808 and was extended by Augustin Cauchy in 1827.

SOLUTION For an arbitrary twice-differentiable function y, we have

$$(D + 1)(D + 3)[y] = (D + 1)\big[(D + 3)[y]\big] = (D + 1)[y' + 3y]$$
$$= D[y' + 3y] + 1[y' + 3y] = (y'' + 3y') + (y' + 3y)$$
$$= y'' + 4y' + 3y = (D^2 + 4D + 3)[y] \ .$$

Hence $(D + 1)(D + 3) = D^2 + 4D + 3$. ■ End

Further discussion of the algebra of differential operators is given in Problems 32–36.

We close this section with a discussion of the existence and uniqueness properties of initial value problems for linear second order differential equations.

EXISTENCE AND UNIQUENESS OF SOLUTION

Theorem 2. Suppose $p(x)$, $q(x)$, and $g(x)$ are continuous on an interval (a, b) that contains the point x_0. Then, for any choice of the initial values y_0 and y_1, there exists a unique solution $y(x)$ on the whole interval (a, b) to the initial value problem

$$\frac{d^2y}{dx^2} + p(x)\frac{dy}{dx} + q(x)y = g(x) \ ;$$

$$y(x_0) = y_0 \ , \qquad y'(x_0) = y_1 \ .$$

Remark. Geometrically, the uniqueness part of Theorem 2 asserts that two solution curves passing through the same point and *with the same slope* must be identical throughout the interval (a, b). For example, if we consider the simple equation $y'' = 0$, we know that the solution curves are just nonvertical lines $(y = c_1x + c_2)$. The uniqueness property then says that if two lines pass through the same point and have the same slope, they must be the same line (a familiar fact!).

The proof of Theorem 2 is more complicated than the proof of Theorem 1 in Section 2.3 for linear first order equations; it is given in Chapter 13.[†]

EXAMPLE 4 Determine the largest interval for which Theorem 2 ensures the existence and uniqueness of a solution to the initial value problem

(13)
$$\frac{d^2y}{dx^2} + \frac{1}{x - 3}\frac{dy}{dx} + \sqrt{x}\, y = \ln x \ ;$$

$$y(1) = 3 \ , \quad y'(1) = -5 \ .$$

SOLUTION Here $p(x) = 1/(x - 3)$, $q(x) = \sqrt{x}$, $g(x) = \ln x$, and $x_0 = 1$. Now $p(x)$ is continuous for $x \neq 3$, and both $q(x)$ and $g(x)$ are continuous for $x > 0$. Hence the largest open interval containing $x_0 = 1$ for which $p(x)$, $q(x)$, and $g(x)$ are simultaneously continuous is the interval $(0, 3)$. From Theorem 2, we can conclude that the initial value problem (13) has a unique solution on the interval $(0, 3)$. ■ End

[†]All references to Chapters 11–13 refer to the expanded text *Fundamentals of Differential Equations and Boundary Value Problems*, 3rd edition.

EXERCISES 4.2

In Problems 1 through 8, determine whether the equation is linear or nonlinear. If it is linear, classify it as being homogeneous or nonhomogeneous, with constant coefficients or with variable coefficients.

1. $6\dfrac{d^2y}{dx^2} + \dfrac{dy}{dx} = xy$. *L, H, NO*

2. $y'' + (1-x)y' + xy = \sin x$. *L, N, NO*

3. $xy'' - yy' = \sin x$. *NonL)*

4. $\dfrac{d^2y}{dx^2} - x\dfrac{dy}{dx} + y = 0$. *L, H, NO*

5. $3t^2\dfrac{d^2x}{dt^2} = t\dfrac{dx}{dt} + 4x - \ln t$. *L, NON, NO*

6. $\dfrac{d^2\theta}{dx^2} = \cos\theta$. *NL*

7. $\dfrac{d^2y}{d\theta^2} + y = \tan\theta$. *L, NON, NO*

8. $\dfrac{d^2s}{dx^2} + 3\dfrac{ds}{dx} - s^{1/2} = x^2$.

9. Let $L[y](x) = x^2y''(x) - 3xy'(x) - 5y(x)$. Compute
 (a) $L[\cos x]$. **(b)** $L[x^{-1}]$.
 (c) $L[x^r]$, r a constant.

10. Let $L[y](x) = y''(x) - 4y'(x) + 3y(x)$. Compute
 (a) $L[x^2]$. **(b)** $L[e^{3x}]$.
 (c) $L[e^{rx}]$, r a constant.

11. Show that T defined by
$$T[y] := y'' - 3y' + y^2 ,$$
where y is any twice-differentiable function, is a nonlinear operator.

12. Show that T defined by
$$T[y] := y'' + \{y'y^2\}^{1/3}$$
is a nonlinear operator. Which property, (6) or (7), fails?

13. Given that $y_1(x) = e^{2x}\cos x$ and $y_2(x) = e^{2x}\sin x$ are solutions to the homogeneous equation
$$y'' - 4y' + 5y = 0 ,$$
find solutions to this equation that satisfy the following initial conditions:

(a) $y(0) = 2$, $y'(0) = 1$.
(b) $y(\pi) = 4e^{2\pi}$, $y'(\pi) = 5e^{2\pi}$.

14. Given that $y_1(x) = e^{2x}$ and $y_2(x) = e^{-x}$ are solutions to the homogeneous equation
$$y'' - y' - 2y = 0 ,$$
find solutions to this equation that satisfy the following initial conditions:

(a) $y(0) = -1$, $y'(0) = 4$.
(b) $y(0) = 3/2$, $y'(0) = 0$.

In Problems 15 through 18, use Theorem 2 to discuss the existence and uniqueness of a solution to the differential equation that satisfies the initial conditions $y(1) = y_0$, $y'(1) = y_1$, where y_0 and y_1 are real constants.

15. $(1 + x^2)y'' + xy' - y = \tan x$.

16. $x(x-3)y'' + 2xy' - y = x^2$.

17. $e^x y'' - \dfrac{y'}{x-3} + y = \ln x$.

18. $x^2y'' + y = \cos x$.

In Problems 19 through 22, determine whether Theorem 2 applies. If it does, then discuss what conclusions can be drawn. If it does not, explain why.

19. $y'' + yy' = x^2 - 1$; $y(0) = 1$, $y'(0) = -1$.

20. $x^2z'' + xz' + z = \cos x$; $z(0) = 1$, $z'(0) = 0$.

21. $(1-t)x'' + tx' - 2x = \sin t$; $x(0) = 1$, $x'(0) = 1$.

22. $y'' + xy' - x^2y = 0$; $y(0) = 0$, $y(1) = 0$.

In Problems 23 through 26, express the given operator L using the differential operator D.

23. $L[y] := y'' + 7y' + 12y$.

24. $L[z] := 2z'' + z' - z$.

25. $L[y] := y'' + xy$.

26. $L[y] := x^2y'' - xy' + y$.

27. Let L be defined by $L[y] := (D^2 - xD + 2)[y]$. Compute
 (a) $L[x^2]$. **(b)** $L[\cos x]$.

28. Complete the proof of Lemma 1 on page 162 by verifying that the differential operator L defined by (5) satisfies property (7).

29. Show that the integro-differential operator L defined by

$$L[y](x) := \int_0^1 (x - t)^2 y(t)\, dt + x^2 y'(x)$$

is linear for any function y having a continuous derivative on $[0, 1]$; that is, L satisfies properties (6) and (7).

30. Boundary Value Problems. When the values of a solution to a differential equation are specified at two different points, these conditions are called **boundary conditions.** (In contrast, initial conditions specify the values of a function and its derivative at the same point.) Given that every solution to

(14) $y'' + y = 0$

is of the form

$$y(x) = c_1 \cos x + c_2 \sin x,$$

where c_1 and c_2 are arbitrary constants, show that

(a) there is a unique solution to (14) that satisfies the boundary conditions $y(0) = 2$ and $y(\pi/2) = 0$;

(b) there is no solution to (14) that satisfies $y(0) = 2$ and $y(\pi) = 0$;

(c) there are infinitely many solutions to (14) that satisfy $y(0) = 2$ and $y(\pi) = -2$.

31. Let $y = \phi(x)$ be the solution to the initial value problem

$$y'' + x^2 y' + e^x y = x - 1 \;;$$
$$y(0) = 1\;, \quad y'(0) = -1\;.$$

Assuming that the differential equation can be repeatedly differentiated, determine $\phi''(0)$, $\phi'''(0)$, and $\phi^{(4)}(0)$.

32. Differential Operators with Constant Coefficients. Let A and B represent two linear differential operators with constant coefficients; that is,

$$A := a_2 D^2 + a_1 D + a_0\;,$$
$$B := b_2 D^2 + b_1 D + b_0\;,$$

where $a_0, a_1, a_2, b_0, b_1,$ and b_2 are constants. We define the **sum** $A + B$ by

$$(A + B)[y] := A[y] + B[y]$$

and the **product** AB by

$$(AB)[y] := A\big[B[y]\big]\;.$$

Notice that the product AB is the composite operator

obtained by applying first B, then A. We say two operators A and B are **equal** if $A[y] = B[y]$ for all functions y with the necessary derivatives. With these definitions, verify the following properties:

(a) Commutative laws:

$$A + B = B + A\;,$$
$$AB = BA\;.$$

(b) Associative laws:

$$(A + B) + C = A + (B + C)\;,$$
$$(AB)C = A(BC)\;.$$

(c) Distributive law: $A(B + C) = AB + AC$.

33. Let $A = (D - 1)$, $B = (D + 2)$, and $y = x^3$. Compute

(a) $A[y]$. **(b)** $B\big[A[y]\big]$. **(c)** $B[y]$.

(d) $A\big[B[y]\big]$. **(e)** AB. **(f)** $(AB)[y]$.

34. Verify, as in Example 3, each of the following operator equations:

(a) $(D - 1)(D + 6) = D^2 + 5D - 6$.

(b) $(D + 1)(2D^2 - 1) = 2D^3 + 2D^2 - D - 1$.

35. Factor the following differential operators:

(a) $D^2 + 3D - 4$. **(b)** $D^2 + D - 6$.

(c) $2D^2 + 9D - 5$. **(d)** $D^2 - 2$.

36. Let $A = (xD - 1)$ and $B = (D + 1)$. Show for the function $y(x) = x$ that $(AB)[y] \neq (BA)[y]$. **Hence, for operators with *variable* coefficients, the commutative law need not hold.**

37. Isolated Zeros. Let $\phi(x)$ be a solution to $y'' + py' + qy = 0$ on (a, b), where p, q are continuous on (a, b). By completing the following steps, prove that if ϕ is not identically zero, then its zeros in (a, b) are *isolated*, i.e., if $\phi(x_0) = 0$, then there exists a $\delta > 0$ such that $\phi(x) \neq 0$ for $0 < |x - x_0| < \delta$.

(a) Suppose $\phi(x_0) = 0$ and assume to the contrary that for each $n = 1, 2, \ldots$, the function ϕ has a zero at x_n, where $0 < |x_0 - x_n| < 1/n$. Show that this implies $\phi'(x_0) = 0$. [*Hint:* Consider the difference quotient for ϕ at x_0.]

(b) With the assumptions of part (a), we have $\phi(x_0) = \phi'(x_0) = 0$. Conclude from this that ϕ must be identically zero, which is a contradiction. Hence there is some integer n_0 such that $\phi(x)$ is not zero for $0 < |x - x_0| < 1/n_0$.

4.3 FUNDAMENTAL SOLUTIONS OF HOMOGENEOUS EQUATIONS

The goal of this section is to show that in order to find *all* solutions to the homogeneous equation

$$y'' + py' + qy = 0 \ ,$$

it is enough to find just two solutions that are not constant multiples of one another. First, let's examine the particular equation

(1) $$L[y] = \frac{d^2y}{dx^2} - y = 0 \ .$$

It is easy to see that the two functions $y_1 = e^x$ and $y_2 = e^{-x}$ are solutions to this homogeneous equation. Further, since L is linear, we know from Theorem 1 in Section 4.2 that $c_1e^x + c_2e^{-x}$ is also a solution for any choice of constants c_1 and c_2. The question is, can all solutions of (1) be represented by

(2) $$c_1e^x + c_2e^{-x} \ ,$$

with appropriate choices for c_1 and c_2? As we now show, the answer lies in interpreting the existence and uniqueness theorem (Theorem 2, page 165).

Let $\phi(x)$ be a solution to (1) and let x_0 be a fixed real number. If we can choose c_1 and c_2 so that

(3) $$\begin{cases} c_1e^{x_0} + c_2e^{-x_0} = \phi(x_0) \ , \\ c_1e^{x_0} - c_2e^{-x_0} = \phi'(x_0) \ , \end{cases}$$

then, because $\phi(x)$ and $c_1e^x + c_2e^{-x}$ satisfy the same initial conditions at x_0, the uniqueness conclusion of Theorem 2 guarantees that $\phi(x) = c_1e^x + c_2e^{-x}$ for all x. Note that we have reduced the question of the general validity of the representation (2), for all solutions to the differential equation, to the simpler question of the existence of solutions to the algebraic system (3).

To solve the system (3) for c_1, we add the two equations and then divide by $2e^{x_0}$ to obtain

(4) $$c_1 = \frac{\phi(x_0) + \phi'(x_0)}{2e^{x_0}} \ .$$

Subtracting the equations and dividing by $2e^{-x_0}$ gives

(5) $$c_2 = \frac{\phi(x_0) - \phi'(x_0)}{2e^{-x_0}} \ .$$

Notice that since e^{x_0} and e^{-x_0} are never zero, we can always carry out the above procedure for finding c_1 and c_2. Thus the algebraic system (3) does indeed possess a solution, and consequently the formula

$$\phi(x) = c_1e^x + c_2e^{-x}$$

is valid with suitable c_1, c_2.

In other words, every solution of (1) can be expressed as a linear combination of the two particular solutions y_1, y_2 of (1). If, on the other hand, we had started with the two solutions $y_1 = e^x$ and $y_3 = 5e^x$ of equation (1), the combination $c_1y_1 + c_2y_3 = c_1e^x + c_25e^x$ would not be rich enough to meet arbitrary initial conditions like (3) (because this combination is simply a multiple of e^x; see Problem 14). The general situation is characterized by the following theorem.

REPRESENTATION OF SOLUTIONS (HOMOGENEOUS CASE)

Theorem 3. Let y_1 and y_2 denote two solutions on (a, b) of

(6) $y'' + p(x)y' + q(x)y = 0$,

where p and q are continuous on (a, b). Suppose at some point x_0 in (a, b) these solutions satisfy

(7) $y_1(x_0)y_2'(x_0) - y_1'(x_0)y_2(x_0) \neq 0$.

Then every solution of (6) on (a, b) can be expressed in the form

(8) $y(x) = c_1y_1(x) + c_2y_2(x)$,

where c_1 and c_2 are constants.

In the proof that follows, we will see that statement (7) is precisely the condition we need to guarantee existence of the solution of the *algebraic* equations determining c_1 and c_2. The linear combination of y_1 and y_2 in (8), written with arbitrary constants c_1 and c_2, is referred to as a **general solution** to (6).

Proof of Theorem 3. Let $\phi(x)$ be any solution to (6). As before, we consider the system

(9) $c_1y_1(x_0) + c_2y_2(x_0) = \phi(x_0)$,

(10) $c_1y_1'(x_0) + c_2y_2'(x_0) = \phi'(x_0)$,

where x_0 is a point at which (7) holds. To solve for c_1, multiply (9) by $y_2'(x_0)$, multiply (10) by $y_2(x_0)$, and subtract to obtain

$$c_1[y_1(x_0)y_2'(x_0) - y_1'(x_0)y_2(x_0)] = \phi(x_0)y_2'(x_0) - \phi'(x_0)y_2(x_0) .$$

Because of condition (7), the expression in brackets is not zero. Hence

$$c_1 = \frac{\phi(x_0)y_2'(x_0) - \phi'(x_0)y_2(x_0)}{y_1(x_0)y_2'(x_0) - y_1'(x_0)y_2(x_0)} .$$

Similarly, we find

$$c_2 = \frac{\phi'(x_0)y_1(x_0) - \phi(x_0)y_1'(x_0)}{y_1(x_0)y_2'(x_0) - y_1'(x_0)y_2(x_0)} .$$

With these choices for c_1 and c_2, the functions $c_1 y_1 + c_2 y_2$ and ϕ are solutions to (6) that satisfy the same initial conditions. Hence they must be the same function on (a, b), according to the uniqueness result in Theorem 2. ■ End

Because the expression on the left-hand side of (7) plays an important role in the theory, we adopt the following terminology.

WRONSKIAN

Definition 1. For any two differentiable functions y_1 and y_2, the function

(11) $W[y_1, y_2](x) := y_1(x)y_2'(x) - y_1'(x)y_2(x)$

is called the **Wronskian** of y_1 and y_2.[†]

A convenient way of writing the Wronskian $W[y_1, y_2]$, which will generalize to higher-order linear equations, is in terms of the determinant

$$W[y_1, y_2](x) = \begin{vmatrix} y_1(x) & y_2(x) \\ y_1'(x) & y_2'(x) \end{vmatrix} .$$

FUNDAMENTAL SOLUTION SET

Definition 2. A pair of solutions $\{y_1, y_2\}$ of $y'' + py' + qy = 0$ on (a, b) is called a **fundamental solution set** if

$$W[y_1, y_2](x_0) \neq 0$$

at some x_0 in (a, b).[††]

Logically speaking, Definition 2 requires only that the Wronskian be nonzero at *one* point. However, it will soon become clear that if the Wronskian of two solutions is nonzero at one point, it will actually be nonzero at *all* points in the interval (a, b). For example, the solutions $y_1 = e^x$, $y_2 = e^{-x}$ of the differential equation (1) form a fundamental solution set for this equation on $(-\infty, \infty)$, since

$$W[y_1, y_2](x) = e^x(-e^{-x}) - e^x e^{-x} = -2 \neq 0$$

for any x.

With the terminology of Definition 2, Theorem 3 can be summarized as follows.

[†]*Historical Footnote:* The Wronskian was named after the Polish mathematician H. Wronski (1778–1853).

[††]*Historical Footnote:* Lazarus Fuchs introduced the terminology *fundamental solution* in 1866.

PROCEDURE FOR SOLVING HOMOGENEOUS EQUATIONS

To determine all solutions to $y'' + py' + qy = 0$:

(a) Find two solutions y_1, y_2 that constitute a fundamental solution set.
(b) Form the linear combination

$$y(x) = c_1 y_1(x) + c_2 y_2(x) \ ,$$

where c_1 and c_2 are arbitrary constants, to obtain a general solution.

EXAMPLE 1 Given that $y_1(x) = \cos 3x$ and $y_2(x) = \sin 3x$ are solutions to

(12) $y'' + 9y = 0$

on $(-\infty, \infty)$, find a general solution to (12).

SOLUTION First we verify that $\{\cos 3x, \sin 3x\}$ is a fundamental solution set. We are given that both $y_1(x) = \cos 3x$ and $y_2(x) = \sin 3x$ are solutions to (12) (the reader can quickly verify this), so we need show only that $W[y_1, y_2](x) \neq 0$ for some x in $(-\infty, \infty)$. Substituting y_1 and y_2 into formula (11) yields

$$
\begin{aligned}
W[y_1, y_2](x) &= (\cos 3x)(3 \cos 3x) - (-3 \sin 3x)(\sin 3x) \\
&= 3(\cos^2 3x + \sin^2 3x) \\
&= 3 \neq 0 \ .
\end{aligned}
$$

Thus $\{\cos 3x, \sin 3x\}$ forms a fundamental solution set, and a general solution to (12) is

$$y(x) = c_1 \cos 3x + c_2 \sin 3x. \quad \blacksquare \text{ End}$$

It is easy to see that a fundamental solution set for $y'' + py' + qy = 0$ always exists. Indeed, let $x_0 \in (a, b)$ and take $y_1(x)$ and $y_2(x)$ to be the solutions that satisfy the initial conditions

$$
\begin{aligned}
y_1(x_0) &= 1 \ , \quad y_2(x_0) = 0 \ , \\
y_1'(x_0) &= 0 \ , \quad y_2'(x_0) = 1 \ .
\end{aligned}
$$

Then $W[y_1, y_2](x_0) = 1 \neq 0$, and so $\{y_1, y_2\}$ is a fundamental solution set.

There is another method that can be used to determine quickly whether a pair of solutions form a fundamental solution set. To describe this test, we need to recall some basic facts about **vectors** (directed line segments) in the plane.

A planar vector is prescribed by an ordered pair of real numbers, say (u_1, u_2). Recall that vector addition is done componentwise, as is scalar multiplication; that is,

$$(u_1, u_2) + (v_1, v_2) = (u_1 + v_1, u_2 + v_2) \ , \qquad c(u_1, u_2) = (cu_1, cu_2) \ .$$

Two vectors are parallel (point in the same or opposite directions) if and only if one vector is a scalar multiple of the other. This is equivalent to saying that there exist two constants c_1, c_2 *not both zero* such that

$$c_1(u_1, u_2) + c_2(v_1, v_2) = (0, 0) \ .$$

Parallel vectors are said to be **linearly dependent,** whereas two nonparallel vectors are **linearly independent.**

These notions also apply to functions.

LINEAR DEPENDENCE OF FUNCTIONS

Definition 3. Two functions y_1 and y_2 are said to be **linearly dependent on an interval I** if there exist constants c_1 and c_2, not both zero, such that

$$c_1 y_1(x) + c_2 y_2(x) = 0$$

for all x in I. If two functions are not linearly dependent, they are said to be **linearly independent.**

The preceding definition extends in a natural way to sets of more than two functions; this is studied in Chapter 6 (see also Problems 24, 25, and 26 in Exercises 4.3). Observe that **in the case of just two functions, linear dependence on I is equivalent to one function being a constant multiple of the other function on I** (see Problem 16).

EXAMPLE 2 Determine whether the following pairs of functions y_1 and y_2 are linearly dependent on $(-5, 5)$:

(a) $y_1(x) = e^{3x}$, $y_2(x) = x + 1$.

(b) $y_1(x) = \sin 2x$, $y_2(x) = \cos x \sin x$.

(c) $y_1(x) = x$, $y_2(x) = |x|$.

SOLUTION (a) A glance at the functions $y_1(x) = e^{3x}$ and $y_2(x) = x + 1$ reveals that neither is a constant multiple of the other. Indeed, if a constant c exists such that

$$e^{3x} = c(x + 1) \quad \text{for all } x \text{ in } (-5, 5) ,$$

then we arrive at a contradiction by setting $x = 0$ and $x = 1$:

$$e^0 = c(0 + 1) \Rightarrow c = 1 ,$$
$$e^3 = c(1 + 1) \Rightarrow c = e^3/2 \neq 1 .$$

Hence e^{3x} and $x + 1$ are linearly independent.

(b) Because $y_1(x) = \sin 2x = 2 \sin x \cos x$, we see that $y_1(x) = 2y_2(x)$. Hence y_1 and y_2 are linearly dependent on $(-5, 5)$.

(c) Here the two functions $y_1(x) = x$ and $y_2(x) = |x|$ are identical on the subinterval $(0, 5)$. $\big($In particular, they are linearly dependent on $(0, 5)$.$\big)$ However, neither function is a fixed constant multiple of the other on the *whole* interval $(-5, 5)$. $\big($On $(0, 5)$, $y_1(x) = 1 \cdot y_2(x)$, but on $(-5, 0)$, $y_1(x) = (-1) \cdot y_2(x)$.$\big)$ Thus y_1 and y_2 are linearly independent on $(-5, 5)$. ■End

Part (c) depicted a situation where $y_1(x)$ and $y_2(x)$ were linearly dependent on a subinterval but not on the whole interval. The next theorem shows (among other things) that this

cannot happen if y_1 and y_2 are both solutions to the same differential equation (on the whole interval).

CONDITION FOR LINEAR DEPENDENCE OF SOLUTIONS

Theorem 4. Let y_1, y_2 be solutions to $y'' + py' + qy = 0$ on (a, b), and let x_0 be any point in (a, b). Then y_1 and y_2 are linearly dependent on (a, b) if and only if the (constant) vectors[†]

$$\begin{bmatrix} y_1(x_0) \\ y_1'(x_0) \end{bmatrix} \text{ and } \begin{bmatrix} y_2(x_0) \\ y_2'(x_0) \end{bmatrix}$$

are linearly dependent.

Remark. Although not explicitly mentioned, we assume in Theorem 4 and in subsequent results that the coefficient functions p and q are continuous on (a, b).

Proof of Theorem 4. (\Rightarrow) Assume that the *functions* y_1 and y_2 are linearly dependent. Then one of these functions is a constant multiple of the other, say $y_1 = cy_2$ on (a, b). Thus $y_1' = cy_2'$, and we have

$$\begin{bmatrix} y_1(x_0) \\ y_1'(x_0) \end{bmatrix} = \begin{bmatrix} cy_2(x_0) \\ cy_2'(x_0) \end{bmatrix} = c\begin{bmatrix} y_2(x_0) \\ y_2'(x_0) \end{bmatrix} .$$

Since one initial vector is a constant multiple of the other, they are linearly dependent.

(\Leftarrow) Assume now that the *initial vectors* are linearly dependent, say

$$\begin{bmatrix} y_1(x_0) \\ y_1'(x_0) \end{bmatrix} = c\begin{bmatrix} y_2(x_0) \\ y_2'(x_0) \end{bmatrix} .$$

This implies that the functions y_1 and cy_2 satisfy the same initial conditions; that is,

$$y_1(x_0) = cy_2(x_0) \quad \text{and} \quad y_1'(x_0) = cy_2'(x_0) .$$

Because y_1 and cy_2 are solutions to the same initial value problem, by the uniqueness property they must be the same function on (a, b). Thus y_1 is a constant multiple of y_2 on (a, b); that is, y_1 and y_2 are linearly dependent on (a, b). ∎ End

A simple algebraic test for the linear independence of two column vectors

(13) $\begin{bmatrix} u_1 \\ u_2 \end{bmatrix}$ and $\begin{bmatrix} v_1 \\ v_2 \end{bmatrix}$

is to compute the determinant

$$\begin{vmatrix} u_1 & v_1 \\ u_2 & v_2 \end{vmatrix} = u_1v_2 - v_1u_2 .$$

[†]Here and in later discussions, we display planar vectors in column rather than row form.

The vectors (13) *are linearly independent if and only if this determinant is not zero* (see Problem 29). Assembling all these considerations, we obtain the following characterization.

FUNDAMENTAL SETS, LINEAR INDEPENDENCE, AND THE WRONSKIAN

Corollary 1. If y_1 and y_2 are solutions to $y'' + py' + qy = 0$ on (a, b), then the following statements are equivalent:

(i) $\{y_1, y_2\}$ is a fundamental solution set on (a, b).
(ii) y_1 and y_2 are linearly independent on (a, b).
(iii) The Wronskian $W[y_1, y_2](x)$ is never zero on (a, b).

Proof. If $\{y_1, y_2\}$ is a fundamental set, then there is *some* point x_0 in (a, b), where

$$(14) \qquad W[y_1, y_2](x_0) = \begin{vmatrix} y_1(x_0) & y_2(x_0) \\ y_1'(x_0) & y_2'(x_0) \end{vmatrix} \neq 0 \ ,$$

by Definition 2. By the determinant test, then, the vectors

$$(15) \qquad \begin{bmatrix} y_1(x_0) \\ y_1'(x_0) \end{bmatrix} \text{ and } \begin{bmatrix} y_2(x_0) \\ y_2'(x_0) \end{bmatrix}$$

are independent. Theorem 4 thus implies the *functions* y_1 and y_2 are linearly independent. Therefore condition (i) implies condition (ii).

But Theorem 4 also states that linear independence of y_1 and y_2 implies that the vectors in (15) are linearly independent, regardless of the choice of x_0. Thus, because of the determinant test (again), the Wronskian of y_1 and y_2 is nonzero everywhere in the interval, and condition (ii) therefore implies condition (iii).

Finally, condition (iii) implies condition (i) trivially because of our definition of fundamental solution sets (Definition 2). ■ End

In particular, the Wronskian of two solutions y_1 and y_2 is either *identically* zero or *never* zero on (a, b), depending on whether the solutions are linearly dependent or linearly independent. In Example 1, for instance, the Wronskian has the constant value 3.

EXAMPLE 3 Show that $y_1(x) = x^{-1}$ and $y_2(x) = x^3$ are solutions to

$$(16) \qquad x^2 y'' - xy' - 3y = 0$$

on the interval $(0, \infty)$ and give a general solution.

SOLUTION The verification that y_1 and y_2 are solutions to (16) is straightforward; substituting $y = x^{-1}$ and $y = x^3$ in (16) gives, respectively, the identities

$$x^2(2x^{-3}) - x(-x^{-2}) - 3(x^{-1}) = 0 \quad \text{and} \quad x^2(6x) - x(3x^2) - 3(x^3) = 0 \ .$$

Furthermore, the solution functions x^{-1} and x^3 are linearly independent on $(0, \infty)$ (neither is a constant multiple of the other on $(0, \infty)$). Hence, by Corollary 1, $\{x^{-1}, x^3\}$ is a fundamental

solution set on $(0, \infty)$, and so a general solution is

$$y(x) = c_1 x^{-1} + c_2 x^3. \quad \blacksquare \text{End}$$

Remark. In the last example we appealed to Corollary 1, which applies only to equations for which the coefficient of y'' is unity and the other coefficients are continuous. But notice that for $x > 0$, equation (16) is equivalent to $y'' - (1/x)y' - (3/x^2)y = 0$, where $p(x) = -1/x$, $q(x) = -3/x^2$ are continuous on $(0, \infty)$. Thus, the application of Corollary 1 is indeed permissible.

EXAMPLE 4 Can the function $w(x) = 3(x - 1)^2$ be the Wronskian on the interval $(0, 2)$ for a pair of solutions to some equation $y'' + py' + qy = 0$ (with p and q continuous)?

SOLUTION Because $w(x) = 0$ only for $x = 1$, then by Corollary 1, $w(x)$ cannot be such a Wronskian on any open interval containing $x = 1$. $\quad \blacksquare \text{End}$

If two differentiable functions are linearly independent on I, can their Wronskian be identically zero? Surprisingly, the answer is yes. The reader can verify (see Problem 18) that the Wronskian of the functions $y_1(x) = x^3$ and $y_2(x) = |x^3|$ is identically zero on $(-\infty, \infty)$. Yet these functions are certainly linearly independent on $(-\infty, \infty)$ because one is not a constant multiple of the other for all x in $(-\infty, \infty)$. Does this example contradict Corollary 1? No, because Corollary 1 refers only to functions $y_1(x)$ and $y_2(x)$ that are solutions to the **same** homogeneous linear second order differential equation. What we *can* conclude from Corollary 1 is that $y_1(x) = x^3$ and $y_2(x) = |x^3|$ cannot be solutions to the same equation on $(-\infty, \infty)$, with continuous $p(x)$ and $q(x)$.

Another representation of the Wronskian for two solutions y_1, y_2 to the equation $y'' + py' + qy = 0$ on (a, b) is **Abel's identity:** [†]

$$(17) \qquad W[y_1, y_2](x) = C \exp\left(-\int_{x_0}^{x} p(t)\,dt\right),$$

where $x_0 \in (a, b)$ and C is a constant that depends on y_1 and y_2. This formula reaffirms our earlier observation that the Wronskian of two solutions is always zero $(C = 0)$ or never zero $(C \neq 0)$, since the exponential factor never vanishes. We leave the proof of (17) as an exercise (see Problem 20).

The properties of the Wronskian and linear independence will be extended to sets of n functions $y_1(x), \ldots, y_n(x)$ when we consider higher-order linear equations in Chapter 6. Although the proofs are a bit more complicated, the ideas are essentially the same as in the case of linear second order equations.

To summarize, we have shown in this section that to find *all* the solutions to a homogeneous second order linear differential equation, it suffices to obtain just two linearly independent solutions and then take linear combinations of them. In Sections 4.5 and 4.6, we concentrate on methods for actually finding such a pair of functions.

[†]*Historical Footnote:*Abel's identity was derived by Niels Abel in 1827.

EXERCISES 4.3

In Problems 1 through 6, use Definition 3 to determine whether the functions y_1 and y_2 are linearly dependent on the interval $(0, 1)$. Also compute the Wronskian $W[y_1, y_2](x)$.

1. $y_1(x) = e^{-x}\cos 2x$, $y_2(x) = e^{-x}\sin 2x$.

2. $y_1(x) = e^{3x}$, $y_2(x) = e^{-4x}$.

3. $y_1(x) = xe^{2x}$, $y_2(x) = e^{2x}$.

4. $y_1(x) = x^2\cos(\ln x)$, $y_2(x) = x^2\sin(\ln x)$.

5. $y_1(x) = \tan^2 x - \sec^2 x$, $y_2(x) = 3$.

6. $y_1(x) = 0$, $y_2(x) = e^x$.

In Problems 7 through 12, (a) verify that the functions y_1 and y_2 are linearly independent solutions of the given differential equation, (b) find a general solution to the given differential equation, and (c) find the solution that satisfies the given initial conditions.

7. $x^2 y'' - 2y = 0$; $y_1(x) = x^2$, $y_2(x) = x^{-1}$;
$y(1) = -2$, $y'(1) = -7$.

8. $y'' - 5y' + 6y = 0$; $y_1(x) = e^{2x}$, $y_2(x) = e^{3x}$;
$y(0) = -1$, $y'(0) = -4$.

9. $y'' - 2y' + 5y = 0$; $y_1(x) = e^x \cos 2x$,
$y_2(x) = e^x \sin 2x$; $y(0) = 2$, $y'(0) = 0$.

10. $y'' - 5y' = 0$; $y_1(x) = 2$, $y_2(x) = e^{5x}$;
$y(0) = 2$, $y'(0) = 5$.

11. $ty'' - (t + 2)y' + 2y = 0$;
$y_1(t) = e^t$, $y_2(t) = t^2 + 2t + 2$;
$y(1) = 0$, $y'(1) = 1$.

12. $y'' - y = 0$; $y_1(x) = \cosh x$, $y_2(x) = \sinh x$;
$y(0) = 1$, $y'(0) = -1$.

13. Consider the differential equation

(18) $y'' + 5y' - 6y = 0$.

 (a) Show that $S_1 := \{e^x, e^x - e^{-6x}\}$ is a fundamental solution set for (18).
 (b) Show that $S_2 := \{e^x, 3e^x + e^{-6x}\}$ is another fundamental solution set for (18).
 (c) Verify that $\phi(x) = e^{-6x}$ is a solution to (18). Then express ϕ as a linear combination of functions in S_1. Likewise, express ϕ as a linear combination of functions in S_2.

14. (a) Verify that $y_1 = e^x$ and $y_3 = 5e^x$ are two solu-

tions to $y'' - y = 0$ (equation (1) of the text), but the solution $\phi(x)$ to this equation that satisfies $\phi(0) = 1$, $\phi'(0) = 0$ cannot be expressed as $c_1 y_1(x) + c_2 y_3(x)$, regardless of the choice of the constants c_1 and c_2.

 (b) Compute the Wronskian of the solutions y_1 and y_3 in part (a).

15. Determine whether the following functions can be Wronskians on $-1 < x < 1$ for a pair of solutions to some equation $y'' + py' + qy = 0$ (with p and q continuous).

 (a) $w(x) = 6e^{4x}$. **(b)** $w(x) = x^3$.
 (c) $w(x) = (x + 1)^{-1}$. **(d)** $w(x) \equiv 0$.

16. Prove that two functions are linearly dependent on an interval I in the sense of Definition 3 if and only if one is a constant times the other on I.

17. Let y_1 and y_2 be two functions defined on $(-\infty, \infty)$.

 (a) (True or False) If y_1 and y_2 are linearly dependent on the interval $[a, b]$, then y_1 and y_2 are linearly dependent on the smaller interval $[c, d] \subset [a, b]$.
 (b) (True or False) If y_1 and y_2 are linearly dependent on the interval $[a, b]$, then y_1 and y_2 are linearly dependent on the larger interval $[C, D] \supset [a, b]$.

18. Let $y_1(x) = x^3$ and $y_2(x) = |x^3|$. Are y_1 and y_2 linearly independent on the following intervals?

 (a) $[0, \infty)$. **(b)** $(-\infty, 0]$. **(c)** $(-\infty, \infty)$.
 (d) Compute the Wronskian $W[y_1, y_2](x)$ on the interval $(-\infty, \infty)$.

19. Use Abel's identity (17) to determine (up to a constant multiple) the Wronskian of two solutions on $(0, \infty)$ to

$$xy'' + (x - 1)y' + 3y = 0 .$$

20. Prove Abel's identity (17) by completing the following steps:

 (a) Show that the Wronskian W satisfies the equation $W' + p(x)W = 0$.
 (b) Solve the separable equation in part (a).

21. Prove that if y_1 and y_2 are linearly independent solutions of $y'' + py' + qy = 0$ on (a, b), then they cannot both be zero at the same point x_0 in (a, b).

22. Show that if y_1 and y_2 are linearly independent solutions of $y'' + py' + qy = 0$ on (a, b), then they cannot both have an extremum at the same point x_0 in (a, b).

23. Normal Form. Show that the substitution $y(x) = u(x)v(x)$, where

$$v(x) := \exp\left(-\frac{1}{2}\int p(x)\,dx\right),$$

transforms the differential equation

$$y'' + p(x)y' + q(x)y = 0$$

into an equation of the form

$$u'' + f(x)u = 0 .$$

The last equation is called the **normal form** of a homogeneous linear second order equation.

24. Linear Dependence of Three Functions. Three functions $y_1(x)$, $y_2(x)$, and $y_3(x)$ are said to be **linearly dependent on an interval I** if, on I, at least one of these functions is a linear combination of the remaining two; that is, there exist constants C_1, C_2, C_3, not all zero, such that

$$C_1 y_1(x) + C_2 y_2(x) + C_3 y_3(x) = 0 ,$$

for all x in I.

Otherwise, we say they are **linearly independent.**

(a) Show that if y_1 and y_2 are two linearly dependent functions on I, then y_1, y_2, and y_3 are linearly dependent on I for any function y_3.

(b) Show that $y_1(x) = e^x$, $y_2(x) = e^{2x}$, and $y_3(x) = e^{-3x}$ are linearly independent on $(-\infty, \infty)$.

(c) Show that $y_1(x) = \cos 2x$, $y_2(x) = \sin^2 x$, and $y_3(x) = \cos^2 x$ are linearly dependent on $(-\infty, \infty)$.

25. Using the definitions in Problem 24, determine whether the given three functions are linearly dependent or linearly independent on $(-\infty, \infty)$:

(a) $y_1(x) = 1$, $y_2(x) = x$, $y_3(x) = x^2$.

(b) $y_1(x) = -3$, $y_2(x) = 5\sin^2 x$, $y_3(x) = \cos^2 x$.

(c) $y_1(x) = e^x$, $y_2(x) = xe^x$, $y_3(x) = x^2 e^x$.

(d) $y_1(x) = e^x$, $y_2(x) = e^{-x}$, $y_3(x) = \cosh x$.

26. Using the definition in Problem 24, prove that $y_1(x) = e^{r_1 x}$, $y_2(x) = e^{r_2 x}$, and $y_3(x) = e^{r_3 x}$ are linearly independent on $(-\infty, \infty)$ if and only if the real numbers r_1, r_2, and r_3 are distinct.

27. Wronskian of Three Functions. The Wronskian of three functions is defined in terms of a determinant as follows:

$$W[y_1, y_2, y_3](x) := \begin{vmatrix} y_1(x) & y_2(x) & y_3(x) \\ y_1'(x) & y_2'(x) & y_3'(x) \\ y_1''(x) & y_2''(x) & y_3''(x) \end{vmatrix} .$$

Find the Wronskian of the three functions given in

(a) part (a) of Problem 25.

(b) part (b) of Problem 25.

(c) part (c) of Problem 25.

(d) What do you think is the connection between the Wronskian of three functions and their linear dependence?

28. Let y_1 and y_2 be solutions to $y'' + py' + qy = 0$ on (a, b). Prove that if they are linearly dependent on a subinterval (c, d), they are linearly dependent on all of (a, b). [*Hint:* Choose $x_0 \in (c, d)$ and apply Theorem 4.]

29. Show that the determinant

$$\begin{vmatrix} u_1 & v_1 \\ u_2 & v_2 \end{vmatrix} \quad (u_1, u_2, v_1, v_2 \text{ constants})$$

is not zero,

(a) if and only if the two column vectors

$$\begin{bmatrix} u_1 \\ u_2 \end{bmatrix} \text{ and } \begin{bmatrix} v_1 \\ v_2 \end{bmatrix}$$

are linearly independent.

(b) if and only if the system with unknowns c_1 and c_2

$$u_1 c_1 + v_1 c_2 = a ,$$
$$u_2 c_1 + v_2 c_2 = b ,$$

has a unique solution for all a, b.

30. Prove that if two functions are linearly dependent, then their zeros must coincide. Demonstrate that this is false for three functions.

31. A Counterexample (?) to Theorem 3. Consider the linear equation

(19) $\qquad x^2 y'' - 3xy' + 3y = 0$
$\qquad\qquad$ for $-\infty < x < \infty$.

(a) Verify that $y_1(x) := x$ and $y_2(x) := x^3$ are two solutions to (19) on $(-\infty, \infty)$. Furthermore, show that $y_1(x_0)y_2'(x_0) - y_1'(x_0)y_2(x_0) \neq 0$ for $x_0 = 1$.

(b) Prove that $y_1(x)$ and $y_2(x)$ are linearly independent on $(-\infty, \infty)$.

(c) Verify that the function $y_3(x) := |x|^3$ is also a solution to (19) on $(-\infty, \infty)$.

(d) Prove that there is *no* choice of constants c_1, c_2 such that $y_3(x) = c_1 y_1(x) + c_2 y_2(x)$ for *all* x in $(-\infty, \infty)$. [*Hint:* Argue that the contrary assumption leads to a contradiction.]

(e) From parts (c) and (d), we see that there is at least one solution to (19) on $(-\infty, \infty)$ that is not expressible as a linear combination of the solutions $y_1(x)$, $y_2(x)$. Does this provide a counterexample to Theorem 3 on page 169? Explain!

4.4 REDUCTION OF ORDER

As we found in the previous section, a general solution to a linear second order homogeneous equation is given by a linear combination of two linearly independent solutions. In the next section, we describe a method for finding such solutions when the equation has constant coefficients. However, the method may produce only a *single* solution (and its multiples), rather than the required *pair* of solutions. So although it may seem like "jumping the gun," we devote the present section to a technique that uses a known solution to reduce any homogeneous linear second order equation to a separable first order equation—a type discussed in Section 2.2—yielding a solution that is independent of the given solution. This procedure is referred to as **reduction of order.** [†]

Let f be a nontrivial (i.e., not identically zero) solution to the homogeneous equation

(1) $y'' + p(x)y' + q(x)y = 0$.

For us to find a second solution that is linearly independent of f, the simplest way would be to use function addition or multiplication to modify f. However, trying a solution of the form $y(x) = v(x) + f(x)$ leads us right back to where we started: to the equation $v'' + pv' + qv = 0$. (Check this!) So, let's try to find a solution of the form

(2) $y(x) = v(x)f(x)$,

where the nonconstant function v is to be determined. Differentiating the product vf, we have

$$y' = vf' + v'f ,$$
$$y'' = vf'' + 2v'f' + v''f .$$

Substituting these expressions into (1) yields

$$(vf'' + 2v'f' + v''f) + p(vf' + v'f) + qvf = 0 ,$$

or, on regrouping,

(3) $(f'' + pf' + qf)v + fv'' + (2f' + pf)v' = 0$.

Because f is a solution to (1), the factor in front of v is zero. Thus (3) reduces to

(4) $fv'' + (2f' + pf)v' = 0$.

[†]*Historical Footnote:* The method for the reduction of order is credited to Jean le Rond d'Alembert circa 1760.

This is a separable (and linear) first order equation in the variable $w = v'$. Indeed, on substituting w for v' in (4), we have

$$fw' + (2f' + pf)w = 0 .$$

Then, separating the variables and integrating gives

$$\int \frac{dw}{w} = -2 \int \frac{f'}{f} dx - \int p\, dx ,$$

$$\ln|w| = \ln(f^{-2}) - \int p\, dx ,$$

$$w = \pm f^{-2} e^{-\int p\, dx} ,$$

which holds on any interval where f does not vanish. If we replace w by v' in the last equation and integrate, we find

$$(5) \qquad v = \pm \int \frac{e^{-\int p(x)dx}}{[f(x)]^2} dx .$$

It is easy to show from this formula that v cannot reduce to a constant function (Problem 19), and so a second linearly independent solution is $y(x) = v(x)f(x)$, where $v(x)$ is given in equation (5). (Either sign will do.)

The above discussion can be summarized as follows.

REDUCTION OF ORDER PROCEDURES

Given a nontrivial solution $f(x)$ to $y'' + py' + qy = 0$, a second linearly independent solution $y(x)$ can be determined in either of the following ways:

1. Set $y(x) = v(x)f(x)$ and substitute for y, y', and y'' in the given equation. This gives a separable equation for v'. Solve for v' and integrate v' to obtain v. The desired second solution is given by $v(x)f(x)$.

2. The solution $y(x)$ can also be obtained by plugging $p(x)$ and $f(x)$ directly into the **reduction of order formula:**

$$(6) \qquad y(x) = f(x) \int \frac{e^{-\int p(x)dx}}{[f(x)]^2} dx .$$

Caution: If the equation is $a_2(x)y'' + a_1(x)y' + a_0(x)y = 0$, then it must first be put in the form $y'' + p(x)y' + q(x)y = 0$, so $p(x) = a_1(x)/a_2(x)$.

The first procedure is a technique for which we only have to remember the substitution $y = v(x)f(x)$. The second procedure is quicker, but we must remember precisely the reduction of order formula. In the next example, we illustrate both of these methods.

EXAMPLE 1 Given that $f(x) = e^x$ is a solution to

$$(7) \qquad y'' - 2y' + y = 0 ,$$

determine a second linearly independent solution.

SOLUTION Let $y = v(x)f(x) = ve^x$. Then

$$y' = ve^x + v'e^x ,$$
$$y'' = ve^x + 2v'e^x + v''e^x .$$

Substituting these representations into (7) gives

$$(ve^x + 2v'e^x + v''e^x) - 2(ve^x + v'e^x) + ve^x = 0 ,$$

which simplifies to $v'' = 0$. Hence $v = c_1 x + c_2$. Because we want y to be linearly independent of e^x, we must take $c_1 \neq 0$, so for simplicity, let $c_1 = 1$ and $c_2 = 0$. Then a second linearly independent solution is

$$y = vf = xe^x .$$

The alternative approach is to use the reduction of order formula given by equation (6). In this example, $p(x) = -2$ and $f(x) = e^x$. Hence

$$-\int p(x)dx = \int 2\, dx = 2x + c .$$

We will take $c = 0$. Now

$$y(x) = f(x) \int \frac{e^{-\int p(x)dx}}{[f(x)]^2} dx = e^x \int \frac{e^{2x}}{(e^x)^2} dx$$

$$= e^x \int dx = e^x(x + C) .$$

Again, we set $C = 0$. This gives us the same linearly independent solution, $y = xe^x$. ∎ End

EXAMPLE 2 Given that $f(x) = x$ is a solution to

(8) $y'' - 2xy' + 2y = 0 ,$

determine a second linearly independent solution.

SOLUTION Here $p(x) = -2x$, and so

$$-\int p(x)dx = \int 2x\, dx = x^2 ,$$

where we have taken the constant of integration to be zero. The reduction of order formula in (6) gives

(9) $$y(x) = x \int \frac{e^{x^2}}{x^2} dx$$

as a second linearly independent solution. Unfortunately, we are not able to evaluate the integral in (9) in terms of elementary functions. One alternative is to write $y(x)$ in terms of a definite integral, say,

$$y(x) = x \int_1^x \frac{e^{t^2}}{t^2} dt .$$

Then $y(x)$ can be approximated for various values of x by using a numerical integration method

such as Simpson's rule (see Problem 17). Another approach is to obtain a **power series expansion** for the solution in (9). This can be derived as follows.

Recall that the Maclaurin series for e^x is

$$e^x = 1 + x + \frac{x^2}{2!} + \frac{x^3}{3!} + \cdots + \frac{x^n}{n!} + \cdots .$$

Hence the expansion for e^{x^2} is just

$$e^{x^2} = 1 + x^2 + \frac{x^4}{2!} + \frac{x^6}{3!} + \cdots + \frac{x^{2n}}{n!} + \cdots .$$

Substituting this series for e^{x^2} into (9), we integrate term by term to obtain

$$y(x) = x \int x^{-2} \left\{ 1 + x^2 + \frac{x^4}{2} + \cdots + \frac{x^{2n}}{n!} + \cdots \right\} dx$$

$$= x \int \left\{ x^{-2} + 1 + \frac{x^2}{2} + \cdots + \frac{x^{2n-2}}{n!} + \cdots \right\} dx$$

$$= x \left\{ -x^{-1} + x + \frac{x^3}{6} + \cdots + \frac{x^{2n-1}}{n!(2n-1)} + \cdots \right\} ,$$

where we have taken the constant of integration to be zero. Hence

$$\textbf{(10)} \qquad y(x) = -1 + x^2 + \frac{x^4}{6} + \cdots + \frac{x^{2n}}{n!(2n-1)} + \cdots . \quad \blacksquare \text{ End}$$

In the last example, we did not bother to justify the use of term-by-term integration to obtain equation (10). Such questions are addressed in Chapter 8.

The reduction of order procedure can also be used to reduce a linear *nonhomogeneous* second order equation to a linear first order equation (see Problem 18 in Exercises 4.7).

EXERCISES 4.4

In Problems 1 through 6, a differential equation and a nontrivial solution f are given. Find a second linearly independent solution.

1. $y'' + 2y' - 15y = 0$; $f(x) = e^{3x}$.

2. $z'' - 3z' + 2z = 0$; $f(x) = e^x$.

3. $x^2 y'' - 2xy' - 4y = 0$, $x > 0$; $f(x) = x^{-1}$.

4. $x^2 y'' + 6xy' + 6y = 0$, $x > 0$; $f(x) = x^{-2}$.

5. $tx'' - (t + 1)x' + x = 0$, $t > 0$; $f(t) = e^t$.

6. $xy'' + (1 - 2x)y' + (x - 1)y = 0$, $x > 0$;
 $f(x) = e^x$.

7. The reduction of order procedure can be used more generally to reduce a homogeneous linear nth order equation to a homogeneous linear $(n - 1)$th order equation. For the equation

$$xy''' - xy'' + y' - y = 0 ,$$

which has $f(x) = e^x$ as a solution, use the substitution $y(x) = v(x)f(x)$ to reduce this third order equation to a homogeneous linear second order equation in the variable $w = v'$.

8. The equation

$$xy''' + (1 - x)y'' + xy' - y = 0$$

has $f(x) = x$ as a solution. Use the substitution $y(x) = v(x)f(x)$ to reduce this third order equation

to a homogeneous linear second order equation in the variable $w = v'$.

9. Given that $f(x) = x$ is a solution to

$$y'' - xy' + y = 0 \; ;$$

(a) Obtain an integral representation for a second linearly independent solution.

(b) Using the integral representation from part (a), obtain a Maclaurin series expansion for this solution.

10. Given that $f(x) = x$ is a solution to

$$(1 - x^2)y'' - 2xy' + 2y = 0 \; , \qquad -1 < x < 1 \; ;$$

(a) Obtain an integral representation for a second linearly independent solution.

(b) Using the integral representation from part (a), obtain a Maclaurin series expansion for this solution. [*Hint:* $(1 - x)^{-1} = 1 + x + x^2 + \cdots$.]

(c) Use partial fractions to evaluate the integral in part (a) and give a general solution to the differential equation.

11. The reduction of order formula (6) can also be derived from Abel's identity [formula (17) of Section 4.3]. Let $f(x)$ be a nontrivial solution to (1) and $y(x)$ a second linearly independent solution. Show that

$$\left(\frac{y}{f}\right)' = \frac{W[f, y]}{f^2} \; ,$$

and then use Abel's identity for the Wronskian $W[f, y]$ to obtain the reduction of order formula.

12. Verify, by direct substitution, that the power series (10) is a solution to the differential equation (8).

13. In quantum mechanics, the study of the Schrödinger equation for the case of a harmonic oscillator leads to a consideration of **Hermite's equation,**

$$y'' - 2xy' + \lambda y = 0 \; ,$$

where λ is a parameter. Use the reduction of order formula to obtain an integral representation of a second linearly independent solution to Hermite's equation for the given value of λ and corresponding solution $f(x)$.

(a) $\lambda = 4$, $\qquad f(x) = 1 - 2x^2$.
(b) $\lambda = 6$, $\qquad f(x) = 3x - 2x^3$.

14. In mathematical physics, many problems with spherical symmetry involve the study of **Legendre's equation,**

$$(1 - x^2)y'' - 2xy' + \lambda(\lambda + 1)y = 0 \; ,$$
$$-1 < x < 1 \; ,$$

where λ is a parameter. Use the reduction of order formula to obtain an integral representation of a second linearly independent solution to Legendre's equation for the given value of λ and corresponding solution $f(x)$.

(a) $\lambda = 1$, $\qquad f(x) = x$.
(b) $\lambda = 2$, $\qquad f(x) = 3x^2 - 1$.
(c) $\lambda = 3$, $\qquad f(x) = 5x^3 - 3x$.

15. In quantum mechanics, the study of the Schrödinger equation for the hydrogen atom leads to a consideration of **Laguerre's equation,**

$$xy'' + (1 - x)y' + \lambda y = 0 \; ,$$

where λ is a parameter. Use the reduction of order formula to obtain an integral representation of a second linearly independent solution to Laguerre's equation for the given value of λ and corresponding solution $f(x)$.

(a) $\lambda = 1$, $\qquad f(x) = x - 1$.
(b) $\lambda = 2$, $\qquad f(x) = x^2 - 4x + 2$.

16. For equation (8), we showed in Example 2 that

$$f(x) = x$$

and

$$h(x) = -1 + x^2 + \frac{x^4}{6} + \cdots$$

$$+ \frac{x^{2n}}{n!(2n - 1)} + \cdots$$

are two linearly independent solutions. Show that the solution to the initial value problem

$$(11) \qquad y'' - 2xy' + 2y = 0 \; ;$$
$$y(0) = -2 \; ,$$
$$y'(0) = -1$$

is given by $y(x) = 2h(x) - f(x)$. Obtain an approximation for the solution to the initial value problem (11) at $x = 1$ by using the first five nonzero terms of the power series for $h(x)$.

17. In Example 2, we showed that a solution to equation

(8) is given by

$$g(x) = x \int_1^x \frac{e^{t^2}}{t^2}\,dt .$$

(a) Show that this solution satisfies $g(1) = 0$ and $g'(1) = e$.

(b) Using some numerical integration algorithm (see Appendix B), approximate the solution to the initial value problem

$$y'' - 2xy' + 2y = 0 ; \quad y(1) = 0 , \quad y'(1) = e ,$$

at $x = 2$.

18. In Example 2, we showed that

$$f(x) = x , \qquad g(x) = x \int_1^x \frac{e^{t^2}}{t^2}\,dt ,$$

and

$$h(x) = -1 + x^2 + \frac{x^4}{6} + \cdots$$

$$+ \frac{x^{2n}}{n!(2n-1)} + \cdots$$

are all solutions to equation (8). Hence, by the representation theorem for homogeneous equations, we know that $h(x)$ can be expressed as a linear combination of $f(x)$ and $g(x)$. This representation can be determined as follows:

(a) Using integration by parts, show that

$$g(x) = ex - e^{x^2} + 2x \int_1^x e^{t^2}\,dt .$$

(b) Prove that

$$h(x) = g(x) + \left(2\int_0^1 e^{t^2}\,dt - e \right) f(x) .$$

19. Prove that the function v given by formula (5) is never identically constant. [*Hint:* Consider v'.]

4.5 HOMOGENEOUS LINEAR EQUATIONS WITH CONSTANT COEFFICIENTS

Let's consider the homogeneous linear second order differential equation with constant coefficients

(1) $ay'' + by' + cy = 0$,

where $a\ (\neq 0)$, b, and c are real constants. This is the type of equation we encountered in Section 4.1 for the mass-spring oscillator when no external forces were present. Because constant functions are everywhere continuous, Theorem 2 on page 165 guarantees that equation (1) has solutions defined for all x in $(-\infty, \infty)$. If we can find two linearly independent solutions to (1), say y_1 and y_2, then we can express a general solution in the form

$$y = c_1 y_1 + c_2 y_2 ,$$

where c_1 and c_2 are arbitrary constants. But how do we find y_1 and y_2? Here the "method" of *judicious guessing* will lead us to our goal. Although there is more inspiration than perspiration involved, making a good guess often serves as an effective starting point for developing rigorous procedures.

A look at equation (1) tells us that a solution of (1) must have the property that its second derivative is expressible as a linear combination of its first and zeroth derivatives. This suggests that we try to find a solution of the form $y = e^{rx}$, since derivatives of e^{rx} are just constants

times e^{rx}. If we substitute $y = e^{rx}$ into (1), we obtain

$$ar^2e^{rx} + bre^{rx} + ce^{rx} = 0 ,$$
$$e^{rx}(ar^2 + br + c) = 0 .$$

Because e^{rx} is never zero, we can divide by it to obtain

(2) $ar^2 + br + c = 0 .$

Consequently, $y = e^{rx}$ is a solution to (1) if and only if r satisfies equation (2). Equation (2) is called the **auxiliary equation** (also known as the **characteristic equation**) associated with the homogeneous equation (1).

Now the auxiliary equation is just a quadratic, and its roots are

$$r_1 = \frac{-b + \sqrt{b^2 - 4ac}}{2a} \quad \text{and} \quad r_2 = \frac{-b - \sqrt{b^2 - 4ac}}{2a} .$$

When the discriminant, $b^2 - 4ac$, is positive, the roots r_1 and r_2 are real and distinct. If $b^2 - 4ac = 0$, the roots are real and equal. And when $b^2 - 4ac < 0$, the roots are complex conjugate numbers. We consider the first two cases here; the complex case is deferred to the next section.

DISTINCT REAL ROOTS

If the auxiliary equation has distinct real roots r_1 and r_2, then e^{r_1x} and e^{r_2x} are linearly independent solutions to (1). Therefore a general solution of (1) is

$$y(x) = c_1e^{r_1x} + c_2e^{r_2x} ,$$

where c_1 and c_2 are arbitrary constants.

Observe that the Wronskian of e^{r_1x} and e^{r_2x}, evaluated at $x = 0$, is $(r_2 - r_1)$. Thus the linear independence is assured if the roots are distinct.

EXAMPLE 1 Find a general solution to

(3) $y'' + 5y' - 6y = 0 .$

SOLUTION The auxiliary equation associated with (3) is

$$r^2 + 5r - 6 = (r - 1)(r + 6) = 0 ,$$

which has the roots $r_1 = 1, r_2 = -6$. Thus $\{e^x, e^{-6x}\}$ is a fundamental solution set for equation (3), and a general solution is

$$y(x) = c_1e^x + c_2e^{-6x} . \quad \blacksquare\text{End}$$

EXAMPLE 2 Solve the initial value problem

(4) $y'' + 2y' - y = 0 ; \quad y(0) = 0 , \quad y'(0) = -1 .$

SOLUTION Here the auxiliary equation is

$$r^2 + 2r - 1 = 0 .$$

Using the quadratic formula, we find that the roots of this equation are

$$r_1 = -1 + \sqrt{2} \quad \text{and} \quad r_2 = -1 - \sqrt{2} \ .$$

Consequently, a general solution to the differential equation in (4) is

(5) $\qquad y(x) = c_1 e^{(-1+\sqrt{2})x} + c_2 e^{(-1-\sqrt{2})x} \ ,$

where c_1 and c_2 are arbitrary constants. To find the specific solution that satisfies the initial conditions given in (4), we first differentiate y as given in (5), then plug y and y' into the initial conditions of (4). This gives

$$y(0) = c_1 e^0 + c_2 e^0 \ ,$$
$$y'(0) = \left(-1 + \sqrt{2}\right)c_1 e^0 + \left(-1 - \sqrt{2}\right)c_2 e^0 \ ,$$

or

$$0 = c_1 + c_2 \ ,$$
$$-1 = \left(-1 + \sqrt{2}\right)c_1 + \left(-1 - \sqrt{2}\right)c_2 \ .$$

Solving this system yields $c_1 = -\sqrt{2}/4$ and $c_2 = \sqrt{2}/4$. Thus

$$y(x) = -\frac{\sqrt{2}}{4}e^{(-1+\sqrt{2})x} + \frac{\sqrt{2}}{4}e^{(-1-\sqrt{2})x}$$

is the desired solution. ■End

When the roots of the auxiliary equation (2) are equal; that is, $r_1 = r_2 = r$, then unlike the previous case, we get only one nontrivial solution, namely $y = e^{rx}$. Of course, constant multiples of this function are also solutions, but they are of no help in finding a second *linearly independent* solution. One way to resolve this shortcoming is to use the reduction of order method discussed in Section 4.4 (see Problem 50). Another method, based on operators, is discussed in Problems 51 and 52. Both procedures ultimately show that in the case of a repeated root r a second linearly independent solution is given simply by xe^{rx}, that is, x times the original solution e^{rx}.

REPEATED ROOT

If the auxiliary equation has a repeated root r, then two linearly independent solutions to (1) are e^{rx} and xe^{rx} and a general solution is

$$y(x) = c_1 e^{rx} + c_2 x e^{rx} \ ,$$

where c_1 and c_2 are arbitrary constants.

EXAMPLE 3 Find a general solution to

(6) $\qquad y'' + 4y' + 4y = 0 \ .$

SOLUTION The auxiliary equation for (6) is

$$r^2 + 4r + 4 = (r + 2)^2 = 0 \ .$$

Because $r = -2$ is a double root, two linearly independent solutions are e^{-2x} and xe^{-2x}. These give

$$y(x) = c_1 e^{-2x} + c_2 x e^{-2x}$$

as a general solution. ■ End

The method we have described for solving homogeneous linear second order equations with constant coefficients applies to any order (even first order) homogeneous linear equations with constant coefficients. We give a detailed treatment of such higher-order equations in Chapter 6. For now, we shall be content to illustrate the method by means of an example. We remark that, in general, a homogeneous linear nth order equation has a fundamental solution set consisting of n linearly independent solutions.

EXAMPLE 4 Find a general solution to

(7) $y''' + 3y'' - y' - 3y = 0$.

SOLUTION If we try to find solutions of the form $y = e^{rx}$, then, as with second order equations, we are led to finding roots of the auxiliary equation

(8) $r^3 + 3r^2 - r - 3 = 0$.

We observe that $r = 1$ is a root of the above equation, and dividing the polynomial on the left-hand side of (8) by $r - 1$ leads to the factorization

$$(r - 1)(r^2 + 4r + 3) = (r - 1)(r + 1)(r + 3) = 0 .$$

Hence the roots of the auxiliary equation are 1, -1, and -3, and so three solutions of (7) are e^x, e^{-x}, and e^{-3x}. The linear independence of these three exponential functions follows from Exercises 4.3, Problem 26, page 177, and indeed they form a fundamental solution set (see also Section 6.1). A general solution to (7) is then

(9) $y(x) = c_1 e^x + c_2 e^{-x} + c_3 e^{-3x}$. ■ End

In the last example, we can actually verify that (9) is a "general solution" by showing, as we did in the proof of Theorem 3 on page 169, that every solution to equation (7) can be written in the form (9) for a suitable choice of constants c_1, c_2, and c_3. For this verification, we need a generalization of the existence and uniqueness theorem to linear third order equations (see Section 6.1).

In the general case of a homogeneous linear nth order equation with constant coefficients, the auxiliary equation is a polynomial equation of order n. The fundamental theorem of algebra[†] states that a polynomial equation of order n has n roots, some of which may be complex. Unfortunately, a result from Galois theory by Abel and Ruffini[††] also states that for arbitrary polynomial equations of order greater than four, there does *not* exist a formula (like the familiar quadratic formula) involving radicals for determining the roots of the equation. Thus,

[†] See, for example, *Classical Abstract Algebra*, by Richard A. Dean (Harper & Row, Publishers, Inc., New York, 1990), Section 13.2.

[††] Ibid., Chapter 14.

when $n > 4$, we will usually have to rely on numerical techniques for approximating the roots (see Problem 49).

Cauchy-Euler Equations

There is a special class of equations with variable coefficients that can be transformed into equations with constant coefficients.

CAUCHY-EULER, OR EQUIDIMENSIONAL, EQUATIONS

Definition 4. A linear second order equation that can be expressed in the form

$$(10) \qquad ax^2 \frac{d^2y}{dx^2} + bx \frac{dy}{dx} + cy = h(x) \;,$$

where a, b, and c are constants is called a **Cauchy-Euler,** or **equidimensional, equation.**[†]

For example, the differential equation

$$3x^2 y'' - 2xy' + 7y = \sin x$$

is a Cauchy-Euler equation, whereas

$$2y'' - 3xy' + 11y = 3x - 1$$

is *not* because the coefficient of y'' is 2, which is not a constant times x^2.

To solve a homogeneous Cauchy-Euler equation (i.e., $h(x) \equiv 0$), we make the substitution $x = e^t$, which transforms equation (10) into an equation with constant coefficients. The technique is illustrated in the next example.

EXAMPLE 5 Find a general solution to

$$(11) \qquad 3x^2 \frac{d^2y}{dx^2} + 11x \frac{dy}{dx} - 3y = 0 \;, \qquad x > 0 \;.$$

SOLUTION We make the substitution $x = e^t$ to transform (11) into an equation in which the new independent variable is t. Because $x = e^t$, it follows by the chain rule that

$$\frac{dy}{dt} = \frac{dy}{dx}\frac{dx}{dt} = \frac{dy}{dx} e^t = x \frac{dy}{dx} \;,$$

and hence

$$(12) \qquad x \frac{dy}{dx} = \frac{dy}{dt} \;.$$

[†]*Historical Footnote:* Although work on this equation was published by Leonhard Euler in 1769 and later by Augustin Cauchy, its solution was known to John Bernoulli prior to 1700.

Differentiating (12) with respect to t, we find from the product rule that

$$\frac{d^2y}{dt^2} = \frac{d}{dt}\left(x\frac{dy}{dx}\right) = \frac{dx}{dt}\frac{dy}{dx} + x\frac{d}{dt}\left(\frac{dy}{dx}\right)$$

$$= \frac{dy}{dt} + x\frac{d^2y}{dx^2}\frac{dx}{dt} = \frac{dy}{dt} + x\frac{d^2y}{dx^2}e^t$$

$$= \frac{dy}{dt} + x^2\frac{d^2y}{dx^2} \ .$$

Thus

(13) $$x^2\frac{d^2y}{dx^2} = \frac{d^2y}{dt^2} - \frac{dy}{dt} \ .$$

Substituting into (11) the expressions for $x\,dy/dx$ and $x^2\,d^2y/dx^2$ given in (12) and (13) yields

$$3\left(\frac{d^2y}{dt^2} - \frac{dy}{dt}\right) + 11\frac{dy}{dt} - 3y = 0 \ ,$$

(14) $$3\frac{d^2y}{dt^2} + 8\frac{dy}{dt} - 3y = 0 \ .$$

The auxiliary equation associated with (14) is

$$3r^2 + 8r - 3 = (3r - 1)(r + 3) = 0 \ .$$

The roots of this equation are $1/3$ and -3, so the general solution to (14) is

$$y(t) = c_1e^{t/3} + c_2e^{-3t} = c_1(e^t)^{1/3} + c_2(e^t)^{-3} \ .$$

Expressing y in terms of the original variable x, we find

(15) $$y(x) = c_1x^{1/3} + c_2x^{-3} \quad \text{for} \quad x > 0. \ \blacksquare \text{ End}$$

It turns out that the solution (15) for the special equation (11) is valid for all $x \neq 0$ even though it was derived only for $x > 0$. In general, it can be shown that if $y(x)$ is a solution to a homogeneous Cauchy-Euler equation for $x > 0$, then $\phi(x) := y(-x)$ is a solution for $x < 0$ (see Problem 43).

Using formulas (12) and (13), the reader can verify that the substitution $x = e^t$ transforms the general Cauchy-Euler equation

(16) $$ax^2\frac{d^2y}{dx^2} + bx\frac{dy}{dx} + cy = h(x) \ , \qquad x > 0$$

into the constant coefficient equation

(17) $$a\frac{d^2y}{dt^2} + (b - a)\frac{dy}{dt} + cy = h(e^t) \ .$$

A method similar to the one discussed here for second order Cauchy-Euler equations can

be derived for higher-order Cauchy-Euler equations of the form

$$a_n x^n y^{(n)}(x) + a_{n-1} x^{n-1} y^{(n-1)}(x) + \cdots + a_0 y(x) = h(x) \, ,$$

where $a_n, a_{n-1}, \ldots, a_0$ are real constants (see Problems 46–48).

Another method for finding solutions to a homogeneous Cauchy-Euler equation involves guessing a solution of the form $y = x^r$. Doing this renders each term xy', $x^2 y''$, \ldots as a constant times x^r, so the result is an auxiliary equation for r. This method is discussed in Section 8.5. However, there is often an advantage in making the substitution $x = e^t$ when trying to solve a *nonhomogeneous* Cauchy-Euler equation by the methods of undetermined coefficients (see Section 4.8) or Laplace transforms (see Chapter 7).

EXERCISES 4.5

In Problems 1 through 12, find a general solution to the given differential equation.

1. $y'' + 5y' + 6y = 0$.

2. $y'' - y' - 2y = 0$.

3. $y'' + 8y' + 16y = 0$.

4. $y'' + 6y' + 9y = 0$.

5. $z'' + z' - z = 0$.

6. $y'' - 5y' + 6y = 0$.

7. $2u'' + 7u' - 4u = 0$.

8. $6y'' + y' - 2y = 0$.

9. $y'' - y' - 11y = 0$.

10. $4y'' - 4y' + y = 0$.

11. $4w'' + 20w' + 25w = 0$. **12.** $3y'' + 11y' - 7y = 0$.

In Problems 13 through 20, solve the given initial value problem.

13. $y'' + 2y' - 8y = 0$; $y(0) = 3$, $y'(0) = -12$.

14. $y'' + y' = 0$; $y(0) = 2$, $y'(0) = 1$.

15. $y'' + 2y' + y = 0$; $y(0) = 1$, $y'(0) = -3$.

16. $y'' - 4y' + 3y = 0$; $y(0) = 1$, $y'(0) = 1/3$.

17. $z'' - 2z' - 2z = 0$; $z(0) = 0$, $z'(0) = 3$.

18. $y'' - 6y' + 9y = 0$; $y(0) = 2$, $y'(0) = 25/3$.

19. $y'' - 4y' - 5y = 0$; $y(-1) = 3$, $y'(-1) = 9$.

20. $y'' - 4y' + 4y = 0$; $y(1) = 1$, $y'(1) = 1$.

21. First Order Constant Coefficient Equations.

 (a) Substituting $y = e^{rx}$, find the auxiliary equation for the first order linear equation

$$ay' + by = 0 \, ,$$

 where a, b, are constants with $a \neq 0$.

 (b) Use the result of part (a) to find the general solution.

In Problems 22 through 25, use the method described in Problem 21 to find a general solution to the given equation.

22. $3y' - 7y = 0$.

23. $5y' + 4y = 0$.

24. $3z' + 11z = 0$.

25. $6w' - 13w = 0$.

In Problems 26 through 31, find a general solution to the given third order differential equation.

26. $y''' + y'' - 6y' + 4y = 0$.

27. $y''' - 6y'' - y' + 6y = 0$.

28. $z''' + 2z'' - 4z' - 8z = 0$.

29. $y''' - 7y'' + 7y' + 15y = 0$.

30. $y''' + 3y'' - 4y' - 12y = 0$.

31. $w''' + w'' - 4w' + 2w = 0$.

32. Solve the initial value problem

$$y''' - y' = 0 \, ;$$
$$y(0) = 2 \, , \quad y'(0) = 3 \, , \quad y''(0) = -1 \, .$$

In Problems 33 through 38, use the method discussed under "Cauchy-Euler Equations" to find a general solution to the given equation for $x > 0$.

33. $x^2 y''(x) + 7xy'(x) - 7y(x) = 0$.

34. $x^2 \dfrac{d^2 y}{dx^2} + 2x \dfrac{dy}{dx} - 6y = 0$.

35. $\dfrac{d^2 w}{dx^2} + \dfrac{6}{x}\dfrac{dw}{dx} + \dfrac{4}{x^2}w = 0$.

36. $x^2 \dfrac{d^2 z}{dx^2} + 5x \dfrac{dz}{dx} + 4z = 0$.

37. $9x^2y''(x) + 15xy'(x) + y(x) = 0$.

38. $x^2y''(x) - 3xy'(x) + 4y(x) = 0$.

In Problems 39 and 40, use the method discussed under "Cauchy-Euler Equations" to solve the given initial value problem.

39. $x^2y''(x) - 4xy'(x) + 4y(x) = 0$;
 $y(1) = -2$, $y'(1) = -11$.

40. $x^2y''(x) + 7xy'(x) + 5y(x) = 0$;
 $y(1) = -1$, $y'(1) = 13$.

In Problems 41 and 42, devise a modification of the method discussed under "Cauchy-Euler Equations" to find a general solution to the given equation.

41. $(x - 2)^2y''(x) - 7(x - 2)y'(x) + 7y(x) = 0$,
 $x > 2$.

42. $(x + 1)^2y''(x) + 10(x + 1)y'(x) + 14y(x) = 0$,
 $x > -1$.

43. Let $y(x)$ be a solution for $x > 0$ to the *homogeneous* Cauchy-Euler equation $ax^2y'' + bxy' + cy = 0$. Show that $\phi(x) := y(-x)$ is a solution to the equation for $x < 0$. Is the analogous statement true for the *non*homogeneous Cauchy-Euler equation?

In Problems 44 and 45, use the result of Problem 43 to find a general solution to the given differential equation for x < 0.

44. $x^2y''(x) + xy'(x) - 2y(x) = 0$.

45. $2x^2y''(x) + 7xy'(x) + 2y(x) = 0$.

46. Use the substitution $x = e^t$ to show that the third order Cauchy-Euler equation

$$ax^3y'''(x) + bx^2y''(x) + cxy'(x) + dy(x) = 0 ,$$
$$x > 0$$

is equivalent to the constant coefficient equation

$$ay'''(t) + (b - 3a)y''(t)$$
$$+ (2a - b + c)y'(t) + dy(t) = 0 .$$

In Problems 47 and 48, use the result of Problem 46 to find a general solution to the given differential equation for x > 0.

47. $x^3y'''(x) - 2x^2y''(x) + 3xy'(x) - 3y(x) = 0$.

48. $x^3y'''(x) + x^2y''(x) - 8xy'(x) - 4y(x) = 0$.

49. By using Newton's method or some other numerical procedure to approximate the roots of the auxiliary equation, find general solutions to the following equations:

(a) $3y''' + 18y'' + 13y' - 19y = 0$.
(b) $y^{iv} - 5y'' + 5y = 0$.
(c) $y^v - 3y^{iv} - 5y''' + 15y'' + 4y' - 12y = 0$.

50. When $b^2 - 4ac = 0$, the roots of the auxiliary equation (2) are equal; that is, $r_1 = r_2 = r = -b/2a$. Hence $f(x) = e^{(-b/2a)x}$ is a solution to (1). Use the reduction of order formula (6) of Section 4.4 to derive $xe^{(-b/2a)x}$ as a second linearly independent solution to (1).

51. Suppose the auxiliary equation (2) has a repeated root r. Then, on dividing by a, equation (1) can be written in the equivalent form

$$(D - r)^2[y] = y'' - 2ry' + r^2y = 0 ,$$

where D is the differential operator d/dx.

(a) Show that for any twice-differentiable function u,

$$(D - r)^2[e^{rx}u] = e^{rx}D^2u ,$$

and hence that the problem of solving $(D - r)^2[y] = 0$ reduces to that of solving $D^2u = 0$, where $y = e^{rx}u$.

(b) Use the result of part (a) to show that e^{rx} and xe^{rx} are linearly independent solutions of $(D - r)^2[y] = 0$.

52. Let $L[y] := ay'' + by' + cy$, where $b^2 - 4ac = 0$.

(a) Show that $L[e^{rx}] = a(r - r_0)^2e^{rx}$, where $r_0 = -b/2a$.

(b) Show that $L[(\partial/\partial r)e^{rx}] = (\partial/\partial r)L[e^{rx}]$, and thus

$$L[xe^{rx}] = 2a(r - r_0)e^{rx} + ax(r - r_0)^2e^{rx} .$$

(c) Use the results of parts (a) and (b) to show that e^{r_0x} and xe^{r_0x} are linearly independent solutions to (1).

53. The suspension in an automobile can be modeled as a vibrating spring with damping due to the shock absorbers (see Figure 4.6). This leads to the equation

$$mx''(t) + bx'(t) + kx(t) = 0 ,$$

where m is the mass of the automobile, b is the damping constant of the shocks, k is the spring constant, and $x(t)$ is the vertical displacement of the automobile at time t. In the next section, we will find that the solution of a homogeneous linear second order equation with constant coefficients oscillates when the auxiliary equation has complex roots. If

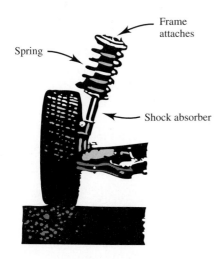

Frame
attaches

Spring

Shock absorber

Figure 4.6 Diagram of an automobile suspension

the mass of an automobile is 1000 kg and the spring constant is 3000 kg/sec², determine the minimum value for the damping constant in kilograms per second that will provide a smooth, oscillation-free (theoretically!) ride. If we replace the springs with heavy duty ones having twice the spring constant, how does this minimum value for b change?

54. Consider the differential equation $y'' - s^2 y = 0$, where s is a positive constant.

 (a) Show that a general solution can be written as $c_1 e^{sx} + c_2 e^{-sx}$.

 (b) Show that the form $d_1 \cosh sx + d_2 \sinh sx$ is also a general solution.

 (c) Use each of these general solution formats to solve the initial value problem $y'' - s^2 y = 0$, $y(0) = a$, $y'(0) = b$. Which is more convenient?

4.6 AUXILIARY EQUATIONS WITH COMPLEX ROOTS

The *simple harmonic equation* $y'' + y = 0$, so called because of its relation to the fundamental vibration of a musical tone, has as solutions $y_1(x) = \cos x$ and $y_2(x) = \sin x$. Notice, however, that the auxiliary equation associated with the harmonic equation is $r^2 + 1 = 0$, which has imaginary roots $r = \pm i$, where i denotes $\sqrt{-1}$.[†] In the previous section, we expressed the solutions to a linear second order equation with constant coefficients in terms of exponential functions. It would appear, then, that one might be able to attribute a meaning to the forms e^{ix} and e^{-ix} and that these "functions" should be related to $\cos x$ and $\sin x$. This matchup is accomplished by Euler's formula, which is discussed in this section.

When $b^2 - 4ac < 0$, the roots of the auxiliary equation

(1) $ar^2 + br + c = 0$

associated with the homogeneous equation

(2) $ay'' + by' + cy = 0$

are the complex conjugate numbers

$$r_1 = \alpha + i\beta \quad \text{and} \quad r_2 = \alpha - i\beta \quad \left(i = \sqrt{-1} \right) ,$$

where α, β are the real numbers

(3) $\alpha = -\dfrac{b}{2a} \quad \text{and} \quad \beta = \dfrac{\sqrt{4ac - b^2}}{2a} .$

[†]Electrical engineers frequently use the symbol j to denote $\sqrt{-1}$.

As in the previous section, we would like to assert that the functions $e^{r_1 x}$ and $e^{r_2 x}$ are solutions to the equation (2). This is, in fact, the case, but before we can proceed, we need to address some fundamental questions. For example, if $r_1 = \alpha + i\beta$ is a complex number, what do we mean by the expression $e^{(\alpha + i\beta)x}$? If we assume that the law of exponents applies to complex numbers, then

(4) $\qquad e^{(\alpha + i\beta)x} = e^{\alpha x + i\beta x} = e^{\alpha x} e^{i\beta x}$.

We now need only clarify the meaning of $e^{i\beta x}$.

For this purpose, let's assume the Maclaurin series for e^z is the same for complex numbers z as it is for real numbers. Observing that $i^2 = -1$, then for θ real we have

$$e^{i\theta} = 1 + (i\theta) + \frac{(i\theta)^2}{2!} + \cdots + \frac{(i\theta)^n}{n!} + \cdots$$

$$= 1 + i\theta - \frac{\theta^2}{2!} - \frac{i\theta^3}{3!} + \frac{\theta^4}{4!} + \frac{i\theta^5}{5!} + \cdots$$

$$= \left(1 - \frac{\theta^2}{2!} + \frac{\theta^4}{4!} + \cdots\right) + i\left(\theta - \frac{\theta^3}{3!} + \frac{\theta^5}{5!} + \cdots\right) .$$

Now recall the Maclaurin series for $\cos\theta$ and $\sin\theta$:

$$\cos\theta = 1 - \frac{\theta^2}{2!} + \frac{\theta^4}{4!} + \cdots ,$$

$$\sin\theta = \theta - \frac{\theta^3}{3!} + \frac{\theta^5}{5!} + \cdots .$$

Recognizing these expansions in the proposed series for $e^{i\theta}$, we make the identification

(5) $\qquad e^{i\theta} = \cos\theta + i\sin\theta$,

which is known as **Euler's formula.**[†]

When Euler's formula (with $\theta = \beta x$) is used in equation (4), we find

(6) $\qquad e^{(\alpha + i\beta)x} = e^{\alpha x}(\cos\beta x + i\sin\beta x)$,

which expresses the complex function $e^{(\alpha + i\beta)x}$ in terms of familiar real functions. Having made sense out of $e^{(\alpha + i\beta)x}$, we can now show (see Problem 30) that

(7) $\qquad \dfrac{d}{dx} e^{(\alpha + i\beta)x} = (\alpha + i\beta)e^{(\alpha + i\beta)x}$,

and, with the choices of α and β as given in (3), the complex function $e^{(\alpha + i\beta)x}$ is indeed a solution to equation (2). In general, if $z(x)$ is a complex-valued function of the real variable x, we can write

$$z(x) = u(x) + iv(x) ,$$

[†]*Historical Footnote:* This formula first appeared in Leonhard Euler's monumental two-volume *Introductio in Analysin Infinitorum* (1748).

where $u(x)$ and $v(x)$ are real-valued functions. The derivatives of $z(x)$ are then given by

$$\frac{dz}{dx} = \frac{du}{dx} + i\frac{dv}{dx} , \qquad \frac{d^2z}{dx^2} = \frac{d^2u}{dx^2} + i\frac{d^2v}{dx^2} .$$

With the following lemma, we show that the complex-valued solution $e^{(\alpha+i\beta)x}$ gives rise to two linearly independent *real-valued* solutions.

REAL SOLUTIONS DERIVED FROM COMPLEX SOLUTIONS

Lemma 2. Let $z(x) = u(x) + iv(x)$ be a solution to equation (2), where a, b, and c are real numbers. Then, the real part $u(x)$ and the imaginary part $v(x)$ are real-valued solutions of (2).[†]

Proof. By assumption, $az'' + bz' + cz = 0$, and hence

$$a(u'' + iv'') + b(u' + iv') + c(u + iv) = 0 ,$$
$$(au'' + bu' + cu) + i(av'' + bv' + cv) = 0 .$$

But a complex number is zero if and only if its real and imaginary parts are both zero. Thus we must have

$$au'' + bu' + cu = 0 \quad \text{and} \quad av'' + bv' + cv = 0 ,$$

which means that both $u(x)$ and $v(x)$ are real-valued solutions of (2). ■ End

When we apply Lemma 2 to the solution

$$e^{(\alpha+i\beta)x} = e^{\alpha x}\cos \beta x + ie^{\alpha x}\sin \beta x ,$$

we obtain the following.

COMPLEX CONJUGATE ROOTS

If the auxiliary equation has complex conjugate roots $\alpha \pm i\beta$, then two linearly independent solutions to (2) are

$$e^{\alpha x}\cos \beta x \quad \text{and} \quad e^{\alpha x}\sin \beta x ,$$

and a general solution is

$$(8) \qquad y(x) = c_1 e^{\alpha x}\cos \beta x + c_2 e^{\alpha x}\sin \beta x ,$$

where c_1 and c_2 are arbitrary constants.

[†]It will be clear from the proof that this property holds for any linear homogeneous differential equation having real-valued coefficients.

In the preceding discussion, we glossed over some important details concerning complex numbers and complex-valued functions. In particular, further analysis is required to justify the use of the law of exponents, Euler's formula, and even the fact that the derivative of e^{rx} is re^{rx} when r is a complex constant.[†] Readers who feel uneasy about our conclusions can readily check the answer by substituting it into (2).

The reader may also be wondering what would have happened if we had worked with the function $e^{(\alpha - i\beta)x}$ instead of $e^{(\alpha + i\beta)x}$. We leave it as an exercise to verify that $e^{(\alpha - i\beta)x}$ gives rise to the same general solution (8).

EXAMPLE 1 Find a general solution to

(9) $y'' + 2y' + 4y = 0$.

SOLUTION The auxiliary equation is

$$r^2 + 2r + 4 = 0 \; ,$$

which has roots

$$r = \frac{-2 \pm \sqrt{4 - 16}}{2} = \frac{-2 \pm \sqrt{-12}}{2} = -1 \pm i\sqrt{3} \; .$$

Hence with $\alpha = -1$, $\beta = \sqrt{3}$, a general solution for (9) is

$$y(x) = c_1 e^{-x} \cos\left(\sqrt{3}x\right) + c_2 e^{-x} \sin\left(\sqrt{3}x\right). \; \blacksquare \text{End}$$

When the auxiliary equation has complex conjugate roots, the (real) solutions oscillate between positive and negative values. This type of behavior is observed in vibrating springs.

EXAMPLE 2 In Section 4.1 we discussed the mechanics of the mass-spring oscillator (Figure 4.1, page 156), and we saw how Newton's second law implies that the position $y(t)$ of the mass m is governed by the second order differential equation

(10) $my''(t) + by'(t) + ky(t) = 0$,

where the terms are physically identified as

$$\left[\text{inertia}\right]y'' + \left[\text{damping}\right]y' + \left[\text{stiffness}\right]y = 0 \; .$$

Determine the equation of motion for a spring system when $m = 36$ kg, $b = 12$ kg/sec (which is equivalent to 12 N-sec/m), $k = 37$ kg/sec^2, $y(0) = 0.7$ m, and $y'(0) = 0.1$ m/sec. Also find $y(10)$, the displacement after 10 sec.

SOLUTION The equation of motion is given by $y(t)$, the solution of the initial value problem for the specified values of m, b, k, $y(0)$, and $y'(0)$. That is, we seek the solution to

(11) $36y'' + 12y' + 37y = 0$, $y(0) = 0.7$, $y'(0) = 0.1$.

[†]For a detailed treatment of these topics see, for example, *Fundamentals of Complex Analysis,* Second Edition, by E. B. Saff and A. D. Snider (Prentice-Hall, Inc., Englewood Cliffs, New Jersey, 1993).

The auxiliary equation for (11) is

$$36r^2 + 12r + 37 = 0 \ ,$$

which has roots

$$r = \frac{-12 \pm \sqrt{144 - 4(36)(37)}}{72} = \frac{-12 \pm 12\sqrt{1-37}}{72} = -\frac{1}{6} \pm i \ .$$

Hence with $\alpha = -1/6$, $\beta = 1$, the displacement $y(t)$ can be expressed in the form

(12) $y(t) = c_1 e^{-t/6}\cos t + c_2 e^{-t/6}\sin t \ .$

We can find c_1 and c_2 by substituting $y(t)$ and $y'(t)$ into the initial conditions given in (11). Differentiating (12), we get a formula for $y'(t)$:

$$y'(t) = \left(-\frac{c_1}{6} + c_2\right)e^{-t/6}\cos t + \left(-c_1 - \frac{c_2}{6}\right)e^{-t/6}\sin t \ .$$

Substituting into the initial conditions now results in the system

$$c_1 = 0.7 \ ,$$
$$-\frac{c_1}{6} + c_2 = 0.1 \ .$$

Upon solving, we find $c_1 = 0.7$ and $c_2 = 1.3/6$. With these values, the equation of motion becomes

$$y(t) = 0.7e^{-t/6}\cos t + \frac{1.3}{6}e^{-t/6}\sin t \ ,$$

and

$$y(10) = 0.7e^{-5/3}\cos 10 + \frac{1.3}{6}e^{-5/3}\sin 10 \approx -0.13 \text{ m.} \quad \blacksquare\text{End}$$

From Example 2 we see that *any* second order constant coefficient differential equation $ay'' + by' + cy = 0$ can be interpreted as describing a mass-spring system with mass a, damping coefficient b, spring stiffness c, and displacement y, if these constants make sense physically; that is, if a is positive and b and c are nonnegative. From the discussion in Section 4.1, then, we expect on physical grounds to see damped oscillatory solutions in such a case. This is consistent with the display in equation (8), which takes the form $c_1 e^{\alpha t}\cos \beta t + c_2 e^{\alpha t}\sin \beta t$ in this context (where the independent variable is time t). With $a = m$ and $c = k$, the exponential decay rate α equals $-b/(2m)$, and the angular frequency β equals $\sqrt{4mk-b^2}/(2m)$, by equation (3).

It is a little surprising, then, that the solutions to the equation $y'' + 4y' + 4y = 0$ do *not* oscillate; the general solution was shown in Example 3 of Section 4.5 (page 185) to be $c_1 e^{-2t} + c_2 te^{-2t}$. The physical significance of this is simply that when the damping coefficient b is too high, the resulting friction prevents the mass from oscillating. Rather than overshoot the spring's equilibrium point, it merely settles in lazily. This could happen if a light mass on a weak spring were submerged in a viscous fluid.

From the above formula for the oscillation frequency β, we can see that the oscillations will not occur for $b > \sqrt{4mk}$. This *overdamping* phenomenon is discussed in more detail in Section 4.11.

It is extremely enlightening to contemplate the predictions of the mass-spring analogy when the coefficients b and c in the equation $ay'' + by' + cy = 0$ are *negative*.

EXAMPLE 3 Interpret the equation

(13) $36y'' - 12y' + 37y = 0$

in terms of the mass-spring system.

SOLUTION Equation (13) is a minor alteration of equation (11) in Example 2; the auxiliary equation $36r^2 - 12r + 37$ has roots $r = (+)\frac{1}{6} \pm i$. Thus its general solution becomes

(14) $y(t) = c_1 e^{+t/6}\cos t + c_2 e^{+t/6}\sin t$.

Comparing equation (13) with the mass-spring model

(15) $\left[\mathbf{inertia}\right]y'' + \left[\mathbf{damping}\right]y' + \left[\mathbf{stiffness}\right]y = 0$,

we have to envision a *negative* damping coefficient $b = -12$, giving rise to a friction force $F_{\text{friction}} = -by'$, which *imparts* energy to the system instead of draining it. The increase in energy over time must then reveal itself in oscillations of ever-greater amplitude—precisely in accordance with formula (14), for which a typical graph is drawn in Figure 4.7. ■End

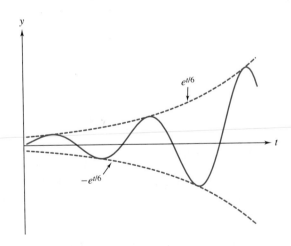

Figure 4.7 Solution graph for Example 3

EXAMPLE 4 Interpret the equation

(16) $y'' + 5y' - 6y = 0$

in terms of the mass-spring system.

SOLUTION Comparing the given equation with (15), we have to envision a spring with a *negative* stiffness $k = -6$. What does this mean? As the mass is moved away from the spring's equilibrium point,

the spring *repels* the mass farther with a force $F_{\text{spring}} = -ky$ that intensifies as the displacement increases. Clearly the spring must "exile" the mass to (plus or minus) infinity, and we expect all solutions $y(t)$ to approach $\pm\infty$ as t increases (except for the equilibrium solution $y(t) \equiv 0$).

In fact, in Example 1 of Section 4.5 we showed the general solution to equation (16) to be

(17) $c_1 e^t + c_2 e^{-6t}$.

Indeed, if we examine the solutions $y(t)$ that start with a unit displacement $y(0) = 1$ and velocity $y'(0) = v_0$, we find

(18) $y(t) = \dfrac{6 + v_0}{7} e^t + \dfrac{1 - v_0}{7} e^{-6t}$,

and the plots in Figure 4.8 confirm our prediction that all (nonequilibrium) solutions diverge— except for the one with $v_0 = -6$.

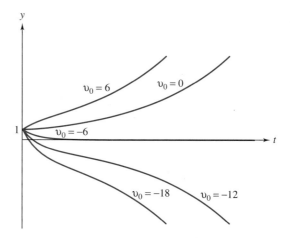

Figure 4.8 Solution graphs for Example 4

What is the physical significance of this isolated bounded solution? Evidently, if the mass is given an initial inwardly directed velocity of -6, it has barely enough energy to overcome the effect of the spring banishing it to $\pm\infty$ but not enough energy to cross the equilibrium point (and get pushed to $-\infty$). So it asymptotically approaches the (extremely delicate) equilibrium position $y = 0$. ■End

In Section 4.10, we will see that taking further liberties with the mass-spring interpretation enables us to predict qualitative features of more complicated equations.

Throughout this section we have assumed that the coefficients a, b, and c in the differential equation were real numbers. If we now allow them to be *complex* constants, then the roots r_1, r_2 of the auxiliary equation (1) are, in general, also complex but not necessarily conjugates of each other. When $r_1 \neq r_2$, a general solution to equation (2) still has the form

$$y(x) = c_1 e^{r_1 x} + c_2 e^{r_2 x} ,$$

but c_1 and c_2 are now arbitrary complex-valued constants.

We also remark that a complex differential equation can be regarded as a system of two real differential equations since we can always work separately with its real and imaginary parts. Systems are discussed in Chapters 5 and 9.

EXERCISES 4.6

In Problems 1 through 8, the auxiliary equation for the given differential equation has complex roots. Find a general solution.

1. $y'' + 9y = 0$. **2.** $y'' + y = 0$.

3. $z'' - 6z' + 10z = 0$. **4.** $y'' - 10y' + 26y = 0$.

5. $w'' + 4w' + 6w = 0$. **6.** $y'' - 4y' + 7y = 0$.

7. $4y'' - 4y' + 26y = 0$. **8.** $4y'' + 4y' + 6y = 0$.

In Problems 9 through 20, find a general solution.

9. $y'' - 8y' + 7y = 0$. **10.** $y'' + 4y' + 8y = 0$.

11. $z'' + 10z' + 25z = 0$. **12.** $u'' + 7u = 0$.

13. $y'' + 2y' + 5y = 0$. **14.** $y'' - 2y' + 26y = 0$.

15. $y'' + 10y' + 41y = 0$. **16.** $y'' - 3y' - 11y = 0$.

17. $y'' - y' + 7y = 0$. **18.** $2y'' + 13y' - 7y = 0$.

T#3 **19.** $y''' + y'' + 3y' - 5y = 0$. **20.** $y''' - y'' + 2y = 0$.

In Problems 21 through 27, solve the given initial value problem.

21. $y'' + 2y' + 2y = 0$; $y(0) = 2$, $y'(0) = 1$.

22. $y'' + 2y' + 17y = 0$; $y(0) = 1$, $y'(0) = -1$.

23. $w'' - 4w' + 2w = 0$; $w(0) = 0$, $w'(0) = 1$.

24. $y'' + 9y = 0$; $y(0) = 1$, $y'(0) = 1$.

25. $y'' - 2y' + 2y = 0$; $y(\pi) = e^{\pi}$, $y'(\pi) = 0$.

26. $y'' - 2y' + y = 0$; $y(0) = 1$, $y'(0) = -2$.

27. $y''' - 4y'' + 7y' - 6y = 0$; $y(0) = 1$, $y'(0) = 0$, $y''(0) = 0$.

28. To see the effect of changing the parameter b in the initial value problem

$$y'' + by' + 4y = 0 ; \qquad y(0) = 1 , \qquad y'(0) = 0 ,$$

solve the problem for $b = 5$, 4, and 2 and sketch the solutions.

29. Find a general solution to the following higher-order equations.

(a) $y''' - y'' + y' + 3y = 0$.

(b) $y''' + 2y'' + 5y' - 26y = 0$.

(c) $y^{iv} + 13y'' + 36y = 0$.

30. Using the representation for $e^{(\alpha + i\beta)x}$ in (6), verify the differentiation formula (7).

31. Using the mass-spring analogy, predict the behavior as $t \to +\infty$ of the solution to the given initial value problem. Then confirm your prediction by actually solving the problem.

(a) $y'' + 16y = 0$; $y(0) = 2$, $y'(0) = 0$.

(b) $y'' + 100y' + y = 0$; $y(0) = 1$, $y'(0) = 0$.

(c) $y'' - 6y' + 8y = 0$; $y(0) = 1$, $y'(0) = 0$.

(d) $y'' + 2y' - 3y = 0$; $y(0) = -2$, $y'(0) = 0$.

(e) $y'' - y' - 6y = 0$; $y(0) = 1$, $y'(0) = 1$.

In Problems 32 through 35, find a general solution to the given Cauchy-Euler equation for $x > 0$ by using the substitution $x = e^t$ as discussed in Section 4.5.

32. $y''(x) - \dfrac{1}{x}y'(x) + \dfrac{5}{x^2}y(x) = 0$.

33. $x^2 y''(x) - 3xy'(x) + 6y(x) = 0$.

34. $x^2 y''(x) + 9xy'(x) + 17y(x) = 0$.

35. $x^2 y''(x) + 3xy'(x) + 5y(x) = 0$.

36. Vibrating Spring without Damping. A vibrating spring without damping can be modeled by the initial value problem (10) in Example 2 by taking $b = 0$.

(a) If $m = 10$ kg, $k = 250$ kg/sec^2, $y(0) = 0.3$ m, and $y'(0) = -0.1$ m/sec, find the equation of motion for this undamped vibrating spring.

(b) When the equation of motion is of the form displayed in (8), the motion is said to be **oscillatory** with **frequency** $\beta/2\pi$. Find the frequency of oscillation for the spring system of part (a).

37. Vibrating Spring with Damping. Using the model for a vibrating spring with damping discussed in Example 2:

(a) Find the equation of motion for the vibrating spring with damping if $m = 10$ kg, $b = 60$ kg/sec, $k = 250$ kg/sec^2, $y(0) = 0.3$ m, and $y'(0) = -0.1$ m/sec.

(b) Find the frequency of oscillation for the spring system of part (a). [*Hint:* See the definition of frequency given in Problem 36(b).]

(c) Compare the results of Problems 36 and 37 and determine what effect the damping has on the frequency of oscillation. What other effects does it have on the solution?

38. *RLC* Series Circuit. In the study of an electrical circuit consisting of a resistor, capacitor, inductor, and an electromotive force (see Figure 4.9), we are led to an initial value problem of the form

$$(19) \qquad L\frac{dI}{dt} + RI + \frac{q}{C} = E(t) \; ;$$

$$q(0) = q_0 \, ,$$
$$I(0) = I_0 \, ,$$

where L is the inductance in henrys, R is the resistance in ohms, C is the capacitance in farads, $E(t)$ is the electromotive force in volts, $q(t)$ is the charge in coulombs on the capacitor at time t, and $I = dq/dt$ is the current in amperes. Find the current at time t if the charge on the capacitor is initially zero, the initial current is zero, $L = 10$ henrys, $R = 20$ ohms, $C = (6260)^{-1}$ farads, and $E(t) = 100$ volts. [*Hint:* Differentiate both sides of the differential equation in (19) to obtain a homogeneous linear second order equation for $I(t)$. Then use (19) to determine dI/dt at $t = 0$.]

Figure 4.9 *RLC* series circuit

39. Swinging Door. The motion of a swinging door with an adjustment screw that controls the amount of friction on the hinges is governed by the initial value problem

$$I\theta'' + b\theta' + k\theta = 0 \; ; \quad \theta(0) = \theta_0 \, , \quad \theta'(0) = v_0 \, ,$$

where θ is the angle that the door is open, I is the moment of inertia of the door about its hinges, $b > 0$ is a damping constant that varies with the amount of friction on the door, $k > 0$ is the spring constant associated with the swinging door, θ_0 is the initial angle that the door is opened, and v_0 is the initial angular velocity imparted to the door (see Figure 4.10). If I and k are fixed, determine for which values of b the door will *not* continually swing back and forth when closing.

Figure 4.10 Top view of swinging door

40. Although the real general solution form (8) is convenient, it is also possible to use the form

$$(20) \qquad d_1 e^{(\alpha + i\beta)x} + d_2 e^{(\alpha - i\beta)x}$$

to solve initial value problems. The coefficients d_1 and d_2 are complex constants.

(a) Use the form (20) to solve Problem 21. Verify that your form is equivalent to the one derived using (8).

(b) Show that, in general, d_1 and d_2 in (20) must be complex conjugates in order that the solution be real.

41. The auxiliary equations for the following differential equations have repeated complex roots. Adapt the "repeated root" procedure of Section 4.5 to find their general solutions:

(a) $y^{iv} + 2y'' + y = 0$.

(b) $y^{iv} + 4y''' + 12y'' + 16y' + 16y = 0$. [*Hint:* The auxiliary equation is $(r^2 + 2r + 4)^2 = 0$.]

4.7 SUPERPOSITION AND NONHOMOGENEOUS EQUATIONS

The differential operator L defined by

(1) $L[y](x) := y'' + p(x)y' + q(x)y$

can be viewed as a "black box" with input the function $y(x)$ and output the function on the right-hand side of (1) (see Figure 4.11).

Figure 4.11 Black box diagram for L

Suppose the input functions $y_1(x)$ and $y_2(x)$ yield, respectively, the output functions $g_1(x)$ and $g_2(x)$; that is,

$$L[y_1](x) = g_1(x) , \quad L[y_2](x) = g_2(x) .$$

Then, since L is a linear operator, an input consisting of the linear combination $c_1 y_1(x) + c_2 y_2(x)$ produces the output $c_1 g_1(x) + c_2 g_2(x)$ in the same combination. Stated in terms of solutions to differential equations, we have the following result.

SUPERPOSITION PRINCIPLE

Theorem 5. Let y_1 be a solution of the differential equation

$$L[y](x) = g_1(x)$$

and let y_2 be a solution of

$$L[y](x) = g_2(x) ,$$

where L is a linear differential operator. Then for any constants c_1 and c_2, the function $c_1 y_1 + c_2 y_2$ is a solution to the differential equation

$$L[y](x) = c_1 g_1(x) + c_2 g_2(x) .$$

EXAMPLE 1 Given that $y_1(x) = -x/3 - 2/9$ is a solution to

$$y'' + 2y' - 3y = x$$

and $y_2(x) = e^{2x}/5$ is a solution to

$$y'' + 2y' - 3y = e^{2x} ,$$

find a solution to

(2) $y'' + 2y' - 3y = 4x - 5e^{2x}$.

SOLUTION Let $L[y] := y'' + 2y' - 3y$. We are given that

$$L[y_1](x) = g_1(x) = x \quad \text{and} \quad L[y_2](x) = g_2(x) = e^{2x} .$$

Since we can express

$$4x - 5e^{2x} = 4g_1(x) - 5g_2(x) ,$$

then, by the superposition principle, the function

$$4y_1(x) - 5y_2(x) = 4\left(\frac{-x}{3} - \frac{2}{9}\right) - 5\left(\frac{e^{2x}}{5}\right)$$

$$= \frac{-4x}{3} - \frac{8}{9} - e^{2x}$$

is a solution to equation (2). ■ End

Now we will show how the solution structure for a nonhomogeneous linear differential equation is determined by solutions to the associated homogeneous equation. By combining the superposition principle with the representation theorem for solutions of homogeneous equations, we can prove that *all* solutions of a *non*homogeneous equation can be found if we know just one of its solutions (called a *particular solution*) and a general solution to the corresponding homogeneous equation.

REPRESENTATION OF SOLUTIONS (NONHOMOGENEOUS CASE)

Theorem 6. Let $y_p(x)$ be a particular solution to the nonhomogeneous equation

(3) $y'' + p(x)y' + q(x)y = g(x)$

on the interval (a, b) and let $y_1(x)$, $y_2(x)$ be linearly independent solutions on (a, b) of the corresponding homogeneous equation

(4) $y'' + p(x)y' + q(x)y = 0$.

Then every solution of (3) on the interval (a, b) can be expressed in the form

(5) $y(x) = y_p(x) + c_1y_1(x) + c_2y_2(x)$.

The linear combination of y_p, y_1, and y_2 in (5), written with arbitrary constants c_1 and c_2, is referred to as a **general solution** to (3).

Proof. Let $\phi(x)$ be any solution to (3). Because both $\phi(x)$ and $y_p(x)$ are solutions to (3), the superposition principle tells us that the difference $\phi(x) - y_p(x)$ is a solution to the homogeneous equation (4). It then follows from Theorem 3 on page 169 that

$$\phi(x) - y_p(x) = c_1y_1(x) + c_2y_2(x)$$

for suitable constants c_1 and c_2. The last equation is equivalent to (5) $\big($with $\phi(x)$ in place of $y(x)\big)$, so the theorem is proved. ■ End

The procedure we advocate for analyzing nonhomogeneous equations is described as follows.

PROCEDURE FOR SOLVING NONHOMOGENEOUS EQUATIONS

To solve $y'' + py' + qy = g$:

(a) Determine a general solution, $c_1 y_1 + c_2 y_2$, of the corresponding homogeneous equation.
(b) Find a particular solution, y_p, of the given nonhomogeneous equation.
(c) Form the sum of the particular solution and a general solution to the homogeneous equation; that is,

$$y = y_p + c_1 y_1 + c_2 y_2 \, ,$$

to obtain a general solution to the given equation.

EXAMPLE 2 Given that $y_p(x) = x^2$ is a particular solution to

(6) $y'' - y = 2 - x^2 \, ,$

find a general solution of (6).

SOLUTION The corresponding homogeneous equation,

$$y'' - y = 0 \, ,$$

has the associated auxiliary equation $r^2 - 1 = 0$. Because $r = \pm 1$ are the roots of this equation, a general solution to the homogeneous equation is $c_1 e^x + c_2 e^{-x}$. Combining this with the particular solution $y_p(x) = x^2$ of the nonhomogeneous equation (6), we find that a general solution is

$$y(x) = x^2 + c_1 e^x + c_2 e^{-x}. \quad ■ \text{End}$$

Obviously, our next job is to learn how to find particular solutions to nonhomogeneous equations. This is the objective of Sections 4.8 and 4.9.

EXERCISES 4.7

In Problems 1 through 10, a nonhomogeneous equation and a particular solution are given. Find a general solution for the equation.

1. $y'' - y = x \, ,$ $y_p(x) = -x \, .$

2. $y'' + y' = 1 \, ,$ $y_p(x) = x \, .$

3. $\theta'' - \theta' - 2\theta = 1 - 2x \, ,$ $\theta_p(x) = x - 1 \, .$

4. $\dfrac{d^2 x}{dt^2} - 4\dfrac{dx}{dt} + 3x = -2e^t \, ,$ $x_p(t) = te^t \, .$

5. $y'' + 2y' + 4y - 4\cos 2x = 0 \, ,$ $y_p(x) = \sin 2x \, .$

6. $\dfrac{d^2 y}{dx^2} + y = 5e^{2x} \, ,$ $y_p(x) = e^{2x} \, .$

7. $\dfrac{d^2\theta}{dt^2} - \dfrac{d\theta}{dt} + \theta = \sin t$, $\theta_p(t) = \cos t$.

8. $y'' + 5y' + 6y = 6x^2 + 10x + 2 + 12e^x$,
$y_p(x) = e^x + x^2$.

9. $y'' = 2y' - y + 2e^x$, $y_p(x) = x^2 e^x$.

10. $y'' = 2y + 2\tan^3 x$, $y_p(x) = \tan x$.

11. Given that $y_1(x) = \cos x$ is a solution to
$$y'' - y' + y = \sin x ,$$
and $y_2(x) = e^{2x}/3$ is a solution to
$$y'' - y' + y = e^{2x} ,$$
find solutions to the following differential equations:

(a) $y'' - y' + y = 5 \sin x$.

(b) $y'' - y' + y = \sin x - 3e^{2x}$.

(c) $y'' - y' + y = 4 \sin x + 18e^{2x}$.

12. Let $L[y] := y'' + 2y' + 4y$, $y_1(x) := (1/4)\sin 2x$, and $y_2(x) := x/4 - 1/8$. Verify that $L[y_1](x) = \cos 2x$ and $L[y_2](x) = x$. Then use the superposition principle to find solutions to the following differential equations:

(a) $L[y](x) = x + \cos 2x$.

(b) $L[y](x) = 2x - 3 \cos 2x$.

(c) $L[y](x) = 11x - 12 \cos 2x$.

(d) $L[y](x) = 0$.

13. All that is known concerning a mysterious differential equation $y'' + p(x)y' + q(x)y = g(x)$ is that the functions x, x^2, and x^3 are solutions.

(a) Determine two linearly independent solutions to the corresponding homogeneous differential equation.

(b) Find the solution to the original equation satisfying the initial conditions $y(2) = 2$, $y'(2) = 5$.

(c) What is $p(x)$? [*Hint:* Use Abel's identity for the Wronskian (Section 4.3, page 175).]

14. Given that $1 + x$, $1 + 2x$, and $1 + 3x^2$ are solutions to the differential equation $y'' + p(x)y' + q(x)y = g(x)$, find the solution to this equation that satisfies $y(1) = 1$, $y'(1) = 0$.

15. Equivalence of General Solutions. Let y_p and \hat{y}_p be two particular solutions to $y'' + py' + qy = g$, and let $\{y_1, y_2\}$ be a fundamental solution set for the corresponding homogeneous equation. Show that any function of the form $y_p + c_1 y_1 + c_2 y_2$ can be

written in the form $\hat{y}_p + k_1 y_1 + k_2 y_2$ for suitable choice of constants k_1, k_2 (and vice versa). Consequently, it makes no difference which particular solution is used in the expression for the general solution of a nonhomogeneous equation.

16. Third Order Superposition Principle. Let $L[y] := y''' + py'' + qy' + ry$. Show that if $L[y_1] = g_1$ and $L[y_2] = g_2$, then for any constants c_1 and c_2, the function $c_1 y_1 + c_2 y_2$ is a solution of the differential equation $L[y] = c_1 g_1 + c_2 g_2$.

17. Let $L[y] := y''' + 2y'' - 13y' + 10y$, $y_1(x) := x^2$, and $y_2(x) := x^3$. Verify that $L[y_1](x) = 10x^2 - 26x + 4$ and $L[y_2](x) = 10x^3 - 39x^2 + 12x + 6$. Using the third order superposition principle (see Problem 16), find a general solution to the following differential equations.

(a) $L[y](x) = -10x^3 + 59x^2 - 64x + 2$.

(b) $L[y](x) = 20x^3 - 68x^2 - 2x + 16$.

18. Reduction of Order. The reduction of order procedure for homogeneous equations (see Section 4.4) can be extended to the nonhomogeneous equation
$$y'' + p(x)y' + q(x)y = g(x) .$$

(a) Show that the substitution $y = v(x)f(x)$, where $f(x)$ is a known nontrivial solution to the corresponding *homogeneous* equation, reduces the nonhomogeneous equation to the linear first order equation
$$fw' + (2f' + pf)w = g ,$$
where $w = v'$. (Such an equation can be solved using the techniques of Section 2.3.)

(b) Use the procedure in part (a) to find a general solution to
$$y'' + x^{-1}y' - 4x^{-2}y = 1 - x^{-3} , x > 0 ,$$
given that $f(x) = x^2$ is a solution to the corresponding homogeneous equation.

19. There is another type of reduction of order procedure that can be applied to equations of the form $L[y](x) = g(x)$, provided L is a differential operator that can be written as the product of first order factors. To illustrate this method, consider the equation

(7) $L[y](x) := y'' - y' - 2y = e^x$.

Notice that $L[y] = (D^2 - D - 2)[y]$ can be

written as $L[y] = (D + 1)(D - 2)[y]$. Let $v = (D - 2)[y]$ and observe that solving (7) is equivalent to solving the two first order equations

(8) $(D + 1)[v] = e^x$,

(9) $(D - 2)[y] = v$,

which, in turn, can be solved using the method discussed in Section 2.3. Carry out this procedure by completing the following steps:

(a) Solve equation (8) for v.

(b) Substitute the v obtained in part (a) into equation (9) and solve for y. This gives a general solution to (7).

20. Use the method of Problem 19 to find a general solution to

$$y'' + 6y' + 5y = 10x + 5 .$$

4.8 METHOD OF UNDETERMINED COEFFICIENTS

In this section, we give a simple procedure for finding a particular solution to a nonhomogeneous linear equation with constant coefficients when the nonhomogeneous term $g(x)$ is of a special type. To motivate the procedure, let's first look at a few examples.

EXAMPLE 1 Find a particular solution to

$$L[y] := y'' + 3y' + 2y = 3x + 1 .$$

SOLUTION We want to find a function y_p such that $L[y_p]$ is a linear function of x, namely $3x + 1$. Now what kind of functions y_p "end up" as linear functions after applications of the operator L? Certainly if y_p is linear, so is $L[y_p]$. So let's try the form

$$y_p(x) = Ax + B$$

and attempt to match up $L[y_p]$ with $3x + 1$. Since

$$L[y_p](x) = 0 + 3A + 2(Ax + B) = 2Ax + (3A + 2B) ,$$

we need to solve the equation

$$2Ax + (3A + 2B) = 3x + 1 .$$

Two polynomials are equal when their corresponding coefficients are equal, so we set

$$2A = 3 ,$$
$$3A + 2B = 1 .$$

Solving this system gives $A = 3/2$ and $B = -7/4$. Thus the function

$$y_p(x) = \frac{3}{2}x - \frac{7}{4}$$

is a particular solution since $L[y_p](x) = 3x + 1$. ∎ End

Throughout this section, L denotes a linear differential operator with *constant coefficients*. For such operators, Example 1 suggests the following method for finding a particular solution to the equation

$$L[y] = p_n ,$$

where $p_n(x)$ is a given polynomial of degree n. Namely, we guess a solution of the indefinite form

$$y_p(x) = A_n x^n + \cdots + A_1 x + A_0$$

and match the coefficients of $L[y_p](x)$ with those of $p_n(x)$:

$$L[y_p](x) = p_n(x) \ .$$

This procedure involves solving $n + 1$ linear equations in the $n + 1$ unknowns A_0, \ldots, A_n and usually has a solution. The technique is called the **method of undetermined coefficients.**[†] The same strategy can be applied to other forms for the nonhomogeneous term.

EXAMPLE 2 Find a particular solution to

$$L[y] := y'' + 3y' + 2y = e^{3x} \ .$$

SOLUTION Here we seek a function y_p such that $L[y_p](x) = e^{3x}$. If we guess $y_p(x) = Ae^{3x}$, where A is a constant, then

$$L[y_p](x) = 9Ae^{3x} + 3(3Ae^{3x}) + 2(Ae^{3x}) = 20Ae^{3x} \ .$$

Setting $20Ae^{3x} = e^{3x}$ and solving for A gives $A = 1/20$. Hence

$$y_p(x) = \frac{e^{3x}}{20}$$

is a particular solution, since $L[y_p](x) = e^{3x}$. ■ End

The method of undetermined coefficients works more generally for equations of the form

$$L[y] = ae^{\alpha x} \ ,$$

where a and α are given constants. Here we would guess

$$y_p(x) = Ae^{\alpha x}$$

and solve $L[y_p](x) = ae^{\alpha x}$ for the unknown coefficient A.

EXAMPLE 3 Find a particular solution to

(1) $L[y] := y'' + 3y' + 2y = \sin x \ .$

SOLUTION This time we seek a function y_p that satisfies $L[y_p](x) = \sin x$. Our initial action might be to guess $y_p(x) = A \sin x$, in which case we get

$$L[y_p](x) = -A \sin x + 3A \cos x + 2A \sin x = A \sin x + 3A \cos x \ .$$

However, since the right-hand side of (1) involves only $\sin x$, this choice of solution would force A (and hence y_p itself) to be zero. This will not work, since $y_p(x) \equiv 0$ is obviously not a solution.
So, to compensate for the $\cos x$ term, let's try

(2) $y_p(x) = A \sin x + B \cos x$

[†]*Historical Footnote:* This method was developed by Leonhard Euler in his 1753 theory of the motion of the moon.

and choose A and B to match up $L[y_p]$ with $\sin x$. We have

$$y_p'(x) = A \cos x - B \sin x , \qquad y_p''(x) = -A \sin x - B \cos x ,$$

so we need to enforce

$$\begin{aligned}
L[y_p](x) &= y_p''(x) + 3y_p'(x) + 2y_p(x) \\
&= -A \sin x - B \cos x + 3A \cos x - 3B \sin x + 2A \sin x + 2B \cos x \\
&= (A - 3B)\sin x + (B + 3A)\cos x \\
&= \sin x .
\end{aligned}$$

The equations $A - 3B = 1, B + 3A = 0$ have the solution $A = 0.1, B = -0.3$. Thus the function

$$y_p(x) = 0.1 \sin x - 0.3 \cos x$$

is a particular solution of $L[y] = \sin x$. ■ End

More generally, for an equation of the form

$$L[y] = a \cos \beta x + b \sin \beta x ,$$

the method of undetermined coefficients suggests that we guess

$$y_p(x) = A \cos \beta x + B \sin \beta x$$

and solve $L[y_p](x) = a \cos \beta x + b \sin \beta x$ for the unknowns A and B.

If we compare equation (1) with the mass-spring system equation

(3) $$[\text{inertia}] \times y'' + [\text{damping}] \times y' + [\text{stiffness}] \times y = F_{\text{external}} ,$$

we can interpret (1) as describing a damped oscillator, shaken with a sinusoidal force. According to our discussion in Section 4.1, then, we would expect the mass ultimately to respond by moving in synchronization with the forcing sinusoid. In other words, the form (2) is suggested by physical, as well as mathematical, experience. A complete description of forced oscillators will be given in Section 4.12.

EXAMPLE 4 Find a particular solution to

(4) $$L[y] := y'' + y' = 5 .$$

SOLUTION The right-hand side is the constant polynomial $p_0(x) \equiv 5$, so Example 1 suggests that we guess $y_p(x) = A$ and set $L[y_p](x) = 5$. But then we get

$$L[y_p](x) = 0 \neq 5 .$$

This unfortunate circumstance occurs because any constant function is a solution to the corresponding homogeneous equation $L[y] = 0$. To circumvent this difficulty, we temporarily abandon the method of undetermined coefficients and observe that we can integrate both sides of equation (4) to obtain the linear first order equation

(5) $$y' + y = 5x + c_1 ,$$

where c_1 is an arbitrary constant. *Now* the method of undetermined coefficients suggests that we approach (5) by trying $y = Ax + B$. This guess leads to

$$A + Ax + B = 5x + c_1$$

or

(6) $A = 5$, $A + B = c_1$,

and so $y_p(x) = 5x + (c_1 - 5)$. In fact, if we take $c_1 = 5$, we get

(7) $y_p(x) = 5x$

as a particular solution to equation (4). This suggests that in our original attempt to use undetermined coefficients, we should have guessed

$$y_p(x) = Ax$$

instead of $y_p(x) = A$. Indeed, with $y_p(x) = Ax$, setting $L[y_p](x) = 5$ immediately gives $A = 5$, and we again obtain (7). ■ End

In the preceding example, the trial choice for y_p did not work because it was a solution to the corresponding homogeneous equation. However, when we replaced y_p by the new function $\hat{y}_p = xy_p$, we were able to find a particular solution to the nonhomogeneous equation. This procedure for "repairing" the method of undetermined coefficients is generalized as follows.

If any term in the trial expression for y_p is a solution to the corresponding homogeneous equation, then replace y_p by xy_p. If the latter contains a term that satisfies the homogeneous equation, try $x^2 y_p$, then $x^3 y_p$, etc., ultimately employing $x^s y_p$, where s is the smallest nonnegative integer such that no term in $x^s y_p$ is a solution to the corresponding homogeneous equation.

EXAMPLE 5 Determine the form of a particular solution to

(8) $y'' - y' - 12y = e^{4x}$.

SOLUTION The auxiliary equation for the corresponding homogeneous equation

(9) $y'' - y' - 12y = 0$

has the roots $r_1 = 4$ and $r_2 = -3$. A general solution for (9) is therefore

$$y_h(x) = c_1 e^{4x} + c_2 e^{-3x}$$.

Now, since the right-hand side of (8) is e^{4x}, we first guess $y_p = Ae^{4x}$. But, because e^{4x} is a solution to equation (9), we replace this choice by $y_p = Axe^{4x}$. Since xe^{4x} is not a solution to equation (9), there exists a particular solution to (8) of the form

$$y_p(x) = Axe^{4x}.$$ ■ End

As our examples illustrate, when the nonhomogeneous term $g(x)$ is a polynomial, an exponential, a sine, or a cosine function, the function $g(x)$ itself suggests the form of the particular solution $y_p(x)$. In fact, we can expand the list of functions $g(x)$ for which the method of undetermined coefficients can be applied to include products of these functions as well. In Table 4.1, we have listed the various forms for $g(x)$ and the corresponding forms for the particular solution

$y_p(x)$. Observe that Type VII includes as special cases all the previous types; we have included the first six for easy reference. In using the table, the reader is reminded to choose s as the smallest nonnegative integer such that no term in y_p is a solution to the corresponding homogeneous equation. For linear second order equations, s will be either 0, 1, or 2. An incorrect choice for s leads to an inconsistent system of equations for the coefficients.

TABLE 4.1 PARTICULAR SOLUTIONS TO $L[y] = g$ (CONSTANT COEFFICIENTS)

Type	$g(x)$	$y_p(x)$
(I)	$p_n(x) = a_n x^n + \cdots + a_1 x + a_0$	$x^s P_n(x) = x^s\{A_n x^n + \cdots + A_1 x + A_0\}^\dagger$
(II)	$ae^{\alpha x}$	$x^s A e^{\alpha x}$
(III)	$a \cos \beta x + b \sin \beta x$	$x^s\{A \cos \beta x + B \sin \beta x\}$
(IV)	$p_n(x)e^{\alpha x}$	$x^s P_n(x)e^{\alpha x}$
(V)	$p_n(x)\cos \beta x + q_m(x)\sin \beta x$, where $q_m(x) = b_m x^m + \cdots + b_1 x + b_0$	$x^s\{P_N(x)\cos \beta x + Q_N(x)\sin \beta x\}$, where $Q_N(x) = B_N x^N + \cdots + B_1 x + B_0$ and $N = \max(n, m)$
(VI)	$ae^{\alpha x}\cos \beta x + be^{\alpha x}\sin \beta x$	$x^s\{A e^{\alpha x}\cos \beta x + B e^{\alpha x}\sin \beta x\}$
(VII)	$p_n(x)e^{\alpha x}\cos \beta x + q_m(x)e^{\alpha x}\sin \beta x$	$x^s e^{\alpha x}\{P_N(x)\cos \beta x + Q_N(x)\sin \beta x\}$, where $N = \max(n, m)$

The nonnegative integer s is chosen to be the smallest integer so that no term in the particular solution $y_p(x)$ is a solution to the corresponding homogeneous equation $L[y] = 0$.

$^\dagger P_n(x)$ must include all its terms even if $p_n(x)$ has some terms that are zero.

EXAMPLE 6 Using Table 4.1, find the form for a particular solution y_p to

(10) $y'' + 2y' - 3y = g(x)$,

where $g(x)$ equals

(a) $7 \cos 3x$, (b) $5e^{-3x}$, (c) $x^2\cos \pi x$,

(d) $2xe^x\sin x - e^x\cos x$, (e) $x^2 e^x + 3xe^x$, (f) $\tan x$.

SOLUTION The auxiliary equation for the corresponding homogeneous equation,

(11) $y'' + 2y' - 3y = 0$,

has roots $r_1 = 1$ and $r_2 = -3$. Thus a general solution for (11) is

$$y_h(x) = c_1 e^x + c_2 e^{-3x} .$$

(a) From Table 4.1, the function $g(x) = 7 \cos 3x$ is of Type III (with $a = 7$, $b = 0$, and $\beta = 3$). Hence y_p has the form

(12) $y_p(x) = A \cos 3x + B \sin 3x$,

since no term in (12) is a solution to (11).

(b) The function $g(x) = 5e^{-3x}$ is of Type II, so y_p has the form $x^s A e^{-3x}$. Here we take $s = 1$, since e^{-3x} is a solution to equation (11) and $x e^{-3x}$ is not. Thus

$$y_p(x) = A x e^{-3x} \ .$$

(c) The function $g(x) = x^2 \cos \pi x$ is of Type V, with $p_2(x) = x^2$, $q_0(x) \equiv 0$. Because $N = \max\{2, 0\} = 2$,

(13) $$y_p(x) = (A_2 x^2 + A_1 x + A_0) \cos \pi x + (B_2 x^2 + B_1 x + B_0) \sin \pi x \ ,$$

where we have taken $s = 0$ because no term in (13) is a solution to the homogeneous equation.

(d) The function $g(x) = 2x e^x \sin x - e^x \cos x$ is of Type VII, with $p_0(x) = -1$ and $q_1(x) = 2x$. For these polynomials, $N = \max\{0, 1\} = 1$, so

$$x^s e^x \{ (A_1 x + A_0) \cos x + (B_1 x + B_0) \sin x \}$$

is the form for the particular solution. None of the terms $e^x x \cos x$, $e^x \cos x$, $e^x x \sin x$, and $e^x \sin x$ is a solution of the homogeneous equation (11), so we set $s = 0$ and obtain

$$y_p(x) = e^x \{ (A_1 x + A_0) \cos x + (B_1 x + B_0) \sin x \} \ .$$

(e) Here $g(x) = x^2 e^x + 3x e^x = (x^2 + 3x) e^x$ is of Type IV with $p_2(x) = x^2 + 3x$. Thus

$$y_p(x) = x(A_2 x^2 + A_1 x + A_0) e^x \ ,$$

where we have taken $s = 1$, since the term $A_0 e^x$ in $(A_2 x^2 + A_1 x + A_0) e^x$ is a solution to equation (11).

(f) Unfortunately, $g(x) = \tan x$ is not one of the forms for which the method of undetermined coefficients can be used. In Section 4.9, we discuss a different method that can handle such nonhomogeneous terms (also see Project C, the Convolution Method, page 254). ■ End

Let's summarize.

METHOD OF UNDETERMINED COEFFICIENTS

To determine a particular solution to $L[y] = g$:

(a) Check that the linear equation has constant coefficients and that the nonhomogeneous term $g(x)$ is one of the types suitable for the method.

(b) Solve the corresponding homogeneous equation $L[y] = 0$.

(c) On the basis of the form of the nonhomogeneous term $g(x)$, determine the appropriate form of the particular solution $y_p(x)$. Remember that no term in the trial expression for $y_p(x)$ may be a solution to the corresponding homogeneous equation.

(d) Since we want $L[y_p](x) = g(x)$, set the corresponding coefficients from both sides of this equation equal to each other to form a system of linear equations.

(e) Solve this system of linear equations for the coefficients appearing in $y_p(x)$.

If we combine the method of undetermined coefficients with the superposition principle, we can solve for $y_p(x)$ when $g(x)$ is a finite *sum* of terms like those listed in Table 4.1.

EXAMPLE 7 Find a general solution to the equation

(14) $y'' + 3y' + 2y = 3x + 1 + 5 \sin x$.

SOLUTION We first solve the associated homogeneous equation $y'' + 3y' + 2y = 0$ and obtain as general solution $y_h(x) = c_1 e^{-x} + c_2 e^{-2x}$.

Notice that the right-hand side of (14) is not one of the forms given in Table 4.1 but is the sum of two such forms. This suggests that we use the superposition principle and consider separately the equations

(15) $y'' + 3y' + 2y = 3x + 1$

and

(16) $y'' + 3y' + 2y = 5 \sin x$.

We solved equation (15) in Example 1 and found the particular solution $\frac{3}{2}x - \frac{7}{4}$. Equation (16) differs only by the factor 5 from equation (1), whose solution was found in Example 3 to be $0.1 \sin x - 0.3 \cos x$. It follows from the superposition principle that a particular solution to the original equation (14) is given by

$$y_p(x) = \frac{3}{2}x - \frac{7}{4} + 5\left[0.1 \sin x - 0.3 \cos x\right] = \frac{3}{2}x - \frac{7}{4} + \frac{1}{2}\sin x - \frac{3}{2}\cos x .$$

Hence a general solution to equation (14), given by the sum of the general solution to the corresponding homogeneous equation and a particular solution, is

$$y(x) = c_1 e^{-x} + c_2 e^{-2x} + \frac{3}{2}x - \frac{7}{4} + \frac{1}{2}\sin x - \frac{3}{2}\cos x . \quad \blacksquare \text{End}$$

Success in applying the method of undetermined coefficients lies in the ability to guess the form of a particular solution. As we have seen, when L is a second order linear differential operator with constant coefficients, there is a class of functions $g(x)$ for which the choice of the form of a particular solution to the equation $L[y] = g$ is quite easy to determine. This class consists of functions of the form

$$p_n(x)e^{\alpha x}\cos \beta x + q_m(x)e^{\alpha x}\sin \beta x ,$$

where $p_n(x)$ and $q_m(x)$ are polynomials. (The use of complex numbers can make the manipulation of this form quite efficient; see Project A.) The method works because when we apply the operator L to a function of this form, we get back a function of the same type. As you might have guessed, these functions are themselves solutions of some (higher-order) homogeneous linear differential equation with constant coefficients. We verify in Chapter 6 that the method of undetermined coefficients applies to all such functions.

EXERCISES 4.8

In Problems 1 through 14, find a particular solution to the differential equation.

1. $y'' + 3y = -9$.

2. $y'' + 2y' - y = 10$.

3. $2z'' + z = 9e^{2x}$.

4. $x''(t) - x(t) = 3e^{-2t}$.

5. $2x' + x = 3t^2 + 10t$.

6. $y'' - y' - 2y = -2x^3 - 3x^2 + 8x + 1$.

7. $y'' - y' + 9y = 3 \sin 3x$.

8. $y'' + y' + y = 2 \cos 2x - 3 \sin 2x$.

9. $\dfrac{d^2\theta}{dr^2} - 5\dfrac{d\theta}{dr} + 6\theta = re^r$.

10. $\dfrac{d^2x}{dt^2} + 2\dfrac{dx}{dt} - 8x = te^{-t} + e^{-t}$.

11. $y'' + y = 2^x$.

12. $\theta''(t) - \theta(t) = t \sin t$.

13. $y'' - 2y' + y = 8e^x$.

14. $z'' - 6z' + 9z = x^2 + e^x$.

In Problems 15 through 24, find a general solution to the differential equation.

15. $y'' - y = -11x + 1$.

16. $y'' - 2y' - 3y = 3x^2 - 5$.

17. $y''(t) - 3y'(t) + 2y(t) = e^t \sin t$.

18. $y''(\theta) + 4y(\theta) = \sin \theta - \cos \theta$.

19. $y''(\theta) + 2y'(\theta) + 2y(\theta) = e^{-\theta}\cos \theta$.

20. $y'' + 6y' + 10y = 10x^4 + 24x^3 + 2x^2 - 12x + 18$.

21. $x''(t) - 4x'(t) + 4x(t) = te^{2t}$.

22. $y'' - 7y' = x^2$.

23. $y'' - 3y = x^2 - e^x$.

24. $y'' - 2y' - 35y = 13 \sin x - e^{3x} + 1$.

In Problems 25 through 32, find the solution to the initial value problem.

25. $y' - y = 1$, $y(0) = 0$.

26. $h'' = 6x$; $h(0) = 3$, $h'(0) = -1$.

27. $z'' + z = 2e^{-x}$; $z(0) = 0$, $z'(0) = 0$.

28. $y'' + 9y = 27$; . $y(0) = 4$, $y'(0) = 6$.

29. $y'' - y' - 2y = \cos x - \sin 2x$;
 $y(0) = -7/20$, $y'(0) = 1/5$.

30. $y'' + y' - 12y = e^x + e^{2x} - 1$;
 $y(0) = 1$, $y'(0) = 3$.

31. $y''(\theta) - y(\theta) = \sin \theta - e^{2\theta}$; $y(0) = 1$,
 $y'(0) = -1$.

32. $y'' + 2y' + y = x^2 + 1 - e^x$; $y(0) = 0$,
 $y'(0) = 2$.

In Problems 33 through 38, determine the form of a particular solution for the differential equation.

33. $y'' + y = \sin x + x \cos x + 10^x$.

34. $y'' - y = e^{2x} + xe^{2x} + x^2e^{2x}$.

35. $x'' - x' - 2x = e^t\cos t - t^2 + t + 1$.

36. $y'' + 5y' + 6y = \sin x - \cos 2x$.

37. $y'' - 4y' + 5y = e^{5x} + x \sin 3x - \cos 3x$.

38. $y'' - 4y' + 4y = x^2e^{2x} - e^{2x}$.

In Problems 39 through 44, decide whether the method of undetermined coefficients (possibly together with superposition) can be applied to find a particular solution of the given equation.

39. $y'' + 2y' - y = x^{-1}e^x$.

40. $y''(\theta) + 3y'(\theta) - y(\theta) = \sec \theta$.

41. $x''(t) - 5x'(t) + 2x(t) = 3^t$.

42. $x''(t) + tx'(t) + x(t) = \sin t$.

43. $xy'' - y' + 2y = \cos x$.

44. $y'' - y' + 2y = x^2 + \sin x$.

In Problems 45 through 48, find a particular solution to the given higher-order equation.

45. $y''' - y'' + y = \sin x$.

46. $2y''' + 3y'' + y' - 4y = e^{-x}$.

47. $y''' + y'' - 2y = xe^x + 1$.

48. $y^{iv} - 3y'' - 8y = \sin x$.

In Problems 49 through 54, use the method discussed under "Cauchy-Euler Equations" in Section 4.5 to find a general solution to the given equation for $x > 0$.

49. $xy''(x) + 3y'(x) - \dfrac{3}{x}y(x) = x^2$.

50. $x^2y''(x) - y(x) = (\ln x)^2 - 1$.

51. $x^4y''(x) - 6x^2y(x) = 1 - 6x^2$.

52. $xy''(x) + y'(x) = 9x^2$.

53. $x^2 y''(x) + x y'(x) + y(x) = (\ln x) \sin(\ln x)$.

54. $x^2 y''(x) + 3 x y'(x) + 4 y(x) = \cos(4 \ln x)$.

55. Discontinuous Forcing Term. In certain physical models, the nonhomogeneous term, or **forcing term,** $g(x)$ in the equation $L[y](x) = g(x)$ may not be continuous but may have a jump discontinuity. If this occurs, we can still obtain a reasonable solution using the following procedure. Consider the initial value problem

$$y'' + 2y' + 5y = g(x) \; ; \quad y(0) = 0 \; , \quad y'(0) = 0 \; ,$$

where

$$g(x) := \begin{cases} 10 \; , & 0 \le x \le 3\pi/2 \; , \\ 0 \; , & x > 3\pi/2 \; . \end{cases}$$

(a) Find a solution to the initial value problem for $0 \le x \le 3\pi/2$.

(b) Find a general solution for $x > 3\pi/2$.

(c) Now choose the constants in the general solution from part (b) so that the solution from part (a) and the solution from part (b) agree, together with their first derivatives, at $x = 3\pi/2$. This gives us a continuously differentiable function that satisfies the differential equation except at $x = 3\pi/2$.

56. Forced Vibrations. A vibrating spring with damping that is under external force can be modeled by

(17) $m x'' + b x' + k x = g(t)$,

where $m > 0$ is the mass of the spring system, $b > 0$ is the damping constant, $k > 0$ is the spring constant, $g(t)$ is the force on the system at time t, and $x(t)$ is the displacement from the equilibrium of the spring system at time t. Assume $b^2 < 4mk$.

(a) Determine the form of the equation of motion for the spring system when $g(t) = \sin \beta t$ by finding a general solution to equation (17).

(b) Discuss the long-term behavior of this system. [*Hint:* Consider what happens to the general solution obtained in part (a) as $t \to +\infty$.]

57. Speed Bumps. Often bumps like the one depicted in Figure 4.12 are built into roads to discourage speeding. The figure suggests that a crude model of the vertical motion $y(t)$ of a car encountering the speed bump with the speed V is given by

$$y(t) = 0 \; , \qquad \text{for } t \le -L/2V \; ,$$

$$m y'' + k y = \begin{cases} \cos\left(\dfrac{\pi V t}{L}\right) , & \text{for } -L/2V < t < L/2V \; , \\ \\ 0 \; , & \text{for } L/2V \le t \; . \end{cases}$$

(The absence of a damping term indicates that the car's shock absorbers are not functioning.)

(a) Taking $m = k = 1$ and $L = \pi$ for convenience, solve this initial value problem. Thereby show that the formula for the oscillatory motion *after the car has traversed the speed bump* is $y(t) = A \sin t$, where the constant A depends on the speed V.

(b) Plot the amplitude $|A|$ of the solution $y(t)$ found in part (a) versus the car's speed V. From the graph, estimate the speed that produces the most violent shaking of the vehicle.

Figure 4.12 Speed bump

Problems 58 through 60 concern justifications for the method of undetermined coefficients.

58. Consider the nonhomogeneous equation

(18) $a y'' + b y' + c y = a_n x^n + \cdots + a_1 x + a_0$,

where $a_n \ne 0$. We want to determine the form of a particular solution y_p to equation (18).

(a) Assume

(19) $y_p(x) = A_n x^n + \cdots + A_1 x + A_0$

and substitute y_p into equation (18) to obtain, after simplifying,

$$c A_n x^n + (c A_{n-1} + n b A_n) x^{n-1} + \cdots$$
$$+ (2 a A_2 + b A_1 + c A_0) = a_n x^n + \cdots + a_0 \; .$$

(b) Show that if $c \ne 0$, then no (nonzero) term in (19) is a solution to the homogeneous equation

(20) $a y'' + b y' + c y = 0$.

(c) Show that if $c \neq 0$, then you can solve for the constants $A_n, A_{n-1}, \ldots, A_0$ in terms of the constants $a_n, a_{n-1}, \ldots, a_0$. Hence there exists a solution to equation (18) of the form (19).

59. Consider the nonhomogeneous solution

(21) $\quad ay'' + by' + cy = e^{\alpha x}(a_n x^n + \cdots + a_0)$.

We want to determine the form of a particular solution y_p to equation (21).

(a) Let $y_p(x) = e^{\alpha x}v(x)$. Substitute for y, y', and y'' in equation (21) and divide by the nonzero factor $e^{\alpha x}$ to obtain

(22) $\quad av'' + (2a\alpha + b)v' + (a\alpha^2 + b\alpha + c)v$
$\qquad = a_n x^n + \cdots + a_0$.

(b) Using the results from Problem 58, show that if $e^{\alpha x}$ is not a solution to the homogeneous equation (20), then equation (21) has a solution of the form
$y_p(x) = e^{\alpha x}(A_n x^n + \cdots + A_1 x + A_0)$.

60. Annihilator Method. The annihilator method, which is discussed in Chapter 6, can be used to justify the method of undetermined coefficients. To illustrate the annihilator approach, consider the equation

(23) $\quad (D-1)(D+1)[y] = y'' - y = e^{3x}$.

(a) Show that applying $(D-3)$ to both sides of equation (23) gives

(24) $\quad (D-3)(D-1)(D+1)[y]$
$\qquad = y''' - 3y'' - y' + 3y = 0$.

(Since $(D-3)[e^{3x}] \equiv 0$, we say that $(D-3)$ "annihilates" e^{3x}.)

(b) If y_p satisfies equation (23), it will also satisfy equation (24). Now verify that

(25) $\quad c_1 e^{3x} + c_2 e^x + c_3 e^{-x}$

is a general solution to (24) and conclude that there exists a particular solution to (23) of the form (25).

(c) Using the fact that $c_2 e^x + c_3 e^{-x}$ is a solution to the homogeneous equation
$(D-1)(D+1)[y] = 0$,
conclude that there must be a particular solution to (23) of the form $y_p(x) = c_1 e^{3x}$.

(d) Use the above procedure to show that there is a particular solution to
$(D-1)(D+1)[y] = e^x$
of the form $y_p(x) = Cxe^x$. [*Hint:* Use the annihilator $(D-1)$.]

4.9 VARIATION OF PARAMETERS

We have seen that the method of undetermined coefficients is a simple procedure for determining a particular solution when the equation has constant coefficients and the nonhomogeneous term is of a special type. Here we present a more general method, called **variation of parameters,**[†] for finding a particular solution. This method applies even when the coefficients of the differential equation are functions of x, *provided* we know a fundamental solution set for the corresponding homogeneous equation.

Consider the nonhomogeneous linear second order equation

(1) $\quad L[y] := y'' + p(x)y' + q(x)y = g(x)$,

where the coefficient of y'' is taken to be 1, and let $\{y_1(x), y_2(x)\}$ be a fundamental solution set for the corresponding homogeneous equation

$\qquad L[y] = 0$.

[†]*Historical Footnote:* The method of variation of parameters was invented by Joseph Lagrange in 1774.

Then we know that the solutions to this homogeneous equation are given by

(2) $y_h(x) = c_1 y_1(x) + c_2 y_2(x)$,

where c_1 and c_2 are constants. To find a particular solution to the nonhomogeneous equation, the idea behind variation of parameters is to replace the constants in (2) by functions of x. That is, we seek a solution of (1) of the form[†]

(3) $y_p(x) = v_1(x) y_1(x) + v_2(x) y_2(x)$.

Because we have introduced two unknown functions, $v_1(x)$ and $v_2(x)$, it is reasonable to expect that we can impose two equations on these functions. Naturally, one of these equations should come from (1). Let's therefore plug $y_p(x)$ given by (3) into (1). To accomplish this, we must first compute $y_p'(x)$ and $y_p''(x)$. From (3), we obtain

$$y_p' = (v_1' y_1 + v_2' y_2) + (v_1 y_1' + v_2 y_2') .$$

To simplify the computation and to avoid second order derivatives for the unknowns v_1, v_2 in the expression for y_p'', we impose the requirement

(4) $v_1' y_1 + v_2' y_2 = 0$;

thus the formula for y_p' becomes

(5) $y_p' = v_1 y_1' + v_2 y_2'$

and so

(6) $y_p'' = v_1' y_1' + v_1 y_1'' + v_2' y_2' + v_2 y_2''$.

Now, substituting y_p, y_p', and y_p'', as given in (3), (5), and (6), into (1), we find

(7) $g = L[y_p]$
$$= (v_1' y_1' + v_1 y_1'' + v_2' y_2' + v_2 y_2'') + p(v_1 y_1' + v_2 y_2') + q(v_1 y_1 + v_2 y_2)$$
$$= (v_1' y_1' + v_2' y_2') + v_1(y_1'' + p y_1' + q y_1) + v_2(y_2'' + p y_2' + q y_2)$$
$$= (v_1' y_1' + v_2' y_2') + v_1 L[y_1] + v_2 L[y_2] .$$

Since y_1 and y_2 are solutions to the homogeneous equation, we have $L[y_1] = L[y_2] = 0$. Thus (7) reduces to

(8) $v_1' y_1' + v_2' y_2' = g$.

To summarize, if we can find v_1 and v_2 that satisfy both (4) and (8), that is,

(9) $y_1 v_1' + y_2 v_2' = 0$,
$y_1' v_1' + y_2' v_2' = g$,

then y_p given by (3) will be a particular solution to (1). To determine v_1 and v_2, we first solve

[†]In Exercises 2.3, Problem 36, we developed this approach for first order linear equations. Because of the similarity of equations (2) and (3), this technique is sometimes known as "variation of constants."

the linear system (9) for v_1' and v_2'. Algebraic manipulation or Cramer's rule immediately gives

$$v_1'(x) = \frac{-g(x)y_2(x)}{W[y_1, y_2](x)} \ , \qquad v_2'(x) = \frac{g(x)y_1(x)}{W[y_1, y_2](x)} \ ,$$

where $W[y_1, y_2](x)$, which occurs in the denominator, is the Wronskian of $y_1(x)$ and $y_2(x)$. Notice that this Wronskian is never zero because $\{y_1, y_2\}$ is a fundamental solution set. Upon integrating these equations, we finally obtain

(10) $$v_1(x) = \int \frac{-g(x)y_2(x)}{W[y_1, y_2](x)} \, dx \ , \qquad v_2(x) = \int \frac{g(x)y_1(x)}{W[y_1, y_2](x)} \, dx \ .$$

Let's review this procedure.

METHOD OF VARIATION OF PARAMETERS

To determine a particular solution to $y'' + py' + qy = g$:

(a) Find a fundamental solution set $\{y_1(x), y_2(x)\}$ for the corresponding homogeneous equation and take

$$y_p(x) = v_1(x)y_1(x) + v_2(x)y_2(x) \ .$$

(b) Determine $v_1(x)$ and $v_2(x)$ by solving the system in (9) for $v_1'(x)$ and $v_2'(x)$ and integrating. (If the given equation is $a_2(x)y'' + a_1(x)y' + a_0(x)y = b(x)$, then it must first be put in the form $y'' + p(x)y' + q(x)y = g(x)$ so that $g(x) = b(x)/a_2(x)$.)

(c) Substitute $v_1(x)$ and $v_2(x)$ into the expression for $y_p(x)$ to obtain a particular solution.

Of course, in step (b) one could use the formulas in (10), but $v_1(x)$ and $v_2(x)$ are so easy to derive that the reader is advised not to memorize them.

EXAMPLE 1 Find a general solution on $(-\pi/2, \pi/2)$ to

(11) $$\frac{d^2y}{dx^2} + y = \tan x \ .$$

SOLUTION Observe that a fundamental solution set for the homogeneous equation $y'' + y = 0$ is $\{\cos x, \sin x\}$. We now set

(12) $$y_p(x) = v_1(x) \cos x + v_2(x) \sin x$$

and, referring to (9), solve the system

$$(\cos x)v_1'(x) + (\sin x)v_2'(x) = 0 \ ,$$
$$(-\sin x)v_1'(x) + (\cos x)v_2'(x) = \tan x$$

for $v_1'(x)$ and $v_2'(x)$. This gives

$$v_1'(x) = -\tan x \sin x \ ,$$
$$v_2'(x) = \tan x \cos x = \sin x \ .$$

Integrating, we obtain

(13) $\displaystyle v_1(x) = -\int \tan x \sin x \, dx = -\int \frac{\sin^2 x}{\cos x} \, dx$

$$= -\int \frac{1 - \cos^2 x}{\cos x} \, dx = \int (\cos x - \sec x) \, dx$$

$$= \sin x - \ln|\sec x + \tan x| + C_1 \ ,$$

(14) $\displaystyle v_2(x) = \int \sin x \, dx = -\cos x + C_2 \ .$

We need only one particular solution, so we take both C_1 and C_2 to be zero for simplicity. Then, substituting $v_1(x)$ and $v_2(x)$ in (12), we obtain

$$y_p(x) = \left[\sin x - \ln|\sec x + \tan x|\right]\cos x - \cos x \sin x \ ,$$

which simplifies to

$$y_p(x) = -(\cos x) \ln|\sec x + \tan x| \ .$$

We may drop the absolute value symbols because $\sec x + \tan x = (1 + \sin x)/\cos x > 0$ for $-\pi/2 < x < \pi/2$.

Recall from Theorem 6 on page 201 that a general solution to a nonhomogeneous equation is given by the sum of a general solution to the homogeneous equation and a particular solution. Consequently, a general solution to equation (11) on the interval $(-\pi/2, \pi/2)$ is

(15) $y(x) = c_1 \cos x + c_2 \sin x - (\cos x) \ln(\sec x + \tan x).$ ■ End

Note that in the above example, the constants C_1 and C_2 appearing in (13) and (14) were chosen to be zero. If we had retained these arbitrary constants, the ultimate effect would be just to add $C_1 \cos x + C_2 \sin x$ to (15), which is clearly redundant.

EXAMPLE 2 Find a particular solution on $(-\pi/2, \pi/2)$ to

(16) $\displaystyle \frac{d^2 y}{dx^2} + y = \tan x + 3x - 1 \ .$

SOLUTION With $g(x) = \tan x + 3x - 1$, the variation of parameters procedure will lead to a solution. But it is simpler in this case to consider separately the equations

(17) $\displaystyle \frac{d^2 y}{dx^2} + y = \tan x \ ,$

(18) $\displaystyle \frac{d^2 y}{dx^2} + y = 3x - 1$

and then use the superposition principle (Theorem 5, page 200).

In Example 1, we found that

$$y_q(x) = -(\cos x)\ln(\sec x + \tan x)$$

is a particular solution for equation (17). For equation (18), the method of undetermined coefficients can be applied. On seeking a solution to (18) of the form $y_r(x) = Ax + B$, we quickly obtain

$$y_r(x) = 3x - 1 \ .$$

Finally, we apply the superposition principle to get

$$y_p(x) = y_q(x) + y_r(x)$$
$$= -(\cos x)\ln(\sec x + \tan x) + 3x - 1$$

as a particular solution for equation (16). ∎ End

One important advantage the method of variation of parameters has over the method of undetermined coefficients is its applicability to linear equations whose coefficients are functions of x (see Problems 23–26). The fly in the ointment is that variation of parameters requires that we know a fundamental solution set for the corresponding homogeneous equation. For equations with variable coefficients, this set may be extremely difficult to determine.

EXERCISES 4.9

In Problems 1 through 12, find a general solution to the differential equation using the method of variation of parameters.

1. $y'' + 4y = \tan 2x$.

2. $y'' + y = \sec x$.

3. $2x'' - 2x' - 4x = 2e^{3t}$.

4. $y'' - y = 2x + 4$.

5. $y'' - 2y' + y = x^{-1}e^x$.

6. $y'' + 2y' + y = e^{-x}$.

7. $y''(\theta) + 16y(\theta) = \sec 4\theta$.

8. $y'' + 9y = \sec^2(3x)$.

9. $y'' + 4y = \csc^2(2x)$.

10. $y'' + 4y' + 4y = e^{-2x}\ln x$.

11. $x^2z'' + xz' + 9z = -\tan(3\ln x)$.

12. $x^2y'' + 3xy' + y = x^{-1}$.

In Problems 13 through 21, find a general solution to the differential equation.

13. $y'' + y = \tan^2 x$.

T#A **14.** $y'' + y = \tan x + e^{3x} - 1$.

15. $v'' + 4v = \sec^4(2x)$.

16. $y''(\theta) + y(\theta) = \sec^3\theta$.

17. $y'' + y = 3\sec x - x^2 + 1$.

18. $y'' + 5y' + 6y = 18x^2$.

19. $\frac{1}{2}y'' + 2y = \tan 2x - \frac{1}{2}e^x$.

20. $y'' - 6y' + 9y = x^{-3}e^{3x}$.

21. $x^2z'' - xz' + z = x\left(1 + \dfrac{3}{\ln x}\right)$.

22. The **Bessel equation** of order one half (cf. page 223),

$$x^2y'' + xy' + \left(x^2 - \frac{1}{4}\right)y = 0 \ , \qquad x > 0 \ ,$$

has two linearly independent solutions,

$$y_1(x) = x^{-1/2}\cos x \ , \qquad y_2(x) = x^{-1/2}\sin x \ .$$

Find a general solution to the nonhomogeneous equation

$$x^2y'' + xy' + \left(x^2 - \frac{1}{4}\right)y = x^{5/2} \ , \qquad x > 0 \ .$$

In Problems 23 through 26, find a general solution to the differential equation given that the functions y_1 and y_2 are linearly independent solutions to the corresponding

homogeneous equation for $x > 0$. Remember to put the equation in the same form as equation (1).

23. $xy'' - (x + 1)y' + y = x^2$;

$y_1 = e^x$, $y_2 = x + 1$.

24. $x^2 y'' - 4xy' + 6y = x^3 + 1$;

$y_1 = x^2$, $y_2 = x^3$.

25. $xy'' + (5x - 1)y' - 5y = x^2 e^{-5x}$;

$y_1 = 5x - 1$, $y_2 = e^{-5x}$.

26. $xy'' + (1 - 2x)y' + (x - 1)y = xe^x$;

$y_1 = e^x$, $y_2 = e^x \ln x$.

27. Express the solution to the initial value problem

$$y'' - y = \frac{1}{x} ; \qquad y(1) = 0 , \qquad y'(1) = -2$$

using definite integrals. Using numerical integration (Appendix B) to approximate the integrals, find an approximation for $y(2)$ to two decimal places.

T#4 **28.** Use the method of variation of parameters to show that

$$y(x) = c_1 \cos x + c_2 \sin x + \int_0^x f(s)\sin(x - s)\,ds$$

is a general solution to the differential equation

$$y'' + y = f(x) ,$$

where $f(x)$ is a continuous function on $(-\infty, \infty)$. [*Hint:* Use the trigonometric identity $\sin(x - s) = \sin x \cos s - \sin s \cos x$.]

4.10 QUALITATIVE CONSIDERATIONS FOR VARIABLE-COEFFICIENT AND NONLINEAR EQUATIONS

In this section, we shall examine some second order linear equations with *variable* coefficients and some *non*linear equations. First we display a few examples that illustrate the profound differences that may occur. Then we show how the mass-spring oscillator analogy, discussed in Section 4.1, can be exploited to predict some of the qualitative features of more complicated equations.

Because we will be making extensive use of the mechanical analogy in this section, it is appropriate to use the notation t, rather than x, for the independent variable (which is interpreted as *time*).

To begin our discussion we display a linear constant-coefficient, a linear variable-coefficient, and two nonlinear equations.

(a) The equation

(1) $3y'' + 2y' + 4y = 0$

is linear and has constant coefficients. We know everything about such equations; the solutions are, at worst, polynomials times exponentials times sinusoids in t, and unique solutions can be found to match any prescribed data $y(a)$, $y'(a)$ at any instant $t = a$.

(b) The equation

(2) $(1 - t^2)y'' - 2ty' + 2y = 0$

is a linear variable-coefficient equation. It is a special case of **Legendre's equation** $(1 - t^2)y'' - 2ty' + \lambda y = 0$, which arises in the analysis of wave and diffusion phenomena in spherical coordinates.

(c) The equations

(3) $y'' - 6y^2 = 0$,

(4) $y'' - 24y^{1/3} = 0$

are nonlinear equations because of the square and the cube root of y terms.

Recall that equations (1) and (2) share the superposition property: If $y_1(t)$ and $y_2(t)$ are solutions, so is $y(t) = y_1(t) + y_2(t)$. But superposition cannot be guaranteed for equation (3) (because, e.g., the quadratic term y^2 does not reduce to $y_1^2 + y_2^2$). (Note that the quadratic t^2 in equation (2) does not prevent Legendre's equation from being linear *in y*.)

The solutions of linear variable-coefficient equations are more complicated than constant-coefficient equations. Typically they are solved using laborious power-series techniques, as will be discussed in Chapter 8. The Legendre equation (2) has one simple solution, $y_1(t) = t$, as can easily be verified by mental calculation. Its second linearly independent solution for $-1 < t < 1$ is usually taken to be

(5) $y_2(t) = \dfrac{t}{2} \ln\left(\dfrac{1 + t}{1 - t}\right) - 1$.

Notice in particular the behavior near $t = \pm 1$; none of the solutions of our *constant*-coefficent ever diverged at a finite point!

We could have anticipated troublesome behavior for Legendre's equation at $t = \pm 1$ if we had rewritten it as

(6) $y'' = \dfrac{2ty' - 2y}{1 - t^2}$.

This form makes it clear that the second derivative "blows up" at, say, $t = 1$ unless $y'(1)$ and $y(1)$ happen to be equal (rendering the fraction in (6) to the indeterminate form $0/0$). Indeed, for our well-behaved solution $y_1(t) = t$, we have $y_1(1) = y_1'(1) = 1$.

It is easy to generalize: A variable-coefficient equation

$$a_2(t)y'' + a_1(t)y' + a_0(t)y = 0$$

cannot be guaranteed to have a solution that matches arbitrarily specified initial data $y(a)$, $y'(a)$ at $t = a$, if the point a is a zero of the leading coefficient $a_2(t)$. Such points are more precisely identified as "singular points" in Chapter 8. *Note that the existence-uniqueness Theorem 2 of Section 4.2 (page 165) was cautiously worded with the differential equation in **standard form** $(a_2 \equiv 1)$ to exclude the case of variable-coefficient equations with singular points, since the latter can generate exceptions to the **existence** statement.*

There are no general solution procedures for solving nonlinear equations. However, the following lemma is very useful in some situations such as equations (3), (4). It has an extremely significant physical interpretation, which we will explore in Project D of Chapter 5; for now we will merely tantalize the reader by giving it a suggestive name.

THE ENERGY INTEGRAL LEMMA

Lemma 3. Let $y(t)$ be a solution to the differential equation

(7) $y'' = f(y)$

where $f(y)$ is a continuous function that does not depend on y' or the independent variable t. Let $F(y)$ be an indefinite integral of $f(y)$, that is,

$$f(y) = \frac{d}{dy} F(y) \ .$$

Then the quantity

(8) $E(t) := \frac{1}{2} y'(t)^2 - F\big(y(t)\big)$

is constant; i.e.,

(9) $\dfrac{d}{dt} E(t) = 0 \ .$

Proof. This is immediate; we insert (8), differentiate, and apply the differential equation (7):

$$\frac{d}{dt} E(t) = \frac{d}{dt}\left[\frac{1}{2} y'(t)^2 - F\big(y(t)\big) \right]$$

$$= 2\frac{y'}{2} y'' - \frac{dF}{dy} y'$$

$$= y'\left[y'' - f(y) \right]$$

$$= 0 \ . \quad \blacksquare \text{End}$$

As a result, an equation of the form (7) can be reduced to

(10) $\dfrac{1}{2}(y')^2 - F(y) = K$

for some constant K, which is equivalent to the *separable* first order equation

$$\frac{dy}{dt} = \pm\sqrt{2\big[F(y) + K\big]}$$

having the implicit two-parameter general solution (Section 2.2)

(11) $t = \pm \displaystyle\int \frac{dy}{\sqrt{2\big[F(y) + K\big]}} + c \ .$

We shall use formula (11) to illustrate some startling features of nonlinear equations.

EXAMPLE 1 Apply the energy integral lemma to explore the solutions of the nonlinear equation $y'' = 6y^2$, given in (3).

SOLUTION Since $6y^2 = \dfrac{d}{dy}(2y^3)$, the solution form (11) becomes

$$t = \pm \int \frac{dy}{\sqrt{2[2y^3 + K]}} + c \ .$$

For simplicity we take the plus sign and focus attention on solutions with $K = 0$. Then we find $t = \int \frac{1}{2} y^{-3/2} \, dy = -y^{-1/2} + c$, or

(12) $y(t) = (c - t)^{-2}$,

for any value of the constant c.

Clearly this equation is an enigma; solutions can blow up at $t = 1$, $t = 2$, $t = \pi$, or anywhere—and there is no clue in equation (3) as to why this should happen! Moreover, we have found an *infinite* family of linearly independent solutions (rather than the expected *two*). Yet we still cannot assemble, out of these, a solution matching (say) $y(0) = 1$, $y'(0) = 3$; all our solutions (12) have $y'(0) = 2y(0)^{3/2}$, and the ploy of using linear combinations $c_1 y_1(t) + c_2 y_2(t)$ will not work for a nonlinear equation.[†] ∎ End

EXAMPLE 2 Apply the energy integral lemma to explore the solutions of the nonlinear equation, $y'' = 24y^{1/3}$, given in (4).

SOLUTION Since $24y^{1/3} = \dfrac{d}{dy}(18y^{4/3})$, formula (11) gives

$$t = \pm \int \frac{dy}{\sqrt{2[18y^{4/3} + K]}} + c \ .$$

Again we take the plus sign and focus attention on solutions with $K = 0$. Then we find $t = y^{1/3}/2 + c$. In particular, $y_1(t) = 8t^3$ is a solution, and it satisfies the initial conditions $y(0) = 0$, $y'(0) = 0$. But note that $y_2(t) \equiv 0$ is also a solution to (4), *and it satisfies the same initial conditions*! Hence the *uniqueness* feature, guaranteed for certain linear equations by Theorem 2 of Section 4.2, can fail in the nonlinear case. ∎ End

So these examples have demonstrated violations of the existence and uniqueness properties (as well as "finiteness") that we have come to expect from the constant-coefficient case. It should not be surprising, then, that the solution techniques for variable-coefficient and nonlinear second order equations are more complicated—when, indeed, they exist. Recall, however, that in Section 4.6 we saw that our familiarity with the mass-spring oscillator equation

(13) $F_{\text{external}} = \big[\text{inertia}\big] y'' + \big[\text{damping}\big] y' + \big[\text{stiffness}\big] y$
 $= my'' + by' + ky$

was helpful in picturing the qualitative features of the solutions of other constant-coefficient equations. (See Figure 4.1, page 156). By pushing these analogies further we can also anticipate

[†]The *existence* of a solution matching these initial conditions is guaranteed by the methods discussed in Chapter 13 of the expanded text *Fundamentals of Differential Equations and Boundary Value Problems,* 3rd edition.

some of the features of the solutions in the variable-coefficient and nonlinear cases. One of the simplest linear second order differential equations with variable coefficients is

(14) $y'' + ty = 0$.

As a matter of fact, equation (14) is a variation of the **Airy equation**

(15) $y'' - ty = 0$.

EXAMPLE 3 Using the mass-spring analogy, predict the nature of the solutions to equation (14) for $t > 0$.

SOLUTION Comparing (13) with (14), we see that the latter equation describes a mass-spring oscillator where the *spring stiffness* "k" varies in time—in fact, it stiffens as time passes ("k" $= t$ in equation (14)). Physically, then, we would expect to see oscillations whose frequency increases with time, while the amplitude of the oscillations diminishes (because the spring gets harder to stretch). The numerically computed solution in Figure 4.13 displays precisely this behavior. ■End

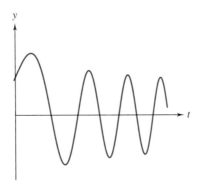

Figure 4.13 Solution to equation (14)

Remark. Schemes for numerically computing solutions to second order equations will be discussed in Section 5.6. If such schemes are available, then why is there a need for the qualitative analysis discussed in this section? The answer is that numerical methods provide only *approximations* to solutions of initial value problems, and their accuracy is sometimes difficult to predict (especially for nonlinear equations). For example, numerical methods are often ineffective near singular points, or over the long time intervals needed to study the asymptotic behavior. And this is precisely when qualitative arguments can lend insight into the reasonableness of the computed solution.

It is easy to verify (Problem 1) that if $y(t)$ is a solution of the *traditional* Airy equation $y'' - ty = 0$, then $y(-t)$ solves $y'' + ty = 0$, so "Airy functions" exhibit the behavior shown in Figure 4.13 for *negative time*. For positive t, the Airy equation has a *negative* stiffness "k" $= -t$, with magnitude increasing in time. As we observed in Section 4.6, negative stiffness tends to reinforce, rather than oppose, displacements, and the solutions $y(t)$ grow rapidly with (positive) time. The solution known as the Airy function of the second kind $Bi(t)$, depicted in Figure 4.14, behaves exactly as expected.

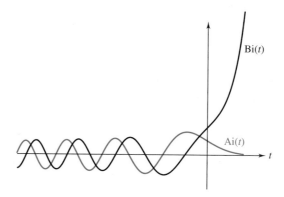

Figure 4.14 Airy functions

In Section 4.6 we also pointed out that mass-spring systems with negative spring stiffness can have isolated *bounded* solutions if the initial displacement $y(0)$ and velocity $y'(0)$ are precisely selected to counteract the repulsive spring force. The Airy function of the first kind $\text{Ai}(t)$, also depicted in Figure 4.14, is such a solution for the "Airy spring"; the initial inwardly directed velocity is just adequate to overcome the outward push of the stiffening spring, and the mass approaches a delicate equilibrium state $y(t) \equiv 0$.

Now let's look at **Bessel's equation.** It arises in the analysis of wave or diffusion phenomena in cylindrical and spherical coordinates. The *Bessel equation of order n* is written

$$(16) \qquad y'' + \frac{1}{t}y' + \left(1 - \frac{n^2}{t^2}\right)y = 0 \ .$$

Clearly there are irregularities at $t = 0$, analogous to those at $t = \pm 1$ for the Legendre equation (2); we will explore those in depth in Chapter 8.

EXAMPLE 4 Apply the mass-spring analogy to predict qualitative features of solutions to Bessel's equation for $t > 0$.

SOLUTION Comparing (16) with the paradigm (13), we observe that

- the inertia "m" $= 1$ is fixed at unity;
- there is positive damping ("b" $= 1/t$), although it weakens with time; and
- the stiffness ("k" $= 1 - n^2/t^2$) is positive—eventually (for $t > n$)—and approaches a constant (one).

Solutions, then, should be expected to oscillate with amplitudes that diminish slowly (due to the damping), and the frequency of the oscillations should settle at a constant value (given, according to the procedures of Section 4.6, by $\sqrt{k/m} = 1$ radian per unit time). The graphs of the Bessel functions $J_n(t)$ and $Y_n(t)$ of the first and second kind of order $n = \frac{1}{2}$ exemplify these qualitative predictions; see Figure 4.15, page 224. The effect of the singularities in the coefficients at $t = 0$ is manifested in the graph of $Y_{1/2}(t)$. ∎End

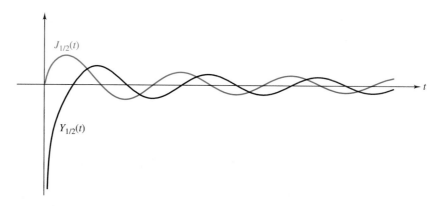

Figure 4.15 Bessel functions

Although most Bessel functions have to be computed by power series methods, if the *order* n is a half-integer then $J_n(t)$ and $Y_n(t)$ have closed-form expressions. In fact, $J_{1/2}(t) = \sqrt{\frac{2}{\pi t}} \sin t$ and $Y_{1/2}(t) = -\sqrt{\frac{2}{\pi t}} \cos t$. The reader can verify directly that these functions solve equation (16).

EXAMPLE 5 Give a qualitative analysis of the **modified Bessel equation** of order n:

$$(17) \qquad y'' + \frac{1}{t}y' - \left(1 - \frac{n^2}{t^2}\right)y = 0 \ .$$

SOLUTION This equation also exhibits unit mass and positive, diminishing, damping. However, the stiffness now converges to *negative* 1. Accordingly, we expect typical solutions to diverge as $t \to +\infty$. The modified Bessel function of the first kind, $I_n(t)$ of order $n = 2$ in Figure 4.16, follows this prediction, while the modified Bessel function of the second kind, $K_n(t)$ of order $n = 2$ in Figure 4.16, exhibits the same sort of balance of initial position and velocity as we saw for the Airy function $\text{Ai}(t)$. Again, the effect of the singularity at $t = 0$ is evident. ∎End

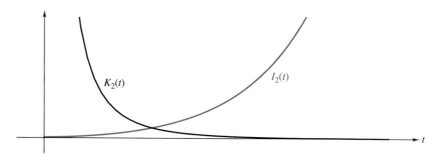

Figure 4.16 Modified Bessel functions

EXAMPLE 6 Use the mass-spring model to predict qualitative features of the solutions to the nonlinear **Duffing equation**

(18) $y'' + y + y^3 = y'' + (1 + y^2)y = 0$.

SOLUTION Although equation (18) is nonlinear, it can be matched with the paradigm (13) if we envision unit mass, no damping, and a (positive) stiffness "k" $= 1 + y^2$, which increases as the *displacement* y gets larger. (This increasing-stiffness effect is built into some popular mattresses for therapeutic reasons.)[†] Such a spring grows stiffer as the mass moves farther away, but it restores to its original value when the mass returns. Thus high-amplitude excursions should oscillate faster than low-amplitude ones, and the sinusoidal shapes in the graphs of $y(t)$ should be "pinched in" somewhat at their peaks. These qualitative predictions are demonstrated by the numerically computed solutions plotted in Figure 4.17. ■ End

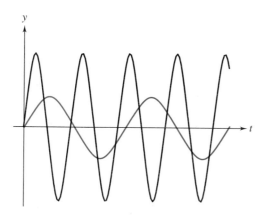

Figure 4.17 Solution graphs for the Duffing equation

Figure 4.18 Solutions to the van der Pol equation

The fascinating **van der Pol equation**

(19) $y'' - (1 - y^2)y' + y = 0$

originated in the study of the electrical oscillations observed in vacuum tubes.

EXAMPLE 7 Predict the behavior of the solutions to equation (19) using the mass-spring model.

SOLUTION By comparison to the paradigm (13), we observe unit mass and stiffness, and positive damping $\left["b" = -(1 - y^2)\right]$ when $|y(t)| > 1$ and negative damping when $|y(t)| < 1$. *Friction thus dampens large-amplitude motions, but energizes small oscillations.* The result, then, is that all (nonzero) solutions tend to the same *limit cycle* whose friction penalty incurred while $|y(t)| > 1$ is balanced by the negative-friction boost received while $|y(t)| < 1$. The computer-generated Figure 4.18 illustrates the convergence to the limit cycle for some solutions to the van der Pol equation. ■ End

[†]Graphic depictions of oscillations on a therapeutic mattress are best left to the reader's imagination, the editors say.

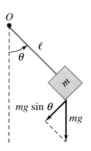

Figure 4.19 A pendulum

Finally, we consider the motion of the pendulum depicted in Figure 4.19. This motion is measured by the angle $\theta(t)$ that the pendulum makes with the vertical line through O at time t. As the diagram shows, the component of gravity, which exerts a torque on the pendulum and thus accelerates the angular velocity $d\theta/dt$, is given by $-mg \sin \theta$. Consequently, the rotational analog of Newton's second law, *torque equals rate of change of angular momentum,* dictates (see Problem 7)

(20) $m\ell^2\theta'' = -\ell mg \sin \theta$,

or

(21) $m\theta'' + m\dfrac{g}{\ell} \sin \theta = 0$.

EXAMPLE 8 Give a qualitative analysis of the motion of the pendulum.

SOLUTION By comparison with the paradigm (13), we see fixed mass m, no damping, and a stiffness given by

$$\text{``}k\text{''} = \frac{mg}{\ell} \frac{\sin \theta}{\theta} \; .$$

This stiffness is plotted in Figure 4.20. Small-amplitude motions are driven by a nearly constant spring stiffness of value mg/ℓ, and the considerations of Section 4.6 dictate the familiar formula

$$\omega = \sqrt{\frac{k}{m}} = \sqrt{\frac{g}{\ell}}$$

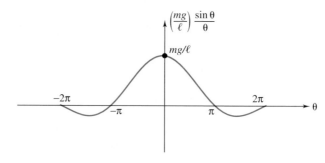

Figure 4.20 Pendulum "stiffness"

for the angular frequency of the nearly sinusoidal oscillations. See Figure 4.21, which compares a computer-generated solution to equation (21) to the solution of the constant-stiffness equation with the same initial conditions.[†]

For larger motions, however, the diminishing stiffness distorts the sinusoidal nature of the graph of $\theta(t)$, and lowers the frequency. This is evident in Figure 4.22.

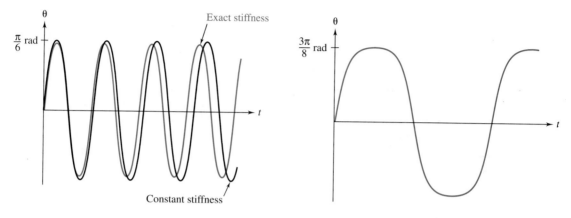

Figure 4.21 Small-amplitude pendulum motion

Figure 4.22 Large-amplitude pendulum motion

Finally, if the motion is so energetic that θ reaches the value π, the stiffness changes sign and abets the displacement; the pendulum passes the apex and gains speed as it falls, and this spinning motion repeats continuously. See Figure 4.23. ∎ End

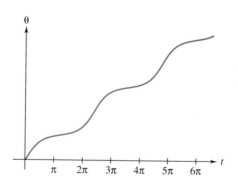

Figure 4.23 Very large amplitude pendulum motion

The computation of the solutions of the Legendre, Bessel, and Airy equations, and the analysis of the nonlinear equations of Duffing, van der Pol, and the pendulum have challenged many of the great mathematicians of the past. It is gratifying, then, to note that so many of their salient features are susceptible to the qualitative reasoning we have used herein.

[†]The latter is identified as the *linearized equation* in Project D.

EXERCISES 4.10

1. Show that if $y(t)$ satisfies $y'' - ty = 0$, then $y(-t)$ satisfies $y'' + ty = 0$.

2. Using the paradigm (13), what are the inertia, damping, and stiffness for the equation $y'' - 6y^2 = 0$? If $y > 0$, what is the sign of the "stiffness constant"? Does your answer help to explain the runaway behavior of the solutions $y(t) = 1/(c - t)^2$?

3. Try to predict the qualitative features of the solution to $y'' - 6y^2 = 0$ that satisfies the initial conditions $y(0) = -1$, $y'(0) = -1$. Compare with the computer-generated Figure 4.24. [*Hint:* Consider the sign of the spring stiffness.]

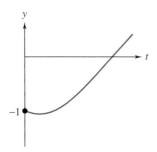

Figure 4.24 Solution for Problem 3

4. Show that the *three* solutions $1/(1 - t)^2$, $1/(2 - t)^2$, and $1/(3 - t)^2$ to $y'' - 6y^2 = 0$ are linearly independent on $(-1, 1)$. (See Problem 24, Exercises 4.3, page 177.)

5. (a) Use the energy integral lemma to derive the family of solutions $y(t) = 1/(t - c)$ to the equation $y'' = 2y^3$.
 (b) For $c \neq 0$, show that these solutions are pairwise linearly independent for different values of c in an appropriate interval around $t = 0$.
 (c) Show that none of these solutions satisfies the initial conditions $y(0) = 1$, $y'(0) = 2$.

6. Use the energy integral lemma to show that motions of the free undamped mass-spring oscillator $my'' + ky = 0$ obey
$$m(y')^2 + ky^2 = \text{constant.}$$

7. **Pendulum Equation.** To derive the pendulum equation (21), complete the following steps.
 (a) The *angular momentum* of the pendulum mass m measured about the support O in Figure 4.19 on page 226 is given by the product of the "lever arm" length ℓ and the component of the vector moment mv perpendicular to the lever arm. Show that this gives
$$\text{angular momentum} = m\ell^2 \frac{d\theta}{dt}.$$
 (b) The *torque* produced by gravity equals the product of the lever arm length ℓ and the component of gravitational (vector) force $m\mathbf{g}$ perpendicular to the lever arm. Show that this gives
$$\text{torque} = -\ell mg \sin\theta.$$
 (c) Now use Newton's law of rotational motion to deduce the pendulum equation (20).

8. Use the energy integral lemma to show that pendulum motions obey
$$\frac{(\theta')^2}{2} - \frac{g}{\ell}\cos\theta = \text{constant.}$$

9. Use the result of Problem 8 to find the value of $\theta'(0)$, the initial velocity that must be imparted to a pendulum at rest, to make it approach (but not cross over) the apex of its motion. Take $\ell = g$.

10. Use the result of Problem 8 to prove that if the pendulum in Figure 4.19 is released *from rest* at the angle α, $0 < \alpha < \pi$, then $|\theta(t)| \leq \alpha$ for all t. [*Hint:* The initial conditions are $\theta(0) = \alpha$, $\theta'(0) = 0$; argue that the *constant* in Problem 8 equals $-(g/\ell)\cos\alpha$.]

11. Use the mass-spring analogy to explain the qualitative nature of the solutions to the **Rayleigh equation**
 (22) $y'' - [1 - (y')^2]y' + y = 0$
 depicted in Figures 4.25 and 4.26.

12. Verify that the function $y_2(t)$ in equation (5) solves Legendre's equation (2).

13. Figure 4.27 contains graphs of solutions to the Duffing, Airy, and van der Pol equations. Try to match the solution to the equation.

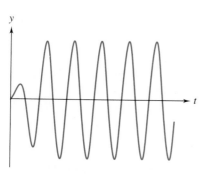

Figure 4.25 Solution to the
Rayleigh equation

Figure 4.26 Solution to the
Rayleigh equation

(a)

(b)

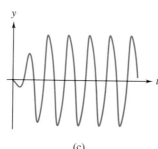

(c)

Figure 4.27 Solution graphs for Problem 13

14. Verify that the formulas for the Bessel functions $J_{1/2}(t)$, $Y_{1/2}(t)$ do indeed solve equation (16).

15. Use the mass-spring oscillator analogy to decide whether all solutions to each of the following differential equations are bounded as $t \rightarrow +\infty$.

(a) $y'' + t^2 y = 0$. **(b)** $y'' - t^2 y = 0$.

(c) $y'' + y^5 = 0$. **(d)** $y'' + y^6 = 0$.

(e) $y'' + (4 + 2\cos t)y = 0$ (**Mathieu's equation**).

(f) $y'' + ty' + y = 0$. **(g)** $y'' - ty' - y = 0$.

16. Use the energy integral lemma to show that every solution to the Duffing equation (18) is bounded; that is, $|y(t)| \leq M$ for some M. [*Hint:* First argue that $y^2/2 + y^4/4 \leq K$ for some K.]

17. Armageddon. Earth revolves around the sun in an approximately circular orbit with radius $r = a$,

completing a revolution in the time $T = 2\pi(a^3/GM)^{1/2}$, which is one Earth year; here M is the mass of the sun and G is the universal gravitational constant. The gravitational force of the sun on Earth is given by GMm/r^2, where m is the mass of Earth. Therefore, if Earth "stood still," losing its orbital velocity, it would fall on a straight line into the sun in accordance with Newton's second law:

$$m\frac{d^2r}{dt^2} = -\frac{GMm}{r^2} \ .$$

If this calamity occurred, what fraction of the normal year T would it take for Earth to splash into the sun (i.e., achieve $r = 0$)? [*Hint:* Use the energy integral lemma and the initial conditions $r(0) = a$, $r'(0) = 0$.]

4.11 A CLOSER LOOK AT FREE MECHANICAL VIBRATIONS

In this section, we return to the mass-spring system depicted in Figure 4.1, page 156, and analyze its motion in more detail. The governing equation is

(1) $F_{\text{external}} = [\text{inertia}]\dfrac{d^2y}{dt^2} + [\text{damping}]\dfrac{dy}{dt} + [\text{stiffness}]y$

$= my'' + by' + ky \ .$

Let's focus on the simple case in which $b = 0$ and $F_{\text{external}} = 0$, the so-called **undamped, free** case. Then equation (1) reduces to

(2) $m\dfrac{d^2y}{dt^2} + ky = 0$

and, when divided by m, becomes

(3) $\dfrac{d^2y}{dt^2} + \omega^2 y = 0 \ ,$

where $\omega = \sqrt{k/m}$. The auxiliary equation associated with (3) is $r^2 + \omega^2 = 0$, which has complex conjugate roots $\pm\omega i$. Hence a general solution to (3) is

(4) $y(t) = C_1 \cos \omega t + C_2 \sin \omega t \ .$

We can express $y(t)$ in the more convenient form

(5) $y(t) = A \sin(\omega t + \phi) \ ,$

with $A \geq 0$, by letting $C_1 = A \sin \phi$ and $C_2 = A \cos \phi$. That is,

$A \sin(\omega t + \phi) = A \cos \omega t \sin \phi + A \sin \omega t \cos \phi$

$= C_1 \cos \omega t + C_2 \sin \omega t \ .$

Solving for A and ϕ in terms of C_1 and C_2, we find

(6) $A = \sqrt{C_1^2 + C_2^2} \quad \text{and} \quad \tan \phi = \dfrac{C_1}{C_2} \ ,$

where the quadrant in which ϕ lies is determined by the signs of C_1 and C_2. This is because $\sin \phi$

has the same sign as C_1 ($\sin\phi = C_1/A$) and $\cos\phi$ has the same sign as C_2 ($\cos\phi = C_2/A$). For example, if $C_1 > 0$ and $C_2 < 0$, then ϕ is in Quadrant II. (Note, in particular, that ϕ is not simply the arctangent of C_1/C_2, which would lie in Quadrant IV.)

It is evident from (5) that, as we predicted in Section 4.1, the motion of a mass in an *undamped, free* system is simply a sine wave, or what is called **simple harmonic motion.** (See Figure 4.28.) The constant A is the amplitude of the motion and ϕ is the phase angle. The motion is periodic with **period** $2\pi/\omega$ and **natural frequency** $\omega/2\pi$, where $\omega = \sqrt{k/m}$. The period is measured in units of time, and the natural frequency has the dimensions of periods (or cycles) per unit time. The constant ω is the **angular frequency** for the sine function in (5) and has dimensions of radians per unit time. To summarize:

angular frequency $= \omega = \sqrt{k/m}$ (rad/sec)

natural frequency $= \omega/2\pi$ (cycles/sec)

period $= 2\pi/\omega$ (sec).

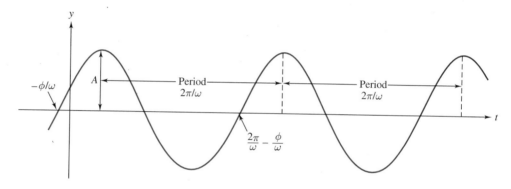

Figure 4.28 Simple harmonic motion of undamped, free vibrations

Observe that the amplitude and phase angle depend on the constants C_1 and C_2, which, in turn, are determined by the initial position and initial velocity of the mass. However, the period and frequency depend only on k and m and not on the initial conditions.

EXAMPLE 1 A 1/8-kg mass is attached to a spring with stiffness $k = 16$ N/m, as depicted in Figure 4.1. The mass is displaced 1/2 m to the right of the equilibrium point and given an outward velocity (to the right) of $\sqrt{2}$ m/sec. Neglecting any damping or external forces that may be present, determine the equation of motion of the mass along with its amplitude, period, and natural frequency. How long after release does the mass pass through the equilibrium position?

SOLUTION Because we have a case of undamped, free vibration, the equation governing the motion is (3). Thus we find the angular frequency to be

$$\omega = \sqrt{\frac{k}{m}} = \sqrt{\frac{16}{1/8}} = 8\sqrt{2} \text{ rad /sec.}$$

Substituting this value for ω into (4) gives

(7) $y(t) = C_1 \cos\left(8\sqrt{2}t\right) + C_2 \sin\left(8\sqrt{2}t\right)$.

Now we use the initial conditions, $y(0) = 1/2$ m and $y'(0) = \sqrt{2}$ m/sec, to solve for C_1 and C_2 in (7). That is,

$$1/2 = y(0) = C_1 \,,$$
$$\sqrt{2} = y'(0) = 8\sqrt{2}\,C_2 \,,$$

and so $C_1 = 1/2$ and $C_2 = 1/8$. Hence the equation of motion of the mass is

(8) $\qquad y(t) = \dfrac{1}{2}\cos\!\left(8\sqrt{2}t\right) + \dfrac{1}{8}\sin\!\left(8\sqrt{2}t\right) \,.$

To express $y(t)$ in the alternative form (5), we set

$$A = \sqrt{C_1^2 + C_2^2} = \sqrt{(1/2)^2 + (1/8)^2} = \frac{\sqrt{17}}{8} \,,$$

$$\tan\phi = \frac{C_1}{C_2} = \frac{1/2}{1/8} = 4 \,.$$

Since both C_1 and C_2 are positive, ϕ is in Quadrant I, so $\phi = \arctan 4 \approx 1.326$. Hence

(9) $\qquad y(t) = \dfrac{\sqrt{17}}{8}\sin\!\left(8\sqrt{2}t + \phi\right) \,.$

Thus the amplitude A is $\sqrt{17}/8$ m and the phase angle ϕ is approximately 1.326 radians. The period is $P = 2\pi/\omega = 2\pi/\left(8\sqrt{2}\right) = \sqrt{2}\pi/8$ sec, and the natural frequency is $1/P = 8/\left(\sqrt{2}\pi\right)$ cycles per sec.

Finally, to determine when the mass will pass through the equilibrium position, $y = 0$, we must solve the trigonometric equation

(10) $\qquad y(t) = \dfrac{\sqrt{17}}{8}\sin\!\left(8\sqrt{2}t + \phi\right) = 0$

for t. Equation (10) will be satisfied whenever

(11) $\qquad 8\sqrt{2}t + \phi = n\pi \quad \text{or} \quad t = \dfrac{n\pi - \phi}{8\sqrt{2}} \approx \dfrac{n\pi - 1.326}{8\sqrt{2}} \,,$

n an integer. Putting $n = 1$ in (11) determines the first time t when the mass crosses its equilibrium position:

$$t = \frac{\pi - \phi}{8\sqrt{2}} \approx 0.16 \text{ sec.} \quad \blacksquare \text{ End}$$

In most applications of vibrational analysis, of course, there is some type of frictional or damping force affecting the vibrations. This force may be due to a component in the system, such as a shock absorber in a car, or to the medium that surrounds the system, such as air or

some liquid. So we turn to a study of the effects of damping on free vibrations, and equation (2) generalizes to

(12) $m\dfrac{d^2y}{dt^2} + b\dfrac{dy}{dt} + ky = 0$.

The auxiliary equation associated with (12) is

(13) $mr^2 + br + k = 0$,

and its roots are

(14) $\dfrac{-b \pm \sqrt{b^2 - 4mk}}{2m} = -\dfrac{b}{2m} \pm \dfrac{1}{2m}\sqrt{b^2 - 4mk}$.

As we found in Sections 4.5 and 4.6, the form of the solution to (12) depends on the nature of these roots and, in particular, on the discriminant $b^2 - 4mk$.

Underdamped or Oscillatory Motion $\left(b^2 < 4mk\right)$

When $b^2 < 4mk$, the discriminant $b^2 - 4mk$ is negative and there are two complex conjugate roots to the auxiliary equation (13). These roots are $\alpha \pm i\beta$, where

(15) $\alpha := -\dfrac{b}{2m}$, $\beta := \dfrac{1}{2m}\sqrt{4mk - b^2}$.

Hence a general solution to (12) is

(16) $y(t) = e^{\alpha t}(C_1 \cos \beta t + C_2 \sin \beta t)$.

As we did with simple harmonic motion, we can express $y(t)$ in the alternate form

(17) $y(t) = Ae^{\alpha t}\sin(\beta t + \phi)$,

where $A = \sqrt{C_1^2 + C_2^2}$ and $\tan \phi = C_1/C_2$. It is now evident that $y(t)$ is the product of an exponential **damping factor,**

$Ae^{\alpha t} = Ae^{-(b/2m)t}$,

and a sine factor, $\sin(\beta t + \phi)$, which accounts for the oscillatory motion. Because the sine factor varies between -1 and 1 with period $2\pi/\beta$, the solution $y(t)$ varies between $-Ae^{\alpha t}$ and $Ae^{\alpha t}$ with **quasiperiod** $P = 2\pi/\beta = 4m\pi/\sqrt{4mk - b^2}$ and **quasifrequency** $1/P$. Moreover, since b and m are positive, $\alpha = -b/2m$ is negative, and thus the exponential factor tends to zero as $t \to +\infty$. A graph of a typical solution $y(t)$ is given in Figure 4.29, page 234. The system is called **underdamped** because there is not enough damping present (b is too small) to prevent the system from oscillating.

It is easily seen that as $b \to 0$, the damping factor approaches the constant A and the quasifrequency approaches the natural frequency of the corresponding undamped harmonic motion. Figure 4.29 demonstrates that the values of t where the graph of $y(t)$ touches the exponential curves $\pm Ae^{\alpha t}$ are *not* the same values of t at which $y(t)$ attains its relative maximum and minimum values (see Problem 13).

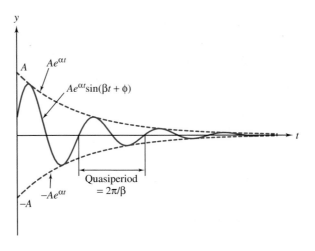

Figure 4.29 Damped oscillatory motion

Overdamped Motion $\left(b^2 > 4mk\right)$

When $b^2 > 4mk$, the discriminant $b^2 - 4mk$ is positive and there are two distinct real roots to the auxiliary equation (13):

(18) $r_1 = -\dfrac{b}{2m} + \dfrac{1}{2m}\sqrt{b^2 - 4mk}$, $r_2 = -\dfrac{b}{2m} - \dfrac{1}{2m}\sqrt{b^2 - 4mk}$.

Hence a general solution to (12) in this case is

(19) $y(t) = C_1 e^{r_1 t} + C_2 e^{r_2 t}$.

Obviously r_2 is negative. And since $b^2 > b^2 - 4mk$ (that is, $b > \sqrt{b^2 - 4mk}$), it follows that r_1 is also negative. Therefore, as $t \to +\infty$, both of the exponentials in (19) decay and $y(t) \to 0$. Moreover, since

$$y'(t) = C_1 r_1 e^{r_1 t} + C_2 r_2 e^{r_2 t} = e^{r_1 t}\left(C_1 r_1 + C_2 r_2 e^{(r_2 - r_1)t}\right) ,$$

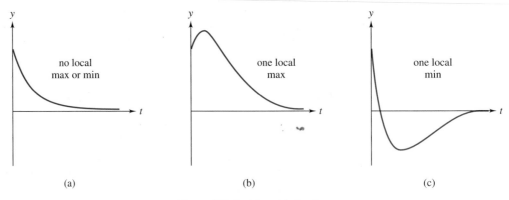

Figure 4.30 Overdamped vibrations

we see that the derivative is either identically zero (when $C_1 = C_2 = 0$) or vanishes for at most one value of t (when the factor in parentheses is zero). If the trivial solution $y(t) \equiv 0$ is ignored, it follows that $y(t)$ has at most one local maximum or minimum for $t > 0$. Therefore $y(t)$ *does not oscillate*. This leaves, qualitatively, only three possibilities for the motion of $y(t)$, depending on the initial conditions. These are illustrated in Figure 4.30. This case where $b^2 > 4mk$ is called **overdamped** motion.

Critically Damped Motion $\left(b^2 = 4mk\right)$

When $b^2 = 4mk$, the discriminant $b^2 - 4mk$ is zero and the auxiliary equation has the repeated root $-b/2m$. Hence a general solution to (12) is now

(20) $y(t) = (C_1 + C_2 t)e^{-(b/2m)t}$.

To understand the motion described by $y(t)$ in (20), we first consider the behavior of $y(t)$ as $t \to +\infty$. By L'Hôpital's rule,

(21) $\displaystyle \lim_{t \to +\infty} y(t) = \lim_{t \to +\infty} \frac{C_1 + C_2 t}{e^{(b/2m)t}} = \lim_{t \to +\infty} \frac{C_2}{(b/2m)e^{(b/2m)t}} = 0$

(recall that $b/2m > 0$). Hence $y(t)$ dies off to zero as $t \to +\infty$. Next, since

$$y'(t) = \left(C_2 - \frac{b}{2m}C_1 - \frac{b}{2m}C_2 t\right)e^{-(b/2m)t} ,$$

we see again that a nontrivial solution can have at most one local maximum or minimum for $t > 0$, so motion is *non*oscillatory. If b were any smaller, oscillation would occur. Thus the special case where $b^2 = 4mk$ is called **critically damped** motion. Qualitatively, critically damped motions are similar to overdamped motions (see Figure 4.30 again).

EXAMPLE 2 Assume the motion of a mass-spring system with damping is governed by

(22) $\displaystyle \frac{d^2 y}{dt^2} + b\frac{dy}{dt} + 25y = 0 ;$ $y(0) = 1 ,$ $y'(0) = 0$.

Find the equation of motion and sketch its graph for the three cases where $b = 6$, 10, and 12.

SOLUTION The auxiliary equation for (22) is

(23) $r^2 + br + 25 = 0$,

whose roots are

(24) $\displaystyle r = -\frac{b}{2} \pm \frac{1}{2}\sqrt{b^2 - 100}$.

Case 1. When $b = 6$, the roots (24) are $-3 \pm 4i$. This is thus a case of underdamping, and the equation of motion has the form

(25) $y(t) = C_1 e^{-3t}\cos 4t + C_2 e^{-3t}\sin 4t$.

Setting $y(0) = 1$ and $y'(0) = 0$ gives the system

$$C_1 = 1 , \quad -3C_1 + 4C_2 = 0 ,$$

whose solution is $C_1 = 1$, $C_2 = 3/4$. To express $y(t)$ as the product of a damping factor and a sine factor (recall equation (17)), we set

$$A = \sqrt{C_1^2 + C_2^2} = \frac{5}{4} , \qquad \tan \phi = \frac{C_1}{C_2} = \frac{4}{3} ,$$

where ϕ is a Quadrant I angle, since C_1 and C_2 are both positive. Then

(26) $$y(t) = \frac{5}{4} e^{-3t} \sin(4t + \phi) ,$$

where $\phi = \arctan(4/3) \approx 0.9273$. The underdamped spring motion is shown in Figure 4.31(a).

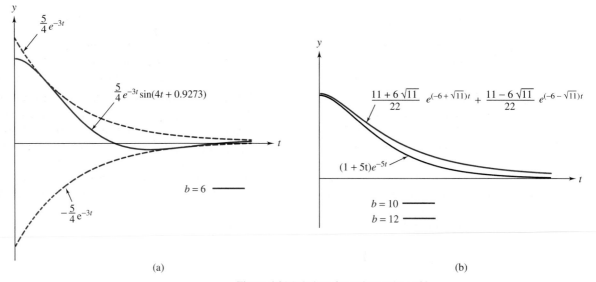

(a) (b)

Figure 4.31 Solutions for various values of b

Case 2. When $b = 10$, there is only one (repeated) root to the auxiliary equation (23), namely $r = -5$. This is a case of critical damping, and the equation of motion has the form

(27) $$y(t) = (C_1 + C_2 t) e^{-5t} .$$

Setting $y(0) = 1$ and $y'(0) = 0$ now gives

$$C_1 = 1 , \qquad C_2 - 5C_1 = 0 ,$$

and so $C_1 = 1$, $C_2 = 5$. Thus

(28) $$y(t) = (1 + 5t) e^{-5t} .$$

The graph of $y(t)$ given in (28) is represented by the dashed line in Figure 4.31(b). Notice that $y(t)$ is zero only for $t = -1/5$ and hence does not cross the t-axis for $t > 0$.

Case 3. When $b = 12$, the roots to the auxiliary equation are $-6 \pm \sqrt{11}$. This is a case of overdamping, and the equation of motion has the form

$$(29) \qquad y(t) = C_1 e^{(-6+\sqrt{11})t} + C_2 e^{(-6-\sqrt{11})t} \; .$$

Setting $y(0) = 1$ and $y'(0) = 0$ gives

$$C_1 + C_2 = 1 \; , \qquad \left(-6 + \sqrt{11}\right)C_1 + \left(-6 - \sqrt{11}\right)C_2 = 0 \; ,$$

from which we find $C_1 = \left(11 + 6\sqrt{11}\right)/22$ and $C_2 = \left(11 - 6\sqrt{11}\right)/22$. Hence

$$(30) \qquad y(t) = \frac{11 + 6\sqrt{11}}{22} e^{(-6+\sqrt{11})t} + \frac{11 - 6\sqrt{11}}{22} e^{(-6-\sqrt{11})t}$$

$$= \frac{e^{(-6+\sqrt{11})t}}{22} \left\{ 11 + 6\sqrt{11} + \left(11 - 6\sqrt{11}\right) e^{-2\sqrt{11}t} \right\} \; .$$

The graph of this overdamped motion is represented by the solid line in Figure 4.31(b). ∎ End

It is interesting to observe in Example 2 that when the system is underdamped ($b = 6$), the solution goes to zero like e^{-3t}; when the system is critically damped ($b = 10$), the solution tends to zero roughly like e^{-5t}; and when the system is overdamped ($b = 12$), the solution goes to zero like $e^{(-6+\sqrt{11})t} \approx e^{-2.68t}$. This means that if the system is underdamped, it not only oscillates, but also dies off slower than if it were critically damped. Moreover, if the system is overdamped, it again dies off more slowly than if it were critically damped. (This agrees with our physical intuition that the damping forces hinder the return to equilibrium.)

EXAMPLE 3 A 1/4-kg mass is attached to a spring with a stiffness 4 N/m as shown in Figure 4.32(a). The damping constant b for the system is 1 N-sec/m. If the mass is displaced 1/2 m to the left and given an initial velocity of 1 m/sec to the left, find the equation of motion. What is the maximum displacement that the mass will attain?

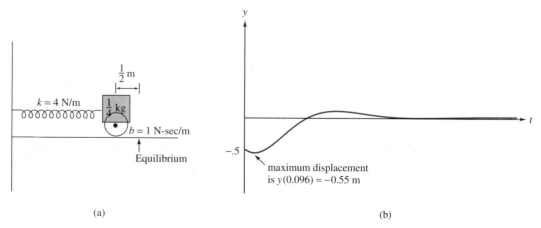

(a)

(b)

Figure 4.32 Mass-spring system and graph of motion for Example 3

SOLUTION Substituting the values for m, b, and k into equation (12) and enforcing the initial conditions, we obtain the initial value problem

(31) $$\frac{1}{4}\frac{d^2y}{dt^2} + \frac{dy}{dt} + 4y = 0 \ ; \qquad y(0) = -\frac{1}{2}, \qquad y'(0) = -1 \ .$$

The negative signs for the initial conditions reflect the facts that the initial displacement and push are to the left.

The reader can verify that the solution to (31) is

(32) $$y(t) = -\frac{1}{2}e^{-2t}\cos\left(2\sqrt{3}t\right) - \frac{1}{\sqrt{3}}e^{-2t}\sin\left(2\sqrt{3}t\right) \ ,$$

or

(33) $$y(t) = \sqrt{\frac{7}{12}}\,e^{-2t}\sin\left(2\sqrt{3}t + \phi\right) \ ,$$

where $\tan\phi = \sqrt{3}/2$ and ϕ lies in Quadrant III because $C_1 = -1/2$ and $C_2 = -1/\sqrt{3}$ are both negative. (See Figure 4.32(b) for a sketch of $y(t)$.)

To determine the maximum displacement from equilibrium, we must determine the maximum value of $|y(t)|$ on the graph in Figure 4.32(b). Because $y(t)$ dies off exponentially, this will occur at the first critical point of $y(t)$. Computing $y'(t)$ from (32), setting it equal to zero, and solving gives

$$y'(t) = e^{-2t}\left\{\frac{5}{\sqrt{3}}\sin\left(2\sqrt{3}t\right) - \cos\left(2\sqrt{3}t\right)\right\} = 0 \ ,$$

$$\frac{5}{\sqrt{3}}\sin\left(2\sqrt{3}t\right) = \cos\left(2\sqrt{3}t\right) \ ,$$

$$\tan\left(2\sqrt{3}t\right) = \frac{\sqrt{3}}{5} \ .$$

Thus the first positive root is

$$t = \frac{1}{2\sqrt{3}}\arctan\frac{\sqrt{3}}{5} \approx 0.096 \ .$$

Substituting this value for t back into equation (32) or (33) gives $y(0.096) \approx -0.55$. Hence the maximum displacement, which occurs to the left of equilibrium, is approximately 0.55 m. ■ End

EXERCISES 4.11

All problems refer to the mass-spring configuration depicted in Figure 4.1, page 156.

 1. A 3-kg mass is attached to a spring with stiffness $k = 48$ N/m. The mass is displaced 1/2 m to the left of the equilibrium point and given a velocity of 2 m/sec to the right. The damping force is negligible. Find the equation of motion of the mass along with the amplitude, period, and frequency. How long

after release does the mass pass through the equilibrium position?

2. A 2-kg mass is attached to a spring with stiffness $k = 50$ N/m. The mass is displaced $1/4$ m to the left of the equilibrium point and given a velocity of 1 m/sec to the left. Neglecting damping, find the equation of motion of the mass along with the amplitude, period, and frequency. How long after release does the mass pass through the equilibrium position?

3. The motion of a mass-spring system with damping is governed by

$$y''(t) + by'(t) + 16y(t) = 0 ;$$
$$y(0) = 1 , \qquad y'(0) = 0 .$$

Find the equation of motion and sketch its graph for $b = 0, 6, 8,$ and 10.

4. The motion of a mass-spring system with damping is governed by

$$y''(t) + by'(t) + 64y(t) = 0 ;$$
$$y(0) = 1 , \qquad y'(0) = 0 .$$

Find the equation of motion and sketch its graph for $b = 0, 10, 16,$ and 20.

5. The motion of a mass-spring system with damping is governed by

$$y''(t) + 10y'(t) + ky(t) = 0 ;$$
$$y(0) = 1 , \qquad y'(0) = 0 .$$

Find the equation of motion and sketch its graph for $k = 20, 25,$ and 30.

6. The motion of a mass-spring system with damping is governed by

$$y''(t) + 4y'(t) + ky(t) = 0 ;$$
$$y(0) = 1 , \qquad y'(0) = 0 .$$

Find the equation of motion and sketch its graph for $k = 2, 4,$ and 6.

7. A $1/8$-kg mass is attached to a spring with stiffness 16 N/m. The damping constant for the system is 2 N-sec/m. If the mass is moved $3/4$ m to the left of equilibrium and given an initial leftward velocity of 2 m/sec, determine the equation of motion of the mass and give its damping factor, quasiperiod, and quasifrequency.

8. A 20-kg mass is attached to a spring with stiffness 200 N/m. The damping constant for the system is 140 N-sec/m. If the mass is pulled 25 cm to the right

of equilibrium and given an initial leftward velocity of 1 m/sec, when will it first return to its equilibrium position?

9. A 2-kg mass is attached to a spring with stiffness 40 N/m. The damping constant for the system is $8\sqrt{5}$ N-sec/m. If the mass is pulled 10 cm to the right of equilibrium and given an initial rightward velocity of 2 m/sec, what is the maximum displacement from equilibrium that it will attain?

10. A $1/4$-kg mass is attached to a spring with stiffness 8 N/m. The damping constant for the system is $1/4$ N-sec/m. If the mass is moved 1 m to the left of equilibrium and released, what is the maximum displacement to the right that it will attain?

11. A 1-kg mass is attached to a spring with stiffness 100 N/m. The damping constant for the system is 0.2 N-sec/m. If the mass is pushed rightward from the equilibrium position with a velocity of 1 m/sec, when will it attain its maximum displacement to the right?

12. A $1/4$-kg mass is attached to a spring with stiffness 8 N/m. The damping constant for the system is 2 N-sec/m. If the mass is pushed 50 cm to the left of equilibrium and given a leftward velocity of 2 m/sec, when will the mass attain its maximum displacement to the left?

13. Show that for the underdamped system of Example 3, the times when the solution curve $y(t)$ in (33) touches the exponential curves $\pm\sqrt{7/12}\,e^{-2t}$ are not the same values of t for which the function $y(t)$ attains its relative extrema.

14. For an underdamped system, verify that as $b \to 0$ the damping factor approaches the constant A and the quasifrequency approaches the natural frequency $\sqrt{k/m}/(2\pi)$.

15. How can one deduce the value of the damping constant b by observing the motion of an underdamped system? Assume that the mass m is known.

16. A mass attached to a spring oscillates with a period of 3 sec. After 2 kg are added, the period becomes 4 sec. Assuming we can neglect any damping or external forces, determine how much mass was originally attached to the spring.

4.12 A CLOSER LOOK AT FORCED MECHANICAL VIBRATIONS

We now consider the vibrations of a mass-spring system when an external force is applied. Of particular interest is the response of the system to a *periodic* forcing term. As a paradigm, let's investigate the effect of a cosine forcing function on the system governed by the differential equation

$$(1) \qquad m\frac{d^2y}{dt^2} + b\frac{dy}{dt} + ky = F_0 \cos \gamma t \ ,$$

where F_0, γ are nonnegative constants and $0 < b^2 < 4mk$ (so the system is *underdamped*).

A solution to (1) has the form $y = y_h + y_p$, where y_p is a particular solution and y_h is a general solution to the corresponding homogeneous equation. We found in equation (17) of Section 4.11 that

$$(2) \qquad y_h(t) = Ae^{-(b/2m)t} \sin\left(\frac{\sqrt{4mk - b^2}}{2m}t + \phi\right) \ ,$$

where A and ϕ are constants.

To determine y_p, we can use the method of undetermined coefficients (Section 4.8). From the form of the nonhomogeneous term, we know that

$$(3) \qquad y_p(t) = A_1 \cos \gamma t + A_2 \sin \gamma t \ ,$$

where A_1, A_2 are constants to be determined. Substituting this expression into equation (1) and simplifying gives

$$(4) \qquad \left[(k - m\gamma^2)A_1 + b\gamma A_2\right]\cos \gamma t + \left[(k - m\gamma^2)A_2 - b\gamma A_1\right]\sin \gamma t = F_0 \cos \gamma t \ .$$

Setting the corresponding coefficients on both sides equal, we have

$$(k - m\gamma^2)A_1 + b\gamma A_2 = F_0 \ ,$$
$$-b\gamma A_1 + (k - m\gamma^2)A_2 = 0 \ .$$

Solving, we obtain

$$(5) \qquad A_1 = \frac{F_0(k - m\gamma^2)}{(k - m\gamma^2)^2 + b^2\gamma^2} \ , \qquad A_2 = \frac{F_0 b\gamma}{(k - m\gamma^2)^2 + b^2\gamma^2} \ .$$

Hence a particular solution to (1) is

$$(6) \qquad y_p(t) = \frac{F_0}{(k - m\gamma^2)^2 + b^2\gamma^2}\left\{(k - m\gamma^2)\cos \gamma t + b\gamma \sin \gamma t\right\} \ .$$

The expression in braces can also be written as

$$\sqrt{(k - m\gamma^2)^2 + b^2\gamma^2} \sin(\gamma t + \theta) \ ,$$

so we can express y_p in the alternative form

$$(7) \qquad y_p(t) = \frac{F_0}{\sqrt{(k - m\gamma^2)^2 + b^2\gamma^2}}\sin(\gamma t + \theta) \ ,$$

where $\tan \theta = A_1/A_2 = (k - m\gamma^2)/(b\gamma)$ and the quadrant in which θ lies is determined by the signs of A_1 and A_2.

Combining equations (2) and (7), we have the following representation of a general solution to (1) in the case $0 < b^2 < 4mk$:

$$(8) \qquad y(t) = Ae^{-(b/2m)t}\sin\left(\frac{\sqrt{4mk - b^2}}{2m}t + \phi\right) + \frac{F_0}{\sqrt{(k - m\gamma^2)^2 + b^2\gamma^2}}\sin(\gamma t + \theta).$$

The solution (8) is the sum of two terms. The first term, y_h, represents damped oscillation and depends only on the parameters of the system and the initial conditions. Because of the damping factor $Ae^{-(b/2m)t}$, this term tends to zero as $t \to +\infty$. Consequently, it is referred to as the **transient** part of the solution. The second term, y_p, in (8) is the offspring of the external forcing function $f(t) = F_0 \cos \gamma t$. Like the forcing function, y_p is a sinusoid with angular frequency γ. It is the synchronous solution that we anticipated in Section 4.1. However, y_p is out of phase with $f(t)$ (by the angle $\theta - \pi/2$), and its magnitude is different by the factor

$$(9) \qquad \frac{1}{\sqrt{(k - m\gamma^2)^2 + b^2\gamma^2}}.$$

As the transient term dies off, the motion of the mass-spring system becomes essentially that of the second term y_p (see Figure 4.33). Hence this term is called the **steady-state** solution. The factor appearing in (9) is referred to as the **frequency gain** or **gain factor** since it represents the ratio of the magnitude of the sinusoidal response to that of the input force. Note that this factor depends on the frequency γ and has units of length/force.

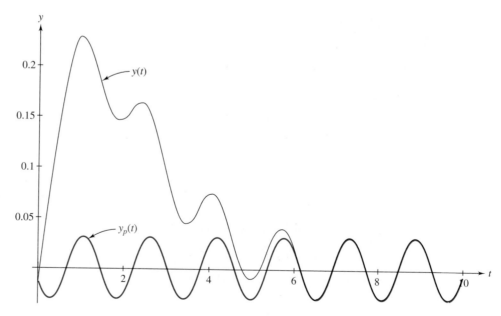

Figure 4.33 Convergence of $y(t)$ to the steady-state solution $y_p(t)$ when $m = 4$, $b = 6$, $k = 3$, $F_0 = 2$, $\gamma = 4$

EXAMPLE 1 A 10-kg mass is attached to a spring with stiffness $k = 49$ N/m. At time $t = 0$, an external force $f(t) = 20 \cos 4t$ N is applied to the system. The damping constant for the system is 3 N-sec/m. Determine the steady-state solution for the system.

SOLUTION Substituting the given parameters into equation (1), we obtain

$$(10) \qquad 10\frac{d^2y}{dt^2} + 3\frac{dy}{dt} + 49y = 20 \cos 4t \ ,$$

where $y(t)$ is the displacement (from equilibrium) of the mass at time t.

To find the steady-state response, we must produce a particular solution to (10) that is a sinusoid. We can do this using the method of undetermined coefficients, guessing a solution of the form $A_1 \cos 4t + A_2 \sin 4t$. But this is precisely how we derived equation (7). Thus we substitute directly into (7) and find

$$(11) \qquad y_p(t) = \frac{20}{\sqrt{(49 - 160)^2 + (9)(16)}} \sin(4t + \theta) \approx (0.18)\sin(4t + \theta) \ ,$$

where $\tan \theta = (49 - 160)/12 \approx -9.25$. Since the numerator, $(49 - 160)$, is negative and the denominator, 12, positive, θ is a Quadrant IV angle. Thus

$$\theta \approx \arctan(-9.25) \approx -1.46 \ ,$$

and the steady-state solution is given (approximately) by

$$(12) \qquad y_p(t) = (0.18)\sin(4t - 1.46). \quad \blacksquare \text{ End}$$

The above example illustrates an important point made earlier: The steady-state response (12) to the sinusoidal forcing function $20 \cos 4t$ is a sinusoid of the same frequency but different amplitude. The gain factor (see (9)) in this case is $(0.18)/20 = 0.009$ m/N.

In general, the amplitude of the steady-state solution to equation (1) depends on the angular frequency γ of the forcing function and is given by $A(\gamma) = F_0 M(\gamma)$, where

$$(13) \qquad M(\gamma) := \frac{1}{\sqrt{(k - m\gamma^2)^2 + b^2\gamma^2}}$$

is the frequency gain (see (9)). This formula is valid even when $b^2 \geq 4mk$. For a given system (m, b, and k fixed), it is often of interest to know how this system reacts to sinusoidal inputs of various frequencies (γ is a variable). For this purpose, the graph of the gain $M(\gamma)$, called the **frequency response curve** or **resonance curve** for the system, is enlightening.

To sketch the frequency response curve, we first observe that for $\gamma = 0$ we find $M(0) = 1/k$. Of course $\gamma = 0$ implies the force $F_0 \cos \gamma t$ is static; there is no motion in the steady-state, so this value of $M(0)$ is appropriate. Also note that as $\gamma \to \infty$, the gain $M(\gamma) \to 0$; the inertia of the system limits the extent to which it can respond to extremely rapid vibrations. As a further aid in describing the graph, we compute from (13)

$$(14) \qquad M'(\gamma) = -\frac{2m^2\gamma\left[\gamma^2 - \left(\frac{k}{m} - \frac{b^2}{2m^2}\right)\right]}{\left[(k - m\gamma^2)^2 + b^2\gamma^2\right]^{3/2}} \ .$$

It follows from (14) that $M'(\gamma) = 0$ if and only if

$$(15) \qquad \gamma = 0 \quad \text{or} \quad \gamma = \gamma_r := \sqrt{\frac{k}{m} - \frac{b^2}{2m^2}} \,.$$

Now when the system is overdamped or critically damped, so $b^2 \geq 4mk > 2mk$, the term inside the radical in (15) is negative and hence $M'(\gamma) = 0$ only when $\gamma = 0$. In this case, as γ increases from 0 to infinity, $M(\gamma)$ decreases from $M(0) = 1/k$ to a limiting value of zero.

When $b^2 < 2mk$ (which implies the system is underdamped), then γ_r is real and positive, and it is easy to verify that $M(\gamma)$ has a *maximum* at γ_r. Substituting γ_r into (13) gives

$$(16) \qquad M(\gamma_r) = \frac{1}{b\sqrt{\dfrac{k}{m} - \dfrac{b^2}{4m^2}}} \,.$$

The value $\gamma_r/2\pi$ is called the **resonance frequency** for the system. When the system is stimulated by an external force at this frequency, it is said to be **at resonance.**

To illustrate the effect of the damping constant b on the resonance curve, we consider a system in which $m = k = 1$. In this case, the frequency response curves are given by

$$(17) \qquad M(\gamma) = \frac{1}{\sqrt{(1 - \gamma^2)^2 + b^2\gamma^2}} \,,$$

and, for $b < \sqrt{2}$, the resonance frequency is $\gamma_r/2\pi = (1/2\pi)\sqrt{1 - b^2/2}$. Figure 4.34

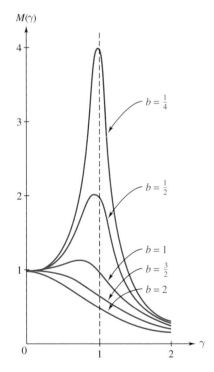

Figure 4.34 Frequency response curves for various values of b

displays the graphs of these frequency response curves for $b = 1/4, 1/2, 1, 3/2$, and 2. Observe that as $b \rightarrow 0$, the maximum magnitude of the frequency gain increases and the resonance frequency $\gamma_r/2\pi$ for the damped system approaches $\sqrt{k/m}/2\pi = 1/2\pi$, the natural frequency for the undamped system.

To understand what is occurring, consider the undamped system $(b = 0)$ with forcing term $F_0 \cos \gamma t$. This system is governed by

$$\textbf{(18)} \qquad m\frac{d^2y}{dt^2} + ky = F_0 \cos \gamma t \ .$$

A general solution to (18) is the sum of a particular solution and a general solution to the homogeneous equation. In Section 4.11, we showed that the latter describes simple harmonic motion:

$$\textbf{(19)} \qquad y_h(t) = A \sin(\omega t + \phi) \ , \qquad \omega := \sqrt{k/m} \ .$$

The formula for the particular solution given in (7) is valid for $b = 0$, provided $\gamma \neq \omega = \sqrt{k/m}$. However, when $b = 0$ and $\gamma = \omega$, then the form we used with undetermined coefficients to derive (7) does not work because $\cos \omega t$ and $\sin \omega t$ are solutions to the corresponding homogeneous equation. The correct form is

$$\textbf{(20)} \qquad y_p(t) = A_1 t \cos \omega t + A_2 t \sin \omega t \ ,$$

which leads to the solution

$$\textbf{(21)} \qquad y_p(t) = \frac{F_0}{2m\omega} t \sin \omega t \ .$$

(The verification of (21) is left to the reader.) Hence, in the *undamped resonant case* (when $\gamma = \omega$), a general solution to (18) is

$$\textbf{(22)} \qquad y(t) = A \sin(\omega t + \phi) + \frac{F_0}{2m\omega} t \sin \omega t \ .$$

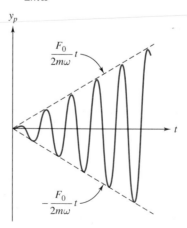

Figure 4.35 Undamped oscillation of the particular solution in (21)

Returning to the question of resonance, observe that the particular solution in (21) oscillates between $\pm(F_0 t)/(2m\omega)$. Hence as $t \to +\infty$, the maximum magnitude of (21) approaches ∞ (see Figure 4.35).

It is obvious from the above discussion that **if the damping constant b is very small, the system is subject to large oscillations when the forcing function has a frequency near the resonance frequency for the system.** It is these large vibrations at resonance that concern engineers. Indeed, resonance vibrations have been known to cause airplane wings to snap, bridges to collapse,[†] and (less catastrophically) wine glasses to shatter.

When the mass-spring system is hung vertically as in Figure 4.36, the force of gravity must be taken into account. This is accomplished very easily. With y measured downward from the unstretched spring position, the governing equation is

$$my'' + by' + ky = mg \ ,$$

and if the right-hand side is recognized as a sinusoidal forcing term with frequency zero $(mg \cos 0t)$, then the synchronous steady-state response is a constant, which is easily seen to be $y_p(t) = mg/k$. Now if we redefine $y(t)$ to be measured *from this (true) equilibrium level,* as indicated in Figure 4.36(c),

$$y_{new}(t) := y(t) - mg/k \ ,$$

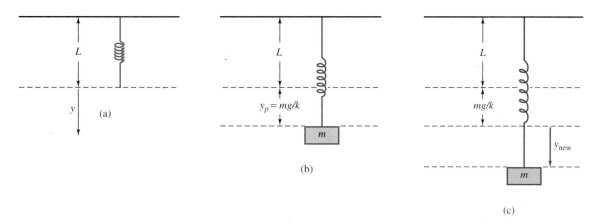

Figure 4.36 Spring (a) in natural position, (b) in equilibrium, and (c) in motion

[†]An interesting discussion of one such disaster, involving the Tacoma Narrows bridge in Washington State, can be found in *Differential Equations and Their Applications,* 4th edition, by M. Braun (Springer-Verlag, New York, 1993). Researchers are still trying to understand this disaster; see the articles "Large-amplitude periodic oscillations in suspension bridges: some new connections with nonlinear analysis," by A. C. Lazer and P. J. McKenna, *SIAM Review,* Vol. 32 (1990), pp. 537–578 or "Still twisting," by Henry Petroski, *American Scientist,* Vol. 19 (1991), pp. 398–401.

then the governing equation

$$my'' + by' + ky = mg + F_{external}(t)$$

simplifies; we find

$$
\begin{aligned}
my''_{new} + by'_{new} + ky_{new} &= m(y - mg/k)'' + b(y - mg/k)' + k(y - mg/k) \\
&= my'' + by' + ky - mg \\
&= mg + F_{external}(t) - mg \ ,
\end{aligned}
$$

or

(23) $my''_{new} + by'_{new} + ky_{new} = F_{external}(t)$.

Thus the gravitational force can be ignored if $y(t)$ is measured from the equilibrium position. Adopting this convention, we drop the "new" subscript hereafter.

EXAMPLE 2 Suppose the mass-spring system in Example 1 is hung vertically. Find the steady-state solution.

SOLUTION This is trivial; the steady-state solution is identical to what we derived before,

$$y_p(t) = \frac{20}{\sqrt{(49 - 160)^2 + (9)(16)}} \sin(4t + \theta)$$

(equation (11)), but now y_p is measured from the equilibrium position, which is $mg/k \approx 10 \times 9.8/49 = 2$ m below the unstretched spring position. ■ End

EXAMPLE 3 A 64-lb weight is attached to a vertical spring, causing it to stretch 3 in. upon coming to rest at equilibrium. The damping constant for the system is 3 lb-sec/ft. An external force $F(t) = 3 \cos 12t$ lb is applied to the weight. Find the steady-state solution for the system.

SOLUTION If a weight of 64 lb stretches a spring by 3 in. (0.25 ft), then the spring stiffness must be $64/0.25 = 256$ lb/ft. Thus if we measured the displacement from the (true) equilibrium level, equation (23) becomes

(24) $my'' + by' + ky = 3 \cos 12t$,

with $b = 3$ and $k = 256$. But recall that the unit of mass in the U.S. Customary System is the *slug*, which equals the weight divided by the gravitational acceleration constant $g \approx 32$ ft/sec^2 (Table 3.2, page 116). Therefore m in equation (24) is $64/32 = 2$ slugs, and we have

$$2y'' + 3y' + 256y = 3 \cos 12t \ .$$

The steady-state solution is given by equation (6) with $F_0 = 3$ and $\gamma = 12$:

$$
\begin{aligned}
y_p(t) &= \frac{3}{(256 - 2 \cdot 12^2)^2 + 3^2 \cdot 12^2}\{(256 - 2 \cdot 12^2) \cos 12t + 3 \cdot 12 \sin 12t\} \\
&= \frac{3}{580}\{-8 \cos 12t + 9 \sin 12t\} \ . \quad ■ End
\end{aligned}
$$

EXERCISES 4.12

In the following problems, take g = 32 ft/sec² for the U.S. Customary System and g = 9.8 m/sec² for the MKS system.

1. Sketch the frequency response curve (13) for the system in which $m = 4$, $k = 1$, $b = 2$.

2. Sketch the frequency response curve (13) for the system in which $m = 2$, $k = 3$, $b = 3$.

3. Determine the equation of motion for an undamped system at resonance governed by

$$\frac{d^2y}{dt^2} + 9y = 2\cos 3t \ ;$$

$$y(0) = 1 \ , \qquad y'(0) = 0 \ .$$

Sketch the solution.

4. Determine the equation of motion for an undamped system at resonance governed by

$$\frac{d^2y}{dt^2} + y = 5\cos t \ ;$$

$$y(0) = 0 \ , \qquad y'(0) = 1 \ .$$

Sketch the solution.

5. An undamped system is governed by

$$m\frac{d^2y}{dt^2} + ky = F_0\cos\gamma t \ ,$$

$$y(0) = y'(0) = 0 \ ,$$

where $\gamma \neq \omega := \sqrt{k/m}$.

 (a) Find the equation of motion of the system.

 (b) Use trigonometric identities to show that the solution can be written in the form

$$y(t) = \frac{2F_0}{m(\omega^2 - \gamma^2)} \sin\left(\frac{\omega + \gamma}{2}t\right)\sin\left(\frac{\omega - \gamma}{2}t\right) \ .$$

 (c) When γ is near ω, then $\omega - \gamma$ is small, while $\omega + \gamma$ is relatively large compared with $\omega - \gamma$. Hence $y(t)$ can be viewed as the product of a slowly varying sine function, $\sin\left[(\omega - \gamma)t/2\right]$, and a rapidly varying sine function, $\sin\left[(\omega + \gamma)t/2\right]$. The net effect is a sine function $y(t)$ with frequency $(\omega - \gamma)/4\pi$, which serves as the time-varying amplitude of a sine function with frequency $(\omega + \gamma)/4\pi$. This vibration phenom-

enon is referred to as **beats** and is used in tuning stringed instruments. This same phenomenon in electronics is called **amplitude modulation.** To illustrate this phenomenon, sketch the curve $y(t)$ for $F_0 = 32$, $m = 2$, $\omega = 9$, and $\gamma = 7$.

6. Derive the formula for $y_p(t)$ given in (21).

7. Shock absorbers in automobiles, aircraft, etc. can be described as forced *overdamped* mass-spring systems. Derive an expression analogous to equation (8) for the general solution to the differential equation (1) when $b^2 > 4mk$.

8. The response of an overdamped system to a constant force is governed by equation (1) with $m = 2$, $b = 8$, $k = 6$, $F_0 = 18$, and $\gamma = 0$. If the system starts from rest $(y(0) = y'(0) = 0)$, compute and sketch the displacement $y(t)$. What is the limiting value of $y(t)$ as $t \to +\infty$? Interpret this physically.

9. An 8-kg mass is attached to a spring hanging from the ceiling, thereby causing the spring to stretch 1.96 m upon coming to rest at equilibrium. At time $t = 0$, an external force $F(t) = \cos 2t$ N is applied to the system. The damping constant for the system is 3 N-sec/m. Determine the steady-state solution for the system.

10. Show that the period of the simple harmonic motion of a mass hanging from a spring is $2\pi\sqrt{l/g}$, where l denotes the amount (beyond its natural length) that the spring is stretched when the mass is at equilibrium.

11. A mass weighing 8 lb is attached to a spring hanging from the ceiling and comes to rest at its equilibrium position. At $t = 0$, an external force $F(t) = 2\cos 2t$ lb is applied to the system. If the spring constant is 10 lb/ft and the damping constant is 1 lb-sec/ft, find the equation of motion of the mass. What is the resonance frequency for the system?

12. A 2-kg mass is attached to a spring hanging from the ceiling, thereby causing the spring to stretch 20 cm upon coming to rest at equilibrium. At time $t = 0$, the mass is displaced 5 cm below the equilibrium position and released. At this same instant, an external force $F(t) = 0.3\cos t$ N is applied to the system.

If the damping constant for the system is 5 N-sec/m, determine the equation of motion for the mass. What is the resonance frequency for the system?

13. A mass weighing 32 lb is attached to a spring hanging from the ceiling and comes to rest at its equilib-

rium position. At time $t = 0$, an external force $F(t) = 3\cos 4t$ lb is applied to the system. If the spring constant is 5 lb/ft and the damping constant is 2 lb-sec/ft, find the steady-state solution for the system.

CHAPTER SUMMARY

In this chapter, we discussed the theory of second order linear equations and presented methods for solving these equations. We also studied the mathematical description of vibrating mechanical systems, and we saw how the mass-spring analogy could be used to predict qualitative features of solutions to some variable-coefficient and nonlinear equations.

The important features and solution techniques for the linear case are listed below. The coefficients $p(x)$ and $q(x)$ are presumed to be continuous on the interval I.

Homogeneous Linear Equations

$$y'' + p(x)y' + q(x)y = 0$$

Fundamental Solution Set: $\{y_1, y_2\}$

Two solutions y_1 and y_2 to the homogeneous equation on the interval I form a fundamental solution set, provided their Wronskian,

$$W[y_1, y_2](x) := y_1(x)y_2'(x) - y_1'(x)y_2(x) ,$$

is different from zero for some (and hence all) x in I. If y_1 and y_2 are linearly independent solutions on I, then $W[y_1, y_2](x) \neq 0$ on I and y_1 and y_2 form a fundamental solution set.

General Solution to Homogeneous Equation: $c_1y_1 + c_2y_2$

If y_1 and y_2 are linearly independent solutions to the homogeneous equation, then a general solution is

$$y(x) = c_1y_1(x) + c_2y_2(x) ,$$

where c_1 and c_2 are arbitrary constants.

Homogeneous Equation with Constant Coefficients: $ay'' + by' + cy = 0$

The form of a general solution for a homogeneous equation with constant coefficients depends on the roots

$$r_1 = \frac{-b + \sqrt{b^2 - 4ac}}{2a} , \qquad r_2 = \frac{-b - \sqrt{b^2 - 4ac}}{2a}$$

of the auxiliary equation

$$ar^2 + br + c = 0 , \qquad a \neq 0 .$$

(a) When $b^2 - 4ac > 0$, the auxiliary equation has two distinct real roots r_1 and r_2 and a

general solution is

$$y(x) = c_1 e^{r_1 x} + c_2 e^{r_2 x} \, .$$

(b) When $b^2 - 4ac = 0$, the auxiliary equation has a repeated real root $r = r_1 = r_2$ and a general solution is

$$y(x) = c_1 e^{rx} + c_2 x e^{rx} \, .$$

(c) When $b^2 - 4ac < 0$, the auxiliary equation has complex conjugate roots $r = \alpha \pm i\beta$ and a general solution is

$$y(x) = c_1 e^{\alpha x} \cos \beta x + c_2 e^{\alpha x} \sin \beta x \, .$$

Reduction of Order Formula: $y(x) = v(x)f(x)$

Let $f(x)$ be a nontrivial solution to the homogeneous equation. The substitution $y(x) = v(x)f(x)$ reduces the homogeneous equation to a first order equation in v'. When solved for v' and integrated to obtain v, a second linearly independent solution $y(x)$ is given by the reduction of order formula

$$y(x) = f(x) \int \frac{e^{-\int p(x)\,dx}}{[f(x)]^2}\,dx \, .$$

Nonhomogeneous Linear Equations

$$y'' + p(x)y' + q(x)y = g(x)$$

General Solution to Nonhomogeneous Equation: $y_p + c_1 y_1 + c_2 y_2$

If y_p is any particular solution to the nonhomogeneous equation and y_1 and y_2 are linearly independent solutions to the corresponding homogeneous equation, then a general solution is

$$y(x) = y_p(x) + c_1 y_1(x) + c_2 y_2(x) \, ,$$

where c_1 and c_2 are arbitrary constants.

Two methods for finding a particular solution y_p are those of undetermined coefficients and variation of parameters.

Undetermined Coefficients: $g(x) = e^{\alpha x} p_n(x) \begin{Bmatrix} \cos \beta x \\ \sin \beta x \end{Bmatrix}$

If the right-hand side $g(x)$ of a nonhomogeneous equation with constant coefficients is a polynomial, an exponential of the form $e^{\alpha x}$, a trigonometric function of the form $\cos \beta x$ or $\sin \beta x$, or any product of these special types of functions, then a particular solution of an appropriate form exists. This special form is given in Table 4.1 on page 208.

Variations of Parameters: $y(x) = v_1(x)y_1(x) + v_2(x)y_2(x)$

If y_1 and y_2 are two linearly independent solutions to the corresponding homogeneous equation, then a particular solution to the nonhomogeneous equation is

$$y(x) = v_1(x)y_1(x) + v_2(x)y_2(x) \, ,$$

where

$$v_1(x) = \int \frac{-g(x)y_2(x)}{W[y_1, y_2](x)}\, dx\ ,\qquad v_2(x) = \int \frac{g(x)y_1(x)}{W[y_1, y_2](x)}\, dx\ .$$

Superposition Principle: $L[y] = c_1g_1 + c_2g_2$

If y_1 and y_2 are solutions to the equations

$$y'' + py' + qy = g_1 \quad\text{and}\quad y'' + py' + qy = g_2\ ,$$

respectively, then $c_1y_1 + c_2y_2$ is a solution to the equation

$$y'' + py' + qy = c_1g_1 + c_2g_2\ .$$

Cauchy-Euler Equations: $ax^2\dfrac{d^2y}{dx^2} + bx\dfrac{dy}{dx} + cy = h(x)\ ,\qquad x > 0$

Let $x = e^t$. This substitution transforms the original equation into the following constant coefficient equation in t:

$$a\frac{d^2y}{dt^2} + (b - a)\frac{dy}{dt} + cy = h(e^t)\ .$$

REVIEW PROBLEMS

In Problems 1 through 28, find a general solution to the given differential equation.

1. $y'' + 8y' - 9y = 0$.

2. $49y'' + 14y' + y = 0$.

3. $4y'' - 4y' + 10y = 0$.

4. $9y'' - 30y' + 25y = 0$.

5. $6y'' - 11y' + 3y = 0$.

6. $y'' + 8y' - 14y = 0$.

7. $36y'' + 24y' + 5y = 0$.

8. $25y'' + 20y' + 4y = 0$.

9. $16z'' - 56z' + 49z = 0$.

10. $u'' + 11u = 0$.

11. $t^2x''(t) + 5x(t) = 0$, $t > 0$.

12. $2y''' - 3y'' - 12y' + 20y = 0$.

13. $y'' + 16y = xe^x$.

14. $v'' - 4v' + 7v = 0$.

15. $3y''' + 10y'' + 9y' + 2y = 0$.

16. $y''' + 3y'' + 5y' + 3y = 0$.

17. $y''' + 10y' - 11y = 0$.

18. $y^{(4)} = 120x$.

19. $4y''' + 8y'' - 11y' + 3y = 0$.

20. $2y'' - y = x\sin x$.

21. $y'' - 3y' + 7y = 7x^2 - e^x$.

22. $y'' - 8y' - 33y = 546\sin x$.

23. $y''(\theta) + 16y(\theta) = \tan 4\theta$.

24. $10y'' + y' - 3y = x - e^{x/2}$.

25. $4y'' - 12y' + 9y = e^{5x} + e^{3x}$.

26. $y'' + 6y' + 15y = e^{2x} + 75$.

27. $x^2y'' + 2xy' - 2y = 6x^{-2} + 3x$, $x > 0$.

28. $y'' = 5x^{-1}y' - 13x^{-2}y$, $x > 0$.

In Problems 29 through 36, find the solution to the given initial value problem.

29. $y'' + 4y' + 7y = 0$;
 $y(0) = 1$, $y'(0) = -2$.

30. $y''(\theta) + 2y'(\theta) + y(\theta) = 2\cos\theta$;
 $y(0) = 3$, $y'(0) = 0$.

31. $y'' - 2y' + 10y = 6\cos 3x - \sin 3x$;
 $y(0) = 2$, $y'(0) = -8$.

32. $4y'' - 4y' + 5y = 0$;
 $y(0) = 1$, $y'(0) = -11/2$.

33. $y''' - 12y'' + 27y' + 40y = 0$;
 $y(0) = -3$, $y'(0) = -6$,
 $y''(0) = -12$.

34. $y'' + 5y' - 14y = 0$;
 $y(0) = 5$, $y'(0) = 1$.

35. $y''(\theta) + y(\theta) = \sec\theta$; $y(0) = 1$, $y'(0) = 2$.

36. $9y'' + 12y' + 4y = 0$;
$$y(0) = -3 , \qquad y'(0) = 3 .$$

In Problems 37 through 39, a differential equation and a nontrivial solution $f(x)$ are given. Find a general solution for the equation.

37. $xy'' + (x-1)y' - y = 0$, $\qquad x > 0$;
$$f(x) = e^{-x} .$$

38. $(x-1)y'' - xy' + y = 0$, $\qquad x > 1$;
$$f(x) = e^{x} .$$

39. $(1-x^2)w'' - 2xw' + 2w = 0$,
$$-1 < x < 1 ; \qquad f(x) = x .$$

40. Determine whether the statement made is *always true* or *sometimes false*.

(a) Two differentiable functions are linearly independent on $[a, b]$ if and only if their Wronskian is not zero on $[a, b]$.

(b) The differential operator
$$L[y](x) := 2y''(x) + x^2y'(x) + \frac{1}{x}y(x)$$
defined for twice-differentiable functions on $(0, \infty)$ is linear.

(c) The initial value problem
$$xy'' + \frac{5}{x-3}y' + 2y = \cos x ;$$
$$y(1) = 6 , \qquad y'(1) = 23$$

has a unique solution on the interval $(0, 3)$.

(d) A general solution of $y'' + y = x$ is given by
$$y = c_1 \cos x + c_2 \sin x + c_3 x ,$$
where c_1, c_2, and c_3 are arbitrary constants.

41. Use the mass-spring oscillator analogy to decide whether all solutions to each of the following differential equations are bounded as $t \to +\infty$.

(a) $y'' + t^4 y = 0$. **(b)** $y'' - t^4 y = 0$.

(c) $y'' + y^7 = 0$. **(d)** $y'' + y^8 = 0$.

(e) $y'' + (3 + \sin t)y = 0$. **(f)** $y'' + t^2 y' + y = 0$.

(g) $y'' - t^2 y' - y = 0$.

42. A 3-kg mass is attached to a spring with stiffness $k = 75$ N/m, as in Figure 4.1, page 156. The mass is displaced $1/4$ m to the left and given a velocity of 1 m/sec to the right. The damping force is negligible. Find the equation of motion of the mass along with the amplitude, period, and frequency. How long after release does the mass pass through the equilibrium position?

43. A 32-lb weight is attached to a vertical spring, causing it to stretch 6 in. upon coming to rest at equilibrium. The damping constant for the system is 2 lb-sec/ft. An external force $F(t) = 4 \cos 8t$ lb is applied to the weight. Find the steady-state solution for the system. What is its resonant frequency?

TECHNICAL WRITING EXERCISES

1. Compare the two methods—undetermined coefficients and variation of parameters—for determining a particular solution to a nonhomogeneous equation. What are the advantages and disadvantages of each?

2. Consider the differential equation
$$\frac{d^2y}{dx^2} + 2b\frac{dy}{dx} + y = 0 ,$$
where b is a constant. Describe how the behavior of solutions to this equation changes as b varies.

3. Consider the differential equation
$$\frac{d^2y}{dx^2} + cy = 0 ,$$

where c is a constant. Describe how the behavior of solutions to this equation changes as c varies.

4. For students with a background in linear algebra: Compare the theory for linear second order equations with that for systems of n linear equations in n unknowns whose coefficient matrix has rank $n - 2$. Use the terminology from linear algebra; for example, subspace, basis, dimension, linear transformation, and kernel. Discuss both homogeneous and nonhomogeneous equations.

A. Undetermined Coefficients Using Complex Arithmetic

The technique of undetermined coefficients described in Section 4.8 can be streamlined with the aid of complex arithmetic and the properties of the complex exponential function. The essential formulas are

$$e^{(\alpha+i\beta)x} = e^{\alpha x}(\cos \beta x + i \sin \beta x) \ ,$$

$$\frac{d}{dx} e^{(\alpha+i\beta)x} = (\alpha + i\beta)e^{(\alpha+i\beta)x} \ ,$$

$$\text{Re } e^{(\alpha+i\beta)x} = e^{\alpha x} \cos \beta x \ , \qquad \text{Im } e^{(\alpha+i\beta)x} = e^{\alpha x} \sin \beta x \ ,$$

and, more generally,

(1) $\qquad \text{Re}\{(a + ib)e^{(\alpha+i\beta)x}\} = e^{\alpha x}(a \cos \beta x - b \sin \beta x) \ ,$

(2) $\qquad \text{Im}\{(a + ib)e^{(\alpha+i\beta)x}\} = e^{\alpha x}(b \cos \beta x + a \sin \beta x) \ .$

Now consider a second order equation of the form

(3) $\qquad L[y] := ay'' + by' + cy = g \ ,$

where a, b, and c are real numbers and g is of the special form

(4) $\qquad g(x) = e^{\alpha x}\{(a_n x^n + \cdots + a_1 x + a_0)\cos \beta x + (b_n x^n + \cdots + b_1 x + b_0)\sin \beta x\} \ ,$

with the a_j's, b_j's, α, and β real numbers. Such a function can always be expressed as the real or imaginary part of a function involving the complex exponential. For example, using equation (1), one can quickly check that

(5) $\qquad g(x) = \text{Re}\{G(x)\} \ ,$

where

(6) $\qquad G(x) = e^{(\alpha-i\beta)x}\left[(a_n + ib_n)x^n + \cdots + (a_1 + ib_1)x + (a_0 + ib_0)\right] \ .$

Now suppose for the moment that we can find a complex-valued solution Y to the equation

(7) $\qquad L[Y] = aY'' + bY' + cY = G \ .$

Then, since a, b, and c are real numbers, we get a real-valued solution y to (3) by simply taking the real part of Y; that is, $y = \text{Re } Y$ solves (3). (Recall that in Lemma 2, page 193, we proved this fact for homogeneous equations.) Thus we need focus only on finding a solution to (7).

The method of undetermined coefficients asserts that any differential equation of the form

(8) $\qquad L[Y] = e^{(\alpha\pm i\beta)x}\left[(a_n + ib_n)x^n + \cdots + (a_1 + ib_1)x + (a_0 + ib_0)\right]$

has a solution of the form

(9) $\qquad Y_p(x) = x^s e^{(\alpha\pm i\beta)x}\left[A_n x^n + \cdots + A_1 x + A_0\right] \ ,$

where A_n, \ldots, A_0 are complex constants and s is the smallest nonnegative integer such that no term

in (9) is a (complex) solution to the corresponding homogeneous equation $L[Y] = 0$. We can solve for the unknown constants A_j by substituting (9) into (8) and equating coefficients of like terms. With these facts in mind, we can (for the small price of using complex arithmetic) dispense with Table 4.1 on page 208 and avoid the unpleasant task of computing derivatives of a function like $e^{3x}(2 + 3x + x^2)\sin(2x)$, which involves both exponential and trigonometric factors.

Carry out this procedure to determine particular solutions to the following equations.

(a) $y'' + y = e^{-x}[\cos 2x - 3\sin 2x]$.
(b) $y'' - 2y' + 10y = xe^x\sin 3x$.

The use of complex arithmetic not only streamlines the computations but proves very useful in analyzing the response of a linear system to a sinusoidal input. Electrical engineers make good use of this in their study of *RLC* circuits by introducing the concept of **impedance.** This will be illustrated in Chapter 5.

B. *An Alternative to the Method of Undetermined Coefficients*[†]

Let

$$L[y] = y'' + by' + cy = g(x)$$

be a general nonhomogeneous constant-coefficient equation. The preceding project demonstrated how any of the forms occurring in Table 4.1—i.e., the permissible forms for the nonhomogeneity $g(x)$ that make the equation susceptible to the undetermined coefficient method—can be expressed in the generic format

$$g(x) = \text{Re}[p_n(x)e^{\gamma x}] ,$$

where $p_n(x)$ is a complex polynomial of degree n and γ is a complex constant. (See equations (5) and (6) of Project A.) This suggests that a change of dependent variable of the form

$$Y(x) = V(x)e^{\gamma x}$$

could simplify the (complex) equation $L[Y](x) = p_n(x)e^{\gamma x}$, because each of the terms Y', Y'' would contain the common exponential factor $e^{\gamma x}$, which would then cancel out of the equation:

$$Y' = V'e^{\gamma x} + \gamma Ve^{\gamma x} ,$$
$$Y'' = V''e^{\gamma x} + 2\gamma V'e^{\gamma x} + \gamma^2 Ve^{\gamma x} ,$$
$$L[Ve^{\gamma x}] = \{V'' + 2\gamma V' + \gamma^2 V\}e^{\gamma x} + b\{V' + \gamma V\}e^{\gamma x} + cVe^{\gamma x} = p_n(x)e^{\gamma x} .$$

Upon rearrangement, then we would be faced with the simpler task of solving a constant-coefficient equation with a polynomial nonhomogeneity:

(10) $V'' + \alpha V' + \beta V = p_n(x) = a_n x^n + a_{n-1}x^{n-1} + \cdots + a_1 x + a_0 ,$

where $\alpha = 2\gamma + b$ and $\beta = \gamma^2 + b\gamma + c$.

Equation (10) can be solved *directly* using the following strategy. Take its first n derivatives,

[†]Suggested in papers by E. R. Love and R. C. Gupta.

reducing the polynomial to a constant:

$$V''' + \alpha V'' + \beta V' = na_n x^{n-1} + (n-1)a_{n-1}x^{n-2} + \cdots + a_1$$
$$V^{(4)} + \alpha V''' + \beta V'' = n(n-1)a_n x^{n-2} + (n-1)(n-2)a_{n-1}x^{n-3} + \cdots + 2a_2$$
$$\vdots$$
$$V^{(n+1)} + \alpha V^{(n)} + \beta V^{(n-1)} = n!a_n x$$
$$V^{(n+2)} + \alpha V^{(n+1)} + \beta V^{(n)} = n!a_n \ .$$

Now if $\beta \neq 0$, the last equation has a solution $V^{(n)}(x) = \text{constant} = n!a_n/\beta$; this can be substituted into the next-to-last equation (together with the fact that $V^{(n+1)} \equiv 0$) to obtain $V^{(n-1)}$, and by continuing this back-substitution we eventually end up with a formula for $V(x)$ (using equation (10)).

On the other hand, if $\beta = 0$ but $\alpha \neq 0$, the last equation has a solution $V^{(n+1)}(x) = \text{constant} = n!a_n/\alpha$. This, in turn, can be substituted into the next-to-last equation to obtain $V^{(n)}$ (remember $\beta = 0$), and so on, until we get to the first equation, which provides us with a formula for $V'(x)$. Simple integration (with the usual additive constant) then yields $V(x)$.

Finally, if α and β are *both* zero, there is no need to implement this process because the equation

$$V'' = p_n(x)$$

can be solved directly with two integrations.

Carry out this procedure to determine particular solutions to the following equations.

(a) $y'' + 3y' + 2y = 3x + 1$.
(b) $y'' + y' = 3x + 1$.
(c) $y'' = 3x + 1$.
(d) $y'' - y' - 12y = e^{4x}$.
(e) $y'' + 2y' - 3y = (x^2 + 3x)e^x$.
(f) $y'' + 2y' - 3y = x^2 \cos \pi x$.
(g) $y'' + 2y' - 3y = 2xe^x\sin x - e^x\cos x$.
(h) $y'' - 2y' + 10y = xe^x\sin 3x$.
(i) $y'' + y = e^{-x}\left[\cos 2x - 3 \sin 2x\right]$.

Explain the significance of the extraneous constant of integration for the case $\beta = 0$. What happened to the factor "x^s" mentioned in Table 4.1?

C. *Convolution Method*

The **convolution** of two functions g and f is the function $g * f$ defined by

$$(g * f)(x) := \int_0^x g(x - v)f(v)\,dv \ .$$

The aim of this project is to show how convolutions can be used to obtain a particular solution to a nonhomogeneous equation of the form

(11) $ay'' + by' + cy = f(x)$,

where a, b, and c are constants, $a \neq 0$.

(a) Use Leibniz's rule,

$$\frac{d}{dx} \int_a^x h(x, v)\, dv = \int_a^x \frac{\partial h}{\partial x}(x, v)\, dv + h(x, x) \ ,$$

to show that

$$(y * f)'(x) = (y' * f)(x) + y(0)f(x)$$

and

$$(y * f)''(x) = (y'' * f)(x) + y'(0)f(x) + y(0)f'(x) \ ,$$

assuming y and f are sufficiently differentiable.

(b) Let $y_s(x)$ be the solution to the homogeneous equation $ay'' + by' + cy = 0$ that satisfies $y_s(0) = 0$, $y_s'(0) = 1/a$. Show that $y_s * f$ is the particular solution to equation (11) satisfying $y(0) = y'(0) = 0$.

(c) Let $y_k(x)$ be the solution to the homogeneous equation $ay'' + by' + cy = 0$ that satisfies $y(0) = y_0$, $y'(0) = y_1$, and let y_s be as defined in part (b). Show that

$$(y_s * f)(x) + y_k(x)$$

is the unique solution to the initial value problem

(12) $ay'' + by' + cy = f(x) \ ;$ $y(0) = y_0 \ ,$ $y'(0) = y_1 \ .$

(d) Use the result of part (c) to determine the solution to each of the following initial value problems. Carry out all integrations and express your answers in terms of elementary functions.

(i) $y'' + y = \tan x \ ;$ $y(0) = 0 \ ,$ $y'(0) = -1.$

(ii) $2y'' + y' - y = e^{-x} \sin x \ ;$ $y(0) = 1 \ ,$ $y'(0) = 1 \ .$

(iii) $y'' + 2y' + y = \sqrt{x}e^x \ ;$ $y(0) = 2 \ ,$ $y'(0) = 0 \ .$

D. *Linearization of Nonlinear Problems*

A useful approach to analyzing a nonlinear equation is to study its **linearized equation,** which is obtained by replacing the nonlinear terms by linear approximations. For example, the nonlinear equation

(13) $\dfrac{d^2\theta}{dt^2} + \sin\theta = 0 \ ,$

which governs the motion of a simple pendulum, has

(14) $\dfrac{d^2\theta}{dt^2} + \theta = 0$

as a linearization for small θ. (The nonlinear term $\sin\theta$ has been replaced by the linear approximation θ.)

A general solution to equation (13) involves Jacobi elliptic functions (see Project G), which are rather complicated, so let's try to approximate the solutions. For this purpose, we consider

two methods: Taylor series and linearization.

(a) Derive the first six terms of the Taylor series about $t = 0$ of the solution to equation (13) with initial conditions $\theta(0) = \pi/12$, $\theta'(0) = 0$. (The Taylor series method is discussed in Project A of Chapter 1 and Section 8.1.)

(b) Solve equation (14) subject to the same initial conditions $\theta(0) = \pi/12$, $\theta'(0) = 0$.

(c) On the same coordinate axes, graph the two approximations found in parts (a) and (b).

(d) Discuss the advantages and disadvantages of the Taylor series method and the linearization method.

(e) Give a linearization for the initial value problem.

$$x''(t) + 0.1\left[1 - x^2(t)\right]x'(t) + x(t) = 0 \; ; \qquad x(0) = 0.4 \; , \qquad x'(0) = 0$$

for x small. Solve this linearized problem to obtain an approximation for the nonlinear problem.

E. *Nonlinear Equations Solvable by First Order Techniques*

Certain *nonlinear* second order equations, namely those with dependent or independent variables missing, can be solved by reducing them to a pair of *first order equations*. This is accomplished by making the substitution $w = dy/dx$.

(a) To solve an equation of the form $y'' = F(x, y')$ in which the dependent variable y is missing, setting $w = y'$ (so that $w' = y''$) yields the pair of equations

$$w' = F(x, w) \; ,$$

$$y' = w \; .$$

Because $w' = F(x, w)$ is a first order equation, we have available the techniques of Chapter 2 to solve it for $w(x)$. Once $w(x)$ is determined, we integrate it to obtain $y(x)$.

Using this method, solve

$$2xy'' - y' + \frac{1}{y'} = 0 \; , \qquad x > 0 \; .$$

(b) To solve an equation of the form $y'' = F(y, y')$ in which the independent variable x is missing, setting $w = dy/dx$ yields, via the chain rule,

$$\frac{d^2y}{dx^2} = \frac{dw}{dx} = \frac{dw}{dy}\frac{dy}{dx} = w\frac{dw}{dy} \; .$$

Thus $y'' = F(y, y')$ is equivalent to the pair of equations

(15) $$w\frac{dw}{dy} = F(y, w) \; ,$$

(16) $$\frac{dy}{dx} = w \; .$$

In equation (15), notice that y plays the role of the *independent* variable; hence solving it yields $w(y)$. Then substituting $w(y)$ into (16), we obtain a separable equation that determines $y(x)$.

Using this method, solve the following equations.

(i) $2y \dfrac{d^2y}{dx^2} = 1 + \left(\dfrac{dy}{dx}\right)^2$.

(ii) $\dfrac{d^2y}{dx^2} + y\dfrac{dy}{dx} = 0$.

(c) Suspended Cable. In the study of a cable suspended between two fixed points (see Figure 4.37), one encounters the initial value problem

$$\frac{d^2y}{dx^2} = \frac{1}{a}\sqrt{1 + \left(\frac{dy}{dx}\right)^2} \; ; \qquad y(0) = a \; , \qquad y'(0) = 0 \; ,$$

where $a \; (\neq 0)$ is a constant. Solve this initial value problem for y. The resulting curve is called a **catenary.**

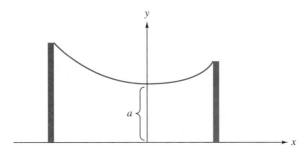

Figure 4.37 Suspended cable

F. *Apollo Reentry*

Alar Toomre, Massachusetts Institute of Technology

Each time the Apollo astronauts returned from the moon circa 1970, they took great care to reenter the Earth's atmosphere along a path that was only a small angle α from the horizontal. (See Figure 4.38.) This was necessary in order to avoid intolerably large "g" forces during their reentry.

Figure 4.38 Reentry path

To appreciate their grounds for concern, consider the idealized problem

$$\frac{d^2s}{dt^2} = -Ke^{s/H}\left(\frac{ds}{dt}\right)^2 ,$$

where K and H are constants and distance s is measured downrange from some reference point on the trajectory, as shown in the figure. This approximate equation pretends that the only force on the capsule during reentry is air drag. For a bluff body such as the Apollo module, drag is proportional to the square of the speed and to the local atmospheric density, which falls off exponentially with height. Intuitively, one might expect that the deceleration predicted by this model would depend heavily on the constant K (which takes into account the vehicle's mass, area, etc.); but, remarkably, for capsules entering the atmosphere (at "$s = -\infty$") with a common speed V_0, the *maximum* deceleration turns out to be independent of K.

(a) Verify this last assertion by demonstrating that this maximum deceleration is just $V_0^2/(2eH)$. [*Hint:* The independent variable t does not appear in the differential equation, so it is helpful to make this substitution $v = ds/dt$; see Project E, part (b).]

(b) Also verify that any such spacecraft at the instant when it is decelerating most fiercely will be traveling exactly with speed V_0/\sqrt{e}, having by then lost almost 40% of its original velocity.

(c) Using the plausible data $V_0 = 11$ km/sec and $H = 10/(\sin \alpha)$ km, estimate how small α had to be chosen so as to inconvenience the returning travelers with no more than 10 g's.

G. *Simple Pendulum*

In Section 4.10, we discussed the simple pendulum consisting of a mass m suspended by a rod of length ℓ having negligible mass and derived the nonlinear initial value problem

(17) $$\frac{d^2\theta}{dt^2} + \frac{g}{\ell}\sin \theta = 0 ; \qquad \theta(0) = \alpha , \qquad \theta'(0) = 0 ,$$

where g is the acceleration due to gravity and $\theta(t)$ is the angle the rod makes with the vertical at time t (see Figure 4.19, page 226). Here it is assumed that the mass is released with zero velocity at an initial angle α, $0 < \alpha < \pi$. We would like to determine the equation of motion for the pendulum and its period of oscillation.

(a) Use equation (17) and the energy integral lemma discussed in Section 4.10 to show that

$$\left(\frac{d\theta}{dt}\right)^2 = \frac{2g}{\ell}(\cos \theta - \cos \alpha)$$

and hence

$$dt = -\sqrt{\frac{\ell}{2g}}\frac{d\theta}{\sqrt{\cos \theta - \cos \alpha}} .$$

(b) Use the trigonometric identity $\cos x = 1 - 2\sin^2(x/2)$ to express dt by

$$dt = -\frac{1}{2}\sqrt{\frac{\ell}{g}}\frac{d\theta}{\sqrt{\sin^2(\alpha/2) - \sin^2(\theta/2)}} .$$

(c) Make the change of variables $\sin(\theta/2) = \sin(\alpha/2)\sin\phi$ to rewrite dt in the form

(18) $$dt = -\sqrt{\frac{\ell}{g}}\frac{d\phi}{\sqrt{1 - k^2\sin^2\phi}} \ , \qquad \text{where } k := \sin(\alpha/2) \ .$$

(d) The period $P(\alpha)$ of the pendulum is defined to be the time required for the pendulum to swing from one extreme to the other and back, that is, from α to $-\alpha$ back to α. Show that the period of oscillation is given by

(19) $$P(\alpha) = 4\sqrt{\frac{\ell}{g}}\int_0^{\pi/2}\frac{d\phi}{\sqrt{1 - k^2\sin^2\phi}} \ ,$$

where $k := \sin(\alpha/2)$. [*Hint:* The period is just four times the time it takes the pendulum to go from $\theta = 0$ to $\theta = \alpha$.]

The integral in (19) is called an **elliptic integral of the first kind** and is denoted by $F(k, \pi/2)$. As you might expect, the period of the simple pendulum depends on the length ℓ of the rod and the initial displacement α. In fact, a check of an elliptic integral table will show that the period nearly doubles as the initial displacement increases from $\pi/8$ to $15\pi/16$ (for fixed ℓ). What happens as α approaches π?

(e) To determine the equation of motion of the pendulum, we observe that as t varies from 0 to t, then θ varies from α to $\theta(t)$ and ϕ varies from $\pi/2$ to $\phi(t)$. Integrating equation (18), show that

(20) $$t + \frac{P(\alpha)}{4} = \sqrt{\frac{\ell}{g}}\int_0^{\phi}\frac{ds}{\sqrt{1 - k^2\sin^2 s}} = \sqrt{\frac{\ell}{g}}F(k, \phi) \ .$$

For fixed k, $F(k, \phi)$ has an "inverse," denoted by $\text{sn}(k, u)$, that satisfies $u = F(k, \phi)$ if and only if $\text{sn}(k, u) = \sin\phi$. The function $\text{sn}(k, u)$ is called a **Jacobi elliptic function** and has many properties that resemble those of the sine function. Using the Jacobi elliptic function $\text{sn}(k, u)$, express the equation of motion for the pendulum in the form

(21) $$\theta = 2\arcsin\left\{k\,\text{sn}\left(k, \sqrt{\frac{g}{\ell}}\left(t + \frac{P(\alpha)}{4}\right)\right)\right\} \ , \qquad 0 \le t \le P(\alpha)/4 \ ,$$

where $k := \sin(\alpha/2)$.

H. *Asymptotic Behavior of Solutions*

In the application of linear systems theory to mechanical problems, we have encountered the equation

(22) $$y'' + py' + qy = f(x) \ ,$$

where p and q are positive constants with $p^2 < 4q$ and $f(x)$ is a forcing function for the system. In many cases, it is important for the design engineer to know that a bounded forcing function gives rise only to bounded solutions. More specifically, how does the behavior of $f(x)$ for large values of x affect the asymptotic behavior of the solution? To answer this question, do the following:

(a) Show that the homogeneous equation associated with equation (22) has two linearly independent solutions given by

$$e^{\alpha x}\cos\beta x \ , \qquad e^{\alpha x}\sin\beta x \ ,$$

where $\alpha = -p/2 < 0$ and $\beta = \frac{1}{2}\sqrt{4q - p^2}$.

(b) Let $f(x)$ be a continuous function defined on the interval $[0, \infty)$. Use the variation of parameters formula to show that any solution to (22) on $[0, \infty)$ can be expressed in the form

(23) $y(x) = c_1 e^{\alpha x} \cos \beta x + c_2 e^{\alpha x} \sin \beta x$

$$-\frac{1}{\beta} e^{\alpha x} \cos \beta x \int_0^x f(t) e^{-\alpha t} \sin \beta t \, dt$$

$$+\frac{1}{\beta} e^{\alpha x} \sin \beta x \int_0^x f(t) e^{-\alpha t} \cos \beta t \, dt \ .$$

(c) Assuming f is bounded on $[0, \infty)$, that is, there exists a constant K such that $|f(t)| \le K$ for all $t \ge 0$, use the triangle inequality and other properties of the absolute value to show that $y(x)$ given in (23) satisfies

$$|y(x)| \le \{|c_1| + |c_2|\} e^{\alpha x} + \frac{2K}{|\alpha|\beta}(1 - e^{\alpha x})$$

for all $x > 0$.

(d) In a similar fashion, show that if $f_1(x)$ and $f_2(x)$ are two bounded continuous functions on $[0, \infty)$ such that $|f_1(x) - f_2(x)| \le \varepsilon$ for all $x > x_0$, and if ϕ_1 is a solution to (22) with $f = f_1$ and ϕ_2 is a solution to (22) with $f = f_2$, then

$$|\phi_1(x) - \phi_2(x)| \le M e^{\alpha x} + \frac{2\varepsilon}{|\alpha|\beta}(1 - e^{\alpha(x - x_0)})$$

for all $x > x_0$, where M is a constant that depends on ϕ_1 and ϕ_2 but not on x.

(e) Now assume $f(x) \to F_0$ as $x \to +\infty$, where F_0 is a constant. Use the result of part (d) to prove that any solution ϕ to (22) must satisfy $\phi(x) \to F_0/q$ as $x \to +\infty$. [*Hint:* Choose $f_1 = f$, $f_2 \equiv F_0$, $\phi_1 = \phi$, $\phi_2 \equiv F_0/q$.]

Introduction to Systems and Phase Plane Analysis

5.1 INTERCONNECTED FLUID TANKS

Two large tanks, each holding 24 liters of a brine solution, are interconnected by pipes as shown in Figure 5.1. Fresh water flows into tank A at a rate of 6 L/min, and fluid is drained out of tank B at the same rate; also 8 L/min of fluid are pumped from tank A to tank B, and 2 L/min from tank B to tank A. The liquids inside each tank are kept well stirred, so that each mixture is homogeneous. If, initially, the brine solution in tank A contains x_0 kg of salt and that in tank B initially contains y_0 kg of salt, determine the mass of salt in each tank at time $t > 0$.

Figure 5.1 Interconnected fluid tanks

Note that the *volume* of liquid in each tank remains constant at 24 L because of the balance between the inflow and outflow volume rates. Hence we have two unknown functions of t: the mass of salt $x(t)$ in tank A, and the mass of salt $y(t)$ in tank B. By focusing attention on one tank at a time, we can derive two equations relating these unknowns. Since the system is being flushed with fresh water, we expect that the salt content of each tank will diminish to zero as $t \to +\infty$.

To formulate the equations for this system, we equate the rate of change of salt in each tank with the *net* rate at which salt is transferred to that tank. The salt *concentration* in tank A is $x(t)/24$ kg/L, so the upper interconnecting pipe carries salt out of tank A at a rate of $8x/24$ kg/min; similarly, the lower interconnecting pipe brings salt into tank A at the rate $2y/24$ kg/min (the concentration of salt in tank B is $y/24$ kg/L). The fresh water inlet, of course, transfers no salt (it simply maintains the volume in tank A at 24 L). From our premise,

$$\frac{dx}{dt} = \text{input rate} - \text{output rate} ,$$

so the rate of change of the mass of salt in tank A is

$$\frac{dx}{dt} = \frac{2}{24}y - \frac{8}{24}x = \frac{1}{12}y - \frac{1}{3}x .$$

The rate of change of salt in tank B is determined by the same interconnecting pipes *and* by the drain pipe, carrying away $6y/24$ kg/min:

$$\frac{dy}{dt} = \frac{8}{24}x - \frac{2}{24}y - \frac{6}{24}y = \frac{1}{3}x - \frac{1}{3}y .$$

The interconnected tanks are thus governed by a *system* of differential equations:

(1)
$$x' = -\frac{1}{3}x + \frac{1}{12}y ,$$
$$y' = \frac{1}{3}x - \frac{1}{3}y .$$

Although both unknowns $x(t)$ and $y(t)$ appear in each of equations (1) (they are "coupled"), the structure is so transparent that we can obtain an equation for y alone by solving the second equation for x,

(2) $x = 3y' + y ,$

and substituting (2) in the first equation to eliminate x:

$$(3y' + y)' = -\frac{1}{3}(3y' + y) + \frac{1}{12}y ,$$

$$3y'' + y' = -y' - \frac{1}{3}y + \frac{1}{12}y ,$$

or

$$3y'' + 2y' + \frac{1}{4}y = 0 .$$

This last equation, which is linear with constant coefficients, is readily solved by the methods of Section 4.5. Since the auxiliary equation

$$3r^2 + 2r + \frac{1}{4} = 0$$

has roots $-1/2, -1/6$, a general solution is given by

(3) $y(t) = c_1 e^{-t/2} + c_2 e^{-t/6}$.

Having determined y, we use equation (2) to deduce a formula for x:

(4) $x(t) = 3\left(-\dfrac{c_1}{2}e^{-t/2} - \dfrac{c_2}{6}e^{-t/6}\right) + c_1 e^{-t/2} + c_2 e^{-t/6} = -\dfrac{1}{2}c_1 e^{-t/2} + \dfrac{1}{2}c_2 e^{-t/6}$.

Formulas (3) and (4) contain two undetermined parameters, c_1 and c_2, which can be adjusted to meet the specified initial conditions:

$$x(0) = -\frac{1}{2}c_1 + \frac{1}{2}c_2 = x_0 , \qquad y(0) = c_1 + c_2 = y_0 ,$$

or

$$c_1 = \frac{y_0 - 2x_0}{2} , \qquad c_2 = \frac{y_0 + 2x_0}{2} .$$

Thus the mass of salt in tanks A and B at time t are, respectively,

(5)
$$x(t) = -\left(\frac{y_0 - 2x_0}{4}\right)e^{-t/2} + \left(\frac{y_0 + 2x_0}{4}\right)e^{-t/6} ,$$

$$y(t) = \left(\frac{y_0 - 2x_0}{2}\right)e^{-t/2} + \left(\frac{y_0 + 2x_0}{2}\right)e^{-t/6} .$$

It is interesting to note from (5) that *all* solutions of (1) tend to the constant solution $x(t) \equiv 0$, $y(t) \equiv 0$ as t approaches (plus) infinity, which is consistent with our expectations based on physical grounds. This constant solution will be identified as a *stable equilibrium solution* in the next section. It turns out that for a general class of systems, equilibria can be identified and classified so as to give qualitative information about the other solutions even when we cannot solve the system explicitly. This is the theme of Section 5.2, in which we introduce phase plane analysis.

The ad hoc elimination procedure that we used to solve the interconnected tanks example can also be generalized and formalized to find solutions of *linear systems with constant coefficients;* this is done in Section 5.3. Furthermore, we will show how to extend numerical algorithms for first order equations to general systems, and we shall consider applications to coupled oscillators and electrical circuits. The final section of this chapter provides an introduction into a newer area of mathematics known as *chaos theory.*

5.2 INTRODUCTION TO THE PHASE PLANE

In this section, we study systems of two first order equations of the form

(1)
$$\frac{dx}{dt} = f(x, y) ,$$

$$\frac{dy}{dt} = g(x, y) .$$

Note that the independent variable t does not appear in the right-hand terms $f(x, y)$ and $g(x, y)$; such systems are called **autonomous.**[†] For example, the system that modeled the interconnected tanks problem in Section 5.1,

$$x' = -\frac{1}{3}x + \frac{1}{12}y \ ,$$

$$y' = \frac{1}{3}x - \frac{1}{3}y \ ,$$

is autonomous. So is the Volterra-Lotka system,

$$x' = Ax - Bxy \ ,$$
$$y' = -Cy + Dxy \ ,$$

(with A, B, C, D constants), which will be discussed in Section 5.6 as a model for population dynamics.

For future reference, we note that the solutions to autonomous systems have a "time-shift immunity," in the sense that if the pair $x(t)$, $y(t)$ solves (1), so does the **time-shifted** pair $x(t + c)$, $y(t + c)$ for any constant c. Specifically, if we let $X(t) := x(t + c)$ and $Y(t) := y(t + c)$, then by the chain rule

$$\frac{dX}{dt}(t) = \frac{dx}{dt}(t + c) = f\big(x(t + c), y(t + c)\big) = f\big(X(t), Y(t)\big) \ ,$$

$$\frac{dY}{dt}(t) = \frac{dy}{dt}(t + c) = g\big(x(t + c), y(t + c)\big) = g\big(X(t), Y(t)\big) \ ,$$

proving that $X(t)$, $Y(t)$ is also a solution to (1).

Since t does not appear explicitly in the system (1), it is certainly tempting to divide the two equations, invoke the chain rule

$$\frac{dy}{dx} = \frac{dy/dt}{dx/dt} \ ,$$

and consider the single first order differential equation

$$(2) \qquad \frac{dy}{dx} = \frac{g(x, y)}{f(x, y)} \ .$$

We shall refer to (2) as the **phase plane equation.** In Chapters 1 and 2 we mastered several approaches to equations like (2): the use of *direction fields* to visualize the solution graphs, and the analytic techniques for the cases of separability, linearity, exactness, etc.

So the form (2) certainly has advantages over (1), but it is important to maintain our perspective by noting these distinctions:

(i) A solution to the original problem (1) is a *pair* of functions of t, namely $x(t)$ and $y(t)$, that satisfies (1) for all t in some interval I. These functions can be visualized as a pair of graphs, as in Figure 5.2. If, in the xy-plane, we plot the points $\big(x(t), y(t)\big)$ as t varies over I, the resulting curve is known as the **trajectory** of the solution pair

[†]The origin of this term is explained in Section 1.4, where the *phase line* is discussed.

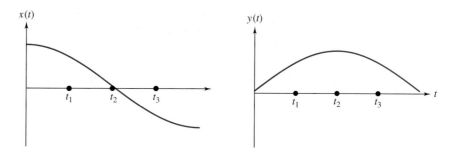

Figure 5.2 Solution pair for system (1)

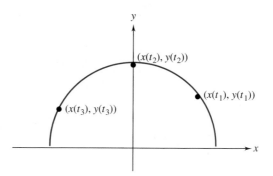

Figure 5.3 Trajectory of the solution pair for system (1)

$x(t)$, $y(t)$, and the xy-plane is called the **phase plane** in this context (see Figure 5.3). Note, however, that the trajectory in this plane contains less information than the original graphs, because the t-dependence has been suppressed. In principle we can construct, point by point, the trajectory from the solution graphs, but we cannot reconstruct the solution graphs from the phase plane trajectory alone (because we would not know what value of t to assign to each point).

(ii) Nonetheless, the slope dy/dx of a trajectory in the phase plane is given by the right-hand side of (2). So, in solving equation (2) we are indeed locating the *trajectories* of the system (1) in the phase plane. More precisely, we have shown that the trajectories satisfy equation (2); therefore the solution curves of (2)—which are known in this context as **integral curves**—contain the trajectories.

(iii) In Chapters 1 and 2 we regarded x as the independent variable and y as the dependent variable, in equations of the form (2). This is no longer true in the context of the system (1); x and y are both dependent variables on an equal footing, and t is the independent variable.

Thus it appears that a phase plane portrait may be a useful, albeit incomplete, tool for analyzing first order autonomous systems like (1).

It is often the case that finding all solutions to the system (1) is an impossible task. But it is relatively easy to find *constant solutions;* if $f(x_0, y_0) = 0$ and $g(x_0, y_0) = 0$, then the constant functions $x(t) \equiv x_0$, $y(t) \equiv y_0$ solve (1). For such solutions the following terminology is used.

CRITICAL POINTS AND EQUILIBRIUM SOLUTIONS

Definition 1. A point (x_0, y_0) where $f(x_0, y_0) = 0$ and $g(x_0, y_0) = 0$ is called a **critical point** or **equilbrium point** of the system $dx/dt = f(x, y)$, $dy/dt = g(x, y)$, and the corresponding constant solution $x(t) \equiv x_0$, $y(t) \equiv y_0$ is called an **equilibrium solution.** The set of all critical points is called the **critical point set.**

Notice that trajectories of equilibrium solutions consist of just single points (the equilibrium points). But what can be said about the other trajectories? Can we predict any of their features from closer examination of the equilibrium points? To explore this we focus on the phase plane equation (2) and exploit its direction field (recall Section 1.3, page 16). However, we shall augment the direction field plot by attaching arrowheads to the line segments, indicating the direction of the "flow" of solutions as t increases. This is easy: When dx/dt is positive, $x(t)$ increases so the trajectory flows to the right. Therefore, according to (1), all direction field segments drawn in a region where $f(x, y)$ is positive should point to the right (and, of course, they point to the left if $f(x, y)$ is negative). If $f(x, y)$ is zero, we can use $g(x, y)$ to decide if the flow is upward $(y(t)$ increases$)$ or downward $(y(t)$ decreases$)$. (*What if both $f(x, y)$ and $g(x, y)$ are zero?*)

In the examples that follow, we advocate the use of computers or calculators for generating these direction fields.

EXAMPLE 1 Sketch the direction field in the phase plane for the system

$$(3) \qquad \begin{aligned} \frac{dx}{dt} &= -x \ , \\ \frac{dy}{dt} &= -2y \end{aligned}$$

and identify its critical point.

SOLUTION Here $f(x, y) = -x$ and $g(x, y) = -2y$ are both zero when $x = y = 0$, so $(0, 0)$ is the critical point. The direction field for the equation

$$(4) \qquad \frac{dy}{dx} = \frac{-2y}{-x} = \frac{2y}{x}$$

is given in Figure 5.4. Since $dx/dt = -x$ in (3), trajectories in the right half-plane (where $x > 0$) flow to the left, and vice versa. From the figure we can see that all solutions "flow into" the critical point $(0, 0)$. Such a critical point is called **asymptotically stable.**[†] ■ End

[†]See Section 12.3 for a rigorous exposition of stability and critical points. All references to Chapters 11–13 refer to the expanded text *Fundamentals of Differential Equations and Boundary Value Problems,* 3rd edition.

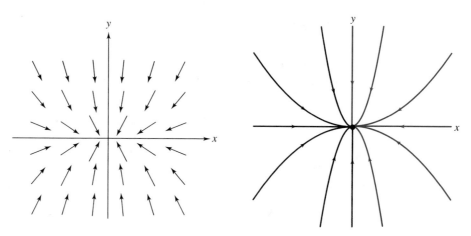

Figure 5.4 Direction field for Example 1

Figure 5.5 Trajectories for Example 1

Remark. For this simple example we can actually solve the system (3) explicitly; indeed, (3) constitutes an uncoupled pair of linear equations whose solutions are $x(t) = c_1 e^{-t}$ and $y(t) = c_2 e^{-2t}$. By elimination of t we obtain the equation $y = c_2 e^{-2t} = c_2[x(t)/c_1]^2 = cx^2$. So the trajectories lie along the parabolas $y = cx^2$. (Alternatively, we could have separated variables in (4) and identified these parabolas as the integral curves.) Notice that each such parabola is made up of three trajectories: an incoming trajectory approaching the origin in the right half-plane; its mirror-image trajectory approaching the origin in the left half-plane; and the origin itself, an equilibrium point. Sample trajectories are indicated in Figure 5.5.

EXAMPLE 2 Sketch the direction field in the phase plane for the system

(5)
$$\frac{dx}{dt} = x ,$$

$$\frac{dy}{dt} = 2y$$

and describe the behavior of solutions near the critical point $(0, 0)$.

SOLUTION This example is almost identical to the previous one; in fact, one could say we have merely "reversed time" in (3). The direction field segments for

(6) $$\frac{dy}{dx} = \frac{2y}{x}$$

are the same as those of (4), but the direction arrows are reversed. Now all solutions flow *away* from the critical point $(0, 0)$; the equilibrium is **unstable.** ■ End

EXAMPLE 3 For the system (7) below, find the critical points, sketch the direction field in the phase plane, and predict the asymptotic nature (i.e., behavior as $t \to +\infty$) of the solution starting

at $x = 2$, $y = 0$ when $t = 0$.

(7)
$$\frac{dx}{dt} = 5x - 3y - 2 \ ,$$

$$\frac{dy}{dt} = 4x - 3y - 1 \ .$$

SOLUTION The only critical point is the solution of the simultaneous equations $f(x, y) = g(x, y) = 0$:

(8)
$$5x_0 - 3y_0 - 2 = 0 \ ,$$

$$4x_0 - 3y_0 - 1 = 0 \ ,$$

from which we find $x_0 = y_0 = 1$. The direction field for the phase plane equation

(9)
$$\frac{dy}{dx} = \frac{4x - 3y - 1}{5x - 3y - 2}$$

is shown in Figure 5.6, with some trajectories rough-sketched in by hand.[†] Note that solutions flow to the right for $5x - 3y - 2 > 0$, i.e., for all points *below* the line $5x - 3y - 2 = 0$.

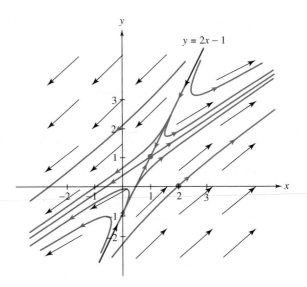

Figure 5.6 Direction field and trajectories for Example 3

The integral curve passing through $(2, 0)$ in Figure 5.6 apparently extends to infinity. Does this imply the corresponding *system* solution $x(t)$, $y(t)$ also approaches infinity in the sense that $|x(t)| + |y(t)| \rightarrow +\infty$ as $t \rightarrow +\infty$, or could its trajectory "stall" at some point along the integral curve, or possibly even "backtrack"? *It cannot backtrack,* because the direction arrows along the integral curve point, unambiguously, to the right. And if $\big(x(t), y(t)\big)$ stalls at some point (x_1, y_1),

[†]The integral curves could be obtained analytically by solving equation (9) using the methods of Section 2.6.

then intuitively we would conclude that (x_1, y_1) was an equilibrium point (since the "speeds" dx/dt and dy/dt would approach zero there). But we have already found the only critical point. So we conclude, with a high degree of confidence,[†] that the system solution does indeed go to infinity.

The critical point $(1, 1)$ is **unstable,** because although many solutions get arbitrarily close to $(1, 1)$, most of them eventually flow away. Solutions that lie on the line $y = 2x - 1$, however, do converge to $(1, 1)$. Such an equilibrium is an example of a **saddle point.** ■ End

In the preceding example we informally argued that endpoints of trajectories of an autonomous system would have to be critical points. This is more carefully stated in the following theorem, whose proof is outlined in Problem 30.

LIMIT POINTS ARE CRITICAL POINTS

Theorem 1. Let the pair $x(t)$, $y(t)$ be a solution on $[0, +\infty)$ to the autonomous system $dx/dt = f(x, y)$, $dy/dt = g(x, y)$, where f and g are continuous in the plane. If the limits

$$x^* := \lim_{t \to +\infty} x(t) \qquad \text{and} \qquad y^* := \lim_{t \to +\infty} y(t)$$

exist and are finite, then the limit point (x^*, y^*) is a critical point for the system.

Some typical trajectory configurations near critical points are displayed and classified in Figure 5.7 on page 270. These phase plane portraits arise from the systems listed in Problem 29, and can be generated by software packages having trajectory-sketching options. A more complete discussion of the nature of various types of equilibrium solutions and their stability is deferred to Chapter 12.[††] For the moment, however, notice that unstable critical points are distinguished by "runaway" trajectories emanating from arbitrarily nearby points, while stable equilibria "trap" all neighboring trajectories. The asymptotically stable critical points *attract* their neighboring trajectories as $t \to +\infty$.

Historically, the phase plane was introduced to facilitate the analysis of mechanical systems governed by Newton's second law, force equals mass times acceleration:

(10) $F = F(t, y, y') = ma = my''$.

(For an *autonomous* mechanical system the force F is independent of time.) The phase plane coordinates were taken to be the position y and the momentum $p = my'$. From (10), we derive the first order differential equation for p:

$$p' = F .$$

[†]These informal arguments are made more rigorous in Chapter 12. All references to Chapters 11–13 refer to the expanded text *Fundamentals of Differential Equations and Boundary Value Problems,* 3rd edition.

[††]All references to Chapters 11–13 refer to the expanded text *Fundamentals of Differential Equations and Boundary Value Problems,* 3rd edition.

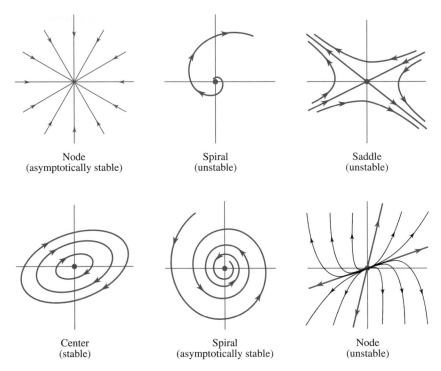

Figure 5.7 Examples of critical points

And the equation for y simply expresses its derivative, the velocity, in terms of the momentum:

$$y' = \frac{p}{m} \ .$$

In general, given any second order differential equation of the form

$$y'' = f(t, y, y') \ ,$$

we will convert it to a first order system by introducing $v = y'$ and simply writing

(11) $\begin{aligned} y' &= v \ , \\ v' &= f(t, y, v) \ . \end{aligned}$

In the yv-plane, trajectories $\big(y(t), v(t)\big)$ flow to the right in the upper half-plane (where $v > 0$), and to the left in the lower half-plane.

EXAMPLE 4 Sketch the direction field in the phase plane for the first order system corresponding to the unforced, undamped mass-spring oscillator described in Section 4.1 (Figure 4.1, page 156). Sketch several trajectories and interpret them physically.

SOLUTION The equation derived in Section 4.1 for this oscillator is $my'' + ky = 0$ or, equivalently, $y'' = -ky/m$. Hence the system (11) takes the form

$$(12) \quad \begin{aligned} y' &= v \ , \\ v' &= -\frac{ky}{m} \ . \end{aligned}$$

The critical point is at the origin $y = v = 0$. The direction field in Figure 5.8 indicates that the trajectories appear to be either closed curves (ellipses?) or spirals that encircle the critical point.

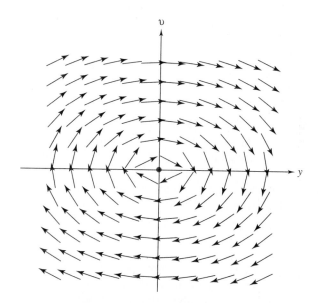

Figure 5.8 Direction field for Example 4

We saw in Section 4.1 that the undamped oscillator motions are periodic; they cycle repeatedly through the same sets of points, with the same velocities. Their trajectories in the phase plane, then, must be closed curves.[†] Let's confirm this mathematically by solving the phase plane equation

$$(13) \quad \frac{dv}{dy} = -\frac{ky}{mv} \ .$$

Equation (13) is separable and we find

$$v \, dv = -\frac{ky}{m} dy \quad \text{or} \quad d\left(\frac{v^2}{2}\right) = -\frac{k}{m} d\left(\frac{y^2}{2}\right) \ ,$$

so its solutions are the ellipses $v^2/2 + ky^2/2m = C$ as shown in Figure 5.9 on page 272. The solutions of (12) are confined to these ellipses and hence flow neither toward nor away from the equilibrium solution. The critical point is thus identified as a **center** in Figure 5.7.

[†]By the same reasoning, *underdamped* oscillations would correspond to spiral trajectories asymptotically approaching the origin as $t \to +\infty$.

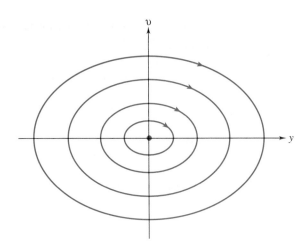

Figure 5.9 Trajectories for Example 4

Furthermore, the system solutions must continually circulate around the ellipses, since there are no critical points to stop them. This confirms that all solutions are periodic. ■ End

Remark. More generally, we argue that if a solution to an autonomous system like (1) passes through a point in the phase plane *twice,* and if it is sufficiently well behaved to satisfy a *uniqueness theorem,* then the second "tour" satisfies the same initial conditions as the first tour and so must replicate it. *Closed trajectories containing no critical points correspond to periodic solutions.*

Through these examples we have seen how, by studying the phase plane, one can often anticipate some of the features (boundedness, periodicity, etc.) of solutions of autonomous systems without solving them explicitly. Much of this information can be predicted simply from the critical points and the direction field (oriented by arrowheads), which are obtainable through standard software packages.[†] The final example ties together several of these ideas.

EXAMPLE 5 Find the critical points and solve the phase plane equation (2) for

(14)
$$\frac{dx}{dt} = -y(y - 2) \ ,$$
$$\frac{dy}{dt} = (x - 2)(y - 2) \ .$$

What is the asymptotic behavior of the solutions starting from $(3, 0)$, $(5, 0)$, and $(2, 3)$?

SOLUTION To find the critical points, we solve the system

$$-y(y - 2) = 0, \qquad (x - 2)(y - 2) = 0 \ .$$

One family of solutions to this system is given by $y = 2$ with x arbitrary; that is, the line $y = 2$. If $y \neq 2$, then the system simplifies to $-y = 0$, and $x - 2 = 0$, which has the solution $x = 2$,

[†]Some packages have advanced features that facilitate phase plane analyses, such as trajectory sketching and zooming.

$y = 0$. Hence the critical point set consists of the isolated point $(2, 0)$ and the horizontal line $y = 2$. The corresponding equilibrium solutions are $x(t) \equiv 2$, $y(t) \equiv 0$, and the family $x(t) \equiv c$, $y(t) \equiv 2$, where c is an arbitrary constant.

The trajectories in the phase plane satisfy the equation

(15)
$$\frac{dy}{dx} = \frac{(x-2)(y-2)}{-y(y-2)} = -\frac{x-2}{y}.$$

Solving (15) by separating variables,

$$y\,dy = -(x-2)\,dx \qquad \text{or} \qquad y^2 + (x-2)^2 = C,$$

demonstrates that the trajectories lie on concentric circles centered at $(2, 0)$. See Figure 5.10.

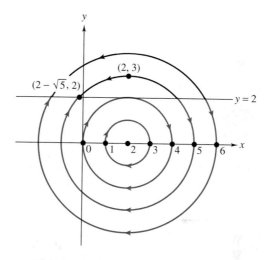

Figure 5.10 Phase plane diagram for Example 5

Next we analyze the flow along each trajectory. From the equation $dx/dt = -y(y-2)$, we see that x is decreasing when $y > 2$. This means the flow is from right to left along the arc of a circle that lies above the line $y = 2$. For $0 < y < 2$, we have $dx/dt > 0$, so in this region the flow is from left to right. Furthermore, for $y < 0$, we have $dx/dt < 0$, and again the flow is from right to left.

We now observe in Figure 5.10 that there are four types of trajectories associated with system (14): (a) those that begin above the line $y = 2$ and follow the arc of a circle counterclockwise back to that line; (b) those that begin below the line $y = 2$ and follow the arc of a circle clockwise back to that line; (c) those that continually move clockwise around a circle centered at $(2, 0)$ with radius less than 2 (i.e., they do not intersect the line $y = 2$); and finally, (d) the critical points $(2, 0)$ and $y = 2$, x arbitrary.

The solution starting at $(3, 0)$ lies on a circle with no critical points; therefore it is a *periodic* solution and the critical point $(2, 0)$ is a center. But the circle containing the solutions starting at $(5, 0)$ and at $(2, 3)$ has critical points at $\left(2 - \sqrt{5}, 2\right)$ and $\left(2 + \sqrt{5}, 2\right)$. The direction arrows indicate that both solutions approach $\left(2 - \sqrt{5}, 2\right)$ asymptotically (as $t \to +\infty$). They lie on the same circle (or integral curve), but they are quite different trajectories. ■End

Note that the critical points on the line $y = 2$ are not isolated, so they do not fit into any of the categories depicted in Figure 5.7. Observe also that all solutions of this system are bounded, since they are confined to circles.

EXERCISES 5.2

In Problems 1 and 2, verify that the pair $x(t)$, $y(t)$ is a solution to the given system. Sketch the trajectory of the given solution in the phase plane.

1. $\dfrac{dx}{dt} = 3y^3$, $\quad \dfrac{dy}{dt} = y$;

$\quad x(t) = e^{3t}$, $\quad y(t) = e^t$.

2. $\dfrac{dx}{dt} = 1$, $\quad \dfrac{dy}{dt} = 3x^2$;

$\quad x(t) = t + 1$, $\quad y(t) = t^3 + 3t^2 + 3t$.

In Problems 3 through 6, find the critical point set for the given system.

3. $\dfrac{dx}{dt} = x - y$,

$\quad \dfrac{dy}{dt} = x^2 + y^2 - 1$.

4. $\dfrac{dx}{dt} = y - 1$,

$\quad \dfrac{dy}{dt} = x + y + 5$.

5. $\dfrac{dx}{dt} = x^2 - 2xy$,

$\quad \dfrac{dy}{dt} = 3xy - y^2$.

6. $\dfrac{dx}{dt} = y^2 - 3y + 2$,

$\quad \dfrac{dy}{dt} = (x - 1)(y - 2)$.

In Problems 7 through 9, determine the integral curves for the given system by solving the phase plane equation given in (2), page 264.

7. $\dfrac{dx}{dt} = y - 1$,

$\quad \dfrac{dy}{dt} = e^{x+y}$.

8. $\dfrac{dx}{dt} = x^2 - 2y^{-3}$,

$\quad \dfrac{dy}{dt} = 3x^2 - 2xy$.

9. $\dfrac{dx}{dt} = 2y - x$,

$\quad \dfrac{dy}{dt} = e^x + y$.

10. Find all the critical points of the system

$$\dfrac{dx}{dt} = x^2 - 1 \ ,$$

$$\dfrac{dy}{dt} = xy \ ,$$

and solve the xy-phase plane equation to determine the integral curves. Thereby prove that there are two trajectories that are genuine semicircles. What are the endpoints of the semicircles?

In Problems 11 through 14, solve the phase plane equation for the given system. Then sketch by hand several representative trajectories (with their flow arrows).

11. $\dfrac{dx}{dt} = 2y$,

$\quad \dfrac{dy}{dt} = 2x$.

12. $\dfrac{dx}{dt} = -8y$,

$\quad \dfrac{dy}{dt} = 18x$.

13. $\dfrac{dx}{dt} = (y - x)(y - 1)$,

$\quad \dfrac{dy}{dt} = (x - y)(x - 1)$.

14. $\dfrac{dx}{dt} = \dfrac{3}{y}$,

$\quad \dfrac{dy}{dt} = \dfrac{2}{x}$.

In Problems 15 through 18, find all of the critical points for the given system. Then use a software package to sketch the direction field in the phase plane, and from this describe the stability of the critical points (i.e., compare with Figure 5.7).

15. $\dfrac{dx}{dt} = 2x + y + 3$,

$\quad \dfrac{dy}{dt} = -3x - 2y - 4$.

16. $\dfrac{dx}{dt} = -5x + 2y$,

$\quad \dfrac{dy}{dt} = x - 4y$.

17. $\dfrac{dx}{dt} = 2x + 13y$,

$\quad \dfrac{dy}{dt} = -x - 2y$.

18. $\dfrac{dx}{dt} = x(7 - x - 2y)$,

$\quad \dfrac{dy}{dt} = y(5 - x - y)$.

 In Problems 19 through 24, convert the given second order equation into a first order system by setting $v = y'$. Then find all the critical points in the yv-plane. Finally, sketch (by hand or software) the direction fields, and describe the stability of the critical points (i.e., compare with Figure 5.7).

19. $\dfrac{d^2y}{dt^2} - y = 0$.

20. $\dfrac{d^2y}{dt^2} + y = 0$.

21. $\dfrac{d^2y}{dt^2} + y + y^5 = 0$.

22. $\dfrac{d^2y}{dt^2} + y^3 = 0$.

23. $y''(t) + y(t) - y(t)^4 = 0$.

24. $y''(t) + y(t) - y(t)^3 = 0$.

 25. Using software, sketch the direction field in the phase plane for the system

$$dx/dt = y ,$$
$$dy/dt = -x + x^3 .$$

From the sketch, conjecture whether the solution passing through each given point is periodic:

(a) $(0.25, 0.25)$, **(b)** $(2, 2)$, **(c)** $(1, 0)$.

 26. Using software, sketch the direction field in the phase plane for the system

$$dx/dt = y ,$$
$$dy/dt = -x - x^3 .$$

From the sketch, conjecture whether all solutions of this system are bounded. Solve the phase plane equation and confirm your conjecture.

27. Using software, sketch the direction field in the phase plane for the system

$$dx/dt = -2x + y ,$$
$$dy/dt = -5x - 4y .$$

From the sketch, predict the asymptotic limit (as $t \to +\infty$) of the solution starting at $(1, 1)$.

28. Figure 5.11 displays some trajectories for the system

$$dx/dt = y ,$$
$$dy/dt = -x + x^2 .$$

What type of critical points (cf. Figure 5.7) occur at $(0, 0)$ and $(1, 0)$?

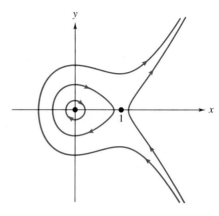

Figure 5.11 Phase plane for Problem 28

29. The phase plane diagrams depicted in Figure 5.7 were derived from the following systems. Use any method (except software) to match the systems to the graphs.

(a) $dx/dt = x ,$
 $dy/dt = 3y .$

(b) $dx/dt = y/2 ,$
 $dy/dt = -2x .$

(c) $dx/dt = -5x + 2y ,$
 $dy/dt = x - 4y .$

(d) $dx/dt = 2x - y ,$
 $dy/dt = x + 2y .$

(e) $dx/dt = 5x - 3y ,$
 $dy/dt = 4x - 3y .$

(f) $dx/dt = -y ,$
 $dy/dt = x - y .$

30. A proof of Theorem 1, page 269, is outlined below. The goal is to show that $f(x^*, y^*) = g(x^*, y^*) = 0$. Justify each step.

(a) From the given hypotheses, deduce that $\lim_{t \to +\infty} x'(t) = f(x^*, y^*)$ and $\lim_{t \to +\infty} y'(t) = g(x^*, y^*)$.

(b) Suppose that $f(x^*, y^*) > 0$. Then, by continuity, $x'(t) > f(x^*, y^*)/2$ for all large t (say, for $t \geq T$). Deduce from this that $x(t) > t f(x^*, y^*)/2 + C$ for $t > T$, where C is some constant.

(c) Conclude from part (b) that $\lim_{t \to +\infty} x(t) = +\infty$, contradicting the fact that this limit is the finite number x^*. Thus $f(x^*, y^*)$ cannot be positive.

(d) Argue similarly that the supposition that $f(x^*, \bar{y}^*) < 0$ also leads to a contradiction; hence $f(x^*, y^*)$ must be zero.

(e) In the same manner, argue that $g(x^*, y^*)$ must be zero.

Therefore, $f(x^*, y^*) = g(x^*, y^*) = 0$, and (x^*, y^*) is a critical point.

31. Phase plane analysis provides a quick derivation of the energy integral lemma of Section 4.10 (page 220). By completing the following steps, prove that solutions of equations of the special form $y'' = f(y)$ satisfy

$$\frac{1}{2}(y')^2 - F(y) = \text{constant} ,$$

where $F(y)$ is an antiderivative of $f(y)$.

(a) Set $v = y'$ and write $y'' = f(y)$ as an equivalent first order system.

(b) Show that the solutions to the yv-phase plane equation for the system in part (a) satisfy $v^2/2 = F(y) + K$. Replacing v by y' then completes the proof.

32. Use the result of Problem 31 to prove that all solutions to the equation

$$y'' + y^3 = 0$$

remain bounded. [*Hint:* Argue that $y^4/4$ is bounded above by the constant appearing in Problem 31.]

33. Epidemic Models. A model for the spread of a disease through a population (an epidemic model) is given by the system

$$\frac{dS}{dt} = -aSI ,$$

$$\frac{dI}{dt} = aSI - bI ,$$

where a and b are positive constants. Here $S(t)$ represents the susceptible population and $I(t)$ the infected population after t days. The constant a is a measure of how fast the disease is transmitted from an infected person to the susceptible population. The constant b represents the rate at which infected people become well, i.e., are removed from the infected population (and are thus presumably immune).

(a) Suppose that for a certain disease it is determined $a = 0.003$ and $b = 0.5$. In the *SI*-phase plane, sketch the trajectory corresponding to the initial condition that one person is infected and 700 persons are susceptible.

(b) From your graph in part (a), estimate the peak (largest possible) number of infected persons.

(c) Using the differential equations that model the spread of disease, argue that the peak number of infected persons occurs when $S = b/a = 1000/6 \approx 167$ persons.

34. (a) For the epidemic model in Problem 33, show that the peak value for I will always occur when $S = b/a = 1000/6$, provided $S(0) > b/a$. (The number b/a is called the **threshold value.**)

(b) The *SI*-phase plane analysis of Problem 33 indicates that an epidemic occurs (the number of infected persons first increases and then decreases to zero) when $S(0) > b/a$. Verify this fact directly from the expression for dI/dt.

(c) As time increases, what happens to the number of infected persons if $S(0) < b/a$?

35. A Problem of Current Interest. The motion of an iron bar attracted by the magnetic field produced by a parallel current wire and restrained by springs (see Figure 5.12) is governed by the equation

$$\frac{d^2x}{dt^2} = -x + \frac{1}{\lambda - x} , \quad \text{for} \quad -x_0 < x < \lambda ,$$

where the constants x_0 and λ are, respectively, the distances from the bar to the wall and to the wire *when the bar is at equilibrium (rest) with the current off.*

(a) Setting $v = dx/dt$, convert the second order equation to an equivalent first order system.

(b) Solve the phase plane equation for the system in part (a) and thereby show that its integral curves are given by

$$v = \pm\sqrt{C - x^2 - 2\ln(\lambda - x)} ,$$

where C is a constant.

(c) Show that if $\lambda < 2$, there are no critical points

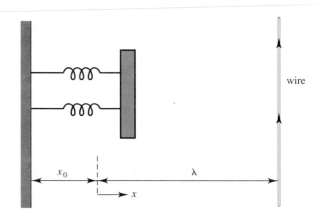

Figure 5.12 Bar restrained by springs and attracted by a parallel current

in the xv-phase plane, whereas if $\lambda > 2$ there are two critical points. For the latter case, determine these critical points.

(d) Physically, the case $\lambda < 2$ corresponds to a current so high that the magnetic attraction completely overpowers the spring. To gain insight into this, use software to plot the phase plane diagrams for the system when $\lambda = 1$ and when $\lambda = 3$.

(e) From your phase plane diagrams in part (d), describe the possible motions of the bar when $\lambda = 1$ and when $\lambda = 3$, under various initial conditions.

36. Falling Object. The motion of an object moving vertically through the air is governed by the equation

$$\frac{d^2y}{dt^2} = -g - \frac{g}{V^2}\frac{dy}{dt}\left|\frac{dy}{dt}\right| ,$$

where y is the upward vertical displacement and V is a constant called the terminal speed. Take $g = 32$ ft/sec^2 and $V = 50$ ft/sec. Sketch trajectories in the yv-phase plane for $-100 \le y \le 100$, $-100 \le v \le 100$, starting from $y = 0$ and $v = -75, -50, -25, 0, 25,$ $50,$ and 75 ft/sec. Interpret the trajectories physically; why is V called the terminal speed?

37. Sticky Friction. An alternative for the damping friction model $F = -by'$ discussed in Section 4.1 is the "sticky friction" model. For a mass sliding on a surface as depicted in Figure 5.13, the contact friction is more complicated than simply $-by'$. We observe, for example, that even if the mass is displaced slightly off the equilibrium location $x = 0$, it may nonetheless remain stationary due to the fact that the spring force

$-ky$ is insufficient to break the static friction's grip. If the maximum force that the friction can exert is denoted by μ, then a feasible model is given by

$$F_{\text{friction}} = \begin{cases} ky , & \text{if } |ky| < \mu \\ & \text{and } \; y' = 0 , \\ \mu\,\text{sign}(y) , & \text{if } |ky| \ge \mu \\ & \text{and } \; y' = 0 , \\ -\mu\,\text{sign}(y') , & \text{if } y' \ne 0 . \end{cases}$$

(The function $\text{sign}(s)$ is $+1$ when $s > 0$, -1 when $s < 0$, and 0 when $s = 0$.) The motion is governed by the equation

(16) $$m\frac{d^2y}{dt^2} = -ky + F_{\text{friction}} .$$

Thus if the mass is at rest, friction *balances* the spring force if $|y| < \mu/k$ but simply *opposes* it with intensity μ if $|y| \ge \mu/k$. If the mass is moving, friction opposes the velocity with the same intensity μ.

(a) Taking $m = \mu = k = 1$, convert (16) into the first order system

$$y' = v ,$$

(17) $$v' = \begin{cases} 0 , & \text{if } |y| < 1 \\ & \text{and } \; v = 0 , \\ -y + \text{sign}(y) , & \text{if } |y| \ge 1 \\ & \text{and } \; v = 0 , \\ -y - \text{sign}(v) , & \text{if } v \ne 0 . \end{cases}$$

(b) Form the phase plane equation for (17) when $v \ne 0$ and solve it to derive the integral curves

$$v^2 + (y \pm 1)^2 = c ,$$

where the plus sign prevails for $v > 0$ and the minus sign for $v < 0$.

(c) Identify the trajectories in the phase plane as two families of concentric semicircles. What is the center of the semicircles in the upper half plane? the lower half plane?

(d) What are the critical points for (17)?

(e) Sketch the trajectory in the phase plane of the mass released from rest at $y = 7.5$. At what value for y does the mass come to rest?

38. Rigid Body Nutation. Euler's equations describe the motion of the principal-axis components of the angular velocity of a freely rotating rigid body (such as a space station), as seen by an observer rotating with the body (the astronauts, for example). This motion is

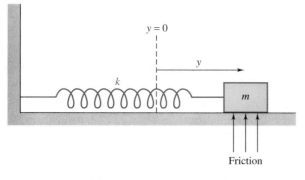

y = 0

y

k

m

Friction

Figure 5.13 Mass-spring system with friction

called *nutation*. If the angular velocity components are denoted by x, y, and z, then an example of Euler's equations is the three-dimensional autonomous system

$$dx/dt = yz \ ,$$
$$dy/dt = -2xz \ ,$$
$$dz/dt = xy \ .$$

The *trajectory* of a solution $x(t)$, $y(t)$, $z(t)$ to these equations is the curve generated by the points $\big(x(t), y(t), z(t)\big)$ in xyz-phase space as t varies over an interval I.

(a) Show that each trajectory of this system lies on the surface of a (possibly degenerate) sphere centered at the origin $(0, 0, 0)$. [*Hint:* Compute $\frac{d}{dt}(x^2 + y^2 + z^2)$.] What does this say about the magnitude of the angular velocity vector?

(b) Find all the critical points of the system, i.e., all points (x_0, y_0, z_0) such that $x(t) \equiv x_0$, $y(t) \equiv y_0$, $z(t) \equiv z_0$ is a solution. For such solutions, the angular velocity vector remains constant in the body system.

(c) Show that the trajectories of the system lie along the intersection of a sphere and an elliptic cylinder of the form $y^2 + 2x^2 = C$, for some constant C. [*Hint:* Consider the expression for dy/dx implied by Euler's equations.]

(d) Using the results of parts (b) and (c), argue that the trajectories of this system are *closed* curves. What does this say about the corresponding solutions?

(e) The front cover of this text displays some typical trajectories for this system. Discuss the stability of the three critical points indicated on the positive axes.[†]

5.3 ELIMINATION METHOD FOR SYSTEMS

The familiar Gauss elimination method used for solving *algebraic* systems such as

$$3x - 2y + z = 4 \ ,$$
$$x + y - z = 0 \ ,$$
$$2x - y + 3z = 6$$

can be adapted to solve systems of *linear differential equations with constant coefficients*. In fact, we used this approach in solving the system that arose in the interconnected tanks problem of Section 5.1. Our goal in this section is to formalize this **elimination method** so that we can tackle more general linear constant coefficient systems.

For this purpose, we begin by rewriting the system in operator notation, utilizing $D := d/dt, D^2 := d^2/dt^2$, etc. We then appeal to the fact that polynomial operators in D, i.e., operators of the form

$$L = c_n D^n + c_{n-1} D^{n-1} + \cdots + c_1 D + c_0$$

with c_0, c_1, \ldots, c_n constants, can be manipulated in much the same way as ordinary polynomials in an algebraic variable; in particular, such operators commute and can be factored (see Section 4.2). For example,

$$D^2 + D - 2 = (D + 2)(D - 1) = (D - 1)(D + 2) \ ,$$

where multiplication denotes the composition of operators.

[†]For further discussion, see the article "The tumbling box," by Susan Jane Colley, *American Math Monthly,* Vol. 94 (1987), pp. 62–68.

We first demonstrate how the method applies to a linear system of two first order differential equations of the form

$$a_1 x'(t) + a_2 x(t) + a_3 y'(t) + a_4 y(t) = f_1(t) ,$$
$$a_5 x'(t) + a_6 x(t) + a_7 y'(t) + a_8 y(t) = f_2(t) ,$$

where a_1, a_2, \ldots, a_8 are constants. In operator notation, this becomes

$$(a_1 D + a_2)[x] + (a_3 D + a_4)[y] = f_1 ,$$
$$(a_5 D + a_6)[x] + (a_7 D + a_8)[y] = f_2 .$$

EXAMPLE 1 Solve the system

(1)
$$x'(t) = 3x(t) - 4y(t) + 1 ,$$
$$y'(t) = 4x(t) - 7y(t) + 10t .$$

SOLUTION The alert reader may observe that since y' is absent from the first equation, we could use the latter to express y in terms of x and x' and substitute into the second equation to derive an "uncoupled" equation containing only x and its derivatives. However, this simple trick will not work on more general systems (see Problem 18).

To utilize the elimination method, we first write the system using the operator notation:[†]

(2)
$$(D - 3)[x] + 4y = 1 ,$$
$$-4x + (D + 7)[y] = 10t .$$

Modeling the elimination procedure for algebraic systems, we can eliminate x from this system by adding 4 times the first equation to $(D - 3)$ applied to the second equation. This gives

$$\big(16 + (D - 3)(D + 7)\big)[y] = 4 \cdot 1 + (D - 3)[10t] = 4 + 10 - 30t ,$$

which simplifies to

(3) $$(D^2 + 4D - 5)[y] = 14 - 30t .$$

Now equation (3) is just a second order linear equation in y with constant coefficients that has the general solution

(4) $$y(t) = C_1 e^{-5t} + C_2 e^t + 6t + 2 ,$$

which can be found using undetermined coefficients.

To find $x(t)$, we have two options.

Method 1. We return to system (2) and eliminate y. This is accomplished by "multiplying" the first equation in (2) by $(D + 7)$ and the second equation by -4 and then adding to obtain

$$(D^2 + 4D - 5)[x] = 7 - 40t .$$

This equation can likewise be solved using undetermined coefficients to yield

(5) $$x(t) = K_1 e^{-5t} + K_2 e^t + 8t + 5 ,$$

[†]Rigorously speaking, we should write $(D - 3)[x] + 4I[y] = 1$, etc., where I is the identity operator.

where we have taken K_1 and K_2 to be the arbitrary constants, which are not necessarily the same as C_1 and C_2 used in formula (4).

It is reasonable to expect that system (1) will involve only *two* arbitrary constants, since it consists of two first order equations. Thus the four constants C_1, C_2, K_1, and K_2 are not independent. To determine the relationships, we substitute the expressions for $x(t)$ and $y(t)$ given in (4) and (5) into one of the equations in (1), say the first one. This yields

$$-5K_1e^{-5t} + K_2e^t + 8 =$$
$$3K_1e^{-5t} + 3K_2e^t + 24t + 15 - 4C_1e^{-5t} - 4C_2e^t - 24t - 8 + 1 \; ,$$

which simplifies to

$$(4C_1 - 8K_1)e^{-5t} + (4C_2 - 2K_2)e^t = 0 \; .$$

Because e^t and e^{-5t} are linearly independent functions on any interval, this last equation holds for all t only if

$$4C_1 - 8K_1 = 0 \quad \text{and} \quad 4C_2 - 2K_2 = 0 \; .$$

Therefore, $K_1 = C_1/2$ and $K_2 = 2C_2$.

A solution to system (1) is then given by the pair

(6) $x(t) = \dfrac{1}{2}C_1e^{-5t} + 2C_2e^t + 8t + 5 \; , \qquad y(t) = C_1e^{-5t} + C_2e^t + 6t + 2 \; .$

As you might expect, this pair is a **general solution** to (1) in the sense that *any* solution to (1) can be expressed in this fashion.

Method 2. A simpler method for determining $x(t)$ once $y(t)$ is known is to use the system to obtain an equation for $x(t)$ in terms of $y(t)$ and $y'(t)$. In this example, solving the second equation in (1) for $x(t)$ gives

$$x(t) = \frac{1}{4}y'(t) + \frac{7}{4}y(t) - \frac{5}{2}t \; .$$

Substituting $y(t)$ as given in (4) yields

$$x(t) = \frac{1}{4}[-5C_1e^{-5t} + C_2e^t + 6] + \frac{7}{4}[C_1e^{-5t} + C_2e^t + 6t + 2] - \frac{5}{2}t$$
$$= \frac{1}{2}C_1e^{-5t} + 2C_2e^t + 8t + 5 \; ,$$

which agrees with (6). ■ End

The above procedure works, more generally, for any linear system of two equations and two unknowns with *constant coefficients* regardless of the order of the equations. For example, if we let L_1, L_2, L_3, and L_4 denote linear differential operators with constant coefficients (i.e., polynomials in D), then the method can be applied to the linear system

$$L_1[x] + L_2[y] = f_1 \; ,$$
$$L_3[x] + L_4[y] = f_2 \; .$$

Because the system has constant coefficients, the operators commute (e.g., $L_2L_4 = L_4L_2$)

and we can eliminate variables in the usual algebraic fashion. Eliminating the variable y gives

(7) $(L_1L_4 - L_2L_3)[x] = g_1$,

where $g_1 := L_4[f_1] - L_2[f_2]$. Similarly, eliminating the variable x yields

(8) $(L_1L_4 - L_2L_3)[y] = g_2$,

where $g_2 := L_1[f_2] - L_3[f_1]$. Now if $L_1L_4 - L_2L_3$ is a differential operator of order n, then a general solution for (7) contains n arbitrary constants, and a general solution for (8) also contains n arbitrary constants. Thus a total of $2n$ constants arise. However, as we saw in Example 1, there are only n of these that are independent for the system; the remaining constants can be expressed in terms of these.[†] The pair of general solutions to (7) and (8) written in terms of the n independent constants is called a **general solution for the system.**

If it turns out that $L_1L_4 - L_2L_3$ is the zero operator, the system is said to be **degenerate.** As with the anomalous problem of solving for the points of intersection of two parallel or coincident lines, a degenerate system may have no solutions, or if it does possess solutions, they may involve any number of arbitrary constants (see Problems 23 and 24).

ELIMINATION PROCEDURE FOR 2 × 2 SYSTEMS

To find a general solution for the system

$$L_1[x] + L_2[y] = f_1 ,$$
$$L_3[x] + L_4[y] = f_2 ,$$

where $L_1, L_2, L_3,$ and L_4 are polynomials in $D = d/dt$:

(a) Make sure the system is written in operator form.
(b) Eliminate one of the variables, say y, and solve the resulting equation for $x(t)$. If the system is degenerate, stop! A separate analysis is required to determine whether or not there are solutions.
(c) (*Shortcut*) If possible, use the system to derive an equation that involves $y(t)$ but not its derivatives. (Otherwise, go to Step (d).) Substitute the found expression for $x(t)$ into this equation to get a formula for $y(t)$. The expressions for $x(t)$, $y(t)$ give the desired general solution.
(d) Eliminate x from the system and solve for $y(t)$. (Solving for $y(t)$ gives more constants—in fact, twice as many as needed.)
(e) Remove the extra constants by substituting the expressions for $x(t)$ and $y(t)$ into one or both of the equations in the system. Write the expressions for $x(t)$ and $y(t)$ in terms of the remaining constants.

EXAMPLE 2 Find a general solution for

(9) $x''(t) + y'(t) - x(t) + y(t) = -1$,
 $x'(t) + y'(t) - x(t) = t^2$.

[†]For a proof of this fact, see *Ordinary Differential Equations,* by M. Tenenbaum and H. Pollard (Dover, New York, 1985), Chapter 7.

SOLUTION We begin by expressing the system in operator notation:

$$\begin{aligned}(D^2 - 1)[x] + (D + 1)[y] &= -1 \ , \\ (D - 1)[x] + D[y] &= t^2 \ . \end{aligned} \tag{10}$$

Here $L_1 := D^2 - 1, L_2 := D + 1, L_3 := D - 1$, and $L_4 := D$.
 Eliminating y gives (cf. (7)):

$$\big((D^2 - 1)D - (D + 1)(D - 1)\big)[x] = D[-1] - (D + 1)[t^2] \ ,$$

which reduces to

$$(D - 1)\big((D + 1)D - (D + 1)\big)[x] = -2t - t^2 \ ,$$
$$(D - 1)^2(D + 1)[x] = -2t - t^2 \ . \tag{11}$$

Since $(D - 1)^2(D + 1)$ is third order, we should expect three arbitrary constants in a general solution to system (9).

 Although the methods of Chapter 4 focused on solving second order equations, we have seen several examples of how they extend in a natural way to higher-order equations. Applying this strategy to the third order equation (11), we observe that the corresponding homogeneous equation has the auxiliary equation $(r - 1)^2(r + 1) = 0$ with roots $r = 1, 1, -1$. Hence a general solution for the homogeneous equation is

$$x_h(t) = C_1 e^t + C_2 t e^t + C_3 e^{-t} \ .$$

To find a particular solution to (11), we use the method of undetermined coefficients with $x_p(t) = At^2 + Bt + C$. Substituting into (11) and solving for A, B, and C yields (after a little algebra)

$$x_p(t) = -t^2 - 4t - 6 \ .$$

Thus a general solution to equation (11) is

$$x(t) = x_h(t) + x_p(t) = C_1 e^t + C_2 t e^t + C_3 e^{-t} - t^2 - 4t - 6 \ . \tag{12}$$

 To find $y(t)$, we take the shortcut described in Step (c) of the elimination procedure box. Subtracting the second equation in (10) from the first, we find

$$(D^2 - D)[x] + y = -1 - t^2 \ ,$$

so that

$$y = (D - D^2)[x] - 1 - t^2 \ .$$

Inserting the expression for $x(t)$, given in (12), we obtain

$$\begin{aligned} y(t) = \ & C_1 e^t + C_2(te^t + e^t) - C_3 e^{-t} - 2t - 4 \\ & - \big[C_1 e^t + C_2(te^t + 2e^t) + C_3 e^{-t} - 2 \big] - 1 - t^2 \ , \end{aligned}$$
$$y(t) = -C_2 e^t - 2C_3 e^{-t} - t^2 - 2t - 3 \ . \tag{13}$$

The formulas for $x(t)$ in (12) and $y(t)$ in (13) give the desired general solution to (9). ■ End

The elimination method also applies to linear systems with three or more equations and unknowns; however, the process becomes more cumbersome as the number of equations and unknowns increases. The matrix methods presented in Chapter 9 are better suited for handling larger systems. Here we illustrate the elimination technique for a 3×3 system.

EXAMPLE 3 Find a general solution to

$$x'(t) = x(t) + 2y(t) - z(t) \ ,$$
(14) $$y'(t) = x(t) + z(t) \ ,$$
$$z'(t) = 4x(t) - 4y(t) + 5z(t) \ .$$

SOLUTION We begin by expressing the system in operator notation:

$$(D - 1)[x] - 2y + z = 0 \ ,$$
(15) $$-x + D[y] - z = 0 \ ,$$
$$-4x + 4y + (D - 5)[z] = 0 \ .$$

Eliminating z from the first two equations (by adding them) and then from the last two equations yields (after some algebra, which we omit)

(16) $$(D - 2)[x] + (D - 2)[y] = 0 \ ,$$
$$-(D - 1)[x] + (D - 1)(D - 4)[y] = 0 \ .$$

On eliminating x from this 2×2 system, we eventually obtain

$$(D - 1)(D - 2)(D - 3)[y] = 0 \ ,$$

which has the general solution

(17) $$y(t) = C_1 e^t + C_2 e^{2t} + C_3 e^{3t} \ .$$

Taking the shortcut approach, we add the two equations in (16) to get an expression for x in terms of y and its derivatives, which simplifies to

$$x = (D^2 - 4D + 2)[y] = y'' - 4y' + 2y \ .$$

When we substitute the expression (17) for $y(t)$ into this equation, we find

(18) $$x(t) = -C_1 e^t - 2C_2 e^{2t} - C_3 e^{3t} \ .$$

Finally, using the second equation in (14) to solve for $z(t)$, we get

$$z(t) = y'(t) - x(t) \ ,$$

and substituting in for $y(t)$ and $x(t)$ yields

(19) $$z(t) = 2C_1 e^t + 4C_2 e^{2t} + 4C_3 e^{3t} \ .$$

The expressions for $x(t)$ in (18), $y(t)$ in (17), and $z(t)$ in (19) give a general solution with C_1, C_2, and C_3 as arbitrary constants. ■End

EXERCISES 5.3

In Problems 1 through 18, use the elimination method to find a general solution for the given linear system, where differentiation is with respect to t.

1. $x' + y' = -2y$,
$\quad x - 2y = y'$.

2. $x' = 3y$,
$\quad y' = 2x - y$.

3. $x' + 2y = 0$,
$\quad x' - y' = 0$.

4. $x' = x - y$,
$\quad y' = y - 4x$.

5. $x' + y' - x = 5$,
$\quad x' + y' + y = 1$.

6. $x' = 3x - 2y + \sin t$,
$\quad y' = 4x - y - \cos t$.

7. $(D + 1)[u] - (D + 1)[v] = e^t$,
$\quad (D - 1)[u] + (2D + 1)[v] = 5$.

8. $(D - 3)[x] + (D - 1)[y] = t$,
$\quad (D + 1)[x] + (D + 4)[y] = 1$.

9. $x' + y' + 2x = 0$,
$\quad x' + y' - x - y = \sin t$.

10. $2x' + y' - x - y = e^{-t}$,
$\quad x' + y' + 2x + y = e^t$.

11. $(D^2 - 1)[u] + 5v = e^t$, **12.** $D^2[u] + D[v] = 2$,
$\quad 2u + (D^2 + 2)[v] = 0$. $\quad 4u + D[v] = 6$.

13. $\dfrac{dx}{dt} = x - 4y$, **14.** $\dfrac{dx}{dt} + y = t^2$,

$\quad \dfrac{dy}{dt} = x + y$. $\quad -x + \dfrac{dy}{dt} = 1$.

15. $\dfrac{dw}{dt} = 5w + 2z + 5t$, **16.** $\dfrac{dx}{dt} + x + \dfrac{dy}{dt} = e^{4t}$,

$\quad \dfrac{dz}{dt} = 3w + 4z + 17t$. $\quad 2x + \dfrac{d^2y}{dt^2} = 0$.

17. $x'' + 5x - 4y = 0$,
$\quad -x + y'' + 2y = 0$.

18. $x'' + y'' - x' = 2t$.
$\quad x'' + y' - x + y = -1$.

In Problems 19 through 21, solve the given initial value problem.

19. $\dfrac{dx}{dt} = 4x + y$; $x(0) = 1$,

$\quad \dfrac{dy}{dt} = -2x + y$; $y(0) = 0$.

20. $\dfrac{dx}{dt} = 2x + y - e^{2t}$; $x(0) = 1$,

$\quad \dfrac{dy}{dt} = x + 2y$; $y(0) = -1$.

21. $\dfrac{d^2x}{dt^2} = y$; $x(0) = 3$, $x'(0) = 1$,

$\quad \dfrac{d^2y}{dt^2} = x$; $y(0) = 1$, $y'(0) = -1$.

22. In Example 3 of Section 5.2 (page 267), phase plane analysis was used to predict that the solution to the initial value problem

$$x' = 5x - 3y - 2 ; x(0) = 2 ,$$
$$y' = 4x - 3y - 1 ; y(0) = 0$$

tends to infinity as $t \to +\infty$, i.e., $|x(t)| + |y(t)| \to +\infty$. Verify this fact directly by solving the initial value problem and taking the limit as $t \to +\infty$.

In Problems 23 and 24, show that the given linear system is degenerate. In attempting to solve the system, determine whether it has no solutions or infinitely many solutions.

23. $(D - 1)[x] + (D - 1)[y] = -3e^{-2t}$,
$\quad (D + 2)[x] + (D + 2)[y] = 3e^t$.

24. $D[x] + (D + 1)[y] = e^t$,
$\quad D[x] + (D + 1)[y] = 0$.

In Problems 25 through 28, use the elimination method to find a general solution for the given system of three equations in the three unknown functions $x(t)$, $y(t)$, $z(t)$.

25. $x' = x + 2y - z$, **26.** $x' = 3x + y - z$,
$\quad y' = x + z$, $\quad y' = x + 2y - z$,
$\quad z' = 4x - 4y + 5z$. $\quad z' = 3x + 3y - z$.

27. $x' = 4x - 4z$, **28.** $x' = x + 2y + z$,
$\quad y' = 4y - 2z$, $\quad y' = 6x - y$,
$\quad z' = -2x - 4y + 4z$. $\quad z' = -x - 2y - z$.

*In Problems 29 and 30, determine the range of values (if any) of the parameter λ that will ensure **all** solutions x(t), y(t) of the given system remain bounded as t → +∞.*

29. $\dfrac{dx}{dt} = \lambda x - y$,

$\dfrac{dy}{dt} = 3x + y$.

30. $\dfrac{dx}{dt} = -x + \lambda y$,

$\dfrac{dy}{dt} = x - y$.

31. Two large tanks, each holding 100 L of liquid, are interconnected by pipes, with the liquid flowing from tank A into tank B at a rate of 3 L/min and from B into A at a rate of 1 L/min (see Figure 5.14). The liquid inside each tank is kept well stirred. A brine solution with a concentration of 2 kg/L of salt flows into tank A at a rate of 6 L/min. The (diluted) solution flows out of the system from tank A at 4 L/min and from tank B at 2 L/min. If, initially, tank A contains pure water and tank B contains 200 kg of salt, determine the mass of salt in each tank at time $t \geq 0$.

32. In Problem 31, 3 L/min of liquid flowed from tank A into tank B and 1 L/min from B into A. Determine the mass of salt in each tank at time $t \geq 0$ if, instead, 5 L/min flows from A into B and 3 L/min flows from B into A, with all other data the same.

33. In Problem 31, assume no solution flows out of the system from tank B, only 1 L/min flows from A into B, and only 4 L/min of brine flows into the system at tank A, other data being the same. Determine the mass of salt in each tank at time $t \geq 0$.

34. Feedback System with Pooling Delay. Many physical and biological systems involve time delays. A pure time delay has the output the same as the input but is shifted in time. A more common type of delay is *pooling delay*. An example of such a feedback system is shown in Figure 5.15. Here the level

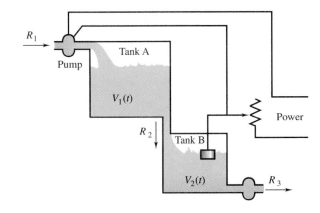

Figure 5.15 Feedback system with pooling delay

of fluid in tank B determines the rate at which fluid enters tank A. Suppose this rate is given by $R_1(t) = \alpha[V - V_2(t)]$, where α and V are positive constants and $V_2(t)$ is the volume of fluid in tank B at time t.

(a) If the outflow rate R_3 from tank B is constant and the flow rate R_2 from tank A into B is $R_2(t) = KV_1(t)$, where K is a positive constant and $V_1(t)$ is the volume of fluid in tank A at time t, then show that this feedback system is governed by the system

$$\frac{dV_1}{dt} = \alpha\left(V - V_2(t)\right) - KV_1(t) ,$$

$$\frac{dV_2}{dt} = KV_1(t) - R_3 .$$

(b) Find a general solution for the system in part (a) when $\alpha = 5$ (min)$^{-1}$, $V = 20$ L, $K = 2$ (min)$^{-1}$, and $R_3 = 10$ L/min.

(c) Using the general solution obtained in part (b), what can be said about the volume of fluid in each of the tanks as $t \to +\infty$?

Figure 5.14 Mixing problem for interconnected tanks

Figure 5.16 Air-conditioned house with attic

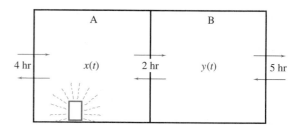

Figure 5.17 Two-zone building with one zone heated

35. A house, for cooling purposes, consists of two zones: the attic area zone A and the living area zone B (see Figure 5.16). The living area is cooled by a 2-ton air conditioning unit that removes 24,000 Btu/hr. The heat capacity of zone B is $1/2°F$ per thousand Btu. The time constant for heat transfer between zone A and the outside is 2 hr, between zone B and the outside is 4 hr, and between the two zones is 4 hr. If the outside temperature stays at 100°F, how warm can it eventually get in the attic zone A? (Heating and cooling of buildings was treated in Section 3.3.)

36. A building consists of two zones A and B (see Figure 5.17). Only zone A is heated by a furnace, which generates 80,000 Btu/hr. The heat capacity of zone A is $1/4°F$ per thousand Btu. The time constant for heat transfer between zone A and the outside is 4 hr, between the unheated zone B and the outside is 5 hr, and between the two zones is 2 hr. If the outside temperature stays at 0°F, how cold can it eventually get in the unheated zone B?

37. In Problem 36, if a small furnace that generates 1000 Btu/hr is placed in zone B, determine the coldest it could eventually get in zone B if zone B has a heat capacity of 2°F per thousand Btu.

38. Arms Race. A simplified mathematical model for an arms race between two countries whose expenditures for defense are expressed by the variables $x(t)$ and $y(t)$ is given by the linear system

$$\frac{dx}{dt} = 2y - x + a \; ; \qquad x(0) = 1 \; ,$$

$$\frac{dy}{dt} = 4x - 3y + b \; ; \qquad y(0) = 4 \; ,$$

where a and b are constants that measure the trust (or distrust) each country has for the other. Determine whether there is going to be disarmament (x and y approach 0 as t increases), a stabilized arms race (x and y approach a constant as $t \to +\infty$), or a runaway arms race (x and y approach $+\infty$ as $t \to +\infty$).

5.4 COUPLED MASS-SPRING SYSTEMS

In this section, we extend our mass-spring model to include situations where coupled springs connect two masses, both of which are free to move. The resulting motions can be very intriguing. For simplicity, we shall neglect the effects of friction, gravity, and external forces. Consider the following experiment.

EXAMPLE 1 On a smooth horizontal surface, a mass $m_1 = 2$ kg is attached to a fixed wall by a spring with spring constant $k_1 = 4$ N/m. Another mass $m_2 = 1$ kg is attached to the first object by a spring with spring constant $k_2 = 2$ N/m. The objects are aligned horizontally so that the springs are their natural lengths (Figure 5.18). If both objects are displaced 3 m to the right of their equilibrium positions (Figure 5.19) and then released, what are the equations of motion for the two objects?

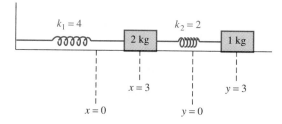

Figure 5.18 Coupled system at equilibrium Figure 5.19 Coupled system at initial displacement

SOLUTION From our assumptions, the only forces we need to consider are those due to the springs them-selves. Recall that Hooke's law asserts that the force acting on an object due to a spring has magnitude proportional to the displacement of the spring from its natural length and has direc-tion opposite to its displacement. That is, if the spring is either stretched or compressed, then it tries to return to its natural length.

Because each mass is free to move, we apply Newton's second law to each object. Let $x(t)$ denote the displacement (to the right) of the 2-kg mass from its equilibrium position, and, similarly, let $y(t)$ denote the corresponding displacement for the 1-kg mass. The 2-kg mass has a force F_1 acting on its left side due to one spring and a force F_2 acting on its right side due to the second spring. Referring to Figure 5.19 and applying Hooke's law, we see that

$$F_1 = -k_1 x , \qquad F_2 = +k_2(y - x) ,$$

since $(y - x)$ is the net displacement of the second spring from its natural length. There is only one force acting on the 1-kg mass: the force due to the second spring, which is

$$F_3 = -k_2(y - x) .$$

Applying Newton's second law to these objects, we obtain the system

(1)
$$m_1 \frac{d^2 x}{dt^2} = F_1 + F_2 = -k_1 x + k_2(y - x) ,$$

$$m_2 \frac{d^2 y}{dt^2} = F_3 = -k_2(y - x) ,$$

or

(2)
$$m_1 \frac{d^2 x}{dt^2} + (k_1 + k_2)x - k_2 y = 0 ,$$

$$m_2 \frac{d^2 y}{dt^2} + k_2 y - k_2 x = 0 .$$

In this problem, we know that $m_1 = 2$, $m_2 = 1$, $k_1 = 4$, and $k_2 = 2$. Substituting these val-ues into system (2) yields

(3) $$2\frac{d^2 x}{dt^2} + 6x - 2y = 0 ,$$

(4) $$\frac{d^2 y}{dt^2} + 2y - 2x = 0 .$$

We shall use the elimination method of Section 5.3 to solve (3)–(4). With $D := d/dt$, we rewrite the system as

(5) $(2D^2 + 6)[x] - 2y = 0$,

(6) $-2x + (D^2 + 2)[y] = 0$.

Adding $(D^2 + 2)$ applied to equation (5) to 2 times equation (6) eliminates y:

$$[(D^2 + 2)(2D^2 + 6) - 4][x] = 0 ,$$

which simplifies to

(7) $2\dfrac{d^4x}{dt^4} + 10\dfrac{d^2x}{dt^2} + 8x = 0$.

Notice that equation (7) is linear with constant coefficients. To solve it, let's proceed as we did with linear second order equations and try to find solutions of the form $x = e^{rt}$. Substituting e^{rt} in equation (7) gives

$$2(r^4 + 5r^2 + 4)e^{rt} = 0 .$$

Thus we get a solution to (7) when r satisfies the auxiliary equation

$$r^4 + 5r^2 + 4 = 0 .$$

From the factorization $r^4 + 5r^2 + 4 = (r^2 + 1)(r^2 + 4)$, we see that the roots of the auxiliary equation are complex numbers $i, -i, 2i, -2i$. Using Euler's formula, it follows that

$$z_1(t) = e^{it} = \cos t + i \sin t \qquad \text{and} \qquad z_2(t) = e^{2it} = \cos 2t + i \sin 2t$$

are complex-valued solutions to equation (7). To obtain real-valued solutions, we take the real and imaginary parts of $z_1(t)$ and $z_2(t)$. Thus four real-valued solutions are

$$x_1(t) = \cos t , \qquad x_2(t) = \sin t , \qquad x_3(t) = \cos 2t , \qquad x_4(t) = \sin 2t ,$$

and a general solution is

(8) $x(t) = a_1 \cos t + a_2 \sin t + a_3 \cos 2t + a_4 \sin 2t$,

where a_1, a_2, a_3, and a_4 are arbitrary constants.[†]
 To obtain a formula for $y(t)$, we use equation (3) to express y in terms of x:

$$y(t) = \dfrac{d^2x}{dt^2} + 3x$$

$$= -a_1 \cos t - a_2 \sin t - 4a_3 \cos 2t - 4a_4 \sin 2t$$
$$+ 3a_1 \cos t + 3a_2 \sin t + 3a_3 \cos 2t + 3a_4 \sin 2t ,$$

and so

(9) $y(t) = 2a_1 \cos t + 2a_2 \sin t - a_3 \cos 2t - a_4 \sin 2t$.

[†]A more detailed discussion of general solutions is given in Chapter 6.

To determine the constants a_1, a_2, a_3, and a_4, let's return to the original problem. We were told that the objects were originally displaced 3 m to the right and then released. Hence

(10) $x(0) = 3$, $\dfrac{dx}{dt}(0) = 0$; $y(0) = 3$, $\dfrac{dy}{dt}(0) = 0$.

On differentiating equations (8) and (9), we find

$$\frac{dx}{dt} = -a_1 \sin t + a_2 \cos t - 2a_3 \sin 2t + 2a_4 \cos 2t ,$$

$$\frac{dy}{dt} = -2a_1 \sin t + 2a_2 \cos t + 2a_3 \sin 2t - 2a_4 \cos 2t .$$

Now, if we put $t = 0$ in the formulas for x, dx/dt, y, and dy/dt, the initial conditions (10) give the four equations

$$x(0) = a_1 + a_3 = 3 , \frac{dx}{dt}(0) = a_2 + 2a_4 = 0 ,$$

$$y(0) = 2a_1 - a_3 = 3 , \frac{dy}{dt}(0) = 2a_2 - 2a_4 = 0 .$$

From this system, we find $a_1 = 2$, $a_2 = 0$, $a_3 = 1$, and $a_4 = 0$. Hence the equations of motion for the two objects are

$$x(t) = 2 \cos t + \cos 2t ,$$
$$y(t) = 4 \cos t - \cos 2t ,$$

which are depicted in Figure 5.20. ■ End

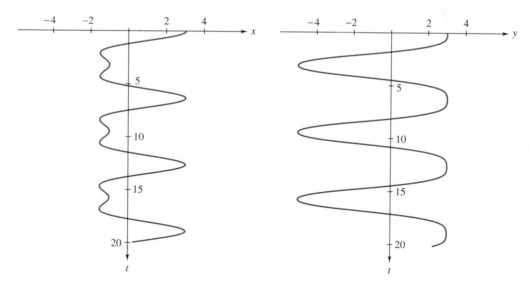

Figure 5.20 Graphs of the motion of the two masses in the coupled mass-spring system

The general solution pair (8), (9) that we have obtained is a combination of sinusoids oscillating at two different angular frequencies: 1 rad/sec and 2 rad/sec. These frequencies extend the notion of the *natural frequency* of the single (free, undamped) mass-spring oscillator (page 231, Section 4.11) and are called the **natural** (or **normal**) **angular frequencies**[†] of the system. A complex system consisting of more masses and springs would have many normal frequencies.

Notice that if the initial conditions were altered so that the constants a_3 and a_4 in (8) and (9) were zero, the motion would be a pure sinusoid oscillating at the single frequency 1 rad/sec. Similarly, if a_1 and a_2 were zero, only the 2 rad/sec oscillation would be "excited." Such solutions, wherein the entire motion is described by a single sinusoid, are called the **normal modes** of the system.[††] The normal modes in the following example are particularly easy to visualize because we shall take all the masses and all the spring constants to be equal.

EXAMPLE 2 Three identical springs with spring constant k and two identical masses m are attached in a straight line with the ends of the outside springs fixed (see Figure 5.21). Determine and interpret the normal modes of the system.

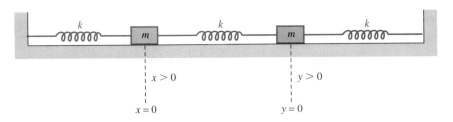

Figure 5.21 Coupled mass-spring system with fixed ends

SOLUTION We define the displacements from equilibrium, x and y, as in Example 1. The equations expressing Newton's second law for the masses are quite analogous to (1), except for the effect of the third spring on the second mass:

(11) $mx'' = -kx + k(y - x)$,

(12) $my'' = -k(y - x) - ky$,

or

$$(mD^2 + 2k)[x] - ky = 0 ,$$
$$-kx + (mD^2 + 2k)[y] = 0 .$$

Eliminating y in the usual manner results in

(13) $\left((mD^2 + 2k)^2 - k^2\right)[x] = 0$.

This has the auxiliary equation

$$(mr^2 + 2k)^2 - k^2 = (mr^2 + k)(mr^2 + 3k) = 0 ,$$

[†]The study of the natural frequencies of oscillations of complex systems is known in engineering as **modal analysis.**

[††]The normal modes are more naturally characterized in terms of *eigenvalues* (see Section 9.5).

with roots $\pm i\sqrt{k/m}$, $\pm i\sqrt{3k/m}$. Setting $\omega := \sqrt{k/m}$, we get the following general solution to (13):

(14) $x(t) = C_1 \cos \omega t + C_2 \sin \omega t + C_3 \cos\left(\sqrt{3}\,\omega t\right) + C_4 \sin\left(\sqrt{3}\,\omega t\right)$.

To obtain $y(t)$, we solve for $y(t)$ in (11) and substitute $x(t)$ as given in (14). Upon simplifying, we get

(15) $y(t) = C_1 \cos \omega t + C_2 \sin \omega t - C_3 \cos\left(\sqrt{3}\,\omega t\right) - C_4 \sin\left(\sqrt{3}\,\omega t\right)$.

From the formulas (14) and (15) we see that the normal angular frequencies are ω and $\sqrt{3}\,\omega$. Indeed, if $C_3 = C_4 = 0$, we have a solution where $y(t) \equiv x(t)$, oscillating at the angular frequency $\omega = \sqrt{k/m}$ rad/sec (equivalent to a frequency $\sqrt{k/m}/2\pi$ periods/sec). Now if $x(t) \equiv y(t)$ in Figure 5.21, the two masses are moving as if they were a single rigid body of mass $2m$, forced by a "double spring" with a spring constant given by $2k$. And indeed, according to equation (4) of Section 4.11 (page 230), we would expect such a system to oscillate at the angular frequency $\sqrt{2k/2m} = \sqrt{k/m}$ (!). This motion is depicted in Figure 5.22(a).

Similarly, if $C_1 = C_2 = 0$, we find the second normal mode where $y(t) = -x(t)$, so that in Figure 5.21 there are two "mirror-image" systems, each with mass m and a "spring and a half" with spring constant $k + 2k = 3k$. (The half-spring will be twice as stiff.) Section 4.11's equation (4) then predicts an angular oscillation frequency for each system of $\sqrt{3k/m} = \sqrt{3}\,\omega$, which again is consistent with (14) and (15). This motion is shown in Figure 5.22(b). ■ End

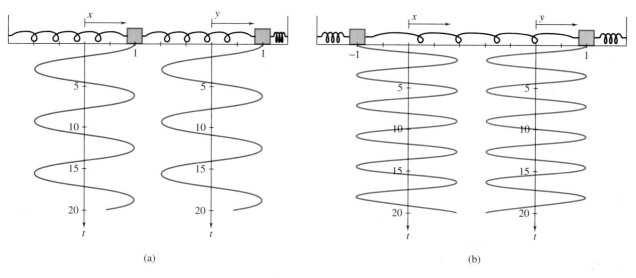

(a) (b)

Figure 5.22 Normal modes for Example 2

EXERCISES 5.4

1. Two springs and two masses are attached in a straight line on a horizontal frictionless surface as illustrated in Figure 5.23 on page 292. The system is set in motion by holding the mass m_2 at its equilibrium position and pulling the mass m_1 to the left of its equilibrium position a distance 1 m and then

Figure 5.23 Coupled mass-spring system with one end free

releasing both masses. Express Newton's law for the system and determine the equations of motion for the two masses if $m_1 = 1$ kg, $m_2 = 2$ kg, $k_1 = 4$ N/m, and $k_2 = 10/3$ N/m.

2. Determine the equations of motion for the two masses described in Problem 1 if $m_1 = 1$ kg, $m_2 = 1$ kg, $k_1 = 3$ N/m, and $k_2 = 2$ N/m.

3. Four springs with the same spring constant and three equal masses are attached in a straight line on a horizontal frictionless surface as illustrated in Figure 5.24. Determine the normal frequencies for the system and describe the three normal modes of vibration.

Figure 5.24 Coupled mass-spring system with three degrees of freedom

4. Two springs, two masses, and a dashpot are attached in a straight line on a horizontal frictionless surface as shown in Figure 5.25. The dashpot provides a damping force on mass m_2, given by $F = -by'$. Derive the system of differential equations for the displacements x and y.

Figure 5.25 Coupled mass-spring system with one end damped

5. Two springs, two masses, and a dashpot are attached in a straight line on a horizontal frictionless surface as shown in Figure 5.26. The system is set in motion by holding the mass m_2 at its equilibrium position and pushing the mass m_1 to the left of its equilibrium position a distance 2 m and then releasing both masses. Determine the equations of motion for the two masses if $m_1 = m_2 = 1$ kg, $k_1 = k_2 = 1$ N/m, and $b = 1$ N-sec/m. [*Hint:* The dashpot damps both m_1 and m_2 with a force whose magnitude equals $b|y' - x'|$.]

Figure 5.26 Coupled mass-spring system with damping between the masses

6. Referring to the coupled mass-spring system discussed in Example 1, suppose an external force $E(t) = 37 \cos 3t$ is applied to the second object of mass 1 kg. The displacement functions $x(t)$ and $y(t)$ now satisfy the system

(16) $2x''(t) + 6x(t) - 2y(t) = 0$,

(17) $y''(t) + 2y(t) - 2x(t) = 37 \cos 3t$.

(a) Show that $x(t)$ satisfies the equation

(18) $x^{(4)}(t) + 5x''(t) + 4x(t) = 37 \cos 3t$.

(b) Find a general solution $x(t)$ to equation (18). [*Hint:* Use undetermined coefficients with $x_p = A \cos 3t + B \sin 3t$.]

(c) Substitute $x(t)$ back into (16) to obtain a formula for $y(t)$.

(d) If both masses are displaced 2 m to the right of their equilibrium positions and then released, find the displacement functions $x(t)$ and $y(t)$.

7. Suppose the displacement functions $x(t)$ and $y(t)$ for a coupled mass-spring system (similar to the one discussed in Problem 6) satisfy the initial value problem

$$x''(t) + 5x(t) - 2y(t) = 0 ,$$
$$y''(t) + 2y(t) - 2x(t) = 3 \sin 2t ;$$
$$x(0) = x'(0) = 0 ,$$
$$y(0) = 1 , \quad y'(0) = 0 .$$

Solve for $x(t)$ and $y(t)$.

Figure 5.27 Double pendulum

Figure 5.28 Coupled pendulums

8. A double pendulum swinging in a vertical plane under the influence of gravity (see Figure 5.27) satisfies the system

$$(m_1 + m_2)l_1^2\theta_1'' + m_2 l_1 l_2 \theta_2'' + (m_1 + m_2)l_1 g\theta_1 = 0 \ ,$$
$$m_2 l_2^2 \theta_2'' + m_2 l_1 l_2 \theta_1'' + m_2 l_2 g\theta_2 = 0 \ ,$$

when θ_1 and θ_2 are small angles. Solve the system when $m_1 = 3$ kg, $m_2 = 2$ kg, $l_1 = l_2 = 5$ m, $\theta_1(0) = \pi/6$, $\theta_2(0) = \theta_1'(0) = \theta_2'(0) = 0$.

9. The motion of a pair of identical pendulums coupled by a spring is modeled by the system

$$mx_1'' = -\frac{mg}{l}x_1 - k(x_1 - x_2) \ ,$$

$$mx_2'' = -\frac{mg}{l}x_2 + k(x_1 - x_2)$$

for small displacements (see Figure 5.28). Determine the two normal frequencies for the system.

10. Suppose the coupled mass-spring system of Problem 1 (Figure 5.23) is hung *vertically* from a support (with mass m_2 above m_1), as in Section 4.12, page 245.

(a) Argue that at equilibrium, the lower spring is stretched a distance l_1 from its natural length L_1, given by $l_1 = m_1 g / k_1$.

(b) Argue that at equilibrium, the upper spring is stretched a distance $l_2 = (m_1 + m_2)g / k_2$.

(c) Show that if x_1 and x_2 are redefined to be displacements from the equilibrium positions of the masses m_1 and m_2, then the equations of motion are identical with those derived in Problem 1.

5.5 ELECTRICAL CIRCUITS

In this section, we consider the application of differential equations to simple electrical circuits consisting of voltage sources (e.g., batteries or generators), resistors, inductors, and capacitors. The so-called *RLC* series circuit is shown schematically in Figure 5.29 on page 294.

The physical principles governing electrical circuits were formulated by G. R. Kirchhoff in 1859. They are the following:

 1. Kirchhoff's current law: The algebraic sum of the currents flowing into any junction point must be zero.

†*Historical Footnote:* Gustav Robert Kirchhoff (1824–1887) was a German physicist noted for his research in spectrum analysis, optics, and electricity.

2. **Kirchhoff's voltage law:** The algebraic sum of the instantaneous changes in potential (voltage drops) around any closed loop must be zero.

Kirchhoff's current law implies that the same current, I, passes through each element in the series circuit of Figure 5.29.

Figure 5.29 Schematic representation of an *RLC* series circuit

To apply Kirchhoff's voltage law, we need to know the voltage drop across each element of the circuit. These voltage formulas are stated below (the reader can consult an introductory physics text for further details).

(a) According to Ohm's law, the voltage drop E_R across a resistor is proportional to the current I passing through the resistor:

$$E_R = RI \ .$$

The proportionality constant R is called the **resistance.**

(b) It can be shown using Faraday's law and Lenz's law that the voltage drop E_L across an inductor is proportional to the instantaneous rate of change of the current I:

$$E_L = L\frac{dI}{dt} \ .$$

The proportionality constant L is called the **inductance.**

(c) The voltage drop E_C across a capacitor is proportional to the electrical charge q on the capacitor:

$$E_C = \frac{1}{C}q \ .$$

The constant C is called the **capacitance.**

A voltage source is assumed to *add* voltage or potential energy to the circuit. If we let $E(t)$ denote the voltage supplied to the circuit at time t, then Kirchhoff's voltage law gives

(1) $E_L + E_R + E_C = E(t) \ .$

Substituting into (1) the expressions for E_L, E_R, and E_C gives

(2) $L\dfrac{dI}{dt} + RI + \dfrac{1}{C}q = E(t) \ .$

Now, the current is just the instantaneous rate of change in charge; that is,

$$I = dq/dt \ .$$

Therefore, we can express (2) in terms of the single variable q as follows:

(3) $$L\frac{d^2q}{dt^2} + R\frac{dq}{dt} + \frac{1}{C}q = E(t) \ .$$

In most applications, we will be interested in determining the current $I(t)$. If we differentiate (3) with respect to t and substitute I for dq/dt, we obtain

(4) $$L\frac{d^2I}{dt^2} + R\frac{dI}{dt} + \frac{1}{C}I = \frac{dE}{dt} \ .$$

For an *RLC* series circuit, one is usually given the initial charge on the capacitor $q(0)$ and the initial current $I(0) = q'(0)$. These constants provide the initial conditions for equation (3). However, for equation (4), we also need to know the initial rate of change in the current, $I'(0)$. This can be obtained from equation (2) by substituting in the initial values $q(0)$, $I(0)$, and $E(0)$.

The common units and symbols used for electrical circuits are listed in Table 5.1.

TABLE 5.1 COMMON UNITS AND SYMBOLS USED WITH ELECTRICAL CIRCUITS

Quantity	Letter Representation	Units	Symbol Representation
Voltage source	E	volt (V)	—O— Generator
			⊣⊢ Battery
Resistance	R	ohm (Ω)	—W—
Inductance	L	henry (H)	—ՇՇՇ—
Capacitance	C	farad (F)	⊣⊢
Charge	q	coulomb	
Current	I	ampere	

EXAMPLE 1 An *RLC* series circuit has a voltage source given by $E(t) = \sin 100t$ volts, a resistor of 0.02 ohms, an inductor of 0.001 henrys, and a capacitor of 2 farads. (These values are selected for convenience; typical values for capacitance are much smaller.) If the initial current and the initial charge on the capacitor are both zero, determine the current in the circuit for $t > 0$.

SOLUTION Here $L = 0.001$, $R = 0.02$, $C = 2$, and $E(t) = \sin 100t$. Substituting these into equation (4) for the current gives

$$(0.001)\frac{d^2I}{dt^2} + (0.02)\frac{dI}{dt} + (0.5)I = 100 \cos 100t \ ,$$

or, equivalently,

(5) $\dfrac{d^2I}{dt^2} + 20\dfrac{dI}{dt} + 500I = 100{,}000 \cos 100t$.

The homogeneous equation associated with (5) has the auxiliary equation

$$r^2 + 20r + 500 = (r + 10)^2 + (20)^2 = 0 \ ,$$

whose roots are $-10 \pm 20i$. Hence the solution to the homogeneous equation is

(6) $I_h(t) = C_1 e^{-10t} \cos 20t + C_2 e^{-10t} \sin 20t$.

To find a particular solution for (5), we can use the method of undetermined coefficients. Setting

$$I_p(t) = A \cos 100t + B \sin 100t$$

and carrying out the procedure in Section 4.8, we ultimately find, to three decimals,

$$A = -10.080 \ , \qquad B = 2.122 \ .$$

Hence a particular solution to (5) is given by

(7) $I_p(t) = -10.080 \cos 100t + 2.122 \sin 100t$.

Since $I = I_h + I_p$, we find from (6) and (7) that

(8) $I(t) = e^{-10t}(C_1 \cos 20t + C_2 \sin 20t) - 10.080 \cos 100t + 2.122 \sin 100t$.

To determine the constants C_1 and C_2, we need the values $I(0)$ and $I'(0)$. We were given $I(0) = q(0) = 0$. To find $I'(0)$, we substitute the values for L, R, and C into equation (2) and equate the two sides at $t = 0$. This gives

$$(0.001)I'(0) + (0.02)I(0) + (0.5)q(0) = \sin 0 \ .$$

Because $I(0) = q(0) = 0$, we find $I'(0) = 0$. Finally, using $I(t)$ in (8) and the initial conditions $I(0) = I'(0) = 0$, we obtain the system

$$I(0) = C_1 - 10.080 = 0 \ ,$$
$$I'(0) = -10C_1 + 20C_2 + 212.2 = 0 \ .$$

Solving this system yields $C_1 = 10.080$ and $C_2 = -5.570$. Hence the current in the *RLC* series circuit is

(9) $I(t) = e^{-10t}(10.080 \cos 20t - 5.570 \sin 20t) - 10.080 \cos 100t + 2.122 \sin 100t$. ■End

Observe that, as was the case with forced mechanical vibrations, the current in (9) is made up of two components. The first, I_h, is a **transient current** that tends to zero as $t \to +\infty$. The second,

$$I_p(t) = -10.080 \cos 100t + 2.122 \sin 100t \ ,$$

is a sinusoidal **steady-state current** that remains.

We leave it for the reader to verify that the steady-state solution $I_p(t)$ that arises from the more general voltage source $E(t) = E_0 \sin \gamma t$ is

(10) $$I_p(t) = \frac{E_0 \sin(\gamma t + \theta)}{\sqrt{R^2 + [\gamma L - 1/(\gamma C)]^2}} \ ,$$

where $\tan \theta = (1/C - L\gamma^2)/(\gamma R)$ (cf. Section 4.11, page 230).

EXAMPLE 2 At time $t = 0$, the charge on the capacitor in the electrical network shown in Figure 5.30 is 2 coulombs, while the current through the capacitor is zero. Determine the charge on the capacitor and the currents in the various branches of the network at any time $t > 0$.

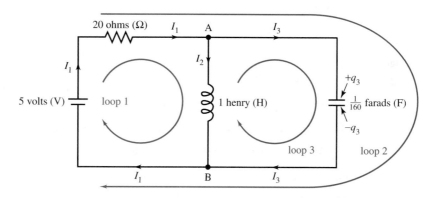

Figure 5.30 Schematic of an electrical network

SOLUTION To determine the charge and currents in the electrical network, we begin by observing that the network consists of three closed circuits: loop 1 through the battery, resistor, and inductor; loop 2 through the battery, resistor, and capacitor; and loop 3 containing the capacitor and inductor. Taking advantage of Kirchhoff's current law, we denote the current passing through the battery and the resistor by I_1, the current through the inductor by I_2, and the current through the capacitor by I_3. For consistency of notation, we denote the charge on the capacitor by q_3. Hence $I_3 = dq_3/dt$.

As discussed at the beginning of this section, the voltage drop at a resistor is RI, at an inductor $L dI/dt$, and at a capacitor q/C. So, applying Kirchhoff's voltage law to the electrical network in Figure 5.30, we find for loop 1,

(11) $$\frac{dI_2}{dt} \quad + \quad 20I_1 \quad = \quad 5 \ ;$$
$\qquad\qquad$ (inductor) \qquad (resistor) \qquad (battery)

for loop 2,

(12) $$20I_1 \quad + \quad 160q_3 \quad = \quad 5 \ ;$$
$\qquad\qquad$ (resistor) \qquad (capacitor) \qquad (battery)

and for loop 3,

$$\textbf{(13)} \qquad -\frac{dI_2}{dt} \quad + \quad 160q_3 = 0 \ .$$

(inductor) (capacitor)

(The minus sign in (13) arises from taking a clockwise path around loop 3 so that the current passing through the inductor is $-I_2$.) Notice that these three equations are not independent: We can obtain equation (13) by subtracting (11) from (12). Hence we have only two equations from which to determine the three unknowns I_1, I_2, and q_3. If we now apply Kirchhoff's current law to the two junction points in the network, we find at point A that $I_1 - I_2 - I_3 = 0$ and at point B that $I_2 + I_3 - I_1 = 0$. In both cases, we get

$$\textbf{(14)} \qquad I_1 - I_2 - \frac{dq_3}{dt} = 0 \ ,$$

since $I_3 = dq_3/dt$. Assembling equations (11), (12), and (14) into a system, we have (with $D = d/dt$)

$$\textbf{(15)} \qquad DI_2 + 20I_1 = 5 \ ,$$
$$\textbf{(16)} \qquad 20I_1 + 160q_3 = 5 \ ,$$
$$\textbf{(17)} \qquad -I_2 + I_1 - Dq_3 = 0 \ .$$

We solve these by the elimination method of Section 5.3. Using equation (16) to eliminate I_1 from the others, we are left with

$$\textbf{(18)} \qquad DI_2 - 160q_3 = 0 \ ,$$
$$\textbf{(19)} \qquad 20I_2 + (20D + 160)q_3 = 5 \ .$$

Elimination of I_2 then leads to

$$\textbf{(20)} \qquad 20D^2 q_3 + 160Dq_3 + 3200q_3 = 0 \ .$$

To obtain the initial conditions for the second order equation (20), recall that at time $t = 0$, the charge on the capacitor is 2 coulombs and the current is zero. Hence

$$\textbf{(21)} \qquad q_3(0) = 2 \ , \qquad \frac{dq_3}{dt}(0) = 0 \ .$$

We can now solve the initial value problem (20)–(21) using the techniques of Chapter 4. Ultimately we find

$$q_3(t) = 2e^{-4t}\cos 12t + \frac{2}{3}e^{-4t}\sin 12t \ ,$$

$$I_3(t) = \frac{dq_3}{dt}(t) = -\frac{80}{3}e^{-4t}\sin 12t \ .$$

Next, to determine I_2, we substitute these expressions into (19) and obtain

$$I_2(t) = \frac{1}{4} - \frac{dq_3}{dt}(t) - 8q_3(t)$$

$$= \frac{1}{4} - 16e^{-4t}\cos 12t + \frac{64}{3}e^{-4t}\sin 12t \ .$$

Finally, from $I_1 = I_2 + I_3$, we get

$$I_1(t) = \frac{1}{4} - 16e^{-4t}\cos 12t - \frac{16}{3}e^{-4t}\sin 12t \ . \quad \blacksquare \text{End}$$

Note that the differential equations that describe mechanical vibrations and *RLC* series circuits are essentially the same. And, in fact, there is a natural identification of the parameters m, b, and k for a mass-spring system with the parameters L, R, and C that describe circuits. This is illustrated in Table 5.2. Moreover, the terms *transient, steady-state, overdamped, critically damped, underdamped,* and *resonance frequency* described in Sections 4.11 and 4.12 apply to electrical circuits as well.

TABLE 5.2 ANALOGY BETWEEN MECHANICAL AND ELECTRICAL SYSTEMS			
Mechanical Mass-Spring System with Damping		**Electrical *RLC* Series Circuit**	
$mx'' + bx' + kx = f(t)$		$Lq'' + Rq' + (1/C)q = E(t)$	
Displacement	x	Charge	q
Velocity	x'	Current	$q' = I$
Mass	m	Inductance	L
Damping constant	b	Resistance	R
Spring constant	k	(Capacitance)$^{-1}$	$1/C$
External force	$f(t)$	Voltage source	$E(t)$

This analogy between a mechanical system and an electrical circuit extends to large-scale systems and circuits. An interesting consequence of this is the use of **analog simulation** and, in particular, analog computers to analyze mechanical systems. Large-scale mechanical systems are modeled by building a corresponding electrical system and then measuring the charges $q(t)$ and currents $I(t)$.

While such analog simulations are important, *both* large-scale mechanical and electrical systems are currently modeled using digital computer simulation. This involves the numerical solution of the initial value problem governing the system. Still, the analogy between mechanical and electrical systems means that basically the same computer software can be used to analyze both systems.

EXERCISES 5.5

1. An *RLC* series circuit has a voltage source given by $E(t) = 20$ volts, a resistor of 100 ohms, an inductor of 4 henrys, and a capacitor of 0.01 farads. If the initial current is zero and the initial charge on the capacitor is 4 coulombs, determine the current in the circuit for $t > 0$.

2. An *RLC* series circuit has a voltage source given by $E(t) = 40 \cos 2t$ volts, a resistor of 2 ohms, an inductor of $1/4$ henrys, and a capacitor of $1/13$ farads. If the initial current is zero and the initial charge on the capacitor is 3.5 coulombs, determine the charge on the capacitor for $t > 0$.

3. An *RLC* series circuit has a voltage source given by $E(t) = 10 \cos 20t$ volts, a resistor of 120 ohms, an inductor of 4 henrys, and a capacitor of $(2200)^{-1}$ farads. Find the steady-state current (solution) for this circuit. What is the resonance frequency of the circuit?

4. An *LC* series circuit has a voltage source given by $E(t) = 30 \sin 50t$ volts, an inductor of 2 henrys, and a capacitor of 0.02 farads (but no resistor). What is the current in this circuit for $t > 0$ if at $t = 0$, $I(0) = q(0) = 0$?

5. An *RLC* series circuit has a voltage source of the form $E(t) = E_0 \cos \gamma t$ volts, a resistor of 10 ohms, an inductor of 4 henrys, and a capacitor of 0.01 farads. Sketch the frequency response curve for this circuit.

6. Show that when the voltage source in (3) is of the form $E(t) = E_0 \sin \gamma t$, then the steady-state solution I_p is as given in equation (10).

7. A mass-spring system with damping consists of a 7-kg mass, a spring with spring constant 3 N/m, a frictional component with damping constant 2 N-sec/m, and an external force given by $f(t) = 10 \cos 10t$ N. Using a 10-ohm resistor, construct an *RLC* series circuit that is the analog of this mechanical system in the sense that the two systems are governed by the same differential equation.

8. A mass-spring system with damping consists of a 16-lb weight, a spring with spring constant 64 lb/ft, a frictional component with damping constant 10

lb-sec/ft, and an external force given by $f(t) = 20 \cos 8t$ lb. Using an inductor of 0.01 henrys, construct an *RLC* series circuit that is the analog of this mechanical system.

9. Because of Euler's formula, $e^{i\theta} = \cos \theta + i \sin \theta$, it is often convenient to treat the voltage sources $E_0 \cos \gamma t$ and $E_0 \sin \gamma t$ simultaneously, using $E(t) = E_0 e^{i\gamma t}$. In this case, equation (3) becomes

(22) $$ L\frac{d^2q}{dt^2} + R\frac{dq}{dt} + \frac{1}{C}q = E_0 e^{i\gamma t}, $$

where q is now complex.

(a) Show that the steady-state solution to (22) is

$$ q_p(t) = \frac{E_0}{1/C - \gamma^2 L + i\gamma R} e^{i\gamma t}. $$

[*Hint:* Use the method of undetermined coefficients with the guess $q_p = Ae^{i\gamma t}$, where A denotes a complex constant. The technique is discussed in detail in Project A at the end of Chapter 4.]

(b) Now show that the steady-state current is

$$ I_p(t) = \frac{E_0}{R + i[\gamma L - 1/(\gamma C)]} e^{i\gamma t}. $$

(c) Use the relation $\alpha + i\beta = \sqrt{\alpha^2 + \beta^2} e^{i\theta}$, where $\tan \theta = \beta/\alpha$, to show that I_p can be expressed in the form

$$ I_p(t) = \frac{E_0}{\sqrt{R^2 + [\gamma L - 1/(\gamma C)]^2}} e^{i(\gamma t + \theta)}, $$

where $\tan \theta = (1/C - L\gamma^2)/(\gamma R)$.

(d) The imaginary part of $e^{i\gamma t}$ is $\sin \gamma t$, so the imaginary part of the solution to (22) must be the solution to equation (3) for $E(t) = E_0 \sin \gamma t$. Verify that this is also the case for the current by showing that the imaginary part of I_p in part (c) is the same as that given in equation (10).

In Problems 10 through 13, find a system of differential equations and initial conditions for the currents in the networks given in the schematic diagram. Assume all

initial currents are zero. Solve for the currents in each branch of the network.

10.

Figure 5.31 *RL* network for Problem 10

11.

Figure 5.32 *RLC* network for Problem 11

12.

Figure 5.33 *RL* network for Problem 12

13.

Figure 5.34 *RLC* network for Problem 13

5.6 NUMERICAL METHODS FOR HIGHER-ORDER EQUATIONS AND SYSTEMS

Many numerical methods for approximating solutions to first order initial value problems of the form

$$(1) \qquad \frac{dx}{dt} = f(t, x) , \qquad x(t_0) = x_0 ,$$

can be readily extended to *systems* of m differential equations in m unknown functions $x_1(t), x_2(t), \ldots, x_m(t)$. This is facilitated by the use of a standard, "normal" form for expressing the system. In this section, we shall address the features of this normal form, show how the classical Runge-Kutta algorithm (discussed in Section 3.6) can be easily carried over to systems in normal form, and consider an application of this methodology to a population dynamics model.

Normal Form

A system of m differential equations expressed in the format

$$
(2) \qquad
\begin{aligned}
x_1'(t) &= f_1(t, x_1, x_2, \ldots, x_m) , \\
x_2'(t) &= f_2(t, x_1, x_2, \ldots, x_m) , \\
&\ \ \vdots \\
x_m'(t) &= f_m(t, x_1, x_2, \ldots, x_m)
\end{aligned}
$$

is said to be in **normal form.** Notice that (2) consists of m first order equations that collectively look like a "vectorized" version of (1). An initial value problem for (2) entails finding a solution to this system that satisfies the initial conditions

$$x_1(t_0) = a_1, \quad x_2(t_0) = a_2, \quad \ldots, \quad x_m(t_0) = a_m$$

for prescribed values $t_0, a_1, a_2, \ldots, a_m$.

The importance of the normal form is underscored by the fact that most professional codes for initial value problems presume that the system is written in this form. Furthermore, for a *linear* system in normal form, the powerful machinery of linear algebra can be readily applied. (Indeed, in Chapter 9 we will show how the solutions $x(t) = ce^{at}$ of the simple equation $x' = ax$ can be generalized to constant coefficient systems in normal form.)

For these reasons, it is gratifying to note that a (single) higher-order equation can be converted to an equivalent system of first order equations. In fact, we have already seen in the phase plane discussion (Section 5.2) how a second order equation $y'' = f(t, y, y')$ can be written as a system of two equations by introducing the variable $v = y'$.

A method for converting an mth order differential equation

(3) $$y^{(m)}(t) = f\left(t, y, y', \ldots, y^{(m-1)}\right)$$

into a first order system is to introduce, as additional unknowns, the sequence of derivatives of y:

$$x_1(t) := y(t), \quad x_2(t) := y'(t), \quad \ldots, \quad x_m(t) := y^{(m-1)}(t) \ .$$

With this scheme, we obtain $m - 1$ first order equations quite trivially:

(4)
$$\begin{aligned} x_1'(t) &= y'(t) = x_2(t) \ , \\ x_2'(t) &= y''(t) = x_3(t) \ , \\ &\vdots \\ x_{m-1}'(t) &= y^{(m-1)}(t) = x_m(t) \ . \end{aligned}$$

The mth and final equation then constitutes a restatement of the original equation (3) in terms of the new unknowns:

(5) $$x_m'(t) = y^{(m)}(t) = f(t, x_1, x_2, \ldots, x_m) \ .$$

If equation (3) has initial conditions $y(t_0) = a_1, y'(t_0) = a_2, \ldots, y^{(m-1)}(t_0) = a_m$, then the system (4)–(5) has initial conditions $x_1(t_0) = a_1, x_2(t_0) = a_2, \ldots, x_m(t_0) = a_m$.

EXAMPLE 1 Convert the initial value problem

(6) $$y''(t) + 3y'(t) + 2y(t) = 0 \ ; \qquad y(0) = 1 \ , \qquad y'(0) = 1$$

into an initial value problem for a system in normal form.

SOLUTION We first express the differential equation in (6) as

$$y''(t) = -3y'(t) - 2y(t) \ .$$

Setting $x_1(t) := y(t)$ and $x_2(t) := y'(t)$, we obtain

$$x_1'(t) = x_2(t) \ ,$$
$$x_2'(t) = -3x_2(t) - 2x_1(t) \ .$$

The initial conditions transform to $x_1(0) = 1, x_2(0) = 1.$ ■ End

Runge-Kutta Algorithm for Systems

Now we will see how the classical Runge-Kutta method of Section 3.6 can be vectorized to provide numerical solutions for systems in normal form. Recall that for the initial value problem (1), the recursive formulas for the classical fourth order Runge-Kutta method are

(7)
$$t_{n+1} = t_n + h \ , \qquad n = 0, 1, \dots,$$
$$x_{n+1} = x_n + \frac{1}{6}(k_1 + 2k_2 + 2k_3 + k_4) \ ,$$

where h is the step size and

(8)
$$k_1 = hf(t_n, x_n) \ , \qquad k_2 = hf\left(t_n + \frac{h}{2}, x_n + \frac{1}{2}k_1\right) \ ,$$
$$k_3 = hf\left(t_n + \frac{h}{2}, x_n + \frac{1}{2}k_2\right) \ , \qquad k_4 = hf(t_n + h, x_n + k_3) \ .$$

The computed value x_n gives an approximation to $x(t_n)$.

Now suppose we want to approximate the solution $x_1(t), x_2(t)$ to the 2×2 system

(9)
$$x_1'(t) = f_1(t, x_1, x_2) \ ,$$
$$x_2'(t) = f_2(t, x_1, x_2)$$

that satisfies the initial conditions

(10) $x_1(t_0) = a_1 \ , \qquad x_2(t_0) = a_2 \ .$

Let $x_{n,1}$ and $x_{n,2}$ denote approximations to $x_1(t_n)$ and $x_2(t_n)$, respectively, where $t_n = t_0 + nh$ for $n = 0, 1, 2, \dots$. The recursive formulas for the classical fourth order Runge-Kutta method for the system (9)–(10) are obtained from the formulas for a single equation given in (7)–(8) by treating each of the quantities $x_n, k_1, k_2, k_3,$ and k_4 as vectors. Namely, we set $\mathbf{x}_n := (x_{n,1}, x_{n,2})$, $\mathbf{k}_1 := (k_{1,1}, k_{1,2})$, $\mathbf{k}_2 := (k_{2,1}, k_{2,2})$, etc. Componentwise, we are implementing the Runge-Kutta method *in parallel* for each of the equations in (9). This approach gives the following recursive formulas for the classical fourth order Runge-Kutta method for system (9)–(10):

(11) $t_{n+1} := t_n + h \ , \qquad n = 0, 1, 2, \dots \ ,$

(12) $x_{n+1,1} := x_{n,1} + \frac{1}{6}(k_{1,1} + 2k_{2,1} + 2k_{3,1} + k_{4,1}) \ ,$

(13) $x_{n+1,2} := x_{n,2} + \frac{1}{6}(k_{1,2} + 2k_{2,2} + 2k_{3,2} + k_{4,2}) \ ,$

where h is the step size and, for $i = 1$ and 2,

$$k_{1,i} := hf_i(t_n, x_{n,1}, x_{n,2}) \;,$$

$$k_{2,i} := hf_i\left(t_n + \frac{h}{2}, x_{n,1} + \frac{1}{2}k_{1,1}, x_{n,2} + \frac{1}{2}k_{1,2}\right) \;,$$

$$k_{3,i} := hf_i\left(t_n + \frac{h}{2}, x_{n,1} + \frac{1}{2}k_{2,1}, x_{n,2} + \frac{1}{2}k_{2,2}\right) \;,$$

$$k_{4,i} := hf_i(t_n + h, x_{n,1} + k_{3,1}, x_{n,2} + k_{3,2}) \;.$$

It is important to note that both $k_{1,1}$ and $k_{1,2}$ must be computed before either $k_{2,1}$ or $k_{2,2}$. Similarly, both $k_{2,1}$ and $k_{2,2}$ are needed to compute $k_{3,1}$ and $k_{3,2}$, etc.

EXAMPLE 2 Use the classical Runge-Kutta method with step size $h = 0.125$ to find an approximation for the solution to the initial value problem

(14) $y''(t) + 3y'(t) + 2y(t) = 0 \;;$ $y(0) = 1 \;,$ $y'(0) = 1 \;,$

on the interval $[0, 1]$.

SOLUTION For the given step size, the method will yield approximations for $y(0.125)$, $y(0.250), \dots, y(1)$. To apply the classical Runge-Kutta method to (14), we first convert it to normal form. This was done in Example 1, where we set $x_1 = y$, $x_2 = y'$ and obtained the system

(15) $\begin{aligned} x_1' &= x_2 \;; & x_1(0) &= 1 \;, \\ x_2' &= -3x_2 - 2x_1 \;; & x_2(0) &= 1 \;. \end{aligned}$

Comparing (15) with (9), we see that $f_1(t, x_1, x_2) = x_2$ and $f_2(t, x_1, x_2) = -3x_2 - 2x_1$. With the starting values of $t_0 = 0$, $x_{0,1} = 1$, and $x_{0,2} = 1$, we compute

$$k_{1,1} = hf_1(t_0, x_{0,1}, x_{0,2}) = hx_{0,2} = (0.125)(1) = 0.125 \;,$$

$$\begin{aligned} k_{1,2} &= hf_2(t_0, x_{0,1}, x_{0,2}) = h(-3x_{0,2} - 2x_{0,1}) \\ &= (0.125)[(-3)(1) - 2(1)] = -0.625 \;, \end{aligned}$$

$$k_{2,1} = hf_1\left(t_0 + \frac{h}{2}, x_{0,1} + \frac{1}{2}k_{1,1}, x_{0,2} + \frac{1}{2}k_{1,2}\right)$$

$$= h\left(x_{0,2} + \frac{1}{2}k_{1,2}\right) = (0.125)(1 - 0.3125) = 0.08594 \;,$$

$$k_{2,2} = hf_2\left(t_0 + \frac{h}{2}, x_{0,1} + \frac{1}{2}k_{1,1}, x_{0,2} + \frac{1}{2}k_{1,2}\right)$$

$$= -3h\left(x_{0,2} + \frac{1}{2}k_{1,2}\right) - 2h\left(x_{0,1} + \frac{1}{2}k_{1,1}\right)$$

$$= -3(0.125)(1 - 0.3125) - 2(0.125)(1 + 0.0625)$$

$$= -0.52344 \;.$$

Similarly, we find

$$k_{3,1} = hf_1\left(t_0 + \frac{h}{2}, x_{0,1} + \frac{1}{2}k_{2,1}, x_{0,2} + \frac{1}{2}k_{2,2}\right) = 0.09229 \ ,$$

$$k_{3,2} = hf_2\left(t_0 + \frac{h}{2}, x_{0,1} + \frac{1}{2}k_{2,1}, x_{0,2} + \frac{1}{2}k_{2,2}\right) = -0.53760 \ ,$$

$$k_{4,1} = hf_1(t_0 + h, x_{0,1} + k_{3,1}, x_{0,2} + k_{3,2}) = 0.05780 \ ,$$

$$k_{4,2} = hf_2(t_0 + h, x_{0,1} + k_{3,1}, x_{0,2} + k_{3,2}) = -0.44647 \ .$$

Hence from (12) and (13), we compute

$$x_{1,1} = x_{0,1} + \frac{1}{6}(k_{1,1} + 2k_{2,1} + 2k_{3,1} + k_{4,1})$$

$$= 1 + \frac{1}{6}\Big(0.125 + 2(0.08594) + 2(0.09229) + 0.05780\Big)$$

$$= 1.08987 \ ,$$

$$x_{1,2} = x_{0,2} + \frac{1}{6}(k_{1,2} + 2k_{2,2} + 2k_{3,2} + k_{4,2})$$

$$= 1 - \frac{1}{6}\Big(0.625 + 2(0.52344) + 2(0.53760) + 0.44647\Big)$$

$$= 0.46774 \ .$$

Continuing the algorithm, we compute $x_{i,1}$ and $x_{i,2}$ for $i = 2, 3, \ldots, 8$. These values, along with the values of the actual solution to (14), $y(t) = x_1(t) = 3e^{-t} - 2e^{-2t}$, are given in Table 5.3 (to five decimal places). Note that the $x_{i,2}$ column gives approximations for $y'(t)$, since $x_2(t) = y'(t)$. ■ End

TABLE 5.3	CLASSICAL 4TH ORDER RUNGE-KUTTA APPROXIMATION OF THE SOLUTION TO (14) USING h = 0.125					
t_i	$x_{i,1}$	$x_{i,2}$	$y(t_i) = x_1(t_i)$	$	y(t_i) - x_{i,1}	$
0.000	1.00000	1.00000	1.00000	0		
0.125	1.08987	0.46774	1.08989	2×10^{-5}		
0.250	1.12332	0.08977	1.12334	2×10^{-5}		
0.375	1.11711	-0.17235	1.11713	2×10^{-5}		
0.500	1.08381	-0.34802	1.08383	2×10^{-5}		
0.625	1.03275	-0.45971	1.03277	2×10^{-5}		
0.750	0.97081	-0.52453	0.97084	3×10^{-5}		
0.875	0.90302	-0.55544	0.90304	2×10^{-5}		
1.000	0.83295	-0.56226	0.83297	2×10^{-5}		

The Runge-Kutta program outlines of Section 3.6 (see pages 139–140) for a single first order equation can be easily modified to handle normal systems. The algorithms for systems of two equations are presented below. (See Appendix E for the general case.) Notice that the algorithm "with tolerance" stops when *each* component of two successive approximations differs by less than a prescribed tolerance ε or after a prescribed maximum number of iterations. The reader is cautioned that the true error in the numerical approximation is only *modeled* by the tolerance factor and may sometimes exceed it.

CLASSICAL 4TH ORDER RUNGE-KUTTA SUBROUTINE (SYSTEMS)

Purpose To approximate the solution to the initial value problem

$$x_1' = f_1(t, x_1, x_2) ; \qquad x_1(t_0) = a_1 ,$$
$$x_2' = f_2(t, x_1, x_2) ; \qquad x_2(t_0) = a_2 ,$$

for $t_0 \leq t \leq c$

INPUT t_0, a_1, a_2, c, N (number of steps), PRNTR (=1 to print table)

Step 1 Set step size $h = (c - t_0)/N$, $x_1 = a_1$, $x_2 = a_2$

Step 2 For $i = 1$ to N, do Steps 3 through 5

Step 3 Set

$$k_{1,1} = hf_1(t, x_1, x_2)$$

$$k_{1,2} = hf_2(t, x_1, x_2)$$

$$k_{2,1} = hf_1\left(t + \frac{h}{2}, x_1 + \frac{1}{2}k_{1,1}, x_2 + \frac{1}{2}k_{1,2}\right)$$

$$k_{2,2} = hf_2\left(t + \frac{h}{2}, x_1 + \frac{1}{2}k_{1,1}, x_2 + \frac{1}{2}k_{1,2}\right)$$

$$k_{3,1} = hf_1\left(t + \frac{h}{2}, x_1 + \frac{1}{2}k_{2,1}, x_2 + \frac{1}{2}k_{2,2}\right)$$

$$k_{3,2} = hf_2\left(t + \frac{h}{2}, x_1 + \frac{1}{2}k_{2,1}, x_2 + \frac{1}{2}k_{2,2}\right)$$

$$k_{4,1} = hf_1(t + h, x_1 + k_{3,1}, x_2 + k_{3,2})$$

$$k_{4,2} = hf_2(t + h, x_1 + k_{3,1}, x_2 + k_{3,2})$$

Step 4 Set

$$t = t + h$$

$$x_1 = x_1 + \frac{1}{6}(k_{1,1} + 2k_{2,1} + 2k_{3,1} + k_{4,1})$$

$$x_2 = x_2 + \frac{1}{6}(k_{1,2} + 2k_{2,2} + 2k_{3,2} + k_{4,2})$$

Step 5 If PRNTR $= 1$, print t, x_1, x_2

CLASSICAL 4TH ORDER RUNGE-KUTTA ALGORITHM WITH TOLERANCE (SYSTEMS)

Purpose To approximate the solution to the initial value problem

$$x_1' = f_1(t, x_1, x_2) \; ; \qquad x_1(t_0) = a_1 \; ,$$
$$x_2' = f_2(t, x_1, x_2) \; ; \qquad x_2(t_0) = a_2 \; ,$$

at $t = c$, with tolerance ε

INPUT $t_0, a_1, a_2, c, \varepsilon, M$ (maximum number of iterations)

Step 1 Set $z_1 = a_1, z_2 = a_2$, PRNTR $= 0$

Step 2 For $m = 0$ to M, do Steps 3 through 7 (Or, to save time, start with $m > 0$.)

Step 3 Set $N = 2^m$

Step 4 Call 4TH ORDER RUNGE-KUTTA SUBROUTINE (SYSTEMS)

Step 5 Print h, x_1, x_2

Step 6 If $|z_1 - x_1| < \varepsilon$ and $|z_2 - x_2| < \varepsilon$, go to Step 10

Step 7 Set $z_1 = x_1, z_2 = x_2$

Step 8 Print "$x_1(c)$ is approximately"; x_1; "and $x_2(c)$ is approximately"; x_2; "but may not be within the tolerance"; ε

Step 9 Go to Step 11

Step 10 Print "$x_1(c)$ is approximately"; x_1; "and $x_2(c)$ is approximately"; x_2; "with tolerance"; ε

Step 11 STOP

OUTPUT Approximations of the solution to the initial value problem at $t = c$, using 2^m steps

For a stopping procedure based on the *relative* error, replace Step 6 by

Step 6' If $\left| \dfrac{z_1 - x_1}{x_1} \right| < \varepsilon$ and $\left| \dfrac{z_2 - x_2}{x_2} \right| < \varepsilon$, go to Step 10.

EXAMPLE 3 Use the preceding Runge-Kutta algorithm with a stopping procedure based on the absolute error and a tolerance of $\varepsilon = 0.001$ to approximate the solution $y(t)$ of the initial value problem

(16) $y'' + ty' + y^3 = 0 \; ; \qquad y(0) = 1 \; , \qquad y'(0) = 0 \; ,$

at $t = 2$.

SOLUTION On converting (16) to the equivalent system

$$x_1' = x_2 \; ; \qquad x_1(0) = 1 \; ,$$
$$x_2' = -tx_2 - x_1^3 \; ; \qquad x_2(0) = 0 \; ,$$

where $x_1 = y$, $x_2 = y'$, we use as inputs for the algorithm $t_0 = 0$, $a_1 = y(0) = 1$, $a_2 = y'(0) = 0$, $c = 2$, $\varepsilon = 0.001$, and $M = 10$ (say). Then, the formulas in Step 3 of the

SUBROUTINE are

$$k_{1,1} = hx_2 \;, \qquad\qquad\qquad k_{1,2} = h(-tx_2 - x_1^3) \;,$$

$$k_{2,1} = h\left(x_2 + \frac{1}{2}k_{1,2}\right) , \qquad k_{2,2} = h\left[-\left(t + \frac{h}{2}\right)\left(x_2 + \frac{1}{2}k_{1,2}\right) - \left(x_1 + \frac{1}{2}k_{1,1}\right)^3\right] ,$$

$$k_{3,1} = h\left(x_2 + \frac{1}{2}k_{2,2}\right) , \qquad k_{3,2} = h\left[-\left(t + \frac{h}{2}\right)\left(x_2 + \frac{1}{2}k_{2,2}\right) - \left(x_1 + \frac{1}{2}k_{2,1}\right)^3\right] ,$$

$$k_{4,1} = h(x_2 + k_{3,2}) \;, \qquad k_{4,2} = h\left[-(t + h)(x_2 + k_{3,2}) - (x_1 + k_{3,1})^3\right] \;.$$

Adopting the notation $x(c; h)$ for the approximation of $x(c)$ obtained using step size h, we give in Table 5.4 the approximations $x_1(2; 2^{-m+1})$ for $x_1(2)$ and $x_2(2; 2^{-m+1})$ for $x_2(2)$ when $m = 0, 1, 2$, and 3. The algorithm stops at $m = 3$ because

$$|x_1(2; 0.5) - x_1(2; 0.25)| = 0.00046 < \varepsilon = 0.001 \;,$$

and

$$|x_2(2; 0.5) - x_2(2; 0.25)| = 0.00040 < \varepsilon = 0.001 \;.$$

Hence $y(2) \approx 0.34400$ with a tolerance of 0.001. ∎ End

TABLE 5.4	RUNGE-KUTTA APPROXIMATION FOR y(2) FROM EXAMPLE 3		
		Approximations	
m	*h*	$y(2) \approx x_1(2; h)$	$y'(2) \approx x_2(2; h)$
0	2	0.33333	−0.66667
1	1	0.33417	−0.13764
2	0.5	0.34446	−0.14822
3	0.25	0.34400	−0.14782
$y(2)$ is approximately 0.34400 with a tolerance of $\varepsilon = 0.001$.			

An Application to Population Dynamics

A mathematical model for the population dynamics of competing species, one a predator with population $x_2(t)$ and the other its prey with population $x_1(t)$, was developed independently in the early 1900s by A. J. Lotka and V. Volterra. It assumes there is plenty of food available for the prey to eat, so the birth rate of the prey should follow the Malthusian or exponential law (see Section 3.2); that is, the birth rate of the prey is Ax_1, where A is a positive constant. The death rate of the prey depends on the number of interactions between the predators and the prey. This is modeled by the expression Bx_1x_2, where B is a positive constant. Therefore the rate of change in the population of the prey per unit time is $dx_1/dt = Ax_1 - Bx_1x_2$. Assuming the predators depend entirely on the prey for their food, it is argued that the birth rate of the predators depends on the number of interactions with the prey; that is, the birth rate of predators is

Dx_1x_2, where D is a positive constant. The death rate of the predators is assumed to be Cx_2 because without food the population would die off at a rate proportional to the population present. Hence the rate of change in the population of predators per unit time is $dx_2/dt = -Cx_2 + Dx_1x_2$. Combining these two equations, we obtain the Volterra-Lotka system for the population dynamics of two competing species:

$$x_1' = Ax_1 - Bx_1x_2 \ ,$$
$$x_2' = -Cx_2 + Dx_1x_2 \ .$$

Such systems are in general not explicitly solvable. In the following example, we obtain an approximate solution for such a system.

EXAMPLE 4 Use the classical fourth order Runge-Kutta algorithm for systems to approximate the solution of the initial value problem

(17)
$$x_1' = 2x_1 - 2x_1x_2 \ ; \qquad x_1(0) = 1 \ ,$$
$$x_2' = x_1x_2 - x_2 \ ; \qquad x_2(0) = 3$$

at $t = 1$ with a tolerance of 0.0001.

SOLUTION Here $f_1(t, x_1, x_2) = 2x_1 - 2x_1x_2$ and $f_2(t, x_1, x_2) = x_1x_2 - x_2$. With the inputs $t_0 = 0$, $a_1 = x_1(0) = 1$, $a_2 = x_2(0) = 3$, and $M = 10$ (say) we proceed with the algorithm to compute $x_1(1; 1)$ and $x_2(1; 1)$, the approximations to $x_1(1)$, $x_2(1)$ using $h = 1$. We find from the formulas in Step 3 of the SUBROUTINE (with $x_1 = 1$, $x_2 = 3$, and $h = 1$)

$$k_{1,1} = hf_1(t, x_1, x_2) = h(2x_1 - 2x_1x_2) = 2(1) - 2(1)(3) = -4 \ ,$$
$$k_{1,2} = hf_2(t, x_1, x_2) = h(x_1x_2 - x_2) = (1)(3) - 3 = 0 \ ,$$
$$k_{2,1} = hf_1\left(t + \frac{h}{2}, x_1 + \frac{1}{2}k_{1,1}, x_2 + \frac{1}{2}k_{1,2}\right)$$
$$= h\left[2\left(x_1 + \frac{1}{2}k_{1,1}\right) - 2\left(x_1 + \frac{1}{2}k_{1,1}\right)\left(x_2 + \frac{1}{2}k_{1,2}\right)\right]$$
$$= 2\left(1 + \frac{1}{2}(-4)\right) - 2\left(1 + \frac{1}{2}(-4)\right)\left(3 + \frac{1}{2}(0)\right)$$
$$= -2 + 2(3) = 4 \ ,$$
$$k_{2,2} = hf_2\left(t + \frac{h}{2}, x_1 + \frac{1}{2}k_{1,1}, x_2 + \frac{1}{2}k_{1,2}\right)$$
$$= h\left[\left(x_1 + \frac{1}{2}k_{1,1}\right)\left(x_2 + \frac{1}{2}k_{1,2}\right) - \left(x_2 + \frac{1}{2}k_{1,2}\right)\right]$$
$$= \left(1 + \frac{1}{2}(-4)\right)\left(3 + \frac{1}{2}(0)\right) - \left(3 + \frac{1}{2}(0)\right)$$
$$= (-1)(3) - 3 = -6 \ .$$

Similarly, we find

$$k_{3,1} = h\left[2\left(x_1 + \frac{1}{2}k_{2,1}\right) - 2\left(x_1 + \frac{1}{2}k_{2,1}\right)\left(x_2 + \frac{1}{2}k_{2,2}\right)\right] = 6 \ ,$$

$$k_{3,2} = h\left[\left(x_1 + \frac{1}{2}k_{2,1}\right)\left(x_2 + \frac{1}{2}k_{2,2}\right) - \left(x_2 + \frac{1}{2}k_{2,2}\right)\right] = 0 \ ,$$

$$k_{4,1} = h\left[2(x_1 + k_{3,1}) - 2(x_1 + k_{3,1})(x_2 + k_{3,2})\right] = -28 \ ,$$

$$k_{4,2} = h\left[(x_1 + k_{3,1})(x_2 + k_{3,2}) - (x_2 + k_{3,2})\right] = 18 \ .$$

Hence in Step 4 of the SUBROUTINE we compute

$$x_1 = x_1 + \frac{1}{6}(k_{1,1} + 2k_{2,1} + 2k_{3,1} + k_{4,1})$$

$$= 1 + \frac{1}{6}\left[-4 + 8 + 12 - 28\right] = -1 \ ,$$

$$x_2 = x_2 + \frac{1}{6}(k_{1,2} + 2k_{2,2} + 2k_{3,2} + k_{4,2})$$

$$= 3 + \frac{1}{6}(0 - 12 + 0 + 18) = 4 \ .$$

Repeating the algorithm with $h = 1/2$ $(N = 2)$ we obtain the approximations $x_1(1; 2^{-1})$ and $x_2(1; 2^{-1})$ for $x_1(1)$ and $x_2(1)$. In Table 5.5, we list the approximations $x_1(1; 2^{-m})$ and $x_2(1; 2^{-m})$ for $x_1(1)$ and $x_2(1)$ using step size $h = 2^{-m}$ for $m = 0, 1, 2, 3$, and 4. We stopped at $m = 4$, since both

$$|x_1(1; 2^{-3}) - x_1(1; 2^{-4})| = 0.00006 < 0.0001$$

and

$$|x_2(1; 2^{-3}) - x_2(1; 2^{-4})| = 0.00001 < 0.0001 \ .$$

Hence $x_1(1) \approx 0.07735$ and $x_2(1) \approx 1.46445$, with tolerance 0.0001. ■ End

TABLE 5.5	APPROXIMATIONS OF THE SOLUTION TO SYSTEM (17) IN EXAMPLE 4		
m	h	$x_1(1; h)$	$x_2(1; h)$
0	1.0	−1.0	4.0
1	0.5	0.14662	1.47356
2	0.25	0.07885	1.46469
3	0.125	0.07741	1.46446
4	0.0625	0.07735	1.46445

To get a better feel for the solution to system (17), we have graphed in Figure 5.35 an approximation of the solution for $0 \le t \le 12$, using linear interpolation to connect the points. This approximation was obtained using step size $h = 0.125$ $(N = 96)$ and the fourth order Runge-Kutta SUBROUTINE for systems. From the graph it appears that the components x_1 and x_2 are periodic in the variable t. Phase plane analysis is used in Project D to show that, indeed, Volterra-Lotka equations have periodic solutions.

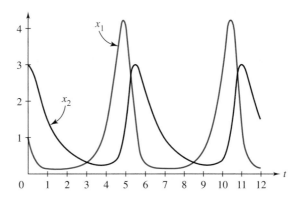

Figure 5.35 Graphs of the components of an approximate solution to the Volterra-Lotka system (17)

EXERCISES 5.6

In Problems 1 through 7, convert the given initial value problem into an initial value problem for a system in normal form.

1. $y''(t) + ty'(t) - 3y(t) = t^2$;
 $y(0) = 3$, $y'(0) = -6$.

2. $y''(t) = \cos(t - y) + y^2(t)$;
 $y(0) = 1$, $y'(0) = 0$.

3. $y^{(4)}(t) - y^{(3)}(t) + 7y(t) = \cos t$;
 $y(0) = y'(0) = 1$, $y''(0) = 0$, $y^{(3)}(0) = 2$.

4. $y^{(6)}(t) = [y'(t)]^3 - \sin\big(y(t)\big) + e^{2t}$;
 $y(0) = y'(0) = \cdots = y^{(5)}(0) = 0$.

5. $x'' + y - x' = 2t$; $x(3) = 5$, $x'(3) = 2$,
 $y'' - x + y = -1$; $y(3) = 1$, $y'(3) = -1$.
 [*Hint:* Set $x_1 = x$, $x_2 = x'$, $x_3 = y$, $x_4 = y'$.]

6. $3x'' + 5x - 2y = 0$; $x(0) = -1$, $x'(0) = 0$,
 $4y'' + 2y - 6x = 0$; $y(0) = 1$, $y'(0) = 2$.

7. $x''' - y = t$; $x(0) = x'(0) = x''(0) = 4$,
 $2x'' + 5y'' - 2y = 1$; $y(0) = y'(0) = 1$.

8. Sturm-Liouville Form. A second order equation is said to be in **Sturm-Liouville form** if it is expressed as

$$[p(t)y'(t)]' + q(t)y(t) = 0 .$$

Show that the substitutions $x_1 = y, x_2 = py'$ result in the normal form

$$x_1' = x_2/p ,$$
$$x_2' = -qx_1 .$$

If $y(0) = a$ and $y'(0) = b$ are the initial values for the Sturm-Liouville problem, what are $x_1(0)$ and $x_2(0)$?

9. In Section 1.5, we discussed Euler's method for approximating the solution to a first order equation. Extend Euler's method to normal systems of 2 equations in 2 unknowns and give the recursive formulas in component form.

10. In Section 3.5, we discussed the improved Euler's method for approximating the solution to a first order equation. Extend this method to normal systems and give the recursive formulas for solving the initial value problem.

You will need a computer and a programmed version of the classical fourth order Runge-Kutta algorithm to solve Problems 11 through 34. (At the instructor's discretion, other software may be used.[†])

In Problems 11 through 14, use the Runge-Kutta SUB-ROUTINE with $h = 0.25$ to find an approximation for the solution to the given initial value problem on the specified interval. (For an interval $t_0 \leq t \leq c$, recall that the input number of steps N and the step size h are related by $Nh = c - t_0$.)

11. $y'' + ty' + y = 0$;

$\quad y(0) = 1$, $\quad y'(0) = 0$ on $[0, 1]$.

12. $(1 + t^2)y'' + y' - y = 0$;

$\quad y(0) = 1$, $\quad y'(0) = -1$ on $[0, 1]$.

13. $t^2 y'' + y = t + 2$;

$\quad y(1) = 1$, $\quad y'(1) = -1$ on $[1, 2]$.

14. $y'' = t^2 - y^2$;

$\quad y(0) = 0$, $\quad y'(0) = 1$ on $[0, 1]$.

(Can you guess the solution?)

15. Using the Runge-Kutta SUBROUTINE with $h = 0.5$, approximate the solution to the initial value problem

$\quad 3t^2 y'' - 5ty' + 5y = 0$;

$\quad y(1) = 0$, $\quad y'(1) = \dfrac{2}{3}$

at $t = 8$. Compare this approximation to the actual solution $y(t) = t^{5/3} - t$.

16. Using the Runge-Kutta algorithm with the stopping procedure based on the absolute error and a tolerance of $\varepsilon = 0.01$, approximate the solution to the initial value problem

$\quad y'' = t^2 + y^2$; $\quad y(0) = 1$, $\quad y'(0) = 0$

at $t = 1$.

17. Using the Runge-Kutta algorithm with the stopping procedure based on the relative error and a tolerance of $\varepsilon = 0.02$, approximate the solution to the initial value problem

$\quad y'' = ty^3$; $\quad y(0) = 1$, $\quad y'(0) = 1$

at $t = 1$.

18. Using the Runge-Kutta SUBROUTINE for systems with $h = 0.125$, approximate the solution to the initial value problem

$\quad x' = 2x - y$; $\quad x(0) = 0$,

$\quad y' = 3x + 6y$; $\quad y(0) = -2$

at $t = 1$. Compare this approximation to the actual solution

$\quad x(t) = e^{5t} - e^{3t}$, $\quad y(t) = e^{3t} - 3e^{5t}$.

19. Using the Runge-Kutta algorithm for systems with the stopping procedure based on the absolute error and a tolerance of $\varepsilon = 0.001$, approximate the solution to the initial value problem

$\quad \dfrac{du}{dx} = 3u - 4v$; $\quad u(0) = 1$,

$\quad \dfrac{dv}{dx} = 2u - 3v$; $\quad v(0) = 1$

at $x = 1$.

In Problems 20 and 21, use the Runge-Kutta SUB-ROUTINE for systems with $h = 0.25$ to find an approximation for the solution to the given initial value problem at the times $t = 0.25$, $t = 0.5$, $t = 0.75$, and $t = 1.0$.

20. $x' = -x^2 + y^2 + 1$; $\quad x(0) = 0$,

$\quad y' = -y$; $\quad y(0) = 1$.

21. $x' = \cos(x + y)$; $\quad x(0) = 1$,

$\quad y' = (x^2 + t)^{-1} + t$; $\quad y(0) = 0.5$.

22. Combat Model. A simplified mathematical model for conventional versus guerrilla combat is given by the system

$\quad x_1' = -(0.1)x_1 x_2$; $\quad x_1(0) = 10$,

$\quad x_2' = -x_1$; $\quad x_2(0) = 15$,

where x_1 and x_2 are the strengths of guerrilla and conventional troops, respectively, and 0.1 and 1 are

[†]When using other software, you may accept the vendor's interpretation of the stopping "tolerance" criterion, even if it differs from the prescription in this section.

the *combat effectiveness coefficients*. Who will win the conflict: the conventional troops or the guerrillas? [*Hint:* Use the Runge-Kutta SUBROUTINE for systems with $h = 0.1$ to approximate the solutions.]

23. Predator-Prey Model. The Volterra-Lotka predator-prey model predicts some rather interesting behavior that is evident in certain biological systems. For example, suppose you fix the initial population of prey but increase the initial population of predators. Then the population cycle for the prey becomes more severe in the sense that there is a long period of time with a reduced population of prey followed by a short period when the population of prey is very large. To demonstrate this behavior, use the 4th order Runge-Kutta SUBROUTINE for systems with $h = 0.5$ to approximate the populations of prey x and of predators y over the period $[0, 5]$ that satisfy the Volterra-Lotka system

$$x' = x(3 - y) ,$$
$$y' = y(x - 3)$$

under each of the following initial conditions:

(a) $x(0) = 2$, $y(0) = 4$.
(b) $x(0) = 2$, $y(0) = 5$.
(c) $x(0) = 2$, $y(0) = 7$.

24. In Project G of Chapter 4, it was shown that the simple pendulum equation

$$\theta''(t) + \sin \theta(t) = 0$$

has periodic solutions when the initial displacement and velocity are small. Show that the period of the solution may depend on the initial conditions by using the Runge-Kutta SUBROUTINE with $h = 0.02$ to approximate the solutions to the simple pendulum problem on $[0, 4]$ for the initial conditions:

(a) $\theta(0) = 0.1$; $\theta'(0) = 0$.
(b) $\theta(0) = 0.5$; $\theta'(0) = 0$.
(c) $\theta(0) = 1.0$; $\theta'(0) = 0$.

[*Hint:* Approximate the length of time it takes to reach $-\theta(0)$.]

25. Fluid Ejection. In the design of a sewage treatment plant, the following equation arises:[†]

$$60 - H = (77.7)H'' + (19.42)(H')^2 ;$$
$$H(0) = H'(0) = 0 ,$$

where H is the level of the fluid in an ejection chamber and t is the time in seconds. Use the Runge-Kutta SUBROUTINE with $h = 0.5$ to approximate $H(t)$ over the interval $[0, 5]$.

26. Oscillations and Nonlinear Equations. For the initial value problem

$$x'' + (0.1)(1 - x^2)x' + x = 0 ;$$
$$x(0) = x_0 , x'(0) = 0 ,$$

use the Runge-Kutta SUBROUTINE with $h = 0.02$ to illustrate that as t increases from 0 to 20, the solution x exhibits damped oscillations when $x_0 = 1$, whereas x exhibits expanding oscillations when $x_0 = 2.1$.

27. Nonlinear Spring. The **Duffing equation**

$$y'' + y + ry^3 = 0 ,$$

where r is a constant, is a model for the vibrations of a mass attached to a *nonlinear* spring. For this model, does the period of vibration vary as the parameter r is varied? Does the period vary as the initial conditions are varied? [*Hint:* Use the Runge-Kutta SUBROUTINE with $h = 0.1$ to approximate the solutions for $r = 1$ and 2, with initial conditions $y(0) = a$, $y'(0) = 0$ for $a = 1, 2$, and 3.]

28. Pendulum with Varying Length. A pendulum is formed by a mass m attached to the end of a wire that is attached to the ceiling. Assume the length $l(t)$ of the wire varies with time in some predetermined fashion. If $\theta(t)$ is the angle between the pendulum and the vertical, then the motion of the pendulum is governed by the initial value problem

$$l^2(t)\theta''(t) + 2l(t)l'(t)\theta'(t) + gl(t)\theta(t) = 0 ;$$
$$\theta(0) = \theta_0 , \theta'(0) = \theta_1 ,$$

where g is the acceleration due to gravity. Assume

$$l(t) = l_0 + l_1 \cos(\omega t - \phi) ,$$

where l_1 is much smaller than l_0. (This might be a model for a person on a swing, where the *pumping* action changes the distance from the center of mass of the swing to the point where the swing is attached.) To simplify the computations, take $g = 1$. Using the Runge-Kutta SUBROUTINE with $h = 0.1$, study the motion of the pendulum when $\theta_0 = 0.5$,

[†]See *Numerical Solution of Differential Equations*, by William Milne (Dover Publications, Inc., New York, 1970), p. 82.

$\theta_1 = 0$, $l_0 = 1$, $l_1 = 0.1$, $\omega = 1$, and $\phi = 0.02$. In particular, does the pendulum ever attain an angle greater in absolute value than the initial angle θ_0? Does the total arc traversed during one half of a swing ever exceed 1?

In Problems 29 through 34, use a software package or the SUBROUTINE in Appendix E.

29. Using the Runge-Kutta algorithm for systems with the stopping procedure based on the relative error and a tolerance of $\varepsilon = 0.01$, approximate the solution to the initial value problem

$$y''' + y'' + y^2 = t \; ;$$
$$y(0) = 1 \; , \qquad y'(0) = 0 \; , \qquad y''(0) = 1$$

at $t = 1$.

30. Use the Runge-Kutta SUBROUTINE for systems with $h = 0.1$ to approximate the solution to the initial value problem

$$x' = yz \; ; \qquad x(0) = 0 \; ,$$
$$y' = -xz \; ; \qquad y(0) = 1 \; ,$$
$$z' = -xy/2 \; ; \qquad z(0) = 1$$

at $t = 1$.

31. Generalized Blasius Equation. H. Blasius, in his study of laminar flow of a fluid, encountered an equation of the form

$$y''' + yy'' = (y')^2 - 1 \; .$$

Use the Runge-Kutta SUBROUTINE for systems with $h = 0.1$ to approximate the solution that satisfies the initial conditions $y(0) = 0$, $y'(0) = 0$, and $y''(0) = 1.32824$. Sketch this solution on the interval $[0, 2]$.

32. Lunar Orbit. The motion of a moon moving in a planar orbit about a planet is governed by the equations

$$\frac{d^2x}{dt^2} = -G\frac{mx}{r^3} \; , \qquad \frac{d^2y}{dt^2} = -G\frac{my}{r^3} \; ,$$

where $r := (x^2 + y^2)^{1/2}$, G is the gravitational constant, and m is the mass of the moon. Assume $Gm = 1$. When $x(0) = 1$, $x'(0) = y(0) = 0$, and $y'(0) = 1$, the motion is a circular orbit of radius 1 and period 2π.

(a) Setting $x_1 = x$, $x_2 = x'$, $x_3 = y$, $x_4 = y'$, express the governing equations as a first order system in normal form.

(b) Using $h = 2\pi/100 \approx 0.0628318$, compute one orbit of this moon (i.e., let $N = 100$). Do your approximations agree with the fact that the orbit is a circle of radius 1?

33. Competing Species. Let $p_i(t)$ denote, respectively, the populations of three competing species S_i, $i = 1$, 2, 3. Suppose these species have the same growth rates and the maximum population that the habitat can support is the same for each species. (We assume it to be one unit.) Also suppose the competitive advantage that S_1 has over S_2 is the same as that of S_2 over S_3 and S_3 over S_1. This situation is modeled by the system

$$p_1' = p_1(1 - p_1 - ap_2 - bp_3) \; ,$$
$$p_2' = p_2(1 - bp_1 - p_2 - ap_3) \; .$$
$$p_3' = p_3(1 - ap_1 - bp_2 - p_3) \; ,$$

where a and b are positive constants. To demonstrate the population dynamics of this system when $a = b = 0.5$, use the Runge-Kutta SUBROUTINE for systems with $h = 0.1$ to approximate the populations p_i over the time interval $[0, 10]$ under each of the following initial conditions:

(a) $p_1(0) = 1.0$, $p_2(0) = 0.1$, $p_3(0) = 0.1$.
(b) $p_1(0) = 0.1$, $p_2(0) = 1.0$, $p_3(0) = 0.1$.
(c) $p_1(0) = 0.1$, $p_2(0) = 0.1$, $p_3(0) = 1.0$.

On the basis of the results of parts (a)–(c), decide what you think will happen to these populations as $t \to +\infty$.

When $a = 0.5$ and $b = 2.0$, the behavior of these populations is entirely different (see Project G).

34. Spring Pendulum. Let a mass be attached to one end of a spring with spring constant k and the other end attached to the ceiling. Let l_0 be the natural length of the spring and let $l(t)$ be its length at time t. If $\theta(t)$ is the angle between the pendulum and the vertical, then the motion of the spring pendulum is governed by the system

$$l''(t) - l(t)\theta'(t) - g\cos\theta(t) + \frac{k}{m}(l - l_0) = 0 \; ,$$

$$l^2(t)\theta''(t) + 2l(t)l'(t)\theta'(t) + gl(t)\sin\theta(t) = 0 \; .$$

Assume $g = 1$, $k = m = 1$, and $l_0 = 4$. When the system is at rest, $l = l_0 + mg/k = 5$.

(a) Describe the motion of the pendulum when $l(0) = 5.5$, $l'(0) = 0$, $\theta(0) = 0$, and $\theta'(0) = 0$. [*Hint:* Use the Runge-Kutta SUBROUTINE for systems with $h = 0.1$ to approximate the solution on the interval $[0, 10]$.]

(b) When the pendulum is both stretched and given an angular displacement, the motion of the pendulum is more complicated. Using the Runge-Kutta SUBROUTINE for systems with $h = 0.1$ to approximate the solution, sketch the graphs of the length l and the angular displacement θ on the interval $[0, 10]$ if $l(0) = 5.5$, $l'(0) = 0$, $\theta(0) = 0.5$, and $\theta'(0) = 0$.

5.7 DYNAMICAL SYSTEMS, POINCARÉ MAPS, AND CHAOS

In this section, we take an excursion through an area of mathematics that is currently receiving a lot of attention both for the interesting mathematical phenomena being observed and for its application to fields such as meteorology, heat conduction, fluid mechanics, lasers, chemical reactions, and nonlinear circuits, among others. The area is that of dynamical systems,[†] especially when these systems are nonlinear.

A **dynamical system** is any system that allows one to determine (at least theoretically) the future states of the system given its present or past state. For example, the recursive formula (difference equation)

$$x_{n+1} = (1.05)x_n , \qquad n = 0, 1, 2, \ldots .$$

is a dynamical system, since we can determine the next state, x_{n+1}, given the previous state, x_n. If we know x_0, then we can compute any future state $\left(\text{indeed } x_{n+1} = x_0(1.05)^{n+1}\right)$.

Another example of a dynamical system is provided by the differential equation

$$\frac{dx}{dt} = -2x ,$$

where the solution $x(t)$ specifies the state of the system at "time" t. If we know $x(t_0) = x_0$, then we can determine the state of the system at any future time $t > t_0$ by solving the initial value problem

$$\frac{dx}{dt} = -2x , \qquad x(t_0) = x_0 .$$

Indeed, a simple calculation yields $x(t) = x_0 e^{-2(t-t_0)}$ for $t \geq t_0$.

For a dynamical system defined by a differential equation, it is often helpful to work with a related dynamical system defined by a difference equation. For example, when we cannot express the solution to a differential equation using elementary functions, we can use a numerical technique such as the improved Euler's method or Runge-Kutta to approximate the

[†]For a more detailed study of dynamical systems, we refer the reader to *An Introduction to Chaotic Dynamical Systems,* Second Edition, by R. L. Devaney (Addison-Wesley Publishing Co., Reading, Mass., 1989) and *Nonlinear Oscillations, Dynamical Systems and Bifurcations of Vector Fields* by J. Guckenheimer and P. J. Holmes (Springer-Verlag, New York, 1983).

solution to an initial value problem. This numerical scheme defines a new (but related) dynamical system that is often easier to study.

In Section 5.2, we used phase plane diagrams to study autonomous systems in the plane. Many important features of the system can be detected just by looking at these diagrams. For example, a closed trajectory corresponds to a periodic solution. The trajectories for *non*autonomous systems in the phase plane are much more complicated to decipher. One technique that is helpful in this regard is the so-called **Poincaré map.** As we shall see, these maps replace the study of a nonautonomous system with the study of a dynamical system defined by the location in the xv-plane $(v = dx/dt)$ of the solution at regularly spaced moments in time such as $t = 2\pi n$, where $n = 0, 1, 2, \ldots$. The advantage in using the Poincaré map will become clear when the method is applied to a nonlinear problem for which no explicit solution is known. In such a case, the trajectories are computed using a numerical scheme such as Runge-Kutta. Several software packages have options that will construct Poincaré maps for a given system.

To illustrate the Poincaré map, consider the equation

(1) $x''(t) + \omega^2 x(t) = F \cos t$,

where F and ω are positive constants. We studied similar equations in Section 4.12 and found that a general solution for $\omega \neq 1$ is given by

(2) $x(t) = A \sin(\omega t + \phi) + \dfrac{F}{\omega^2 - 1} \cos t$,

where the amplitude A and the phase angle ϕ are arbitrary constants. Since $v = x'$,

$$v(t) = \omega A \cos(\omega t + \phi) - \frac{F}{\omega^2 - 1}\sin t .$$

Because the forcing function $F \cos t$ is 2π-periodic, it is natural to seek 2π-periodic solutions to (1). For this purpose, we define the Poincaré map as

(3) $x_n := x(2\pi n) = A \sin(2\pi\omega n + \phi) + F/(\omega^2 - 1)$,
 $v_n := v(2\pi n) = \omega A \cos(2\pi\omega n + \phi)$,

for $n = 0, 1, 2, \ldots$. In Figure 5.36, we plotted the first 100 values of (x_n, v_n) in the xv-plane for different choices of ω. For simplicity, we have taken $A = F = 1$ and $\phi = 0$. These graphs are called **Poincaré sections.** We will interpret them shortly.

Now let's play the following game. We agree to ignore the fact that we already know the formula for $x(t)$ for all $t \geq 0$. We want to see what information about the solution we can glean just from the Poincaré section and the form of the differential equation.

Notice that the first two Poincaré sections in Figure 5.36, corresponding to $\omega = 2$ and 3, consist of a single point. This tells us that, starting with $t = 0$, every increment 2π of t returns us to the same point in the phase plane. This in turn implies that equation (1) has a 2π-periodic solution, which can be proved as follows: For $\omega = 2$, let $x(t)$ be the solution to (1) with $\big(x(0), v(0)\big) = (1/3, 2)$ and let $X(t) := x(t + 2\pi)$. Since the Poincaré section is just the point $(1/3, 2)$, we have $X(0) = x(2\pi) = 1/3$ and $X'(0) = x'(2\pi) = 2$. Thus $x(t)$ and $X(t)$ have the same initial values at $t = 0$. Further, because $\cos t$ is 2π-periodic, we also have

$$X''(t) + \omega^2 X(t) = x''(t + 2\pi) + \omega^2 x(t + 2\pi) = \cos(t + 2\pi) = \cos t .$$

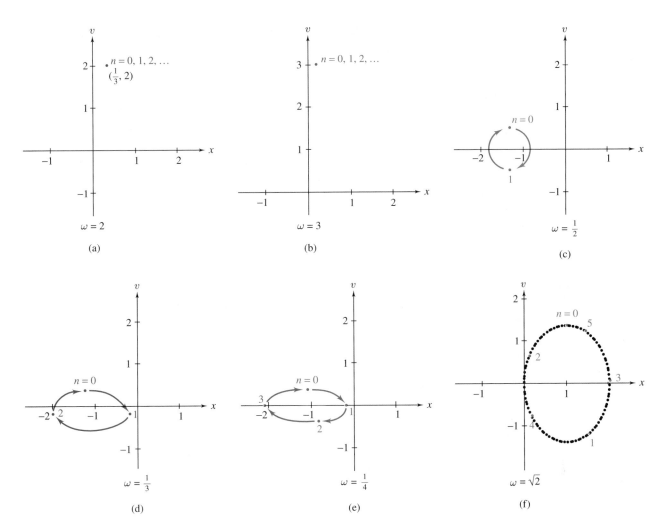

Figure 5.36 Poincaré sections for equation (1) for various values of ω

Consequently, $x(t)$ and $X(t)$ satisfy the same initial value problem. By the existence and uniqueness theorem (see Theorem 2 on page 165), these functions must agree on the interval $[0, \infty)$. Hence $x(t) = X(t) = x(t + 2\pi)$ for all $t \geq 0$; that is, $x(t)$ is 2π-periodic. (The same reasoning works for $\omega = 3$.) With a similar argument, it follows from the Poincaré section for $\omega = 1/2$ that there is a solution of period 4π that alternates between the two points displayed in Figure 5.36(c) as t is incremented by 2π. For the case $\omega = 1/3$, we deduce that there is a solution of period 6π rotating among three points, and for $\omega = 1/4$, there is an 8π-periodic solution rotating among four points. We call these last three solutions **subharmonics.**

The case $\omega = \sqrt{2}$ is different. So far, in Figure 5.36(f), none of the points has repeated. Did we stop too soon? Will the points ever repeat? Here, the fact that $\sqrt{2}$ is irrational plays a crucial role. It turns out that every integer n yields a distinct point in the Poincaré section (see Problem 8). However, there is a pattern developing. The points all appear to lie on a simple

curve, possibly an ellipse. To see that this is indeed the case, notice that when $\omega = \sqrt{2}$, $A = F = 1$, and $\phi = 0$, we have

$$x_n = \sin\left(2\sqrt{2}\pi n\right) + 1 \; , \qquad v_n = \sqrt{2}\cos\left(2\sqrt{2}\pi n\right) \; , \qquad n = 0, 1, 2, \ldots \; .$$

It is then an easy computation to show that each (x_n, v_n) lies on the ellipse

$$(x - 1)^2 + \frac{v^2}{2} = 1 \; .$$

In our investigation of equation (1), we concentrated on 2π-periodic solutions because the forcing term $F \cos t$ has period 2π. (We observed subharmonics when $\omega = 1/2, 1/3$, and $1/4$, that is, solutions with periods $2(2\pi), 3(2\pi)$, and $4(2\pi)$.) When a damping term is introduced into the differential equation, the Poincaré map displays a different behavior. Recall that the solution will now be the sum of a transient and a steady-state term. For example, let's consider the equation

$$(4) \qquad x''(t) + bx'(t) + \omega^2 x(t) = F \cos t \; ,$$

where b, F, and ω are positive constants.

When $b^2 < 4\omega^2$, the solution to (4) can be expressed as

$$(5) \qquad x(t) = Ae^{-(b/2)t}\sin\left(\frac{\sqrt{4\omega^2 - b^2}}{2}t + \phi\right) + \frac{F}{\sqrt{(\omega^2 - 1)^2 + b^2}}\sin(t + \theta) \; ,$$

where $\tan\theta = (\omega^2 - 1)/b$ and A and ϕ are arbitrary constants (cf. equations (7) and (8) in Section 4.12). The first term on the right-hand side of (5) is the transient and the second term, the steady-state solution. Let's construct the Poincaré map using $t = 2\pi n, n = 0, 1, 2, \ldots$. We will take $b = 0.22$, $\omega = A = F = 1$, and $\phi = 0$ to simplify the computations. Because $\tan\theta = (\omega^2 - 1)/b = 0$, we will take $\theta = 0$ as well. Then we have

$$x(2\pi n) = x_n = e^{-0.22\pi n}\sin\left(\sqrt{0.9879}\; 2\pi n\right) \; ,$$
$$x'(2\pi n) = v_n = -0.11e^{-0.22\pi n}\sin\left(\sqrt{0.9879}\; 2\pi n\right)$$
$$+ \sqrt{0.9879}\; e^{-0.22\pi n}\cos\left(\sqrt{0.9879}\; 2\pi n\right) + \frac{1}{(0.22)} \; .$$

The Poincaré section for $n = 0, 1, 2, \ldots, 10$ is shown in Figure 5.37 (black points). After just a few iterations, we observe that $x_n \approx 0$ and $v_n \approx 1/(0.22) \approx 4.545$; that is, the points of the Poincaré section are approaching a single point in the xv-plane (colored point). Thus we might expect that there is a 2π-periodic solution corresponding to a particular choice of A and ϕ. (In this example, where we can explicitly represent the solution, we see that indeed a 2π-periodic solution arises when we take $A = 0$ in (5).)

There is an important difference between the Poincaré sections for equation (1) and those for equation (4). In Figure 5.36, the location of all of the points in (a)–(e) depends on the initial value selected (here $A = 1$ and $\phi = 0$). (See Problem 10.) However, in Figure 5.37, the first few points (black points) depend on the initial conditions, while the limit point (colored point) does *not* (see Problem 6). The latter behavior is typical for equations that have a "damping" term (i.e., $b > 0$); namely, the Poincaré section has a limit set[†] that is essentially independent of the initial conditions.

[†]The **limit set** for a map $(x_n, v_n), n = 1, 2, 3, \ldots$, is the set of points (p, q) such that $\lim_{k \to \infty}(x_{n_k}, v_{n_k}) = (p, q)$, where $n_1 < n_2 < n_3 < \cdots$ is some subsequence of the positive integers.

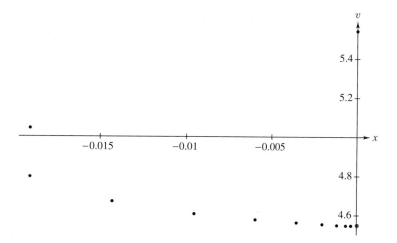

Figure 5.37 Poincaré section for equation (4) with $F = 1$, $b = 0.22$, and $\omega = 1$

For equations with damping, the limit set may be more complicated than just a point. For example, the Poincaré map for the equation

$$(6) \qquad x''(t) + (0.22)x'(t) + x(t) = \cos t + \cos\left(\sqrt{2}\,t\right)$$

has a limit set consisting of an ellipse (see Problem 11). This is illustrated in Figure 5.38 for the initial values $x_0 = 2$, $v_0 = 4$ and $x_0 = 2$, $v_0 = 6$.

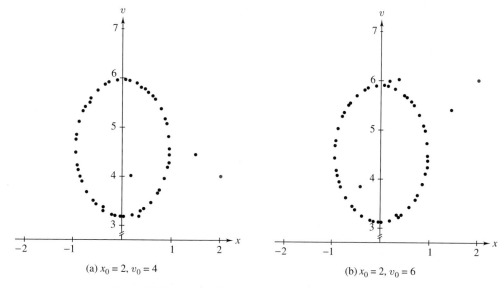

(a) $x_0 = 2$, $v_0 = 4$

(b) $x_0 = 2$, $v_0 = 6$

Figure 5.38 Poincaré section for equation (6) with initial values x_0, v_0

So far we have seen limit sets for the Poincaré map that were either a single point or an ellipse—independent of the initial values. These particular limit sets are **attractors.** In general,

an attractor is a set A with the property that there exists an open set[†] B containing A such that whenever the Poincaré map enters B, its points remain in B *and* the limit set of the Poincaré map is a subset of A. Further, we assume A has the *invariance property:* Whenever the Poincaré map starts at a point A, it remains in A.

In the previous examples, the attractors of the dynamical system (Poincaré map) were easy to describe. In recent years, however, many investigators, working on a variety of applications, have encountered dynamical systems that do *not* behave orderly—their attractor sets are very complicated (not just isolated points or familiar geometric objects such as ellipses). The behavior of such systems is called **chaotic,** and the corresponding limit sets are referred to as **strange attractors.**

To illustrate chaotic behavior and what is meant by a strange attractor, we discuss two nonlinear differential equations and a simple difference equation. First, let's consider the **forced Duffing equation**

$$(7) \qquad x''(t) + bx'(t) - x(t) + x^3(t) = F \sin \gamma t \ .$$

We cannot express the solution to (7) in any explicit form, so we must obtain the Poincaré map by numerically approximating the solution to (7) for fixed initial values and then plot the approximations for $x(2\pi n/\gamma)$ and $v(2\pi n/\gamma) = x'(2\pi n/\gamma)$. (Because the forcing term $F \sin \gamma t$ has period $2\pi/\gamma$, we seek $2\pi/\gamma$-periodic solutions and subharmonics.) In Figure 5.39, we display the limit sets (attractors) when $b = 0.3$ and $\gamma = 1.2$ in the cases (a) $F = 0.2$, (b) $F = 0.28$, (c) $F = 0.29$, and (d) $F = 0.37$.

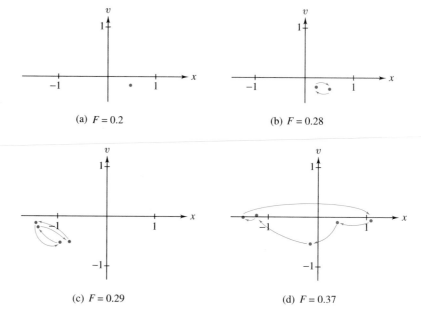

(a) $F = 0.2$ (b) $F = 0.28$

(c) $F = 0.29$ (d) $F = 0.37$

Figure 5.39 Poincaré sections for the Duffing equation (7) with $b = 0.3$ and $\gamma = 1.2$

[†]A set $B \subset \mathbf{R}^2$ is an **open set** if for each point $p \in B$ there is an open disk V containing p such that $V \subset B$.

Notice that as the constant F increases, the Poincaré map changes character. When $F = 0.2$, the Poincaré section tells us that there is a $2\pi/\gamma$-periodic solution. For $F = 0.28$, there is a subharmonic of period $4\pi/\gamma$, and for $F = 0.29$ and 0.37, there are subharmonics with periods $8\pi/\gamma$ and $10\pi/\gamma$, respectively.

Things are dramatically different when $F = 0.5$: The solution is neither $2\pi/\gamma$-periodic nor subharmonic. The Poincaré section for $F = 0.5$ is illustrated in Figure 5.40. This section was generated by numerically approximating the solution to (7) when $\gamma = 1.2$, $b = 0.3$, and $F = 0.5$, for fixed initial values.[†] Not all of the approximations $x(2\pi n/\gamma)$ and $v(2\pi n/\gamma)$ that were calculated are graphed; because of the presence of a transient solution, the first few points were omitted. It turns out that the plotted set is essentially independent of the initial values and has the property that once a point is in the set, all subsequent points lie in the set. Because of the complicated shape of the set, it is indeed a strange attractor. While the shape of the strange attractor does not depend on the initial values, the picture does change if we consider different sections; for example, $t = (2\pi n + \pi/2)/\gamma$, $n = 0, 1, 2, \ldots$ yields a different configuration.

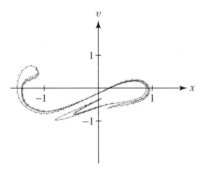

Figure 5.40 Poincaré section for the Duffing equation (7) with $b = 0.3$, $\gamma = 1.2$, and $F = 0.5$

Another example of a strange attractor occurs when we consider the **forced pendulum equation**

$$(8) \qquad x''(t) + bx'(t) + \sin\big(x(t)\big) = F \cos t ,$$

where the $x(t)$ term in (4) has been replaced by $\sin\big(x(t)\big)$. Here $x(t)$ is the angle between the pendulum and the vertical rest position, b is related to damping, and F represents the strength of the forcing function (see Figure 5.41 on page 322). For $F = 2.7$ and $b = 0.22$, we have graphed in Figure 5.42 on page 322 approximately 90,000 points in the Poincaré map. Since we cannot express the solution to (8) in any explicit form, the Poincaré map was obtained by numerically approximating the solution to (8) for fixed initial values and plotting the approximations for $x(2\pi n)$ and $v(2\pi n) = x'(2\pi n)$.

[†]*Historical Footnote:* When researchers first encountered these strange-looking Poincaré sections, they would check their computations using different computers and different numerical schemes (see Hénon and Heiles, "The applicability of the third integral of motion: some numerical experiments," *Astronomical Journal,* Vol. 69 (1964), p. 75). For special types of dynamical systems, such as the Hénon map, it can be shown that there exists a true trajectory that *shadows* the numerical trajectory (see M. Hammel, J. A. Yorke, and C. Grebogi, "Numerical orbits of chaotic processes represent true orbits," *Bulletin American Mathematical Society,* Vol. 19 (1988), pp. 466–469).

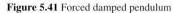

Figure 5.41 Forced damped pendulum

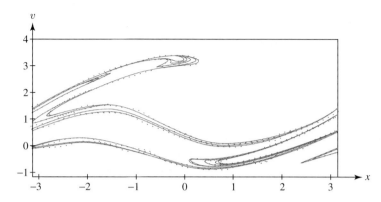

Figure 5.42 Poincaré section for the forced pendulum equation (8) with $b = 0.22$ and $F = 2.7$

The Poincaré maps for the forced Duffing equation and for the forced pendulum equation not only illustrate the idea of a strange attractor; they also exhibit another peculiar behavior called **chaos.** Chaos occurs when small changes in the initial conditions lead to major changes in the behavior of the solution. Henri Poincaré described the situation as follows:

"... it may happen that small differences in the initial conditions will produce very large ones in the final phenomena. A small error in the former produces an enormous error in the latter. Prediction becomes impossible"

In a physical experiment, we can never *exactly* (with infinite accuracy) reproduce the same initial conditions. Consequently, if the behavior is chaotic, even a slight difference in the initial conditions may lead to quite different values for the corresponding Poincaré map when n is large. Such behavior does not occur for solutions to either equation (4) or equation (1) (see Problems 6 and 7). However, two solutions to the Duffing equation (7) with $F = 0.5$ that correspond to two different, but close, initial values have Poincaré maps that do *not* remain close together. While they both are attracted to the same set, their locations with respect to this set may be relatively far apart.

The phenomenon of chaos can also be illustrated by the following simple map. Let x_0 lie in $[0, 1)$ and define

(9) $x_{n+1} = 2x_n \pmod 1$,

where by (mod 1) we mean the decimal part of the number if it is greater than or equal to 1; that is

$$x_{n+1} = \begin{cases} 2x_n , & \text{for } 0 \le x_n < 1/2 , \\ 2x_n - 1 , & \text{for } 1/2 \le x_n < 1 . \end{cases}$$

When $x_0 = 1/3$, we find

$$x_1 = 2 \cdot (1/3) \pmod 1 = 2/3 ,$$
$$x_2 = 2 \cdot (2/3) \pmod 1 = 4/3 \pmod 1 = 1/3 ,$$
$$x_3 = 2 \cdot (1/3) \pmod 1 = 2/3 ,$$
$$x_4 = 2 \cdot (2/3) \pmod 1 = 1/3 ,$$
 etc.

Written as a sequence, we get $\{1/3, 2/3, \overline{1/3, 2/3}, \ldots\}$, where the overbar denotes the repeated pattern.

What happens when we pick a starting value x_0 near $1/3$? Does the sequence cluster about $1/3$ and $2/3$ as does the mapping when $x_0 = 1/3$? For example, when $x_0 = 0.3$, we get the sequence

$$\{0.3, 0.6, 0.2, 0.4, 0.8, \overline{0.6, 0.2, 0.4, 0.8}, \ldots\} \ .$$

In Figure 5.43, we have plotted the values of x_n for $x_0 = 0.3$, 0.33, and 0.333. We have not plotted the first few terms, but only those that repeat. (This omission of the first few terms parallels the situation depicted in Figure 5.40, where transient solutions arise.)

It is clear from Figure 5.43 that while the values for x_0 are getting closer to $1/3$, the corresponding maps are spreading out over the whole interval $[0, 1]$ and *not* clustering near $1/3$ and $2/3$. This behavior is chaotic, since the Poincaré maps for initial values near $1/3$ behave quite differently from the map for $x_0 = 1/3$. If we had selected x_0 to be irrational (which we can't do with a calculator), the sequence would *not* repeat and would be dense in $[0, 1]$.

Figure 5.43 Plots of the map $x_{n+1} = 2x_n(\text{mod } 1)$ for $x_0 = 0.3$, 0.33, and 0.333

Systems that exhibit chaotic behavior arise in many applications. The challenge to engineers is to design systems that avoid this chaos and, instead, enjoy the property of **stability.** The topic of stable systems is discussed at length in Chapter 12.[†]

EXERCISES 5.7

A software package that supports the construction of Poincaré maps is required for the problems in this section.

1. Compute and graph the points of the Poincaré map with $t = 2\pi n$, $n = 0, 1, \ldots, 20$ for equation (1) taking $A = F = 1$, $\phi = 0$, and $\omega = 3/2$. Repeat, taking $\omega = 3/5$. Do you think the equation has a

2π-periodic solution for either choice of ω? A subharmonic solution?

2. Compute and graph the points of the Poincaré map with $t = 2\pi n$, $n = 0, 1, \ldots, 20$ for equation (1) taking $A = F = 1$, $\phi = 0$, and $\omega = 1/\sqrt{3}$. Describe the limit set for this system.

[†]All references to Chapters 11–13 refer to the expanded text *Fundamentals of Differential Equations and Boundary Value Problems,* 3rd edition.

3. Compute and graph the points of the Poincaré map with $t = 2\pi n$, $n = 0, 1, \ldots, 20$ for equation (4) taking $A = F = 1$, $\phi = 0$, $\omega = 1$, and $b = -0.1$. What is happening to these points as $n \to \infty$?

4. Compute and graph the Poincaré map with $t = 2\pi n$, $n = 0, 1, \ldots, 20$ for equation (4) taking $A = F = 1$, $\phi = 0$, $\omega = 1$, and $b = 0.1$. Describe the attractor for this system.

5. Compute and graph the Poincaré map with $t = 2\pi n$, $n = 0, 1, \ldots, 20$ for equation (4) taking $A = F = 1$, $\phi = 0$, $\omega = 1/3$, and $b = 0.22$. Describe the attractor for this system.

6. Show that for $b > 0$ the Poincaré map for equation (4) is not chaotic by showing that as t gets large

$$x_n = x(2\pi n) \approx \frac{F}{\sqrt{(\omega^2 - 1)^2 + b^2}} \sin(2\pi n + \theta) ,$$

$$v_n = x'(2\pi n) \approx \frac{F}{\sqrt{(\omega^2 - 1)^2 + b^2}} \cos(2\pi n + \theta)$$

independent of the initial values $x_0 = x(0)$ and $v_0 = x'(0)$.

7. Show that the Poincaré map for equation (1) is not chaotic by showing that if (x_0, v_0) and (x_0^*, v_0^*) are two initial values that define the Poincaré maps $\{(x_n, v_n)\}$ and $\{(x_n^*, v_n^*)\}$, respectively, using the recursive formulas in (3), then one can make the distance between (x_n, v_n) and (x_n^*, v_n^*) small by making the distance between (x_0, v_0) and (x_0^*, v_0^*) small. [*Hint:* Let (A, ϕ) and (A^*, ϕ^*) be the polar coordinates of two points in the plane. From the law of cosines, it follows that the distance d between them is given by $d^2 = (A - A^*)^2 + 2AA^*[1 - \cos(\phi - \phi^*)]$.]

8. Consider the Poincaré maps defined in (3) with $\omega = \sqrt{2}$, $A = F = 1$, and $\phi = 0$. If this map were ever to repeat, then, for two distinct positive integers n and m, $\sin(2\sqrt{2}\pi n) = \sin(2\sqrt{2}\pi m)$. Using basic properties of the sine function, show that this would imply that $\sqrt{2}$ is rational. It follows from this contradiction that the points of the Poincaré map do not repeat.

9. The doubling modulo 1 map defined by equation (9) exhibits some fascinating behavior. Compute the sequence obtained when

(a) $x_0 = k/7$ for $k = 1, 2, \ldots, 6$.
(b) $x_0 = k/15$ for $k = 1, 2, \ldots, 14$.
(c) $x_0 = k/2^j$, where j is a positive integer and $k = 1, 2, \ldots, 2^j - 1$.

Numbers of the form $k/2^j$ are called **dyadic numbers** and are dense in $[0, 1]$. That is, there is a dyadic number arbitrarily close to any real number (rational or irrational).

10. To show that the limit set of the Poincaré map given in (3) depends on the initial values, do the following:
(a) Show that when $\omega = 2$ or 3, the Poincaré map consists of the single point
$$(x, v) = \left(A \sin \phi + \frac{F}{\omega^2 - 1} , \quad \omega A \cos \phi \right) .$$

(b) Show that when $\omega = 1/2$, the Poincaré map alternates between the two points
$$\left(\frac{F}{\omega^2 - 1} \pm A \sin \phi , \quad \pm \omega A \cos \phi \right) .$$

(c) Use the results of parts (a) and (b) to show that when $\omega = 2$, 3, or $1/2$, the Poincaré map (3) depends on the initial values (x_0, v_0).

11. To show that the limit set for the Poincaré map $x_n := x(2\pi n)$, $v_n := x'(2\pi n)$, where $x(t)$ is a solution to equation (6), is an ellipse and that this ellipse is the same for any initial values x_0, v_0, do the following:
(a) Argue that since the initial values affect only the transient solution to (6), the limit set for the Poincaré map is independent of the initial values.
(b) Now show that for n large,
$$x_n \approx a \sin(2\sqrt{2}\pi n + \psi) ,$$
$$v_n \approx c + \sqrt{2}a \cos(2\sqrt{2}\pi n + \psi) ,$$
where $a = (1 + 2(0.22)^2)^{-1/2}$, $c = (0.22)^{-1}$, and $\psi = \arctan\{-[(0.22)\sqrt{2}]^{-1}\}$.
(c) Use the result of part (b) to conclude that the ellipse
$$x^2 + \frac{(v - c)^2}{2} = a^2$$
contains the limit set of the Poincaré map.

12. Using a numerical scheme such as Runge-Kutta or a software package, calculate the Poincaré map for

equation (7) when $b = 0.3$, $\gamma = 1.2$, and $F = 0.2$. (Notice that the closer you start to the limiting point, the sooner the transient part will die out.) Compare your map with Figure 5.39(a) on page 320. Redo for $F = 0.28$.

13. Redo Problem 12 with $F = 0.31$. What kind of behavior does the solution exhibit?

14. Redo Problem 12 with $F = 0.65$. What kind of behavior does the solution exhibit?

15. Chaos Machine. Chaos can be illustrated using a long ruler, a short ruler, a pin, and a tie tack (pivot). Construct the double pendulum as shown in Figure 5.44(a). The pendulum is set in motion by releasing it from a position such as the one shown in Figure 5.44(b). Repeatedly set the pendulum in motion, each time trying to release it from the same position. Record the number of times the short ruler flips over and the direction in which it was moving. If the pendulum was released in *exactly* the same position each time, then the motion would be the same. However, from your experiments you will observe that even beginning close to the same position leads to very different motions. This double pendulum exhibits chaotic behavior.

(a) double pendulum

(b) release position

Figure 5.44 Double pendulum as a chaos machine

CHAPTER SUMMARY

Systems of differential equations arise in the study of coupled mass-spring oscillators, electrical circuits, ecological models, and many other areas. A technique for studying the qualitative behavior of solutions to the **autonomous system**

$$\textbf{(1)} \qquad \frac{dx}{dt} = f(x, y) , \qquad \frac{dy}{dt} = g(x, y)$$

is **phase plane analysis.** We begin by finding the **critical points** of (1), namely points (x_0, y_0) where

$$f(x_0, y_0) = 0 \quad \text{and} \quad g(x_0, y_0) = 0 .$$

The corresponding constant solution pairs $x(t) \equiv x_0, y(t) \equiv y_0$ are called **equilibrium solutions** to (1). We then sketch the direction field for the phase plane equation

$$\textbf{(2)} \qquad \frac{dy}{dx} = \frac{g(x, y)}{f(x, y)}$$

with appropriate direction arrows (oriented by the sign of dx/dt or dy/dt). From this we can

usually conjecture qualitative features of the solutions and of the critical points, such as stability and asymptotic behavior. Software is typically used to visualize the solution curves to (2), which contain the **trajectories** of the system (1).

Linear systems with constant coefficients can be solved explicitly using a variant of the **Gauss elimination process.** For this purpose, we begin by writing the system with operator notation, using $D := d/dt$, $D^2 := d^2/dt^2$, etc. A system of two equations in two unknown functions then takes the form

$$L_1[x] + L_2[y] = f_1 \ ,$$
$$L_3[x] + L_4[y] = f_2 \ ,$$

where L_1, L_2, L_3, and L_4 are polynomial operator expressions in D. Applying L_4 to the first equation, L_2 to the second, and subtracting, we get a single (typically higher-order) equation in $x(t)$, namely

$$(L_4 L_1 - L_2 L_3)[x] = L_4[f_1] - L_2[f_2] \ .$$

We then solve this constant coefficient equation for $x(t)$. Similarly, we can eliminate x from the system to obtain a single equation for $y(t)$, which we can also solve. This procedure introduces some extraneous constants, but by substituting the expressions for x and y back into one of the original equations we can determine the relationships among these constants.

A preliminary step for the application of numerical algorithms for solving systems or single equations of higher order is to rewrite them as an equivalent system of first order equations in **normal form:**

$$\begin{aligned} x_1'(t) &= f_1(t, x_1, x_2, \ldots, x_m) \ , \\ x_2'(t) &= f_2(t, x_1, x_2, \ldots, x_m) \ , \\ &\ \vdots \\ x_m'(t) &= f_m(t, x_1, x_2, \ldots, x_m) \ . \end{aligned}$$

(3)

For example, by setting $v = y'$, we can rewrite the second order equation $y'' = f(t, y, y')$ as the normal system

$$y' = v \ ,$$
$$v' = f(t, y, v) \ .$$

The normal system (3) has the outward appearance of a vectorized version of a single first order equation, and as such it suggests how to generalize numerical algorithms such as those of Euler and Runge-Kutta.

Nonautonomous systems can be studied by considering a **Poincaré map** for the system. A Poincaré map can be used to detect periodic and subharmonic solutions and to study systems whose solutions exhibit chaotic behavior.

REVIEW PROBLEMS

1. Explain the distinction between the *integral curves* and the *trajectories* for an autonomous system.

In Problems 2 and 3, solve the phase plane equation for the given system. Then sketch by hand several representative trajectories (with their flow arrows) and describe the stability of the critical points (i.e., compare with Figure 5.7, page 270).

2. $x' = y - 2$,
$y' = 2 - x$.

3. $x' = 4 - 4y$,
$y' = -4x$.

4. Find all the critical points and determine an equation for the integral curves of the system

$$x' = \sin x \cos y ,$$
$$y' = \cos x \sin y .$$

In Problems 5 and 6, sketch some typical trajectories for the given system, and by comparing with Figure 5.7, page 270, identify the type of critical point at the origin.

5. $x' = -2x - y$,
$y' = 3x - y$.

6. $x' = -x + 2y$,
$y' = x + y$.

In Problems 7 through 10, find a general solution $x(t)$, $y(t)$ for the given system.

7. $x' + y'' + y = 0$,
$x'' + y' = 0$.

8. $x' = x + 2y$,
$y' = -4x - 3y$.

9. $2x' - y' = y + 3x + e^t$,
$3y' - 4x' = y - 15x + e^{-t}$.

10. $x'' + x - y'' = 2e^{-t}$,
$x'' - x + y'' = 0$.

In Problems 11 and 12, solve the given initial value problem.

11. $x' = z - y$; $x(0) = 0$,
$y' = z$; $y(0) = 0$,
$z' = z - x$; $z(0) = 2$.

12. $x' = y + z$; $x(0) = 2$,
$y' = x + z$; $y(0) = 2$,
$z' = x + y$; $z(0) = -1$.

13. For the interconnected tanks problem of Section 5.1, page 261, suppose that instead of pure water being fed into tank A, a brine solution with concentration 2 kg/L is used; all other data remain the same. Determine the mass of salt in each tank at time t if the initial masses are $x_0 = 1$ kg and $y_0 = 3$ kg.

14. In the coupled mass-spring system depicted in Figure 5.18, page 287, take each mass to be 1 kg and let $k_1 = 8$ N/m, while $k_2 = 3$ N/m. What are the natural angular frequencies of the system? What is the general solution?

15. In the electrical circuit of Figure 5.45, take $R_1 = R_2 = 1$ ohm, $C = 1$ farad, and $L = 1$ henry. Derive three equations for the unknown currents I_1, I_2, and I_3 by writing Kirchhoff's voltage law for loops 1 and 2, and Kirchhoff's current for the top juncture. Find the general solution.

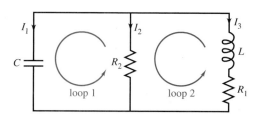

Figure 5.45 Electrical circuit for Problem 15

In Problems 16 through 19, write the given higher-order equation or system in an equivalent normal form (cf. Section 5.6).

16. $2y'' - ty' + 8y = \sin t$.

17. $3y''' + 2y' - e^t y = 5$.

18. $x'' - x + y = 0$,
$x' - y + y'' = 0$.

19. $x''' + y' + y'' = t$,
$x'' - x' + y''' = 0$.

A. *Designing a Landing System for Interplanetary Travel*

Alfred Clark, Jr., University of Rochester

You are a second-year Starfleet Academy Cadet aboard the U.S.S. Enterprise on a continuing study of the starsystem Glia. The object of study on the present expedition is the large airless planet Glia-4. A Class 1 Sensor Probe of mass m is to be sent to the planet's surface to collect data. The probe has a modifiable landing system so that it can be used on planets of different gravity. The system consists of a linear spring (force $= -kx$, where x is displacement), a nonlinear spring (force $= -ax^3$), and a shock-damper (force $= -b\dot{x}$),[†] all in parallel. Figure 5.46 shows a schematic of the system. During the landing process, the probe's thrusters are used to create a constant rate of descent. The velocity at impact varies; the symbol V_L is used to denote the largest velocity likely to happen in practice. At the instant of impact, (1) the thrust is turned off, and (2) the suspension springs are at their unstretched natural length.

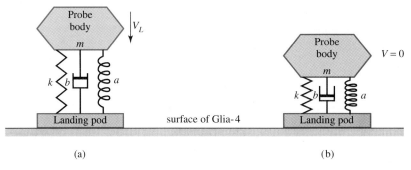

(a) (b)

Figure 5.46 Schematic of the probe landing system. (a) The system at the instant of impact. The springs are not stretched or compressed, the thrusters have been turned off, and the velocity is V_L downward. (b) The probe has reached a state of rest on the surface, and the springs are compressed enough to support the weight. Between states (a) and (b), the probe oscillates relative to the landing pod.

(a) Let the displacement x be measured from the unstretched length of the springs and be taken negative downward (i.e., compression gives a negative x). Show that the equation governing the oscillations after impact is

$$m\ddot{x} + b\dot{x} + kx + ax^3 = -mg \ .$$

(b) The probe has a mass $m = 1220$ kg. The linear spring is permanently installed and has a stiffness $k = 35{,}600$ N/m. The gravity on the surface of Glia-4 is $g = 17.5$ m/sec². The nonlinear spring is removable; an appropriate spring must be chosen for each mission. These nonlinear springs are made of coralidium, a rare and difficult-to-fabricate alloy. Therefore the Enterprise stocks only four different strengths: $a = 150{,}000$, $300{,}000$, $450{,}000$, and $600{,}000$ N/m³. Determine which springs give a compression as close as

[†]The symbol \dot{x} denotes dx/dt.

possible to 0.3 m without exceeding 0.3 m, when the ship is resting on the surface of Glia-4. (The limit of 0.3 m is imposed by unloading clearance requirements.)

(c) The other adjustable component on the landing system is the linear shock-damper, which may be adjusted in increments of $\Delta b = 500$ N-sec/m, from a low value of 1000 N-sec/m to a high value of 10,000 N-sec/m. It is desirable to make b as small as possible because a large b produces large forces at impact. However, if b is too small, there is some danger that the probe will rebound after impact. To minimize the chance of this, find the smallest value of b such that the springs are always in compression during the oscillations after impact. Use a minimum impact velocity $V_L = 5$ m/sec downward. To find this value of b, you will need to use a software package to integrate the differential equation.

B. Things That Bob

Richard Bernatz, Luther College

The motion of various-shaped objects that bob in a pool of water can be modeled by a second order differential equation derived from Newton's second law of motion, $F = ma$. The forces acting on the object include the force due to gravity, a frictional force due to the motion of the object in the water, and a buoyant force based on Archimedes' principle: An object that is completely or partially submerged in a fluid is acted on by an upward (buoyant) force equal to the weight of the water it displaces.

(a) The first step is to write down the governing differential equation. The dependent variable is the depth z of the object's lowest point in the water. Take z to be negative downward so that $z = -1$ means 1 ft of the object has submerged. Let $V(z)$ be the submerged volume of the object, m be the mass of the object, ρ be the density of water (in pounds per cubic foot), g be the acceleration due to gravity, and γ_w be the coefficient of friction for water. Assuming the frictional force is proportional to the vertical velocity of the object, write down the governing second order ODE.

(b) For the time being, neglect the effect of friction and assume the object is a cube measuring L feet on a side. Write down the governing differential equation for this case. Next, designate $z = l$ to be the depth of submersion such that the buoyant force is equal and opposite the gravitational force. Introduce a new variable, ζ, that gives the displacement of the object from its equilibrium position l (that is, $z = \zeta + l$). You can now write the ODE in a more familiar form. [*Hint:* Recall the mass-spring system and the equilibrium case.] Now you should recognize the type of solution for this problem. What is the natural frequency?

(c) In this task, you will consider the effect of friction. The bobbing object is a cube, one foot on a side, that weighs 32 lb. Let $\gamma_w = 3$ lb-sec/ft, $\rho = 62.57$ lb/ft^3, and suppose the object is initially placed on the surface of the water. Solve the governing ODE by hand to find the general solution. Next, find the particular solution for the case in which the cube is initially placed on the surface of the water and is given no initial velocity. Provide a plot of the position of the object as a function of time t.

(d) In this step of the project, you will develop a numerical solution to the same problem presented in part (c). The numerical solution will be useful (indeed necessary) for sub-

sequent parts of the project. This case will provide a trial to verify that your numerical solution is correct. Go back to the initial ODE you developed in part (a). Using parameter values given in part (c), solve the initial value problem for the cube starting on the surface with no initial velocity. To solve this problem numerically, you will have to write the second order ODE as a system of two first order ODEs, one for vertical position z and one for vertical velocity w. Plot your results for vertical position as a function of time t for the first 3 or 4 sec and compare with the analytical solution you found in part (c). Are they in close agreement? What might you have to do in order to compare these solutions? Provide a plot of both your analytical and numerical solutions on the same graph.

(e) Suppose a sphere of radius R is allowed to bob in the water. Derive the governing second order equation for the sphere using Archimedes' principle and allowing for friction due to its motion in the water. Suppose a sphere weighs 32 lb, has a radius of $1/2$ ft, and $\gamma_w = 3.0$ lb-sec/ft. Determine the limiting value of the position of the sphere without solving the ODE. Next, solve the governing ODE for the velocity and position of the sphere as a function of time for a sphere placed on the surface of the water. You will need to write the governing second order ODE as a system of two ODEs, one for velocity and one for position. What is the limiting position of the sphere for your solution? Does it agree with the equilibrium solution you found above? How does it compare with the equilibrium position of the cube? If it is different, explain why.

(f) Suppose the sphere in part (d) is a volleyball. Calculate the position of the sphere as a function of time t for the first 3 sec if the ball is submerged so that its lowest point is 5 ft under water. Will the ball leave the water? How high will it go? Next, calculate the ball's trajectory for initial depths lower than 5 ft, and higher than 5 ft. Provide plots of velocity and position for each case and comment on what you see. Specifically, comment on the relationship between the initial depth of the ball and the maximum height the ball eventually attains.

You might consider taking a volleyball into a swimming pool to gather real data in order to verify and improve on your model. If you do so, report the data you found and explain how you used it for verification and improvement of your model.

C. *Effects of Hunting on Predator-Prey Systems*

As discussed in Section 5.6 (page 308), cyclic variations in the populations of predators and their prey have been studied using the Volterra-Lotka predator-prey model given by the system

(1) $\qquad \dfrac{dx}{dt} = Ax - Bxy \ ,$

(2) $\qquad \dfrac{dy}{dt} = -Cy + Dxy \ ,$

where A, B, C, and D are positive constants, $x(t)$ is the population of prey at time t, and $y(t)$ is the population of predators. It can be shown that such a system has a periodic solution (see Project D). That is, there exists some constant T such that $x(t) = x(t + T)$ and $y(t) = y(t + T)$ for all t. This periodic or cyclic variation in the populations has been observed in various systems such as sharks-food fish, lynx-rabbits, and ladybird beetles-cottony cushion scale. Because of this periodic behavior,

it is useful to consider the average populations \bar{x} and \bar{y} defined by

$$\bar{x} := \frac{1}{T} \int_0^T x(t)\,dt \;, \qquad \bar{y} := \frac{1}{T} \int_0^T y(t)\,dt \;.$$

(a) Show that $\bar{x} = C/D$ and $\bar{y} = A/B$. [*Hint:* Use equation (1) and the fact that $x(0) = x(T)$ to show that

$$\int_0^T \Big(A - By(t)\Big)\,dt = \int_0^T \frac{x'(t)}{x(t)}\,dt = 0.]$$

(b) To determine the effect of indiscriminate hunting on the populations, assume hunting reduces the rate of change in a population by a constant times the population. Then, the predator-prey system satisfies the new set of equations

(3) $$\frac{dx}{dt} = Ax - Bxy - \varepsilon x = (A - \varepsilon)x - Bxy \;,$$

(4) $$\frac{dy}{dt} = -Cy + Dxy - \delta y = -(C + \delta)y + Dxy \;,$$

where ε and δ are positive constants with $\varepsilon < A$. What effect does this have on the average population of prey? On the average population of predators?

(c) Assume the hunting is done selectively, as in shooting only rabbits (or shooting only lynx). Then we have $\varepsilon > 0$ and $\delta = 0$ (or $\varepsilon = 0$ and $\delta > 0$) in (3)–(4). What effect does this have on the average populations of predator and prey?

(d) In a rural county, foxes prey mainly on rabbits but occasionally include a chicken in their diet. The farmers decide to put a stop to the chicken killing by hunting the foxes. What do you predict will happen? What will happen to the farmers' gardens?

For a discussion of other differential equation models in population biology, see *Differential Equation Models,* by Martin Braun, Courtney S. Coleman, and Donald A. Drew (eds.) (Springer-Verlag, New York, 1983), Part V.

D. *Periodic Solutions to Volterra-Lotka Systems*

The Volterra-Lotka predator-prey model is given by the system (1)–(2) in Project C, where $x(t)$ and $y(t)$ are the populations of the prey and the predators—say, rabbits and foxes—respectively. For simplicity, let's assume units have been selected so that $A = B = C = D = 1$ in the model, and write it as

(5) $$\begin{aligned} x' &= x - xy \;, \\ y' &= -y + xy \;. \end{aligned}$$

(a) Show that $(0, 0)$ and $(1, 1)$ are the only critical points of (5).

(b) Plot the direction field in the phase plane (cf. Section 5.2) for the trajectories of (5). [*Hint:* The direction field for the phase plane equation

(6) $$\frac{dy}{dx} = \frac{y(x - 1)}{x(1 - y)}$$

is easy to plot along the lines $x = 0$, $y = 0$, $x = 1$, and $y = 1$. You may use software if you wish to fill in some of the others.]

From the direction field plot it appears that the equilibrium at $(1, 1)$ is either a center or a (stable or unstable) spiral. The distinction is very important to biologists. If it is a stable spiral, then we can expect any initial population distribution to settle, eventually, at $(1, 1)$ (in these units). If it is an unstable spiral, then we will be up to our ears in rabbits and foxes! And if it is a stable center, then the populations will be periodic, with lots of rabbits some years and only a few rabbits other years, perpetually changing back and forth. So it is worth taking some trouble to establish the exact nature of the equilibrium.

(c) Solve equation (6) by separating variables and show that the result can be manipulated into the form

(7) $$\left(xe^{-x}\right)\left(ye^{-y}\right) = C \ .$$

Note that formula (7) for the integral curves implies that they are symmetric about the line $y = x$.

(d) We need to decide whether these curves are spirals approaching $(1, 1)$ as $t \to +\infty$ or $t \to -\infty$. From the graph of ue^{-u} in Figure 5.47, argue that the graph of equation (7) contains no points if $C > e^{-2}$ and that it contains one point, $(1, 1)$, if $C = e^{-2}$. Thus the trajectories in question arise only for $C < e^{-2}$.

(e) Using Figure 5.47, argue that the point $(1, 1)$ *cannot* be a limit point of the trajectory for fixed C $(C < e^{-2})$, because when $y = 1$ equation (7) has only two solutions x_{\min} and x_{\max} satisfying $xe^{-x} = Ce$, both distinct from $x = 1$.

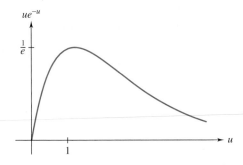

Figure 5.47 Graph of ue^{-u}

The phase plane sketch thus leads us to conclude that the critical point $(1, 1)$ is a center, the trajectories are closed orbits, and the populations vary periodically.

E. *Hamiltonian Systems*

The problems in this project explore the **Hamiltonian**[†] **formulation** of the laws of motion of a system, and its phase plane implications. This formulation replaces Newton's second law

[†]*Historical Footnote:* Sir William Rowan Hamilton (1805–1865) was an Irish mathematical physicist. Besides his work in mechanics, he invented quaternions and discovered the anticommutative law for vector products.

$F = ma = my''$ and is based on three mathematical manipulations:

 (i) It is presumed that the force $F(t, y, y')$ depends only on y and has an antiderivative $-V(y)$: $F = F(y) = -dV(y)/dy$.

 (ii) The velocity variable y' is replaced throughout by the momentum $p = my'$ (so $y' = p/m$).

 (iii) The **Hamiltonian** of the system is defined as

$$H = H(y, p) = \frac{p^2}{2m} + V(y) \ .$$

 (a) Express Newton's law $F = my''$ as an equivalent first order system in the manner prescribed in Section 5.2.

 (b) Show that this system is equivalent to **Hamilton's equations**

(8) $$\frac{dy}{dt} = \frac{\partial H}{\partial p} \quad \left(= \frac{p}{m} \right)$$

(9) $$\frac{dp}{dt} = -\frac{\partial H}{\partial y} \quad \left(= -\frac{dV}{dy} \right) \ .$$

 (c) Using Hamilton's equations and the chain rule, show that the Hamiltonian remains constant along the solution curves:

$$\frac{d}{dt} H(y, p) = 0 \ .$$

In the formula for the Hamiltonian function $H(y, p)$, the first term, $p^2/(2m) = m(y')^2/2$, is the **kinetic energy** of the mass. By analogy, then, the second term $V(y)$ is known as the **potential energy** of the mass, and the Hamiltonian is the total energy. The total (mechanical[†]) energy is constant—hence "conserved"—when the forces $F(y)$ do not depend on time t or velocity y'; such forces are called **conservative.** The energy integral lemma of Section 4.10 (page 220) is simply an alternate statement of the conservation of energy.

Hamilton's formulation for mechanical systems and the conservation of energy principle imply that the phase plane trajectories of conservative systems lie on the curves where the Hamiltonian $H(y, p)$ is constant, and plotting these curves may be considerably easier than solving for the trajectories directly (which, in turn, is easier than solving the original system!).

 (d) For the mass-spring oscillator of Section 4.1, the spring force is given by $F = -ky$ (where k is the spring constant). Find the Hamiltonian, express Hamilton's equations, and show that the phase plane trajectories $H(y, p) = $ constant for this system are the ellipses given by $p^2/(2m) + ky^2/2 = $ constant. See Figure 5.9, page 272.

The **damping** force $-by'$ considered in Section 4.1 is not conservative, of course. Physically speaking, we know that damping drains the energy from a system until it grinds to a halt at an equilibrium point. In the phase plane, we can qualitatively describe the trajectory as continuously migrating to successively lower constant-energy orbits; stable centers become asymptotically stable spiral points when damping is taken into consideration.

[†]Physics states that when all forms of energy, such as heat and radiation, are taken into account, energy is conserved even when the forces are not conservative.

(e) The second Hamiltonian equation (9), which effectively states $p' = my'' = F$, has to be changed to

$$p' = -\frac{\partial H}{\partial y} - by' = -\frac{\partial H}{\partial y} - \frac{bp}{m}$$

when damping is present. Show that the Hamiltonian decreases along trajectories in this case (for $b > 0$):

$$\frac{d}{dt}H(y, p) = -b\left(\frac{p}{m}\right)^2 = -b(y')^2 \; .$$

(f) The force on a mass-spring system suspended vertically in a gravitational field was shown in Section 4.12 (page 245) to be $F = -ky + mg$. Derive the Hamiltonian and sketch the phase plane trajectories. Sketch the trajectories when damping is present.

(g) As indicated in Section 4.10 (page 225), the Duffing spring force is modeled by $F = -y - y^3$. Derive the Hamiltonian and sketch the phase plane trajectories. Sketch the trajectories when damping is present.

(h) For the pendulum system studied in Section 4.10, Example 8, the force is given by (Figure 4.19, page 226)

$$F = -\ell mg \sin \theta = -\frac{\partial}{\partial \theta}(-\ell mg \cos \theta) = -\frac{\partial}{\partial \theta}V(\theta)$$

(where ℓ is the length of the pendulum). For angular variables, the Hamiltonian formulation dictates the replacement of the *angular velocity variable θ' by the angular momentum $p = m\ell^2\theta'$*; the kinetic energy, mass \times velocity2/2, is expressed as $m(\ell\theta')^2/2 = p^2/(2m\ell^2)$. Derive the Hamiltonian for the pendulum and sketch the phase plane trajectories. Sketch the trajectories when damping is present.

(i) The Coulomb force field is a force that varies as the reciprocal square of the distance from the origin: $F = k/y^2$. The force is *attractive* if $k < 0$ and *repulsive* if $k > 0$. Sketch the phase plane trajectories for this motion. Sketch the trajectories when damping is present.

(j) For an attractive Coulomb force field, what is the *escape velocity* for a particle situated at a position y? That is, what is the minimal (outward-directed) velocity required for the trajectory to reach $y = \infty$?

F. Limit Cycles

In the study of triode vacuum tubes, one encounters the van der Pol equation:[†]

$$y'' - \mu(1 - y^2)y' + y = 0 \; ,$$

where the constant μ is regarded as a parameter. In Section 4.10 (page 225), we used the mass-spring

[†]*Historical Footnote:* Experimental research by **E. V. Appleton** and **B. van der Pol** in 1921 on the oscillations of an electrical circuit containing a triode generator (vacuum tube) led to the nonlinear equation now called **van der Pol's equation.** Methods of solution were developed by van der Pol in 1926–1927. **Mary L. Cartwright** continued research into nonlinear oscillation theory and together with **J. E. Littlewood** obtained existence results for forced oscillations in nonlinear systems in 1945.

oscillator analogy to argue that the nonzero solutions to the van der Pol equation with $\mu = 1$ should approach a periodic limit cycle. The same argument applies for any positive value of μ.

(a) Recast the van der Pol equation as a system in normal form and use software to plot some typical trajectories for $\mu = 0.1$, 1, and 10. Rescale the plots if necessary until you can discern the limit cycle trajectory; find trajectories that spiral in, and ones that spiral out, to the limit cycle.

(b) Now let $\mu = -0.1$, -1, and -10. Try to predict the nature of the solutions using the mass-spring analogy. Then use the software to check your predictions. Are there limit cycles? Do the neighboring trajectories spiral into, or spiral out from, the limit cycles?

(c) Repeat parts (a) and (b) for the Rayleigh equation

$$y'' - \mu\big[1 - (y')^2\big]y' + y = 0 \ .$$

G. Strange Behavior of Competing Species—Part I

Let $p_i(t)$ denote, respectively, the populations of three competing species S_i, $i = 1, 2, 3$. Suppose that these species have the same growth rates and that the maximum population of the species the habitat can support is the same for each. (We assume it to be one unit.) Moreover, suppose the competitive advantage that S_1 has over S_2 is the same as that of S_2 over S_3 and S_3 over S_1. This situation is modeled by the system

$$
\begin{aligned}
p_1' &= p_1(1 - p_1 - ap_2 - bp_3) \ ,\\
p_2' &= p_2(1 - bp_1 - p_2 - ap_3) \ ,\\
p_3' &= p_3(1 - ap_1 - bp_2 - p_3) \ ,
\end{aligned}
$$

(10)

where a and b are positive constants.

(a) Show that system (10) has five equilibrium populations (solutions) corresponding to the critical points $(0, 0, 0)$, $(1, 0, 0)$, $(0, 1, 0)$, $(0, 0, 1)$, and (δ, δ, δ), where $\delta := (1 + a + b)^{-1}$.

(b) Study the system when $a = b = 0.5$. Since the system is *not* linear, we cannot apply the technique discussed in Section 5.2. The system has three unknowns, so the phase analysis is much more complex than that presented in Section 5.2. Instead, we perform numerical experiments on the system to see if we can determine the behavior of its solutions. Use a scheme such as the classical fourth order Runge-Kutta method for systems (see Appendix E) to perform these experiments. Use various initial values, in particular, values near the five equilibrium points. Describe the behavior of the populations as $t \to +\infty$.

(c) The system behaves quite differently when $a = 0.5$ and $b = 2.0$. Perform numerical experiments to show that the solution for this system approaches one of the equilibrium solutions—$(1, 0, 0)$, $(0, 1, 0)$, $(0, 0, 1)$—and then moves toward another, then toward the third, then back toward the first, and so on. Furthermore, the solutions linger longer and longer about these points before rapidly moving to the next one.

(d) Use the results of part (c) to argue that "most of the time,"

$$p_1p_2 \approx 0 \ , \qquad p_2p_3 \approx 0 \ , \qquad p_1p_3 \approx 0 \ .$$

Moreover, $p_1 + p_2 + p_3 \approx 1$.

(e) From system (10), derive the equations

$$\frac{dT}{dt} = T(1 - T) - \eta(p_1 p_2 + p_2 p_3 + p_1 p_3) \ ,$$

$$\frac{d \ln P}{dt} = -\eta + (3 + \eta)(1 - T) \ ,$$

where $T := p_1 + p_2 + p_3$, $P := p_1 p_2 p_3$, and $\eta := a + b - 2 = 0.5$.

(f) Using the results of parts (d) and (e), show that

$$\frac{dT}{dt} = T(1 - T) \ , \qquad \frac{d \ln P}{dt} = -\eta$$

are reasonable simplifications.

(g) Solve the equations in part (f) and verify that as $t \to +\infty$, the sum T approaches 1 and the product P approaches zero.

This system is studied in more detail in Project D in Chapter 9.

H. *Cleaning Up the Great Lakes*

A simple mathematical model that can be used to determine the time it would take to clean up the Great Lakes can be developed using a multiple compartmental analysis approach. In particular, we can view each lake as a tank that contains a liquid in which is dissolved a particular pollutant (DDT, phosphorus, mercury). Schematically, we view the lakes as consisting of five tanks connected as indicated in Figure 5.48.

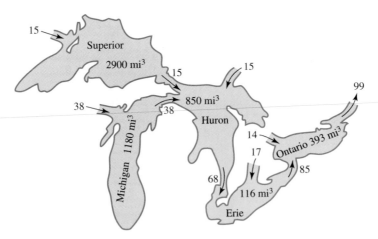

Figure 5.48 Compartmental model of the Great Lakes with flow rates (mi^3/year) and volumes (mi^3)

For our model, we make the following assumptions:

1. The volume of each lake remains constant.
2. The flow rates are constant throughout the year.

3. When a liquid enters the lake, perfect mixing occurs and the pollutants are uniformly distributed.

4. Pollutants are dissolved in the water and enter or leave by inflow or outflow of solution.

Before using this model to obtain estimates on the cleanup times for the lakes, we consider some simpler models:

(a) Use the outflow rates given in Figure 5.48 to determine the time it would take to "drain" each lake. This gives a lower bound on how long it would take to remove all the pollutants.

(b) A better estimate is obtained by assuming each lake is a separate tank with *only* clean water flowing in. Use this approach to determine how long it would take the pollution level in each lake to be reduced to 50% of its original level. How long would it take to reduce the pollution to 5% of its original level?

(c) Finally, to take into account the fact that pollution from one lake flows into the next lake in the chain, use the entire multiple compartment model given in Figure 5.48 to determine when the pollution level in each lake has been reduced to 50% of its original level, assuming pollution has ceased (that is, inflows not from a lake are clean water). Assume all the lakes initially have the same pollution concentration p. How long would it take for the pollution to be reduced to 5% of its original level?

For a detailed discussion of this model, see *An Introduction to Mathematical Modeling* by Edward A. Bender (Krieger, New York, 1991), Chapter 8.

6

Theory of Higher-Order Linear Differential Equations

In this chapter, we discuss the basic theory of linear higher-order differential equations. The material is a generalization of the results we obtained in Chapter 4 for second order equations. In the statements and proofs of these results, we use concepts usually covered in an elementary linear algebra course; namely, linear dependence, determinants, and methods for solving systems of linear equations. These concepts also arise in the matrix approach for solving systems of differential equations and are discussed in Chapter 9.

6.1 BASIC THEORY OF LINEAR DIFFERENTIAL EQUATIONS

Recall that a *linear* differential equation is an equation that can be written in the form

(1) $a_n(x)y^{(n)}(x) + a_{n-1}(x)y^{(n-1)}(x) + \cdots + a_0(x)y(x) = b(x)$,

where $a_0(x), a_1(x), \ldots, a_n(x)$ and $b(x)$ depend only on x, not y. When a_0, a_1, \ldots, a_n are constants, we say equation (1) has **constant coefficients;** otherwise it has **variable coefficients.** If $b(x) \equiv 0$, equation (1) is called **homogeneous;** otherwise it is **nonhomogeneous.**

In developing a basic theory, we assume $a_0(x), a_1(x), \ldots, a_n(x)$ and $b(x)$ are continuous on an interval I and $a_n(x) \neq 0$ on I. Then, on dividing by $a_n(x)$, we can rewrite (1) in the **standard form**

(2) $y^{(n)}(x) + p_1(x)y^{(n-1)}(x) + \cdots + p_n(x)y(x) = g(x)$,

where the functions $p_1(x), \ldots, p_n(x)$, and $g(x)$ are continuous on I.

For a linear higher-order differential equation, the initial value problem always has a unique solution.

EXISTENCE AND UNIQUENESS

Theorem 1. Suppose that $p_1(x), \ldots, p_n(x)$ and $g(x)$ are each continuous on an interval (a, b) that contains the point x_0. Then, for any choice of the initial values $\gamma_0, \gamma_1, \ldots, \gamma_{n-1}$, there exists a unique solution $y(x)$ on the whole interval (a, b) to the initial value problem

(3) $y^{(n)}(x) + p_1(x)y^{(n-1)}(x) + \cdots + p_n(x)y(x) = g(x)$,

(4) $y(x_0) = \gamma_0,\; y'(x_0) = \gamma_1,\; \ldots,\; y^{(n-1)}(x_0) = \gamma_{n-1}$.

The proof of Theorem 1 can be found in Chapter 13.[†]

EXAMPLE 1 For the initial value problem

(5) $x(x - 1)y''' - 3xy'' + 6x^2y' - (\cos x)y = \sqrt{x + 5}$;

(6) $y(x_0) = 1$, $y'(x_0) = 0$, $y''(x_0) = 7$,

determine the values of x_0 and the intervals (a, b) containing x_0 for which Theorem 1 guarantees the existence of a unique solution on (a, b).

SOLUTION Putting equation (5) in standard form, we find that $p_1(x) = -3/(x - 1)$, $p_2(x) = 6x/(x - 1)$, $p_3(x) = -(\cos x)/[x(x - 1)]$, and $g(x) = \sqrt{x + 5}/[x(x - 1)]$. Now $p_1(x)$ and $p_2(x)$ are continuous on every interval not containing $x = 1$, while $p_3(x)$ is continuous on every interval not containing $x = 0$ or $x = 1$. The function $g(x)$ is not defined for $x < -5$, $x = 0$, and $x = 1$, but is continuous on $(-5, 0)$, $(0, 1)$, and $(1, \infty)$. Hence the functions p_1, p_2, p_3, and g are *simultaneously* continuous on the intervals $(-5, 0)$, $(0, 1)$, and $(1, \infty)$. From Theorem 1, it follows that if we choose $x_0 \in (-5, 0)$, then there exists a unique solution to the initial value problem (5)–(6) on the whole interval $(-5, 0)$. Similarly, for $x_0 \in (0, 1)$, there is a unique solution on $(0, 1)$ and for $x_0 \in (1, \infty)$, a unique solution on $(1, \infty)$. ■ End

If we let the left-hand side of equation (3) define the differential operator L,

(7) $L[y] := \dfrac{d^n y}{dx^n} + p_1 \dfrac{d^{n-1}y}{dx^{n-1}} + \cdots + p_n y$,

then we can express equation (3) in the operator form

(8) $L[y](x) = g(x)$.

[†]All references to Chapters 11–13 refer to the expanded text *Fundamentals of Differential Equations and Boundary Value Problems,* 3rd edition.

It is essential to keep in mind that L is a *linear* operator; it satisfies

(9) $L[y_1 + y_2] = L[y_1] + L[y_2]$,

(10) $L[cy] = cL[y]$, c any constant

(see Problem 25).

As a consequence of this linearity, if y_1, \ldots, y_m are solutions to the *homogeneous* equation

(11) $L[y](x) = 0$,

then any linear combination of these functions, $C_1 y_1 + \cdots + C_m y_m$, is also a solution. Imagine now that we have found n solutions y_1, \ldots, y_n to the nth order linear equation (11). Is it true that *every* solution to (11) can be represented by

(12) $C_1 y_1 + \cdots + C_n y_n$

for appropriate choices of the constants C_1, \ldots, C_n? The answer is yes, provided the solutions y_1, \ldots, y_n satisfy a certain property of independence that we shall now derive.

Let $\phi(x)$ be a solution to (11) on the interval (a, b) and let x_0 be a fixed number in (a, b). If it is possible to choose the constants C_1, \ldots, C_n so that

(13)
$$
\begin{aligned}
C_1 y_1(x_0) \quad + \cdots + C_n y_n(x_0) &= \phi(x_0) , \\
C_1 y_1'(x_0) \quad + \cdots + C_n y_n'(x_0) &= \phi'(x_0) , \\
\vdots \qquad\qquad\qquad \vdots \qquad\qquad &\quad \vdots \\
C_1 y_1^{(n-1)}(x_0) + \cdots + C_n y_n^{(n-1)}(x_0) &= \phi^{(n-1)}(x_0) ,
\end{aligned}
$$

then, since $\phi(x)$ and $C_1 y_1(x) + \cdots + C_n y_n(x)$ are two solutions satisfying the same initial conditions at x_0, the uniqueness conclusion of Theorem 1 gives

(14) $\phi(x) = C_1 y_1(x) + \cdots + C_n y_n(x)$

for all x in (a, b).

The system (13) consists of n linear equations in the n unknowns C_1, \ldots, C_n. It has a unique solution for all possible values of $\phi(x_0), \phi'(x_0), \ldots, \phi^{(n-1)}(x_0)$ if and only if the determinant[†] of the coefficients is different from zero; that is, if and only if

(15)
$$
\begin{vmatrix}
y_1(x_0) & y_2(x_0) & \cdots & y_n(x_0) \\
y_1'(x_0) & y_2'(x_0) & \cdots & y_n'(x_0) \\
\vdots & \vdots & & \vdots \\
y_1^{(n-1)}(x_0) & y_2^{(n-1)}(x_0) & \cdots & y_n^{(n-1)}(x_0)
\end{vmatrix} \neq 0 .
$$

Hence, if y_1, \ldots, y_n are solutions to equation (11) and there is some point x_0 in (a, b) such that (15) holds, then every solution $\phi(x)$ to (11) is a linear combination of y_1, \ldots, y_n.

Before putting this observation in the form of a theorem, it is convenient for us to extend the Wronskian concept to a set of n functions.

[†]Determinants are discussed in Section 9.3.

WRONSKIAN

Definition 1. Let f_1, \ldots, f_n be any n functions that are $(n-1)$ times differentiable. The function

$$
(16) \qquad W[f_1, \ldots, f_n](x) := \begin{vmatrix} f_1(x) & f_2(x) & \cdots & f_n(x) \\ f_1'(x) & f_2'(x) & \cdots & f_n'(x) \\ \vdots & \vdots & & \vdots \\ f_1^{(n-1)}(x) & f_2^{(n-1)}(x) & \cdots & f_n^{(n-1)}(x) \end{vmatrix}
$$

is called the **Wronskian** of f_1, \ldots, f_n.

We now state the representation theorem that we proved above for solutions to homogeneous linear differential equations.

REPRESENTATION OF SOLUTIONS (HOMOGENEOUS CASE)

Theorem 2. Let y_1, \ldots, y_n be n solutions on (a, b) of

$$
(17) \qquad y^{(n)}(x) + p_1(x)y^{(n-1)}(x) + \cdots + p_n(x)y(x) = 0 \ ,
$$

where p_1, \ldots, p_n are continuous on (a, b). If at some point x_0 in (a, b) these solutions satisfy

$$
(18) \qquad W[y_1, \ldots, y_n](x_0) \neq 0 \ ,
$$

then every solution of (17) on (a, b) can be expressed in the form

$$
(19) \qquad y(x) = C_1 y_1(x) + \cdots + C_n y_n(x) \ ,
$$

where C_1, \ldots, C_n are constants.

A set of solutions $\{y_1, \ldots, y_n\}$ satisfying (18) for some x_0 in (a, b) is called a **fundamental solution set** for (17) on (a, b). (The *existence* of fundamental solution sets is easy to prove; see Problem 26.) The linear combination of y_1, \ldots, y_n in (19), written with arbitrary constants C_1, \ldots, C_n, is referred to as a **general solution** to (17).

EXAMPLE 2 Given that $y_1(x) = x$, $y_2(x) = x^2$, and $y_3(x) = x^{-1}$ are solutions to

$$
(20) \qquad x^3 y''' + x^2 y'' - 2xy' + 2y = 0 \ , \qquad x > 0 \ ,
$$

find a general solution.

SOLUTION We first show that $\{y_1, y_2, y_3\}$ is a fundamental solution set for equation (20) on $(0, \infty)$. Since we

are told that y_1, y_2, and y_3 satisfy (20) (the reader can easily verify this), we need only consider

$$W[y_1, y_2, y_3](x) = \begin{vmatrix} y_1(x) & y_2(x) & y_3(x) \\ y_1'(x) & y_2'(x) & y_3'(x) \\ y_1''(x) & y_2''(x) & y_3''(x) \end{vmatrix} = \begin{vmatrix} x & x^2 & x^{-1} \\ 1 & 2x & -x^{-2} \\ 0 & 2 & 2x^{-3} \end{vmatrix}.$$

Evaluating the determinant, we find after a little algebra that

$$W[y_1, y_2, y_3](x) = 6x^{-1},$$

which is not zero for $x > 0$. Thus $\{y_1, y_2, y_3\}$ is a fundamental solution set, and hence a general solution is

(21) $y(x) = C_1 x + C_2 x^2 + C_3 x^{-1}, \qquad x > 0.$ ■End

The important condition (18) concerning the nonvanishing of the Wronskian can also be described in terms of the linear independence of the solutions y_1, \ldots, y_n (recall Definition 3, Section 4.3, page 172).

LINEAR DEPENDENCE OF FUNCTIONS

Definition 2. The m functions f_1, \ldots, f_m are said to be **linearly dependent on an interval I** if there exist constants c_1, \ldots, c_m, not all zero, such that

(22) $c_1 f_1(x) + \cdots + c_m f_m(x) = 0$

for all x in I. If the functions f_1, \ldots, f_m are not linearly dependent on I, they are said to be **linearly independent on I.**

Remark. Linear dependence of f_1, \ldots, f_m on I means that at least one of these functions is expressible as a linear combination of the remaining functions (on I).

EXAMPLE 3 Show that the functions $f_1(x) = e^x$, $f_2(x) = e^{-2x}$, and $f_3(x) = 3e^x - 2e^{-2x}$ are linearly dependent on $(-\infty, \infty)$.

SOLUTION Notice that f_3 is a linear combination of f_1 and f_2:

$$f_3(x) = 3e^x - 2e^{-2x} = 3f_1(x) - 2f_2(x).$$

Therefore we have

(23) $3f_1(x) - 2f_2(x) - f_3(x) = 0$

for all x in $(-\infty, \infty)$. Consequently, f_1, f_2, and f_3 are linearly dependent on $(-\infty, \infty)$. ■End

To prove that functions f_1, f_2, \ldots, f_m are linearly *independent* on (a, b), a convenient approach is the following: *Assume* equation (22) holds on (a, b) and show that this forces $c_1 = c_2 = \cdots = c_m = 0$.

EXAMPLE 4 Show that the functions $f_1(x) = x$, $f_2(x) = x^2$, and $f_3(x) = 1 - 2x^2$ are linearly independent on $(-\infty, \infty)$.

SOLUTION Assume c_1, c_2, and c_3 are constants for which

(24) $c_1 x + c_2 x^2 + c_3(1 - 2x^2) = 0$

holds at every x. If we can prove that (24) implies $c_1 = c_2 = c_3 = 0$, then linear independence follows. Let's set $x = 0$, 1, and -1 in equation (24). These x values are, essentially, "picked out of a hat" but will get the job done. Substituting in (24) gives

$$
\begin{aligned}
c_3 &= 0 & (x = 0)\ , \\
\textbf{(25)} \quad c_1 + c_2 - c_3 &= 0 & (x = 1)\ , \\
-c_1 + c_2 - c_3 &= 0 & (x = -1)\ .
\end{aligned}
$$

When we solve this system (or compute the determinant of the coefficients), we find that the only possible solution is $c_1 = c_2 = c_3 = 0$. Consequently, the functions f_1, f_2, and f_3 are linearly independent on $(-\infty, \infty)$. ∎ End

The connection between linear independence and fundamental solution sets is stated in the next theorem.

FUNDAMENTAL SETS, LINEAR INDEPENDENCE, AND THE WRONSKIAN

Theorem 3. If y_1, \ldots, y_n are n solutions to $y^{(n)} + p_1 y^{(n-1)} + \cdots + p_n y = 0$ on the interval (a, b), with p_1, \ldots, p_n continuous on (a, b), then the following statements are equivalent:

(i) $\{y_1, \ldots, y_n\}$ is a fundamental solution set on (a, b).
(ii) y_1, \ldots, y_n are linearly independent on (a, b).
(iii) The Wronskian $W[y_1, \ldots, y_n](x)$ is never zero on (a, b).

Theorem 3 is a generalization of Corollary 1 in Section 4.3. Its proof is similar, so we leave it as an exercise (see Problem 30).

As with linear second order equations, *the Wronskian of solutions is either identically zero or never zero on (a, b)*. This fact follows from Theorem 3 or from **Abel's identity**

(26) $W[y_1, \ldots, y_n](x) = W[y_1, \ldots, y_n](x_0) \exp\left(-\int_{x_0}^{x} p_1(t)\,dt\right)\ ,$

which holds for any n solutions y_1, \ldots, y_n to $y^{(n)} + p_1 y^{(n-1)} + \cdots + p_n y = 0$ on (a, b). In (26), the point x_0 can be anywhere in (a, b). For a proof of Abel's identity when $n = 3$, see Problem 36.

Notice that Theorem 3 provides a simpler approach to proving that the solutions $y_1(x) = x$, $y_2(x) = x^2$, $y_3(x) = x^{-1}$ in Example 2 form a fundamental solution set. For, without having to compute a determinant, we can see that no one of these functions is a linear combination of the remaining two (e.g., by comparing their growth rates at infinity). So these functions are linearly independent on $(0, \infty)$.

It is useful to keep in mind that the following sets consist of functions that are linearly

independent on every open interval (a, b):

$$\{1, x, x^2, \ldots, x^n,\}$$
$$\{1, \cos x, \sin x, \cos 2x, \sin 2x, \ldots, \cos nx, \sin nx\} \ ,$$
$$\{e^{\alpha_1 x}, e^{\alpha_2 x}, \ldots, e^{\alpha_n x}\} \ , \qquad \alpha_i\text{'s distinct constants.}$$

(See Problems 27 and 28, and Section 6.2 (page 348).)

Recall that the **superposition principle** (Section 4.7, page 200) for linear differential operators states that if $L[y_1] = g_1$ and $L[y_2] = g_2$, then the linear combination $C_1 y_1 + C_2 y_2$ satisfies

$$L[C_1 y_1 + C_2 y_2] = C_1 g_1 + C_2 g_2 \ .$$

If we combine this principle with the representation theorem for solutions of the homogeneous equation, we obtain the following representation theorem for nonhomogeneous equations.

REPRESENTATION OF SOLUTIONS (NONHOMOGENEOUS CASE)

Theorem 4. Let $y_p(x)$ be a particular solution to the nonhomogeneous equation

(27) $y^{(n)}(x) + p_1(x)y^{(n-1)}(x) + \cdots + p_n(x)y(x) = g(x)$

on the interval (a, b) with p_1, p_2, \ldots, p_n continuous on (a, b), and let $\{y_1, \ldots, y_n\}$ be a fundamental solution set for the corresponding homogeneous equation

(28) $y^{(n)}(x) + p_1(x)y^{(n-1)}(x) + \cdots + p_n(x)y(x) = 0 \ .$

Then every solution of (27) on the interval (a, b) can be expressed in the form

(29) $y(x) = y_p(x) + C_1 y_1(x) + \cdots + C_n y_n(x) \ .$

Proof. Let $\phi(x)$ be any solution to (27). Because both $\phi(x)$ and $y_p(x)$ are solutions to (27), the superposition principle states that the difference $\phi(x) - y_p(x)$ is a solution to the homogeneous equation (28). It then follows from Theorem 2 that

$$\phi(x) - y_p(x) = C_1 y_1(x) + \cdots + C_n y_n(x)$$

for suitable constants C_1, \ldots, C_n. The last equation is equivalent to (29) $\left(\text{with } \phi(x) \text{ in place of } y(x)\right)$, so the theorem is proved. ■ End

The linear combination of y_p, y_1, \ldots, y_n in (29) written with arbitrary constants C_1, \ldots, C_n is, for obvious reasons, referred to as a **general solution** to (27). Theorem 4 can be easily generalized. For example, if L denotes the operator appearing as the left-hand side in equation (27), and if $L[y_{p1}] = g_1(x)$ and $L[y_{p2}] = g_2(x)$, then any solution of $L[y] = c_1 g_1 + c_2 g_2$ can be expressed as

$$y(x) = c_1 y_{p1}(x) + c_2 y_{p2}(x) + C_1 y_1(x) + C_2 y_2(x) + \cdots + C_n y_n(x) \ ,$$

for a suitable choice of the constants C_1, C_2, \ldots, C_n.

EXAMPLE 5 Find a general solution on the interval $(-\infty, \infty)$ to

(30) $L[y] := y''' - 2y'' - y' + 2y = 2x^2 - 2x - 4 - 24e^{-2x}$,

given that $y_{p1}(x) = x^2$ is a particular solution to $L[y] = 2x^2 - 2x - 4$, that $y_{p2}(x) = e^{-2x}$ is a particular solution to $L[y] = -12e^{-2x}$, and that $y_1(x) = e^{-x}$, $y_2(x) = e^x$, and $y_3(x) = e^{2x}$ are solutions to the corresponding homogeneous equation.

SOLUTION We previously remarked that the functions e^{-x}, e^x, e^{2x} are linearly independent because the exponents $-1, 1$, and 2 are distinct. Since each of these functions is a solution to the corresponding homogeneous equation, then $\{e^{-x}, e^x, e^{2x}\}$ is a fundamental solution set. It now follows from the remarks above for nonhomogeneous equations that a general solution to (30) is

(31) $y(x) = y_{p1} + 2y_{p2} + C_1 y_1 + C_2 y_2 + C_3 y_3$
$= x^2 + 2e^{-2x} + C_1 e^{-x} + C_2 e^x + C_3 e^{2x}$. ■ End

EXERCISES 6.1

In Problems 1 through 6, determine the largest interval (a, b) for which Theorem 1 guarantees the existence of a unique solution on (a, b) to the given initial value problem.

1. $xy''' - 3y' + e^x y = x^2 - 1$;
$y(-2) = 1$, $y'(-2) = 0$, $y''(-2) = 2$.

2. $y''' - \sqrt{x}\,y = \sin x$;
$y(\pi) = 0$, $y'(\pi) = 11$, $y''(\pi) = 3$.

3. $y''' - y'' + \sqrt{x - 1}\,y = \tan x$;
$y(5) = y'(5) = y''(5) = 1$.

4. $x(x + 1)y''' - 3xy' + y = 0$;
$y(-1/2) = 1$, $y'(-1/2) = y''(-1/2) = 0$.

5. $x\sqrt{x + 1}\,y''' - y' + xy = 0$;
$y(1/2) = y'(1/2) = -1$, $y''(1/2) = 1$.

6. $(x^2 - 1)y''' + e^x y = \ln x$;
$y(3/4) = 1$, $y'(3/4) = y''(3/4) = 0$.

In Problems 7 through 14, determine whether the given functions are linearly dependent or linearly independent on the specified interval. Justify your decisions.

7. $\{e^{3x}, e^{5x}, e^{-x}\}$ on $(-\infty, \infty)$.

8. $\{x^2, x^2 - 1, 5\}$ on $(-\infty, \infty)$.

9. $\{\sin^2 x, \cos^2 x, 1\}$ on $(-\infty, \infty)$.

10. $\{\sin x, \cos x, \tan x\}$ on $(-\pi/2, \pi/2)$.

11. $\{x^{-1}, x^{1/2}, x\}$ on $(0, \infty)$.

12. $\{\cos 2x, \cos^2 x, \sin^2 x\}$ on $(-\infty, \infty)$.

13. $\{x, x^2, x^3, x^4\}$ on $(-\infty, \infty)$.

14. $\{x, xe^x, 1\}$ on $(-\infty, \infty)$.

Using the Wronskian, in Problems 15 through 18 verify that the given functions form a fundamental solution set for the given differential equation, and find a general solution.

15. $y''' + 2y'' - 11y' - 12y = 0$;
$\{e^{3x}, e^{-x}, e^{-4x}\}$.

16. $y''' - y'' + 4y' - 4y = 0$;
$\{e^x, \cos 2x, \sin 2x\}$.

17. $x^3 y''' - 3x^2 y'' + 6xy' - 6y = 0$, $x > 0$;
$\{x, x^2, x^3\}$.

18. $y^{(4)} - y = 0$; $\{e^x, e^{-x}, \cos x, \sin x\}$.

In Problems 19 through 22, a particular solution and a fundamental solution set are given for a nonhomogeneous equation and its corresponding homogeneous equation. (a) Find a general solution to the nonhomogeneous equation. (b) Find the solution that satisfies the specified initial conditions.

19. $y''' + y'' + 3y' - 5y = 2 + 6x - 5x^2$;
$y(0) = -1$, $y'(0) = 1$, $y''(0) = -3$;
$y_p = x^2$; $\{e^x, e^{-x}\cos 2x, e^{-x}\sin 2x\}$.

20. $xy''' - y'' = -2$; $y(1) = 2$, $y'(1) = -1$,
$y''(1) = -4$; $y_p = x^2$; $\{1, x, x^3\}$.

21. $x^3y''' + xy' - y = 3 - \ln x$, $x > 0$;
$y(1) = 3$, $y'(1) = 3$, $y''(1) = 0$;
$y_p = \ln x$; $\{x, x \ln x, x(\ln x)^2\}$.

22. $y^{(4)} + 4y = 5 \cos x$;
$y(0) = 2$, $y'(0) = 1$, $y''(0) = -1$,
$y'''(0) = -2$; $y_p = \cos x$;
$\{e^x\cos x, e^x\sin x, e^{-x}\cos x, e^{-x}\sin x\}$.

23. Let $L[y] := y''' + y' + xy$, $y_1(x) := \sin x$ and
$y_2(x) := x$. Verify that $L[y_1](x) = x \sin x$ and
$L[y_2](x) = x^2 + 1$. Then use the superposition principle to find a solution to the differential equation:
(a) $L[y] = 2x \sin x - x^2 - 1$.
(b) $L[y] = 4x^2 + 4 - 6x \sin x$.

24. Let $L[y] := y''' - xy'' + 4y' - 3xy$, $y_1(x) := \cos 2x$,
and $y_2(x) := -1/3$. Verify that $L[y_1](x) = x \cos 2x$
and $L[y_2](x) = x$. Then use the superposition principle to find a solution to the differential equation:
(a) $L[y] = 7x \cos 2x - 3x$.
(b) $L[y] = -6x \cos 2x + 11x$.

25. Prove that L defined in (7) is a linear operator by verifying that properties (9) and (10) hold for any n-times differentiable functions y, y_1, y_2 on (a, b).

26. Existence of Fundamental Solution Sets. By Theorem 1, for each $j = 1, 2, \ldots, n$ there is a unique solution $y_j(x)$ to equation (17) satisfying the initial conditions

$$y_j^{(k)}(x_0) = \begin{cases} 1 , & \text{for } k = j - 1 , \\ 0 , & \text{for } k \neq j - 1, \quad 0 \leq k \leq n - 1 . \end{cases}$$

(a) Show that $\{y_1, y_2, \ldots, y_n\}$ is a fundamental solution set for (17). [*Hint:* Write out the Wronskian at x_0.]
(b) For given initial values $\gamma_0, \gamma_1, \ldots, \gamma_{n-1}$, express the solution $y(x)$ to (17) satisfying $y^{(k)}(x_0) = \gamma_k$, $k = 0, \ldots, n - 1$, (as in equations (4)) in terms of this fundamental solution set.

27. Show that the set of functions $\{1, x, x^2, \ldots, x^n\}$, where n is a positive integer, is linearly independent on every open interval (a, b). [*Hint:* Use the fact that

a polynomial of degree at most n has no more than n zeros unless it is identically zero.]

28. The set of functions

$$\{1, \cos x, \sin x, \ldots, \cos nx, \sin nx\},$$

where n is a positive integer, is linearly independent on every interval (a, b). Prove this in the special case $n = 2$ and $(a, b) = (-\infty, \infty)$.

29. (a) Show that if f_1, \ldots, f_m are linearly independent on $(-1, 1)$, then they are linearly independent on $(-\infty, \infty)$.
(b) Give an example to show that if f_1, \ldots, f_m are linearly independent on $(-\infty, \infty)$, then they need not be linearly independent on $(-1, 1)$.

30. To prove Theorem 3, proceed as follows:
(a) Let $x_0 \in (a, b)$. Prove that the solutions y_1, \ldots, y_n are linearly independent on (a, b) if and only if the initial vectors $\left(y_k(x_0), y_k'(x_0), \ldots, y_k^{(n-1)}(x_0)\right)$, $k = 1, 2, \ldots, n$, are linearly independent.
(b) To complete the proof, use the result of part (a) and the fact that *the column vectors of a matrix are linearly independent if and only if the determinant of the matrix is not zero.*

31. Reduction of Order. If a nontrivial solution $f(x)$ is known for the *homogeneous* equation

$$y^{(n)} + p_1(x)y^{(n-1)} + \cdots + p_n(x)y = 0 ,$$

the substitution $y(x) = v(x)f(x)$ can be used to reduce the order of the equation. (See Section 4.4 for the case of second order linear equations.) By completing the following steps, demonstrate the method for the third order equation

(32) $y''' - 2y'' - 5y' + 6y = 0$,

given that $f(x) = e^x$ is a solution.
(a) Set $y(x) = v(x)e^x$ and compute y', y'', and y'''.
(b) Substitute your expressions from (a) into (32) to obtain a *second order* equation in $w = v'$.
(c) Solve the second order equation in part (b) for w and integrate to find v. Determine two linearly independent choices for v, say v_1 and v_2.
(d) By part (c), the functions $y_1(x) = v_1(x)e^x$ and $y_2(x) = v_2(x)e^x$ are two solutions to (32). Verify that the three solutions e^x, $y_1(x)$, and $y_2(x)$ are linearly independent on $(-\infty, \infty)$.

32. Given that the function $f(x) = x$ is a solution to $y''' - x^2 y' + xy = 0$, show that the substitution $y(x) = v(x)f(x) = v(x)x$ reduces this equation to $xw'' + 3w' - x^3 w = 0$, where $w = v'$.

33. Use the reduction of order method described in Problem 31 to find three linearly independent solutions to $y''' - 2y'' + y' - 2y = 0$, given that $f(x) = e^{2x}$ is a solution.

34. Constructing Differential Equations. Given three functions $f_1(x)$, $f_2(x)$, $f_3(x)$ that are each three times differentiable and whose Wronskian is never zero on $[a, b]$, show that the equation

$$\begin{vmatrix} f_1(x) & f_2(x) & f_3(x) & y \\ f_1'(x) & f_2'(x) & f_3'(x) & y' \\ f_1''(x) & f_2''(x) & f_3''(x) & y'' \\ f_1'''(x) & f_2'''(x) & f_3'''(x) & y''' \end{vmatrix} = 0$$

is a third order linear differential equation for which $\{f_1, f_2, f_3\}$ is a fundamental solution set. What is the coefficient of y''' in this equation?

35. Use the result of Problem 34 to construct a third order differential equation for which $\{x, \sin x, \cos x\}$ is a fundamental solution set.

36. To prove Abel's identity for $n = 3$, proceed as follows:

(a) Let $W(x) := W[y_1, y_2, y_3](x)$. Use the product rule for differentiation to show

$$W'(x) = \begin{vmatrix} y_1' & y_2' & y_3' \\ y_1' & y_2' & y_3' \\ y_1'' & y_2'' & y_3'' \end{vmatrix} + \begin{vmatrix} y_1 & y_2 & y_3 \\ y_1'' & y_2'' & y_3'' \\ y_1'' & y_2'' & y_3'' \end{vmatrix}$$

$$+ \begin{vmatrix} y_1 & y_2 & y_3 \\ y_1' & y_2' & y_3' \\ y_1''' & y_2''' & y_3''' \end{vmatrix} .$$

(b) Show that the above expression reduces to

$$(33) \qquad W'(x) = \begin{vmatrix} y_1 & y_2 & y_3 \\ y_1' & y_2' & y_3' \\ y_1''' & y_2''' & y_3''' \end{vmatrix} .$$

(c) Since each y_i satisfies (17), show that

$$(34) \qquad y_i^{(3)}(x) = -\sum_{k=1}^{3} p_k(x) y_i^{(3-k)}(x) ,$$

$$i = 1, 2, 3 .$$

(d) Substituting the expressions in (34) into (33), show that

$$(35) \qquad W'(x) = -p_1(x) W(x) .$$

(e) Deduce Abel's identity by solving the first order differential equation (35).

6.2 HOMOGENEOUS LINEAR EQUATIONS WITH CONSTANT COEFFICIENTS

Our goal in this section is to obtain a general solution to an nth order linear differential equation with constant coefficients. Based on the experience gained with second order equations in Section 4.5, the reader should have little trouble guessing the form of such a solution. However, our interest here is to help the reader understand *why* these techniques work. This is done using an operator approach—a technique that is useful in tackling many other problems in analysis such as solving partial differential equations.

Let's consider the homogeneous linear nth order differential equation

$$(1) \qquad a_n y^{(n)}(x) + a_{n-1} y^{(n-1)}(x) + \cdots + a_1 y'(x) + a_0 y(x) = 0 ,$$

where $a_n (\neq 0), a_{n-1}, \ldots, a_0$ are real constants.[†] Since constant functions are everywhere continuous, equation (1) has solutions defined for all x in $(-\infty, \infty)$ (recall Theorem 1 in

[†]*Historical Footnote:* In a letter to John Bernoulli dated September 15, 1739, Leonhard Euler claimed to have solved the general case of the homogeneous linear nth order equation with constant coefficients.

Section 6.1). If we can find n linearly independent solutions to (1) on $(-\infty, \infty)$, say y_1, \ldots, y_n, then we can express a general solution to (1) in the form

(2) $y(x) = C_1 y_1(x) + \cdots + C_n y_n(x)$,

with C_1, \ldots, C_n as arbitrary constants.

To find these n linearly independent solutions, we capitalize on our previous success with second order equations. Namely, experience suggests that we begin by trying a function of the form $y = e^{rx}$.

If we let L be the differential operator defined by the left-hand side of (1), that is,

(3) $L[y] := a_n y^{(n)} + a_{n-1} y^{(n-1)} + \cdots + a_1 y' + a_0 y$,

then we can write (1) in the operator form

(4) $L[y](x) = 0$.

For $y = e^{rx}$, we find

(5) $L[e^{rx}](x) = a_n r^n e^{rx} + a_{n-1} r^{n-1} e^{rx} + \cdots + a_0 e^{rx}$
$= e^{rx}(a_n r^n + a_{n-1} r^{n-1} + \cdots + a_0) = e^{rx} P(r)$,

where $P(r)$ is the polynomial $a_n r^n + a_{n-1} r^{n-1} + \cdots + a_0$. Thus e^{rx} is a solution to equation (4), provided r is a root of the **auxiliary** (or **characteristic**) **equation**

(6) $P(r) = a_n r^n + a_{n-1} r^{n-1} + \cdots + a_0 = 0$.

According to the fundamental theorem of algebra, the auxiliary equation has n roots (counting multiplicities), which may be either real or complex. As we mentioned in Chapter 4, there are no formulas for determining the zeros of an arbitrary polynomial of degree greater than four. However, if we can determine one zero r_1, then we can divide out the factor $r - r_1$ and be left with a polynomial of lower degree. (For convenience, we have chosen most of our examples and exercises so that 0, ± 1, or ± 2 are zeros of any polynomial of degree greater than two that we must factor.) When a zero cannot be exactly determined, numerical algorithms such as Newton's method or the quotient-difference algorithm can be used to compute approximate roots of the polynomial equation.[†] Some pocket calculators even have these algorithms built in.

We proceed to discuss the various possibilities.

Distinct Real Roots

If the roots r_1, \ldots, r_n of the auxiliary equation (6) are real and distinct, then n solutions to equation (1) are

(7) $y_1(x) = e^{r_1 x}, \ y_2(x) = e^{r_2 x}, \ \ldots, \ y_n(x) = e^{r_n x}$.

As stated in the previous section, these functions are linearly independent on $(-\infty, \infty)$, a fact

[†]See, for example, *Applied and Computational Complex Analysis,* by P. Henrici (Wiley-Interscience, New York, 1974), Volume 1, or *Numerical Analysis,* 6th edition, by R. L. Burden and J. D. Faires (Prindle, Weber & Schmidt, Boston, 1997).

that we shall now officially verify. Let's assume that c_1, \ldots, c_n are constants such that

(8) $c_1 e^{r_1 x} + \cdots + c_n e^{r_n x} = 0$

for all x in $(-\infty, \infty)$. Our goal is to prove that $c_1 = c_2 = \cdots = c_n = 0$.

One way to show this is to construct a linear operator L_k that annihilates (maps to zero) everything on the left-hand side of (8) except the kth term. For this purpose, we note that since r_1, \ldots, r_n are the zeros of the auxiliary polynomial $P(r)$, then $P(r)$ can be factored as

(9) $P(r) = a_n(r - r_1) \cdots (r - r_n)$.

Consequently, the operator $L[y] = a_n y^{(n)} + a_{n-1} y^{(n-1)} + \cdots + a_0 y$ can be expressed as the following composition:[†]

(10) $L = P(D) = a_n(D - r_1) \cdots (D - r_n)$.

We now construct the polynomial $P_k(r)$ by deleting the factor $(r - r_k)$ from $P(r)$. Then we set $L_k := P_k(D)$; that is,

(11) $L_k := P_k(D) = a_n(D - r_1) \cdots (D - r_{k-1})(D - r_{k+1}) \cdots (D - r_n)$.

Applying L_k to both sides of (8), we get, via linearity,

(12) $c_1 L_k[e^{r_1 x}] + \cdots + c_n L_k[e^{r_n x}] = 0$.

Also, since $L_k = P_k(D)$, we find (just as in equation (5)) that $L_k[e^{rx}](x) = e^{rx} P_k(r)$ for all r. Thus (12) can be written as

$$c_1 e^{r_1 x} P_k(r_1) + \cdots + c_n e^{r_n x} P_k(r_n) = 0 ,$$

which simplifies to

(13) $c_k e^{r_k x} P_k(r_k) = 0$,

because $P_k(r_i) = 0$ for $i \neq k$. Since r_k is not a root of $P_k(r)$, then $P_k(r_k) \neq 0$. It now follows from (13) that $c_k = 0$. But as k is arbitrary, all the constants c_1, \ldots, c_n must be zero. Thus $y_1(x), \ldots, y_n(x)$ as given in (7) are linearly independent. (See Problem 26 for an alternative proof.)

We have proved that, in the case of n distinct real roots, a general solution to (1) is

(14) $y(x) = C_1 e^{r_1 x} + \cdots + C_n e^{r_n x}$,

where C_1, \ldots, C_n are arbitrary constants.

EXAMPLE 1 Find a general solution to

(15) $y''' - 2y'' - 5y' + 6y = 0$.

SOLUTION The auxiliary equation is

(16) $r^3 - 2r^2 - 5r + 6 = 0$.

[†]*Historical Footnote:* The symbolic notation $P(D)$ was introduced by Augustin Cauchy in 1827.

By inspection, we find that $r = 1$ is a root. Then, using polynomial division, we get

$$r^3 - 2r^2 - 5r + 6 = (r - 1)(r^2 - r - 6) ,$$

which further factors into $(r - 1)(r + 2)(r - 3)$. Hence the roots of equation (16) are $r_1 = 1$, $r_2 = -2$, $r_3 = 3$. Since these roots are real and distinct, a general solution to (15) is

$$y(x) = C_1 e^x + C_2 e^{-2x} + C_3 e^{3x} . \quad \blacksquare \text{ End}$$

Complex Roots

If $\alpha + i\beta$ (α, β real) is a complex root of the auxiliary equation (6), then so is its complex conjugate $\alpha - i\beta$, since the coefficients of $P(r)$ are real-valued (see Problem 24). If we accept complex-valued functions as solutions, then both $e^{(\alpha+i\beta)x}$ and $e^{(\alpha-i\beta)x}$ are solutions to (1). Moreover, if there are no repeated roots, then a general solution to (1) is again given by (14). To find two real-valued solutions corresponding to the roots $\alpha \pm i\beta$, we can just take the real and imaginary parts of $e^{(\alpha+i\beta)x}$. That is, since

(17) $\quad e^{(\alpha+i\beta)x} = e^{\alpha x}\cos \beta x + ie^{\alpha x}\sin \beta x ,$

then two linearly independent solutions to (1) are

(18) $\quad e^{\alpha x}\cos \beta x , \qquad e^{\alpha x}\sin \beta x .$

In fact, using these solutions in place of $e^{(\alpha+i\beta)x}$ and $e^{(\alpha-i\beta)x}$ in (14) preserves the linear independence of the set of n solutions. Thus, treating each of the conjugate pairs of roots in this manner, we obtain a real-valued general solution to (1).

EXAMPLE 2 Find a general solution to

(19) $\quad y''' + y'' + 3y' - 5y = 0 .$

SOLUTION The auxiliary equation is

(20) $\quad r^3 + r^2 + 3r - 5 = (r - 1)(r^2 + 2r + 5) = 0 ,$

which has distinct roots $r_1 = 1$, $r_2 = -1 + 2i$, $r_3 = -1 - 2i$. Thus a general solution is

(21) $\quad y(x) = C_1 e^x + C_2 e^{-x}\cos 2x + C_3 e^{-x}\sin 2x . \quad \blacksquare \text{ End}$

Repeated Roots

If r_1 is a root of multiplicity m, then the n solutions given in (7) are not even distinct, let alone linearly independent. Recall that for a second order equation, when we had a repeated root r_1 to the auxiliary equation, we obtained two linearly independent solutions by taking $e^{r_1 x}$ and $xe^{r_1 x}$. So if r_1 is a root of (6) of multiplicity m, we might expect that m linearly independent solutions are

(22) $\quad e^{r_1 x} , \quad xe^{r_1 x} , \quad x^2 e^{r_1 x} , \quad \ldots , \quad x^{m-1} e^{r_1 x} .$

To see that this is the case, observe that if r_1 is a root of multiplicity m, then the auxiliary equa-

tion can be written in the form

(23) $a_n(r - r_1)^m(r - r_{m+1}) \cdots (r - r_n) = (r - r_1)^m \tilde{P}(r) = 0$,

where $\tilde{P}(r) := a_n(r - r_{m+1}) \cdots (r - r_n)$ and $\tilde{P}(r_1) \neq 0$. With this notation, we have the identity

(24) $L[e^{rx}](x) = e^{rx}(r - r_1)^m \tilde{P}(r)$

(see (5)). Setting $r = r_1$ in (24), we again see that $e^{r_1 x}$ is a solution to $L[y] = 0$.

To find other solutions, we take the kth partial derivative with respect to r of both sides of (24):

(25) $\dfrac{\partial^k}{\partial r^k} L[e^{rx}](x) = \dfrac{\partial^k}{\partial r^k}[e^{rx}(r - r_1)^m \tilde{P}(r)]$.

Carrying out the differentiation on the right-hand side of (25), we find that the resulting expression will still have $(r - r_1)$ as a factor, provided $k \leq m - 1$. Thus setting $r = r_1$ in (25) gives

(26) $\left. \dfrac{\partial^k}{\partial r^k} L[e^{rx}](x) \right|_{r=r_1} = 0$ if $k \leq m - 1$.

Now notice that the function e^{rx} has continuous partial derivatives of all orders with respect to r and x. Hence for mixed partial derivatives of e^{rx}, it makes no difference whether the differentiation is done first with respect to x, then with respect to r, or vice versa. Since L involves derivatives with respect to x, this means we can interchange the order of differentiation in (26) to obtain

$$L\left[\left. \dfrac{\partial^k}{\partial r^k}(e^{rx}) \right|_{r=r_1} \right](x) = 0 .$$

Thus

(27) $\left. \dfrac{\partial^k}{\partial r^k}(e^{rx}) \right|_{r=r_1} = x^k e^{r_1 x}$

will be a solution to (1) for $k = 0, 1, \ldots, m - 1$. So m distinct solutions to (1), due to the root $r = r_1$ of multiplicity m, are indeed given by (22). We leave it as an exercise to show that the m functions in (22) are linearly independent on $(-\infty, \infty)$ (see Problem 25).

If $\alpha + i\beta$ is a repeated complex root of multiplicity m, then we can replace the $2m$ complex-valued functions

$$e^{(\alpha + i\beta)x} , \quad xe^{(\alpha + i\beta)x} , \quad \ldots, \quad x^{m-1}e^{(\alpha + i\beta)x} ,$$
$$e^{(\alpha - i\beta)x} , \quad xe^{(\alpha - i\beta)x} , \quad \ldots, \quad x^{m-1}e^{(\alpha - i\beta)x}$$

by the $2m$ linearly independent real-valued functions

(28) $e^{\alpha x}\cos \beta x , \quad xe^{\alpha x}\cos \beta x , \quad \ldots, \quad x^{m-1}e^{\alpha x}\cos \beta x ,$
 $e^{\alpha x}\sin \beta x , \quad xe^{\alpha x}\sin \beta x , \quad \ldots, \quad x^{m-1}e^{\alpha x}\sin \beta x .$

Using the results of the three cases discussed above, we can obtain a set of n linearly independent solutions that yield a real-valued general solution for (1).

EXAMPLE 3 Find a general solution to

(29) $y^{(4)} - y^{(3)} - 3y'' + 5y' - 2y = 0$.

SOLUTION The auxiliary equation is

$$r^4 - r^3 - 3r^2 + 5r - 2 = (r - 1)^3(r + 2) = 0 ,$$

which has roots $r_1 = 1, r_2 = 1, r_3 = 1, r_4 = -2$. Because the root at 1 has multiplicity 3, a general solution is

(30) $y(x) = C_1 e^x + C_2 x e^x + C_3 x^2 e^x + C_4 e^{-2x}$. ■End

EXAMPLE 4 Find a general solution to

(31) $y^{(4)} - 8y^{(3)} + 26y'' - 40y' + 25y = 0$,

whose auxiliary equation can be factored as

(32) $r^4 - 8r^3 + 26r^2 - 40r + 25 = (r^2 - 4r + 5)^2 = 0$.

SOLUTION The auxiliary equation (32) has repeated complex roots: $r_1 = 2 + i, r_2 = 2 + i, r_3 = 2 - i$, and $r_4 = 2 - i$. Hence a general solution is

$$y(x) = C_1 e^{2x} \cos x + C_2 x e^{2x} \cos x + C_3 e^{2x} \sin x + C_4 x e^{2x} \sin x .$$ ■End

EXERCISES 6.2

In Problems 1 through 14, find a general solution for the differential equation with x as the independent variable.

1. $y''' + 2y'' - 8y' = 0$.
2. $y''' - 3y'' - y' + 3y = 0$.
3. $6z''' + 7z'' - z' - 2z = 0$.
4. $y''' + 2y'' - 19y' - 20y = 0$.
5. $y''' + 3y'' + 28y' + 26y = 0$.
6. $y''' - y'' + 2y = 0$.
7. $2y''' - y'' - 10y' - 7y = 0$.
8. $y''' + 5y'' - 13y' + 7y = 0$.
9. $u''' - 9u'' + 27u' - 27u = 0$.
10. $y''' + 3y'' - 4y' - 6y = 0$.
11. $y^{(4)} + 4y''' + 6y'' + 4y' + y = 0$.
12. $y''' + 5y'' + 3y' - 9y = 0$.
13. $y^{(4)} + 4y'' + 4y = 0$.
14. $\theta''' + 5\theta'' + 5\theta' - 11\theta = 0$.

In Problems 15 through 18, find a general solution to the given homogeneous equation.

15. $(D - 1)^2(D + 3)(D^2 + 2D + 5)^2[y] = 0$.
16. $(D + 1)^2(D - 6)^3(D + 5)(D^2 + 1) \cdot$
$$(D^2 + 4)[y] = 0 .$$
17. $(D + 4)(D - 3)(D + 2)^3(D^2 + 4D + 5)^2 \cdot$
$$D^5[y] = 0 .$$
18. $(D - 1)^3(D - 2)(D^2 + D + 1) \cdot$
$$(D^2 + 6D + 10)^3[y] = 0 .$$

In Problems 19 through 21, solve the given initial value problem.

19. $y''' - y'' - 4y' + 4y = 0$;
$$y(0) = -4 , \qquad y'(0) = -1 , \qquad y''(0) = -19 .$$
20. $y''' + 7y'' + 14y' + 8y = 0$;
$$y(0) = 1 , \qquad y'(0) = -3 , \qquad y''(0) = 13 .$$

21. $y''' - 4y'' + 7y' - 6y = 0$;

$y(0) = 1$, $y'(0) = 0$, $y''(0) = 0$.

In Problems 22 and 23, find a general solution for the given linear system using the elimination method of Section 5.3.

22. $\dfrac{d^2x}{dt^2} - x + 5y = 0$,

$2x + \dfrac{d^2y}{dt^2} + 2y = 0$.

23. $\dfrac{d^3x}{dt^3} - x + \dfrac{dy}{dt} + y = 0$,

$\dfrac{dx}{dt} - x + y = 0$.

24. Let $P(r) = a_n r^n + \cdots + a_1 r + a_0$ be a polynomial with real coefficients a_n, \ldots, a_0. Prove that if r_1 is a zero of $P(r)$, then so is its complex conjugate \bar{r}_1. [*Hint:* Show that $\overline{P(r)} = P(\bar{r})$, where the bar denotes complex conjugation.]

25. Show that the m functions $e^{rx}, xe^{rx}, \ldots, x^{m-1}e^{rx}$ are linearly independent on $(-\infty, \infty)$. [*Hint:* Show that these functions are linearly independent if and only if $1, x, \ldots, x^{m-1}$ are linearly independent.]

26. As an alternative proof that the functions $e^{r_1 x}, e^{r_2 x}, \ldots, e^{r_n x}$ are linearly independent on $(-\infty, \infty)$ when r_1, r_2, \ldots, r_n are distinct, assume

(33) $C_1 e^{r_1 x} + C_2 e^{r_2 x} + \cdots + C_n e^{r_n x} = 0$

holds for all x in $(-\infty, \infty)$ and proceed as follows:

(a) Because the r_i's are distinct we can (if necessary) relabel them so that

$r_1 > r_2 > \cdots > r_n$.

Divide equation (33) by $e^{r_1 x}$ to obtain

$C_1 + C_2 \dfrac{e^{r_2 x}}{e^{r_1 x}} + \cdots + C_n \dfrac{e^{r_n x}}{e^{r_1 x}} = 0$.

Now let $x \to +\infty$ on the left-hand side to obtain $C_1 = 0$.

(b) Since $C_1 = 0$, equation (33) becomes

$C_2 e^{r_2 x} + C_3 e^{r_3 x} + \cdots + C_n e^{r_n x} = 0$

for all x in $(-\infty, \infty)$. Divide this equation by $e^{r_2 x}$ and let $x \to +\infty$ to conclude that $C_2 = 0$.

(c) Continuing in the manner of (b), argue that all the coefficients, C_1, C_2, \ldots, C_n are zero and hence $e^{r_1 x}, e^{r_2 x}, \ldots, e^{r_n x}$ are linearly independent on $(-\infty, \infty)$.

27. Find a general solution to

$$y^{(4)} + 2y''' - 3y'' - y' + \frac{1}{2}y = 0$$

by using Newton's method (Appendix A) or some other numerical procedure to approximate the roots of the auxiliary equation.

28. Find a general solution to $y''' - 3y' - y = 0$ by using Newton's method or some other numerical procedure to approximate the roots of the auxiliary equation.

29. Find a general solution to

$$y^{(4)} + 2y^{(3)} + 4y'' + 3y' + 2y = 0$$

by using Newton's method to approximate numerically the roots of the auxiliary equation. [*Hint:* To find complex roots, use the Newton recursion formula $z_{n+1} = z_n - f(z_n)/f'(z_n)$ and start with a *complex* initial guess z_0.]

30. (a) Derive the form

$$y(x) = A_1 e^x + A_2 e^{-x} + A_3 \cos x + A_4 \sin x$$

for the general solution to the equation $y^{(4)} = y$, from the observation that the fourth roots of unity are $1, -1, i,$ and $-i$.

(b) Derive the form

$$y(x) = A_1 e^x + A_2 e^{-x/2} \cos\left(\sqrt{3}x/2\right) + A_3 e^{-x/2} \sin\left(\sqrt{3}x/2\right)$$

for the general solution to the equation $y^{(3)} = y$, from the observation that the cube roots of unity are $1, e^{i2\pi/3},$ and $e^{-i2\pi/3}$.

31. Higher-Order Cauchy-Euler Equations. A differential equation that can be expressed in the form

$$a_n x^n y^{(n)}(x) + a_{n-1} x^{n-1} y^{(n-1)}(x) + \cdots + a_0 y(x) = 0 ,$$

where $a_n, a_{n-1}, \ldots, a_0$ are constants, is called a homogeneous **Cauchy-Euler** equation. (The second order case is discussed in Sections 4.5 and 8.5.) Use the substitution $y = x^r$ to help determine a fundamental solution set for the following Cauchy-Euler equations:

(a) $x^3y''' + x^2y'' - 2xy' + 2y = 0$, $x > 0$.

(b) $x^4y^{(4)} + 6x^3y''' + 2x^2y'' - 4xy' + 4y = 0$, $x > 0$.

(c) $x^3y''' - 2x^2y'' + 13xy' - 13y = 0$, $x > 0$.

[*Hint:* $x^{\alpha+i\beta} = e^{(\alpha+i\beta)\ln x}$
$= x^\alpha \{\cos(\beta \ln x) + i \sin(\beta \ln x)\}$.]

32. Let $y(x) = Ce^{rx}$, where $C\ (\neq 0)$ and r are real numbers, be a solution to a differential equation. Suppose we can not determine r exactly but can only approximate it by \tilde{r}. Let $\tilde{y}(x) := Ce^{\tilde{r}x}$ and consider the error $|y(x) - \tilde{y}(x)|$.

(a) If r and \tilde{r} are positive, $r \neq \tilde{r}$, show that the error grows exponentially large as x approaches $+\infty$.

(b) If r and \tilde{r} are negative, $r \neq \tilde{r}$, show that the error goes to zero exponentially as x approaches $+\infty$.

33. On a smooth horizontal surface, a mass of m_1 kg is attached to a fixed wall by a spring with spring constant k_1 N/m. Another mass of m_2 kg is attached to the first object by a spring with spring constant k_2 N/m. The objects are aligned horizontally so that the springs are their natural lengths. As we showed in Section 5.4, this coupled mass-spring system is governed by the system of differential equations

(34) $\quad m_1 \dfrac{d^2x}{dt^2} + (k_1 + k_2)x - k_2y = 0$,

(35) $\quad m_2 \dfrac{d^2y}{dt^2} - k_2x + k_2y = 0$.

Let's assume that $m_1 = m_2 = 1, k_1 = 3$, and

$k_2 = 2$. If both objects are displaced 1 m to the right of their equilibrium positions (cf. Figure 5.18, page 287) and then released, determine the equations of motion for the objects as follows:

(a) Show that $x(t)$ satisfies the equation

(36) $\quad x^{(4)}(t) + 7x''(t) + 6x(t) = 0$.

(b) Find a general solution $x(t)$ to (36).

(c) Substitute $x(t)$ back into (34) to obtain a general solution for $y(t)$.

(d) Use the initial conditions to determine the solutions, $x(t)$ and $y(t)$, which are the equations of motion.

34. Suppose the two springs in the coupled mass-spring system discussed in Problem 33 are switched, giving the new data $m_1 = m_2 = 1, k_1 = 2$, and $k_2 = 3$. If both objects are now displaced 1 m to the right of their equilibrium positions and then released, determine the equations of motion of the two objects.

35. Vibrating Beam. In studying the transverse vibrations of a beam, one encounters the homogeneous equation

$$EI\frac{d^4y}{dx^4} - ky = 0 ,$$

where $y(x)$ is related to the displacement of the beam at position x, the constant E is Young's modulus, I is the area moment of inertia, and k is a parameter. Assuming E, I, and k are positive constants, find a general solution in terms of sines, cosines, hyperbolic sines, and hyperbolic cosines.

6.3 UNDETERMINED COEFFICIENTS AND THE ANNIHILATOR METHOD

When a higher-order linear equation has constant coefficients, the method of undetermined coefficients discussed in Section 4.8 can still be used to obtain a particular solution y_p, provided the forcing term is a sum of functions of the type given in Table 4.1 on page 208 (or inside back cover).

EXAMPLE 1 Use the method of undetermined coefficients to find a general solution to

(1) $\qquad y'''(x) - 3y''(x) + 4y(x) = xe^{2x}$.

SOLUTION We first solve the corresponding homogeneous equation

(2) $\qquad y'''(x) - 3y''(x) + 4y(x) = 0$.

The auxiliary equation

$$r^3 - 3r^2 + 4 = (r + 1)(r - 2)^2 = 0$$

has a single root at -1 and a double root at 2, and hence a general solution to (2) is

(3) $y_h(x) = C_1 e^{-x} + C_2 e^{2x} + C_3 x e^{2x}$.

We can find the form for a particular solution y_p to equation (1) by referring to Table 4.1. Since $g(x) = xe^{2x}$ is of Type IV, y_p has the form

$$y_p(x) = x^s(A_1 x + A_0)e^{2x} .$$

Because e^{2x} and xe^{2x} are solutions to the corresponding homogeneous equation (2), we take $s = 2$. That is, y_p has the form

(4) $y_p(x) = (A_1 x^3 + A_0 x^2)e^{2x}$.

Substituting y_p for y in (1) gives

$$\left[8A_1 x^3 + (36A_1 + 8A_0)x^2 + (36A_1 + 24A_0)x + (6A_1 + 12A_0)\right]e^{2x}$$
$$- 3\left[4A_1 x^3 + (12A_1 + 4A_0)x^2 + (6A_1 + 8A_0)x + 2A_0\right]e^{2x}$$
$$+ 4\left[A_1 x^3 + A_0 x^2\right]e^{2x} = xe^{2x} ,$$

which reduces to

$$\left[18A_1 x + (6A_1 + 6A_0)\right]e^{2x} = xe^{2x} .$$

This leads to the system of equations

$$18A_1 = 1 , \qquad 6A_1 + 6A_0 = 0 ,$$

which yields $A_1 = 1/18$ and $A_0 = -1/18$. Thus

(5) $y_p(x) = \dfrac{1}{18}(x^3 - x^2)e^{2x}$.

A general solution to (1) is $y_h + y_p$, that is,

(6) $y(x) = C_1 e^{-x} + C_2 e^{2x} + C_3 x e^{2x} + \dfrac{1}{18}(x^3 - x^2)e^{2x}$. ■ End

For a better understanding of why the method of undetermined coefficients works for linear equations with constant coefficients, we will study a procedure for solving nonhomogeneous equations that begins by transforming the nonhomogeneous equation into a homogeneous one. For this purpose, we introduce the concept of an annihilator.

ANNIHILATOR

Definition 3. A linear differential operator A is said to **annihilate** a function f if

(7) $A[f](x) = 0 ,$ for all x .

That is, A annihilates f if f is a solution to the homogeneous linear differential equation (7) on $(-\infty, \infty)$.

For example, $A = D - 3$ annihilates $f(x) = e^{3x}$, since

$$(D - 3)[e^{3x}] = 3e^{3x} - 3e^{3x} = 0 .$$

Similarly, $D^2 + \beta^2$ annihilates both $\cos \beta x$ and $\sin \beta x$, since each of these functions satisfies the differential equation $y'' + \beta^2 y = 0$. Observe also that functions like $e^{\alpha x}\cos \beta x$ and $e^{\alpha x}\sin \beta x$ solve homogeneous differential equations whose auxiliary equation has roots $\alpha \pm i\beta$. The simplest such is

$$[r - (\alpha + i\beta)][r - (\alpha - i\beta)] = (r - \alpha)^2 + \beta^2 = r^2 - 2\alpha r + \alpha^2 + \beta^2 ,$$

so $(D - \alpha)^2 + \beta^2$ annihilates these functions.

From what we learned in the previous section about auxiliary equations with repeated roots, it follows that the differential operator $(D - r)^m$, m a positive integer, annihilates each of the functions

(8) $e^{rx}, xe^{rx}, \dots, x^{m-1}e^{rx} .$

Moreover, the differential operator $[(D - \alpha)^2 + \beta^2]^m$ annihilates each of the functions

(9)
$$e^{\alpha x}\cos \beta x, \quad xe^{\alpha x}\cos \beta x, \quad \dots, \quad x^{m-1}e^{\alpha x}\cos \beta x ,$$
$$e^{\alpha x}\sin \beta x, \quad xe^{\alpha x}\sin \beta x, \quad \dots, \quad x^{m-1}e^{\alpha x}\sin \beta x ,$$

which are $2m$ linearly independent solutions to $[(D - \alpha)^2 + \beta^2]^m[y] = 0$.

In the next example, we make use of the fact that linear differential operators with *constant* coefficients commute (see Exercises 4.2, Problem 32).

EXAMPLE 2 Find a differential operator that annihilates

(10) $6xe^{-4x} + 5e^x\sin 2x .$

SOLUTION Consider the two functions whose sum appears in (10). Observe that $(D + 4)^2$ annihilates the function $f_1(x) := 6xe^{-4x}$. Further, $f_2(x) := 5e^x\sin 2x$ is annihilated by the operator $(D - 1)^2 + 4$. Hence the composite operator

$$A := (D + 4)^2[(D - 1)^2 + 4] ,$$

which is the same as the operator

$$[(D - 1)^2 + 4](D + 4)^2 ,$$

annihilates both f_1 and f_2. But then, by linearity, A also annihilates the sum $f_1 + f_2$. ■ End

We now show how annihilators can be used to determine particular solutions to certain *non*homogeneous equations. Consider the nth order differential equation with constant coefficients

(11) $a_n y^{(n)}(x) + a_{n-1}y^{(n-1)}(x) + \cdots + a_0 y(x) = g(x) ,$

which can be written in the operator form

(12) $L[y](x) = g(x) ,$

where

$$L[y] := a_n y^{(n)} + a_{n-1} y^{(n-1)} + \cdots + a_0 y .$$

In this section, we restrict our attention to nonhomogeneous terms that can be annihilated by a linear differential operator with *constant* coefficients. Assuming A is such an annihilator of g, then applying A to both sides of (12) yields

$$A\big[L[y]\big](x) = A[g](x) = 0 .$$

Therefore, if y is a solution to (12), then y is also a solution to the *homogeneous* linear differential equation

(13) $AL[y](x) = 0$

involving the composition of the operators A and L. Because equation (13) has constant coefficients (why?), we can apply the methods of Section 6.2 to obtain its general solution. Comparing this with a general solution of $L[y] = 0$, it is then possible to determine the *form* of a particular solution to equation (12). This procedure, known as the **annihilator method,** is illustrated in the following examples.

EXAMPLE 3 Find a general solution to

(14) $y''(x) + y(x) = e^{2x} + 1 .$

SOLUTION Since $D - 2$ annihilates e^{2x} and D annihilates 1, then $A := D(D - 2)$ annihilates $e^{2x} + 1$. Therefore applying A to both sides of (14) yields

$$A[y'' + y] = A[e^{2x} + 1]$$

(15) $D(D - 2)(D^2 + 1)[y] = 0 .$

The auxiliary equation associated with (15) is

$$r(r - 2)(r^2 + 1) = 0 ,$$

which has the roots $i, -i, 2, 0$. Hence a general solution to (15) is

(16) $y(x) = C_1 \cos x + C_2 \sin x + C_3 e^{2x} + C_4 .$

Now recall that a general solution to (14) is of the form $y_h + y_p$, where y_p is a particular solution to (14) and y_h is a general solution to

$$y''(x) + y(x) = 0 .$$

Since every solution to (14) is also a solution to (15), then $y_h + y_p$ must have the form displayed on the right-hand side of (16). But we recognize that $y_h(x) = C_1 \cos x + C_2 \sin x$, so there exists a particular solution of the form

(17) $y_p(x) = C_3 e^{2x} + C_4 .$

To determine the constants C_3 and C_4, we substitute y_p given by (17) into (14):

$$4C_3 e^{2x} + C_3 e^{2x} + C_4 = e^{2x} + 1 ,$$
$$5C_3 e^{2x} + C_4 = e^{2x} + 1 .$$

Equating the coefficients of e^{2x} and setting the constant terms equal yields $C_3 = 1/5$ and $C_4 = 1$. Thus

$$y_p(x) = \frac{1}{5}e^{2x} + 1 \ ,$$

so a general solution to (14) is

$$y(x) = C_1 \cos x + C_2 \sin x + \frac{1}{5}e^{2x} + 1 \ . \quad \blacksquare \text{End}$$

EXAMPLE 4 Use the annihilator method to determine the form of a particular solution to

(18) $y'' - y = e^{-2x} \sin x$.

SOLUTION The function $g(x) = e^{-2x} \sin x$ is annihilated by the operator

$$A := (D + 2)^2 + 1^2 = D^2 + 4D + 5 \ .$$

If we apply A to both sides of (18), we obtain

$$A\big[y'' - y\big] = A\big[e^{-2x} \sin x\big]$$

(19) $(D^2 + 4D + 5)(D^2 - 1)\big[y\big] = 0$.

Now the auxiliary equation associated with (19) is

$$(r^2 + 4r + 5)(r - 1)(r + 1) = 0 \ ,$$

which has roots $1, -1, -2 + i, -2 - i$. Hence a general solution to (19) is

(20) $y(x) = C_1 e^x + C_2 e^{-x} + C_3 e^{-2x} \cos x + C_4 e^{-2x} \sin x$.

A general solution to the corresponding homogeneous equation $y'' - y = 0$ is $y_h(x) = C_1 e^x + C_2 e^{-x}$, so we see that a particular solution for (18) has the form

(21) $y_p(x) = C_3 e^{-2x} \cos x + C_4 e^{-2x} \sin x$. $\quad \blacksquare$ End

Of course the forms of the particular solutions derived in Examples 3 and 4 using the annihilator method can be quickly obtained from Table 4.1, page 208, for the method of undetermined coefficients. The essential point is that the annihilator method can be used to *justify* the method of undetermined coefficients; that is, to derive the entries that appear in Table 4.1. This applies not only to second order, but also to higher-order linear differential equations with constant coefficients.

The connection between the two methods is even more evident once one realizes that those functions for which the method of undetermined coefficients works are *exactly* those functions that are solutions to higher-order linear differential equations with constant coefficients, that is, those functions $g(x)$ that are annihilated by a linear differential operator with constant coefficients. In particular, these methods work for polynomials, exponentials, sines, and cosines, and for any product or sum of these functions. (Why?)

We leave it to the reader to show how the annihilator method yields various entries in Table 4.1 (see Problems 34–37). The derivation for the general case, where

$$g(x) = p_n(x)e^{\alpha x}\cos \beta x + q_m(x)e^{\alpha x}\sin \beta x \ ,$$

p_n, q_m polynomials, is discussed in Project A at the end of this chapter.

Another procedure for solving linear constant coefficient equations, using the *Laplace transform*, will be introduced in Chapter 7.

EXERCISES 6.3

In Problems 1 through 4, use the method of undetermined coefficients (Table 4.1, page 208) to determine the form of a particular solution for the given equation.

1. $y''' - 2y'' - 5y' + 6y = e^x + x^2$.

2. $y''' + y'' - 5y' + 3y = e^{-x} + \sin x$.

3. $y''' + 3y'' - 4y = e^{-2x}$.

4. $y''' + y'' - 2y = xe^x + 1$.

In Problems 5 through 10, find a general solution to the given equation.

5. $y''' - 2y'' - 5y' + 6y = e^x + x^2$.

6. $y''' + y'' - 5y' + 3y = e^{-x} + \sin x$.

7. $y''' + 3y'' - 4y = e^{-2x}$.

8. $y''' + y'' - 2y = xe^x + 1$.

9. $y''' - 3y'' + 3y' - y = e^x$.

10. $y''' + 4y'' + y' - 26y = e^{-3x}\sin 2x + x$.

In Problems 11 through 20, find a differential operator that annihilates the given function.

11. $x^4 - x^2 + 11$.

12. $3x^2 - 6x + 1$.

13. e^{-7x} .

14. e^{5x} .

15. $e^{2x} - 6e^x$.

16. $x^2 - e^x$.

17. $x^2e^{-x}\sin 2x$.

18. $xe^{3x}\cos 5x$.

19. $xe^{-2x} + xe^{-5x}\sin 3x$. **20.** $x^2e^x - x\sin 4x + x^3$.

In Problems 21 through 30, use the annihilator method to determine the form of a particular solution for the given equation.

21. $u'' - 5u' + 6u = \cos 2x + 1$.

22. $y'' + 6y' + 8y = e^{3x} - \sin x$.

23. $y'' - 5y' + 6y = e^{3x} - x^2$.

24. $\theta'' - \theta = xe^x$.

25. $y'' - 6y' + 9y = \sin 2x + x$.

26. $y'' + 2y' + y = x^2 - x + 1$.

27. $y'' + 2y' + 2y = e^{-x}\cos x + x^2$.

28. $y'' - 6y' + 10y = e^{3x} - x$.

29. $z''' - 2z'' + z' = x - e^x$.

30. $y''' + 2y'' - y' - 2y = e^x - 1$.

In Problems 31 through 33, solve the given initial value problem.

31. $y''' + 2y'' - 9y' - 18y = -18x^2 - 18x + 22$;
$y(0) = -2$, $y'(0) = -8$, $y''(0) = -12$.

32. $y''' - 2y'' + 5y' = -24e^{3x}$;
$y(0) = 4$, $y'(0) = -1$, $y''(0) = -5$.

33. $y''' - 2y'' - 3y' + 10y$
$= 34xe^{-2x} - 16e^{-2x} - 10x^2 + 6x + 34$;
$y(0) = 3$, $y'(0) = 0$, $y''(0) = 0$.

34. Use the annihilator method to show that if $a_0 \neq 0$ in equation (11) and $g(x)$ has the form

(22) $g(x) = b_m x^m + b_{m-1}x^{m-1} + \cdots + b_1x + b_0$,

then

$$y_p(x) = B_m x^m + B_{m-1}x^{m-1} + \cdots + B_1x + B_0$$

is the form of a particular solution to equation (11).

35. Use the annihilator method to show that if $a_0 = 0$ and $a_1 \neq 0$ in (11) and $g(x)$ has the form given in (22), then equation (11) has a particular solution of the form
$$y_p(x) = x\{B_m x^m + B_{m-1}x^{m-1} + \cdots + B_1x + B_0\} .$$

36. Use the annihilator method to show that if $g(x)$ in (11) has the form $g(x) = Be^{\alpha x}$, then equation (11) has a particular solution of the form $y_p(x) = x^sBe^{\alpha x}$, where s is chosen to be the smallest nonnegative integer such that $x^se^{\alpha x}$ is not a solution to the corresponding homogeneous equation.

37. Use the annihilator method to show that if $g(x)$ in

(11) has the form

$$g(x) = a \cos \beta x + b \sin \beta x \ ,$$

then equation (11) has a particular solution of the form

(23) $y_p(x) = x^s\{A \cos \beta x + B \sin \beta x\} \ ,$

where s is chosen to be the smallest nonnegative integer such that $x^s\cos \beta x$ and $x^s\sin \beta x$ are not solutions to the corresponding homogeneous equation.

In Problems 38 and 39, use the elimination method of Section 5.3 to find a general solution to the given system.

38. $x - \dfrac{d^2y}{dt^2} = t + 1 \ ,$

$\dfrac{dx}{dt} + \dfrac{dy}{dt} - 2y = e^t \ .$

39. $\dfrac{d^2x}{dt^2} - x + y = 0 \ ,$

$x + \dfrac{d^2y}{dt^2} - y = e^{3t} \ .$

40. The currents in the electrical network in Figure 6.1 satisfy the system

$$\frac{1}{9}I_1 + 64I_2'' = -2 \sin\frac{t}{24} \ ,$$

$$\frac{1}{64}I_3 + 9I_3'' - 64I_2'' = 0 \ ,$$

$$I_1 = I_2 + I_3 \ ,$$

where I_1, I_2, and I_3 are the currents through the different branches of the network. Using the elimination method of Section 5.3, determine the currents if initially $I_1(0) = I_2(0) = I_3(0) = 0$, $I_1'(0) = 73/12$, $I_2'(0) = 3/4$, and $I_3'(0) = 16/3$.

Figure 6.1 An electrical network

6.4 METHOD OF VARIATION OF PARAMETERS

In the previous section, we discussed the method of undetermined coefficients and the annihilator method. These methods work only for linear equations with constant coefficients *and* when the nonhomogeneous term is a solution to some homogeneous linear equation with constant coefficients. In this section, we show how the method of **variation of parameters** discussed in Section 4.9 generalizes to higher-order linear equations with variable coefficients.

Our goal is to find a particular solution to

(1) $L[y](x) = g(x) \ ,$

where $L[y] := y^{(n)} + p_1y^{(n-1)} + \cdots + p_ny$ and the coefficient functions p_1, \ldots, p_n, as well as g, are continuous on (a, b). The method to be described requires that we already know a fundamental solution set $\{y_1, \ldots, y_n\}$ for the corresponding homogeneous equation

(2) $L[y](x) = 0 \ .$

A general solution to (2) is then

(3) $y_h(x) = C_1y_1(x) + \cdots + C_ny_n(x) \ ,$

where C_1, \ldots, C_n are arbitrary constants. In the method of variation of parameters, we assume there exists a particular solution to (1) of the form

(4) $y_p(x) = v_1(x)y_1(x) + \cdots + v_n(x)y_n(x)$

and try to determine the functions v_1, \ldots, v_n.

There are n unknown functions, so we will need n conditions (equations) to determine them. These conditions are obtained as follows. Differentiating y_p in (4) gives

(5) $y_p' = (v_1 y_1' + \cdots + v_n y_n') + (v_1' y_1 + \cdots + v_n' y_n)$.

To prevent second derivatives of the unknowns v_1, \ldots, v_n from entering the formula for y_p'', we impose the condition

$$v_1' y_1 + \cdots + v_n' y_n = 0 .$$

In a like manner, as we compute y_p'', y_p''', \ldots, $y_p^{(n-1)}$, we impose $(n-2)$ additional conditions involving v_1', \ldots, v_n'; namely,

$$v_1' y_1' + \cdots + v_n' y_n' = 0, \ldots, v_1' y_1^{(n-2)} + \cdots + v_n' y_n^{(n-2)} = 0 .$$

Finally, the nth condition that we impose is that y_p satisfy the given equation (1). Using the previous conditions and the fact that y_1, \ldots, y_n are solutions to the homogeneous equation, then $L[y_p] = g$ reduces to

(6) $v_1' y_1^{(n-1)} + \cdots + v_n' y_n^{(n-1)} = g$

(see Problem 12). We therefore seek n functions v_1', \ldots, v_n' that satisfy the system

(7)
$$\begin{aligned}
y_1 v_1' + \cdots + y_n v_n' &= 0 , \\
&\;\;\vdots \qquad\qquad \vdots \;\; \vdots \\
y_1^{(n-2)} v_1' + \cdots + y_n^{(n-2)} v_n' &= 0 , \\
y_1^{(n-1)} v_1' + \cdots + y_n^{(n-1)} v_n' &= g .
\end{aligned}$$

A sufficient condition for the existence of a solution to system (7) for x in (a, b) is that the determinant of the matrix made up of the coefficients of v_1', \ldots, v_n' be different from zero for all x in (a, b). But this determinant is just the Wronskian:

(8)
$$\begin{vmatrix}
y_1 & \cdots & y_n \\
\vdots & & \vdots \\
y_1^{(n-2)} & \cdots & y_n^{(n-2)} \\
y_1^{(n-1)} & \cdots & y_n^{(n-1)}
\end{vmatrix} = W[y_1, \ldots, y_n](x) ,$$

which is never zero on (a, b) because $\{y_1, \ldots, y_n\}$ is a fundamental solution set. Solving (7) via Cramer's rule (Appendix C), we find

(9) $v_k'(x) = \dfrac{g(x) W_k(x)}{W[y_1, \ldots, y_n](x)} , \qquad k = 1, \ldots, n ,$

where $W_k(x)$ is the determinant of the matrix obtained from the Wronskian $W[y_1, \ldots y_n](x)$ by replacing the kth column by $\text{col}[0, \ldots, 0, 1]$. Using a cofactor expansion about this column, we can express $W_k(x)$ in terms of an $(n-1)$th order Wronskian:

(10) $W_k(x) = (-1)^{n-k} W[y_1, \ldots, y_{k-1}, y_{k+1}, \ldots, y_n](x) , \qquad k = 1, \ldots, n .$

Integrating $v_k'(x)$ in (9) gives

(11) $v_k(x) = \displaystyle\int \dfrac{g(x) W_k(x)}{W[y_1, \ldots, y_n](x)} dx , \qquad k = 1, \ldots, n .$

Finally, substituting the v_k's back into (4), we obtain a particular solution to equation (1):

(12) $$y_p(x) = \sum_{k=1}^{n} y_k(x) \int \frac{g(x)W_k(x)}{W[y_1, \ldots, y_n](x)} \, dx \ .$$

Note that in equation (1) we presumed that the coefficient of the leading term, $y^{(n)}$, was unity. If, instead, it is $p_0(x)$, we must replace $g(x)$ by $g(x)/p_0(x)$ in (12).

Although equation (12) gives a neat formula for a particular solution to (1), its implementation requires one to evaluate $n + 1$ determinants and then perform n integrations. This may entail several tedious computations. However, the method does work in cases when the technique of undetermined coefficients does not apply (provided, of course, we know a fundamental solution set).

EXAMPLE 1 Find a general solution to the Cauchy-Euler equation

(13) $x^3 y''' + x^2 y'' - 2xy' + 2y = x^3 \sin x \ , \qquad x > 0 \ ,$

given that $\{x, x^{-1}, x^2\}$ is a fundamental solution set to the corresponding homogeneous equation.

SOLUTION An important first step is to divide (13) by x^3 to obtain the standard form

(14) $y''' + \dfrac{1}{x} y'' - \dfrac{2}{x^2} y' + \dfrac{2}{x^3} y = \sin x \ , \qquad x > 0 \ ,$

from which we see that $g(x) = \sin x$. Since $\{x, x^{-1}, x^2\}$ is a fundamental solution set, we can obtain a particular solution of the form

(15) $y_p(x) = v_1(x)x + v_2(x)x^{-1} + v_3(x)x^2 \ .$

To use formula (12), we must first evaluate the four determinants:

$$W[x, x^{-1}, x^2](x) = \begin{vmatrix} x & x^{-1} & x^2 \\ 1 & -x^{-2} & 2x \\ 0 & 2x^{-3} & 2 \end{vmatrix} = -6x^{-1} \ ,$$

$$W_1(x) = (-1)^{(3-1)} W[x^{-1}, x^2](x) = (-1)^2 \begin{vmatrix} x^{-1} & x^2 \\ -x^{-2} & 2x \end{vmatrix} = 3 \ ,$$

$$W_2(x) = (-1)^{(3-2)} \begin{vmatrix} x & x^2 \\ 1 & 2x \end{vmatrix} = -x^2 \ ,$$

$$W_3(x) = (-1)^{(3-3)} \begin{vmatrix} x & x^{-1} \\ 1 & -x^{-2} \end{vmatrix} = -2x^{-1} \ .$$

Substituting the above expressions into (12), we find

$$y_p(x) = x \int \frac{(\sin x)3}{-6x^{-1}} \, dx + x^{-1} \int \frac{(\sin x)(-x^2)}{-6x^{-1}} \, dx + x^2 \int \frac{(\sin x)(-2x^{-1})}{-6x^{-1}} \, dx$$

$$= x \int \left(-\frac{1}{2} x \sin x \right) dx + x^{-1} \int \frac{1}{6} x^3 \sin x \, dx + x^2 \int \frac{1}{3} \sin x \, dx \ ,$$

which after some labor simplifies to

(16) $y_p(x) = \cos x - x^{-1}\sin x + C_1 x + C_2 x^{-1} + C_3 x^2 ,$

where $C_1, C_2,$ and C_3 denote the constants of integration. Since $\{x, x^{-1}, x^2\}$ is a fundamental solution set for the homogeneous equation, we can take $C_1, C_2,$ and C_3 to be arbitrary constants. The right-hand side of (16) then gives the desired general solution. ■ End

In the preceding example, the fundamental solution set $\{x, x^{-1}, x^2\}$ can be derived by substituting $y = x^r$ into the homogeneous equation corresponding to (13) (see Problem 31, Exercises 6.2). However, in dealing with other equations that have variable coefficients, the determination of a fundamental set may be extremely difficult. In Chapter 8 we tackle this problem using power series methods.

EXERCISES 6.4

In Problems 1 through 6, use the method of variation of parameters to determine a particular solution to the given equation.

1. $y''' - 3y'' + 4y = e^{2x}$.

2. $y''' - 2y'' + y' = x$.

3. $z''' + 3z'' - 4z = e^{2x}$.

4. $y''' - 3y'' + 3y' - y = e^x$.

5. $y''' + y' = \tan x ,\qquad 0 < x < \pi/2$.

6. $y''' + y' = \sec\theta\tan\theta ,\qquad 0 < \theta < \pi/2$.

7. Find a general solution to the Cauchy-Euler equation
$$x^3 y''' - 3x^2 y'' + 6xy' - 6y = x^{-1} ,$$
$$x > 0 ,$$
given that $\{x, x^2, x^3\}$ is a fundamental solution set for the corresponding homogeneous equation.

8. Find a general solution to the Cauchy-Euler equation
$$x^3 y''' - 2x^2 y'' + 3xy' - 3y = x^2 ,$$
$$x > 0 ,$$
given that $\{x, x\ln x, x^3\}$ is a fundamental solution set for the corresponding homogeneous equation.

9. Given that $\{e^x, e^{-x}, e^{2x}\}$ is a fundamental solution set for the homogeneous equation corresponding to the equation
$$y''' - 2y'' - y' + 2y = g(x) ,$$

determine a formula involving integrals for a particular solution.

10. Given that $\{x, x^{-1}, x^4\}$ is a fundamental solution set for the homogeneous equation corresponding to the equation
$$x^3 y''' - x^2 y'' - 4xy' + 4y = g(x) ,$$
$$x > 0 ,$$
determine a formula involving integrals for a particular solution.

11. Find a general solution to the Cauchy-Euler equation
$$x^3 y''' - 3xy' + 3y = x^4\cos x ,\qquad x > 0 .$$

12. Derive the system (7) in the special case when $n = 3$. [*Hint:* To determine the last equation, require that $L[y_p] = g$ and use the fact that $y_1, y_2,$ and y_3 satisfy the corresponding homogeneous equation.]

13. Show that
$$W_k(x) = (-1)^{(n-k)}W[y_1, \dots, y_{k-1}, y_{k+1}, \dots, y_n](x) .$$

14. **Deflection of a Beam under Axial Force.** A uniform beam under a load and subject to a constant axial force is governed by the differential equation
$$y^{(4)}(x) - k^2 y''(x) = q(x) ,\qquad 0 < x < L,$$
where $y(x)$ is the deflection of the beam, L is the length of the beam, k^2 is proportional to the axial force, and $q(x)$ is proportional to the load (see Figure 6.2, page 364).

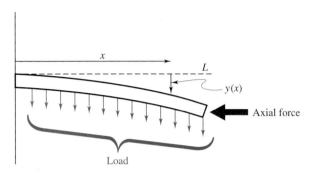

Figure 6.2 Deformation of a beam under axial force and load

(a) Show that a general solution can be written in the form

$$y(x) = C_1 + C_2 x + C_3 e^{kx} + C_4 e^{-kx}$$

$$+ \frac{1}{k^2} \int q(x) x \, dx - \frac{x}{k^2} \int q(x) \, dx$$

$$+ \frac{e^{kx}}{2k^3} \int q(x) e^{-kx} dx - \frac{e^{-kx}}{2k^3} \int q(x) e^{kx} dx \ .$$

(b) Show that the general solution in part (a) can be rewritten in the form

$$y(x) = c_1 + c_2 x + c_3 e^{kx} + c_4 e^{-kx}$$

$$+ \int_0^x q(s) G(s, x) ds \ ,$$

where

$$G(s, x) := \frac{s - x}{k^2} - \frac{\sinh[k(s - x)]}{k^3} \ .$$

(c) Let $q(x) \equiv 1$. First compute the general solution using the formula in part (a) and then using the formula in part (b). Compare these two general solutions with the general solution

$$y(x) = B_1 + B_2 x + B_3 e^{kx} + B_4 e^{-kx} - \frac{1}{2k^2} x^2 \ ,$$

which one would obtain using the method of undetermined coefficients.

(d) What are some advantages of the formula in part (b)?

CHAPTER SUMMARY

The theory and techniques for solving an nth order linear differential equation

(1) $\qquad y^{(n)} + p_1(x) y^{(n-1)} + \cdots + p_n(x) y = g(x)$

are natural extensions of the development for linear second order equations given in Chapter 4. Assuming p_1, \ldots, p_n and g are continuous functions on an open interval I, there is a unique solution to (1) on I that satisfies the n initial conditions: $y(x_0) = \gamma_0$, $y'(x_0) = \gamma_1, \ldots, y^{(n-1)}(x_0) = \gamma_{n-1}$, where $x_0 \in I$.

For the corresponding homogeneous equation

(2) $\qquad y^{(n)} + p_1(x) y^{(n-1)} + \cdots + p_n(x) y = 0 \ ,$

there exists a set of n **linearly independent** solutions $\{y_1, \ldots, y_n\}$ on I. Such functions are said to form a **fundamental solution set,** and every solution to (2) can be written as a linear combination of these functions:

$$y(x) = C_1 y_1(x) + C_2 y_2(x) + \cdots + C_n y_n(x) \ .$$

The linear independence of solutions to (2) is equivalent to the nonvanishing on I of the **Wronskian**

$$W[y_1, \ldots, y_n](x) := \det \begin{bmatrix} y_1(x) & \cdots & y_n(x) \\ y_1'(x) & \cdots & y_n'(x) \\ \vdots & & \vdots \\ y_1^{(n-1)}(x) & \cdots & y_n^{(n-1)}(x) \end{bmatrix} .$$

When equation (2) has (real) constant coefficients so that it is of the form

(3) $a_n y^{(n)} + a_{n-1} y^{(n-1)} + \cdots + a_0 y = 0 , \qquad a_n \neq 0 ,$

then the problem of determining a fundamental solution set can be reduced to the algebraic problem of solving the **auxiliary** or **characteristic equation**

(4) $a_n r^n + a_{n-1} r^{n-1} + \cdots + a_0 = 0 .$

If the n roots of (4), say r_1, r_2, \ldots, r_n, are all distinct, then

(5) $\{e^{r_1 x}, e^{r_2 x}, \ldots, e^{r_n x}\}$

is a fundamental solution set for (3). If some real root, say r_1, occurs with multiplicity m (e.g., $r_1 = r_2 = \cdots = r_m$), then m of the functions in (5) are replaced by

$$e^{r_1 x}, xe^{r_1 x}, \ldots, x^{m-1} e^{r_1 x} .$$

When a complex root $\alpha + i\beta$ to (4) occurs with multiplicity m, then so does its conjugate and $2m$ members of the set (5) are replaced by the real-valued functions

$$e^{\alpha x} \sin \beta x, \ xe^{\alpha x} \sin \beta x, \ \ldots, \ x^{m-1} e^{\alpha x} \sin \beta x ,$$
$$e^{\alpha x} \cos \beta x, \ xe^{\alpha x} \cos \beta x, \ \ldots, \ x^{m-1} e^{\alpha x} \cos \beta x .$$

A general solution to the nonhomogeneous equation (1) can be written as

$$y(x) = y_p(x) + y_h(x) ,$$

where y_p is some particular solution to (1) and y_h is a general solution to the corresponding homogeneous equation. Two useful techniques for finding particular solutions are the **annihilator method** (undetermined coefficients) and the method of **variation of parameters.**

The annihilator method applies to equations of the form

(6) $L[y] = g(x) ,$

where L is a linear differential operator with constant coefficients and the forcing term $g(x)$ is a polynomial, exponential, sine, or cosine, or a linear combination of products of these. Such a function $g(x)$ is annihilated (mapped to zero) by a linear differential operator A that also has constant coefficients. Every solution to the nonhomogeneous equation (6) is then a solution to the homogeneous equation $AL[y] = 0$, and, by comparing the solutions of the latter equation with a general solution to $L[y] = 0$, we can obtain the *form* of a particular solution to (6). These forms have previously been compiled in Table 4.1, page 208, for the method of undetermined coefficients.

The method of variation of parameters is more general in that it applies to arbitrary equations of the form (1). The idea is, starting with a fundamental solution set $\{y_1, \ldots, y_n\}$ for (2), to determine functions v_1, \ldots, v_n such that

(7) $y_p(x) = v_1(x) y_1(x) + \cdots + v_n(x) y_n(x)$

satisfies (1). This method leads to the formula

(8) $\displaystyle y_p(x) = \sum_{k=1}^{n} y_k(x) \int \frac{g(x) W_k(x)}{W[y_1, \ldots, y_n](x)} \, dx ,$

where

$$W_k(x) = (-1)^{n-k} W[y_1, \ldots, y_{k-1}, y_{k+1}, \ldots, y_n](x) , \qquad k = 1, \ldots, n .$$

REVIEW PROBLEMS

1. Determine the intervals for which Theorem 1 on page 339 guarantees the existence of a solution in that interval.
 (a) $y^{(4)} - (\ln x)y'' + xy' + 2y = \cos 3x$.
 (b) $(x^2 - 1)y''' + (\sin x)y'' + \sqrt{x + 4}\, y' + e^x y = x^2 + 3$.

2. Determine whether the given functions are linearly dependent or linearly independent on the interval $(0, \infty)$.
 (a) $\{e^{2x}, x^2 e^{2x}, e^{-x}\}$.
 (b) $\{e^x \sin 2x, xe^x \sin 2x, e^x, xe^x\}$.
 (c) $\{2e^{2x} - e^x, e^{2x} + 1, e^{2x} - 3, e^x + 1\}$.

3. Show that the set of functions $\{\sin x, x \sin x, x^2 \sin x, x^3 \sin x\}$ is linearly independent on $(-\infty, \infty)$.

4. Find a general solution for the given differential equation
 (a) $y^{(4)} + 2y''' - 4y'' - 2y' + 3y = 0$.
 (b) $y''' + 3y'' - 5y' + y = 0$.
 (c) $y^{(5)} - y^{(4)} + 2y''' - 2y'' + y' - y = 0$.
 (d) $y''' - 2y'' - y' + 2y = e^x + x$.

5. Find a general solution for the homogeneous linear differential equation with constant coefficients whose auxiliary equation is
 (a) $(r + 5)^2 (r - 2)^3 (r^2 + 1)^2 = 0$.
 (b) $r^4 (r - 1)^2 (r^2 + 2r + 4)^2 = 0$.

6. Given that $y_p = \sin(x^2)$ is a particular solution to
 $$y^{(4)} + y = (16x^4 - 11)\sin(x^2) - 48x^2 \cos(x^2)$$
 on $(0, \infty)$, find a general solution.

7. Find a differential operator that annihilates the given function.
 (a) $x^2 - 2x + 5$. (b) $e^{3x} + x - 1$.
 (c) $x \sin 2x$. (d) $x^2 e^{-2x} \cos 3x$.
 (e) $x^2 - 2x + xe^{-x} + \sin 2x - \cos 3x$.

8. Use the annihilator method to determine the form of a particular solution for the given equation.
 (a) $y'' + 6y' + 5y = e^{-x} + x^2 - 1$.
 (b) $y''' + 2y'' - 19y' - 20y = xe^{-x}$.
 (c) $y^{(4)} + 6y'' + 9y = x^2 - \sin 3x$.
 (d) $y''' - y'' + 2y = x \sin x$.

9. Find a general solution to the Cauchy-Euler equation
 $$x^3 y''' - 2x^2 y'' - 5xy' + 5y = x^{-2},$$
 $$x > 0,$$
 given that $\{x, x^5, x^{-1}\}$ is a fundamental solution set to the corresponding homogeneous equation.

10. Find a general solution to the given Cauchy-Euler equation.
 (a) $4x^3 y''' + 8x^2 y'' - xy' + y = 0$, $x > 0$.
 (b) $x^3 y''' + 2x^2 y'' + 2xy' + 4y = 0$, $x > 0$.

TECHNICAL WRITING EXERCISES

1. Describe the differences and similarities between second order and higher-order linear differential equations. Include in your comparisons both theoretical results and the methods of solution. For example, what complications arise in solving higher-order equations that are not present for the second order case?

2. Explain the relationship between the method of undetermined coefficients and the annihilator method. What difficulties would you encounter in applying the annihilator method if the linear equation did not have constant coefficients?

3. For students with a background in linear algebra: Compare the theory for kth order linear differential equations with that for systems of n linear equations in n unknowns whose coefficient matrix has rank $n - k$. Use the terminology from linear algebra; for example, subspaces, basis, dimension, linear transformation, and kernel. Discuss both homogeneous and nonhomogeneous equations.

A. *Justifying the Method of Undetermined Coefficients*

The annihilator method discussed in Section 6.3 can be used to derive the entries in Table 4.1, page 208, for the method of undetermined coefficients. To show this, it suffices to work with Type VII functions—that is, functions of the form

(1) $\qquad g(x) = p_n(x)e^{\alpha x}\cos \beta x + q_m(x)e^{\alpha x}\sin \beta x$,

where p_n and q_m are polynomials of degrees n and m, respectively—since the other types listed in Table 4.1 are just special cases of (1).

Consider the nonhomogeneous equation

(2) $\qquad L[y](x) = g(x)$,

where L is the linear operator

(3) $\qquad L[y] := a_n y^{(n)} + a_{n-1}y^{(n-1)} + \cdots + a_0 y$,

with $a_n, a_{n-1}, \ldots, a_0$ constants and $g(x)$ as given in equation (1). Let $N := \max(n, m)$.

(a) Show that

$$A := \left[(D - \alpha)^2 + \beta^2 \right]^{N+1}$$

is an annihilator for g.

(b) Show that the auxiliary equation associated with $AL[y] = 0$ is of the form

(4) $\qquad a_n \left[(r - \alpha)^2 + \beta^2 \right]^{s+N+1}(r - r_{2s+1}) \cdots (r - r_n) = 0$,

where s (≥ 0) is the multiplicity of $\alpha \pm i\beta$ as roots of the auxiliary equation associated with $L[y] = 0$, and r_{2s+1}, \ldots, r_n are the remaining roots of this equation.

(c) Find a general solution for $AL[y] = 0$ and compare it with a general solution for $L[y] = 0$ to verify that equation (2) has a particular solution of the form

$$y_p(x) = x^s e^{\alpha x}\{P_N(x)\cos \beta x + Q_N(x)\sin \beta x\} ,$$

where P_N and Q_N are polynomials of degree N.

B. *Transverse Vibrations of a Beam*

In applying elasticity theory to study the transverse vibrations of a beam, one encounters the equation

$$EIy^{(4)}(x) - \gamma\lambda y(x) = 0 ,$$

where $y(x)$ is related to the displacement of the beam at position x; the constant E is Young's modulus; I is the area moment of inertia, which we assume is constant; γ is the constant mass per unit length of the beam; and λ is a positive parameter to be determined. We can simplify the equation by letting $r^4 := \gamma\lambda/EI$; that is, we consider

(5) $\qquad y^{(4)}(x) - r^4 y(x) = 0$.

When the beam is clamped at each end, we seek a solution to (5) that satisfies the boundary conditions

(6) $\qquad y(0) = y'(0) = 0 \quad \text{and} \quad y(L) = y'(L) = 0$,

where L is the length of the beam. The problem is to determine those nonnegative values of r for which equation (5) has a nontrivial solution $(y(x) \not\equiv 0)$ that satisfies (6). To do this, proceed as follows:

 (a) Show that there are no nontrivial solutions to the boundary value problem (5)–(6) when $r = 0$.

 (b) Represent the general solution to (5) for $r > 0$ in terms of sines, cosines, hyperbolic sines, and hyperbolic cosines.

 (c) Substitute the general solution obtained in part (b) into the equations (6) to obtain four linear algebraic equations for the four coefficients appearing in the general solution.

 (d) Show that the system of equations in part (c) has nontrivial solutions only for those values of r satisfying

(7) $\cosh(rL) = \sec(rL)$.

 (e) On the same coordinate system, sketch the graphs of $\cosh(rL)$ and $\sec(rL)$ versus r for $L = 1$, and argue that equation (7) has an infinite number of positive solutions.

 (f) For $L = 1$, determine the first two positive solutions to (7) numerically, and plot the corresponding solutions to the boundary value problem (5)–(6). [*Hint:* You may want to use Newton's method in Appendix A.]

Laplace Transforms

7.1 INTRODUCTION: A MIXING PROBLEM

Figure 7.1 depicts a mixing problem with valved input feeders. At time $t = 0$, valve A is opened, delivering 6 L/min of a brine solution containing 4 kg of salt per liter. At $t = 10$ minutes, valve A is closed and valve B is opened, delivering 6 L/min of brine at a concentration of 2 kg/L. Initially there are 300 kg of salt dissolved in 1000 L of water in the tank. The exit valve C, which empties the tank at 6 L/min, maintains the contents of the tank at constant volume. Assuming the solution is kept well stirred, determine the amount of salt in the tank at all times $t > 0$.

Figure 7.1 Mixing tank with valve A open

We analyzed a simpler version of this problem in Example 1 of Section 3.2. Let $x(t)$ be the amount of salt (in kilograms) in the tank at time t. Then of course $x(t)/1000$ is the concentration, in kg/L. The salt content is depleted at the rate (6 L/min) \times $(x(t)/1000 \text{ kg/L}) = 3x(t)/500$ kg/min through the exit valve. Simultaneously it is enriched through valves A and B at the rate

$g(t)$, given by

(1) $g(t) = \begin{cases} 4 \text{ kg/L} \times 6 \text{ L/min} = 24 \text{ kg/min} , & 0 < t < 10 \text{ (valve } A) , \\ 2 \text{ kg/L} \times 6 \text{ L/min} = 12 \text{ kg/min} , & t > 10 \text{ (valve } B) . \end{cases}$

Thus $x(t)$ changes at a rate

$$\frac{d}{dt}x(t) = g(t) - \frac{3x(t)}{500} ,$$

or

(2) $\dfrac{dx}{dt} + \dfrac{3}{500}x = g(t) ,$

with initial condition

(3) $x(0) = 300 .$

To solve the initial value problem (2)–(3) using the techniques of Chapter 4, we would have to break up the time interval $(0, \infty)$ into two subintervals $(0, 10)$ and $(10, \infty)$. On these subintervals, the nonhomogeneous term $g(t)$ is constant, and the method of undetermined coefficients could be applied to equation (2) to determine general solutions for each subinterval, each containing one arbitrary constant (in the associated homogeneous solutions). The initial condition (3) would fix this constant for $0 < t < 10$, but then we would need to evaluate $x(10)$ and use it to reset the constant in the general solution for $t > 10$.

Our purpose here is to illustrate a new approach using Laplace transforms. As we shall see, this method offers several advantages over the previous techniques. For one thing, it is much more convenient in solving initial value problems for linear constant coefficient equations when the forcing term contains jump discontinuities.

The **Laplace transform** of a function $f(t)$, defined on $[0, \infty)$, is given by[†]

(4) $F(s) := \displaystyle\int_0^\infty e^{-st}f(t)\,dt .$

Thus we multiply $f(t)$ by e^{-st} and integrate with respect to t from 0 to ∞. This takes a function of t and produces a function of s. For example, the Laplace transform $G(s)$ of the discontinuous function $g(t)$ in (1) is given by[††]

$$G(s) = \int_{t=0}^\infty e^{-st}g(t)\,dt = \int_{t=0}^{10} e^{-st}24\,dt + \int_{t=10}^\infty e^{-st}12\,dt$$

$$= 24\frac{e^{-st}}{-s}\Big|_{t=0}^{t=10} + 12\frac{e^{-st}}{-s}\Big|_{t=10}^{t=\infty} .$$

[†] **Historical Footnote:** The Laplace transform was first introduced by Pierre Laplace in 1779 in his research on probability. G. Doetsch helped develop the use of Laplace transforms to solve differential equations. His work in the 1930s served to justify the operational calculus procedures earlier used by Oliver Heaviside.

[††] In computing the integral we treat s as a constant.

Now for $s > 0$, the function e^{-st} tends to zero as $t \to +\infty$, so

(5) $\qquad G(s) = \dfrac{24}{-s}(e^{-10s} - 1) + \dfrac{12}{-s}(0 - e^{-10s}) = \dfrac{24}{s} - \dfrac{12}{s}e^{-10s}, \quad$ for $\quad s > 0$.

Note that although the discontinuous function $g(t)$ is specified "in pieces," its transform $G(s)$ is continuous and is represented by a single formula. This is one benefit of the Laplace transform; it handles many discontinuous functions conveniently.

But the biggest advantage of working in the "s-domain" instead of the "t-domain" arises when solving initial value problems such as (2)–(3). *The Laplace transform replaces these differential equations in the t-domain by (simpler) algebraic equations in the s-domain!* To see how this works, let's denote the Laplace transform of the unknown function $x(t)$ by $X(s)$:

$$X(s) = \int_0^\infty e^{-st}x(t)\,dt \ .$$

Multiplying both sides of the equation (2) by e^{-st} and integrating from $t = 0$ to $t = \infty$, we get

$$\int_0^\infty e^{-st}\left\{ \frac{dx}{dt} + \frac{3}{500}x \right\}dt = \int_0^\infty e^{-st}g(t)\,dt$$

or

(6) $\qquad \displaystyle\int_0^\infty e^{-st}\frac{dx}{dt}\,dt + \frac{3}{500}X(s) = G(s)$.

Next we rewrite the integral in (6) using the integration by parts formula, $\int u\,dv = uv - \int v\,du$, with $u = e^{-st}$ and $dv = x'(t)dt$ to obtain

(7) $\qquad \displaystyle\int_0^\infty e^{-st}\frac{dx}{dt}(t)\,dt = e^{-st}x(t)\Big|_{t=0}^{t=\infty} + s\int_0^\infty e^{-st}x(t)\,dt$

$$= \lim_{t\to\infty} e^{-st}x(t) - e^0 x(0) + sX(s) \ .$$

Making the physically reasonable assumption that $x(t)$ remains bounded as $t \to \infty$ (in fact, the salt concentration in the tank, $x(t)/1000$ kg/L, should eventually level off at the incoming value 2 kg/L from valve B), we conclude that for $s > 0$ the product $e^{-st}x(t)$ tends to zero as $t \to \infty$. Hence (7) becomes

(8) $\qquad \displaystyle\int_0^\infty e^{-st}\frac{dx}{dt}(t)\,dt = -e^0 x(0) + sX(s) = -300 + sX(s)$,

where we have inserted the given initial value $x(0) = 300$. We have thus found a simple formula that relates the Laplace transform of the derivative $x'(t)$ to the Laplace transform of $x(t)$. In essence, this formula says that we replace the operation "differentiation with respect to t" by "multiplication by s" when we move from the t-domain to the s-domain.

Insertion of (8) into (6) yields

$$-300 + sX(s) + \frac{3}{500}X(s) = G(s) \ ,$$

which is just an algebraic equation for $X(s)$. Solving it, we find

$$X(s) = \frac{300}{s + 3/500} + \frac{G(s)}{s + 3/500} \, ,$$

which together with the formula (5) for $G(s)$ gives

(9) $$X(s) = \frac{300}{s + 3/500} + \frac{24}{s(s + 3/500)} - \frac{12e^{-10s}}{s(s + 3/500)} \, .$$

With $X(s)$ determined, the final step consists of returning to the t-domain to recover $x(t)$ from its Laplace transform—that is, taking the *inverse transform*. This is not always easy, but in Sections 7.4 and 7.6 we will see how to use the table of Laplace transforms (on the inside back cover of this book) to derive from (9) the formula for the salt content $x(t)$:

(10) $$x(t) = 4000 - 3700e^{-3t/500} - 2000 \cdot \begin{cases} 0 \, , & t \le 10 \, , \\ \left[1 - e^{-3(t-10)/500} \right] , & t \ge 10 \, . \end{cases}$$

See Figure 7.2.

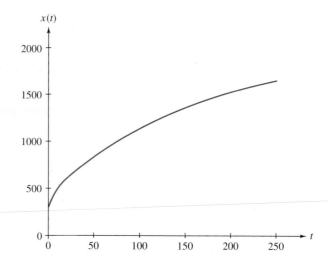

Figure 7.2 Solution to mixing tank example

Note that to arrive at (10), we did not have to take derivatives of trial solutions, break up intervals, or evaluate constants through initial data. The Laplace transform machinery replaces all of these operations by basic algebra: addition, subtraction, multiplication, division—and, of course, the judicious use of the table. Figure 7.3 depicts the advantages of the transform method.

Unfortunately the Laplace transform method is less helpful with equations containing variable coefficients or nonlinear equations (and sometimes determining inverse transforms can be a Herculean task!). But it is ideally suited for many problems arising in applications. Thus we devote this chapter to this important topic.

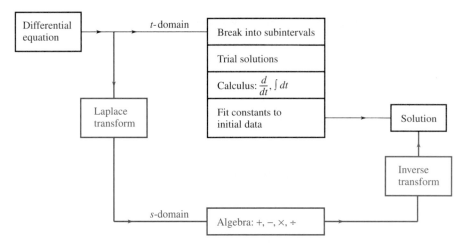

Figure 7.3 Comparison of solution methods

7.2 DEFINITION OF THE LAPLACE TRANSFORM

In earlier chapters we studied differential operators. These operators took a function and mapped or transformed it (via differentiation) into another function. The Laplace transform, which is an integral operator, is another such transformation.

LAPLACE TRANSFORM

Definition 1. Let $f(t)$ be a function on $[0, \infty)$. The **Laplace transform** of f is the function F defined by the integral

(1) $$F(s) := \int_0^\infty e^{-st}f(t)\,dt \ .$$

The domain of $F(s)$ is all the values of s for which the integral in (1) exists.[†] The Laplace transform of f is denoted by both F and $\mathscr{L}\{f\}$.

Notice that the integral in (1) is an **improper** integral. More precisely,

$$\int_0^\infty e^{-st}f(t)\,dt := \lim_{N\to\infty} \int_0^N e^{-st}f(t)\,dt$$

whenever the limit exists.

[†]We treat s as real-valued, but in certain applications s may be a complex variable. For a detailed treatment of complex-valued Laplace transforms, see *Complex Variables and the Laplace Transform for Engineers,* by Wilbur R. LePage (Dover Publications, New York, 1980), or *Fundamentals of Complex Analysis for Mathematics, Science, and Engineering* (2nd edition), by E. B. Saff and A. D. Snider (Prentice Hall, 1993).

EXAMPLE 1 Determine the Laplace transform of the constant function $f(t) = 1, t \geq 0$.

SOLUTION Using the definition of the transform, we compute

$$F(s) = \int_0^\infty e^{-st} \cdot 1 \, dt = \lim_{N \to \infty} \int_0^N e^{-st} dt$$

$$= \lim_{N \to \infty} \frac{-e^{-st}}{s} \Big|_{t=0}^{t=N} = \lim_{N \to \infty} \left[\frac{1}{s} - \frac{e^{-sN}}{s} \right].$$

Since $e^{-sN} \to 0$ when $s > 0$ is fixed and $N \to \infty$, we get

$$F(s) = \frac{1}{s} \quad \text{for} \quad s > 0 .$$

When $s \leq 0$, the integral $\int_0^\infty e^{-st} dt$ diverges. (Why?) Hence $F(s) = 1/s$, with the domain of $F(s)$ being all $s > 0$. ∎ End

EXAMPLE 2 Determine the Laplace transform of $f(t) = e^{at}$, where a is a constant.

SOLUTION Using the definition of the transform,

$$F(s) = \int_0^\infty e^{-st} e^{at} dt = \int_0^\infty e^{-(s-a)t} dt$$

$$= \lim_{N \to \infty} \int_0^N e^{-(s-a)t} dt = \lim_{N \to \infty} \frac{-e^{-(s-a)t}}{s-a} \Big|_0^N$$

$$= \lim_{N \to \infty} \left[\frac{1}{s-a} - \frac{e^{-(s-a)N}}{s-a} \right]$$

$$= \frac{1}{s-a} \quad \text{for} \quad s > a .$$

Again, if $s \leq a$ the integral diverges, and hence the domain of $F(s)$ is all $s > a$. ∎ End

It is comforting to note from Example 2 that the transform of the constant function $f(t) = 1 = e^{0t}$ is $1/(s - 0) = 1/s$, which agrees with the solution in Example 1.

EXAMPLE 3 Find $\mathcal{L}\{\sin bt\}$, where b is a nonzero constant.

SOLUTION We need to compute

$$\mathcal{L}\{\sin bt\}(s) = \int_0^\infty e^{-st} \sin bt \, dt = \lim_{N \to \infty} \int_0^N e^{-st} \sin bt \, dt .$$

Referring to the table of integrals on the inside front cover, we see that

$$\mathscr{L}\{\sin bt\}(s) = \lim_{N \to \infty} \left[\frac{e^{-st}}{s^2 + b^2}(-s \sin bt - b \cos bt) \Big|_0^N \right]$$

$$= \lim_{N \to \infty} \left[\frac{b}{s^2 + b^2} - \frac{e^{-sN}}{s^2 + b^2}(s \sin bN + b \cos bN) \right]$$

$$= \frac{b}{s^2 + b^2} \quad \text{for} \quad s > 0$$

(since for such s we have $\lim_{N \to \infty} e^{-sN}(s \sin bN + b \cos bN) = 0$; see Problem 32). ■ End

EXAMPLE 4 Determine the Laplace transform of

$$f(t) = \begin{cases} 2, & 0 < t < 5, \\ 0, & 5 < t < 10, \\ e^{4t}, & 10 < t. \end{cases}$$

SOLUTION Since $f(t)$ is defined by a different formula on different intervals, we begin by breaking up the integral in (1) into three separate parts.† Thus

$$F(s) = \int_0^\infty e^{-st} f(t) \, dt$$

$$= \int_0^5 e^{-st} \cdot 2 \, dt + \int_5^{10} e^{-st} \cdot 0 \, dt + \int_{10}^\infty e^{-st} e^{4t} \, dt$$

$$= 2 \int_0^5 e^{-st} \, dt + \lim_{N \to \infty} \int_{10}^N e^{-(s-4)t} \, dt$$

$$= \frac{2}{s} - \frac{2e^{-5s}}{s} + \lim_{N \to \infty} \left[\frac{e^{-10(s-4)}}{s - 4} - \frac{e^{-(s-4)N}}{s - 4} \right]$$

$$= \frac{2}{s} - \frac{2e^{-5s}}{s} + \frac{e^{-10(s-4)}}{s - 4} \quad \text{for} \quad s > 4. \quad ■ \text{End}$$

Notice that the function $f(t)$ of Example 4 has jump discontinuities at $t = 5$ and $t = 10$. These values are reflected in the exponential terms e^{-5s} and e^{-10s} that appear in the formula for $F(s)$. We shall make this connection more precise when we discuss the unit step function in Section 7.6.

An important property of the Laplace transform is its **linearity.** That is, the Laplace transform \mathscr{L} is a linear operator.

†Notice that $f(t)$ is not defined at the points $t = 0$, 5, and 10. Nevertheless, the integral in (1) is still meaningful and unaffected by the function's values at finitely many points.

LINEARITY OF THE TRANSFORM

Theorem 1. Let $f, f_1,$ and f_2 be functions whose Laplace transforms exist for $s > \alpha$ and let c be a constant. Then, for $s > \alpha$,

(2) $\mathscr{L}\{f_1 + f_2\} = \mathscr{L}\{f_1\} + \mathscr{L}\{f_2\}$,

(3) $\mathscr{L}\{cf\} = c\mathscr{L}\{f\}$.

Proof. Using the linearity properties of integration, we have for $s > \alpha$

$$\mathscr{L}\{f_1 + f_2\}(s) = \int_0^\infty e^{-st}\big[f_1(t) + f_2(t)\big]\,dt$$

$$= \int_0^\infty e^{-st}f_1(t)\,dt + \int_0^\infty e^{-st}f_2(t)\,dt$$

$$= \mathscr{L}\{f_1\}(s) + \mathscr{L}\{f_2\}(s) .$$

Hence equation (2) is satisfied. In a similar fashion, we see that

$$\mathscr{L}\{cf\}(s) = \int_0^\infty e^{-st}\big[cf(t)\big]\,dt = c\int_0^\infty e^{-st}f(t)\,dt$$

$$= c\mathscr{L}\{f\}(s) .\quad \blacksquare\text{End}$$

EXAMPLE 5 Determine $\mathscr{L}\{11 + 5e^{4t} - 6\sin 2t\}$.

SOLUTION From the linearity property, we know that the Laplace transform of the sum of any finite number of functions is the sum of their Laplace transforms. Thus

$$\mathscr{L}\{11 + 5e^{4t} - 6\sin 2t\} = \mathscr{L}\{11\} + \mathscr{L}\{5e^{4t}\} + \mathscr{L}\{-6\sin 2t\}$$
$$= 11\mathscr{L}\{1\} + 5\mathscr{L}\{e^{4t}\} - 6\mathscr{L}\{\sin 2t\} .$$

In Examples 1, 2, and 3, we determined that

$$\mathscr{L}\{1\}(s) = \frac{1}{s} , \qquad \mathscr{L}\{e^{4t}\}(s) = \frac{1}{s - 4} , \qquad \mathscr{L}\{\sin 2t\}(s) = \frac{2}{s^2 + 2^2} .$$

Using these results, we find

$$\mathscr{L}\{11 + 5e^{4t} - 6\sin 2t\}(s) = 11\left(\frac{1}{s}\right) + 5\left(\frac{1}{s - 4}\right) - 6\left(\frac{2}{s^2 + 4}\right)$$

$$= \frac{11}{s} + \frac{5}{s - 4} - \frac{12}{s^2 + 4} .$$

Since $\mathscr{L}\{1\}, \mathscr{L}\{e^{4t}\},$ and $\mathscr{L}\{\sin 2t\}$ are all defined for $s > 4$, so is the transform $\mathscr{L}\{11 + 5e^{4t} - 6\sin 2t\}$. \blacksquareEnd

Existence of the Transform

There are functions for which the improper integral in (1) fails to converge for any value of s. For example, this is the case for the function $f(t) = 1/t$, which grows too fast near zero. Likewise, no Laplace transform exists for the function $f(t) = e^{t^2}$, which increases too rapidly as $t \to \infty$. Fortunately, the set of functions for which the Laplace transform is defined includes many of the functions that arise in applications involving linear differential equations. We now discuss some properties that will (collectively) ensure the existence of the Laplace transform.

A function $f(t)$ on $[a, b]$ is said to have a **jump discontinuity** at $t_0 \in (a, b)$ if $f(t)$ is discontinuous at t_0, but the one-sided limits

$$\lim_{t \to t_0^-} f(t) \quad \text{and} \quad \lim_{t \to t_0^+} f(t)$$

exist as finite numbers. If the discontinuity occurs at an endpoint, $t_0 = a$ (or b), a jump discontinuity occurs if the one-sided limit of $f(t)$ as $t \to a^+ (t \to b^-)$ exists as a finite number. We can now define piecewise continuity.

PIECEWISE CONTINUITY

Definition 2. A function $f(t)$ is said to be **piecewise continuous on a finite interval** $[a, b]$ if $f(t)$ is continuous at every point in $[a, b]$, except possibly for a finite number of points at which $f(t)$ has a jump discontinuity.

A function $f(t)$ is said to be **piecewise continuous on $[0, \infty)$** if $f(t)$ is piecewise continuous on $[0, N]$ for all $N > 0$.

EXAMPLE 6 Show that

$$f(t) = \begin{cases} t , & 0 < t < 1 , \\ 2 , & 1 < t < 2 , \\ (t - 2)^2 , & 2 \leq t \leq 3 , \end{cases}$$

whose graph is sketched in Figure 7.4, is piecewise continuous on $[0, 3]$.

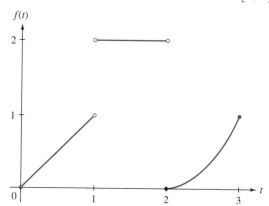

Figure 7.4 Graph of $f(t)$ in Example 6

SOLUTION From the graph of $f(t)$ we see that $f(t)$ is continuous on the intervals $(0, 1)$, $(1, 2)$, and $(2, 3]$. More-over, at the points of discontinuity, $t = 0$, 1, and 2, the function has jump discontinuities, since the one-sided limits exist as finite numbers. In particular, at $t = 1$, the left-hand limit is 1 and the right-hand limit is 2. Therefore $f(t)$ is piecewise continuous on $[0, 3]$. ■ End

Observe that the function $f(t)$ of Example 4 is piecewise continuous on $[0, \infty)$ because it is piecewise continuous on every finite interval of the form $[0, N]$, with $N > 0$. In contrast, the function $f(t) = 1/t$ is not piecewise continuous on any interval containing the origin, since it has an "infinite jump" at the origin (see Figure 7.5).

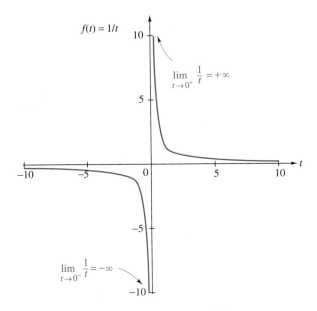

Figure 7.5 Infinite jump at origin

A function that is piecewise continuous on a *finite* interval is necessarily integrable over that interval. However, piecewise continuity on $[0, \infty)$ is not enough to guarantee the existence (as a finite number) of the improper integral over $[0, \infty)$; we also need to consider the growth of the integrand for large t. Roughly speaking, we shall show that the Laplace transform of a piecewise continuous function exists, provided the function does not grow "faster than an exponential."

EXPONENTIAL ORDER α

Definition 3. A function $f(t)$ is said to be of **exponential order** α if there exist positive constants T and M such that

(4) $|f(t)| \leq Me^{\alpha t}$, for all $t \geq T$.

For example, $f(t) = e^{5t} \sin 2t$ is of exponential order $\alpha = 5$ since

$$|e^{5t} \sin 2t| \leq e^{5t} ,$$

and hence (4) holds with $M = 1$ and T any positive constant.

We use the phrase $f(t)$ *is of exponential order* to mean that for *some* value of α, the function $f(t)$ satisfies the conditions of Definition 3; that is, $f(t)$ grows no faster than a function of the form $Me^{\alpha t}$. The function e^{t^2} is *not* of exponential order. To see this, observe that

$$\lim_{t \to \infty} \frac{e^{t^2}}{e^{\alpha t}} = \lim_{t \to \infty} e^{t(t-\alpha)} = +\infty$$

for any α. Consequently, e^{t^2} grows faster than $e^{\alpha t}$ for every choice of α.

The functions usually encountered in solving linear differential equations with constant coefficients (e.g., polynomials, exponentials, sines, and cosines) are both piecewise continuous and of exponential order. As we now show, the Laplace transforms of such functions exist for large enough values of s.

CONDITIONS FOR EXISTENCE OF THE TRANSFORM

Theorem 2. If $f(t)$ is piecewise continuous on $[0, \infty)$ and of exponential order α, then $\mathscr{L}\{f\}(s)$ exists for $s > \alpha$.

Proof. We need to show that the integral

$$\int_0^\infty e^{-st} f(t)\,dt$$

converges for $s > \alpha$. We begin by breaking up this integral into two separate integrals:

$$\textbf{(5)} \qquad \int_0^T e^{-st} f(t)\,dt + \int_T^\infty e^{-st} f(t)\,dt ,$$

where T is chosen so that inequality (4) holds. The first integral in (5) exists because $f(t)$ and hence $e^{-st} f(t)$ are piecewise continuous on the interval $[0, T]$ for any fixed s. To see that the second integral in (5) converges, we use the **comparison test for improper integrals.**

Since $f(t)$ is of exponential order α, we have for $t \geq T$

$$|f(t)| \leq Me^{\alpha t} ,$$

and hence

$$|e^{-st} f(t)| = e^{-st}|f(t)| \leq Me^{-(s-\alpha)t} ,$$

for all $t \geq T$. Now for $s > \alpha$.

$$\int_T^\infty Me^{-(s-\alpha)t}\,dt = M \int_T^\infty e^{-(s-\alpha)t}\,dt = \frac{Me^{-(s-\alpha)T}}{s - \alpha} < \infty .$$

Since $|e^{-st} f(t)| \leq Me^{-(s-\alpha)t}$ for $t \geq T$ and the improper integral of the larger function

converges for $s > \alpha$, then, by the comparison test, the integral

$$\int_T^\infty e^{-st} f(t)\, dt$$

converges for $s > \alpha$. Finally, because the two integrals in (5) exist, the Laplace transform $\mathscr{L}\{f\}(s)$ exists for $s > \alpha$. ∎End

In Table 7.1 we have listed the Laplace transforms of some of the elementary functions. The reader should become familiar with these, since they are frequently encountered in solving linear differential equations with constant coefficients. The entries in the table can be derived from the definition of the Laplace transform. A more elaborate table of transforms is given on the inside back cover of this book.

TABLE 7.1 BRIEF TABLE OF LAPLACE TRANSFORMS

$f(t)$	$F(s) = \mathscr{L}\{f\}(s)$
1	$\dfrac{1}{s}, \quad s > 0$
e^{at}	$\dfrac{1}{s-a}, \quad s > a$
$t^n, \quad n = 1, 2, \ldots$	$\dfrac{n!}{s^{n+1}}, \quad s > 0$
$\sin bt$	$\dfrac{b}{s^2 + b^2}, \quad s > 0$
$\cos bt$	$\dfrac{s}{s^2 + b^2}, \quad s > 0$
$e^{at} t^n, \quad n = 1, 2, \ldots$	$\dfrac{n!}{(s-a)^{n+1}}, \quad s > a$
$e^{at} \sin bt$	$\dfrac{b}{(s-a)^2 + b^2}, \quad s > a$
$e^{at} \cos bt$	$\dfrac{s-a}{(s-a)^2 + b^2}, \quad s > a$

EXERCISES 7.2

In Problems 1 through 12, use Definition 1 to determine the Laplace transform of the given function.

1. t .
2. t^2 .
3. e^{6t} .
4. te^{3t} .
5. $\cos 2t$.
6. $\cos bt$, b a constant.
7. $e^{2t} \cos 3t$.
8. $e^{-t} \sin 2t$.

9. $f(t) = \begin{cases} 0, & 0 < t < 2, \\ t, & 2 < t. \end{cases}$

10. $f(t) = \begin{cases} 1 - t, & 0 < t < 1, \\ 0, & 1 < t. \end{cases}$

11. $f(t) = \begin{cases} \sin t, & 0 < t < \pi, \\ 0, & \pi < t. \end{cases}$

12. $f(t) = \begin{cases} e^{2t}, & 0 < t < 3, \\ 1, & 3 < t. \end{cases}$

In Problems 13 through 20, use the Laplace transform table and the linearity of the Laplace transform to determine the following transforms.

13. $\mathcal{L}\{6e^{-3t} - t^2 + 2t - 8\}$.

14. $\mathcal{L}\{5 - e^{2t} + 6t^2\}$.

15. $\mathcal{L}\{t^3 - te^t + e^{4t}\cos t\}$.

16. $\mathcal{L}\{t^2 - 3t - 2e^{-t}\sin 3t\}$.

17. $\mathcal{L}\{e^{3t}\sin 6t - t^3 + e^t\}$.

18. $\mathcal{L}\{t^4 - t^2 - t + \sin\sqrt{2}t\}$.

19. $\mathcal{L}\{t^4 e^{5t} - e^t\cos\sqrt{7}t\}$.

20. $\mathcal{L}\{e^{-2t}\cos\sqrt{3}t - t^2 e^{-2t}\}$.

In Problems 21 through 28, determine whether $f(t)$ is continuous, piecewise continuous, or neither on $[0, 10]$ and sketch the graph of $f(t)$.

21. $f(t) = \begin{cases} 1, & 0 \le t \le 1, \\ (t - 2)^2, & 1 < t \le 10, \end{cases}$

22. $f(t) = \begin{cases} 0, & 0 \le t < 2, \\ t, & 2 \le t \le 10. \end{cases}$

23. $f(t) = \begin{cases} 1, & 0 \le t < 1, \\ t - 1, & 1 < t < 3, \\ t^2 - 4, & 3 < t \le 10. \end{cases}$

24. $f(t) = \dfrac{t^2 - 3t + 2}{t^2 - 4}$.

25. $f(t) = \dfrac{t^2 - t - 20}{t^2 + 7t + 10}$.

26. $f(t) = \dfrac{t}{t^2 - 1}$.

27. $f(t) = \begin{cases} 1/t, & 0 < t < 1, \\ 1, & 1 \le t \le 2, \\ 1 - t, & 2 < t \le 10. \end{cases}$

28. $f(t) = \begin{cases} \dfrac{\sin t}{t}, & t \ne 0, \\ 1, & t = 0. \end{cases}$

29. Which of the following functions are of exponential order?

(a) $t^3\sin t$. (b) $100e^{49t}$.
(c) e^{t^3} (d) $t \ln t$.
(e) $\cosh(t^2)$ (f) $\dfrac{1}{t^2 + 1}$.
(g) $\sin(t^2) + t^4 e^{6t}$. (h) $3 - e^{t^2} + \cos 4t$.
(i) $\exp\{t^2/(t + 1)\}$. (j) $\sin(e^{t^2}) + e^{\sin t}$.

30. For the transforms $F(s)$ in Table 7.1, what can be said about $\lim_{s\to\infty} F(s)$?

31. Thanks to Euler's formula (page 192) and the algebraic properties of complex numbers, several of the entries of Table 7.1 can be derived from a single formula; namely,

(6) $\mathcal{L}\{e^{(a+ib)t}\}(s) = \dfrac{s - a + ib}{(s - a)^2 + b^2}$, $s > a$.

(a) By computing the integral in the definition of the Laplace transform on page 373 with $f(t) = e^{(a+ib)t}$, show that

$$\mathcal{L}\{e^{(a+ib)t}\}(s) = \frac{1}{s - (a + ib)}, \quad s > a.$$

(b) Deduce (6) from part (a) by showing that

$$\frac{1}{s - (a + ib)} = \frac{s - a + ib}{(s - a)^2 + b^2}.$$

(c) By equating the real and imaginary parts in formula (6), deduce the last two entries in Table 7.1.

32. Prove that for fixed $s > 0$, we have

$$\lim_{N\to\infty} e^{-sN}(s \sin bN + b \cos bN) = 0.$$

33. Prove that if f is piecewise continuous on $[a, b]$ and g is continuous on $[a, b]$, then the product fg is piecewise continuous on $[a, b]$.

7.3 PROPERTIES OF THE LAPLACE TRANSFORM

In the previous section we defined the Laplace transform of a function $f(t)$ as

$$\mathcal{L}\{f\}(s) := \int_0^\infty e^{-st} f(t)\, dt \ .$$

Using this definition to get an explicit expression for $\mathcal{L}\{f\}$ requires the evaluation of the improper integral—frequently a tedious task! We have already seen how the linearity property of the transform can help relieve this burden. In this section, we discuss some further properties of the Laplace transform that simplify its computation. These new properties will also enable us to use the Laplace transform to solve initial value problems.

TRANSLATION IN s

Theorem 3. If the Laplace transform $\mathcal{L}\{f\}(s) = F(s)$ exists for $s > \alpha$, then

$$(1) \qquad \mathcal{L}\{e^{at}f(t)\}(s) = F(s - a)$$

for $s > \alpha + a$.

Proof. We simply compute

$$\mathcal{L}\{e^{at}f(t)\}(s) = \int_0^\infty e^{-st} e^{at} f(t)\, dt$$

$$= \int_0^\infty e^{-(s-a)t} f(t)\, dt$$

$$= F(s - a) \ . \quad \blacksquare \text{ End}$$

Theorem 3 illustrates the effect on the Laplace transform of multiplication of a function $f(t)$ by e^{at} .

EXAMPLE 1 Determine the Laplace transform of $e^{at}\sin bt$.

SOLUTION In Example 3 in Section 7.2, we found that

$$\mathcal{L}\{\sin bt\}(s) = F(s) = \frac{b}{s^2 + b^2} \ .$$

Thus, by the translation property of $F(s)$, we have

$$\mathcal{L}\{e^{at}\sin bt\}(s) = F(s - a) = \frac{b}{(s - a)^2 + b^2} \ . \quad \blacksquare \text{ End}$$

LAPLACE TRANSFORM OF THE DERIVATIVE

Theorem 4. Let $f(t)$ be continuous on $[0, \infty)$ and $f'(t)$ be piecewise continuous on $[0, \infty)$, with both of exponential order α. Then, for $s > \alpha$,

(2) $\mathcal{L}\{f'\}(s) = s\mathcal{L}\{f\}(s) - f(0)$.

Proof. Since $\mathcal{L}\{f'\}$ exists, we can use integration by parts $\left(\text{with } u = e^{-st} \text{ and } dv = f'(t)dt\right)$ to obtain

$$(3) \qquad \mathcal{L}\{f'\}(s) = \int_0^\infty e^{-st} f'(t)\,dt = \lim_{N\to\infty} \int_0^N e^{-st} f'(t)\,dt$$

$$= \lim_{N\to\infty}\left[e^{-st}f(t)\bigg|_0^N + s\int_0^N e^{-st}f(t)\,dt \right]$$

$$= \lim_{N\to\infty} e^{-sN}f(N) - f(0) + s\lim_{N\to\infty}\int_0^N e^{-st}f(t)\,dt$$

$$= \lim_{N\to\infty} e^{-sN}f(N) - f(0) + s\mathcal{L}\{f\}(s)$$.

To evaluate $\lim_{N\to\infty} e^{-sN}f(N)$, we observe that since $f(t)$ is of exponential order α, there exists a constant M such that for N large,

$$\left| e^{-sN}f(N) \right| \le e^{-sN}Me^{\alpha N} = Me^{-(s-\alpha)N}$$.

Hence, for $s > \alpha$,

$$0 \le \lim_{N\to\infty} \left| e^{-sN}f(N) \right| \le \lim_{N\to\infty} Me^{-(s-\alpha)N} = 0$$,

so

$$\lim_{N\to\infty} e^{-sN}f(N) = 0$$

for $s > \alpha$. Equation (3) now reduces to

$$\mathcal{L}\{f'\}(s) = s\mathcal{L}\{f\}(s) - f(0)$$. ■ End

Using induction, we can extend the last theorem to higher-order derivatives of $f(t)$. For example,

$$\mathcal{L}\{f''\}(s) = s\mathcal{L}\{f'\}(s) - f'(0)$$
$$= s\left[s\mathcal{L}\{f\}(s) - f(0) \right] - f'(0)$$,

which simplifies to

$$\mathcal{L}\{f''\}(s) = s^2\mathcal{L}\{f\}(s) - sf(0) - f'(0)$$.

In general we obtain the following result.

LAPLACE TRANSFORM OF HIGHER-ORDER DERIVATIVES

Theorem 5. Let $f(t), f'(t), \ldots, f^{(n-1)}(t)$ be continuous on $[0, \infty)$ and let $f^{(n)}(t)$ be piecewise continuous on $[0, \infty)$, with all these functions of exponential order α. Then, for $s > \alpha$,

$$(4) \qquad \mathscr{L}\left\{ f^{(n)} \right\}(s) = s^n \mathscr{L}\{f\}(s) - s^{n-1}f(0) - s^{n-2}f'(0) - \cdots - f^{(n-1)}(0) \ .$$

The last two theorems shed light on the reason why the Laplace transform is such a useful tool in solving initial value problems. Roughly speaking, they tell us that by using the Laplace transform we can replace "differentiation with respect to t" with "multiplication by s," thereby converting a differential equation into an algebraic one. This idea is explored in Section 7.5. For now, we show how Theorem 4 can be helpful in computing a Laplace transform.

EXAMPLE 2 Using Theorem 4 and the fact that

$$\mathscr{L}\{\sin bt\}(s) = \frac{b}{s^2 + b^2} \ ,$$

determine $\mathscr{L}\{\cos bt\}$.

SOLUTION Let $f(t) = \sin bt$. Then $f(0) = 0$ and $f'(t) = b \cos bt$. Substituting into equation (2), we have

$$\mathscr{L}\{f'\}(s) = s\mathscr{L}\{f\}(s) - f(0) \ ,$$
$$\mathscr{L}\{b \cos bt\}(s) = s\mathscr{L}\{\sin bt\}(s) - 0 \ ,$$
$$b\mathscr{L}\{\cos bt\}(s) = \frac{sb}{s^2 + b^2} \ .$$

Dividing by b gives

$$\mathscr{L}\{\cos bt\}(s) = \frac{s}{s^2 + b^2} \ . \quad \blacksquare \text{ End}$$

EXAMPLE 3 Prove the following identity for continuous functions $f(t)$ (assuming the transforms exist):

$$(5) \qquad \mathscr{L}\left\{ \int_0^t f(\tau)d\tau \right\}(s) = \frac{1}{s}\mathscr{L}\{f(t)\}(s) \ .$$

Use it to verify the solution to Example 2.

SOLUTION Define the function $g(t)$ by the integral

$$g(t) := \int_0^t f(\tau)d\tau \ .$$

Observe that $g(0) = 0$ and $g'(t) = f(t)$. Thus if we apply Theorem 4 to $g(t)$ $\big($instead of $f(t)\big)$,

equation (2) on page 383 reads

$$\mathcal{L}\{f(t)\}(s) = s\mathcal{L}\left\{\int_0^t f(\tau)\,d\tau\right\}(s) - 0 \ ,$$

which is equivalent to equation (5).

Now since

$$\sin bt = \int_0^t b\cos b\tau\,d\tau \ ,$$

equation (5) predicts

$$\mathcal{L}\{\sin bt\}(s) = \frac{1}{s}\mathcal{L}\{b\cos bt\}(s) = \frac{b}{s}\mathcal{L}\{\cos bt\}(s).$$

This identity is indeed valid for the transforms in Example 2. ■ End

Another question arises concerning the Laplace transform. If $F(s)$ is the Laplace transform of $f(t)$, is $F'(s)$ also a Laplace transform of some function of t? The answer is yes:

$$F'(s) = \mathcal{L}\{-tf(t)\}(s) \ .$$

In fact, the following more general assertion holds.

DERIVATIVES OF THE LAPLACE TRANSFORM

Theorem 6. Let $F(s) = \mathcal{L}\{f\}(s)$ and assume $f(t)$ is piecewise continuous on $[0, \infty)$ and of exponential order α. Then, for $s > \alpha$,

(6) $$\mathcal{L}\{t^n f(t)\}(s) = (-1)^n \frac{d^n F}{ds^n}(s) \ .$$

Proof. Consider the identity

$$\frac{dF}{ds}(s) = \frac{d}{ds}\int_0^\infty e^{-st}f(t)\,dt \ .$$

Because of the assumptions on $f(t)$, we can apply a theorem from advanced calculus (sometimes called **Leibniz's rule**) to interchange the order of integration and differentiation:

$$\frac{dF}{ds}(s) = \int_0^\infty \frac{d}{ds}(e^{-st})f(t)\,dt$$

$$= -\int_0^\infty e^{-st}tf(t)\,dt = -\mathcal{L}\{tf(t)\}(s) \ .$$

Thus

$$\mathcal{L}\{tf(t)\}(s) = (-1)\frac{dF}{ds}(s) \ .$$

The general result (6) now follows by induction on n. ■ End

A consequence of the above theorem is that if $f(t)$ is piecewise continuous and of exponential order, then its transform $F(s)$ has derivatives of all orders.

EXAMPLE 4 Determine $\mathcal{L}\{t \sin bt\}$.

SOLUTION We already know that

$$\mathcal{L}\{\sin bt\}(s) = F(s) = \frac{b}{s^2 + b^2} \ .$$

Differentiating $F(s)$, we obtain

$$\frac{dF}{ds}(s) = \frac{-2bs}{(s^2 + b^2)^2} \ .$$

Hence, using formula (6), we have

$$\mathcal{L}\{t \sin bt\}(s) = -\frac{dF}{ds}(s) = \frac{2bs}{(s^2 + b^2)^2} \ . \quad \blacksquare \ \text{End}$$

For easy reference, we have listed in Table 7.2 some of the basic properties of the Laplace transform derived so far.

TABLE 7.2 PROPERTIES OF LAPLACE TRANSFORMS

$\mathcal{L}\{f + g\} = \mathcal{L}\{f\} + \mathcal{L}\{g\}$.

$\mathcal{L}\{cf\} = c\mathcal{L}\{f\}$ for any constant c.

$\mathcal{L}\{e^{at}f(t)\}(s) = \mathcal{L}\{f\}(s - a)$.

$\mathcal{L}\{f'\}(s) = s\mathcal{L}\{f\}(s) - f(0)$.

$\mathcal{L}\{f''\}(s) = s^2\mathcal{L}\{f\}(s) - sf(0) - f'(0)$.

$\mathcal{L}\left\{f^{(n)}\right\}(s) = s^n\mathcal{L}\{f\}(s) - s^{n-1}f(0) - s^{n-2}f'(0) - \cdots - f^{(n-1)}(0)$.

$\mathcal{L}\{t^n f(t)\}(s) = (-1)^n \dfrac{d^n}{ds^n}\left(\mathcal{L}\{f\}(s)\right)$.

EXERCISES 7.3

In Problems 1 through 20, determine the Laplace transform of the given function using Table 7.1, and the properties of the transform given in Table 7.2. [Hint: In Problems 12 through 20, use an appropriate trigonometric identity.]

1. $t^2 + e^t \sin 2t$.

2. $3t^2 - e^{2t}$.

3. $e^{-t}\cos 3t + e^{6t} - 1$.

4. $3t^4 - 2t^2 + 1$.

5. $2t^2 e^{-t} - t + \cos 4t$.

6. $e^{-2t}\sin 2t + e^{3t}t^2$.

7. $(t - 1)^4$.

8. $(1 + e^{-t})^2$.

9. $e^{-t}t \sin 2t$.

10. $te^{2t}\cos 5t$.

11. $\cosh bt$.

12. $\sin 3t \cos 3t$.

13. $\sin^2 t$.

14. $e^{7t}\sin^2 t$.

15. $\cos^3 t$.

16. $t \sin^2 t$.

17. $\sin 2t \sin 5t$.

18. $\cos nt \cos mt$,
 $m \neq n$.

19. $\cos nt \sin mt$,
 $m \neq n$.

20. $t \sin 2t \sin 5t$.

21. Given that $\mathcal{L}\{\cos bt\}(s) = s/(s^2 + b^2)$, use the translation property to compute $\mathcal{L}\{e^{at}\cos bt\}$.

22. Starting with the transform $\mathcal{L}\{1\}(s) = 1/s$, use formula (6) for the derivatives of the Laplace transform to show that $\mathcal{L}\{t^n\}(s) = n!/s^{n+1}$.

23. Use Theorem 4 to show how entry 31 follows from entry 30 in the Laplace transform table on the inside back cover of the text.

24. Show that $\mathcal{L}\{e^{at}t^n\}(s) = n!/(s - a)^{n+1}$ in two ways:
 (a) Use the translation property for $F(s)$.
 (b) Use formula (6) for the derivatives of the Laplace transform.

25. Use formula (6) to help determine
 (a) $\mathcal{L}\{t \cos bt\}$. (b) $\mathcal{L}\{t^2\cos bt\}$.

26. Let $f(t)$ be piecewise continuous on $[0, \infty)$ and of exponential order.
 (a) Show that there exist constants K and α such that
 $$|f(t)| \leq Ke^{\alpha t} \quad \text{for all } t \geq 0 .$$
 (b) By using the definition of the transform and estimating the integral with the help of part (a), prove that
 $$\lim_{s \to \infty} \mathcal{L}\{f\}(s) = 0 .$$

27. Let $f(t)$ be piecewise continuous on $[0, \infty)$ and of exponential order α and assume $\lim_{t \to 0^+}[f(t)/t]$ exists. Show that
 $$\mathcal{L}\left\{\frac{f(t)}{t}\right\}(s) = \int_s^\infty F(u)\,du ,$$
 where $F(s) = \mathcal{L}\{f\}(s)$. [*Hint:* First show that $\frac{d}{ds}\mathcal{L}\{f(t)/t\}(s) = -F(s)$ and then use the result of Problem 26.]

28. Verify the identity in Problem 27 for the following functions. (Use the table of Laplace transforms on the inside back cover.)
 (a) $f(t) = t^5$. (b) $f(t) = t^{3/2}$.

29. The **transfer function** of a linear system is defined as the ratio of the Laplace transform of the output function $y(t)$ to the Laplace transform of the input function $g(t)$, when all initial conditions are zero. If a linear system is governed by the differential equation
 $$y''(t) + 6y'(t) + 10y(t) = g(t) , \qquad t > 0 ,$$
 use the linearity property of the Laplace transform and Theorem 5 on the Laplace transform of higher-order derivatives to determine the transfer function $H(s) = Y(s)/G(s)$ for this system.

30. Find the transfer function, as defined in Problem 29, for the linear system governed by
 $$y''(t) + 5y'(t) + 6y(t) = g(t) , \qquad t > 0 .$$

31. **Translation in t.** Show that for $c > 0$, the translated function
 $$g(t) = \begin{cases} 0 , & 0 < t < c , \\ f(t - c) , & c < t \end{cases}$$
 has Laplace transform
 $$\mathcal{L}\{g\}(s) = e^{-cs}\mathcal{L}\{f\}(s) .$$

In Problems 32 through 35, let $g(t)$ be the given function $f(t)$ translated to the right by c units. Sketch $f(t)$ and $g(t)$ and find $\mathcal{L}\{g(t)\}(s)$. (See Problem 31.)

32. $f(t) \equiv 1$, $c = 2$.

33. $f(t) = t$, $c = 1$.

34. $f(t) = \sin t$, $c = \pi$.

35. $f(t) = \sin t$, $c = \pi/2$.

36. Use equation (5) to derive the formula $\mathcal{L}\{t^n\}(s) = n!/s^{n+1}$. [*Hint:* Start with $\mathcal{L}\{1\}(s)$ and use induction.]

37. **Initial Value Theorem.** Apply the relation
 $$(7) \quad \mathcal{L}\{f'\}(s) = \int_0^\infty e^{-st}f'(t)\,dt = s\mathcal{L}\{f\}(s) - f(0)$$
 to argue that for any function $f(t)$ whose derivative is piecewise continuous and of exponential order on $[0, \infty)$,
 $$f(0) = \lim_{s \to \infty} s\mathcal{L}\{f\}(s) .$$

38. Verify the initial value theorem (Problem 37) for the following functions. (Use the table of Laplace transforms on the inside back cover.)
 (a) 1 . (b) e^t . (c) e^{-t} . (d) $\cos t$.
 (e) $\sin t$. (f) t^2 . (g) $t \cos t$.

7.4 INVERSE LAPLACE TRANSFORM

In Section 7.2, we defined the Laplace transform as an integral operator that maps a function $f(t)$ into a function $F(s)$. In this section we consider the problem of finding the function $f(t)$ when we are given the transform $F(s)$. That is, we seek an **inverse mapping** for the Laplace transform.

To see the usefulness of such an inverse, let's consider the simple initial value problem

(1) $y'' - y = -t$; $y(0) = 0$, $y'(0) = 1$.

If we take the transform of both sides of equation (1) and use the linearity property of the transform, we find

$$\mathscr{L}\{y''\}(s) - Y(s) = -\frac{1}{s^2} \ ,$$

where $Y(s) := \mathscr{L}\{y\}(s)$. We know the initial values of the solution $y(t)$, so we can use Theorem 5, page 384, on the Laplace transform of higher-order derivatives to express

$$\mathscr{L}\{y''\}(s) = s^2 Y(s) - sy(0) - y'(0) = s^2 Y(s) - 1 \ .$$

Substituting for $\mathscr{L}\{y''\}(s)$ yields

$$s^2 Y(s) - 1 - Y(s) = -\frac{1}{s^2} \ .$$

Solving this algebraic equation for $Y(s)$ gives

$$Y(s) = \frac{1 - \left(\dfrac{1}{s^2}\right)}{s^2 - 1} = \frac{s^2 - 1}{s^2(s^2 - 1)} = \frac{1}{s^2} \ .$$

We now recall that $\mathscr{L}\{t\}(s) = 1/s^2$, and since $Y(s) = \mathscr{L}\{y\}(s)$, we have

$$\mathscr{L}\{y\}(s) = 1/s^2 = \mathscr{L}\{t\}(s) \ .$$

It therefore seems reasonable to conclude that $y(t) = t$ is the solution to the initial value problem (1). A quick check confirms this!

Notice that in the above procedure, a crucial step is to determine $y(t)$ from its Laplace transform $Y(s) = 1/s^2$. As we noted, $y(t) = t$ is such a function, but it is *not* the only function whose Laplace function is $1/s^2$. For example, the transform of

$$g(t) := \begin{cases} t \ , & t \neq 6 \ , \\ 0 \ , & t = 6 \end{cases}$$

is also $1/s^2$. This is because the transform is an integral, and integrals are not affected by changing a function's values at isolated points. The significant difference between $y(t)$ and $g(t)$ as far as we are concerned is that $y(t)$ is continuous on $[0, \infty)$, whereas $g(t)$ is not. Naturally, we prefer to work with continuous functions, since solutions to differential equations are continuous.

Fortunately, it can be shown that if two different functions have the same Laplace transform, at most one of them can be continuous.[†] With this in mind we give the following definition.

INVERSE LAPLACE TRANSFORM

Definition 4. Given a function $F(s)$, if there is a function $f(t)$ that is continuous on $[0, \infty)$ and satisfies

$$(2) \qquad \mathscr{L}\{f\} = F \ ,$$

then we say that $f(t)$ is the **inverse Laplace transform** of $F(s)$ and employ the notation $f = \mathscr{L}^{-1}\{F\}$.

In case every function $f(t)$ satisfying (2) is discontinuous (and hence not a solution of a differential equation), one could choose any one of them to be the inverse transform; the distinction among them has no physical significance. (Indeed, two *piecewise* continuous functions satisfying (2) can only differ at their points of discontinuity.)

Naturally the Laplace transform tables will be a great help in determining the inverse Laplace transform of a given function $F(s)$.

EXAMPLE 1 Determine $\mathscr{L}^{-1}\{F\}$, where

(a) $F(s) = \dfrac{2}{s^3}$. (b) $F(s) = \dfrac{3}{s^2 + 9}$. (c) $F(s) = \dfrac{s - 1}{s^2 - 2s + 5}$.

SOLUTION To compute $\mathscr{L}^{-1}\{F\}$, we refer to the Laplace transform table on page 380.

(a) $\mathscr{L}^{-1}\left\{\dfrac{2}{s^3}\right\}(t) = \mathscr{L}^{-1}\left\{\dfrac{2!}{s^3}\right\}(t) = t^2$.

(b) $\mathscr{L}^{-1}\left\{\dfrac{3}{s^2 + 9}\right\}(t) = \mathscr{L}^{-1}\left\{\dfrac{3}{s^2 + 3^2}\right\}(t) = \sin 3t$.

(c) $\mathscr{L}^{-1}\left\{\dfrac{s - 1}{s^2 - 2s + 5}\right\}(t) = \mathscr{L}^{-1}\left\{\dfrac{s - 1}{(s - 1)^2 + 2^2}\right\}(t) = e^t \cos 2t$.

In part (c), we used the technique of completing the square to rewrite the denominator in a form that we could find in the table. ■ End

In practice, we do not always encounter a transform $F(s)$ that exactly corresponds to an entry in the second column of the Laplace transform table. To handle more complicated functions $F(s)$, we use properties of \mathscr{L}^{-1}, just as we used properties of \mathscr{L}. One such tool is the linearity of the inverse Laplace transform, a property that is inherited from the linearity of the operator \mathscr{L}.

[†]For this result and further properties of the Laplace transform and its inverse, we refer the reader to *Operational Mathematics,* 3rd edition, by R. V. Churchill (McGraw-Hill, New York, 1972).

LINEARITY OF THE INVERSE TRANSFORM

Theorem 7. Assume $\mathcal{L}^{-1}\{F\}$, $\mathcal{L}^{-1}\{F_1\}$, and $\mathcal{L}^{-1}\{F_2\}$ exist and are continuous on $[0, \infty)$ and let c be any constant. Then

(3) $\mathcal{L}^{-1}\{F_1 + F_2\} = \mathcal{L}^{-1}\{F_1\} + \mathcal{L}^{-1}\{F_2\}$,

(4) $\mathcal{L}^{-1}\{cF\} = c\mathcal{L}^{-1}\{F\}$.

The proof of Theorem 7 is outlined in Problem 37. We illustrate the usefulness of this theorem in the next example.

EXAMPLE 2 Determine $\mathcal{L}^{-1}\left\{\dfrac{5}{s - 6} - \dfrac{6s}{s^2 + 9} + \dfrac{3}{2s^2 + 8s + 10}\right\}$.

SOLUTION We begin by using the linearity property. Thus

$$\mathcal{L}^{-1}\left\{\frac{5}{s - 6} - \frac{6s}{s^2 + 9} + \frac{3}{2(s^2 + 4s + 5)}\right\}$$

$$= 5\mathcal{L}^{-1}\left\{\frac{1}{s - 6}\right\} - 6\mathcal{L}^{-1}\left\{\frac{s}{s^2 + 9}\right\} + \frac{3}{2}\mathcal{L}^{-1}\left\{\frac{1}{s^2 + 4s + 5}\right\} .$$

Referring to the Laplace transform tables, we see that

$$\mathcal{L}^{-1}\left\{\frac{1}{s - 6}\right\}(t) = e^{6t} \quad \text{and} \quad \mathcal{L}^{-1}\left\{\frac{s}{s^2 + 3^2}\right\}(t) = \cos 3t .$$

This gives us the first two terms. To determine $\mathcal{L}^{-1}\{1/(s^2 + 4s + 5)\}$, we complete the square of the denominator to obtain $s^2 + 4s + 5 = (s + 2)^2 + 1$. We now recognize from the tables that

$$\mathcal{L}^{-1}\left\{\frac{1}{(s + 2)^2 + 1^2}\right\}(t) = e^{-2t}\sin t .$$

Hence

$$\mathcal{L}^{-1}\left\{\frac{5}{s - 6} - \frac{6s}{s^2 + 9} + \frac{3}{2s^2 + 8s + 10}\right\}(t) = 5e^{6t} - 6\cos 3t + \frac{3e^{-2t}}{2}\sin t . \quad \blacksquare\text{End}$$

EXAMPLE 3 Determine $\mathcal{L}^{-1}\left\{\dfrac{5}{(s + 2)^4}\right\}$.

SOLUTION The $(s + 2)^4$ in the denominator suggests that we work with the formula

$$\mathcal{L}^{-1}\left\{\frac{n!}{(s - a)^{n+1}}\right\}(t) = e^{at}t^n .$$

Here we have $a = -2$ and $n = 3$, so $\mathcal{L}^{-1}\{6/(s + 2)^4\}(t) = e^{-2t}t^3$. Using the linearity

property, we find

$$\mathcal{L}^{-1}\left\{\frac{5}{(s+2)^4}\right\}(t) = \frac{5}{6}\mathcal{L}^{-1}\left\{\frac{3!}{(s+2)^4}\right\}(t) = \frac{5}{6}e^{-2t}t^3 . \quad \blacksquare\,\text{End}$$

EXAMPLE 4 Determine $\mathcal{L}^{-1}\left\{\dfrac{3s+2}{s^2+2s+10}\right\}$.

SOLUTION By completing the square, the quadratic in the denominator can be written as

$$s^2 + 2s + 10 = s^2 + 2s + 1 + 9 = (s+1)^2 + 3^2 .$$

The form of $F(s)$ now suggests that we use one or both of the formulas

$$\mathcal{L}^{-1}\left\{\frac{s-a}{(s-a)^2+b^2}\right\}(t) = e^{at}\cos bt ,$$

$$\mathcal{L}^{-1}\left\{\frac{b}{(s-a)^2+b^2}\right\}(t) = e^{at}\sin bt .$$

In this case, $a = -1$ and $b = 3$. The next step is to express

(5) $\qquad \dfrac{3s+2}{s^2+2s+10} = A\dfrac{s+1}{(s+1)^2+3^2} + B\dfrac{3}{(s+1)^2+3^2} ,$

where A, B are constants to be determined. Multiplying both sides of (5) by $s^2 + 2s + 10$ leaves

$$3s + 2 = A(s+1) + 3B = As + (A+3B) ,$$

which is an identity between two polynomials in s. Equating the coefficients of like terms gives

$$A = 3 , \qquad A + 3B = 2 ,$$

so $A = 3$ and $B = -1/3$. Finally, from (5) and the linearity property, we find

$$\mathcal{L}^{-1}\left\{\frac{3s+2}{s^2+2s+10}\right\}(t) = 3\mathcal{L}^{-1}\left\{\frac{s+1}{(s+1)^2+3^2}\right\}(t) - \frac{1}{3}\mathcal{L}^{-1}\left\{\frac{3}{(s+1)^2+3^2}\right\}(t)$$

$$= 3e^{-t}\cos 3t - \frac{1}{3}e^{-t}\sin 3t . \quad \blacksquare\,\text{End}$$

Given the choice of finding the inverse Laplace transform of

$$F_1(s) = \frac{7s^2 + 10s - 1}{s^3 + 3s^2 - s - 3}$$

or of

$$F_2(s) = \frac{2}{s-1} + \frac{1}{s+1} + \frac{4}{s+3} ,$$

which would you select? No doubt $F_2(s)$ is the easier one. Actually, the two functions $F_1(s)$ and $F_2(s)$ are identical. This can be checked by combining the simple fractions that form $F_2(s)$. Thus, if we are faced with the problem of computing \mathcal{L}^{-1} of a rational function such as $F_1(s)$, we will first express it, as we did $F_2(s)$, as a sum of simple rational functions. This is accomplished by the **method of partial fractions.**

We briefly review this method. Recall from calculus that a rational function of the form $P(s)/Q(s)$, where $P(s)$ and $Q(s)$ are polynomials with the degree of P less than the degree of Q, has a partial fraction expansion whose form is based on the linear and quadratic factors of $Q(s)$. (We assume the coefficients of the polynomials to be real numbers.) There are three cases to consider:

1. Nonrepeated linear factors.

2. Repeated linear factors.

3. Quadratic factors.

1. Nonrepeated Linear Factors

If $Q(s)$ can be factored into a product of distinct linear factors,

$$Q(s) = (s - r_1)(s - r_2) \cdots (s - r_n) ,$$

where the r_i's are all distinct real numbers, then the partial fraction expansion has the form

$$\frac{P(s)}{Q(s)} = \frac{A_1}{s - r_1} + \frac{A_2}{s - r_2} + \cdots + \frac{A_n}{s - r_n} ,$$

where the A_i's are real numbers. There are various ways of determining the constants A_1, \ldots, A_n. In the next example, we demonstrate two such methods.

EXAMPLE 5 Determine $\mathcal{L}^{-1}\{F\}$, where

$$F(s) = \frac{7s - 1}{(s + 1)(s + 2)(s - 3)} .$$

SOLUTION We begin by finding the partial fraction expansion for $F(s)$. The denominator consists of three distinct linear factors, so the expansion has the form

(6) $$\frac{7s - 1}{(s + 1)(s + 2)(s - 3)} = \frac{A}{s + 1} + \frac{B}{s + 2} + \frac{C}{s - 3} ,$$

where A, B, and C are real numbers to be determined.

One procedure that works for all partial fraction expansions is first to multiply the expansion equation by the denominator of the given rational function. This leaves us with two identical polynomials. Equating the coefficients of s^k leads to a system of linear equations that we can solve to determine the unknown constants. In this example, we

multiply (6) by $(s + 1)(s + 2)(s - 3)$ and find

(7) $7s - 1 = A(s + 2)(s - 3) + B(s + 1)(s - 3) + C(s + 1)(s + 2)$,†

which reduces to

$$7s - 1 = (A + B + C)s^2 + (-A - 2B + 3C)s + (-6A - 3B + 2C).$$

Equating the coefficients of s^2, s, and 1 gives the system of linear equations

$$A + B + C = 0 ,$$
$$-A - 2B + 3C = 7 ,$$
$$-6A - 3B + 2C = -1 .$$

Solving this system yields $A = 2$, $B = -3$, and $C = 1$. Hence

(8) $$\frac{7s - 1}{(s + 1)(s + 2)(s - 3)} = \frac{2}{s + 1} - \frac{3}{s + 2} + \frac{1}{s - 3} .$$

An alternative method for finding the constants A, B, and C from (7) is to choose three values for s and substitute them into (7) to obtain three linear equations in the three unknowns. If we are careful in our choice of the values for s, the system is easy to solve. In this case, equation (7) obviously simplifies if $s = -1$, -2, or 3. Putting $s = -1$ gives

$$-7 - 1 = A(1)(-4) + B(0) + C(0) ,$$
$$-8 = -4A .$$

Hence $A = 2$. Next, setting $s = -2$ gives

$$-14 - 1 = A(0) + B(-1)(-5) + C(0) ,$$
$$-15 = 5B ,$$

and so $B = -3$. Finally, letting $s = 3$, we similarly find that $C = 1$. In the case of nonrepeated linear factors, the alternative method is easier to use.

Now that we have obtained the partial fraction expansion (8), we use linearity to compute

$$\mathcal{L}^{-1}\left\{\frac{7s - 1}{(s + 1)(s + 2)(s - 3)}\right\}(t) = \mathcal{L}^{-1}\left\{\frac{2}{s + 1} - \frac{3}{s + 2} + \frac{1}{s - 3}\right\}(t)$$

$$= 2\mathcal{L}^{-1}\left\{\frac{1}{s + 1}\right\}(t) - 3\mathcal{L}^{-1}\left\{\frac{1}{s + 2}\right\}(t)$$

$$+ \mathcal{L}^{-1}\left\{\frac{1}{s - 3}\right\}(t)$$

$$= 2e^{-t} - 3e^{-2t} + e^{3t} .\quad ∎\,\text{End}$$

†Rigorously speaking, equation (7) was derived for s different from -1, -2, and 3, but by continuity it holds for these values as well.

2. Repeated Linear Factors

Let $s - r$ be a factor of $Q(s)$ and suppose $(s - r)^m$ is the highest power of $s - r$ that divides $Q(s)$. Then the portion of the partial fraction expansion of $P(s)/Q(s)$ that corresponds to the term $(s - r)^m$ is

$$\frac{A_1}{s - r} + \frac{A_2}{(s - r)^2} + \cdots + \frac{A_m}{(s - r)^m} \, ,$$

where the A_i's are real numbers.

EXAMPLE 6 Determine $\mathscr{L}^{-1}\left\{\dfrac{s^2 + 9s + 2}{(s - 1)^2(s + 3)}\right\}$.

SOLUTION Since $s - 1$ is a repeated linear factor with multiplicity two and $s + 3$ is a nonrepeated linear factor, the partial fraction expansion has the form

$$\frac{s^2 + 9s + 2}{(s - 1)^2(s + 3)} = \frac{A}{s - 1} + \frac{B}{(s - 1)^2} + \frac{C}{s + 3} \, .$$

We begin by multiplying both sides by $(s - 1)^2(s + 3)$ to obtain

(9) $s^2 + 9s + 2 = A(s - 1)(s + 3) + B(s + 3) + C(s - 1)^2$.

Now observe that when we set $s = 1$ (or $s = -3$), two terms on the right-hand side of (9) vanish, leaving a linear equation that we can solve for B (or C). Setting $s = 1$ in (9) gives

$$1 + 9 + 2 = A(0) + 4B + C(0) \, ,$$
$$12 = 4B \, ,$$

and hence $B = 3$. Similarly, setting $s = -3$ in (9) gives

$$9 - 27 + 2 = A(0) + B(0) + 16C$$
$$-16 = 16C \, .$$

Thus $C = -1$. Finally, to find A, we pick a different value for s, say $s = 0$. Then, since $B = 3$ and $C = -1$, plugging $s = 0$ into (9) yields

$$2 = -3A + 3B + C = -3A + 9 - 1$$

so that $A = 2$. Hence

(10) $\dfrac{s^2 + 9s + 2}{(s - 1)^2(s + 3)} = \dfrac{2}{s - 1} + \dfrac{3}{(s - 1)^2} - \dfrac{1}{s + 3}$.

We could also have determined the constants A, B, and C by first rewriting equation (9) in the form

$$s^2 + 9s + 2 = (A + C)s^2 + (2A + B - 2C)s + (-3A + 3B + C) \, .$$

Then, equating the corresponding coefficients of s^2, s, and 1 and solving the resulting system, we again find $A = 2$, $B = 3$, and $C = -1$.

Now that we have derived the partial fraction expansion (10) for the given rational function, we can determine its inverse Laplace transform:

$$\mathcal{L}^{-1}\left\{\frac{s^2 + 9s + 2}{(s-1)^2(s+3)}\right\}(t) = \mathcal{L}^{-1}\left\{\frac{2}{s-1} + \frac{3}{(s-1)^2} - \frac{1}{s+3}\right\}(t)$$

$$= 2\mathcal{L}^{-1}\left\{\frac{1}{s-1}\right\}(t) + 3\mathcal{L}^{-1}\left\{\frac{1}{(s-1)^2}\right\}(t)$$

$$- \mathcal{L}^{-1}\left\{\frac{1}{s+3}\right\}(t)$$

$$= 2e^t + 3te^t - e^{-3t} . \quad \blacksquare \text{ End}$$

3. Quadratic Factors

Let $(s-\alpha)^2 + \beta^2$ be a quadratic factor of $Q(s)$ that cannot be reduced to linear factors with real coefficients. Suppose m is the highest power of $(s-\alpha)^2 + \beta^2$ that divides $Q(s)$. Then the portion of the partial fraction expansion that corresponds to $(s-\alpha)^2 + \beta^2$ is

$$\frac{C_1 s + D_1}{(s-\alpha)^2 + \beta^2} + \frac{C_2 s + D_2}{[(s-\alpha)^2 + \beta^2]^2} + \cdots + \frac{C_m s + D_m}{[(s-\alpha)^2 + \beta^2]^m} .$$

As we saw in Example 4, it is more convenient to express $C_i s + D_i$ in the form $A_i(s-\alpha) + \beta B_i$ when we look up the Laplace transforms. So let's agree to write this portion of the partial fraction expansion in the equivalent form

$$\frac{A_1(s-\alpha) + \beta B_1}{(s-\alpha)^2 + \beta^2} + \frac{A_2(s-\alpha) + \beta B_2}{[(s-\alpha)^2 + \beta^2]^2} + \cdots + \frac{A_m(s-\alpha) + \beta B_m}{[(s-\alpha)^2 + \beta^2]^m} .$$

EXAMPLE 7 Determine $\mathcal{L}^{-1}\left\{\dfrac{2s^2 + 10s}{(s^2 - 2s + 5)(s+1)}\right\}$.

SOLUTION We first observe that the quadratic factor $s^2 - 2s + 5$ is irreducible (check the sign of the discriminant in the quadratic formula). Next we write the quadratic in the form $(s-\alpha)^2 + \beta^2$ by completing the square:

$$s^2 - 2s + 5 = (s-1)^2 + 2^2 .$$

Since $s^2 - 2s + 5$ and $s + 1$ are nonrepeated factors, the partial fraction expansion has the form

$$\frac{2s^2 + 10s}{(s^2 - 2s + 5)(s+1)} = \frac{A(s-1) + 2B}{(s-1)^2 + 2^2} + \frac{C}{s+1} .$$

When we multiply both sides by the common denominator, we obtain

(11) $$2s^2 + 10s = [A(s-1) + 2B](s+1) + C(s^2 - 2s + 5) .$$

In equation (11), let's put $s = -1, 1,$ and 0. With $s = -1$, we find

$$2 - 10 = [A(-2) + 2B](0) + C(8) ,$$
$$-8 = 8C ,$$

and hence $C = -1$. With $s = 1$ in (11), we obtain

$$2 + 10 = [A(0) + 2B](2) + C(4) ,$$

and since $C = -1$, the last equation becomes $12 = 4B - 4$. Thus $B = 4$. Finally, setting $s = 0$ in (11) and using $C = -1$ and $B = 4$ gives

$$0 = [A(-1) + 2B](1) + C(5) ,$$
$$0 = -A + 8 - 5 ,$$
$$A = 3 .$$

Hence $A = 3$, $B = 4$, and $C = -1$ so that

$$\frac{2s^2 + 10s}{(s^2 - 2s + 5)(s + 1)} = \frac{3(s - 1) + 2(4)}{(s - 1)^2 + 2^2} - \frac{1}{s + 1} .$$

With this partial fraction expansion in hand, we can immediately determine the inverse Laplace transform:

$$\mathcal{L}^{-1}\left\{\frac{2s^2 + 10s}{(s^2 - 2s + 5)(s + 1)}\right\}(t) = \mathcal{L}^{-1}\left\{\frac{3(s - 1) + 2(4)}{(s - 1)^2 + 2^2} - \frac{1}{s + 1}\right\}(t)$$

$$= 3\mathcal{L}^{-1}\left\{\frac{s - 1}{(s - 1)^2 + 2^2}\right\}(t)$$

$$+ 4\mathcal{L}^{-1}\left\{\frac{2}{(s - 1)^2 + 2^2}\right\}(t) - \mathcal{L}^{-1}\left\{\frac{1}{s + 1}\right\}(t)$$

$$= 3e^t\cos 2t + 4e^t\sin 2t - e^{-t} . \quad \blacksquare \text{End}$$

In Section 7.7, we discuss a different method (involving convolutions) for computing inverse transforms that does not require partial fraction decompositions. Moreover, the convolution method is convenient in the case of a rational function with a repeated quadratic factor in the denominator. Other helpful tools are described in Problems 33–36 and 38–43.

EXERCISES 7.4

In Problems 1 through 10, determine the inverse Laplace transform of the given function.

1. $\dfrac{6}{(s - 1)^4}$.

2. $\dfrac{2}{s^2 + 4}$.

3. $\dfrac{s + 1}{s^2 + 2s + 10}$.

4. $\dfrac{4}{s^2 + 9}$.

5. $\dfrac{1}{s^2 + 4s + 8}$.

6. $\dfrac{3}{(2s + 5)^3}$.

7. $\dfrac{2s + 16}{s^2 + 4s + 13}$.

8. $\dfrac{1}{s^5}$.

9. $\dfrac{3s - 15}{2s^2 - 4s + 10}$.

10. $\dfrac{s - 1}{2s^2 + s + 6}$.

In Problems 11 through 20, determine the partial fraction expansions for the given rational function.

11. $\dfrac{s^2 - 26s - 47}{(s-1)(s+2)(s+5)}$.

12. $\dfrac{-s-7}{(s+1)(s-2)}$.

13. $\dfrac{-2s^2 - 3s - 2}{s(s+1)^2}$.

14. $\dfrac{-8s^2 - 5s + 9}{(s+1)(s^2 - 3s + 2)}$.

15. $\dfrac{8s - 2s^2 - 14}{(s+1)(s^2 - 2s + 5)}$.

16. $\dfrac{-5s - 36}{(s+2)(s^2 + 9)}$.

17. $\dfrac{3s+5}{s(s^2 + s - 6)}$.

18. $\dfrac{3s^2 + 5s + 3}{s^4 + s^3}$.

19. $\dfrac{1}{(s-3)(s^2 + 2s + 2)}$.

20. $\dfrac{s}{(s-1)(s^2 - 1)}$.

In Problems 21 through 30, determine $\mathcal{L}^{-1}\{F\}$.

21. $F(s) = \dfrac{6s^2 - 13s + 2}{s(s-1)(s-6)}$.

22. $F(s) = \dfrac{s + 11}{(s-1)(s+3)}$.

23. $F(s) = \dfrac{5s^2 + 34s + 53}{(s+3)^2(s+1)}$.

24. $F(s) = \dfrac{7s^2 - 41s + 84}{(s-1)(s^2 - 4s + 13)}$.

25. $F(s) = \dfrac{7s^2 + 23s + 30}{(s-2)(s^2 + 2s + 5)}$.

26. $F(s) = \dfrac{7s^3 - 2s^2 - 3s + 6}{s^3(s-2)}$.

27. $s^2 F(s) - 4F(s) = \dfrac{5}{s+1}$.

28. $s^2 F(s) + sF(s) - 6F(s) = \dfrac{s^2 + 4}{s^2 + s}$.

29. $sF(s) + 2F(s) = \dfrac{10s^2 + 12s + 14}{s^2 - 2s + 2}$.

30. $sF(s) - F(s) = \dfrac{2s + 5}{s^2 + 2s + 1}$.

31. Determine the Laplace transform of each of the following functions:

(a) $f_1(t) = \begin{cases} 0, & t = 2, \\ t, & t \neq 2. \end{cases}$

(b) $f_2(t) = \begin{cases} 5, & t = 1, \\ 2, & t = 6, \\ t, & t \neq 1, 6. \end{cases}$

(c) $f_3(t) = t$.

Which of the preceding functions is the inverse Laplace transform of $1/s^2$?

32. Determine the Laplace transform of each of the following functions:

(a) $f_1(t) = \begin{cases} t, & t = 1, 2, 3, \ldots, \\ e^t, & t \neq 1, 2, 3, \ldots. \end{cases}$

(b) $f_2(t) = \begin{cases} e^t, & t \neq 5, 8, \\ 6, & t = 5, \\ 0, & t = 8. \end{cases}$

(c) $f_3(t) = e^t$.

Which of the preceding functions is the inverse Laplace transform of $1/(s-1)$?

Theorem 6 in Section 7.3 can be expressed in terms of the inverse Laplace transform as

$$\mathcal{L}^{-1}\left\{\frac{d^n F}{ds^n}\right\}(t) = (-t)^n f(t),$$

where $f = \mathcal{L}^{-1}\{F\}$. Use this equation in Problems 33 through 36 to compute $\mathcal{L}^{-1}\{F\}$.

33. $F(s) = \ln\left(\dfrac{s+2}{s-5}\right)$.

34. $F(s) = \ln\left(\dfrac{s-4}{s-3}\right)$.

35. $F(s) = \ln\left(\dfrac{s^2 + 9}{s^2 + 1}\right)$.

36. $F(s) = \arctan(1/s)$.

37. Prove Theorem 7 on the linearity of the inverse transform. [*Hint:* Show that the right-hand side of equation (3) is a continuous function on $[0, \infty)$ whose Laplace transform is $F_1(s) + F_2(s)$.]

38. Residue Computation. Let $P(s)/Q(s)$ be a rational function with $\deg P < \deg Q$ and suppose $s - r$ is a nonrepeated linear factor of $Q(s)$. Prove that the portion of the partial fraction expansion of $P(s)/Q(s)$ corresponding to $s - r$ is

$$\frac{A}{s-r},$$

where A (called the **residue**) is given by the formula

$$A = \lim_{s \to r} \frac{(s - r)P(s)}{Q(s)} .$$

39. Use the residue computation formula derived in Problem 38 to determine quickly the partial fraction expansion for

$$F(s) = \frac{2s + 1}{s(s - 1)(s + 2)} .$$

40. Heaviside's Expansion Formula.[†] Let $P(s)$ and $Q(s)$ be polynomials with the degree of $P(s)$ less than the degree of $Q(s)$. Let

$$Q(s) = (s - r_1)(s - r_2) \cdots (s - r_n) ,$$

where the r_i's are distinct real numbers. Show that

$$\mathscr{L}^{-1}\left\{\frac{P}{Q}\right\}(t) = \sum_{i=1}^{n} \frac{P(r_i)}{Q'(r_i)} e^{r_i t} .$$

41. Use Heaviside's expansion formula derived in Problem 40 to determine the inverse Laplace transform of

$$F(s) = \frac{3s^2 - 16s + 5}{(s + 1)(s - 3)(s - 2)} .$$

42. Complex Residues. Let $P(s)/Q(s)$ be a rational function with $\deg P < \deg Q$ and suppose $(s - \alpha)^2 + \beta^2$ is a nonrepeated quadratic factor of Q. (That is, $\alpha \pm i\beta$ are complex conjugate zeros of Q.) Prove that the portion of the partial fraction expansion of $P(s)/Q(s)$ corresponding to $(s - \alpha)^2 + \beta^2$ is

$$\frac{A(s - \alpha) + \beta B}{(s - \alpha)^2 + \beta^2} ,$$

where the **complex residue** $\beta B + i\beta A$ is given by the formula

$$\beta B + i\beta A = \lim_{s \to \alpha + i\beta} \frac{[(s - \alpha)^2 + \beta^2]P(s)}{Q(s)} .$$

(Thus we can determine B and A by taking the real and imaginary parts of the limit and dividing them by β.)

43. Use the residue formulas derived in Problems 38 and 42 to determine the partial fraction expansion for

$$F(s) = \frac{6s^2 + 28}{(s^2 - 2s + 5)(s + 2)} .$$

7.5 SOLVING INITIAL VALUE PROBLEMS

Our goal is to show how Laplace transforms can be used to solve initial value problems for linear differential equations. Recall that we have already studied ways of solving such initial value problems in Chapter 4. These previous methods required that we first find a *general solution* of the differential equation and then use the initial conditions to determine the desired solution. As we shall see, the method of Laplace transforms leads to the solution of the initial value problem *without* first finding a general solution.

Other advantages to the transform method are worth noting. For example, the technique can easily handle equations involving forcing functions having jump discontinuities, as illustrated in Section 7.1. Further, the method can be used for certain linear differential equations with variable coefficients, a special class of integral equations, systems of differential equations, and partial differential equations.

[†]*Historical Footnote:* This formula played an important role in the "operational solution" to ordinary differential equations developed by Oliver Heaviside in the 1890s.

METHOD OF LAPLACE TRANSFORMS

To solve an initial value problem:

(a) Take the Laplace transform of both sides of the equation.
(b) Use the properties of the Laplace transform and the initial conditions to obtain an equation for the Laplace transform of the solution and then solve this equation for the transform.
(c) Determine the inverse Laplace transform of the solution by looking it up in a table or by using a suitable method (such as partial fractions) in combination with the table.

In step (a), we are tacitly assuming the solution is piecewise continuous on $[0, \infty)$ and of exponential order. Once we have obtained the inverse Laplace transform in step (c), we can verify that these tacit assumptions are satisfied.

EXAMPLE 1 Solve the initial value problem

(1) $y'' - 2y' + 5y = -8e^{-t} \; ; \qquad y(0) = 2 \, , \qquad y'(0) = 12 \, .$

SOLUTION The differential equation in (1) is an identity between two functions of t. Hence equality holds for the Laplace transforms of these functions:

$$\mathscr{L}\{y'' - 2y' + 5y\} = \mathscr{L}\{-8e^{-t}\} \, .$$

Using the linearity property of \mathscr{L} and the previously computed transform of the exponential function, we can write

(2) $\mathscr{L}\{y''\}(s) - 2\mathscr{L}\{y'\}(s) + 5\mathscr{L}\{y\}(s) = \dfrac{-8}{s + 1} \, .$

Now let $Y(s) := \mathscr{L}\{y\}(s)$. From the formulas for the Laplace transform of higher-order derivatives (see Section 7.3) and the initial conditions in (1), we find

$$\mathscr{L}\{y'\}(s) = sY(s) - y(0) = sY(s) - 2 \, .$$
$$\mathscr{L}\{y''\}(s) = s^2 Y(s) - sy(0) - y'(0) = s^2 Y(s) - 2s - 12 \, .$$

Substituting these expressions into (2) and solving for $Y(s)$ yields

$$\left[s^2 Y(s) - 2s - 12 \right] - 2\left[sY(s) - 2 \right] + 5Y(s) = \frac{-8}{s + 1} \, ,$$

$$(s^2 - 2s + 5)Y(s) = 2s + 8 - \frac{8}{s + 1} \, ,$$

$$(s^2 - 2s + 5)Y(s) = \frac{2s^2 + 10s}{s + 1} \, ,$$

$$Y(s) = \frac{2s^2 + 10s}{(s^2 - 2s + 5)(s + 1)} \, .$$

Our remaining task is to compute the inverse transform of the rational function $Y(s)$. This was done in Example 7 of Section 7.4, where, using a partial fraction expansion, we found

(3) $y(t) = 3e^t\cos 2t + 4e^t\sin 2t - e^{-t}$,

which is the solution to the initial value problem (1). ■ End

As a quick check on the accuracy of our computations, the reader is advised to verify that the computed solution satisfies the given initial conditions.

The reader is probably questioning the wisdom of using the Laplace transform method to solve an initial value problem that can be easily handled by the methods discussed in Chapter 4. The objective of the first few examples in this section is simply to make the reader familiar with the Laplace transform procedure. We will see in Example 4 and in later sections that the method is applicable to problems that cannot be readily handled by the techniques discussed in the previous chapters.

EXAMPLE 2 Solve the initial value problem

(4) $y'' + 4y' - 5y = te^t$; $y(0) = 1$, $y'(0) = 0$.

SOLUTION Let $Y(s) := \mathcal{L}\{y\}(s)$. Taking the Laplace transform of both sides of the differential equation in (4) gives

(5) $\mathcal{L}\{y''\}(s) + 4\mathcal{L}\{y'\}(s) - 5Y(s) = \dfrac{1}{(s-1)^2}$.

Using the initial conditions, we can express $\mathcal{L}\{y'\}(s)$ and $\mathcal{L}\{y''\}(s)$ in terms of $Y(s)$. That is,

$$\mathcal{L}\{y'\}(s) = sY(s) - y(0) = sY(s) - 1 \ ,$$
$$\mathcal{L}\{y''\}(s) = s^2Y(s) - sy(0) - y'(0) = s^2Y(s) - s \ .$$

Substituting back into (5) and solving for $Y(s)$ gives

$$\left[s^2Y(s) - s\right] + 4\left[sY(s) - 1\right] - 5Y(s) = \frac{1}{(s-1)^2}$$

$$(s^2 + 4s - 5)Y(s) = s + 4 + \frac{1}{(s-1)^2}$$

$$(s + 5)(s - 1)Y(s) = \frac{s^3 + 2s^2 - 7s + 5}{(s-1)^2}$$

$$Y(s) = \frac{s^3 + 2s^2 - 7s + 5}{(s+5)(s-1)^3} \ .$$

The partial fraction expansion for $Y(s)$ has the form

(6) $\dfrac{s^3 + 2s^2 - 7s + 5}{(s+5)(s-1)^3} = \dfrac{A}{s+5} + \dfrac{B}{s-1} + \dfrac{C}{(s-1)^2} + \dfrac{D}{(s-1)^3}$.

Solving for the numerators, we ultimately obtain $A = 35/216$, $B = 181/216$, $C = -1/36$,

and $D = 1/6$. Substituting these values into (6) gives

$$Y(s) = \frac{35}{216}\left(\frac{1}{s+5}\right) + \frac{181}{216}\left(\frac{1}{s-1}\right) - \frac{1}{36}\left(\frac{1}{(s-1)^2}\right) + \frac{1}{12}\left(\frac{2}{(s-1)^3}\right) ,$$

where we have written $D = 1/6 = (1/12)2$ to facilitate the final step of taking the inverse transform. From the tables, we now obtain

(7) $$y(t) = \frac{35}{216}e^{-5t} + \frac{181}{216}e^t - \frac{1}{36}te^t + \frac{1}{12}t^2e^t$$

as the solution to the initial value problem (4). ∎ End

EXAMPLE 3 Solve the initial value problem

(8) $$w''(t) - 2w'(t) + 5w(t) = -8e^{\pi - t} ; \qquad w(\pi) = 2 , \qquad w'(\pi) = 12 .$$

SOLUTION To use the method of Laplace transforms, we first move the initial conditions to $t = 0$. This can be done by setting $y(t) := w(t + \pi)$. Then

$$y'(t) = w'(t + \pi) , \qquad y''(t) = w''(t + \pi) .$$

Replacing t by $t + \pi$ in the differential equation in (8), we have

(9) $$w''(t + \pi) - 2w'(t + \pi) + 5w(t + \pi) = -8e^{\pi - (t+\pi)} = -8e^{-t} .$$

Substituting $y(t) = w(t + \pi)$ in (9), the initial value problem in (8) becomes

$$y''(t) - 2y'(t) + 5y(t) = -8e^{-t} ; \qquad y(0) = 2 , \qquad y'(0) = 12 .$$

Because the initial conditions are now given at the origin, the Laplace transform method is applicable. In fact, we carried out the procedure in Example 1, where we found

(10) $$y(t) = 3e^t\cos 2t + 4e^t\sin 2t - e^{-t} .$$

Since $w(t + \pi) = y(t)$, then $w(t) = y(t - \pi)$. Hence, replacing t by $t - \pi$ in (10) gives

$$w(t) = y(t - \pi) = 3e^{t-\pi}\cos\left[2(t - \pi)\right] + 4e^{t-\pi}\sin\left[2(t - \pi)\right] - e^{-(t-\pi)}$$
$$= 3e^{t-\pi}\cos 2t + 4e^{t-\pi}\sin 2t - e^{\pi-t} . ∎ End$$

Thus far we have applied the Laplace transform method only to linear equations with constant coefficients. Yet several important equations in mathematical physics involve linear equations whose coefficients are polynomials in t. To solve such equations using Laplace transforms, we shall apply Theorem 6, page 385, where we proved that

(11) $$\mathcal{L}\{t^n f(t)\}(s) = (-1)^n\frac{d^n F}{ds^n}(s) .$$

If we let $n = 1$ and $f(t) = y'(t)$, we find

$$\mathcal{L}\{ty'(t)\}(s) = -\frac{d}{ds}\mathcal{L}\{y'\}(s)$$

$$= -\frac{d}{ds}\left[sY(s) - y(0)\right] = -sY'(s) - Y(s) .$$

Similarly, with $n = 1$ and $f(t) = y''(t)$, we obtain from (11)

$$\mathcal{L}\{ty''(t)\}(s) = -\frac{d}{ds}\mathcal{L}\{y''\}(s)$$

$$= -\frac{d}{ds}[s^2Y(s) - sy(0) - y'(0)]$$

$$= -s^2Y'(s) - 2sY(s) + y(0) .$$

Thus we see that for a linear differential equation in $y(t)$ whose coefficients are polynomials in t, the method of Laplace transforms will convert the given equation into a linear differential equation in $Y(s)$ whose coefficients are polynomials in s. Moreover, if the coefficients of the given equation are polynomials of degree one in t, then (regardless of the order of the given equation) the differential equation for $Y(s)$ is just a linear *first order* equation. Since we know how to solve this first order equation, the only serious obstacle we may encounter is obtaining the inverse Laplace transform of $Y(s)$. (This problem may be insurmountable, since the solution $y(t)$ may *not* have a Laplace transform.)

In illustrating the technique, we make use of the following fact. *If $f(t)$ is piecewise continuous on $[0, \infty)$ and of exponential order, then*

(12) $$\lim_{s \to \infty} \mathcal{L}\{f\}(s) = 0 .$$

(The reader may have already guessed this from the entries in Table 7.1, page 380.) An outline of the proof of (12) is given in Exercises 7.3, Problem 26.

EXAMPLE 4 Solve the initial value problem

(13) $$y'' + 2ty' - 4y = 1 , \qquad y(0) = y'(0) = 0 .$$

SOLUTION Let $Y(s) = \mathcal{L}\{y\}(s)$ and take the Laplace transform of both sides of the equation in (13):

(14) $$\mathcal{L}\{y''\}(s) + 2\mathcal{L}\{ty'(t)\}(s) - 4Y(s) = \frac{1}{s} .$$

Using the initial conditions, we find

$$\mathcal{L}\{y''\}(s) = s^2Y(s) - sy(0) - y'(0) = s^2Y(s)$$

and

$$\mathcal{L}\{ty'(t)\}(s) = -\frac{d}{ds}\mathcal{L}\{y'\}(s)$$

$$= -\frac{d}{ds}[sY(s) - y(0)] = -sY'(s) - Y(s) .$$

Substituting these expressions into (14) gives

$$s^2Y(s) + 2[-sY'(s) - Y(s)] - 4Y(s) = \frac{1}{s} ,$$

$$-2sY'(s) + (s^2 - 6)Y(s) = \frac{1}{s} ,$$

(15) $$Y'(s) + \left(\frac{3}{s} - \frac{s}{2}\right)Y(s) = \frac{-1}{2s^2} .$$

Equation (15) is a linear first order equation and has the integrating factor

$$\mu(s) = e^{\int(3/s - s/2)ds} = e^{\ln s^3 - s^2/4} = s^3 e^{-s^2/4}$$

(see Section 2.3). Multiplying (15) by $\mu(s)$, we obtain

$$\frac{d}{ds}\{\mu(s)Y(s)\} = \frac{d}{ds}\left\{s^3 e^{-s^2/4}Y(s)\right\} = -\frac{s}{2}e^{-s^2/4} .$$

Integrating and solving for $Y(s)$ yields

$$s^3 e^{-s^2/4}Y(s) = -\int \frac{s}{2}e^{-s^2/4}\, ds = e^{-s^2/4} + C ,$$

(16) $$Y(s) = \frac{1}{s^3} + C\frac{e^{s^2/4}}{s^3} .$$

Now if $Y(s)$ is the Laplace transform of a piecewise continuous function of exponential order, then it follows from equation (12) that

$$\lim_{s\to\infty} Y(s) = 0 .$$

For this to occur, the constant C in equation (16) must be zero. Hence $Y(s) = 1/s^3$, and taking the inverse transform gives $y(t) = t^2/2$. We can easily verify that $y(t) = t^2/2$ is the solution to the given initial value problem by substituting it into (13). ■End

We end this section with an application from **control theory.** Let's consider a servomechanism that models an automatic pilot. Such a mechanism applies a torque to the steering control shaft so that a plane or boat will follow a prescribed course. If we let $y(t)$ be the true direction (angle) of the craft at time t and $g(t)$ be the desired direction at time t, then

$$e(t) := y(t) - g(t)$$

denotes the **error** or **deviation** between the desired direction and the true direction.

Let's assume the servomechanism can measure the error $e(t)$ and feed back to the steering shaft a component of torque that is proportional to $e(t)$ but opposite in sign (see Figure 7.6). Newton's second law, expressed in terms of torques, states that

$$(\textbf{moment of inertia}) \times (\textbf{angular acceleration}) = \textbf{total torque.}$$

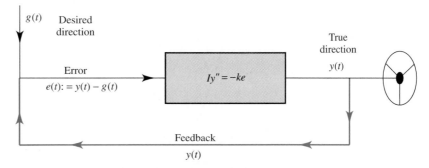

Figure 7.6 Servomechanism with feedback

For the servomechanism described, this becomes

(17) $Iy''(t) = -ke(t)$,

where I is the moment of inertia of the steering shaft and k is a positive proportionality constant.

EXAMPLE 5 Determine the error $e(t)$ for the automatic pilot if the steering shaft is initially at rest in the zero direction and the desired direction is given by $g(t) = at$, where a is a constant.

SOLUTION Based on the discussion leading to equation (17), a model for the mechanism is given by the initial value problem

(18) $Iy''(t) = -ke(t)$; $y(0) = 0$, $y'(0) = 0$,

where $e(t) = y(t) - g(t) = y(t) - at$. We begin by taking the Laplace transform of both sides of (18):

$$I\mathscr{L}\{y''\}(s) = -k\mathscr{L}\{e\}(s) ,$$
$$I[s^2Y(s) - sy(0) - y'(0)] = -kE(s) ,$$
(19) $$s^2IY(s) = -kE(s) ,$$

where $Y(s) = \mathscr{L}\{y\}(s)$ and $E(s) = \mathscr{L}\{e\}(s)$. Since

$$E(s) = \mathscr{L}\{y(t) - at\}(s) = Y(s) - \mathscr{L}\{at\}(s) = Y(s) - as^{-2} ,$$

we find from (19) that

$$s^2IE(s) + aI = -kE(s) .$$

Solving this equation for $E(s)$ gives

$$E(s) = -\frac{aI}{s^2I + k} = \frac{-a}{\sqrt{k/I}}\frac{\sqrt{k/I}}{s^2 + k/I} .$$

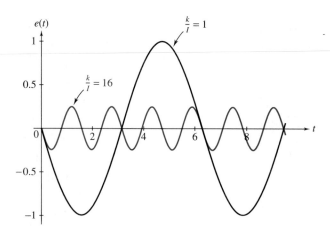

Figure 7.7 Error for automatic pilot when $k/I = 1$ and when $k/I = 16$

Hence, on taking the inverse Laplace transform, we obtain the error

(20) $e(t) = -\dfrac{a}{\sqrt{k/I}} \sin\left(\sqrt{k/I}\, t\right)$. ■ End

As we can see from equation (20), the automatic pilot will oscillate back and forth about the desired course, always "oversteering" by the factor $a\sqrt{k/I}$. Clearly we can make the error small by making k large relative to I, but then the term $\sqrt{k/I}$ becomes large, causing the error to oscillate more rapidly. (See Figure 7.7.) As with vibrations, the oscillations or oversteering can be controlled by introducing a damping torque proportional to $e'(t)$ but opposite in sign (see Problem 40).

EXERCISES 7.5

In Problems 1 through 14, solve the given initial value problem using the method of Laplace transforms.

1. $y'' - 2y' + 5y = 0$;
$\quad y(0) = 2$, $y'(0) = 4$.

2. $y'' - y' - 2y = 0$;
$\quad y(0) = -2$, $y'(0) = 5$.

3. $y'' + 6y' + 9y = 0$;
$\quad y(0) = -1$, $y'(0) = 6$.

4. $y'' + 6y' + 5y = 12e^t$;
$\quad y(0) = -1$, $y'(0) = 7$.

5. $w'' + w = t^2 + 2$;
$\quad w(0) = 1$, $w'(0) = -1$.

6. $y'' - 4y' + 5y = 4e^{3t}$;
$\quad y(0) = 2$, $y'(0) = 7$.

7. $y'' - 7y' + 10y = 9\cos t + 7\sin t$;
$\quad y(0) = 5$, $y'(0) = -4$.

8. $y'' + 4y = 4t^2 - 4t + 10$;
$\quad y(0) = 0$, $y'(0) = 3$.

9. $z'' + 5z' - 6z = 21e^t$;
$\quad z(0) = -1$, $z'(0) = 9$.

10. $y'' - 4y = 4t - 8e^{-2t}$;
$\quad y(0) = 0$, $y'(0) = 5$.

11. $y'' - y = t - 2$; $y(2) = 3$, $y'(2) = 0$.

12. $w'' - 2w' + w = 6t - 2$;
$\quad w(-1) = 3$; $w'(-1) = 7$.

13. $y'' - y' - 2y = -8\cos t - 2\sin t$;
$\quad y(\pi/2) = 1$, $y'(\pi/2) = 0$.

14. $y'' + y = t$; $y(\pi) = 0$, $y'(\pi) = 0$.

In Problems 15 through 24, solve for $Y(s)$, the Laplace transform of the solution $y(t)$ to the given initial value problem.

15. $y'' - 3y' + 2y = \cos t$; [Don't find actual soln.]
$\quad y(0) = 0$, $y'(0) = -1$.

16. $y'' + 6y = t^2 - 1$;
$\quad y(0) = 0$, $y'(0) = -1$.

17. $y'' + y' - y = t^3$;
$\quad y(0) = 1$, $y'(0) = 0$.

18. $y'' - 2y' - y = e^{2t} - e^t$;
$\quad y(0) = 1$, $y'(0) = 3$.

19. $y'' + 5y' - y = e^t - 1$;
$\quad y(0) = 1$, $y'(0) = 1$.

20. $y'' + 3y = t^3$; $y(0) = 0$, $y'(0) = 0$.

21. $y'' - 2y' + y = \cos t - \sin t$;
$\quad y(0) = 1$, $y'(0) = 3$.

22. $y'' - 6y' + 5y = te^t$;
$\quad y(0) = 2$, $y'(0) = -1$.

23. $y'' + 4y = g(t)$; $y(0) = -1$; $y'(0) = 0$,
where
$$g(t) = \begin{cases} t , & t < 2 , \\ 5 , & t > 2 . \end{cases}$$

24. $y'' - y = g(t)$; $\quad y(0) = 1$, $\quad y'(0) = 2$,
 where

$$g(t) = \begin{cases} 1 , & t < 3 , \\ t , & t > 3 . \end{cases}$$

In Problems 25 through 28, solve the given third order initial value problem for $y(t)$ using the method of Laplace transforms.

25. $y''' - y'' + y' - y = 0$;
 $y(0) = 1$, $\quad y'(0) = 1$, $\quad y''(0) = 3$.

26. $y''' + 4y'' + y' - 6y = -12$;
 $y(0) = 1$, $\quad y'(0) = 4$, $\quad y''(0) = -2$.

27. $y''' + 3y'' + 3y' + y = 0$;
 $y(0) = -4$, $\quad y'(0) = 4$, $\quad y''(0) = -2$.

28. $y''' + y'' + 3y' - 5y = 16e^{-t}$;
 $y(0) = 0$, $\quad y'(0) = 2$, $\quad y''(0) = -4$.

In Problems 29 through 32, use the method of Laplace transforms to find a general solution to the given differential equation by assuming $y(0) = a$ and $y'(0) = b$, where a and b are arbitrary constants.

29. $y'' - 4y' + 3y = 0$. **30.** $y'' + 6y' + 5y = t$.

31. $y'' + 2y' + 2y = 5$.

32. $y'' - 5y' + 6y = -6te^{2t}$.

33. Use Theorem 6 in Section 7.3 to show that
$$\mathcal{L}\{t^2 y'(t)\}(s) = sY''(s) + 2Y'(s) ,$$
 where $Y(s) = \mathcal{L}\{y\}(s)$.

34. Use Theorem 6 in Section 7.3 to show that
$$\mathcal{L}\{t^2 y''(t)\}(s) = s^2 Y''(s) + 4sY'(s) + 2Y(s) ,$$
 where $Y(s) = \mathcal{L}\{y\}(s)$.

In Problems 35 through 38, find solutions to the given initial value problem.

35. $y'' + 3ty' - 6y = 1$;
 $y(0) = 0$, $\quad y'(0) = 0$.

36. $ty'' - ty' + y = 2$;
 $y(0) = 2$, $\quad y'(0) = -1$.

37. $ty'' - 2y' + ty = 0$;
 $y(0) = 1$, $\quad y'(0) = 0$.
 [*Hint:* $\mathcal{L}^{-1}\{1/(s^2 + 1)^2\}(t) = (\sin t - t \cos t)/2.$]

38. $y'' + ty' - y = 0$;
 $y(0) = 0$, $\quad y'(0) = 3$.

39. Determine the error $e(t)$ for the automatic pilot in Example 5 if the shaft is initially at rest in the zero direction and the desired direction is $g(t) = a$, where a is a constant.

40. In Example 5, assume that in order to control oscillations, a component of torque proportional to $e'(t)$, but opposite in sign, is also fed back to the steering shaft. Show that equation (17) is now replaced by

$$Iy''(t) = -ke(t) - \mu e'(t) ,$$

 where μ is a positive constant. Determine the error $e(t)$ for the automatic pilot with mild damping (i.e., $\mu < 2\sqrt{Ik}$) if the steering shaft is initially at rest in the zero direction and the desired direction is given by $g(t) = a$, where a is a constant.

41. In Problem 40, determine the error $e(t)$ when the desired direction is given by $g(t) = at$, where a is a constant.

7.6 TRANSFORMS OF DISCONTINUOUS AND PERIODIC FUNCTIONS

In this section we study special functions that often arise when the method of Laplace transforms is applied to physical problems. Of particular interest are methods for handling functions with jump discontinuities. Jump discontinuities occur naturally in physical problems such as electric circuits with on/off switches. To handle such behavior, Oliver Heaviside introduced the following step function.

UNIT STEP FUNCTION

Definition 5. The **unit step function** $u(t)$ is defined by

$$(1) \qquad u(t) := \begin{cases} 0, & t < 0, \\ 1, & 0 < t. \end{cases}$$

By shifting the argument of $u(t)$, the jump can be moved to a different location. That is,

$$(2) \qquad u(t-a) = \begin{cases} 0, & t-a < 0, \\ 1, & 0 < t-a \end{cases} = \begin{cases} 0, & t < a, \\ 1, & a < t \end{cases}$$

has its jump at $t = a$. By multiplying by a constant M, the height of the jump can also be modified (see Figure 7.8):

$$Mu(t-a) = \begin{cases} 0, & t < a, \\ M, & a < t. \end{cases}$$

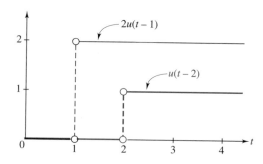

Figure 7.8 Two-step functions expressed using the unit step function

Any piecewise continuous function can be expressed in terms of unit step functions.

EXAMPLE 1 Write the function

$$(3) \qquad f(t) = \begin{cases} 3, & t < 2, \\ 1, & 2 < t < 5, \\ t, & 5 < t < 8, \\ t^2/10, & 8 < t \end{cases}$$

(see Figure 7.9 on page 408) in terms of unit step functions.

SOLUTION From the graph in Figure 7.9, we see that the function $f(t)$ equals 3 until t reaches 2, at which point it jumps, by -2 units, to the value 1. We can express this jump with the formula $3 - 2u(t-2)$, since $u(t-2)$ is zero until t reaches 2, after which it has the value 1.

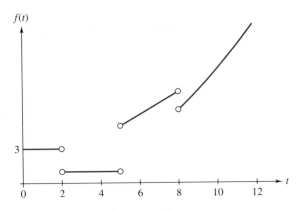

Figure 7.9 Graph of $f(t)$ in equation (3)

At $t = 5$, the figure shows that the function changes from the (constant) value 1 to the value t. Effectively, one subtracts the constant 1 from $f(t)$ at that point, and adds the function t. We can accomplish this in the formula for f by adding the term $u(t - 5)$ times $[t] - [1]$, or $(t - 1)u(t - 5)$. Similarly, we change the formula for f from t to $t^2/10$ at $t = 8$ by adding $[t^2/10 - t]u(t - 8)$, and the result is

$$f(t) = 3 - 2u(t - 2) + (t - 1)u(t - 5) + (t^2/10 - t)u(t - 8) .$$ ■ End

The Laplace transform of $u(t - a)$ with $a \geq 0$ is

(4) $$\mathscr{L}\{u(t - a)\}(s) = \frac{e^{-as}}{s} ,$$

since, for $s > 0$,

$$\mathscr{L}\{u(t - a)\}(s) = \int_0^\infty e^{-st}u(t - a)\,dt = \int_a^\infty e^{-st}\,dt$$

$$= \lim_{N \to \infty} \frac{-e^{-st}}{s}\bigg|_a^N = \frac{e^{-as}}{s} .$$

Conversely, for $a > 0$, we say that the piecewise continuous function $u(t - a)$ is an inverse Laplace transform for e^{-as}/s and we write[†]

$$\mathscr{L}^{-1}\left\{\frac{e^{-as}}{s}\right\}(t) = u(t - a) .$$

The translation property of $F(s)$ discussed in Section 7.3 described the effect on the Laplace transform of multiplying a function by e^{at}. The next theorem illustrates an analogous effect of multiplying the Laplace transform of a function by e^{-as}.

[†]The absence of a specific value for $u(0)$ in Definition 5 is a reflection of the ambiguity of the inverse Laplace transform, when no continuous inverse transform exists.

TRANSLATION IN t

Theorem 8. Let $F(s) = \mathcal{L}\{f\}(s)$ exist for $s > \alpha \geq 0$. If a is a positive constant, then

(5) $\mathcal{L}\{f(t - a)u(t - a)\}(s) = e^{-as}F(s)$,

and, conversely, an inverse Laplace transform† of $e^{-as}F(s)$ is given by

(6) $\mathcal{L}^{-1}\{e^{-as}F(s)\}(t) = f(t - a)u(t - a)$.

Proof. By the definition of the Laplace transform, we have

(7) $\mathcal{L}\{f(t - a)u(t - a)\}(s) = \displaystyle\int_0^\infty e^{-st}f(t - a)u(t - a)\,dt$

$\qquad\qquad\qquad\qquad\quad = \displaystyle\int_a^\infty e^{-st}f(t - a)\,dt$,

where, in the last equation, we used the fact that $u(t - a)$ is zero for $t < a$ and equals 1 for $t > a$. Now let $v = t - a$. Then we have $dv = dt$, and equation (7) becomes

$\qquad \mathcal{L}\{f(t - a)u(t - a)\}(s) = \displaystyle\int_0^\infty e^{-as}e^{-sv}f(v)\,dv$

$\qquad\qquad\qquad\qquad\qquad\qquad = e^{-as}\displaystyle\int_0^\infty e^{-sv}f(v)\,dv = e^{-as}F(s)$. ■ End

Notice that formula (5) includes as a special case the formula for $\mathcal{L}\{u(t - a)\}$; indeed, if we take $f(t) \equiv 1$, then $F(s) = 1/s$ and (5) becomes $\mathcal{L}\{u(t - a)\}(s) = e^{-as}/s$.

In practice it is more common to be faced with the problem of computing the transform of a function expressed as $g(t)u(t - a)$ rather than $f(t - a)u(t - a)$. To compute $\mathcal{L}\{g(t)u(t - a)\}$, we simply identify $g(t)$ with $f(t - a)$ so that $f(t) = g(t + a)$. Equation (5) then gives

(8) $\mathcal{L}\{g(t)u(t - a)\}(s) = e^{-as}\mathcal{L}\{g(t + a)\}(s)$.

EXAMPLE 2 Determine the Laplace transform of $t^2 u(t - 1)$.

SOLUTION To apply equation (8), we take $g(t) = t^2$ and $a = 1$. Then

$\qquad g(t + a) = g(t + 1) = (t + 1)^2 = t^2 + 2t + 1$.

Now the Laplace transform of $g(t + a)$ is

$\qquad \mathcal{L}\{g(t + a)\}(s) = \mathcal{L}\{t^2 + 2t + 1\}(s) = \dfrac{2}{s^3} + \dfrac{2}{s^2} + \dfrac{1}{s}$.

†This inverse transform is, in fact, a *continuous* function of t if $f(0) = 0$ and $f(t)$ is continuous for $t \geq 0$; the values of $f(t)$ for $t < 0$ are of no consequence, since the factor $u(t - a)$ is zero there.

So, by formula (8), we have

$$\mathscr{L}\{t^2 u(t-1)\}(s) = e^{-s}\left\{\frac{2}{s^3} + \frac{2}{s^2} + \frac{1}{s}\right\} . \quad \blacksquare \text{End}$$

EXAMPLE 3 Determine $\mathscr{L}\{(\cos t)u(t-\pi)\}$.

SOLUTION Here $g(t) = \cos t$ and $a = \pi$. Hence

$$g(t+a) = g(t+\pi) = \cos(t+\pi) = -\cos t ,$$

and so the Laplace transform of $g(t+a)$ is

$$\mathscr{L}\{g(t+a)\}(s) = -\mathscr{L}\{\cos t\}(s) = -\frac{s}{s^2+1} .$$

Thus, from formula (8), we get

$$\mathscr{L}\{(\cos t)u(t-\pi)\}(s) = -e^{-\pi s}\frac{s}{s^2+1} . \quad \blacksquare \text{End}$$

In Examples 2 and 3, we could also have computed the Laplace transform directly from the definition. In dealing with inverse transforms, however, we do not have a simple alternative formula[†] upon which to rely, and so formula (6) is especially useful whenever the transform has e^{-as} as a factor.

EXAMPLE 4 Determine $\mathscr{L}^{-1}\left\{\dfrac{e^{-2s}}{s^2}\right\}$ and sketch its graph.

SOLUTION To use the translation property (6), we first express e^{-2s}/s^2 as the product $e^{-as}F(s)$. For this purpose, we put $e^{-as} = e^{-2s}$ and $F(s) = 1/s^2$. Thus $a = 2$ and

$$f(t) = \mathscr{L}^{-1}\left\{\frac{1}{s^2}\right\}(t) = t .$$

It now follows from the translation property that

$$\mathscr{L}^{-1}\left\{\frac{e^{-2s}}{s^2}\right\}(t) = f(t-2)u(t-2) = (t-2)u(t-2) .$$

See Figure 7.10. \blacksquare End

As illustrated by the next example, step functions arise in the modeling of on/off switches,

[†]Under certain conditions, the inverse transform is given by the contour integral

$$\mathscr{L}^{-1}\{F\}(t) = \frac{1}{2\pi i}\int_{a-i\infty}^{a+i\infty} e^{st}F(s)\,ds .$$

See, for example, *Complex Variables and the Laplace Transform for Engineers,* by Wilbur R. LePage (Dover Publications, New York, 1980) or *Fundamentals of Complex Analysis for Mathematics, Science, and Engineering* (2nd edition), by E. B. Saff and A. D. Snider (Prentice Hall, Englewood Cliffs, 1993).

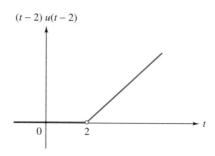

Figure 7.10 Graph of solution to Example 4

changes in polarity, etc.

EXAMPLE 5 The current I in an LC series circuit is governed by the initial value problem

(9) $I''(t) + 4I(t) = g(t)$; $I(0) = 0$, $I'(0) = 0$,

where

$$g(t) := \begin{cases} 1 , & 0 < t < 1 , \\ -1 , & 1 < t < 2 , \\ 0 , & 2 < t . \end{cases}$$

Determine the current as a function of time t.

SOLUTION Let $J(s) := \mathcal{L}\{I\}(s)$. Then we have $\mathcal{L}\{I''\}(s) = s^2 J(s)$. Expressing

$$g(t) = u(t) - 2u(t - 1) + u(t - 2) ,$$

we find that

$$\mathcal{L}\{g\}(s) = \frac{1}{s} - \frac{2e^{-s}}{s} + \frac{e^{-2s}}{s} .$$

Thus, when we take the Laplace transform of both sides of (9), we obtain

$$\mathcal{L}\{I''\}(s) + 4\mathcal{L}\{I\}(s) = \mathcal{L}\{g\}(s) ,$$

$$s^2 J(s) + 4J(s) = \frac{1}{s} - \frac{2e^{-s}}{s} + \frac{e^{-2s}}{s} ,$$

$$J(s) = \frac{1}{s(s^2 + 4)} - \frac{2e^{-s}}{s(s^2 + 4)} + \frac{e^{-2s}}{s(s^2 + 4)} .$$

To find $I = \mathcal{L}^{-1}\{J\}$, we first observe that

$$J(s) = F(s) - 2e^{-s}F(s) + e^{-2s}F(s) ,$$

where

$$F(s) := \frac{1}{s(s^2 + 4)} = \frac{1}{4}\left(\frac{1}{s}\right) - \frac{1}{4}\left(\frac{s}{s^2 + 4}\right) .$$

Computing the inverse transform of $F(s)$ gives

$$f(t) := \mathscr{L}^{-1}\{F\}(t) = \frac{1}{4} - \frac{1}{4}\cos 2t \ .$$

Hence, via the translation property (6), we find

$$\begin{aligned}
I(t) &= \mathscr{L}^{-1}\{F(s) - 2e^{-s}F(s) + e^{-2s}F(s)\}(t) \\
&= f(t) - 2f(t-1)u(t-1) + f(t-2)u(t-2) \\
&= \left(\frac{1}{4} - \frac{1}{4}\cos 2t\right) - \left[\frac{1}{2} - \frac{1}{2}\cos 2(t-1)\right]u(t-1) \\
&\quad + \left[\frac{1}{4} - \frac{1}{4}\cos 2(t-2)\right]u(t-2) \ .
\end{aligned}$$

The current is graphed in Figure 7.11. Note that $I(t)$ is smoother than $g(t)$; the former has discontinuities in its second derivative at the points where the latter has jumps. ■ End

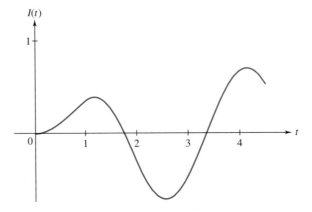

Figure 7.11 Solution to Example 5

Periodic functions are another class of functions that occur frequently in applications.

PERIODIC FUNCTION

Definition 6. A function $f(t)$ is said to be **periodic of period T** if

$$f(t + T) = f(t)$$

for all t in the domain of f.

As we know, the sine and cosine functions are periodic with period 2π and the tangent function is periodic with period π.[†] To specify a periodic function, it is sufficient to give its val-

[†]A function that has period T will also have period $2T$, $3T$, etc. For example, the sine function has periods 2π, 4π, 6π, etc. Some authors refer to the smallest period as the **fundamental period** or just the period of the function.

ues over one period. For example, the square wave function in Figure 7.12 can be expressed as

(10) $f(t) := \begin{cases} 1 , & 0 < t < 1 , \\ 0 , & 1 < t < 2 , \end{cases}$ and $f(t)$ has period 2.

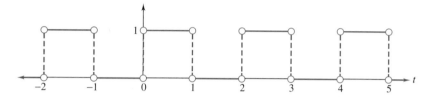

Figure 7.12 Graph of square wave function $f(t)$

It is convenient to introduce a notation for the "windowed" version of a periodic function $f(t)$:

(11) $f_T(t) := \begin{cases} f(t) , & 0 < t < T , \\ 0 , & \text{otherwise.} \end{cases}$

Figure 7.13 Windowed version of periodic function

(See Figure 7.13.) The Laplace transform of $f_T(t)$ is given by

$$F_T(s) = \int_0^\infty e^{-st} f_T(t)\, dt = \int_0^T e^{-st} f(t)\, dt \ .$$

It is related to the Laplace transform of $f(t)$ as follows.

TRANSFORM OF PERIODIC FUNCTION

Theorem 9. If f has period T and is piecewise continuous on $\left[0, T\right]$, then

(12) $\mathcal{L}\{f\}(s) = \dfrac{F_T(s)}{1 - e^{-sT}} = \dfrac{\displaystyle\int_0^T e^{-st} f(t)\, dt}{1 - e^{-sT}} \ .$

Proof. Observe that $f(t)$ can be written as a sum of translated copies of its windowed version:

(13) $f(t) = f_T(t) + f_T(t - T)u(t - T) + f_T(t - 2T)u(t - 2T) + \cdots \ .$

Therefore

$$\mathcal{L}\{f\} = \mathcal{L}\{f_T\} + \mathcal{L}\{f_T(t - T)u(t - T)\} + \mathcal{L}\{f_T(t - 2T)u(t - 2T)\} + \cdots ,$$

so, by the translation property expressed in Theorem 8 (page 409),

$$\mathcal{L}\{f\} = \mathcal{L}\{f_T\} + \mathcal{L}\{f_T\}e^{-sT} + \mathcal{L}\{f_T\}e^{-2sT} + \cdots$$
$$= F_T(s)\left[1 + e^{-sT} + e^{-2sT} + \cdots\right] .$$

Now the series in brackets is a geometric series with ratio $r = e^{-sT}$, which sums to $1/(1 - r)$ or $1/(1 - e^{-sT})$ for $s > 0$. Hence

$$\mathcal{L}\{f\} = F_T(s)/(1 - e^{-sT}) . \quad \blacksquare \text{End}$$

EXAMPLE 6 Determine $\mathcal{L}\{f\}$, where f is the square wave function in (10).

SOLUTION Here $T = 2$. From the formula (11), we express the windowed version of f as

$$f_T(t) = 1 - u(t - 1) ,$$

so

$$F_T(s) = \mathcal{L}\{1 - u(t - 1)\} = \frac{1}{s} - \frac{1}{s}e^{-s} = \frac{1}{s}(1 - e^{-s}) .$$

Therefore, from formula (12), we obtain

$$\mathcal{L}\{f\} = \frac{1}{s}(1 - e^{-s})\frac{1}{1 - e^{-2s}} = \frac{1}{s(1 + e^{-s})} . \quad \blacksquare \text{End}$$

In the proof of Theorem 9, we determined the Laplace transform of $f(t)$ by writing $f(t)$ as a series of functions with known Laplace transforms. This approach can also be used for functions that have a power series expansion, since we know that

$$\mathcal{L}\{t^n\}(s) = \frac{n!}{s^{n+1}} , \qquad n = 0, 1, 2, \ldots .$$

EXAMPLE 7 Determine $\mathcal{L}\{f\}$, where

$$f(t) := \begin{cases} \dfrac{\sin t}{t} , & t \neq 0 , \\ 1 , & t = 0 . \end{cases}$$

SOLUTION We begin by expressing $f(t)$ in a Taylor series[†] about $t = 0$. Since

$$\sin t = t - \frac{t^3}{3!} + \frac{t^5}{5!} - \frac{t^7}{7!} + \cdots ,$$

[†]For a discussion of Taylor series, see Sections 8.1 and 8.2.

then dividing by t, we obtain

$$f(t) = \frac{\sin t}{t} = 1 - \frac{t^2}{3!} + \frac{t^4}{5!} - \frac{t^6}{7!} + \cdots$$

for $t > 0$. This representation also holds at $t = 0$ since

$$\lim_{t\to 0} f(t) = \lim_{t\to 0} \frac{\sin t}{t} = 1 \ .$$

Observe that $f(t)$ is continuous on $[0, \infty)$ and of exponential order. Hence its Laplace transform exists for all s sufficiently large. Because of the linearity of the Laplace transform, we would expect that

$$\mathcal{L}\{f\}(s) = \mathcal{L}\{1\}(s) - \frac{1}{3!}\mathcal{L}\{t^2\}(s) + \frac{1}{5!}\mathcal{L}\{t^4\}(s) + \cdots$$

$$= \frac{1}{s} - \frac{2!}{3!s^3} + \frac{4!}{5!s^5} - \frac{6!}{7!s^7} + \cdots$$

$$= \frac{1}{s} - \frac{1}{3s^3} + \frac{1}{5s^5} - \frac{1}{7s^7} + \cdots \ .$$

Indeed, using tools from analysis, it can be verified that this series representation is valid for all $s > 1$. Moreover, one can show that the series converges to the function $\arctan(1/s)$ (see Problem 54). Thus

(14) $\mathcal{L}\left\{\dfrac{\sin t}{t}\right\}(s) = \arctan\dfrac{1}{s} \ .$ ■ End

A similar procedure involving the series expansion for $F(s)$ in powers of $1/s$ can be used to compute $f(t) = \mathcal{L}^{-1}\{F\}(t)$ (see Problems 55–57).

We have previously shown, for every nonnegative integer n, that $\mathcal{L}\{t^n\}(s) = n!/s^{n+1}$. But what if the power of t is *not* an integer? Is this formula still valid? To answer this question, we need to extend the idea of "factorial." This is accomplished by the gamma function.[†]

GAMMA FUNCTION

Definition 7. The **gamma function** $\Gamma(t)$ is defined by

(15) $\Gamma(t) := \displaystyle\int_0^\infty e^{-u}u^{t-1}\,du \ , \qquad t > 0 \ .$

It can be shown that the integral in (15) converges for $t > 0$. A useful property of the

[†]*Historical Footnote:* The gamma function was introduced by Leonhard Euler.

gamma function is the recursive relation

(16) $\Gamma(t + 1) = t\Gamma(t)$.

This identity follows from the definition (15) after performing an integration by parts:

$$\Gamma(t + 1) = \int_0^\infty e^{-u} u^t \, du = \lim_{N \to \infty} \int_0^N e^{-u} u^t \, du$$

$$= \lim_{N \to \infty} \left\{ -e^{-u} u^t \Big|_0^N + \int_0^N t e^{-u} u^{t-1} \, du \right\}$$

$$= \lim_{N \to \infty} (-e^{-N} N^t) + t \lim_{N \to \infty} \int_0^N e^{-u} u^{t-1} \, du$$

$$= 0 + t\Gamma(t) = t\Gamma(t) .$$

When t is a positive integer, say $t = n$, then the recursive relation (16) can be repeatedly applied to obtain

$$\Gamma(n + 1) = n\Gamma(n) = n(n - 1)\Gamma(n - 1) = \cdots$$
$$= n(n - 1)(n - 2) \cdots 2\Gamma(1) .$$

It follows from the definition (15) that $\Gamma(1) = 1$, so we find

$$\Gamma(n + 1) = n! .$$

Thus the gamma function extends the notion of factorial!

As an application of the gamma function, let's return to the problem of determining the Laplace transform of an arbitrary power of t. We shall verify that the formula

(17) $\mathcal{L}\{t^r\}(s) = \dfrac{\Gamma(r + 1)}{s^{r+1}}$

holds for every constant $r > -1$.

By definition,

$$\mathcal{L}\{t^r\}(s) = \int_0^\infty e^{-st} t^r \, dt .$$

Let's make the substitution $u = st$. Then $du = s \, dt$, and we find

$$\mathcal{L}\{t^r\}(s) = \int_0^\infty e^{-u} \left(\frac{u}{s}\right)^r \left(\frac{1}{s}\right) du$$

$$= \frac{1}{s^{r+1}} \int_0^\infty e^{-u} u^r \, du = \frac{\Gamma(r + 1)}{s^{r+1}} .$$

Notice that when $r = n$ is a nonnegative integer, then $\Gamma(n + 1) = n!$, and so formula (17) reduces to the familiar formula for $\mathcal{L}\{t^n\}$.

EXERCISES 7.6

In Problems 1 through 4, sketch the graph of the given function and determine its Laplace transform.

1. $(t - 1)^2 u(t - 1)$.

2. $u(t - 1) - u(t - 4)$.

3. $t^2 u(t - 2)$.

4. $tu(t - 1)$.

In Problems 5 through 10, express the given function using unit step functions and compute its Laplace transform.

5. $g(t) = \begin{cases} 0 , & 0 < t < 1 , \\ 2 , & 1 < t < 2 , \\ 1 , & 2 < t < 3 , \\ 3 , & 3 < t . \end{cases}$

6. $g(t) = \begin{cases} 0 , & 0 < t < 2 , \\ t + 1 , & 2 < t . \end{cases}$

7.

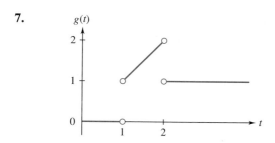

Figure 7.14 Function in Problem 7

8.

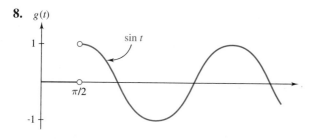

Figure 7.15 Function in Problem 8

9.

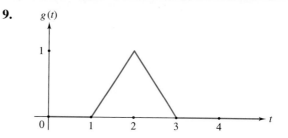

Figure 7.16 Function in Problem 9

10.

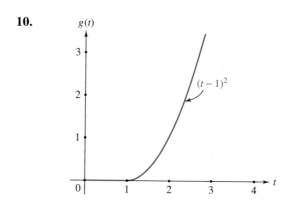

Figure 7.17 Function in Problem 10

In Problems 11 through 18, determine an inverse Laplace transform of the given function.

11. $\dfrac{e^{-2s}}{s - 1}$.

12. $\dfrac{e^{-3s}}{s^2}$.

13. $\dfrac{e^{-2s} - 3e^{-4s}}{s + 2}$.

14. $\dfrac{e^{-3s}}{s^2 + 9}$.

15. $\dfrac{se^{-3s}}{s^2 + 4s + 5}$.

16. $\dfrac{e^{-s}}{s^2 + 4}$.

17. $\dfrac{e^{-3s}(s - 5)}{(s + 1)(s + 2)}$.

18. $\dfrac{e^{-s}(3s^2 - s + 2)}{(s - 1)(s^2 + 1)}$.

19. The current $I(t)$ in an RLC series circuit is governed by the initial value problem

$$I''(t) + 2I'(t) + 2I(t) = g(t) ;$$
$$I(0) = 10 , \qquad I'(0) = 0 ,$$

where

$$g(t) := \begin{cases} 20 , & 0 < t < 3\pi , \\ 0 , & 3\pi < t < 4\pi , \\ 20 , & 4\pi < t . \end{cases}$$

Determine the current as a function of time t. Sketch $I(t)$ for $0 < t < 8\pi$.

20. The current $I(t)$ in an LC series circuit is governed by the initial value problem

$$I''(t) + 4I(t) = g(t) ;$$

$$I(0) = 1 , \qquad I'(0) = 3 ,$$

where

$$g(t) := \begin{cases} 3 \sin t , & 0 \le t \le 2\pi , \\ 0 , & 2\pi < t . \end{cases}$$

Determine the current as a function of time t.

In Problems 21 through 24, determine $\mathcal{L}\{f\}$, where $f(t)$ is periodic with the given period. Also graph $f(t)$.

21. $f(t) = t ,\quad 0 < t < 2 ,\quad$ and $f(t)$ has period 2.

22. $f(t) = e^t ,\quad 0 < t < 1 ,\quad$ and $f(t)$ has period 1.

23. $f(t) = \begin{cases} e^{-t} , & 0 < t < 1 , \\ 1 , & 1 < t < 2 , \end{cases}$

and $f(t)$ has period 2.

24. $f(t) = \begin{cases} t , & 0 < t < 1 , \\ 1 - t , & 1 < t < 2 , \end{cases}$

and $f(t)$ has period 2.

In Problems 25 through 28, determine $\mathcal{L}\{f\}$, where the periodic function is described by its graph.

25.

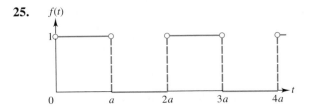

Figure 7.18 Square wave

26.

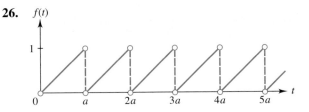

Figure 7.19 Sawtooth wave

27.

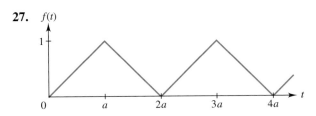

Figure 7.20 Triangular wave

28.

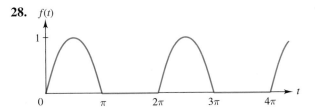

Figure 7.21 Half-rectified sine wave

In Problems 29 through 32, solve the given initial value problem using the method of Laplace transforms. Sketch the graph of the solution.

29. $y'' + y = u(t - 3) ;$
$y(0) = 0 , \qquad y'(0) = 1 .$

30. $w'' + w = u(t - 2) - u(t - 4) ;$
$w(0) = 1 , \qquad w'(0) = 0 .$

31. $y'' + y = t - (t - 4)u(t - 2) ;$
$y(0) = 0 , \qquad y'(0) = 1 .$

32. $y'' + y = 3 \sin 2t - 3(\sin 2t)u(t - 2\pi) ;$
$y(0) = 1 , \qquad y'(0) = -2 .$

In Problems 33 through 40, solve the given initial value problem using the method of Laplace transforms.

33. $y'' + 2y' + 2y = u(t - 2\pi) - u(t - 4\pi) ;$
$y(0) = 1 , \qquad y'(0) = 1 .$

34. $y'' + 4y' + 4y = u(t - \pi) - u(t - 2\pi)$;
$y(0) = 0$, $\quad y'(0) = 0$.

35. $z'' + 3z' + 2z = e^{-3t}u(t - 2)$;
$z(0) = 2$, $\quad z'(0) = -3$.

36. $y'' + 5y' + 6y = tu(t - 2)$;
$y(0) = 0$, $\quad y'(0) = 1$.

37. $y'' + 4y = g(t)$; $\quad y(0) = 1$, $\quad y'(0) = 3$,

where $g(t) = \begin{cases} \sin t , & 0 \le t \le 2\pi , \\ 0 , & 2\pi < t . \end{cases}$

38. $y'' + 2y' + 10y = g(t)$;
$y(0) = -1$, $\quad y'(0) = 0$,

where $g(t) = \begin{cases} 10 , & 0 \le t \le 10 , \\ 20 , & 10 < t < 20 , \\ 0 , & 20 < t . \end{cases}$

39. $y'' + 5y' + 6y = g(t)$;
$y(0) = 0$, $\quad y'(0) = 2$,

where $g(t) = \begin{cases} 0 , & 0 \le t < 1 , \\ t , & 1 < t < 5 , \\ 1 , & 5 < t . \end{cases}$

40. $y'' + 3y' + 2y = g(t)$;
$y(0) = 2$, $\quad y'(0) = -1$,

where $g(t) = \begin{cases} e^{-t} , & 0 \le t < 3 , \\ 1 , & 3 < t . \end{cases}$

41. Show that if $\mathcal{L}\{g\}(s) = [(s + \alpha)(1 - e^{-Ts})]^{-1}$, where $T > 0$ is fixed, then

(18) $g(t) = e^{-\alpha t} + e^{-\alpha(t-T)}u(t - T)$
$\qquad + e^{-\alpha(t-2T)}u(t - 2T)$
$\qquad + e^{-\alpha(t-3T)}u(t - 3T) + \cdots$.

[*Hint:* Use the fact that
$1 + x + x^2 + \cdots = 1/(1 - x)$.]

42. The function $g(t)$ in (18) can be expressed in a more convenient form as follows:

(a) Show that for each $n = 0, 1, 2, \ldots$,

$$g(t) = e^{-\alpha t}\left[\frac{e^{(n+1)\alpha T} - 1}{e^{\alpha T} - 1}\right]$$

for $nT < t < (n + 1)T$.

[*Hint:* Use the fact that $1 + x + x^2 + \cdots + x^n = (x^{n+1} - 1)/(x - 1)$.]

(b) Let $v = t - (n + 1)T$. Show that when $nT < t < (n + 1)T$, then $-T < v < 0$ and

(19) $g(t) = \dfrac{e^{-\alpha v}}{e^{\alpha T} - 1} - \dfrac{e^{-\alpha t}}{e^{\alpha T} - 1}$.

(c) Use the facts that the first term in (19) is periodic with period T and the second term is independent of n to sketch the graph of $g(t)$ in (19) for $\alpha = 1$ and $T = 2$.

43. Show that if $\mathcal{L}\{g\}(s) = \beta[(s^2 + \beta^2)(1 - e^{-Ts})]^{-1}$, then

$g(t) = \sin \beta t + [\sin \beta(t - T)]u(t - T)$
$\qquad + [\sin \beta(t - 2T)]u(t - 2T)$
$\qquad + [\sin \beta(t - 3T)]u(t - 3T) + \cdots$.

44. Use the result of Problem 43 to show that

$$\mathcal{L}^{-1}\left\{\frac{1}{(s^2 + 1)(1 - e^{-\pi s})}\right\}(t) = g(t) ,$$

where $g(t)$ is periodic with period 2π and

$g(t) := \begin{cases} \sin t , & 0 \le t \le \pi , \\ 0 , & \pi \le t \le 2\pi . \end{cases}$

In Problems 45 and 46, use the method of Laplace transforms and the results of Problems 41 and 42 to solve the initial value problem.

$$y'' + 3y' + 2y = f(t) ;$$
$$y(0) = 0 , \quad y'(0) = 0 ,$$

where $f(t)$ is the periodic function defined in the stated problem.

45. Problem 22 **46.** Problem 25 with $a = 1$.

In Problems 47 through 50, find a Taylor series for $f(t)$ about $t = 0$. Assuming the Laplace transform of $f(t)$ can be computed term by term, find an expansion for $\mathcal{L}\{f\}(s)$ in powers of $1/s$. If possible, sum the series.

47. $f(t) = e^t$. **48.** $f(t) = \sin t$.

49. $f(t) = \dfrac{1 - \cos t}{t}$. **50.** $f(t) = e^{-t^2}$.

51. Using the recursive relation (16) and the fact that $\Gamma(1/2) = \sqrt{\pi}$, determine

(a) $\mathcal{L}\{t^{-1/2}\}$. **(b)** $\mathcal{L}\{t^{7/2}\}$.

52. Use the recursive relation (16) and the fact that $\Gamma(1/2) = \sqrt{\pi}$ to show that

$$\mathcal{L}^{-1}\left\{s^{-(n+1/2)}\right\}(t) = \frac{2^n t^{n-1/2}}{1 \cdot 3 \cdot 5 \cdots (2n-1)\sqrt{\pi}} \ ,$$

where n is a positive integer.

53. Explain why, for each fixed t, the series (13) in the proof of Theorem 9 has only a *finite* number of nonzero terms.

54. By replacing s by $1/s$ in the Maclaurin series expansion for arctan s, show that

$$\arctan\frac{1}{s} = \frac{1}{s} - \frac{1}{3s^3} + \frac{1}{5s^5} - \frac{1}{7s^7} + \cdots .$$

55. Find an expansion for $e^{-1/s}$ in powers of $1/s$. Use the expansion for $e^{-1/s}$ to obtain an expansion for $s^{-1/2}e^{-1/s}$ in terms of $1/s^{n+1/2}$. Assuming the inverse Laplace transform can be computed term by term, show that

$$\mathcal{L}^{-1}\left\{s^{-1/2}e^{-1/s}\right\}(t) = \frac{1}{\sqrt{\pi t}}\cos 2\sqrt{t} \ .$$

[*Hint:* Use the result of Problem 52.]

56. Use the procedure discussed in Problem 55 to show that

$$\mathcal{L}^{-1}\left\{s^{-3/2}e^{-1/s}\right\}(t) = \frac{1}{\sqrt{\pi}}\sin 2\sqrt{t} \ .$$

57. Find an expansion for $\ln\left[1 + (1/s^2)\right]$ in powers of $1/s$. Assuming the inverse Laplace transform can be computed term by term, show that

$$\mathcal{L}^{-1}\left\{\ln\left(1 + \frac{1}{s^2}\right)\right\}(t) = \frac{2}{t}(1 - \cos t) \ .$$

58. The **unit gate function** $G_a(t)$ is defined by

$$G_a(t) := \begin{cases} 0, & t < 0 , \\ 1, & 0 < t < a , \\ 0, & a < t . \end{cases}$$

(a) Show that $G_a(t) = u(t) - u(t-a)$.

(b) Verify that $\mathcal{L}\{G_a\}(s) = (1/s)(1 - e^{-as})$.
(c) Show that

$$\mathcal{L}\{G_a(t-b)\}(s) = (1/s)\left(e^{-bs} - e^{-(a+b)s}\right) .$$

In Problems 59 and 60, use the results of Problem 58 and the method of Laplace transforms to solve the given initial value problem.

59. $y'' - y = G_3(t-1)$;
$\quad y(0) = 0 , \qquad y'(0) = 2$.

60. $y'' - y = G_4(t-3)$;
$\quad y(0) = 1 , \qquad y'(0) = -1$.

61. The mixing tank in Figure 7.22 initially holds 500 L of a brine solution with a salt concentration of 2 kg/L. For the first 10 min of operation, valve A is open, adding 12 L/min of brine containing a 4 kg/L salt concentration. After 10 min, valve B is switched in, adding a 6 kg/L concentration at 12 L/min. The exit valve C removes 12 L/min, thereby keeping the volume constant. Find the concentration of salt in the tank as a function of time.

62. Suppose that in Problem 61 valve B is initially opened for 10 min and then valve A is switched in for 10 min. Finally valve B is switched back in. Find the concentration of salt in the tank as a function of time.

63. Suppose valve C removes only 6 L/min in Problem 61. Can Laplace transforms be used to solve the problem? Discuss.

Figure 7.22 Mixing tank

7.7 CONVOLUTION

Consider the initial value problem

(1) $y'' + y = g(t)$; $y(0) = 0 ,$ $y'(0) = 0$.

If we let $Y(s) = \mathcal{L}\{y\}(s)$ and $G(s) = \mathcal{L}\{g\}(s)$, then taking the Laplace transform of both sides of (1) yields

$$s^2 Y(s) + Y(s) = G(s) \ ,$$

and hence

(2) $Y(s) = \left(\dfrac{1}{s^2 + 1}\right) G(s) \ .$

That is, the Laplace transform of the solution to (1) is the product of the Laplace transform of sin t and the Laplace transform of the forcing term $g(t)$. What we would now like to have is a simple formula for $y(t)$ in terms of sin t and $g(t)$. Just as the integral of a product is not the product of the integrals, $y(t)$ is not the product of sin t and $g(t)$. However, we can express $y(t)$ as the "convolution" of sin t and $g(t)$.

CONVOLUTION

Definition 8. Let $f(t)$ and $g(t)$ be piecewise continuous on $[0, \infty)$. The **convolution** of $f(t)$ and $g(t)$, denoted $f * g$, is defined by

(3) $(f * g)(t) := \displaystyle\int_0^t f(t - v)g(v)\,dv \ .$

For example, the convolution of t and t^2 is

$$t * t^2 = \int_0^t (t - v)v^2\,dv = \int_0^t (tv^2 - v^3)\,dv$$

$$= \left(\frac{tv^3}{3} - \frac{v^4}{4}\right)\Bigg|_0^t = \frac{t^4}{3} - \frac{t^4}{4} = \frac{t^4}{12} \ .$$

Convolution is certainly different from ordinary multiplication. For example, $1 * 1 = t \neq 1$ and in general $1 * f \neq f$. However, convolution does satisfy some of the same properties as multiplication.

PROPERTIES OF CONVOLUTION

Theorem 10. Let $f(t)$, $g(t)$, and $h(t)$ be piecewise continuous on $[0, \infty)$. Then

(4) $f * g = g * f \ ,$
(5) $f * (g + h) = (f * g) + (f * h) \ ,$
(6) $(f * g) * h = f * (g * h) \ ,$
(7) $f * 0 = 0 \ .$

Proof. To prove equation (4), we begin with the definition

$$(f * g)(t) := \int_0^t f(t - v)g(v)\,dv \ .$$

Using the change of variables $w = t - v$, we have

$$(f * g)(t) = \int_t^0 f(w)g(t - w)(-dw) = \int_0^t g(t - w)f(w)dw = (g * f)(t) \ ,$$

which proves (4). The proofs of equations (5) and (6) are left to the exercises (see Problems 33 and 34). Equation (7) is obvious, since $f(t - v) \cdot 0 \equiv 0$. ∎ End

Returning to our original goal, we now prove that if $Y(s)$ is the product of the Laplace transforms $F(s)$ and $G(s)$, then $y(t)$ is the convolution $(f * g)(t)$.

CONVOLUTION THEOREM

Theorem 11. Let $f(t)$ and $g(t)$ be piecewise continuous on $[0, \infty)$ and of exponential order α and set $F(s) = \mathscr{L}\{f\}(s)$ and $G(s) = \mathscr{L}\{g\}(s)$. Then

(8) $\mathscr{L}\{f * g\}(s) = F(s)G(s) \ ,$

or, equivalently,

(9) $\mathscr{L}^{-1}\{F(s)G(s)\}(t) = (f * g)(t) \ .$

Proof. Starting with the left-hand side of (8), we use the definition of convolution to write for $s > \alpha$

$$\mathscr{L}\{f * g\}(s) = \int_0^\infty e^{-st}\left(\int_0^t f(t - v)g(v)dv\right)dt \ .$$

To simplify the evaluation of this iterated integral, we introduce the unit step function $u(t - v)$ and write

$$\mathscr{L}\{f * g\}(s) = \int_0^\infty e^{-st}\left(\int_0^\infty u(t - v)f(t - v)g(v)dv\right)dt \ ,$$

where we have used the fact that $u(t - v) = 0$ if $v > t$. Reversing the order of integration[†] gives

(10) $\mathscr{L}\{f * g\}(s) = \int_0^\infty g(v)\left(\int_0^\infty e^{-st}u(t - v)f(t - v)dt\right)dv \ .$

Recall from the translation property in Section 7.6 that the integral in parentheses in equation (10) equals $e^{-sv}F(s)$. Hence

$$\mathscr{L}\{f * g\}(s) = \int_0^\infty g(v)e^{-sv}F(s)dv = F(s)\int_0^\infty e^{-sv}g(v)dv = F(s)G(s) \ .$$

This proves formula (8). ∎ End

[†]This is permitted since, for each $s > \alpha$, the absolute value of the integrand is integrable on $(0, \infty) \times (0, \infty)$.

For the initial value problem (1), recall that we found

$$Y(s) = \left(\frac{1}{s^2 + 1}\right) G(s) = \mathscr{L}\{\sin t\}(s)\, \mathscr{L}\{g\}(s) \ .$$

It now follows from the convolution theorem that

$$y(t) = \sin t * g(t) = \int_0^t \sin(t - v) g(v)\, dv \ .$$

Thus we have obtained an integral representation for the solution to the initial value problem (1) for any forcing function $g(t)$ that is piecewise continuous on $[0, \infty)$ and of exponential order.

EXAMPLE 1 Use the convolution theorem to solve the initial value problem

(11) $y'' - y = g(t) \ ; \qquad y(0) = 1 \ , \qquad y'(0) = 1 \ ,$

where $g(t)$ is piecewise continuous on $[0, \infty)$ and of exponential order.

SOLUTION Let $Y(s) = \mathscr{L}\{y\}(s)$ and $G(s) = \mathscr{L}\{g\}(s)$. Taking the Laplace transform of both sides of the differential equation in (11) and using the initial conditions gives

$$s^2 Y(s) - s - 1 - Y(s) = G(s) \ .$$

Solving for $Y(s)$, we have

$$Y(s) = \frac{s + 1}{s^2 - 1} + \left(\frac{1}{s^2 - 1}\right) G(s) = \frac{1}{s - 1} + \left(\frac{1}{s^2 - 1}\right) G(s) \ .$$

Hence

$$y(t) = \mathscr{L}^{-1}\left\{\frac{1}{s - 1}\right\}(t) + \mathscr{L}^{-1}\left\{\frac{1}{s^2 - 1}\, G(s)\right\}(t)$$

$$= e^t + \mathscr{L}^{-1}\left\{\frac{1}{s^2 - 1}\, G(s)\right\}(t) \ .$$

Referring to the table of Laplace transforms on the inside back cover, we find

$$\mathscr{L}\{\sinh t\}(s) = \frac{1}{s^2 - 1} \ ,$$

so we can now express

$$\mathscr{L}^{-1}\left\{\frac{1}{s^2 - 1}\, G(s)\right\}(t) = \sinh t * g(t) \ .$$

Thus

$$y(t) = e^t + \int_0^t \sinh(t - v) g(v)\, dv$$

is the solution to the initial value problem (11). ∎ End

EXAMPLE 2 Use the convolution theorem to find $\mathscr{L}^{-1}\{1/(s^2 + 1)^2\}$.

SOLUTION Write

$$\frac{1}{(s^2 + 1)^2} = \left(\frac{1}{s^2 + 1}\right)\left(\frac{1}{s^2 + 1}\right) .$$

Since $\mathscr{L}\{\sin t\}(s) = 1/(s^2 + 1)$, it follows from the convolution theorem that

$$\mathscr{L}^{-1}\left\{\frac{1}{(s^2 + 1)^2}\right\}(t) = \sin t * \sin t = \int_0^t \sin(t - v)\sin v\, dv$$

$$= \frac{1}{2}\int_0^t \left[\cos(2v - t) - \cos t\right]dv^\dagger$$

$$= \frac{1}{2}\left[\frac{\sin(2v - t)}{2}\right]_0^t - \frac{1}{2}t\cos t$$

$$= \frac{1}{2}\left[\frac{\sin t}{2} - \frac{\sin(-t)}{2}\right] - \frac{1}{2}t\cos t$$

$$= \frac{\sin t - t\cos t}{2} . \quad \blacksquare \text{End}$$

As the preceding example attests, the convolution theorem is useful in determining the inverse transforms of rational functions of s. In fact, it provides an alternative to the method of partial fractions. For example,

$$\mathscr{L}^{-1}\left\{\frac{1}{(s - a)(s - b)}\right\}(t) = \mathscr{L}^{-1}\left\{\left(\frac{1}{s - a}\right)\left(\frac{1}{s - b}\right)\right\}(t) = e^{at} * e^{bt} ,$$

and all that remains in finding the inverse is to compute the convolution $e^{at} * e^{bt}$.

In the early 1900s, V. Volterra introduced **integro-differential** equations in his study of population growth. These equations enabled him to take into account "hereditary influences." In certain cases, these equations involved a convolution. As the next example shows, the convolution theorem helps to solve such integro-differential equations.

EXAMPLE 3 Solve the integro-differential equation

(12) $y'(t) = 1 - \int_0^t y(t - v)e^{-2v}\, dv , \qquad y(0) = 1 .$

SOLUTION Equation (12) can be written as

(13) $y'(t) = 1 - y(t) * e^{-2t} .$

†Here we used the identity $\sin \alpha \sin \beta = \frac{1}{2}\left[\cos(\beta - \alpha) - \cos(\beta + \alpha)\right]$.

Let $Y(s) = \mathcal{L}\{y\}(s)$. Taking the Laplace transform of (13) (with the help of the convolution theorem) and solving for $Y(s)$, we obtain

$$sY(s) - 1 = \frac{1}{s} - Y(s)\left(\frac{1}{s + 2}\right)$$

$$sY(s) + \left(\frac{1}{s + 2}\right)Y(s) = 1 + \frac{1}{s}$$

$$\left(\frac{s^2 + 2s + 1}{s + 2}\right)Y(s) = \frac{s + 1}{s}$$

$$Y(s) = \frac{(s + 1)(s + 2)}{s(s + 1)^2} = \frac{s + 2}{s(s + 1)}$$

$$Y(s) = \frac{2}{s} - \frac{1}{s + 1} .$$

Hence $y(t) = 2 - e^{-t}$. ∎ End

The **transfer function** $H(s)$ of a linear system is defined as the ratio of the Laplace transform of the output function $y(t)$ to the Laplace transform of the input function $g(t)$, under the assumption that all initial conditions are zero. That is, $H(s) = Y(s)/G(s)$. If the linear system is governed by the differential equation

(14) $ay'' + by' + cy = g(t)$, $t > 0$,

where a, b, and c are constants, we can compute the transfer function as follows. Take the Laplace transform of both sides of (14) to get

$$as^2Y(s) - asy(0) - ay'(0) + bsY(s) - by(0) + cY(s) = G(s) .$$

Because the initial conditions are assumed to be zero, the equation reduces to

$$(as^2 + bs + c)Y(s) = G(s) .$$

Thus the transfer function for equation (14) is

(15) $H(s) = \dfrac{Y(s)}{G(s)} = \dfrac{1}{as^2 + bs + c}$.

The reader may note the similarity of these calculations to those for finding the auxiliary equation for the homogeneous equation associated with (14) (recall Section 4.5, page 184). Indeed, the first step in inverting $Y(s) = G(s)/(as^2 + bs + c)$ would be to find the roots of the denominator $as^2 + bs + c$, which is identical to solving the characteristic equation for (14).

The function $h(t) := \mathcal{L}^{-1}\{H\}(t)$ is called the **impulse response function** for the system because, physically speaking, it describes the solution when a mass-spring system is struck by a hammer (see Section 7.8). We can also characterize $h(t)$ as the unique solution to the homogeneous problem

(16) $ah'' + bh' + ch = 0$; $h(0) = 0$, $h'(0) = 1/a$.

Indeed, observe that taking the Laplace transform of the equation in (16) gives

(17) $a\left[s^2H(s) - sh(0) - h'(0)\right] + b\left[sH(s) - h(0)\right] + cH(s) = 0$.

Substituting in $h(0) = 0$ and $h'(0) = 1/a$ and solving for $H(s)$ yields

$$H(s) = \frac{1}{as^2 + bs + c} \; ,$$

which is the same as the formula for the transfer function given in equation (15).

One nice feature of the impulse response function h is that it can help us describe the solution to the *general* initial value problem

(18) $ay'' + by' + cy = g(t) \; ; \qquad y(0) = y_0 \; , \qquad y'(0) = y_1 \; .$

From the discussion of equation (14), we can see that the convolution $h * g$ is the solution to (18) in the special case when the initial conditions are zero (i.e., $y_0 = y_1 = 0$). To deal with nonzero initial conditions, let y_k denote the solution to the corresponding *homogeneous* initial value problem; that is, y_k solves

(19) $ay'' + by' + cy = 0 \; ; \qquad y(0) = y_0 \; , \qquad y'(0) = y_1 \; .$

Then, the desired solution to the general initial value problem (18) must be $h * g + y_k$. Indeed, it follows from the superposition principle (see Theorem 5 in Section 4.7) that since $h * g$ is a solution to equation (14) and y_k is a solution to the corresponding homogeneous equation, then $h * g + y_k$ is a solution to equation (14). Moreover, since $h * g$ has initial conditions zero,

$$(h * g)(0) + y_k(0) = 0 + y_0 = y_0 \; ,$$
$$(h * g)'(0) + y_k'(0) = 0 + y_1 = y_1 \; .$$

We summarize these observations in the following theorem.

SOLUTION USING IMPULSE RESPONSE FUNCTION

Theorem 12. Let I be an interval containing the origin. The unique solution to the initial value problem

$$ay'' + by' + cy = g \; ; \qquad y(0) = y_0 \; , \qquad y'(0) = y_1 \; ,$$

where a, b, and c are constants and g is continuous on I, is given by

(20) $y(t) = (h * g)(t) + y_k(t) = \displaystyle\int_0^t h(t - v)g(v)\, dv + y_k(t) \; ,$

where h is the impulse response function for the system and y_k is the unique solution to (19).

Equation (20) is instructive in that it highlights how the value of y at time t depends on the initial conditions (through $y_k(t)$) and on the nonhomogeneity $g(t)$ (through the convolution integral). It even displays the *causal* nature of the dependence, in that the value of $g(v)$ cannot influence $y(t)$ until $t \geq v$.

A proof of Theorem 12 that does not involve Laplace transforms is outlined in Project C in Chapter 4.

In the next example we use Theorem 12 to find a formula for the solution to an initial value problem.

EXAMPLE 4 A linear system is governed by the differential equation

(21) $y'' + 2y' + 5y = g(t)$; $y(0) = 2$, $y'(0) = -2$.

Find the transfer function for the system, the impulse response function, and a formula for the solution.

SOLUTION According to formula (15), the transfer function for (21) is

$$H(s) = \frac{1}{as^2 + bs + c} = \frac{1}{s^2 + 2s + 5} = \frac{1}{(s + 1)^2 + 2^2} .$$

The inverse Laplace transform of $H(s)$ is the impulse response function

$$h(t) = \mathcal{L}^{-1}\{H\}(t) = \frac{1}{2}\mathcal{L}^{-1}\left\{\frac{2}{(s + 1)^2 + 2^2}\right\}(t)$$

$$= \frac{1}{2}e^{-t}\sin 2t .$$

To solve the initial value problem, we need the solution to the corresponding homogeneous problem. The auxiliary equation for the homogeneous equation is $r^2 + 2r + 5 = 0$, which has roots $r = -1 \pm 2i$. Thus a general solution is $C_1 e^{-t}\cos 2t + C_2 e^{-t}\sin 2t$. Choosing C_1 and C_2 so that the initial conditions in (21) are satisfied, we obtain $y_k(t) = 2e^{-t}\cos 2t$.

Hence, a formula for the solution to the initial value problem (21) is

$$(h * g)(t) + y_k(t) = \frac{1}{2}\int_0^t e^{-(t-v)}\sin\left[2(t - v)\right]g(v)dv + 2e^{-t}\cos 2t . \quad \blacksquare \text{ End}$$

EXERCISES 7.7

In Problems 1 through 4, use the convolution theorem to obtain a formula for the solution to the given initial value problem, where $g(t)$ is piecewise continuous on $[0, \infty)$ and of exponential order.

1. $y'' - 2y' + y = g(t)$;
$y(0) = -1$, $y'(0) = 1$.

2. $y'' + 9y = g(t)$; $y(0) = 1$, $y'(0) = 0$.

3. $y'' + 4y' + 5y = g(t)$;
$y(0) = 1$, $y'(0) = 1$.

4. $y'' + y = g(t)$; $y(0) = 0$, $y'(0) = 1$.

In Problems 5 through 12, use the convolution theorem to find the inverse Laplace transform of the given function.

5. $\dfrac{1}{s(s^2 + 1)}$.

6. $\dfrac{1}{(s + 1)(s + 2)}$.

7. $\dfrac{14}{(s + 2)(s - 5)}$.

8. $\dfrac{1}{(s^2 + 4)^2}$.

9. $\dfrac{s}{(s^2 + 1)^2}$.

10. $\dfrac{1}{s^3(s^2 + 1)}$.

11. $\dfrac{s}{(s - 1)(s + 2)}$. $\left[\text{Hint: } \dfrac{s}{s - 1} = 1 + \dfrac{1}{s - 1} .\right]$

12. $\dfrac{s + 1}{(s^2 + 1)^2}$.

13. Find the Laplace transform of

$$f(t) := \int_0^t (t - v)e^{3v}dv .$$

14. Find the Laplace transform of

$$f(t) := \int_0^t e^v\sin(t - v)dv .$$

In Problems 15 through 22, solve the given integral equation or integro-differential equation for $y(t)$.

15. $y(t) + 3 \int_0^t y(v)\sin(t - v)dv = t$.

16. $y(t) + \int_0^t e^{t-v}y(v)dv = \sin t$.

17. $y(t) + \int_0^t (t - v)y(v)dv = 1$.

18. $y(t) + \int_0^t (t - v)y(v)dv = t^2$.

19. $y(t) + \int_0^t (t - v)^2 y(v)dv = t^3 + 3$.

20. $y'(t) + \int_0^t (t - v)y(v)dv = t$, $y(0) = 0$.

21. $y'(t) + y(t) - \int_0^t y(v)\sin(t - v)dv = -\sin t$,

$y(0) = 1$.

22. $y'(t) - 2\int_0^t e^{t-v}y(v)dv = t$, $y(0) = 2$.

In Problems 23 through 28, a linear system is governed by the given initial value problem. Find the transfer function $H(s)$ for the system and the impulse response function $h(t)$ and give a formula for the solution to the initial value problem.

23. $y'' + 9y = g(t)$;
 $y(0) = 2$, $y'(0) = -3$.

24. $y'' - 9y = g(t)$; $y(0) = 2$, $y'(0) = 0$.

25. $y'' - y' - 6y = g(t)$;
 $y(0) = 1$, $y'(0) = 8$.

26. $y'' + 2y' - 15y = g(t)$;
 $y(0) = 0$, $y'(0) = 8$.

27. $y'' - 2y' + 5y = g(t)$;
 $y(0) = 0$, $y'(0) = 2$.

28. $y'' - 4y' + 5y = g(t)$;
 $y(0) = 0$, $y'(0) = 1$.

In Problems 29 and 30, the current $I(t)$ in an RLC circuit with voltage source $E(t)$ is governed by the initial value problem

$$LI''(t) + RI'(t) + \frac{1}{C}I(t) = e(t) ,$$

$$I(0) = a , \qquad I'(0) = b ,$$

where $e(t) = E'(t)$ (see Figure 7.23). For the given constants R, L, C, a, and b, find a formula for the solution $I(t)$ in terms of $e(t)$.

Figure 7.23 Schematic representation of an *RLC* series circuit

29. $R = 20$ ohms, $L = 5$ henrys, $C = 0.005$ farads, $a = -1$ amps, $b = 8$ amps/sec.

30. $R = 80$ ohms, $L = 10$ henrys, $C = 1/410$ farads, $a = 2$ amps, $b = -8$ amps/sec.

31. Use the convolution theorem and Laplace transforms to compute $1 * 1 * 1$.

32. Use the convolution theorem and Laplace transforms to compute $1 * t * t^2$.

33. Prove property (5) in Theorem 10.

34. Prove property (6) in Theorem 10.

35. Use the convolution theorem to show that

$$\mathcal{L}^{-1}\left\{\frac{F(s)}{s}\right\}(t) = \int_0^t f(v)dv ,$$

where $F(s) = \mathcal{L}\{f\}(s)$.

36. Using Theorem 5 in Section 7.3 and the convolution theorem, show that

$$\int_0^t \int_0^v f(z) \, dz \, dv = \mathcal{L}^{-1}\left\{\frac{F(s)}{s^2}\right\}(t)$$

$$= t\int_0^t f(v) \, dv - \int_0^t vf(v) \, dv ,$$

where $F(s) = \mathcal{L}\{f\}(s)$.

37. Prove directly that if $h(t)$ is the impulse response function characterized by equation (16), then for any continuous $g(t)$, we have $(h * g)(0) = (h * g)'(0) = 0$.

[*Hint:* Use Leibniz's rule, described in Project C of Chapter 4.]

7.8 IMPULSES AND THE DIRAC DELTA FUNCTION

In mechanical systems, electrical circuits, bending of beams, and other applications, one encounters functions that have a very large value over a very short interval. For example, the strike of a hammer exerts a relatively large force over a relatively short time, and a heavy weight concentrated at a spot on a suspended beam exerts a large force over a very small section of the beam. To deal with violent forces of short duration, physicists and engineers use the delta function introduced by Paul A. M. Dirac. Relaxing our standards of rigor for the moment, we present the following somewhat informal defintion.

DIRAC DELTA FUNCTION

Definition 9. The **Dirac delta function** $\delta(t)$ is characterized by the following two properties:

$$(1) \qquad \delta(t) = \begin{cases} 0 , & t \neq 0 , \\ \text{``infinite,''} & t = 0 , \end{cases}$$

and

$$(2) \qquad \int_{-\infty}^{\infty} f(t)\delta(t)\, dt = f(0)$$

for any function $f(t)$ that is continuous on an open interval containing $t = 0$.

By shifting the argument of $\delta(t)$, we have $\delta(t - a) = 0, t \neq a$, and

$$(3) \qquad \int_{-\infty}^{\infty} f(t)\delta(t - a)\, dt = f(a)$$

for any function $f(t)$ that is continuous on an interval containing $t = a$.

It is obvious that $\delta(t - a)$ is not a function in the usual sense; instead it is an example of what is called a **generalized function** or a **distribution.** In spite of this shortcoming, the Dirac delta function was successfully used for several years to solve various physics and engineering problems before Laurent Schwartz mathematically justified its use!

A heuristic argument for the existence of the Dirac delta function can be made by considering the impulse of a force over a short interval. If a force $\mathscr{F}(t)$ is applied from time t_0 to time t_1, then the **impulse** due to \mathscr{F} is the integral

$$\text{Impulse} := \int_{t_0}^{t_1} \mathscr{F}(t)\, dt .$$

By Newton's second law, we see that

(4) $$\int_{t_0}^{t_1} \mathscr{F}(t)\,dt = \int_{t_0}^{t_1} m\frac{dv}{dt}\,dt = mv(t_1) - mv(t_0) \;,$$

where m denotes mass and v denotes velocity. Since mv represents the momentum, we can interpret equation (4) as saying: **The impulse equals the change in momentum.**

When a hammer strikes an object, it transfers momentum to the object. This change in momentum takes place over a very short period of time, say $[t_0, t_1]$. If we let $\mathscr{F}_1(t)$ represent the force due to the hammer, then the *area* under the curve $\mathscr{F}_1(t)$ is the impulse or change in momentum (see Figure 7.24). If, as is illustrated in Figure 7.25, the same change in momentum takes place over shorter and shorter time intervals, say $[t_0, t_2]$ or $[t_0, t_3]$, then the average force must get greater and greater in order for the impulses (the areas under the curves \mathscr{F}_n) to remain the same. In fact, if the forces \mathscr{F}_n having the same impulse act, respectively, over the intervals $[t_0, t_n]$, where $t_n \rightarrow t_0$ as $n \rightarrow \infty$, then \mathscr{F}_n approaches a function that is zero for $t \neq t_0$ but has an infinite value for $t = t_0$. Moreover, the areas under the \mathscr{F}_n's have a common value. Normalizing this value to be 1 gives

$$\int_{-\infty}^{\infty} \mathscr{F}_n(t)\,dt = 1 \qquad \text{for all } n.$$

When $t_0 = 0$, we derive from the limiting properties of the \mathscr{F}_n's a "function" δ that satisfies property (1) and the integral condition

(5) $$\int_{-\infty}^{\infty} \delta(t)\,dt = 1 \;.$$

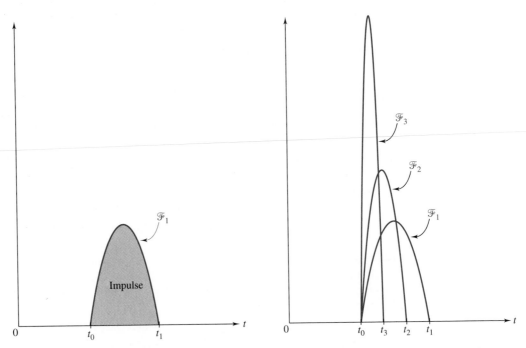

Figure 7.24 Force due to a blow from a hammer **Figure 7.25** Forces with the same impulse

Notice that (5) is a special case of property (2) that is obtained by taking $f(t) \equiv 1$. It is interesting to note that (1) and (5) actually imply the general property (2) (see Problem 33).

The Laplace transform of the Dirac delta function can be quickly derived from property (3). Since $\delta(t - a) = 0$ for $t \neq a$, then setting $f(t) = e^{-st}$ in (3), we find for $a \geq 0$

$$\int_0^\infty e^{-st}\delta(t - a)dt = \int_{-\infty}^\infty e^{-st}\delta(t - a)dt = e^{-as} \ .$$

Thus, for $a \geq 0$,

(6) $\mathcal{L}\{\delta(t - a)\}(s) = e^{-as} \ .$

An interesting connection exists between the unit step function and the Dirac delta function. Observe that as a consequence of equation (5) and the fact that $\delta(x - a)$ is zero for $x < a$ and for $x > a$, we have

(7) $\displaystyle\int_{-\infty}^t \delta(x - a)dx = \begin{cases} 0 \ , & t < a \ , \\ 1 \ , & t > a \end{cases}$

$$= u(t - a) \ .$$

If we formally differentiate both sides of (7) with respect to t (in the spirit of the fundamental theorem of calculus), we find

$$\delta(t - a) = u'(t - a) \ .$$

Thus it appears that the Dirac delta function is the derivative of the unit step function. That is, in fact, the case if we consider "differentiation" in a more general sense.[†]

The Dirac delta function is used in modeling mechanical vibration problems involving an impulse. For example, a mass-spring system at rest that is struck by a hammer exerting an impulse on the mass might be governed by the *symbolic* initial value problem

(8) $x'' + x = \delta(t) \ ; \qquad x(0) = 0 \ , \qquad x'(0) = 0 \ ,$

where $x(t)$ denotes the displacement from equilibrium at time t. We refer to this as a symbolic problem because while the left-hand side of equation (8) represents an ordinary function, the right-hand side does *not*. Consequently, it is not clear what we mean by a solution to problem (8). Because $\delta(t)$ is zero everywhere except at $t = 0$, one might be tempted to treat (8) as a homogeneous equation with zero initial conditions. But the solution to the latter is zero everywhere, which certainly does not describe the motion of the spring after the mass is struck by the hammer.

To define what is meant by a solution to (8), recall that $\delta(t)$ is depicted as the limit of forces $\mathcal{F}_n(t)$ having unit impulse and acting over shorter and shorter intervals. If we let $y_n(t)$ be the solution to the initial value problem

(9) $y_n'' + y_n = \mathcal{F}_n(t) \ ; \qquad y_n(0) = 0 \ , \qquad y_n'(0) = 0 \ ,$

where δ is replaced by \mathcal{F}_n, then we can think of the solution $x(t)$ to (8) as the limit (as $n \to \infty$) of the solutions $y_n(t)$.

[†]See *Distributions, Complex Variables, and Fourier Transforms,* by H. J. Bremermann (Addison-Wesley, Reading, Mass., 1965).

For example, let

$$\mathscr{F}_n(t) := n - nu(t - 1/n) = \begin{cases} n\,, & 0 < t < 1/n\,, \\ 0\,, & \text{otherwise.} \end{cases}$$

Taking the Laplace transform of equation (9), we find

$$(s^2 + 1)Y_n(s) = \frac{n}{s}\left(1 - e^{-s/n}\right)\,,$$

and so

$$Y_n(s) = \frac{n}{s(s^2 + 1)} - e^{-s/n}\frac{n}{s(s^2 + 1)}\,.$$

Now

$$\frac{n}{s(s^2 + 1)} = \frac{n}{s} - \frac{ns}{s^2 + 1} = \mathscr{L}\{n - n\cos t\}(s)\,.$$

Hence

(10) $y_n(t) = n - n\cos t - \left(n - n\cos(t - 1/n)\right)u(t - 1/n)$
$$= \begin{cases} n - n\cos t\,, & 0 < t < 1/n\,, \\ n\cos(t - 1/n) - n\cos t\,, & 1/n < t\,. \end{cases}$$

Fix $t > 0$. Then for n large enough, we have $1/n < t$. Thus

$$\lim_{n\to\infty} y_n(t) = \lim_{n\to\infty}\left[n\cos(t - 1/n) - n\cos t\right]$$
$$= -\lim_{n\to\infty}\frac{\cos(t - 1/n) - \cos t}{-1/n}$$
$$= -\lim_{h\to 0}\frac{\cos(t + h) - \cos t}{h}\,, \quad (\text{where } h = -1/n)\,,$$
$$= -\frac{d}{dt}(\cos t) = \sin t\,.$$

Also, for $t = 0$, $\lim_{n\to\infty}y_n(0) = 0 = \sin 0$. Therefore

$$\lim_{n\to\infty} y_n(t) = \sin t\,.$$

Hence the solution to the symbolic initial value problem (8) is $x(t) = \sin t$.

Fortunately, we do not have to go through the tedious process of solving for each y_n in order to find the solution x of the symbolic problem. It turns out that the Laplace transform method when applied directly to (8) yields the derived solution $x(t)$. Indeed, simply taking the Laplace transform of both sides of (8), we obtain from (6) (with $a = 0$)

$$(s^2 + 1)X(s) = 1\,,$$
$$X(s) = \frac{1}{s^2 + 1}\,,$$

which gives

$$x(t) = \mathcal{L}^{-1}\left\{\frac{1}{s^2 + 1}\right\}(t) = \sin t \ .$$

A peculiarity of using the Dirac delta function is that the solution $x(t) = \sin t$ of the symbolic initial value problem (8) does not satisfy both initial conditions, that is, $x'(0) = 1 \neq 0$. This reflects the fact that the impulse $\delta(t)$ is applied at $t = 0$. Thus the momentum x' (observe that in equation (8) the mass equals 1) jumps abruptly from $x'(0) = 0$ to $x'(0^+) = 1$.

In the next example, the Dirac delta function is used in modeling a mechanical vibration problem.

EXAMPLE 1 A mass attached to a spring is released from rest 1 m below the equilibrium position for the mass-spring system and begins to vibrate. After π seconds, the mass is struck by a hammer exerting an impulse on the mass. The system is governed by the symbolic initial value problem

$$(11) \qquad \frac{d^2x}{dt^2} + 9x = 3\delta(t - \pi) \ ; \qquad x(0) = 1 \ , \qquad \frac{dx}{dt}(0) = 0 \ ,$$

where $x(t)$ denotes the displacement from equilibrium at time t. Determine $x(t)$.

SOLUTION Let $X(s) = \mathcal{L}\{x\}(s)$. Since

$$\mathcal{L}\{x''\}(s) = s^2X(s) - s \quad \text{and} \quad \mathcal{L}\{\delta(t - \pi)\}(s) = e^{-\pi s} \ ,$$

taking the Laplace transform of both sides of (11) and solving for $X(s)$ yields

$$s^2X(s) - s + 9X(s) = 3e^{-\pi s}$$

$$X(s) = \frac{s}{s^2 + 9} + e^{-\pi s}\frac{3}{s^2 + 9}$$

$$= \mathcal{L}\{\cos 3t\}(s) + e^{-\pi s}\mathcal{L}\{\sin 3t\}(s) \ .$$

Using the translation property (cf. page 409) to determine the inverse Laplace transform of $X(s)$, we find

$$x(t) = \cos 3t + \left[\sin 3(t - \pi)\right]u(t - \pi)$$

$$= \begin{cases} \cos 3t \ , & t < \pi \ , \\ \cos 3t - \sin 3t \ , & \pi < t \end{cases}$$

$$= \begin{cases} \cos 3t \ , & t < \pi \ , \\ \sqrt{2}\cos\left(3t + \dfrac{\pi}{4}\right) \ , & \pi < t \ . \end{cases}$$

The graph of $x(t)$ is given in color in Figure 7.26 on page 434. For comparison, the dashed curve depicts the displacement of an undisturbed vibrating spring. Note that the impulse effectively adds 3 units to the momentum at time $t = \pi$. ∎ End

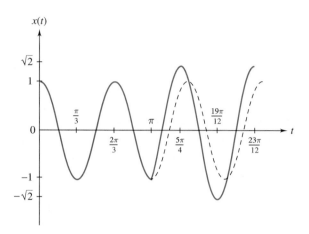

Figure 7.26 Displacement of a vibrating spring that is struck by a hammer at $t = \pi$

In Section 7.7, we defined the **impulse response function** for

(12) $ay'' + by' + cy = g(t)$

as the function $h(t) := \mathscr{L}^{-1}\{H\}(t)$, where $H(s)$ is the **transfer function.** Recall that $H(s)$ is the ratio

$$H(s) := \frac{Y(s)}{G(s)} \; ,$$

where $Y(s)$ is the Laplace transform of the solution to (12) with zero initial conditions and $G(s)$ is the Laplace transform of $g(t)$. It is important to note that $H(s)$, and hence $h(t)$, does not depend on the choice of the function $g(t)$ in (12) (see equation (15) in Section 7.7). However, it is useful to think of the impulse response function as the solution of the symbolic initial value problem

(13) $ay'' + by' + cy = \delta(t) \; ; \qquad y(0) = 0 \; , \qquad y'(0) = 0 \; .$

Indeed, with $g(t) = \delta(t)$, we have $G(s) = 1$, and hence $H(s) = Y(s)$. Consequently $h(t) = y(t)$. So we see that the function $h(t)$ is the response to the impulse $\delta(t)$ for a mechanical system governed by the symbolic initial value problem (13).

EXERCISES 7.8

In Problems 1 through 6, evaluate the given integral.

1. $\displaystyle\int_{-\infty}^{\infty} (t^2 - 1)\delta(t)\,dt$.

2. $\displaystyle\int_{-\infty}^{\infty} e^{3t}\delta(t)\,dt$.

3. $\displaystyle\int_{-\infty}^{\infty} (\sin 3t)\delta\left(t - \frac{\pi}{2}\right)dt$.

4. $\displaystyle\int_{-\infty}^{\infty} e^{-2t}\delta(t + 1)\,dt$.

5. $\int_0^\infty e^{-2t}\delta(t-1)\,dt$.

6. $\int_{-1}^1 (\cos 2t)\delta(t)\,dt$.

In Problems 7 through 12, determine the Laplace transform of the given generalized function.

7. $\delta(t-1) - \delta(t-3)$. **8.** $3\delta(t-1)$.

9. $t\delta(t-1)$. **10.** $t^3\delta(t-3)$.

11. $\delta(t-\pi)\sin t$. **12.** $e^t\delta(t-3)$.

In Problems 13 through 20, solve the given symbolic initial value problem.

13. $w'' + w = \delta(t-\pi)$;
 $w(0) = 0$, $w'(0) = 0$.

14. $y'' + 2y' + 2y = \delta(t-\pi)$;
 $y(0) = 1$, $y'(0) = 1$.

15. $y'' + 2y' - 3y = \delta(t-1) - \delta(t-2)$;
 $y(0) = 2$, $y'(0) = -2$.

16. $y'' - 2y' - 3y = 2\delta(t-1) - \delta(t-3)$;
 $y(0) = 2$, $y'(0) = 2$.

17. $y'' - y = 4\delta(t-2) + t^2$;
 $y(0) = 0$, $y'(0) = 2$.

18. $y'' - y' - 2y = 3\delta(t-1) + e^t$;
 $y(0) = 0$, $y'(0) = 3$.

19. $w'' + 6w' + 5w = e^t\delta(t-1)$;
 $w(0) = 0$, $w'(0) = 4$.

20. $y'' + 5y' + 6y = e^{-t}\delta(t-2)$;
 $y(0) = 2$, $y'(0) = -5$.

In Problems 21 through 24, solve the given symbolic initial value problem and sketch a graph of the solution.

21. $y'' + y = \delta(t-2\pi)$;
 $y(0) = 0$, $y'(0) = 1$.

22. $y'' + y = \delta(t-\pi/2)$;
 $y(0) = 0$, $y'(0) = 1$.

23. $y'' + y = -\delta(t-\pi) + \delta(t-2\pi)$;
 $y(0) = 0$, $y'(0) = 1$.

24. $y'' + y = \delta(t-\pi) - \delta(t-2\pi)$;
 $y(0) = 0$, $y'(0) = 1$.

In Problems 25 through 28, find the impulse response function $h(t)$ by using the fact that $h(t)$ is the solution to the symbolic initial value problem with $g(t) = \delta(t)$ and zero initial conditions.

25. $y'' + 4y' + 8y = g(t)$.

26. $y'' - 6y' + 13y = g(t)$.

27. $y'' - 2y' + 5y = g(t)$. **28.** $y'' - y = g(t)$.

29. A mass attached to a spring is released from rest 1 m below the equilibrium position for the mass-spring system and begins to vibrate. After $\pi/2$ sec, the mass is struck by a hammer exerting an impulse on the mass. The system is governed by the symbolic initial value problem

$$\frac{d^2x}{dt^2} + 9x = -3\delta\left(t - \frac{\pi}{2}\right) ;$$

$$x(0) = 1 ,$$

$$\frac{dx}{dt}(0) = 0 ,$$

where $x(t)$ denotes the displacement from equilibrium at time t. What happens to the mass after it is struck?

30. You have probably heard that soldiers are told not to march in cadence when crossing a bridge. By solving the symbolic initial value problem

$$y'' + y = \sum_{k=1}^\infty \delta(t - 2k\pi) ;$$

$$y(0) = 0 , y'(0) = 0 ,$$

explain why soldiers are so instructed. [*Hint:* See Section 4.12.]

31. A linear system is said to be **stable** if its impulse response function $h(t)$ remains bounded as $t \to +\infty$. If the linear system is governed by

$$ay'' + by' + cy = g(t) ,$$

where b and c are not both zero, show that the system is stable if and only if the real parts of the roots to

$$ar^2 + br + c = 0$$

are less than or equal to zero.

32. A linear system is said to be **asymptotically stable** if its impulse response function satisfies $h(t) \to 0$ as $t \to +\infty$. If the linear system is governed by

$$ay'' + by' + cy = g(t) ,$$

show that the system is asymptotically stable if and only if the real parts of the roots to

$$ar^2 + br + c = 0$$

are strictly less than zero.

33. The Dirac delta function may also be characterized by the properties

$$\delta(t) = \begin{cases} 0 , & t \neq 0 , \\ \text{"infinite,"} & t = 0 , \end{cases}$$

and

$$\int_{-\infty}^{\infty} \delta(t)\,dt = 1 .$$

Formally using the mean value theorem for definite integrals, verify that if $f(t)$ is continuous, then the above properties imply

$$\int_{-\infty}^{\infty} f(t)\delta(t)\,dt = f(0) .$$

34. Formally using integration by parts, show that

$$\int_{-\infty}^{\infty} f(t)\delta'(t)\,dt = -f'(0) .$$

Also show that, in general,

$$\int_{-\infty}^{\infty} f(t)\delta^{(n)}(t)\,dt = (-1)^n f^{(n)}(0) .$$

35. Figure 7.27 shows a beam of length 2λ that is imbedded in a support on the left side and free on the right. The vertical deflection of the beam a distance x from the support is denoted by $y(x)$. If the beam has a concentrated load L acting on it in the center of the beam, then the deflection must satisfy the symbolic boundary value problem

$$EIy^{(4)}(x) = L\delta(x - \lambda) ;$$
$$y(0) = y'(0) = y''(2\lambda) = y'''(2\lambda) = 0 ,$$

where E, the modulus of elasticity, and I, the moment of inertia, are constants. Find a formula for the displacement $y(x)$ in terms of the constants $\lambda, L, E,$ and I. [*Hint:* Let $y''(0) = A$ and $y'''(0) = B$. First solve the fourth order symbolic initial value problem and then use the conditions $y''(2\lambda) = y'''(2\lambda) = 0$ to determine A and B.]

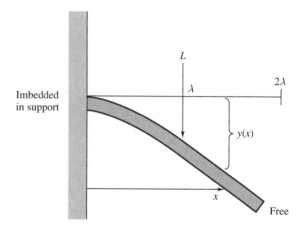

Figure 7.27 Beam imbedded in a support under a concentrated load at $x = \lambda$

7.9 SOLVING LINEAR SYSTEMS WITH LAPLACE TRANSFORMS

We can use the Laplace transform to reduce certain *systems* of linear differential equations with initial conditions to a system of linear algebraic equations, where again the unknowns are the transforms of the functions that make up the solution. Solving for these unknowns and taking their inverse Laplace transforms, we can then obtain the solution to the initial value problem for the system.

EXAMPLE 1 Solve the initial value problem

$$\textbf{(1)} \qquad \begin{aligned} x'(t) - 2y(t) &= 4t ; & x(0) &= 4 , \\ y'(t) + 2y(t) - 4x(t) &= -4t - 2 ; & y(0) &= -5 . \end{aligned}$$

SOLUTION Taking the Laplace transform of both sides of the differential equations gives

$$\mathscr{L}\{x'\}(s) - 2\mathscr{L}\{y\}(s) = \frac{4}{s^2} \, ,$$

(2)

$$\mathscr{L}\{y'\}(s) + 2\mathscr{L}\{y\}(s) - 4\mathscr{L}\{x\}(s) = -\frac{4}{s^2} - \frac{2}{s} \, .$$

Let $X(s) := \mathscr{L}\{x\}(s)$ and $Y(s) := \mathscr{L}\{y\}(s)$. Then, we have, by Theorem 4 on page 383,

$$\mathscr{L}\{x'\}(s) = sX(s) - x(0) = sX(s) - 4 \, ,$$
$$\mathscr{L}\{y'\}(s) = sY(s) - y(0) = sY(s) + 5 \, .$$

Substituting these expressions into system (2) and simplifying, we find

$$sX(s) - 2Y(s) = \frac{4s^2 + 4}{s^2} \, ,$$

(3)

$$-4X(s) + (s + 2)Y(s) = -\frac{5s^2 + 2s + 4}{s^2} \, .$$

To eliminate $Y(s)$ from the system, we multiply the first equation by $(s + 2)$ and the second by 2 and then add to obtain

$$\left[s(s + 2) - 8\right]X(s) = \frac{(s + 2)(4s^2 + 4)}{s^2} - \frac{10s^2 + 4s + 8}{s^2} \, .$$

This simplifies to

$$X(s) = \frac{4s - 2}{(s + 4)(s - 2)} \, .$$

To compute the inverse transform, we first write $X(s)$ in the partial fraction form

$$X(s) = \frac{3}{s + 4} + \frac{1}{s - 2} \, .$$

Hence, from the Laplace transform table on the inside back cover, we find that

(4) $$x(t) = 3e^{-4t} + e^{2t} \, .$$

To determine $y(t)$, we could solve system (3) for $Y(s)$ and then compute its inverse Laplace transform. However, it is easier just to solve the first equation in system (1) for $y(t)$ in terms of $x(t)$. Thus

$$y(t) = \frac{1}{2}x'(t) - 2t \, .$$

Substituting $x(t)$ from equation (4), we find that

(5) $$y(t) = -6e^{-4t} + e^{2t} - 2t \, .$$

The solution to the initial value problem (1) consists of the pair of functions $x(t)$, $y(t)$ given by equations (4) and (5). ∎ End

EXERCISES 7.9

In Problems 1 through 19, use the method of Laplace transforms to solve the given initial value problem. Here x', y', etc., denotes differentiation with respect to t; so does the symbol D.

1. $x' = 3x - 2y$; $x(0) = 1$,
 $y' = 3y - 2x$; $y(0) = 1$.

2. $x' = x - y$; $x(0) = -1$,
 $y' = 2x + 4y$; $y(0) = 0$.

3. $z' + w' = z - w$; $z(0) = 1$,
 $z' - w' = z - w$; $w(0) = 0$.

4. $x' - 3x + 2y = \sin t$; $x(0) = 0$,
 $4x - y' - y = \cos t$; $y(0) = 0$.

5. $x' = y + \sin t$; $x(0) = 2$,
 $y' = x + 2\cos t$; $y(0) = 0$.

6. $x' - x - y = 1$; $x(0) = 0$,
 $-x + y' - y = 0$; $y(0) = -5/2$.

7. $(D - 4)[x] + 6y = 9e^{-3t}$; $x(0) = -9$,
 $x - (D - 1)[y] = 5e^{-3t}$; $y(0) = 4$.

8. $D[x] + y = 0$; $x(0) = 7/4$,
 $4x + D[y] = 3$; $y(0) = 4$.

9. $x'' + 2y' = -x$; $x(0) = 2$, $x'(0) = -7$,
 $-3x'' + 2y'' = 3x - 4y$; $y(0) = 4$,
 $y'(0) = -9$.

10. $x'' + y = 1$; $x(0) = 1$, $x'(0) = 1$,
 $x + y'' = -1$; $y(0) = 1$, $y'(0) = -1$.

11. $x' + y = 1 - u(t - 2)$; $x(0) = 0$,
 $x + y' = 0$; $y(0) = 0$.

12. $x' + y = x$; $x(0) = 0$, $y(0) = 1$,
 $2x' + y'' = u(t - 3)$; $y'(0) = -1$.

13. $x' - y' = (\sin t)u(t - \pi)$; $x(0) = 1$,
 $x + y' = 0$; $y(0) = 1$.

14. $x'' = y + u(t - 1)$; $x(0) = 1$, $x'(0) = 0$,
 $y'' = x + 1 - u(t - 1)$; $y(0) = 0$,
 $y'(0) = 0$.

15. $x' - 2y = 2$; $x(1) = 1$,
 $x' + x - y' = t^2 + 2t - 1$; $y(1) = 0$.

16. $x' - 2x + y' = -(\cos t + 4\sin t)$; $x(\pi) = 0$,
 $2x + y' + y = \sin t + 3\cos t$; $y(\pi) = 3$.

17. $x' + x - y' = 2(t - 2)e^{t-2}$; $x(2) = 0$,
 $x'' - x' - 2y = -e^{t-2}$; $x'(2) = 1$,
 $y(2) = 1$.

18. $x' - 2y = 0$; $x(0) = 0$,
 $x' - z' = 0$; $y(0) = 0$,
 $x + y' - z = 3$; $z(0) = -2$.

19. $x' = 3x + y - 2z$; $x(0) = -6$,
 $y' = -x + 2y + z$; $y(0) = 2$,
 $z' = 4x + y - 3z$; $z(0) = -12$.

20. Use the method of Laplace transforms to solve
 $x'' + y' = 2$; $x(0) = 3$, $x'(0) = 0$,
 $4x + y' = 6$; $y(1) = 4$.
 [*Hint:* Let $y(0) = c$ and then solve for c.]

21. For the interconnected tanks problem of Section 5.1, page 261, suppose that the input to tank A is now controlled by a valve which for the first five minutes delivers 6 L/min of pure water, but thereafter delivers 6 L/min of brine at a concentration of 2 kg/L. Assuming that all other data remain the same (see Figure 5.1), determine the mass of salt in each tank for $t > 0$ if $x_0 = 0$ and $y_0 = 4$.

22. Recompute the coupled mass-spring oscillator motion in Problem 1, Exercises 5.4 (page 291), using Laplace transforms.

In Problems 23 and 24, find a system of differential equations and initial conditions for the currents in the networks given by the schematic diagrams; the initial currents are all assumed to be zero. Solve for the currents in each branch of the network (see Section 5.5 for a discussion of electrical networks).

23.

Figure 7.28 *RL* network for Problem 23

24.

Figure 7.29 *RL* network for Problem 24

CHAPTER SUMMARY

The use of the Laplace transform helps to simplify the process of solving initial value problems for certain differential and integral equations, especially when a forcing function with jump discontinuities is involved. The Laplace transform $\mathscr{L}\{f\}$ of a function $f(t)$ is defined by

$$\mathscr{L}\{f\}(s) := \int_0^\infty e^{-st} f(t)\, dt$$

for all values of s for which the improper integral exists. If $f(t)$ is piecewise continuous on $[0, \infty)$ and of exponential order α (that is, $|f(t)|$ grows no faster than a constant times $e^{\alpha t}$ as $t \to \infty$), then $\mathscr{L}\{f\}(s)$ exists for all $s > \alpha$.

The Laplace transform can be interpreted as an integral operator that maps a function $f(t)$ to a function $F(s)$. The transforms of commonly occurring functions appear in Table 7.1, page 380, and on the inside back cover of this book. The use of these tables is enhanced by several important properties of the operator \mathscr{L}.

Linearity: $\mathscr{L}\{af + bg\} = a\mathscr{L}\{f\} + b\mathscr{L}\{g\}$.

Translation in s: $\mathscr{L}\{e^{at}f(t)\}(s) = F(s - a)$, where $F = \mathscr{L}\{f\}$.

Translation in t: $\mathscr{L}\{g(t)u(t - a)\}(s) = e^{-as}\mathscr{L}\{g(t + a)\}(s)$, where $u(t - a)$ is the step function that equals 1 for $t > a$ and 0 for $t < a$. If $f(t)$ is continuous and $f(0) = 0$, then

$$\mathscr{L}^{-1}\{e^{-as}F(s)\}(t) = f(t - a)u(t - a) \ ,$$

where $f = \mathscr{L}^{-1}\{F\}$.

Convolution Property: $\mathscr{L}\{f * g\} = \mathscr{L}\{f\}\mathscr{L}\{g\}$, where $f * g$ denotes the convolution function

$$(f * g)(t) := \int_0^t f(t - v)g(v)\, dv \ .$$

One reason for the usefulness of the Laplace transform lies in the simple formula for the transform of the derivative f':

(1) $\mathscr{L}\{f'\}(s) = sF(s) - f(0) \ ,$ where $F = \mathscr{L}\{f\} \ .$

This formula shows that by using the Laplace transform, "differentiation with respect to t" can essentially be replaced by the simple operation of "multiplication by s." The extension of (1) to higher-order derivatives is

(2) $\mathcal{L}\{f^{(n)}\}(s) = s^n F(s) - s^{n-1}f(0) - s^{n-2}f'(0) - \cdots - f^{(n-1)}(0)$.

To solve an initial value problem of the form

(3) $ay'' + by' + cy = f(t)$; $y(0) = \alpha$, $y'(0) = \beta$

via the Laplace transform method, one takes the transform of both sides of the differential equation in (3). Using the linearity of \mathcal{L} and formula (2) leads to an equation involving the Laplace transform $Y(s)$ of the (unknown) solution $y(t)$. The next step is to solve this simpler equation for $Y(s)$. Finally, one computes the *inverse* Laplace transform of $Y(s)$ to obtain the desired solution. This last step of finding $\mathcal{L}^{-1}\{Y\}$ is often the most difficult. Sometimes it requires a partial fractions decomposition, a judicious use of the properties of the transform, an excursion through the Laplace transform tables, or the evaluation of a contour integral in the complex plane.

For the special problem in (3), where a, b, and c are constants, the differential equation is transformed to a simple *algebraic* equation for $Y(s)$. Another nice feature of this latter equation is that it incorporates the initial conditions. When the coefficients of the equation in (3) depend on t, the following formula may be helpful in taking the transform:

$$\mathcal{L}\{t^n f(t)\}(s) = (-1)^n \frac{d^n F}{ds^n}(s) , \quad \text{where} \quad F = \mathcal{L}\{f\} .$$

If the forcing function $f(t)$ in equation (3) has jump discontinuities, it is often convenient to write $f(t)$ in terms of unit step functions $u(t-a)$ before proceeding with the Laplace transform method. The transform of a periodic forcing function $f(t)$ with period T is given by

$$\mathcal{L}\{f\}(s) = \frac{\displaystyle\int_0^T e^{-st}f(t)\,dt}{1 - e^{-sT}} .$$

The Dirac delta function $\delta(t)$ is useful in modeling a system that is excited by a large force applied over a short time interval. It is not a function in the usual sense but can be roughly interpreted as the derivative of a unit step function. The transform of $\delta(t-a)$ is

$$\mathcal{L}\{\delta(t-a)\}(s) = e^{-as} , \quad a \geq 0 .$$

REVIEW PROBLEMS

In Problems 1 and 2, use the definition of the Laplace transform to determine $\mathcal{L}\{f\}$.

1. $f(t) = \begin{cases} 3 , & 0 \leq t \leq 2 , \\ 6 - t , & 2 < t . \end{cases}$

2. $f(t) = \begin{cases} e^{-t} , & 0 \leq t \leq 5 , \\ -1 , & 5 < t . \end{cases}$

In Problems 3 through 10, determine the Laplace transform of the given function.

3. $t^2 e^{-9t}$.

4. $e^{3t} \sin 4t$.

5. $e^{2t} - t^3 + t^2 - \sin 5t$.

6. $7e^{2t}\cos 3t - 2e^{7t}\sin 5t$.

7. $t \cos 6t$.

8. $(t + 3)^2 - (e^t + 3)^2$. **9.** $t^2 u(t - 4)$.

10. $f(t) = \cos t, -\pi/2 \le t \le \pi/2$, and $f(t)$ has period π.

In Problems 11 through 17, determine the inverse Laplace transform of the given function.

11. $\dfrac{7}{(s + 3)^3}$.

12. $\dfrac{2s - 1}{s^2 - 4s + 6}$.

13. $\dfrac{4s^2 + 13s + 19}{(s - 1)(s^2 + 4s + 13)}$.

14. $\dfrac{s^2 + 16s + 9}{(s + 1)(s + 3)(s - 2)}$.

15. $\dfrac{2s^2 + 3s - 1}{(s + 1)^2(s + 2)}$.

16. $\dfrac{1}{(s^2 + 9)^2}$.

17. $\dfrac{e^{-2s}(4s + 2)}{(s - 1)(s + 2)}$.

18. Find the Taylor series for $f(t) = e^{-t^2}$ about $t = 0$. Then, assuming the Laplace transform of $f(t)$ can be computed term by term, find an expansion for $\mathcal{L}\{f\}(s)$ in powers of $1/s$.

In Problems 19 through 24, solve the given initial value problem for $y(t)$ using the method of Laplace transforms.

19. $y'' - 7y' + 10y = 0$;
$y(0) = 0$, $y'(0) = -3$.

20. $y'' + 6y' + 9y = 0$;
$y(0) = -3$, $y'(0) = 10$.

21. $y'' + 2y' + 2y = t^2 + 4t$;
$y(0) = 0$, $y'(0) = -1$.

22. $y'' + 9y = 10e^{2t}$;
$y(0) = -1$, $y'(0) = 5$.

23. $y'' + 3y' + 4y = u(t - 1)$;
$y(0) = 0$, $y'(0) = 1$.

24. $y'' - 4y' + 4y = t^2 e^t$;
$y(0) = 0$, $y'(0) = 0$.

In Problems 25 and 26, find solutions to the given initial value problem.

25. $ty'' + 2(t - 1)y' - 2y = 0$;
$y(0) = 0$, $y'(0) = 0$.

26. $ty'' + 2(t - 1)y' + (t - 2)y = 0$;
$y(0) = 1$, $y'(0) = -1$.

In Problems 27 and 28, solve the given equation for $y(t)$.

27. $y(t) + \displaystyle\int_0^t (t - v)y(v)dv = e^{-3t}$.

28. $y'(t) - 2\displaystyle\int_0^t y(v)\sin(t - v)dv = 1$;

$y(0) = -1$.

29. A linear system is governed by
$y'' - 5y' + 6y = g(t)$.
Find the transfer function and the impulse response function.

30. Solve the symbolic initial value problem
$$y'' + 4y = \delta\left(t - \frac{\pi}{2}\right) ;$$
$$y(0) = 0 , y'(0) = 1 .$$

In Problems 31 and 32, use Laplace transforms to solve the given system.

31. $x' + y = 0$; $x(0) = 0$,
$x + y' = 1 - u(t - 2)$; $y(0) = 0$.

32. $x'' + 2y' = u(t - 3)$; $x(0) = 1$, $x'(0) = -1$,
$x + y' = y$; $y(0) = 0$.

TECHNICAL WRITING EXERCISES

1. Compare the use of Laplace transforms in solving linear differential equations with constant coefficients with the use of logarithms in solving algebraic equations of the form $x^r = a$.

2. Explain why the method of Laplace transforms works so well for linear differential equations with constant coefficients *and* integro-differential equations involving a convolution.

3. Discuss several examples of initial value problems in which the method of Laplace transforms cannot be applied.

4. A linear system is said to be **asymptotically stable** if its impulse response function $h(t) \to 0$ as $t \to +\infty$. Assume $H(s)$, the Laplace transform of $h(t)$, is a rational function in reduced form with the degree of its numerator less than the degree of its denominator. Explain in detail how the asymptotic stability of the linear system can be characterized in terms of the zeros of the denominator of $H(s)$. Give examples.

5. Compare and contrast the solution of initial value problems by Laplace transforms versus the methods of Chapter 4.

A. Duhamel's Formulas

For a linear system governed by the equation

(1) $ay'' + by' + cy = g(t)$,

where a, b, c are real constants, the function

(2) $H(s) := \dfrac{\mathscr{L}\{y\}(s)}{\mathscr{L}\{g\}(s)} = \dfrac{\mathscr{L}\{\text{output}\}}{\mathscr{L}\{\text{input}\}}$,

where all initial conditions are taken to be zero, is called the **transfer function** for the system. (As mentioned in Section 7.7, the transfer function $H(s)$ depends only on the constants a, b, c of the system; it is not affected by the choice of g.) If the output function $g(t)$ is the unit step function $u(t)$, then equation (2) yields

$$\mathscr{L}\{y\}(s) = \mathscr{L}\{u\}(s)H(s) = \frac{H(s)}{s} \; .$$

The solution (output function) in this special case is called the **indicial admittance** and is denoted by $A(t)$. Hence $\mathscr{L}\{A\}(s) = H(s)/s$.

It is possible to express the response $y(t)$ of the system to a general input function $g(t)$ in terms of $g(t)$ and the indicial admittance $A(t)$. To derive these relations, proceed as follows:

(a) Show that

(3) $\mathscr{L}\{y\}(s) = s\mathscr{L}\{A\}(s)\mathscr{L}\{g\}(s)$.

(b) Now apply the convolution theorem to (3) and show that

(4) $y(t) = \dfrac{d}{dt}\left[\displaystyle\int_0^t A(t - v)g(v)dv \right] = \dfrac{d}{dt}\left[\displaystyle\int_0^t A(v)g(t - v)dv \right]$.

(c) To perform the differentiation indicated in (4), one can use **Leibniz's rule:**

$$\frac{d}{dt}\left[\int_{a(t)}^{b(t)} f(v, t)dv \right] = \int_{a(t)}^{b(t)} \frac{\partial f}{\partial t}(v, t)dv + f\big(b(t), t\big)\frac{db}{dt}(t) - f\big(a(t), t\big)\frac{da}{dt}(t) \; ,$$

where f and $\partial f/\partial t$ are assumed continuous in v and t, and $a(t), b(t)$ are differentiable functions of t. Applying this rule to (4), derive the formulas

(5) $y(t) = \displaystyle\int_0^t A'(t - v)g(v)dv$,

(6) $y(t) = \displaystyle\int_0^t A(v)g'(t - v)dv + A(t)g(0)$.

[*Hint:* Recall that the initial conditions on A are $A(0) = A'(0) = 0$.]

443

(d) In equations (5) and (6), make the change of variables $w = t - v$ and show that

$$(7) \qquad y(t) = \int_0^t A'(w)g(t - w)dw \; ,$$

$$(8) \qquad y(t) = \int_0^t A(t - w)g'(w)dw + A(t)g(0) \; .$$

Equations (5)–(8) are referred to as **Duhamel's formulas,** in honor of the French mathematician J. M. C. Duhamel. These formulas are helpful in determining the response of the system to a general input $g(t)$, since the indicial admittance of the system can be determined experimentally by measuring the response of the system to a unit step function.

(e) The impulse response function $h(t)$ is defined as $h(t) := \mathcal{L}^{-1}\{H\}(t)$, where $H(s)$ is the transfer function. Show that $h(t) = A'(t)$, so that equations (5) and (7) can be written in the form

$$(9) \qquad y(t) = \int_0^t h(t - v)g(v)dv = \int_0^t h(v)g(t - v)dv \; .$$

We remark that the indicial admittance is the response of the system to a unit step function, and the impulse response function is the response to the unit impulse or delta function (see Section 7.8). But the delta function is the derivative (in a generalized sense) of the unit step function. Therefore the fact that $h(t) = A'(t)$ is not really surprising.

B. *Frequency Response Modeling*

Frequency response modeling of a linear system is based on the premise that the dynamics of a linear system can be recovered from a knowledge of how the system responds to sinusoidal inputs. (This will be made mathematically precise in Theorem 13.) In other words, to determine (or identify) a linear system, all one has to do is observe how the system reacts to sinusoidal inputs.

Let's assume that we have a linear system governed by

$$(10) \qquad y'' + py' + qy = g(t) \; ,$$

where p and q are real constants. The function $g(t)$ is called the **forcing function** or **input function.** When $g(t)$ is a sinusoid, the particular solution to (10) obtained by the method of undetermined coefficients is the **steady-state solution** or **output function** $y_{ss}(t)$ corresponding to $g(t)$. We can think of a linear system as a compartment or block into which goes an input function g and out of which comes the output function y_{ss} (see Figure 7.30). To **identify** a linear system means to determine the coefficients p and q in equation (10).

Figure 7.30 Block diagram depicting a linear system

It will be convenient for us to work with complex variables. A complex number z is usually expressed in the form $z = \alpha + i\beta$, with α, β real numbers and i denoting $\sqrt{-1}$. We can also express z in polar form, $z = re^{i\theta}$, where $r^2 = \alpha^2 + \beta^2$ and $\tan\theta = \beta/\alpha$. Here r (≥ 0) is called the **magnitude** and θ the **phase angle** of z.

The following theorem gives the relationship between the linear system and its response to sinusoidal inputs in terms of the **transfer function** $H(s)$ (see Project A, equation (2)).

STEADY-STATE SOLUTIONS TO SINUSOIDAL INPUTS

Theorem 13. Let $H(s)$ be the transfer function for equation (10). If $H(s)$ is finite at $s = i\omega$, with ω real, then the steady-state solution to (10) for $g(t) = e^{i\omega t}$ is

(11) $y_{ss}(t) = H(i\omega)e^{i\omega t} = H(i\omega)\{\cos\omega t + i\sin\omega t\}$.

(a) Prove Theorem 13. $\left[\textit{Hint: Guess } y_{ss}(t) = Ae^{i\omega t} \text{ and show that } A = H(i\omega).\right]$
(b) Use Theorem 13 to show that if $g(t) = \sin\omega t$, then the steady-state solution to (10) is
$M(\omega)\sin\left(\omega t + N(\omega)\right)$, where $H(i\omega) = M(\omega)e^{iN(\omega)}$ is the polar form for $H(i\omega)$.
(c) Solve for $M(\omega)$ and $N(\omega)$ in terms of p and q.
(d) Experimental results for modeling done by frequency response methods are usually presented in **frequency response**[†] or **Bode plots.** There are two types of Bode plots. The first is the log of the magnitude $M(\omega)$ of $H(i\omega)$ versus the angular frequency ω using a log scale for ω. The second is a plot of the phase angle or argument $N(\omega)$ of $H(i\omega)$ versus the angular frequency using a log scale for ω. The Bode plots for the transfer function $H(s) = (1 + 0.2s + s^2)^{-1}$ are given in Figure 7.31.

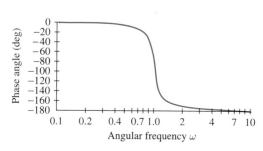

Figure 7.31 Bode plots for $H(i\omega) = \left[1 + 0.2(i\omega) + (i\omega)^2\right]^{-1}$

Sketch the Bode plots of the linear system governed by equation (10) with $p = 0.4$ and $q = 1.0$. Use $\omega = 0.3, 0.6, 0.9, 1.2,$ and 1.5 for the plot of $M(\omega)$ and $\omega = 0.5, 0.8, 1, 2,$ and 5 for the plot of $N(\omega)$.

(e) Assume we know that $q = 1$. When we input a sine wave with $\omega = 2$, the system settles

[†]Frequency response curves are also discussed in Section 4.12.

into a steady-state sinusoidal output with magnitude $M(2) = 0.333$. Find p and thus identify the linear system.

(f) Suppose a sine wave input with $\omega = 2$ produces a steady-state sinusoidal output with magnitude $M(2) = 0.5$ and that when $\omega = 4$, then $M(4) = 0.1$. Find p and q and thus identify the system.

In most applications, there are some inaccuracies in the measurement of the magnitudes and frequencies. To compensate for these errors, sinusoids with several different frequencies are used as input. A least squares approximation for p and q is then found. For a discussion of frequency response modeling as a mathematical modeling tool, see the chapter by W. F. Powers, "Modeling Linear Systems by Frequency Response Methods," in *Differential Equations Models*, by M. Braun, C. Coleman, and D. Drew (eds.) (Springer-Verlag, New York, 1983), Chapter 9. Additional examples may be found in *Schaum's Outline on Feedback and Control Systems*, by J. J. DiStefano, A. R. Stubberud, and I. J. Williams (McGraw-Hill, New York, 1990, 2nd edition), Chapter 15.

C. *Determining System Parameters*

In mechanical design, one sometimes must determine system parameters even though information on the system forces is incomplete. For example, the differential equation governing the motion $x(t)$ of an externally forced damped mass-spring oscillator was shown in Section 4.1 (page 156) to be

$$m\frac{d^2x}{dt^2} + b\frac{dx}{dt} + kx = f(t) \ ,$$

where $f(t)$ is the external force; the other parameters are defined in Section 4.1. Assume the system is underdamped $(b^2 < 4mk$; see Section 4.11) and starts from rest $\left(x(0) = 0, x'(0) = 0\right)$ and that the force is bounded: $|f(t)| \leq A$ for all t.

(a) Show that the transforms $X(s)$ and $F(s)$ of $x(t)$ and $f(t)$ are related by

$$X(s) = \frac{1}{ms^2 + bs + k}F(s) \ .$$

(b) Use the convolution theorem to derive the formula

$$x(t) = \frac{1}{\beta m}\int_0^t f(t - v)e^{-bv/2m}\sin \beta v \ dv \ ,$$

where (as in Section 4.11, equation (15), page 233)

$$\beta := \frac{1}{2m}\sqrt{4mk - b^2} \ .$$

(c) Show that the motion $x(t)$ is bounded under these circumstances by

$$|x(t)| \leq At/\beta m$$

and also by

$$|x(t)| \leq 2A/\beta b \ .$$

(d) Suppose the mass $m = 5$ kg, the spring constant $k = 3000$ N/m, and the force is bounded by $A = 10$ N. What range of values for the damping constant b will ensure a displacement of 1 cm (.01 m) or less?

Series Solutions of Differential Equations

Probably the best tool for numerically approximating a function $f(x)$ near a particular point x_0 is the *Taylor polynomial*. The formula for the Taylor polynomial of degree n centered at x_0, approximating a function $f(x)$ possessing n derivatives at x_0, is given by

$$(1) \qquad p_n(x) = f(x_0) + f'(x_0)(x - x_0) + \frac{f''(x_0)}{2!}(x - x_0)^2$$

$$+ \frac{f'''(x_0)}{3!}(x - x_0)^3 + \cdots + \frac{f^{(n)}(x_0)}{n!}(x - x_0)^n$$

$$= \sum_{j=0}^{n} \frac{f^{(j)}(x_0)}{j!}(x - x_0)^j \ .$$

This polynomial matches the value of f and the values of its derivatives, up to the order of the polynomial, at the point x_0:

$$p_n(x_0) = f(x_0) \ ,$$
$$p_n'(x_0) = f'(x_0) \ ,$$
$$p_n''(x_0) = f''(x_0) \ ,$$
$$\vdots$$
$$p_n^{(n)}(x_0) = f^{(n)}(x_0) \ .$$

For example, the first four Taylor polynomials for e^x, expanded around $x_0 = 0$, are

$$p_0(x) = 1 \ ,$$
$$p_1(x) = 1 + x \ ,$$

(2)
$$p_2(x) = 1 + x + \frac{x^2}{2} \ ,$$

$$p_3(x) = 1 + x + \frac{x^2}{2} + \frac{x^3}{6} \ .$$

Their efficacy in approximating the exponential function is demonstrated in Figure 8.1.

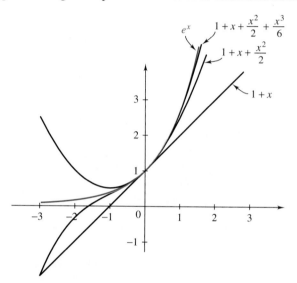

Figure 8.1 Graphs of Taylor polynomials for e^x

The Taylor polynomial of degree n differs from the polynomial of the next lower degree only in the addition of a single term:

$$p_n(x) = p_{n-1}(x) + \frac{f^{(n)}(x_0)}{n!}(x - x_0)^n \ ,$$

so a listing like (2) is clearly redundant—one can read off $p_0(x)$, $p_1(x)$, and $p_2(x)$ from the formula for $p_3(x)$. In fact, if f is infinitely differentiable, $p_n(x)$ is just the $(n + 1)$th partial sum of the **Taylor series**[†]

(3)
$$\sum_{j=0}^{\infty} \frac{f^{(j)}(x_0)}{j!}(x - x_0)^j \ .$$

EXAMPLE 1 Determine the fourth degree Taylor polynomials matching the functions e^x, $\cos x$, and $\sin x$ at $x_0 = 2$.

[†]Truncated Taylor series were introduced in Section 3.6 (page 135) as a tool for constructing recursive formulas for approximate solutions of differential equations.

SOLUTION For $f(x) = e^x$ we have $f^{(j)}(2) = e^2$ for each $j = 0, 1, \ldots$, so from (1) we obtain

$$e^x \approx e^2 + e^2(x - 2) + \frac{e^2}{2!}(x - 2)^2 + \frac{e^2}{3!}(x - 2)^3 + \frac{e^2}{4!}(x - 2)^4 \ .$$

For $f(x) = \cos x$, we have $f'(x) = -\sin x$, $f''(x) = -\cos x$, $f'''(x) = \sin x$, $f^{(4)}(x) = \cos x$, so that

$$\cos x \approx \cos 2 - (\sin 2)(x - 2) - \frac{\cos 2}{2!}(x - 2)^2 + \frac{\sin 2}{3!}(x - 2)^3 + \frac{\cos 2}{4!}(x - 2)^4 \ .$$

In a similar fashion we find

$$\sin x \approx \sin 2 + (\cos 2)(x - 2) - \frac{\sin 2}{2!}(x - 2)^2 - \frac{\cos 2}{3!}(x - 2)^3$$

$$+ \frac{\sin 2}{4!}(x - 2)^4 \ . \quad \blacksquare \text{End}$$

To relate this approximation scheme to our theme (the solution of differential equations), we alter our point of view; we regard a differential equation *not* as a "condition to be satisfied," but as a prescription for constructing the Taylor polynomials for its solutions. Besides providing a very general method for computing accurate approximate solutions to the equation near any particular "starting" point, this interpretation also provides insight into the role of the initial conditions. The following example illustrates the method.

EXAMPLE 2 Find the first few Taylor polynomials approximating the solution around $x_0 = 0$ of the initial value problem

$$y'' = 3y' + x^{7/3}y \ , \qquad y(0) = 10 \ , \qquad y'(0) = 5 \ .$$

SOLUTION To construct

$$p_n(x) = y(0) + y'(0)x + \frac{y''(0)}{2!}x^2 + \frac{y'''(0)}{3!}x^3 + \cdots + \frac{y^{(n)}(0)}{n!}x^n \ ,$$

we need the values of $y(0), y'(0), y''(0), y'''(0)$, etc. The first two are provided by the given initial conditions. The value of $y''(0)$ can be deduced from the differential equation itself and the values of the lower derivatives:

$$y''(0) = 3y'(0) + 0^{7/3}y(0) = 3 \cdot 5 + 0 \cdot 10 = 15 \ .$$

Now since $y'' = 3y' + x^{7/3}y$ holds for some interval around $x_0 = 0$, we can differentiate both sides to derive

$$y''' = 3y'' + \frac{7}{3}x^{4/3}y + x^{7/3}y' \ ,$$

$$y^{(4)} = 3y''' + \frac{28}{9}x^{1/3}y + \frac{14}{3}x^{4/3}y' + x^{7/3}y'' \ ,$$

$$y^{(5)} = 3y^{(4)} + \frac{28}{27}x^{-2/3}y + (\cdots) \ .$$

Thus on substituting $x = 0$ we deduce, in turn, that

$$y'''(0) = 3 \cdot 15 + \frac{7}{3} \cdot 0^{4/3} \cdot 10 + 0^{7/3} \cdot 5 = 45 \ ,$$

$$y^{(4)}(0) = 3 \cdot 45 + \frac{28}{9} \cdot 0^{1/3} \cdot 10 + \frac{14}{3} \cdot 0^{4/3} \cdot 5 + 0^{7/3} \cdot 15 = 135 \ ,$$

$$y^{(5)}(0) = 3 \cdot 135 + \frac{28}{27} \cdot 0^{-2/3} \cdot 10 + (\cdots) \text{ does not exist!}$$

Consequently, we can only construct the Taylor polynomials of degree 0 through 4 for the solution, and $p_4(x)$ is given by

$$p_4(x) = 10 + 5x + \frac{15}{2}x^2 + \frac{45}{6}x^3 + \frac{135}{24}x^4$$

$$= 10 + 5x + \frac{15}{2}x^2 + \frac{15}{2}x^3 + \frac{45}{8}x^4 \ . \ \blacksquare \text{End}$$

The next example demonstrates the application of the Taylor polynomial method to a *nonlinear* equation.

EXAMPLE 3 Determine the Taylor polynomial of degree 3 for the solution to the initial value problem

(4) $$y' = \frac{1}{x + y + 1} \ , \qquad y(0) = 0 \ .$$

SOLUTION Using $y(0) = 0$, we substitute $x = 0$ and $y = 0$ into equation (4) and find that $y'(0) = 1$. To determine $y''(0)$, we differentiate both sides of the equation in (4) with respect to x, thereby getting an expression for $y''(x)$ in terms of x, $y(x)$, and $y'(x)$. That is,

(5) $$y''(x) = (-1)[x + y(x) + 1]^{-2}[1 + y'(x)] \ .$$

Substituting $x = 0$, $y(0) = 0$, and $y'(0) = 1$ in (5), we obtain

$$y''(0) = (-1)(1)^{-2}(1 + 1) = -2 \ .$$

Similarly, differentiating (5) and substituting, we obtain

$$y^{(3)}(x) = 2[x + y(x) + 1]^{-3}[1 + y'(x)]^2 - [x + y(x) + 1]^{-2}y''(x) \ ,$$
$$y^{(3)}(0) = 2(1)^{-3}(1 + 1)^2 - (1)^{-2}(-2) = 10 \ .$$

Thus, the Taylor polynomial of degree 3 is

$$p_3(x) = 0 + x - x^2 + \frac{10}{6}x^3 = x - x^2 + \frac{5}{3}x^3 \ . \ \blacksquare \text{End}$$

In a theoretical sense we can estimate the accuracy to which a Taylor polynomial $p_n(x)$ approximates its target function $f(x)$ for x near x_0. Indeed, if we let $\varepsilon_n(x)$ measure the accuracy of the approximation,

$$\varepsilon_n(x) := f(x) - p_n(x) \ ,$$

then calculus provides us with several formulas for estimating ε_n. The most transparent is due to Lagrange: *if the $(n + 1)$st derivative of f exists and is continuous on an interval containing x_0 and x, then*

(6) $$\varepsilon_n(x) = \frac{f^{(n+1)}(\xi)}{(n + 1)!}(x - x_0)^{n+1} ,$$

where ξ, although unknown, is guaranteed to lie between x_0 and x.[†]

Figure 8.1 and equation (6) suggest that one might control the error in the Taylor polynomial approximation by increasing the degree n of the polynomial (i.e., taking more terms), thereby increasing the factor $(n + 1)!$ in the denominator. This possibility is limited, of course, by the number of times f can be differentiated. In Example 2, for instance, the solution did not have a fifth derivative at $x_0 = 0$ ($f^{(5)}(0)$ is "infinite"). Thus we could not construct $p_5(x)$, nor could we conclude anything about the accuracy of $p_4(x)$ from the Lagrange formula.

However, for Example 3 we could, in theory, compute *every* derivative of the solution $y(x)$ at $x_0 = 0$, and speculate on the *convergence* of the Taylor series

$$\sum_{j=0}^{\infty} \frac{y^{(j)}(x_0)}{j!}(x - x_0)^j = \lim_{n \to \infty} \sum_{j=0}^{n} \frac{y^{(j)}(x_0)}{j!}(x - x_0)^j$$

to the solution $y(x)$. Now for nonlinear equations such as (4), the factor $f^{(n+1)}(\xi)$ in the Lagrange error formula may grow too rapidly with n, and the convergence can be thwarted. But if the differential equation is linear and its coefficients and nonhomogeneous term enjoy a feature known as *analyticity,* our wish is granted; the error does indeed diminish to zero as the degree n goes to infinity, and the sequence of Taylor polynomials can be guaranteed to converge to the actual solution on a certain (known) interval. For instance, the exponential, sine, and cosine functions in Example 1 all satisfy linear differential equations with constant coefficients, and their Taylor series converge to the corresponding function values. (Indeed, *all* their derivatives are bounded on any interval of finite length, so the Lagrange formula for their approximation errors approaches zero as n increases, for each value of x.) This topic is the theme for the first sections of this chapter.

In Sections 8.5 and 8.6 we shall see that solutions to second order linear equations can exhibit very wild behavior near points x_0 where the coefficient of y'' is zero; so wild, in fact, that no Euler or Runge-Kutta algorithm could hope to keep up with them. But a clever modification of the Taylor polynomial method, due to Frobenius, provides very accurate approximations to the solutions in such regions. It is this latter feature, perhaps, that underscores the value of the Taylor methodology in the current practice of applied mathematics.

[†]Equation (6) is proved by invoking the mean value theorem; see, e.g., *Principles of Mathematical Analysis,* 3rd edition, by Walter Rudin (McGraw-Hill, New York, 1976).

EXERCISES 8.1

In Problems 1 through 8, determine the first three nonzero terms in the Taylor polynomial approximations for the given initial value problem.

1. $y' = x^2 + y^2$; $y(0) = 1$.

2. $y' = y^2$; $y(0) = 2$.

3. $y' = \sin y + e^x$; $y(0) = 0$.

4. $y' = \sin(x + y)$; $y(0) = 0$.

5. $x'' + tx = 0$; $x(0) = 1$, $x'(0) = 0$.

6. $y'' + y = 0$; $y(0) = 0$, $y'(0) = 1$.

7. $y''(\theta) + y(\theta)^3 = \sin\theta$;
$y(0) = 0$, $y'(0) = 0$.

8. $y'' + \sin y = 0$; $y(0) = 1$, $y'(0) = 0$.

9. (a) Construct the Taylor polynomial $p_3(x)$ of degree 3 for the function $f(x) = \ln x$ around $x = 1$.

(b) Using the error formula (6), show that
$$\left| \ln(1.5) - p_3(1.5) \right| \le \frac{(0.5)^4}{4} = 0.015625 .$$

(c) Compare the estimate in part (b) with the actual error by calculating $\left| \ln(1.5) - p_3(1.5) \right|$.

(d) Sketch the graphs of $\ln x$ and $p_3(x)$ (on the same axes) for $0 < x < 2$.

10. (a) Construct the Taylor polynomial $p_3(x)$ of degree 3 for the function $f(x) = 1/(2 - x)$ around $x = 0$.

(b) Using the error formula (6), show that
$$\left| f\!\left(\tfrac{1}{2}\right) - p_3\!\left(\tfrac{1}{2}\right) \right| = \left| \tfrac{2}{3} - p_3\!\left(\tfrac{1}{2}\right) \right| \le \frac{2}{3^5} .$$

(c) Compare the estimate in (b) with the actual error
$$\left| \tfrac{2}{3} - p_3\!\left(\tfrac{1}{2}\right) \right| .$$

(d) Sketch the graphs of $1/(2 - x)$ and $p_3(x)$ (on the same axes) for $-2 < x < 2$.

11. Argue that if $y = \phi(x)$ is a solution to the differential equation $y'' + p(x)y' + q(x)y = g(x)$ on the interval (a, b), where p, q, and g are each twice-differentiable, then the fourth derivative of $\phi(x)$ exists on (a, b).

12. Argue that if $y = \phi(x)$ is a solution to the differential equation $y'' + p(x)y' + q(x)y = g(x)$ on the interval (a, b), where p, q, and g possess derivatives of all

orders, then ϕ has derivatives of all orders on (a, b).

13. Duffing's Equation. In the study of a nonlinear spring with periodic forcing, the following equation arises:
$$y'' + ky + ry^3 = A \cos \omega t .$$
Let $k = r = A = 1$ and $\omega = 10$. Find the first three nonzero terms in the Taylor polynomial approximations to the solution with initial values $y(0) = 0$, $y'(0) = 1$.

14. Soft versus Hard Springs. For Duffing's equation given in Problem 13, the behavior of the solutions changes as r changes sign. When $r > 0$, the restoring force $ky + ry^3$ becomes stronger than for the linear spring $(r = 0)$. Such a spring is called **hard.** When $r < 0$, the restoring force becomes weaker than the linear spring and the spring is called **soft.** Pendulums act like soft springs.

(a) Redo Problem 13 with $r = -1$. Notice that for the initial conditions $y(0) = 0$, $y'(0) = 1$, the soft and hard springs appear to respond in the same way for t small.

(b) Keeping $k = A = 1$ and $\omega = 10$, change the initial conditions to $y(0) = 1$ and $y'(0) = 0$. Now redo Problem 13 with $r = \pm 1$.

(c) Based on the results of part (b), is there a difference between the behavior of soft and hard springs for t small? Describe.

15. The solution to the initial value problem
$$xy''(x) + 2y'(x) + xy(x) = 0 ,$$
$$y(0) = 1 ,\quad y'(0) = 0$$
has derivatives of all orders at $x = 0$ (although this is far from obvious). Use L'Hôpital's rule to compute the Taylor polynomial of degree 2 approximating this solution.

16. van der Pol Equation. In the study of the vacuum tube, the following equation is encountered:
$$y'' + (0.1)(y^2 - 1)y' + y = 0 .$$
Find the Taylor polynomial of degree 4 approximating the solution with the initial values $y(0) = 1$, $y'(0) = 0$.

8.2 POWER SERIES AND ANALYTIC FUNCTIONS

The differential equations studied in earlier sections often possessed solutions $y(x)$ that could be written in terms of elementary functions such as polynomials, exponentials, sines, and cosines. However, many important equations arise whose solutions cannot be so expressed. In the previous chapters, when we encountered such an equation we either settled for expressing the solution as an integral (see Exercises 2.2, Problem 27, page 51) or as a numerical approximation (Sections 3.5–3.7 and 5.6). However, the Taylor polynomial approximation scheme of the preceding section suggests another possibility. Suppose the differential equation (and initial conditions) permit the computation of *every* derivative $y^{(n)}$ at the expansion point x_0. Are there any conditions that would guarantee that the sequence of Taylor polynomials would *converge* to the solution $y(x)$ as the degree of the polynomials tends to infinity:

$$\lim_{n\to\infty}\sum_{j=0}^{n}\frac{y^{(j)}(x_0)}{j!}(x-x_0)^j = \sum_{j=0}^{\infty}\frac{y^{(j)}(x_0)}{j!}(x-x_0)^j = y(x)\ ?$$

In other words, when can we be sure that a solution to a differential equation is represented by its *Taylor series*? As we shall see, the answer is quite favorable, and it enables a powerful new technique for solving equations.

The most efficient way to begin an exploration of this issue is by investigating the algebraic and convergence properties of generic expressions that include Taylor series—"long polynomials" so to speak, or more conventionally, *power series*.

Power Series

A **power series** about the point x_0 is an expression of the form

(1) $\quad \sum_{n=0}^{\infty} a_n(x-x_0)^n = a_0 + a_1(x-x_0) + a_2(x-x_0)^2 + \cdots,$

where x is a variable and the a_n's are constants. We say that (1) **converges** at the point $x = c$ if the infinite series (of real numbers) $\sum_{n=0}^{\infty} a_n(c-x_0)^n$ converges; that is, the limit of the partial sums,

$$\lim_{N\to\infty}\sum_{n=0}^{N} a_n(c-x_0)^n,$$

exists (as a finite number). If this limit does not exist, the power series is said to **diverge** at $x = c$. Observe that (1) converges at $x = x_0$, since

$$\sum_{n=0}^{\infty} a_n(x_0-x_0)^n = a_0 + 0 + 0 + \cdots = a_0.$$

But what about convergence for other values of x? As stated in the following Theorem 1, a power series of the form (1) converges for all values of x in some "interval" centered at x_0 and diverges for x outside this interval. Moreover, at the interior points of this interval, the power series **converges absolutely** in the sense that $\sum_{n=0}^{\infty}|a_n(x-x_0)^n|$ converges. (Recall that absolute convergence of a series implies (ordinary) convergence of the series.)

RADIUS OF CONVERGENCE

Theorem 1. For each power series of the form (1), there is a number ρ $(0 \le \rho \le \infty)$, called the **radius of convergence** of the power series, such that (1) converges absolutely for $|x - x_0| < \rho$ and diverges for $|x - x_0| > \rho$. (See Figure 8.2.)

If the series (1) converges for all values of x, then $\rho = \infty$. When the series (1) converges only at x_0, then $\rho = 0$.

Figure 8.2 Interval of convergence

Notice that Theorem 1 settles the question of convergence except at the endpoints $x_0 \pm \rho$. Thus these two points require separate analysis. To determine the radius of convergence ρ, one method that is often easy to apply is the ratio test.

RATIO TEST

Theorem 2. If, for n large, the coefficients a_n are nonzero and satisfy

$$\lim_{n \to \infty} \left| \frac{a_{n+1}}{a_n} \right| = L \qquad (0 \le L \le \infty) \, ,$$

then the radius of convergence of the power series $\sum_{n=0}^{\infty} a_n(x - x_0)^n$ is $\rho = 1/L$, with $\rho = \infty$ if $L = 0$ and $\rho = 0$ if $L = \infty$.

Remark. We caution the reader that if the ratio $|a_{n+1}/a_n|$ does not have a limit, then methods other than the ratio test (e.g., root test) must be used to determine ρ. In particular, if infinitely many of the a_n's are zero, then the ratio test cannot be directly applied. (However, Problem 7 demonstrates how to apply the result for series containing only "even order" or "odd order" terms.)

EXAMPLE 1 Determine the convergence set of

(2) $$\sum_{n=0}^{\infty} \frac{(-2)^n}{n + 1}(x - 3)^n \, .$$

SOLUTION Since $a_n = (-2)^n/(n + 1)$, we have

$$\lim_{n \to \infty} \left| \frac{a_{n+1}}{a_n} \right| = \lim_{n \to \infty} \left| \frac{(-2)^{n+1}(n + 1)}{(-2)^n(n + 2)} \right|$$

$$= \lim_{n \to \infty} \frac{2(n + 1)}{(n + 2)} = 2 = L \, .$$

By the ratio test, the radius of convergence is $\rho = 1/2$. Hence the series (2) converges absolutely for $|x - 3| < 1/2$ and diverges when $|x - 3| > 1/2$. It remains only to determine what happens when $|x - 3| = 1/2$, that is, when $x = 5/2$ and $x = 7/2$.

Set $x = 5/2$, and the series (2) becomes the harmonic series $\Sigma_{n=0}^{\infty} (n + 1)^{-1}$, which is known to diverge. When $x = 7/2$, the series (2) becomes an **alternating** harmonic series, which is known to converge. Thus the power series converges for each x in the half-open interval $(5/2, 7/2]$; outside this interval it diverges. ∎ End

For each value of x for which the power series $\Sigma_{n=0}^{\infty} a_n (x - x_0)^n$ converges, we get a number that is the sum of the series. It is appropriate to denote this sum by $f(x)$, since its value depends on the choice of x. Thus we write

$$f(x) = \sum_{n=0}^{\infty} a_n(x - x_0)^n$$

for all numbers x in the convergence interval. For example, the **geometric series** $\Sigma_{n=0}^{\infty} x^n$ has the radius of convergence $\rho = 1$ and the sum function $f(x) = 1/(1 - x)$; that is,

$$(3) \qquad \frac{1}{1 - x} = 1 + x + x^2 + \cdots = \sum_{n=0}^{\infty} x^n \quad \text{for} \quad -1 < x < 1.$$

In this chapter we shall frequently appeal to the following basic property of power series.

POWER SERIES VANISHING ON AN INTERVAL

Theorem 3. If $\Sigma_{n=0}^{\infty} a_n(x - x_0)^n = 0$ for all x in some open interval, then each coefficient a_n equals zero.

Given two power series

$$(4) \qquad f(x) = \sum_{n=0}^{\infty} a_n(x - x_0)^n \, , \qquad g(x) = \sum_{n=0}^{\infty} b_n(x - x_0)^n \, ,$$

with nonzero radii of convergence, we want to find power series representations for the sum, product, and quotient of the functions $f(x)$ and $g(x)$. The sum is simply obtained by termwise addition:

$$f(x) + g(x) = \sum_{n=0}^{\infty} (a_n + b_n)(x - x_0)^n$$

for all x in the common interval of convergence of the power series in (4). The power series representation for the product $f(x)g(x)$ is a bit more complicated. To provide motivation for the formula, we treat the power series for $f(x)$ and $g(x)$ as "long polynomials," apply the distributive law, and group the terms in powers of $(x - x_0)$:

$$\left[a_0 + a_1(x - x_0) + a_2(x - x_0)^2 + \cdots \right] \cdot \left[b_0 + b_1(x - x_0) + b_2(x - x_0)^2 + \cdots \right]$$
$$= a_0 b_0 + (a_0 b_1 + a_1 b_0)(x - x_0) + (a_0 b_2 + a_1 b_1 + a_2 b_0)(x - x_0)^2 + \cdots .$$

The general formula for the product is

(5) $\qquad f(x)g(x) = \sum_{n=0}^{\infty} c_n(x - x_0)^n$,

where

(6) $\qquad c_n := \sum_{k=0}^{n} a_k b_{n-k}$.

The power series in (5) is called the **Cauchy product,** and it will converge for all x in the common *open* interval of convergence for the power series of f and g.[†]

The quotient $f(x)/g(x)$ will also have a power series expansion about x_0, provided $g(x_0) \neq 0$. However, the radius of convergence for this quotient series may be smaller than that for $f(x)$ or $g(x)$. Unfortunately, there is no nice formula for obtaining the coefficients in the power series for $f(x)/g(x)$. However, we can use the Cauchy product to divide power series indirectly (see Problem 15). The quotient series can also be obtained by formally carrying out polynomial long division (see Problem 16).

The next theorem explains, in part, why power series are so useful.

DIFFERENTIATION AND INTEGRATION OF POWER SERIES

Theorem 4. If the series $f(x) = \sum_{n=0}^{\infty} a_n(x - x_0)^n$ has a positive radius of convergence ρ, then f is differentiable in the interval $|x - x_0| < \rho$ and termwise differentiation gives the power series for the derivative:

$$f'(x) = \sum_{n=1}^{\infty} n a_n(x - x_0)^{n-1} \quad \text{for} \quad |x - x_0| < \rho .$$

Furthermore, termwise integration gives the power series for the integral of f:

$$\int f(x)\,dx = \sum_{n=0}^{\infty} \frac{a_n}{n+1}(x - x_0)^{n+1} + C \quad \text{for} \quad |x - x_0| < \rho .$$

EXAMPLE 2 Starting with the geometric series (3) for $1/(1 - x)$, find a power series for each of the following functions:

(a) $\dfrac{1}{1 + x^2}$. (b) $\dfrac{1}{(1 - x)^2}$. (c) $\arctan x$.

SOLUTION (a) Replacing x by $-x^2$ in (3) immediately gives

(7) $\qquad \dfrac{1}{1 + x^2} = 1 - x^2 + x^4 - x^6 + \cdots + (-1)^n x^{2n} + \cdots$.

(b) Notice that $1/(1 - x)^2$ is the derivative of the function $f(x) = 1/(1 - x)$. Hence on

[†]Actually, it may happen that the radius of convergence of the power series for $f(x)g(x)$ or $f(x) + g(x)$ is larger than that for the power series of f or g.

differentiating (3) term by term, we obtain

(8) $f'(x) = \dfrac{1}{(1-x)^2} = 1 + 2x + 3x^2 + 4x^3 + \cdots + nx^{n-1} + \cdots$.

(c) Since

$$\arctan x = \int_0^x \frac{1}{1+t^2}\,dt \ ,$$

we can integrate the series in (7) termwise to obtain the series for arctan x. Thus

$$\int_0^x \frac{1}{1+t^2}\,dt = \int_0^x \left\{ 1 - t^2 + t^4 - t^6 + \cdots + (-1)^n t^{2n} + \cdots \right\} dt$$

(9) $\arctan x = x - \dfrac{1}{3}x^3 + \dfrac{1}{5}x^5 - \dfrac{1}{7}x^7 + \cdots + \dfrac{(-1)^n x^{2n+1}}{2n+1} + \cdots$. ∎ End

It is important to keep in mind that since the geometric series (3) has the (open) interval of convergence $(-1, 1)$, the representations (7), (8), and (9) are at least valid in this interval. (Actually, the series (9) for arctan x converges for all $|x| \leq 1$.)

Shifting the Summation Index

The index of summation in a power series is a dummy index just like the variable of integration in a definite integral. Hence,

$$\sum_{n=0}^{\infty} a_n(x - x_0)^n = \sum_{k=0}^{\infty} a_k(x - x_0)^k = \sum_{i=0}^{\infty} a_i(x - x_0)^i \ .$$

Just as there are times when we want to change the variable of integration, there are situations (and we will encounter many in this chapter) when it is desirable to change or shift the index of summation. This is particularly important when one has to combine two different power series.

EXAMPLE 3 Express the series

$$\sum_{n=2}^{\infty} n(n-1)a_n x^{n-2}$$

as a series where the generic term is x^k instead of x^{n-2}.

SOLUTION Setting $k = n - 2$, we have $n = k + 2$ and $n - 1 = k + 1$. Note that when $n = 2$, then $k = 0$. Hence, substituting into the given series, we find

$$\sum_{n=2}^{\infty} n(n-1)a_n x^{n-2} = \sum_{k=0}^{\infty} (k+2)(k+1)a_{k+2} x^k \ .$$ ∎ End

EXAMPLE 4 Show that

$$x^3 \sum_{n=0}^{\infty} n^2(n-2)a_n x^n = \sum_{n=3}^{\infty} (n-3)^2(n-5)a_{n-3} x^n \ .$$

SOLUTION We start by taking the x^3 inside the summation on the left-hand side:

$$x^3 \sum_{n=0}^{\infty} n^2(n-2)a_n x^n = \sum_{n=0}^{\infty} n^2(n-2)a_n x^{n+3} .$$

To rewrite this with generic term x^k, we set $k = n + 3$. Thus $n = k - 3$, and $n = 0$ corresponds to $k = 3$. Straightforward substitution thus yields

$$\sum_{n=0}^{\infty} n^2(n-2)a_n x^{n+3} = \sum_{k=3}^{\infty} (k-3)^2(k-5)a_{k-3} x^k .$$

By replacing k by n, we obtain the desired form. ∎ End

EXAMPLE 5 Show that the identity

$$\sum_{n=1}^{\infty} n a_{n-1} x^{n-1} + \sum_{n=2}^{\infty} b_n x^{n+1} = 0$$

implies that $a_0 = a_1 = a_2 = 0$ and $a_n = -b_{n-1}/(n+1)$ for $n \geq 3$.

SOLUTION First we rewrite both series in terms of x^k. For the first series, we set $k = n - 1$, and hence $n = k + 1$, to write

$$\sum_{n=1}^{\infty} n a_{n-1} x^{n-1} = \sum_{k=0}^{\infty} (k+1)a_k x^k .$$

Then with $k = n + 1$, $n = k - 1$, the second series becomes

$$\sum_{n=2}^{\infty} b_n x^{n+1} = \sum_{k=3}^{\infty} b_{k-1} x^k .$$

The identity thus states

$$\sum_{k=0}^{\infty} (k+1)a_k x^k + \sum_{k=3}^{\infty} b_{k-1} x^k = 0 ,$$

and so we have a power series that sums to zero; consequently, by Theorem 3, each of its coefficients equals zero. For $k = 3, 4, \ldots$, both series contribute to the coefficient of x^k, and thus we confirm that

$$(k+1)a_k + b_{k-1} = 0$$

or $a_k = -b_{k-1}/(k+1)$ for $k \geq 3$. For $k = 0, 1,$ or 2, only the first series contributes, and we find, in turn,

$$(0+1)a_0 = 0 ,$$
$$(1+1)a_1 = 0 ,$$
$$(2+1)a_2 = 0 .$$

Hence $a_0 = a_1 = a_2 = 0$. ∎ End

Analytic Functions

Not all functions are expressible as power series. Those distinguished functions that can be so represented are called **analytic.**

ANALYTIC FUNCTION

Definition 1. A function f is said to be **analytic at x_0** if, in an open interval about x_0, this function is the sum of a power series $\sum_{n=0}^{\infty} a_n(x - x_0)^n$ that has a positive radius of convergence.

For example, a polynomial function $b_0 + b_1 x + \cdots + b_n x^n$ is analytic at every x_0, since we can always rewrite it in the form $a_0 + a_1(x - x_0) + \cdots + a_n(x - x_0)^n$. A rational function $P(x)/Q(x)$, where $P(x)$ and $Q(x)$ are polynomials without a common factor, is an analytic function except at those x_0 for which $Q(x_0) = 0$. As the reader may recall from calculus, the elementary functions e^x, $\sin x$, and $\cos x$ are analytic for all x, while $\ln x$ is analytic for $x > 0$. Some familiar representations are

$$(10) \qquad e^x = 1 + x + \frac{x^2}{2!} + \frac{x^3}{3!} + \cdots \qquad\qquad = \sum_{n=0}^{\infty} \frac{x^n}{n!} \; \left[= 1 + x + \frac{x^2}{2!} + \right]$$

$$(11) \qquad \sin x = x - \frac{x^3}{3!} + \frac{x^5}{5!} - \cdots \qquad\qquad = \sum_{n=0}^{\infty} \frac{(-1)^n}{(2n + 1)!} x^{2n+1} \; ,$$

$$(12) \qquad \cos x = 1 - \frac{x^2}{2!} + \frac{x^4}{4!} - \cdots \qquad\qquad = \sum_{n=0}^{\infty} \frac{(-1)^n}{(2n)!} x^{2n} \; ,$$

$$(13) \qquad \ln x = (x - 1) - \frac{1}{2}(x - 1)^2 + \frac{1}{3}(x - 1)^3 - \cdots = \sum_{n=1}^{\infty} \frac{(-1)^{n-1}}{n}(x - 1)^n \; ,$$

where (10), (11), and (12) are valid for all x, whereas (13) is valid for x in the half-open interval $(0, 2]$.

From Theorem 4 on the differentiation of power series, we see that a function f analytic at x_0 is differentiable in a neighborhood of x_0. Moreover, because f' has a power series representation in this neighborhood, it too is analytic at x_0. Repeating this argument, we see that f'', $f^{(3)}$, etc., exist and are analytic at x_0. Consequently, if a function does not have derivatives of all orders at x_0, then it cannot be analytic at x_0. The function $f(x) = |x - 1|$ is not analytic at $x_0 = 1$ because $f'(1)$ does not exist; and $f(x) = x^{7/3}$ is not analytic at $x_0 = 0$ because $f'''(0)$ does not exist.

Now we can deduce something very specific about the possible power series that can represent an analytic function. If $f(x)$ is analytic at x_0, then (by definition) it is the sum of *some* power series that converges in a neighborhood of x_0:

$$f(x) = \sum_{n=0}^{\infty} a_n(x - x_0)^n \; .$$

By the reasoning in the previous paragraph, the derivatives of f have convergent power series

representations

$$f'(x) = \sum_{n=0}^{\infty} na_n(x - x_0)^{n-1} = 0 + \sum_{n=1}^{\infty} na_n(x - x_0)^{n-1}$$

$$f''(x) = \sum_{n=0}^{\infty} n(n-1)a_n(x - x_0)^{n-2} = 0 + 0 + \sum_{n=2}^{\infty} n(n-1)a_n(x - x_0)^{n-2}$$

$$\vdots$$

$$f^{(j)}(x) = \sum_{n=0}^{\infty} n(n-1) \cdots \left(n - [j-1]\right) a_n(x - x_0)^{n-j}$$

$$= \sum_{n=j}^{\infty} n(n-1) \cdots \left(n - [j-1]\right) a_n(x - x_0)^{n-j}$$

$$\vdots$$

But if we evaluate these series at $x = x_0$, we learn that

$$f(x_0) = a_0 \ ,$$

$$f'(x_0) = 1 \cdot a_1 \ ,$$

$$f''(x_0) = 2 \cdot 1 \cdot a_2 \ ,$$

$$\vdots$$

$$f^{(j)}(x_0) = j! \cdot a_j \ ,$$

$$\vdots$$

that is, $a_j = f^{(j)}(x_0)/j!$ and the power series *must coincide* with the Taylor series[†]

$$\sum_{j=0}^{\infty} \frac{f^{(j)}(x_0)}{j!}(x - x_0)^j$$

about x_0. *Any power series—regardless of how it is derived—that converges in some neighborhood of x_0 to a function has to be the Taylor series of that function.* For example, the expansion for arctan x given in (9) of Example 2 must be its Taylor expansion.

With these facts in mind we are ready to turn to the study of the effectiveness of power series techniques for solving differential equations. In the next sections the reader will find it helpful to keep in mind that if f and g are analytic at x_0, then so are $f + g$, cf, fg, and f/g if $g(x_0) \neq 0$. These facts follow from the algebraic properties of power series discussed earlier.

EXERCISES 8.2

In Problems 1 through 6, determine the convergence set of the given power series.

1. $\displaystyle\sum_{n=0}^{\infty} \frac{2^{-n}}{n+1}(x-1)^n$.

2. $\displaystyle\sum_{n=0}^{\infty} \frac{3^n}{n!}x^n$.

3. $\displaystyle\sum_{n=0}^{\infty} \frac{n^2}{2^n}(x+2)^n$.

4. $\displaystyle\sum_{n=1}^{\infty} \frac{4}{n^2 + 2n}(x-3)^n$.

5. $\displaystyle\sum_{n=1}^{\infty} \frac{3}{n^3}(x-2)^n$.

6. $\displaystyle\sum_{n=0}^{\infty} \frac{(n+2)!}{n!}(x+2)^n$.

7. Sometimes the ratio test (Theorem 2) can be applied to a power series containing an infinite number of zero coefficients, provided the zero pattern is regular. Use Theorem 2 to show, for

[†]When the expansion point x_0 is zero, the Taylor series is also known as the **Maclaurin series**.

example, that the series

$$a_0 + a_2x^2 + a_4x^4 + a_6x^6 + \cdots = \sum_{k=0}^{\infty} a_{2k}x^{2k}$$

has a radius of convergence $\rho = 1/\sqrt{L}$, if

$$\lim_{k\to\infty} \left| \frac{a_{2k+2}}{a_{2k}} \right| = L \ ,$$

and that

$$a_1x + a_3x^3 + a_5x^5 + a_7x^7 + \cdots$$

$$= \sum_{k=0}^{\infty} a_{2k+1}x^{2k+1}$$

has a radius of convergence $\rho = 1/\sqrt{M}$, if

$$\lim_{k\to\infty} \left| \frac{a_{2k+3}}{a_{2k+1}} \right| = M \ .$$

[*Hint: Let $z = x^2$.*]

8. Determine the convergence set of the given power series.

(a) $\displaystyle\sum_{k=0}^{\infty} 2^{2k}x^{2k}$. **(b)** $\displaystyle\sum_{k=0}^{\infty} 2^{2k+1}x^{2k+1}$.

(c) $\sin x$ (equation (11)). **(d)** $\cos x$ (equation (12)).

(e) $(\sin x)/x = \displaystyle\sum_{n=0}^{\infty} (-1)^n x^{2n}/[(2n + 1)!]$.

(f) $\displaystyle\sum_{k=0}^{\infty} 2^{2k}x^{4k}$.

In Problems 9 and 10, find the power series expansion $\sum_{n=0}^{\infty} a_n x^n$ for $f(x) + g(x)$, given the expansions for $f(x)$ and $g(x)$.

9. $f(x) = \displaystyle\sum_{n=0}^{\infty} \frac{1}{n+1}x^n$, $g(x) = \displaystyle\sum_{n=1}^{\infty} 2^{-n}x^{n-1}$.

10. $f(x) = \displaystyle\sum_{n=3}^{\infty} \frac{2^n}{n!}(x-1)^{n-2}$,

$g(x) = \displaystyle\sum_{n=1}^{\infty} \frac{n^2}{2^n}(x-1)^{n-1}$.

In Problems 11 through 14, find the first three nonzero terms in the power series expansion for the product $f(x)g(x)$.

11. $f(x) = e^x = \displaystyle\sum_{n=0}^{\infty} \frac{1}{n!}x^n$,

$g(x) = \sin x = \displaystyle\sum_{k=0}^{\infty} \frac{(-1)^k}{(2k+1)!}x^{2k+1}$.

12. $f(x) = \sin x = \displaystyle\sum_{k=0}^{\infty} \frac{(-1)^k}{(2k+1)!}x^{2k+1}$,

$g(x) = \cos x = \displaystyle\sum_{k=0}^{\infty} \frac{(-1)^k}{(2k)!}x^{2k}$.

13. $f(x) = e^{-x} = \displaystyle\sum_{n=0}^{\infty} \frac{(-1)^n}{n!}x^n$,

$g(x) = (1 + x)^{-1} = \displaystyle\sum_{n=0}^{\infty} (-1)^n x^n$.

14. $f(x) = e^x = \displaystyle\sum_{n=0}^{\infty} \frac{1}{n!}x^n$,

$g(x) = e^{-x} = \displaystyle\sum_{n=0}^{\infty} \frac{(-1)^n}{n!}x^n$.

15. Find the first few terms of the power series for the quotient

$$q(x) = \left(\sum_{n=0}^{\infty} \frac{1}{2^n}x^n \right) \Big/ \left(\sum_{n=0}^{\infty} \frac{1}{n!}x^n \right)$$

by completing the following:

(a) Let $q(x) = \sum_{n=0}^{\infty} a_n x^n$, where the coefficients a_n are to be determined. Argue that $\sum_{n=0}^{\infty} x^n/2^n$ is the Cauchy product of $q(x)$ and $\sum_{n=0}^{\infty} x^n/n!$.

(b) Use formula (6) of the Cauchy product to deduce the equations

$$\frac{1}{2^0} = a_0 \ , \quad \frac{1}{2} = a_0 + a_1 \ ,$$

$$\frac{1}{2^2} = \frac{a_0}{2} + a_1 + a_2 \ ,$$

$$\frac{1}{2^3} = \frac{a_0}{6} + \frac{a_1}{2} + a_2 + a_3 , \dots \ .$$

(c) Solve the equations in part (b) to determine the constants a_0, a_1, a_2, a_3.

16. To find the first few terms in the power series for the quotient $q(x)$ in Problem 15, treat the power series in the numerator and denominator as "long polynomials" and carry out long division. That is, perform

$$1 + x + \frac{1}{2}x^2 + \cdots \quad \Big| \; 1 + \frac{1}{2}x + \frac{1}{4}x^2 + \cdots \ .$$

In Problems 17 through 20, find a power series expansion for $f'(x)$, given the expansion for $f(x)$.

17. $f(x) = (1 + x)^{-1} = \displaystyle\sum_{n=0}^{\infty} (-1)^n x^n$.

18. $f(x) = \sin x = \displaystyle\sum_{k=0}^{\infty} \frac{(-1)^k}{(2k+1)!}x^{2k+1}$.

19. $f(x) = \displaystyle\sum_{n=0}^{\infty} a_n x^n$.

20. $f(x) = \sum_{n=1}^{\infty} n a_n x^{n-1}$.

In Problems 21 and 22, find a power series expansion for $g(x) := \int_0^x f(t)\,dt$, given the expansion for $f(x)$.

21. $f(x) = (1 + x)^{-1} = \sum_{n=0}^{\infty} (-1)^n x^n$.

22. $f(x) = \dfrac{\sin x}{x} = \sum_{k=0}^{\infty} \dfrac{(-1)^k}{(2k+1)!} x^{2k}$.

In Problems 23 through 26, express the given power series as a series with generic term x^k.

23. $\sum_{n=1}^{\infty} n a_n x^{n-1}$.

24. $\sum_{n=2}^{\infty} n(n-1) a_n x^{n+2}$.

25. $\sum_{n=0}^{\infty} a_n x^{n+1}$.

26. $\sum_{n=1}^{\infty} \dfrac{a_n}{n+3} x^{n+3}$.

27. Show that

$$x^2 \sum_{n=0}^{\infty} n(n+1) a_n x^n$$

$$= \sum_{n=2}^{\infty} (n-2)(n-1) a_{n-2} x^n .$$

28. Show that

$$2 \sum_{n=0}^{\infty} a_n x^{n+1} + \sum_{n=1}^{\infty} n b_n x^{n-1}$$

$$= b_1 + \sum_{n=1}^{\infty} \left[2a_{n-1} + (n+1) b_{n+1} \right] x^n .$$

In Problems 29 through 34, determine the Taylor series about the point x_0 for the given functions and values of x_0.

29. $f(x) = \cos x$, $x_0 = \pi$.

30. $f(x) = x^{-1}$, $x_0 = 1$.

31. $f(x) = \dfrac{1+x}{1-x}$, $x_0 = 0$.

32. $f(x) = \ln(1 + x)$, $x_0 = 0$.

33. $f(x) = x^3 + 3x - 4$, $x_0 = 1$.

34. $f(x) = \sqrt{x}$, $x_0 = 1$.

35. The Taylor series for $f(x) = \ln x$ about $x_0 = 1$ given in equation (13) can also be obtained as follows:

(a) Starting with the expansion $1/(1 - s) = \sum_{n=0}^{\infty} s^n$ and observing that

$$\frac{1}{x} = \frac{1}{1 + (x - 1)} ,$$

obtain the Taylor series for $1/x$ about $x_0 = 1$.

(b) Since $\ln x = \int_1^x 1/t\,dt$, use the result of part (a) and termwise integration to obtain the Taylor series for $f(x) = \ln x$ about $x_0 = 1$.

36. Let $f(x)$ and $g(x)$ be analytic at x_0. Determine whether the following statements are always true or sometimes false:

(a) $3f(x) + g(x)$ is analytic at x_0 .
(b) $f(x)/g(x)$ is analytic at x_0 .
(c) $f'(x)$ is analytic at x_0 .
(d) $\left[f(x) \right]^3 - \int_{x_0}^x g(t)\,dt$ is analytic at x_0 .

37. Let

$$f(x) = \begin{cases} e^{-1/x^2} , & x \neq 0 , \\ 0 , & x = 0 . \end{cases}$$

Show that $f^{(n)}(0) = 0$ for $n = 0, 1, 2, \ldots$ and hence that the Maclaurin series for $f(x)$ is $0 + 0 + 0 + \cdots$, which converges for all x but is equal to $f(x)$ only when $x = 0$. This is an example of a function possessing derivatives of all orders (at $x_0 = 0$), whose Taylor series converges, but the Taylor series (about $x_0 = 0$) does not converge to the original function! Consequently, this function is not analytic at $x = 0$.

38. Compute the Taylor series for $f(x) = \ln(1 + x^2)$ about $x_0 = 0$. [*Hint:* Multiply the series for $(1 + x^2)^{-1}$ by $2x$ and integrate.]

8.3 POWER SERIES SOLUTIONS TO LINEAR DIFFERENTIAL EQUATIONS

In this section we demonstrate a method for obtaining a power series solution to a linear differential equation with polynomial coefficients. This method is easier to use than the Taylor series

method discussed in Section 8.1 and sometimes gives a nice expression for the general term in the power series expansion. Knowing the form of the general term also allows us to test for the radius of convergence of the power series.

We begin by writing the linear differential equation

(1) $a_2(x)y'' + a_1(x)y' + a_0(x)y = 0$

in the standard form

(2) $y'' + p(x)y' + q(x)y = 0$,

where $p(x) := a_1(x)/a_2(x)$ and $q(x) := a_0(x)/a_2(x)$.

ORDINARY AND SINGULAR POINTS

Definition 2. A point x_0 is called an **ordinary point** of equation (1) if both $p = a_1/a_2$ and $q = a_0/a_2$ are analytic at x_0. If x_0 is not an ordinary point, it is called a **singular point** of the equation.

EXAMPLE 1 Determine all the singular points of

$$xy'' + x(1 - x)^{-1}y' + (\sin x)y = 0 .$$

SOLUTION Dividing the equation by x, we find that

$$p(x) = \frac{x}{x(1 - x)} , \qquad q(x) = \frac{\sin x}{x} .$$

The singular points are those points where $p(x)$ *or* $q(x)$ fails to be analytic. Observe that $p(x)$ and $q(x)$ are the ratios of functions that are everywhere analytic. Hence $p(x)$ and $q(x)$ are analytic except, *perhaps,* when their denominators are zero. For $p(x)$ this occurs at $x = 0$ and $x = 1$. But since we can cancel an x in the numerator and denominator of $p(x)$, that is,

$$p(x) = \frac{x}{x(1 - x)} = \frac{1}{1 - x} ,$$

we see that $p(x)$ is actually analytic at $x = 0.^\dagger$ Therefore $p(x)$ is analytic except at $x = 1$. For $q(x)$, the denominator is zero at $x = 0$. Just as with $p(x)$, this zero is removable since $q(x)$ has the power series expansion

$$q(x) = \frac{\sin x}{x} = \frac{x - \dfrac{x^3}{3!} + \dfrac{x^5}{5!} - \cdots}{x} = 1 - \frac{x^2}{3!} + \frac{x^4}{5!} - \cdots .$$

Thus $q(x)$ is everywhere analytic. Consequently, the only singular point of the given equation is $x = 1$. ∎End

At an ordinary point x_0 of equation (1) (or (2)), the coefficient functions $p(x)$ and $q(x)$ are

†Such points are called **removable singularities.** In this chapter we assume in such cases that the function has been defined (or redefined) so that it is analytic at the point.

analytic. Hence we might expect that the solutions to these equations inherit this property. From the discussion in Section 4.3 on linear second order equations, the continuity of p and q in a neighborhood of x_0 is sufficient to imply that equation (2) has two linearly independent solutions defined in that neighborhood. But analytic functions are not merely continuous—they possess derivatives of all orders in a neighborhood of x_0. Thus we can differentiate equation (2) to show that $y^{(3)}$ exists and, by a "bootstrap" argument, prove that solutions to (2) must likewise possess derivatives of all orders. Although we cannot conclude by this reasoning that the solutions enjoy the stronger property of analyticity, this is nonetheless the case (see Theorem 5 in Section 8.4). Hence, in a neighborhood of an ordinary point x_0, the solutions to (1) (or (2)) can be expressed as a power series about x_0.

To illustrate the power series method about an ordinary point, let's look at a simple *first order* linear differential equation.

EXAMPLE 2 Find a power series solution about $x = 0$ to

$$(3) \qquad y' + 2xy = 0 \ .$$

SOLUTION The coefficient of y is the polynomial $2x$, which is analytic everywhere, so $x = 0$ is an ordinary point[†] of equation (3). Thus we expect to find a power series solution of the form

$$(4) \qquad y(x) = a_0 + a_1 x + a_2 x^2 + \cdots = \sum_{n=0}^{\infty} a_n x^n \ .$$

Our task is to determine the coefficients a_n.

For this purpose we need the expansion for $y'(x)$ that is given by termwise differentiation of (4):

$$y'(x) = 0 + a_1 + 2a_2 x + 3a_3 x^2 + \cdots = \sum_{n=1}^{\infty} n a_n x^{n-1} \ .$$

We now substitute the series expansions for y and y' into (3) and obtain

$$\sum_{n=1}^{\infty} n a_n x^{n-1} + 2x \sum_{n=0}^{\infty} a_n x^n = 0 \ ,$$

which simplifies to

$$(5) \qquad \sum_{n=1}^{\infty} n a_n x^{n-1} + \sum_{n=0}^{\infty} 2 a_n x^{n+1} = 0 \ .$$

To add the two power series in (5), we add the coefficients of like powers of x. If we write out the first few terms of these summations and add, we get

$$(a_1 + 2a_2 x + 3a_3 x^2 + 4a_4 x^3 + \cdots) + (2a_0 x + 2a_1 x^2 + 2a_2 x^3 + \cdots) = 0 \ ,$$

$$(6) \qquad a_1 + (2a_2 + 2a_0)x + (3a_3 + 2a_1)x^2 + (4a_4 + 2a_2)x^3 + \cdots = 0 \ .$$

For the power series on the left-hand side of equation (6) to be identically zero, we must have

[†]By an ordinary point of a first order equation $y' + q(x)y = 0$, we mean a point where $q(x)$ is analytic.

all the coefficients equal to zero. Thus

$$a_1 = 0 \, , \qquad \left(2a_2 + 2a_0\right) = 0 \, ,$$
$$\left(3a_3 + 2a_1\right) = 0 \, , \qquad \left(4a_4 + 2a_2\right) = 0 \, , \qquad \text{etc.}$$

Solving the preceding system, we find

$$a_1 = 0 \, , \qquad a_2 = -a_0 \, , \qquad a_3 = -\frac{2}{3}a_1 = 0 \, ,$$

$$a_4 = -\frac{1}{2}\,a_2 = -\frac{1}{2}(-a_0) = \frac{1}{2}a_0 \, .$$

Hence the power series for the solution takes the form

$$(7) \qquad y(x) = a_0 - a_0 x^2 + \frac{1}{2}\,a_0 x^4 + \cdots \, .$$

While the first few terms displayed in (7) are useful, it is sometimes advantageous to have a formula for the *general term* in the power series expansion for the solution. To achieve this goal, let's return to equation (5). This time, instead of just writing out a few terms, let's shift the indices in the two power series so that they sum over the same powers of x, say x^k. To do this, we shift the index in the first summation in (5) by letting $k = n - 1$. Then $n = k + 1$ and $k = 0$ when $n = 1$. Hence the first summation in (5) becomes

$$(8) \qquad \sum_{n=1}^{\infty} na_n x^{n-1} = \sum_{k=0}^{\infty} (k + 1)a_{k+1}x^k \, .$$

In the second summation of (5), we put $k = n + 1$ so that $n = k - 1$ and $k = 1$ when $n = 0$. This gives

$$(9) \qquad \sum_{n=0}^{\infty} 2a_n x^{n+1} = \sum_{k=1}^{\infty} 2a_{k-1}x^k \, .$$

Substituting (8) and (9) into (5) yields

$$(10) \qquad \sum_{k=0}^{\infty} (k + 1)a_{k+1}x^k + \sum_{k=1}^{\infty} 2a_{k-1}x^k = 0 \, .$$

Since the first summation in (10) begins at $k = 0$ and the second at $k = 1$, we break up the first into

$$\sum_{k=0}^{\infty} (k + 1)a_{k+1}x^k = a_1 + \sum_{k=1}^{\infty} (k + 1)a_{k+1}x^k \, .$$

Then (10) becomes

$$(11) \qquad a_1 + \sum_{k=1}^{\infty} \left[(k + 1)a_{k+1} + 2a_{k-1}\right]x^k = 0 \, .$$

When we set all the coefficients in (11) equal to zero, we find

$$a_1 = 0 \, ,$$

and, for all $k \geq 1$,

(12) $(k + 1)a_{k+1} + 2a_{k-1} = 0$.

Equation (12) is a **recurrence relation** that we can use to determine the coefficient a_{k+1} in terms of a_{k-1}, that is,

$$a_{k+1} = -\frac{2}{k+1}a_{k-1} \ .$$

Setting $k = 1, 2, \ldots, 8$ and using the fact that $a_1 = 0$, we find

$$a_2 = -\frac{2}{2}a_0 = -a_0 \qquad (k = 1) , \qquad a_3 = -\frac{2}{3}a_1 = 0 \qquad (k = 2) ,$$

$$a_4 = -\frac{2}{4}a_2 = \frac{1}{2}a_0 \qquad (k = 3) , \qquad a_5 = -\frac{2}{5}a_3 = 0 \qquad (k = 4) ,$$

$$a_6 = -\frac{2}{6}a_4 = -\frac{1}{3!}a_0 \qquad (k = 5) , \qquad a_7 = -\frac{2}{7}a_5 = 0 \qquad (k = 6) ,$$

$$a_8 = -\frac{2}{8}a_6 = \frac{1}{4!}a_0 \qquad (k = 7) , \qquad a_9 = -\frac{2}{9}a_7 = 0 \qquad (k = 8) .$$

After a moment's reflection, we realize that

$$a_{2n} = \frac{(-1)^n}{n!}a_0 , \qquad n = 1, 2, \ldots ,$$

$$a_{2n+1} = 0 , \qquad n = 0, 1, 2, \ldots .$$

Substituting back into the expression (4), we obtain the power series solution

(13) $y(x) = a_0 - a_0 x^2 + \frac{1}{2!}a_0 x^4 + \cdots = a_0 \sum_{n=0}^{\infty} \frac{(-1)^n}{n!}x^{2n}$.

Since a_0 is left undetermined, it serves as an arbitrary constant, and hence (13) gives a general solution to equation (3). ∎ End

Applying the ratio test as described in Problem 7, Exercises 8.2 (page 460), we can verify that the power series in (13) has radius of convergence $\rho = \infty$. Moreover, (13) is reminiscent of the expansion for the exponential function; the reader should check that it converges to

$$y(x) = a_0 e^{-x^2} \ .$$

This general solution to the simple equation (3) can also be obtained by the method of separation of variables.

The convergence of the partial sums of (13), with $a_0 = 1$, to the actual solution e^{-x^2} is depicted in Figure 8.3. Notice that taking more terms results in better approximations around $x = 0$. Note also, however, that every partial sum is a *polynomial,* and hence must diverge at $x = \pm\infty$, while e^{-x^2} converges to zero, of course. Thus each partial sum approximation eventually deteriorates for large $|x|$. This is a typical feature of power series approximations.

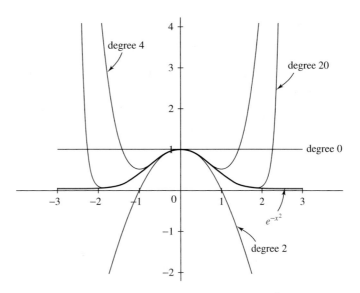

Figure 8.3 Partial sum approximations to e^{-x^2}

In the next example we use the power series method to obtain a general solution to a **linear second order differential equation.**

EXAMPLE 3 Find a general solution to

(14) $2y'' + xy' + y = 0$

in the form of a power series about the ordinary point $x = 0$.

SOLUTION Writing

(15) $y(x) = a_0 + a_1 x + a_2 x^2 + \cdots = \displaystyle\sum_{n=0}^{\infty} a_n x^n$,

we differentiate termwise to obtain

$$y'(x) = a_1 + 2a_2 x + 3a_3 x^2 + \cdots = \sum_{n=1}^{\infty} n a_n x^{n-1} ,$$

$$y''(x) = 2a_2 + 6a_3 x + 12a_4 x^2 + \cdots = \sum_{n=2}^{\infty} n(n-1) a_n x^{n-2} .$$

Substituting these power series into equation (14), we find

(16) $\displaystyle\sum_{n=2}^{\infty} 2n(n-1)a_n x^{n-2} + \sum_{n=1}^{\infty} n a_n x^n + \sum_{n=0}^{\infty} a_n x^n = 0$.

To simplify the addition of the three summations in (16), let's shift the indices so that the general term in each is a constant times x^k. For the first summation, we substitute $k = n - 2$ and get

$$\sum_{n=2}^{\infty} 2n(n-1)a_n x^{n-2} = \sum_{k=0}^{\infty} 2(k+2)(k+1)a_{k+2} x^k .$$

In the second and third summations, we simply substitute k for n. With these changes of indices, equation (16) becomes

$$\sum_{k=0}^{\infty} 2(k+2)(k+1)a_{k+2}x^k + \sum_{k=1}^{\infty} ka_k x^k + \sum_{k=0}^{\infty} a_k x^k = 0 \ .$$

Next, we separate the x^0 terms from the others and then combine the like powers of x in the three summations to get

$$4a_2 + a_0 + \sum_{k=1}^{\infty} \left[2(k+2)(k+1)a_{k+2} + ka_k + a_k \right] x^k = 0 \ .$$

Setting the coefficients of this power series equal to zero yields

(17) $4a_2 + a_0 = 0$

and the recurrence relation

(18) $2(k+2)(k+1)a_{k+2} + (k+1)a_k = 0 \ , \qquad k \geq 1 \ .$

We can now use (17) and (18) to determine all the coefficients a_k of the solution in terms of a_0 and a_1. Solving (18) for a_{k+2} gives

(19) $a_{k+2} = \dfrac{-1}{2(k+2)} a_k \ , \qquad k \geq 1 \ .$

Thus

$$a_2 = \frac{-1}{2^2} a_0 \ ,$$

$$a_3 = \frac{-1}{2 \cdot 3} a_1 \qquad\qquad (k = 1) \ ,$$

$$a_4 = \frac{-1}{2 \cdot 4} a_2 = \frac{1}{2^2 \cdot 2 \cdot 4} a_0 \qquad (k = 2) \ ,$$

$$a_5 = \frac{-1}{2 \cdot 5} a_3 = \frac{1}{2^2 \cdot 3 \cdot 5} a_1 \qquad (k = 3) \ ,$$

$$a_6 = \frac{-1}{2 \cdot 6} a_4 = \frac{-1}{2^3 \cdot 2 \cdot 4 \cdot 6} a_0 = \frac{-1}{2^6 \cdot 3!} a_0 \qquad (k = 4) \ ,$$

$$a_7 = \frac{-1}{2 \cdot 7} a_5 = \frac{-1}{2^3 \cdot 3 \cdot 5 \cdot 7} a_1 \qquad (k = 5) \ ,$$

$$a_8 = \frac{-1}{2 \cdot 8} a_6 = \frac{1}{2^4 \cdot 2 \cdot 4 \cdot 6 \cdot 8} a_0 = \frac{1}{2^8 \cdot 4!} a_0 \qquad (k = 6) \ .$$

The pattern for the coefficients is now apparent. Since a_0 and a_1 are not restricted, we find

$$a_{2n} = \frac{(-1)^n}{2^{2n} n!} a_0 \ , \qquad n \geq 1 \ ,$$

and

$$a_{2n+1} = \frac{(-1)^n}{2^n \left[1 \cdot 3 \cdot 5 \cdots (2n + 1) \right]} a_1 , \qquad n \geq 1.$$

From this, two linearly independent solutions emerge; namely,

(20) $y_1(x) = \displaystyle\sum_{n=0}^{\infty} \frac{(-1)^n}{2^{2n}n!} x^{2n}$ $\left(\text{take } a_0 = 1, a_1 = 0 \right)$,

(21) $y_2(x) = \displaystyle\sum_{n=0}^{\infty} \frac{(-1)^n}{2^n \left[1 \cdot 3 \cdot 5 \cdots (2n + 1) \right]} x^{2n+1}$ $\left(\text{take } a_0 = 0, a_1 = 1 \right)$.

Hence a general solution to (14) is $a_0 y_1(x) + a_1 y_2(x)$. Approximations to the solutions $y_1(x), y_2(x)$ are depicted in Figure 8.4. ∎ End

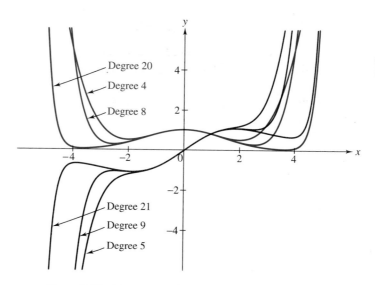

Figure 8.4 Partial sum approximations to solutions for Example 3

The method illustrated in Example 3 can also be used to solve initial value problems. Suppose we are given the values of $y(0)$ and $y'(0)$; then, from equation (15), we see that $a_0 = y(0)$ and $a_1 = y'(0)$. Knowing these two coefficients leads to a unique power series solution for the initial value problem.

The recurrence relation (18) in Example 3 involved just two of the coefficients, a_{k+2} and a_k, and we were fortunate in being able to deduce from this relation the general form for the coefficient a_n. However, many cases arise that lead to more complicated two-term or even to many-term recurrence relations. When this occurs, it may be impossible to determine the general form for the coefficients a_n. In the next example we consider an equation that gives rise to a three-term recurrence relation.

EXAMPLE 4 Find the first few terms in a power series expansion about $x = 0$ for a general solution to

(22) $(1 + x^2)y'' - y' + y = 0$.

SOLUTION Since $p(x) = -(1 + x^2)^{-1}$ and $q(x) = (1 + x^2)^{-1}$ are analytic at $x = 0$, then $x = 0$ is an ordinary point for equation (22). Hence we can express its general solution in the form

$$y(x) = \sum_{n=0}^{\infty} a_n x^n \ .$$

Substituting this expansion into (22) yields

$$(1 + x^2) \sum_{n=2}^{\infty} n(n-1)a_n x^{n-2} - \sum_{n=1}^{\infty} n a_n x^{n-1} + \sum_{n=0}^{\infty} a_n x^n = 0 \ ,$$

(23) $\sum_{n=2}^{\infty} n(n-1)a_n x^{n-2} + \sum_{n=2}^{\infty} n(n-1)a_n x^n - \sum_{n=1}^{\infty} n a_n x^{n-1} + \sum_{n=0}^{\infty} a_n x^n = 0$.

To sum over like powers x^k, we put $k = n - 2$ in the first summation of (23), $k = n - 1$ in the third, and $k = n$ in the second and fourth. This gives

$$\sum_{k=0}^{\infty} (k+2)(k+1)a_{k+2} x^k + \sum_{k=2}^{\infty} k(k-1)a_k x^k - \sum_{k=0}^{\infty} (k+1)a_{k+1} x^k + \sum_{k=0}^{\infty} a_k x^k = 0 \ .$$

Separating the terms corresponding to $k = 0$ and $k = 1$ and combining the rest under one summation, we have

$$(2a_2 - a_1 + a_0) + (6a_3 - 2a_2 + a_1)x$$
$$+ \sum_{k=2}^{\infty} \left[(k+2)(k+1)a_{k+2} - (k+1)a_{k+1} + \left(k(k-1) + 1 \right)a_k \right] x^k = 0 \ .$$

Setting the coefficients equal to zero gives

(24) $2a_2 - a_1 + a_0 = 0$,

(25) $6a_3 - 2a_2 + a_1 = 0$

and the recurrence relation

(26) $(k+2)(k+1)a_{k+2} - (k+1)a_{k+1} + (k^2 - k + 1)a_k = 0$, $k \geq 2$.

We can solve (24) for a_2 in terms of a_0 and a_1:

$$a_2 = \frac{a_1 - a_0}{2} \ .$$

Now that we have a_2, we can use (25) to express a_3 in terms of a_0 and a_1:

$$a_3 = \frac{2a_2 - a_1}{6} = \frac{(a_1 - a_0) - a_1}{6} = \frac{-a_0}{6} \ .$$

Solving the recurrence relation (26) for a_{k+2}, we obtain

(27) $a_{k+2} = \dfrac{(k+1)a_{k+1} - (k^2 - k + 1)a_k}{(k+2)(k+1)}$, $k \geq 2$.

For $k = 2$, 3, and 4, this gives

$$a_4 = \frac{3a_3 - 3a_2}{4 \cdot 3} = \frac{a_3 - a_2}{4}$$

$$= \frac{\dfrac{-a_0}{6} - \left(\dfrac{a_1 - a_0}{2}\right)}{4} = \frac{2a_0 - 3a_1}{24} \qquad (k = 2) \ ,$$

$$a_5 = \frac{4a_4 - 7a_3}{5 \cdot 4} = \frac{3a_0 - a_1}{40} \qquad (k = 3) \ ,$$

$$a_6 = \frac{5a_5 - 13a_4}{6 \cdot 5} = \frac{36a_1 - 17a_0}{720} \qquad (k = 4) \ .$$

We can now express a general solution in terms up to order 6, using a_0 and a_1 as the arbitrary constants. Thus

$$(28) \qquad y(x) = a_0 + a_1 x + \left(\frac{a_1 - a_0}{2}\right)x^2 - \frac{a_0}{6}x^3$$

$$+ \left(\frac{2a_0 - 3a_1}{24}\right)x^4 + \left(\frac{3a_0 - a_1}{40}\right)x^5 + \left(\frac{36a_1 - 17a_0}{720}\right)x^6 + \cdots$$

$$= a_0\left(1 - \frac{1}{2}x^2 - \frac{1}{6}x^3 + \frac{1}{12}x^4 + \frac{3}{40}x^5 - \frac{17}{720}x^6 + \cdots\right)$$

$$+ a_1\left(x + \frac{1}{2}x^2 - \frac{1}{8}x^4 - \frac{1}{40}x^5 + \frac{1}{20}x^6 + \cdots\right) . \quad \blacksquare \text{ End}$$

Given specific values for a_0 and a_1, will the partial sums of the power series representation (28) yield useful approximations to the solution when $x = 0.5$? What about when $x = 2.3$ or $x = 7.8$? The answers to these questions certainly depend on the radius of convergence of the power series in (28). But since we were not able to determine a general form for the coefficients a_n in this example, we cannot use the ratio test (or other methods such as the root test, integral test, or comparison test) to compute the radius ρ. In the next section we remedy this situation by giving a simple procedure that determines a lower bound for the radius of convergence of power series solutions.

EXERCISES 8.3

In Problems 1 through 10, determine all the singular points of the given differential equation.

1. $(x + 1)y'' - x^2y' + 3y = 0$.

2. $x^2y'' + 3y' - xy = 0$.

3. $(\theta^2 - 2)y'' + 2y' + (\sin \theta)y = 0$.

4. $(x^2 + x)y'' + 3y' - 6xy = 0$.

5. $(t^2 - t - 2)x'' + (t + 1)x' - (t - 2)x = 0$.

6. $(x^2 - 1)y'' + (1 - x)y' + (x^2 - 2x + 1)y = 0$.

7. $(\sin x)y'' + (\cos x)y = 0$.

8. $e^x y'' - (x^2 - 1)y' + 2xy = 0$.

9. $(\sin \theta)y'' - (\ln \theta)y = 0$.

10. $[\ln(x - 1)]y'' + (\sin 2x)y' - e^x y = 0$.

In Problems 11 through 18, find at least the first four nonzero terms in a power series expansion about $x = 0$ for a general solution to the given differential equation.

11. $y' + (x + 2)y = 0$. **12.** $y' - y = 0$.

13. $z'' - x^2 z = 0$. **14.** $(x^2 + 1)y'' + y = 0$.

15. $y'' + (x - 1)y' + y = 0$.

16. $y'' - 2y' + y = 0$.

17. $w'' - x^2 w' + w = 0$.

18. $(2x - 3)y'' - xy' + y = 0$.

In Problems 19 through 24, find a power series expansion about $x = 0$ for a general solution to the given differential equation. Your answer should include a general formula for the coefficients.

19. $y' - 2xy = 0$. **20.** $y'' + y = 0$.

21. $y'' - xy' + 4y = 0$. **22.** $y'' - xy = 0$ (FE # 10

23. $z'' - x^2 z' - xz = 0$.

24. $(x^2 + 1)y'' - xy' + y = 0$.

In Problems 25 through 28, find at least the first four nonzero terms in a power series expansion about $x = 0$ for the solution to the given initial value problem.

25. $w'' + 3xw' - w = 0$;
$\quad w(0) = 2$, $w'(0) = 0$.

26. $(x^2 - x + 1)y'' - y' - y = 0$;
$\quad y(0) = 0$, $y'(0) = 1$.

27. $(x + 1)y'' - y = 0$;
$\quad y(0) = 0$, $y'(0) = 1$.

28. $y'' + (x - 2)y' - y = 0$;
$\quad y(0) = -1$, $y'(0) = 0$.

In Problems 29 through 31, use the first few terms of the power series expansion to find a cubic polynomial approximation for the solution to the given initial value problem. Graph the linear, quadratic, and cubic polynomial approximations for $-5 \leq x \leq 5$.

29. $y'' + y' - xy = 0$;
$\quad y(0) = 1$, $y'(0) = -2$.

30. $y'' - 4xy' + 5y = 0$;
$\quad y(0) = -1$, $y'(0) = 1$.

31. $(x^2 + 2)y'' + 2xy' + 3y = 0$;
$\quad y(0) = 1$, $y'(0) = 2$.

32. Consider the initial value problem
$$y'' - 2xy' - 2y = 0 ;$$
$$y(0) = a_0 , \qquad y'(0) = a_1 ,$$
where a_0 and a_1 are constants.
 (a) Show that if $a_0 = 0$, then the solution will be an odd function (that is, $y(-x) = -y(x)$ for all x). What happens when $a_1 = 0$?
 (b) Show that if a_0 and a_1 are positive, then the solution is increasing on $(0, \infty)$.
 (c) Show that if a_0 is negative and a_1 is positive, then the solution is increasing on $(-\infty, 0)$.
 (d) What conditions on a_0 and a_1 would guarantee that the solution is increasing on $(-\infty, \infty)$?

33. Use the ratio test to show that the radius of convergence of the series in equation (13) is infinite. [*Hint:* See Problem 7, Exercises 8.2, page 460.]

34. **Emden's Equation.** A classical nonlinear equation that occurs in the study of the thermal behavior of a spherical cloud is **Emden's equation**
$$y'' + \frac{2}{x}y' + y^n = 0 ,$$
with initial conditions $y(0) = 1, y'(0) = 0$. Even though $x = 0$ is *not* an ordinary point for this equation (which is nonlinear for $n \neq 1$), it turns out that there does exist a solution analytic at $x = 0$. Assuming n is a positive integer, show that the first few terms in a power series solution are
$$y = 1 - \frac{x^2}{3!} + n\frac{x^4}{5!} + \cdots .$$
[*Hint:* Substitute $y = 1 + c_2 x^2 + c_3 x^3 + c_4 x^4 + c_5 x^5 + \cdots$ into the equation and carefully compute the first few terms in the expansion for y^n.]

35. **Variable Resistor.** In Section 5.5, we showed that the charge q on the capacitor in a simple *RLC* circuit is governed by the equation
$$Lq''(t) + Rq'(t) + \frac{1}{C}q(t) = E(t) ,$$
where L is the inductance, R the resistance, C the capacitance, and E the voltage source. Since the resistance of a resistor increases with temperature, let's assume the resistor is heated so that the resistance at time t is $R(t) = 1 + t/10$ ohms (see Figure 8.5). If $L = 0.1$ henrys, $C = 2$ farads, $E(t) \equiv 0, q(0) = 10$

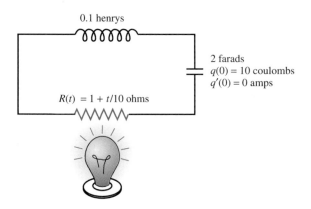

Figure 8.5 An *RLC* circuit whose resistor is being heated

Figure 8.6 A mass-spring system whose spring is being heated

coulombs, and $q'(0) = 0$ amps, find at least the first four nonzero terms in a power series expansion about $t = 0$ for the charge on the capacitor.

36. Variable Spring Constant. As a spring is heated, its spring "constant" decreases. Suppose the spring is heated so that the spring "constant" at time t is $k(t) = 6 - t$ N/m (see Figure 8.6). If the unforced

mass-spring system has mass $m = 2$ kg and a damping constant $b = 1$ N-sec/m with initial conditions $x(0) = 3$ m and $x'(0) = 0$ m/sec, then the displacement $x(t)$ is governed by the initial value problem

$$2x''(t) + x'(t) + (6 - t)x(t) = 0 \; ;$$
$$x(0) = 3 \; , \qquad x'(0) = 0 \; .$$

Find at least the first four nonzero terms in a power series expansion about $t = 0$ for the displacement.

8.4 EQUATIONS WITH ANALYTIC COEFFICIENTS

In Section 8.3 we introduced a method for obtaining a power series solution about an ordinary point. In this section we continue the discussion of this procedure. We begin by stating a basic existence theorem for the equation

(1) $y''(x) + p(x)y'(x) + q(x)y(x) = 0 \; ,$

which justifies the power series method.

EXISTENCE OF ANALYTIC SOLUTIONS

Theorem 5. Suppose x_0 is an ordinary point for equation (1). Then (1) has two linearly independent analytic solutions of the form

(2) $y(x) = \displaystyle\sum_{n=0}^{\infty} a_n(x - x_0)^n \; .$

Moreover, the radius of convergence of any power series solution of the form given by (2) is at least as large as the distance from x_0 to the nearest singular point (real or complex-valued) of equation (1).

The key element in the proof of Theorem 5 is the construction of a convergent geometric

series that dominates the series expansion (2) of a solution to equation (1). The convergence of the series in (2) then follows by the comparison test. The details of the proof can be found in more advanced books on differential equations.[†]

As we saw in Section 8.3, the power series method gives us a general solution in the same form as (2), with a_0 and a_1 as arbitrary constants. The two linearly independent solutions referred to in Theorem 5 can be obtained by taking $a_0 = 1$, $a_1 = 0$ for the first and $a_0 = 0$, $a_1 = 1$ for the second. Thus we can extend Theorem 5 by saying that *equation (1) has a general solution of the form* (2) *with a_0 and a_1 as the arbitrary constants.*

The second part of Theorem 5 gives a simple way of determining a minimum value for the radius of convergence of the power series. We need only find the singular points of equation (1) and then determine the distance between the ordinary point x_0 and the nearest singular point.

EXAMPLE 1 Find a minimum value for the radius of convergence of a power series solution about $x = 0$ to

$$(3) \qquad 2y'' + xy' + y = 0 \ .$$

SOLUTION For this equation, $p(x) = x/2$ and $q(x) = 1/2$. Both of these functions are analytic for all real or complex values of x. Since equation (3) has no singular points, the distance between the ordinary point $x = 0$ and the nearest singular point is infinite. Hence the radius of convergence is infinite. ■ End

The next example helps to answer the questions posed at the end of the last section.

EXAMPLE 2 Find a minimum value for the radius of convergence of a power series solution about $x = 0$ to

$$(4) \qquad (1 + x^2)y'' - y' + y = 0 \ .$$

SOLUTION Here $p(x) = -1/(1 + x^2)$, $q(x) = 1/(1 + x^2)$, and so the singular points of equation (4) occur when $1 + x^2 = 0$; that is, when $x = \pm\sqrt{-1} = \pm i$. Since the only singular points of equation (4) are the complex numbers $\pm i$, we see that $x = 0$ is an ordinary point. Moreover, the distance[††] from 0 to either $\pm i$ is 1. Thus the radius of convergence of a power series solution about $x = 0$ is at least 1. ■ End

In equation (28) of Section 8.3, we found the first few terms of a power series solution to equation (4). Because we now know that this series has radius of convergence at least 1, the partial sums of this series will converge to the solution for $|x| < 1$. However, when $|x| \geq 1$, we have no basis on which to decide whether we can use the series to approximate the solution.

Power series expansions about $x_0 = 0$ are somewhat easier to manipulate than expansions about nonzero points. As the next example shows, a simple shift in variable enables us always to expand about the origin.

[†]See, for example, *Ordinary Differential Equations,* 4th edition, by G. Birkhoff and G.-C. Rota (John Wiley & Sons, New York, 1989).

[††]Recall that the distance between the two complex numbers $z = a + bi$ and $w = c + di$ is given by $\sqrt{(a - c)^2 + (b - d)^2}$.

EXAMPLE 3 Find the first few terms in a power series expansion about $x = 1$ for a general solution to

(5) $2y'' + xy' + y = 0$.

Also determine the radius of convergence of the series.

SOLUTION As seen in Example 1, there are no singular points for equation (5). Thus $x = 1$ is an ordinary point, and, as a consequence of Theorem 5, equation (5) has a general solution of the form

(6) $y(x) = \displaystyle\sum_{n=0}^{\infty} a_n(x - 1)^n$.

Moreover, the radius of convergence of the series in (6) must be infinite.

We can simplify the computation of the coeffients a_n by shifting the center of the expansion (6) from $x_0 = 1$ to $t_0 = 0$. This is accomplished by the substitution $x = t + 1$. Setting $Y(t) = y(t + 1)$, we find via the chain rule

$$\frac{dy}{dx} = \frac{dY}{dt} \ , \qquad \frac{d^2y}{dx^2} = \frac{d^2Y}{dt^2} \ ,$$

and hence equation (5) is changed into

(7) $2\dfrac{d^2Y}{dt^2} + (t + 1)\dfrac{dY}{dt} + Y = 0$.

We now seek a general solution of the form

(8) $Y(t) = \displaystyle\sum_{n=0}^{\infty} a_n t^n$,

where the a_n's in equations (6) and (8) are the same. Proceeding as usual, we substitute the power series for $Y(t)$ into (7), derive a recurrence relation for the coefficients, and ultimately find that

$$Y(t) = a_0\left\{1 - \frac{1}{4}t^2 + \frac{1}{24}t^3 + \cdots\right\} + a_1\left\{t - \frac{1}{4}t^2 - \frac{1}{8}t^3 + \cdots\right\}$$

(the details are left for the reader). Thus restoring $t = x - 1$ we have

(9) $y(x) = a_0\left\{1 - \frac{1}{4}(x - 1)^2 + \frac{1}{24}(x - 1)^3 + \cdots\right\}$

$$+ a_1\left\{(x - 1) - \frac{1}{4}(x - 1)^2 - \frac{1}{8}(x - 1)^3 + \cdots\right\} . \quad \blacksquare \text{ End}$$

When the coefficients of a linear equation are not polynomials in x, but are analytic functions, we can still find analytic solutions by essentially the same method.

EXAMPLE 4 Find a power series expansion for the solution to

(10) $y''(x) + e^x y'(x) + (1 + x^2)y(x) = 0 \ ; \qquad y(0) = 1 \ , \qquad y'(0) = 0$.

SOLUTION Here $p(x) = e^x$ and $q(x) = 1 + x^2$, and both are analytic for all x. Thus by Theorem 5, the

initial value problem (10) has a power series solution

(11) $$y(x) = \sum_{n=0}^{\infty} a_n x^n$$

that converges for all x ($\rho = \infty$). To find the first few terms of this series, we first expand $p(x) = e^x$ in its Maclaurin series:

$$e^x = 1 + x + \frac{x^2}{2!} + \frac{x^3}{3!} + \cdots \;.$$

Substituting the expansions for $y(x)$, $y'(x)$, $y''(x)$, and e^x into (10) gives

(12) $$\sum_{n=2}^{\infty} n(n-1)a_n x^{n-2} + \left(1 + x + \frac{x^2}{2} + \frac{x^3}{6} + \frac{x^4}{24} + \cdots\right)\sum_{n=1}^{\infty} n a_n x^{n-1}$$

$$+ (1 + x^2)\sum_{n=0}^{\infty} a_n x^n = 0 \;.$$

Because of the computational difficulties due to the appearance of the product of the power series for e^x and $y'(x)$, we concern ourselves with just those terms up to order 4. Writing out (12) and keeping track of all such terms, we find

(13) $(2a_2 + 6a_3 x + 12a_4 x^2 + 20a_5 x^3 + 30a_6 x^4 + \cdots)$

$+ (a_1 + 2a_2 x + 3a_3 x^2 + 4a_4 x^3 + 5a_5 x^4 + \cdots)$ $1 \cdot \Sigma\, n a_n x^{n-1}$

$+ (a_1 x + 2a_2 x^2 + 3a_3 x^3 + 4a_4 x^4 + \cdots)$ $x \cdot \Sigma\, n a_n x^{n-1}$

$+ \left(\frac{1}{2}a_1 x^2 + a_2 x^3 + \frac{3}{2}a_3 x^4 + \cdots\right)$ $\frac{x^2}{2}\cdot \Sigma\, n a_n x^{n-1}$

$+ \left(\frac{1}{6}a_1 x^3 + \frac{1}{3}a_2 x^4 + \cdots\right)$

$+ \left(\frac{1}{24}a_1 x^4 + \cdots\right)$

\vdots

$+ (a_0 + a_1 x + a_2 x^2 + a_3 x^3 + a_4 x^4 + \cdots)$ $1 \cdot \Sigma\, a_n x^n$

$+ (a_0 x^2 + a_1 x^3 + a_2 x^4 + \cdots) = 0 \;.$ $x^2 \cdot \Sigma\, a_n x^n$

Grouping the like powers of x in equation (13) (for example, the x^2 terms are shown in color) and then setting the coefficients equal to zero yields the system of equations

$$2a_2 + a_1 + a_0 = 0 \qquad (x^0 \text{ term}) \;,$$

$$6a_3 + 2a_2 + 2a_1 = 0 \qquad (x^1 \text{ term}) \;,$$

$$12a_4 + 3a_3 + 3a_2 + \frac{1}{2}a_1 + a_0 = 0 \qquad (x^2 \text{ term}) \;,$$

$$20a_5 + 4a_4 + 4a_3 + a_2 + \frac{7}{6}a_1 = 0 \qquad (x^3 \text{ term}) \;,$$

$$30a_6 + 5a_5 + 5a_4 + \frac{3}{2}a_3 + \frac{4}{3}a_2 + \frac{1}{24}a_1 = 0 \qquad (x^4 \text{ term}) \;.$$

The initial conditions in (10) imply that $y(0) = a_0 = 1$ and $y'(0) = a_1 = 0$. Using these

values for a_0 and a_1, we can solve the above system first for a_2, then a_3, and so on:

$$2a_2 + 0 + 1 = 0 \Rightarrow a_2 = -\frac{1}{2} \ ,$$

$$6a_3 - 1 + 0 = 0 \Rightarrow a_3 = \frac{1}{6} \ ,$$

$$12a_4 + \frac{1}{2} - \frac{3}{2} + 0 + 1 = 0 \Rightarrow a_4 = 0 \ ,$$

$$20a_5 + 0 + \frac{2}{3} - \frac{1}{2} + 0 = 0 \Rightarrow a_5 = -\frac{1}{120} \ ,$$

$$30a_6 - \frac{1}{24} + 0 + \frac{1}{4} - \frac{2}{3} + 0 = 0 \Rightarrow a_6 = \frac{11}{720} \ .$$

Thus the solution to the initial value problem in (10) is

(14) $y(x) = 1 - \frac{1}{2}x^2 + \frac{1}{6}x^3 - \frac{1}{120}x^5 + \frac{11}{720}x^6 + \cdots \ .$ ■ End

Thus far we have used the power series method only for homogeneous equations. But the same method applies, with obvious modifications, to nonhomogeneous equations of the form

(15) $y''(x) + p(x)y'(x) + q(x)y(x) = g(x) \ ,$

provided the forcing term $g(x)$ and the coefficient functions are analytic at x_0. For example, to find a power series about $x = 0$ for a general solution to

(16) $y''(x) - xy'(x) - y(x) = \sin x \ ,$

we use the substitution $y(x) = \Sigma\, a_n x^n$ to obtain a power series expansion for the left-hand side of (16). We then equate the coefficients of this series with the corresponding coefficients of the Maclaurin expansion for $\sin x$:

$$\sin x = \sum_{n=0}^{\infty} \frac{(-1)^n}{(2n + 1)!} x^{2n+1} \ .$$

Carrying out the details (see Problem 20), we ultimately find that an expansion for a general solution to (16) is

(17) $y(x) = a_0 y_1(x) + a_1 y_2(x) + y_p(x) \ ,$

where

(18) $y_1(x) = 1 + \frac{1}{2}x^2 + \frac{1}{8}x^4 + \frac{1}{48}x^6 + \cdots \ ,$

(19) $y_2(x) = x + \frac{1}{3}x^3 + \frac{1}{15}x^5 + \frac{1}{105}x^7 + \cdots$

are the solutions to the homogeneous equation associated with equation (16) and

(20) $y_p(x) = \frac{1}{6}x^3 + \frac{1}{40}x^5 + \frac{19}{5040}x^7 + \cdots$

is a particular solution to equation (16).

EXERCISES 8.4

In Problems 1 through 6, find a minimum value for the radius of convergence of a power series solution about x_0.

1. $(x + 1)y'' - 3xy' + 2y = 0$; $x_0 = 1$.

2. $y'' - xy' - 3y = 0$; $x_0 = 2$.

3. $(1 + x + x^2)y'' - 3y = 0$; $x_0 = 1$.

4. $(x^2 - 5x + 6)y'' - 3xy' - y = 0$; $x_0 = 0$.

5. $y'' - (\tan x)y' + y = 0$; $x_0 = 0$.

6. $(1 + x^3)y'' - xy' + 3x^2y = 0$; $x_0 = 1$.

In Problems 7 through 12, find at least the first four nonzero terms in a power series expansion about x_0 for a general solution to the given differential equation with the given value for x_0.

7. $y' + 2(x - 1)y = 0$; $x_0 = 1$.

8. $y' - 2xy = 0$; $x_0 = -1$.

9. $(x^2 - 2x)y'' + 2y = 0$; $x_0 = 1$.

10. $x^2y'' - xy' + 2y = 0$; $x_0 = 2$.

11. $x^2y'' - y' + y = 0$; $x_0 = 2$.

12. $y'' + (3x - 1)y' - y = 0$; $x_0 = -1$.

In Problems 13 through 19, find at least the first four nonzero terms in a power series expansion of the solution to the given initial value problem.

13. $x' + (\sin t)x = 0$; $x(0) = 1$.

14. $y' - e^x y = 0$; $y(0) = 1$.

15. $(x^2 + 1)y'' - e^x y' + y = 0$;
 $y(0) = 1$, $y'(0) = 1$.

16. $y'' + ty' + e^t y = 0$;
 $y(0) = 0$, $y'(0) = -1$.

17. $y'' - (\sin x)y = 0$;
 $y(\pi) = 1$, $y'(\pi) = 0$.

18. $y'' - (\cos x)y' - y = 0$;
 $y(\pi/2) = 1$, $y'(\pi/2) = 1$.

19. $y'' - e^{2x}y' + (\cos x)y = 0$;
 $y(0) = -1$, $y'(0) = 1$.

20. To derive the general solution given by equations (17)–(20) for the nonhomogeneous equation (16), complete the following steps:

 (a) Substitute $y(x) = \sum_{n=0}^{\infty} a_n x^n$ and the Maclaurin

series for $\sin x$ into equation (16) to obtain

$$(2a_2 - a_0) + \sum_{k=1}^{\infty} \left[(k + 2)(k + 1)a_{k+2} - (k + 1)a_k \right]x^k$$

$$= \sum_{n=0}^{\infty} \frac{(-1)^n}{(2n + 1)!} x^{2n + 1} \ .$$

 (b) Equate the coefficients of like powers of x on both sides of the equation in part (a) and thereby deduce the equations

$$a_2 = \frac{a_0}{2} \ , \quad a_3 = \frac{1}{6} + \frac{a_1}{3} \ , \quad a_4 = \frac{a_0}{8} \ ,$$

$$a_5 = \frac{1}{40} + \frac{a_1}{15} \ , \quad a_6 = \frac{a_0}{48} \ ,$$

$$a_7 = \frac{19}{5040} + \frac{a_1}{105} \ .$$

 (c) Show that the relations in part (b) yield the general solution to (16) given in equations (17)–(20).

In Problems 21 through 28, use the procedure illustrated in Problem 20 to find at least the first four nonzero terms in a power series expansion about $x = 0$ of a general solution to the given differential equation.

21. $y' - xy = \sin x$.

22. $w' + xw = e^x$.

23. $z'' + xz' + z = x^2 + 2x + 1$.

24. $y'' - 2xy' + 3y = x^2$.

25. $(1 + x^2)y'' - xy' + y = e^{-x}$.

26. $y'' - xy' + 2y = \cos x$.

27. $(1 - x^2)y'' - y' + y = \tan x$.

28. $y'' - (\sin x)y = \cos x$.

29. The equation

$$(1 - x^2)y'' - 2xy' + n(n + 1)y = 0 \ ,$$

where n is an unspecified parameter, is called **Legendre's equation.** This equation occurs in applications of differential equations to physics and engineering.

 (a) Find a power series expansion about $x = 0$ for a solution to Legendre's equation.

 (b) Show that for n a nonnegative integer, there exists an nth degree polynomial that is a solution

to Legendre's equation. These polynomials, up to a constant multiple, are called **Legendre polynomials.**

(c) Determine the first three Legendre polynomials (up to a constant multiple).

30. Aging Spring. As a spring ages, its spring "constant" decreases in value. One such model for a mass-spring system with an aging spring is

$$mx''(t) + bx'(t) + ke^{-\eta t}x(t) = 0 \; ,$$

where m is the mass, b the damping constant, k and η positive constants, and $x(t)$ the displacement of the spring from its equilibrium position. Let $m = 1$ kg, $b = 2$ N-sec/m, $k = 1$ N/m, and $\eta = 1(\text{sec})^{-1}$. The system is set in motion by displacing the mass one

meter from its equilibrium position and then releasing it $(x(0) = 1, x'(0) = 0)$. Find at least the first four nonzero terms in a power series expansion about $t = 0$ for the displacement.

31. Aging Spring without Damping. In the mass-spring system for an aging spring discussed in Problem 30, assume there is no damping (i.e., $b = 0$), $m = 1$, and $k = 1$. To see the effect of aging, consider η as a positive parameter.

(a) Redo Problem 30 with $b = 0$ and η arbitrary but fixed.

(b) Set $\eta = 0$ in the expansion obtained in part (a). Does this expansion agree with the expansion for the solution to the problem with $\eta = 0$? [*Hint:* When $\eta = 0$, the solution is $x(t) = \cos t.$]

8.5 CAUCHY-EULER (EQUIDIMENSIONAL) EQUATIONS REVISITED

In the previous sections we considered methods for obtaining power series solutions about an ordinary point for a linear second order equation. However, in certain cases we may want a series expansion about a *singular point* of the equation. To motivate a procedure for finding such expansions, we return to the class of **Cauchy-Euler equations.** In Chapter 4 we solved these equations by making the change of variables $x = e^t$, which transforms a Cauchy-Euler equation into an equation with constant coefficients. However, it is more instructive for our study of series expansions about singular points to work directly in the variable x.

Recall that a second order homogeneous Cauchy-Euler equation has the form

(1) $$ax^2y''(x) + bxy'(x) + cy(x) = 0 \; , \qquad x > 0,$$

where $a(\neq 0)$, b, and c are (real) constants. Since here $p(x) = b/(ax)$ and $q(x) = c/(ax^2)$, it follows that $x = 0$ is a singular point for (1).

As we found in Chapter 4, equation (1) has solutions of the form $y = x^r$. To determine the values for r, we can proceed as follows. Let L be the differential operator defined by the left-hand side of equation (1), that is,

(2) $$L[y](x) := ax^2y''(x) + bxy'(x) + cy(x) \; ,$$

and set

(3) $$w(r, x) := x^r \; .$$

When we substitute $w(r, x)$ for $y(x)$ in (2), we find

$$\begin{aligned} L[w](x) &= ax^2 r(r - 1)x^{r-2} + bxrx^{r-1} + cx^r \\ &= \{ar^2 + (b - a)r + c\}x^r \; . \end{aligned}$$

From this, we see that $w = x^r$ is a solution to (1) if and only if r satisfies

(4) $\qquad ar^2 + (b - a)r + c = 0$.

Equation (4) is referred to as the **auxiliary, or indicial, equation** for (1).

When the indicial equation has two distinct roots, we have

$$L[w](x) = a(r - r_1)(r - r_2)x^r ,$$

from which it follows that equation (1) has the two solutions

(5) $\qquad y_1(x) = w(r_1, x) = x^{r_1} , \qquad x > 0 ,$

(6) $\qquad y_2(x) = w(r_2, x) = x^{r_2} , \qquad x > 0 ,$

which are linearly independent.

When r_1 and r_2 are complex conjugates, $\alpha \pm i\beta$, we can use Euler's formula to express

$$x^{\alpha + i\beta} = e^{(\alpha + i\beta)\ln x} = e^{\alpha \ln x}\cos(\beta \ln x) + ie^{\alpha \ln x}\sin(\beta \ln x)$$
$$= x^\alpha\cos(\beta \ln x) + ix^\alpha\sin(\beta \ln x) .$$

Since the real and imaginary parts of $x^{\alpha + i\beta}$ must also be solutions to (1), we can replace (5) and (6) by the two linearly independent real-valued solutions

$$y_1(x) = x^\alpha\cos(\beta \ln x) , \qquad y_2(x) = x^\alpha\sin(\beta \ln x) .$$

If the indicial equation (4) has a repeated real root r_0, then it turns out that x^{r_0} and $x^{r_0} \ln x$ are two linearly independent solutions. This can be deduced using the reduction of order approach of Section 4.4. However, it is more instructive for later applications to see how these two linearly independent solutions can be obtained via the operator approach. If r_0 is a repeated root, then

(7) $\qquad L[w](x) = a(r - r_0)^2 x^r$.

Setting $r = r_0$ gives the solution

(8) $\qquad y_1(x) = w(r_0, x) = x^{r_0} , \qquad x > 0$.

To find a second linearly independent solution, we make the following observation. The right-hand side of (7) has the factor $(r - r_0)^2$, so taking the partial derivative of (7) with respect to r and setting $r = r_0$, we get *zero*. That is,

(9) $\qquad \dfrac{\partial}{\partial r}\{L[w](x)\}\bigg|_{r=r_0} = \{a(r - r_0)^2 x^r \ln x + 2a(r - r_0)x^r\}\bigg|_{r=r_0} = 0$.

While it may not appear that we have made any progress toward finding a second solution, a closer look at the expression on the left in (9) will soon vindicate our efforts.

First note that $w(r, x) = x^r$ has continuous partial derivatives of all orders with respect to both r and x. Hence the mixed partial derivatives are equal:

$$\frac{\partial^3 w}{\partial r\partial x^2} = \frac{\partial^3 w}{\partial x^2\partial r} , \qquad \frac{\partial^2 w}{\partial r\partial x} = \frac{\partial^2 w}{\partial x\partial r} .$$

Consequently, for the differential operator L, we have

$$\frac{\partial}{\partial r}L[w] = \frac{\partial}{\partial r}\left\{ax^2\frac{\partial^2 w}{\partial x^2} + bx\frac{\partial w}{\partial x} + cw\right\}$$

$$= ax^2\frac{\partial^3 w}{\partial r\partial x^2} + bx\frac{\partial^2 w}{\partial r\partial x} + c\frac{\partial w}{\partial r}$$

$$= ax^2\frac{\partial^3 w}{\partial x^2\partial r} + bx\frac{\partial^2 w}{\partial x\partial r} + c\frac{\partial w}{\partial r}$$

$$= L\left[\frac{\partial w}{\partial r}\right] ;$$

that is, the operators $\partial/\partial r$ and L commute. With this fact, (9) can be written as

$$L\left[\frac{\partial w}{\partial r}\right]\Bigg|_{r=r_0} = 0 .$$

Thus for the case of a repeated root r_0, a second linearly independent solution to (1) is

$$\textbf{(10)} \qquad y_2(x) = \frac{\partial w}{\partial r}(r_0, x) = \frac{\partial}{\partial r}(x^r)\Bigg|_{r=r_0} = x^{r_0}\ln x , \qquad x > 0 .$$

EXAMPLE 1 Find a general solution to

$$\textbf{(11)} \qquad 4x^2 y''(x) + y(x) = 0 , \qquad x > 0 .$$

SOLUTION Let $w(r, x) = x^r$ and let L denote the left-hand side of (11). A short calculation gives

$$L[w](x) = (4r^2 - 4r + 1)x^r .$$

Solving the indicial equation

$$4r^2 - 4r + 1 = (2r - 1)^2 = 4\left(r - \frac{1}{2}\right)^2 = 0$$

yields the repeated root $r_0 = 1/2$. Thus a general solution to (11) is obtained from equations (8) and (10) by setting $r_0 = 1/2$. That is,

$$y(x) = C_1\sqrt{x} + C_2\sqrt{x}\ln x , \qquad x > 0 . \quad \blacksquare \text{ End}$$

In closing we note that the solutions to Cauchy-Euler equations can exhibit some extraordinary features near their singular points—features unlike anything we have encountered in the case of constant coefficient equations. Observe that:

(i) x^r and $x^r \ln x$, for $0 < r < 1$, have infinite slope at the origin;

(ii) x^r, for $r < 0$, diverges at the origin; and

(iii) $x^\alpha \cos(\beta \ln x)$ and $x^\alpha \sin(\beta \ln x)$ oscillate "infinitely rapidly" near the origin. (See Figure 8.7 on page 482.)

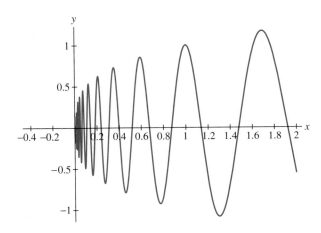

Figure 8.7 Graph of $x^{0.3}\cos(12 \ln x)$

Numerical solvers based on Euler or Runge-Kutta procedures would be completely frustrated in trying to chart such behavior; analytic methods are essential for characterizing solutions near singular points.

In Section 8.7 we discuss the problem of finding a second linearly independent *series* solution to certain differential equations. As we shall see, operator methods similar to those described in this section will lead to the desired second solution.

EXERCISES 8.5

In Problems 1 through 10, use the substitution $y = x^r$ to find a general solution to the given equation for $x > 0$.

1. $x^2 y''(x) + 6xy'(x) + 6y(x) = 0$.

2. $2x^2 y''(x) + 13xy'(x) + 15y(x) = 0$.

3. $x^2 y''(x) - xy'(x) + 17y(x) = 0$.

4. $x^2 y''(x) + 2xy'(x) - 3y(x) = 0$.

5. $\dfrac{d^2 y}{dx^2} = \dfrac{5}{x}\dfrac{dy}{dx} - \dfrac{13}{x^2}y$.

6. $\dfrac{d^2 y}{dx^2} = \dfrac{1}{x}\dfrac{dy}{dx} - \dfrac{4}{x^2}y$.

7. $x^3 y'''(x) + 4x^2 y''(x) + 10xy'(x) - 10y(x) = 0$.

8. $x^3 y'''(x) + 4x^2 y''(x) + xy'(x) = 0$.

9. $x^3 y'''(x) + 3x^2 y''(x) + 5xy'(x) - 5y(x) = 0$.

10. $x^3 y'''(x) + 9x^2 y''(x) + 19xy'(x) + 8y(x) = 0$.

In Problems 11 and 12, use a substitution of the form $y = (x - c)^r$ to find a general solution to the given

equation for $x > c$.

11. $2(x - 3)^2 y''(x) + 5(x - 3)y'(x) - 2y(x) = 0$.

12. $4(x + 2)^2 y''(x) + 5y(x) = 0$.

In Problems 13 and 14, use variation of parameters to find a general solution to the given equation for $x > 0$.

13. $x^2 y''(x) - 2xy'(x) + 2y(x) = x^{-1/2}$.

14. $x^2 y''(x) + 2xy'(x) - 2y(x) = 6x^{-2} + 3x$.

In Problems 15 through 17, solve the given initial value problem.

15. $t^2 x''(t) - 12x(t) = 0$;
$\quad x(1) = 3$, $x'(1) = 5$.

16. $x^2 y''(x) + 5xy'(x) + 4y(x) = 0$;
$\quad y(1) = 3$, $y'(1) = 7$.

17. $x^3 y'''(x) + 6x^2 y''(x) + 29xy'(x) - 29y(x) = 0$;
$\quad y(1) = 2$, $y'(1) = -3$, $y''(1) = 19$.

18. When r_0 is a repeated root of the auxiliary equation $ar^2 + br + c = 0$, then $y_1(t) = e^{r_0 t}$ is a solution to the equation $ay'' + by' + cy = 0$, where a, b, and c are constants. Use a derivation similar to the one given in this section for the case when the indicial equation has a repeated root to show that a second linearly independent solution is $y_2(t) = te^{r_0 t}$.

19. Let $L[y](x) := x^3 y'''(x) + xy'(x) - y(x)$.
 (a) Show that $L[x^r](x) = (r - 1)^3 x^r$.
 (b) Using an extension of the argument given in this section for the case when the indicial equation has a double root, show that $L[y] = 0$ has the general solution

$$y(x) = C_1 x + C_2 x \ln x + C_3 x (\ln x)^2 \ .$$

8.6 METHOD OF FROBENIUS

In the previous section we showed that a homogeneous Cauchy-Euler equation has a solution of the form $y(x) = x^r$, $x > 0$, where r is a certain constant. Cauchy-Euler equations have, of course, a very special form with only one singular point (at $x = 0$). In this section we show how the theory for Cauchy-Euler equations generalizes to other equations that have a special type of singularity.

To motivate the procedure, let's rewrite the Cauchy-Euler equation,

(1) $ax^2 y''(x) + bxy'(x) + cy(x) = 0$, $x > 0$,

in the standard form

(2) $y''(x) + p(x)y'(x) + q(x)y(x) = 0$, $x > 0$,

where

$$p(x) = \frac{p_0}{x} , \qquad q(x) = \frac{q_0}{x^2} ,$$

and p_0, q_0 are the constants b/a and c/a, respectively. When we substitute $w(r, x) = x^r$ for y into equation (2), we get

$$\left[r(r - 1) + p_0 r + q_0 \right] x^{r-2} = 0 ,$$

which yields the indicial equation

(3) $r(r - 1) + p_0 r + q_0 = 0$.

Thus, if r_1 is a root of (3), then $w(r_1, x) = x^{r_1}$ is a solution to equations (1) and (2).
Let's now assume, more generally, that (2) is an equation for which $xp(x)$ and $x^2 q(x)$, instead of being constants, are *analytic functions*. That is, in some open interval about $x = 0$,

(4) $xp(x) = p_0 + p_1 x + p_2 x^2 + \cdots = \displaystyle\sum_{n=0}^{\infty} p_n x^n$,

(5) $x^2 q(x) = q_0 + q_1 x + q_2 x^2 + \cdots = \displaystyle\sum_{n=0}^{\infty} q_n x^n$.

It follows from (4) and (5) that

(6) $\displaystyle\lim_{x \to 0} xp(x) = p_0$ and $\displaystyle\lim_{x \to 0} x^2 q(x) = q_0$,

and hence, for x near 0 we have $xp(x) \approx p_0$ and $x^2 q(x) \approx q_0$. Therefore it is reasonable to expect that the solutions to (2) will behave (for x near 0) like the solutions to the Cauchy-Euler equation

$$x^2 y'' + p_0 x y' + q_0 y = 0 \ .$$

When $p(x)$ and $q(x)$ satisfy (4) and (5), we say that the singular point at $x = 0$ is regular. More generally, we state the following.

REGULAR SINGULAR POINT

Definition 3. A singular point x_0 of

(7) $y''(x) + p(x)y'(x) + q(x)y(x) = 0$

is said to be a **regular singular point** if both $(x - x_0)p(x)$ and $(x - x_0)^2 q(x)$ are analytic at x_0.[†] Otherwise x_0 is called an **irregular singular point.**

EXAMPLE 1 Classify the singular points of the equation

(8) $(x^2 - 1)^2 y''(x) + (x + 1)y'(x) - y(x) = 0 \ .$

SOLUTION Here

$$p(x) = \frac{x + 1}{(x^2 - 1)^2} = \frac{1}{(x + 1)(x - 1)^2} \ ,$$

$$q(x) = \frac{-1}{(x^2 - 1)^2} = \frac{-1}{(x + 1)^2 (x - 1)^2} \ ,$$

from which we see that ± 1 are the singular points of (8). For the singularity at 1, we have

$$(x - 1)p(x) = \frac{1}{(x + 1)(x - 1)} \ ,$$

which is not analytic at $x = 1$. Therefore $x = 1$ is an irregular singular point.
For the singularity at -1, we have

$$(x + 1)p(x) = \frac{1}{(x - 1)^2} \ , \qquad (x + 1)^2 q(x) = \frac{-1}{(x - 1)^2} \ ,$$

both of which are analytic at $x = -1$. Hence $x = -1$ is a regular singular point. ■ End

Let's assume $x = 0$ is a regular singular point for equation (7) so that $p(x)$ and $q(x)$ satisfy (4) and (5); that is,

(9) $p(x) = \displaystyle\sum_{n=0}^{\infty} p_n x^{n-1} \ , \qquad q(x) = \sum_{n=0}^{\infty} q_n x^{n-2} \ .$

The idea of the mathematician Frobenius was that since Cauchy-Euler equations have solu-

[†]In the terminology of complex variables, p has a pole of order at most 1 and q has a pole of order at most 2 at x_0.

tions of the form x^r, then for the regular singular point $x = 0$, there should be solutions to (7) of the form x^r *times an analytic function.*[†] Hence we seek solutions to (7) of the form

(10) $\quad w(r, x) = x^r \sum_{n=0}^{\infty} a_n x^n = \sum_{n=0}^{\infty} a_n x^{n+r}$, $\quad x > 0$.

In writing (10), we have assumed a_0 is the first nonzero coefficient, so we are left with determining r and the coefficients a_n, $n \geq 1$. Differentiating $w(r, x)$ with respect to x, we have

(11) $\quad w'(r, x) = \sum_{n=0}^{\infty} (n + r)a_n x^{n+r-1}$,

(12) $\quad w''(r, x) = \sum_{n=0}^{\infty} (n + r)(n + r - 1)a_n x^{n+r-2}$.

If we substitute the above expansions for $w(r, x)$, $w'(r, x)$, $w''(r, x)$, $p(x)$, and $q(x)$ into (7), we obtain

(13) $\quad \sum_{n=0}^{\infty} (n + r)(n + r - 1)a_n x^{n+r-2} + \left(\sum_{n=0}^{\infty} p_n x^{n-1} \right) \left(\sum_{n=0}^{\infty} (n + r)a_n x^{n+r-1} \right)$

$\quad + \left(\sum_{n=0}^{\infty} q_n x^{n-2} \right) \left(\sum_{n=0}^{\infty} a_n x^{n+r} \right) = 0$.

Now we use the Cauchy product to perform the series multiplications and then group like powers of x, starting with the lowest power, x^{r-2}. This gives

(14) $\quad [r(r - 1) + p_0 r + q_0]a_0 x^{r-2}$

$\quad\quad + [(r + 1)ra_1 + (r + 1)p_0 a_1 + p_1 r a_0 + q_0 a_1 + q_1 a_0]x^{r-1} + \cdots = 0$.

For the expansion on the left-hand side of equation (14) to sum to zero, each coefficient must be zero. Considering the first term, x^{r-2}, we find

(15) $\quad [r(r - 1) + p_0 r + q_0]a_0 = 0$.

We have assumed $a_0 \neq 0$, so the quantity in brackets must be zero. This gives the indicial equation; it is the same as the one we derived for Cauchy-Euler equations.

INDICIAL EQUATION

Definition 4. If x_0 is a regular singular point of $y'' + py' + qy = 0$, then the **indicial equation** for this point is

(16) $\quad r(r - 1) + p_0 r + q_0 = 0$,

where

$$p_0 := \lim_{x \to x_0} (x - x_0)p(x) , \quad\quad q_0 := \lim_{x \to x_0} (x - x_0)^2 q(x) .$$

The roots of the indicial equation are called the **exponents (indices)** of the singularity x_0.

[†]*Historical Footnote:* George Frobenius (1848–1917) developed this method in 1873. He is also known for his research on group theory.

EXAMPLE 2 Find the indicial equation and the exponents at the singularity $x = -1$ of

(17) $(x^2 - 1)^2 y''(x) + (x + 1)y'(x) - y(x) = 0$.

SOLUTION In Example 1, we showed that $x = -1$ is a regular singular point. Since $p(x) = (x + 1)^{-1}(x - 1)^{-2}$ and $q(x) = -(x + 1)^{-2}(x - 1)^{-2}$, we find

$$p_0 = \lim_{x \to -1} (x + 1)p(x) = \lim_{x \to -1} (x - 1)^{-2} = \frac{1}{4} ,$$

$$q_0 = \lim_{x \to -1} (x + 1)^2 q(x) = \lim_{x \to -1} \left[-(x - 1)^{-2} \right] = -\frac{1}{4} .$$

Substituting these values for p_0 and q_0 into (16), we obtain the indicial equation

(18) $r(r - 1) + \frac{1}{4}r - \frac{1}{4} = 0$.

Multiplying by 4 and factoring gives $(4r + 1)(r - 1) = 0$. Hence $r = 1, -1/4$ are the exponents. ∎ End

As we have seen, we can use the indicial equation to determine those values of r for which the coefficient of x^{r-2} in (14) is zero. If we set the coefficient of x^{r-1} in (14) equal to zero, we have

(19) $\left[(r + 1)r + (r + 1)p_0 + q_0 \right]a_1 + (p_1 r + q_1)a_0 = 0$.

Since a_0 is arbitrary and we know the p_i's, q_i's, and r, we can solve equation (19) for a_1, provided the coefficient of a_1 in (19) is not zero. This will be the case if we take r to be the larger of the two roots of the indicial equation (see Problem 43).[†] Similarly, when we set the coefficient of x^r equal to zero, we can solve for a_2 in terms of the p_i's, q_i's, r, a_0, and a_1. Continuing in this manner, we can recursively solve for the a_n's. The procedure is illustrated in the following example.

EXAMPLE 3 Find a series expansion about the regular singular point $x = 0$ for a solution to

(20) $(x + 2)x^2 y''(x) - xy'(x) + (1 + x)y(x) = 0$, $x > 0$.

SOLUTION Here $p(x) = -x^{-1}(x + 2)^{-1}$ and $q(x) = x^{-2}(x + 2)^{-1}(1 + x)$, so

$$p_0 = \lim_{x \to 0} xp(x) = \lim_{x \to 0} \left[-(x + 2)^{-1} \right] = -\frac{1}{2} ,$$

$$q_0 = \lim_{x \to 0} x^2 q(x) = \lim_{x \to 0} (x + 2)^{-1}(1 + x) = \frac{1}{2} .$$

Since $x = 0$ is a regular singular point, we seek a solution to (20) of the form

(21) $w(r, x) = x^r \sum_{n=0}^{\infty} a_n x^n = \sum_{n=0}^{\infty} a_n x^{n+r}$.

[†]Several interesting possibilities arise if we take r to be the smaller root. This will be discussed in the next section.

By the previous discussion, r must satisfy the indicial equation (16). Substituting for p_0 and q_0 in (16), we obtain

$$r(r-1) - \frac{1}{2}r + \frac{1}{2} = 0 ,$$

which simplifies to $2r^2 - 3r + 1 = (2r-1)(r-1) = 0$. Thus $r = 1$ and $r = 1/2$ are the roots of the indicial equation associated with $x = 0$.

Let's use the larger root $r = 1$ and solve for a_1, a_2, etc., to obtain the solution $w(1, x)$. We can simplify the computations by substituting $w(r, x)$ directly into equation (20), where the coefficients are polynomials in x, rather than dividing by $(x + 2)x^2$ and having to work with the rational functions $p(x)$ and $q(x)$. Inserting $w(r, x)$ in (20) and recalling the formulas for $w'(r, x)$ and $w''(r, x)$ in (11) and (12) gives (with $r = 1$)

$$(22) \quad (x+2)x^2 \sum_{n=0}^{\infty} (n+1)na_nx^{n-1} - x \sum_{n=0}^{\infty} (n+1)a_nx^n$$

$$+ (1+x) \sum_{n=0}^{\infty} a_nx^{n+1} = 0 ,$$

which we can write as

$$(23) \quad \sum_{n=0}^{\infty} (n+1)na_nx^{n+2} + \sum_{n=0}^{\infty} 2(n+1)na_nx^{n+1} - \sum_{n=0}^{\infty} (n+1)a_nx^{n+1}$$

$$+ \sum_{n=0}^{\infty} a_nx^{n+1} + \sum_{n=0}^{\infty} a_nx^{n+2} = 0 .$$

Next we shift the indices so that each summation in (23) is in powers x^k. With $k = n + 2$ in the first and last summations and $k = n + 1$ in the rest, (23) becomes

$$(24) \quad \sum_{k=2}^{\infty} \left[(k-1)(k-2) + 1\right]a_{k-2}x^k + \sum_{k=1}^{\infty} \left[2k(k-1) - k + 1\right]a_{k-1}x^k = 0 .$$

Separating off the $k = 1$ term and combining the rest under one summation yields

$$(25) \quad \left[2(1)(0) - 1 + 1\right]a_0x + \sum_{k=2}^{\infty} \left[(k^2 - 3k + 3)a_{k-2} + (2k-1)(k-1)a_{k-1}\right]x^k = 0 .$$

Notice that the coefficient of x in (25) is zero. This is because $r = 1$ is a root of the indicial equation, which is the equation we obtained by setting the coefficient of the lowest power of x equal to zero.

We can now determine the a_k's in terms of a_0 by setting the coefficients of x^k in equation (25) equal to zero for $k = 2, 3$, etc. This gives the recurrence relation

$$(26) \quad (k^2 - 3k + 3)a_{k-2} + (2k-1)(k-1)a_{k-1} = 0 ,$$

or, equivalently,

$$(27) \quad a_{k-1} = -\frac{k^2 - 3k + 3}{(2k-1)(k-1)}a_{k-2} , \quad k \geq 2 .$$

Setting $k = 2, 3$, and 4 in (27), we find

$$a_1 = -\frac{1}{3}a_0 \qquad\qquad (k = 2) \ ,$$

$$a_2 = -\frac{3}{10}a_1 = \frac{1}{10}a_0 \qquad (k = 3) \ ,$$

$$a_3 = -\frac{1}{3}a_2 = -\frac{1}{30}a_0 \qquad (k = 4) \ .$$

Substituting these values for r, a_1, a_2, and a_3 into (21) gives

$$\textbf{(28)} \qquad w(1, x) = a_0 x^1 \left(1 - \frac{1}{3}x + \frac{1}{10}x^2 - \frac{1}{30}x^3 + \cdots \right) ,$$

where a_0 is arbitrary. In particular, for $a_0 = 1$, we get the solution

$$y_1(x) = x - \frac{1}{3}x^2 + \frac{1}{10}x^3 - \frac{1}{30}x^4 + \cdots \ , \qquad x > 0 \ .$$

See Figure 8.8. ■End

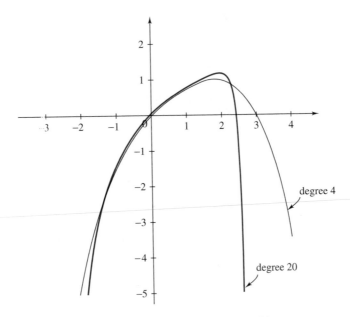

Figure 8.8 Partial sums approximating the solution $y_1(x)$ of Example 3

To find a second linearly independent solution to equation (20), we could try setting $r = 1/2$ and solving for a_1, a_2, \ldots to obtain a solution $w(1/2, x)$ (see Problem 44). In this particular case, the approach would work. However, if we encounter an indicial equation that has a

repeated root, then the method of Frobenius would yield just one solution (apart from constant multiples). To find the desired second solution, we must use another technique, such as the reduction of order procedure discussed in Section 4.4. We tackle the problem of finding a second linearly independent solution in the next section.

The method of Frobenius can be summarized as follows.

METHOD OF FROBENIUS

To derive a series solution about the singular point x_0 of

(29) $a_2(x)y''(x) + a_1(x)y'(x) + a_0(x)y(x) = 0$, $x > x_0$:

(a) Set $p(x) := a_1(x)/a_2(x)$, $q(x) := a_0(x)/a_2(x)$. If both $(x - x_0)p(x)$ and $(x - x_0)^2 q(x)$ are analytic at x_0, then x_0 is a regular singular point and the remaining steps apply.

(b) Let

(30) $$w(r, x) = (x - x_0)^r \sum_{n=0}^{\infty} a_n(x - x_0)^n = \sum_{n=0}^{\infty} a_n(x - x_0)^{n+r} \ ,$$

and, using termwise differentiation, substitute $w(r, x)$ into equation (29) to obtain an equation of the form

$$A_0(x - x_0)^{r+J} + A_1(x - x_0)^{r+J+1} + \cdots = 0 \ .$$

(c) Set the coefficients A_0, A_1, A_2, \ldots equal to zero. (Notice that the equation $A_0 = 0$ is just a constant multiple of the indicial equation $r(r - 1) + p_0 r + q_0 = 0$.)

(d) Use the system of equations

$$A_0 = 0 , \quad A_1 = 0 , \quad \ldots , \quad A_k = 0$$

to find a recurrence relation involving a_k and $a_0, a_1, \ldots, a_{k-1}$.

(e) Take $r = r_1$, the larger root of the indicial equation, and use the relation obtained in step (d) to determine a_1, a_2, \ldots recursively in terms of a_0 and r_1.

(f) A series expansion of a solution to (29) is

(31) $$w(r_1, x) = (x - x_0)^{r_1} \sum_{n=0}^{\infty} a_n(x - x_0)^n \ , \quad x > x_0 \ ,$$

where a_0 is arbitrary and the a_n's are defined in terms of a_0 and r_1.

One important question that remains concerns the radius of convergence of the power series that appears in (31). The following theorem contains an answer.[†]

[†]For a proof of this theorem, see *Ordinary Differential Equations,* by E. L. Ince (Dover Publications, Inc., New York, 1956), Chapter XVI.

FROBENIUS'S THEOREM

Theorem 6. If x_0 is a regular singular point of equation (29), then there exists at least one series solution of the form (30), where $r = r_1$ is the larger root of the associated indicial equation. Moreover, this series converges for all x such that $0 < x - x_0 < R$, where R is the distance from x_0 to the nearest other singular point (real or complex) of (29).

For simplicity, in the examples that follow we consider only series expansions about the regular singular point $x = 0$ and only those equations for which the associated indicial equation has real roots.

The following three examples not only illustrate the method of Frobenius but are important models to which we refer in later sections.

EXAMPLE 4 Find a series solution about the regular singular point $x = 0$ of

(32) $x^2 y''(x) - xy'(x) + (1 - x)y(x) = 0$, $x > 0$.

SOLUTION Here $p(x) = -x^{-1}$ and $q(x) = (1 - x)x^{-2}$. It is easy to check that $x = 0$ is a regular singular point of (32), so we compute

$$p_0 = \lim_{x \to 0} xp(x) = \lim_{x \to 0}(-1) = -1 \ ,$$
$$q_0 = \lim_{x \to 0} x^2 q(x) = \lim_{x \to 0}(1 - x) = 1 \ .$$

Then the indicial equation is

$$r(r - 1) - r + 1 = r^2 - 2r + 1 = (r - 1)^2 = 0 \ .$$

which has the roots $r_1 = r_2 = 1$.

Next we substitute

(33) $w(r, x) = x^r \sum_{n=0}^{\infty} a_n x^n = \sum_{n=0}^{\infty} a_n x^{n+r}$

into (32) and obtain

(34) $x^2 \sum_{n=0}^{\infty} (n + r)(n + r - 1)a_n x^{n+r-2} - x \sum_{n=0}^{\infty} (n + r)a_n x^{n+r-1}$

$$+ (1 - x) \sum_{n=0}^{\infty} a_n x^{n+r} = 0 \ ,$$

which we write as

(35) $\sum_{n=0}^{\infty} (n + r)(n + r - 1)a_n x^{n+r} - \sum_{n=0}^{\infty} (n + r)a_n x^{n+r}$

$$+ \sum_{n=0}^{\infty} a_n x^{n+r} - \sum_{n=0}^{\infty} a_n x^{n+r+1} = 0 \ .$$

Shifting the indices so that each summation in (35) is in powers x^{k+r}, we take $k = n + 1$ in the

last summation and $k = n$ in the rest. This gives

(36) $\displaystyle\sum_{k=0}^{\infty} \big[(k+r)(k+r-1) - (k+r) + 1 \big] a_k x^{k+r} - \sum_{k=1}^{\infty} a_{k-1} x^{k+r} = 0$.

Singling out the term corresponding to $k = 0$ and combining the rest under one summation yields

(37) $\big[r(r-1) - r + 1 \big] a_0 x^r$

$\displaystyle\qquad\qquad + \sum_{k=1}^{\infty} \big\{ [(k+r)(k+r-1) - (k+r) + 1] a_k - a_{k-1} \big\} x^{k+r} = 0$.

When we set the coefficients equal to zero, we recover the indicial equation

(38) $\big[r(r-1) - r + 1 \big] a_0 = 0$,

and obtain for $k \geq 1$, the recurrence relation

(39) $\big[(k+r)^2 - 2(k+r) + 1 \big] a_k - a_{k-1} = 0$,

which reduces to

(40) $(k+r-1)^2 a_k - a_{k-1} = 0$.

Relation (40) can be used to solve for a_k in terms of a_{k-1}:

(41) $a_k = \dfrac{1}{(k+r-1)^2} a_{k-1} , \qquad k \geq 1$.

Setting $r = r_1 = 1$ in (38) gives (as expected) $0 \cdot a_0 = 0$, and in (41) gives

(42) $a_k = \dfrac{1}{k^2} a_{k-1} , \qquad k \geq 1$.

For $k = 1, 2,$ and 3, we now find

$$a_1 = \frac{1}{1^2} a_0 = a_0 \qquad\qquad\qquad\qquad (k = 1) ,$$

$$a_2 = \frac{1}{2^2} a_1 = \frac{1}{(2 \cdot 1)^2} a_0 = \frac{1}{4} a_0 \qquad\qquad (k = 2) ,$$

$$a_3 = \frac{1}{3^2} a_2 = \frac{1}{(3 \cdot 2 \cdot 1)^2} a_0 = \frac{1}{36} a_0 \qquad (k = 3) .$$

In general, we have

(43) $a_k = \dfrac{1}{(k!)^2} a_0$.

Hence equation (32) has a series solution given by

(44) $w(1, x) = a_0 x \left\{ 1 + x + \dfrac{1}{4} x^2 + \dfrac{1}{36} x^3 + \cdots \right\}$

$$= a_0 x \sum_{k=0}^{\infty} \frac{1}{(k!)^2} x^k , \qquad x > 0 . \quad \blacksquare \text{ End}$$

Since $x = 0$ is the only singular point for equation (32), it follows from Frobenius's theorem or directly by the ratio test that the series solution (44) converges for all $x > 0$.

In the next two examples, we only outline the method; the reader is invited to furnish the intermediate steps.

EXAMPLE 5 Find a series solution about the regular singular point $x = 0$ of

(45) $xy''(x) + 4y'(x) - xy(x) = 0 , \qquad x > 0 .$

SOLUTION Since $p(x) = 4/x$ and $q(x) = -1$, we see that $x = 0$ is indeed a regular singular point and

$$p_0 = \lim_{x \to 0} xp(x) = 4 , \qquad q_0 = \lim_{x \to 0} x^2 q(x) = 0 .$$

The indicial equation is

$$r(r - 1) + 4r = r^2 + 3r = r(r + 3) = 0 ,$$

with roots $r_1 = 0$ and $r_2 = -3$.

Now we substitute

(46) $w(r, x) = x^r \sum_{n=0}^{\infty} a_n x^n = \sum_{n=0}^{\infty} a_n x^{n+r}$

into (45). After a little algebra and a shift in indices, we get

(47) $\left[r(r - 1) + 4r \right] a_0 x^{r-1} + \left[(r + 1)r + 4(r + 1) \right] a_1 x^r$

$\qquad + \sum_{k=1}^{\infty} \left[(k + r + 1)(k + r + 4) a_{k+1} - a_{k-1} \right] x^{k+r} = 0 .$

Next we set the coefficients equal to zero and find

(48) $\left[r(r - 1) + 4r \right] a_0 = 0 ,$

(49) $\left[(r + 1)r + 4(r + 1) \right] a_1 = 0 ,$

and, for $k \geq 1$, the recurrence relation

(50) $(k + r + 1)(k + r + 4) a_{k+1} - a_{k-1} = 0 .$

For $r = r_1 = 0$, equation (48) becomes $0 \cdot a_0 = 0$ and (49) becomes $4 \cdot a_1 = 0$. Hence, while a_0 is arbitrary, a_1 must be zero. Setting $r = r_1 = 0$ in (50), we find

(51) $a_{k+1} = \dfrac{1}{(k + 1)(k + 4)} a_{k-1} , \qquad k \geq 1 ,$

from which it follows (after a few experimental computations) that $a_{2k+1} = 0$ for $k = 0, 1, \ldots ,$ and

(52) $a_{2k} = \dfrac{1}{\left[2 \cdot 4 \cdots (2k) \right] \left[5 \cdot 7 \cdots (2k + 3) \right]} a_0$

$\qquad = \dfrac{1}{2^k k! \left[5 \cdot 7 \cdots (2k + 3) \right]} a_0 , \qquad k \geq 1 .$

Hence equation (45) has a power series solution

$$
(53) \qquad w(0, x) = a_0 \left\{ 1 + \sum_{k=1}^{\infty} \frac{1}{2^k k! [5 \cdot 7 \cdots (2k + 3)]} x^{2k} \right\} , \qquad x > 0 . \quad \blacksquare \text{End}
$$

If in Example 5 we had worked with the root $r = r_2 = -3$, then we would actually have obtained *two* linearly independent solutions (see Problem 45).

EXAMPLE 6 Find a series solution about the regular singular point of $x = 0$ of

$$
(54) \qquad xy''(x) + 3y'(x) - xy(x) = 0 , \qquad x > 0 .
$$

SOLUTION Since $p(x) = 3/x$ and $q(x) = -1$, we see that $x = 0$ is a regular singular point. Moreover,

$$
p_0 = \lim_{x \to 0} xp(x) = 3 , \qquad q_0 = \lim_{x \to 0} x^2 q(x) = 0 .
$$

So the indicial equation is

$$
(55) \qquad r(r - 1) + 3r = r^2 + 2r = r(r + 2) = 0 ,
$$

with roots $r_1 = 0$ and $r_2 = -2$.

Substituting

$$
(56) \qquad w(r, x) = x^r \sum_{n=0}^{\infty} a_n x^n = \sum_{n=0}^{\infty} a_n x^{n+r}
$$

into (54) ultimately gives

$$
(57) \qquad \left[r(r - 1) + 3r \right] a_0 x^{r-1} + \left[(r + 1)r + 3(r + 1) \right] a_1 x^r
$$

$$
+ \sum_{k=1}^{\infty} \left[(k + r + 1)(k + r + 3) a_{k+1} - a_{k-1} \right] x^{k+r} = 0 .
$$

Setting the coefficients equal to zero, we have

$$
(58) \qquad \left[r(r - 1) + 3r \right] a_0 = 0 ,
$$
$$
(59) \qquad \left[(r + 1)r + 3(r + 1) \right] a_1 = 0 ,
$$

and, for $k \geq 1$, the recurrence relation

$$
(60) \qquad (k + r + 1)(k + r + 3) a_{k+1} - a_{k-1} = 0 .
$$

With $r = r_1 = 0$, these equations lead to the following formulas: $a_{2k+1} = 0$, $k = 0, 1, \ldots$, and

$$
(61) \qquad a_{2k} = \frac{1}{[2 \cdot 4 \cdots (2k)][4 \cdot 6 \cdots (2k + 2)]} a_0 = \frac{1}{2^{2k} k! (k + 1)!} a_0 , \qquad k \geq 0 .
$$

Hence equation (54) has a power series solution

$$
(62) \qquad w(0, x) = a_0 \sum_{k=0}^{\infty} \frac{1}{2^{2k} k! (k + 1)!} x^{2k} , \qquad x > 0 . \quad \blacksquare \text{End}
$$

Unlike in Example 5, if we work with the second root $r = r_2 = -2$ in Example 6, then we do *not* obtain a second linearly independent solution (see Problem 46).

In the preceding examples we were able to use the method of Frobenius to find a series solution valid to the right ($x > 0$) of the regular singular point $x = 0$. For $x < 0$, we can use the change of variables $x = -t$ and then solve the resulting equation for $t > 0$.

The method of Frobenius also applies to higher-order linear equations (see Problems 35–38).

EXERCISES 8.6

In Problems 1 through 10, classify each singular point (real or complex) of the given equation as regular or irregular.

1. $(x^2 - 1)y'' + xy' + 3y = 0$.

2. $x^2 y'' + 8xy' - 3xy = 0$.

3. $(x^2 + 1)z'' + 7x^2 z' - 3xz = 0$.

4. $x^2 y'' - 5xy' + 7y = 0$.

5. $(x^2 - 1)^2 y'' - (x - 1)y' + 3y = 0$.

6. $(x^2 - 4)y'' + (x + 2)y' + 3y = 0$.

7. $(t^2 - t - 2)^2 x'' + (t^2 - 4)x' - tx = 0$.

8. $(x^2 - x)y'' + xy' + 7y = 0$.

9. $(x^2 + 2x - 8)^2 y'' + (3x + 12)y' - x^2 y = 0$.

10. $x^3(x - 1)y'' + (x^2 - 3x)(\sin x)y' - xy = 0$.

In Problems 11 through 18, find the indicial equation and the exponents for the specified singularity of the given differential equations.

11. $x^2 y'' - 2xy' - 10y = 0$, at $x = 0$.

12. $x^2 y'' + 4xy' + 2y = 0$, at $x = 0$.

13. $(x^2 - x - 2)^2 z'' + (x^2 - 4)z' - 6xz = 0$, at $x = 2$.

14. $(x^2 - 4)y'' + (x + 2)y' + 3y = 0$, at $x = -2$.

15. $\theta^3 y'' + \theta(\sin \theta)y' - (\tan \theta)y = 0$, at $\theta = 0$.

16. $(x^2 - 1)y'' - (x - 1)y' - 3y = 0$, at $x = 1$.

17. $(x - 1)^2 y'' + (x^2 - 1)y' - 12y = 0$, at $x = 1$.

18. $4x(\sin x)y'' - 3y = 0$, at $x = 0$.

In Problems 19 through 24, use the method of Frobenius to find at least the first four nonzero terms in the series expansion about $x = 0$ for a solution to the given equation for $x > 0$.

19. $9x^2 y'' + 9x^2 y' + 2y = 0$.

20. $2x(x - 1)y'' + 3(x - 1)y' - y = 0$.

21. $x^2 y'' + xy' + x^2 y = 0$.

22. $xy'' + y' - 4y = 0$.

23. $x^2 z'' + (x^2 + x)z' - z = 0$.

24. $3xy'' + (2 - x)y' - y = 0$.

In Problems 25 through 30, use the method of Frobenius to find a general formula for the coefficient a_n in a series expansion about $x = 0$ for a solution to the given equation for $x > 0$.

25. $4x^2 y'' + 2x^2 y' - (x + 3)y = 0$.

26. $x^2 y'' + (x^2 - x)y' + y = 0$.

27. $xw'' - w' - xw = 0$.

28. $3x^2 y'' + 8xy' + (x - 2)y = 0$.

29. $xy'' + (x - 1)y' - 2y = 0$.

30. $x(x + 1)y'' + (x + 5)y' - 4y = 0$.

In Problems 31 through 34, first determine a recurrence formula for the coefficients in the (Frobenius) series expansion of the solution about $x = 0$. Use this recurrence formula to determine if there exists a solution to the differential equation that is decreasing for $x > 0$.

31. $xy'' + (1 - x)y' - y = 0$.

32. $x^2 y'' - x(1 + x)y' + y = 0$.

33. $3xy'' + 2(1 - x)y' - 4y = 0$.

34. $xy'' + (x + 2)y' - y = 0$.

In Problems 35 through 38, use the method of Frobenius to find at least the first four nonzero terms in the series expansion about $x = 0$ for a solution to the given linear third order equation for $x > 0$.

35. $6x^3 y''' + 13x^2 y'' + (x + x^2)y' + xy = 0$.

36. $6x^3 y''' + 11x^2 y'' - 2xy' - (x - 2)y = 0$.

37. $6x^3y''' + 13x^2y'' - (x^2 + 3x)y' - xy = 0$.

38. $6x^3y''' + (13x^2 - x^3)y'' + xy' - xy = 0$.

In Problems 39 and 40, try to use the method of Frobenius to find a series expansion about the irregular singular point $x = 0$ for a solution to the given differential equation. If the method works, give at least the first four nonzero terms in the expansion. If the method does not work, explain what went wrong.

39. $x^2y'' + (3x - 1)y' + y = 0$.

40. $x^2y'' + y' - 2y = 0$.

In certain applications, it is desirable to have an expansion about the point at infinity. To obtain such an expansion, we use the change of variables $z = 1/x$ and expand about $z = 0$. In Problems 41 and 42, show that infinity is a regular singular point of the given differential equation by showing that $z = 0$ is a regular singular point for the transformed equation in z. Also find at least the first four nonzero terms in the series expansion about infinity of a solution to the original equation in x.

41. $x^3y'' - x^2y' - y = 0$.

42. $18(x - 4)^2(x - 6)y'' + 9x(x - 4)y' - 32y = 0$.

43. Show that if r_1 and r_2 are roots of the indicial equation (16), with r_1 the larger root (Re $r_1 \geq$ Re r_2), then the coefficient of a_1 in equation (19) is not zero when $r = r_1$.

44. To obtain a second linearly independent solution to equation (20):

 (a) Substitute $w(r, x)$ given in (21) into (20) and conclude that the coefficients a_k, $k \geq 1$, must

satisfy the recurrence relation
$$(k + r - 1)(2k + 2r - 1)a_k$$
$$+ \big[(k + r - 1)(k + r - 2) + 1\big]a_{k-1} = 0 .$$

 (b) Use the recurrence relation with $r = 1/2$ to derive the second series solution
$$w\left(\frac{1}{2}, x\right) = a_0\left(x^{1/2} - \frac{3}{4}x^{3/2} + \frac{7}{32}x^{5/2}\right.$$
$$\left. - \frac{133}{1920}x^{7/2} + \cdots\right) .$$

 (c) Use the recurrence relation with $r = 1$ to obtain $w(1, x)$ in (28).

45. In Example 5, show that if we choose $r = r_2 = -3$, then we obtain *two* linearly independent solutions to equation (45). [*Hint:* a_0 and a_3 are arbitrary constants.]

46. In Example 6, show that if we choose $r = r_2 = -2$, then we obtain a solution that is a constant multiple of the solution given in (62). [*Hint:* Show that a_0 and a_1 must be zero while a_2 is arbitrary.]

47. In applying the method of Frobenius, the following recurrence relation arose: $a_{k+1} = 15^7a_k/(k + 1)^9$, $k = 0, 1, 2, \ldots$.

 (a) Show that the coefficients are given by the formula $a_k = 15^{7k}a_0/(k!)^9$, $k = 0, 1, 2, \ldots$.

 (b) Use the formula obtained in part (a) with $a_0 = 1$ to compute $a_5, a_{10}, a_{15}, a_{20}$, and a_{25} on your computer or calculator. What goes wrong?

 (c) Now use the recurrence relation to compute a_k for $k = 1, 2, 3, \ldots, 25$, assuming $a_0 = 1$.

 (d) What advantage does the recurrence relation have over the formula?

8.7 FINDING A SECOND LINEARLY INDEPENDENT SOLUTION

In the previous section we showed that if $x = 0$ is a regular singular point of

(1) $y''(x) + p(x)y'(x) + q(x)y(x) = 0$, $x > 0$,

then the method of Frobenius can be used to find a series solution valid for x near zero. The first step in the method is to find the roots r_1 and r_2 (Re $r_1 \geq$ Re r_2) of the associated indicial equation

(2) $r(r - 1) + p_0 r + q_0 = 0$.

Then, utilizing the larger root r_1, equation (1) has a series solution of the form

$$(3) \qquad w(r_1, x) = x^{r_1} \sum_{n=0}^{\infty} a_n x^n = \sum_{n=0}^{\infty} a_n x^{n+r_1} \ ,$$

where $a_0 \neq 0$. To find a second linearly independent solution, our first inclination is to set $r = r_2$ and seek a solution of the form

$$(4) \qquad w(r_2, x) = x^{r_2} \sum_{n=0}^{\infty} a_n x^n = \sum_{n=0}^{\infty} a_n x^{n+r_2} \ .$$

We shall see that this procedure works, *provided $r_1 - r_2$ is not an integer*. However, when $r_1 - r_2$ *is* an integer, the Frobenius method with $r = r_2$ may just lead to the same solution that we obtained using the root r_1. (This is obviously true when $r_1 = r_2$.)

EXAMPLE 1 Find the first few terms in the series expansion about the regular singular point $x = 0$ for a general solution to

$$(5) \qquad (x + 2)x^2 y''(x) - xy'(x) + (1 + x)y(x) = 0 \ , \qquad x > 0 \ .$$

SOLUTION In Example 3 of Section 8.6, we used the method of Frobenius to find a series solution for (5). In the process we determined that $p_0 = -1/2$, $q_0 = 1/2$ and that the indicial equation has roots $r_1 = 1$, $r_2 = 1/2$. Since these roots do not differ by an integer $(r_1 - r_2 = 1/2)$, the method of Frobenius will give two linearly independent solutions of the form

$$(6) \qquad w(r, x) := x^r \sum_{n=0}^{\infty} a_n x^n = \sum_{n=0}^{\infty} a_n x^{n+r} \ .$$

In Problem 44 of Exercises 8.6, the reader was requested to show that substituting $w(r, x)$ into (5) leads to the recurrence relation

$$(7) \qquad (k + r - 1)(2k + 2r - 1)a_k + \big[(k + r - 1)(k + r - 2) + 1 \big] a_{k-1} = 0 \ , \qquad k \geq 1 \ .$$

With $r = r_1 = 1$ and $a_0 = 1$, we find

$$(8) \qquad y_1(x) = x - \frac{1}{3}x^2 + \frac{1}{10}x^3 - \frac{1}{30}x^4 + \cdots$$

as obtained in the previous section. Moreover, taking $r = r_2 = 1/2$ and $a_0 = 1$ in (7) leads to the second solution,

$$(9) \qquad y_2(x) = x^{1/2} - \frac{3}{4}x^{3/2} + \frac{7}{32}x^{5/2} - \frac{133}{1920}x^{7/2} + \cdots \ .$$

Consequently a general solution to equation (5) is

$$(10) \qquad y(x) = c_1 y_1(x) + c_2 y_2(x) \ , \qquad x > 0 \ ,$$

where $y_1(x)$ and $y_2(x)$ are the series solutions given in equations (8) and (9). See Figure 8.9. ■ End

When the indicial equation has repeated roots, $r_1 = r_2$, obviously substituting $r = r_2$ just gives us back the first solution and gets us nowhere. However, in such a case, we can use the reduction of order formula discussed in Section 4.4 to obtain a second linearly independent solution.

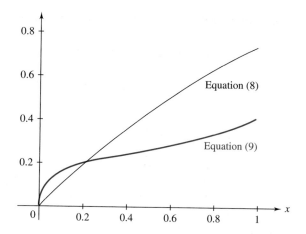

Figure 8.9 Partial sums approximating solutions to Example 1

Recall that if $f(x)$ is a nontrivial solution to equation (1), then a second linearly independent solution is given by the reduction of order formula

(11) $$y(x) = f(x) \int \frac{e^{-\int p(x)dx}}{[f(x)]^2} \, dx \ .$$

In the next example we use this formula to hunt for a second series solution.

EXAMPLE 2 Find the first few terms in the series expansion about the regular singular point $x = 0$ for a general solution to

(12) $$x^2 y''(x) - xy'(x) + (1 - x)y(x) = 0 , \qquad x > 0 \ .$$

SOLUTION In Example 4 of Section 8.6, we used the method of Frobenius to obtain a series solution to equation (12). In the process, we found the indicial equation to be $r^2 - 2r + 1 = 0$, which has roots $r_1 = r_2 = 1$. Working with $r_1 = 1$, we derived the series solution

(13) $$y_1(x) = x + x^2 + \frac{1}{4}x^3 + \frac{1}{36}x^4 + \frac{1}{576}x^5 + \cdots = \sum_{k=0}^{\infty} \frac{1}{(k!)^2} x^{k+1}$$

(see equation (44) in Section 8.6 with $a_0 = 1$).

Since $p(x) = -x^{-1}$ for equation (12), we can substitute $f(x) = y_1(x)$ and $p(x) = -x^{-1}$ into the reduction of order formula (11) to obtain

(14) $$y_2(x) = y_1(x) \int \frac{e^{\int x^{-1}dx}}{[y_1(x)]^2} dx = y_1(x) \int \frac{e^{\ln x}}{[y_1(x)]^2} dx$$

$$= y_1(x) \int \frac{x}{\left[x + x^2 + \frac{1}{4}x^3 + \frac{1}{36}x^4 + \frac{1}{576}x^5 + \cdots \right]^2} dx \ .$$

Using the Cauchy product to square the series in the denominator and canceling out an x gives

(15) $$y_2(x) = y_1(x) \int \frac{x}{\left[x^2 + 2x^3 + \frac{3}{2}x^4 + \frac{5}{9}x^5 + \frac{35}{288}x^6 + \cdots \right]} dx$$

$$= y_1(x) \int \left\{ \frac{1}{x + 2x^2 + \frac{3}{2}x^3 + \frac{5}{9}x^4 + \frac{35}{288}x^5 + \cdots} \right\} dx \ .$$

Next, we can use long division to compute the power series for the quotient in the braces, thereby obtaining

(16) $$y_2(x) = y_1(x) \int \left\{ \frac{1}{x} - 2 + \frac{5}{2}x - \frac{23}{9}x^2 + \cdots \right\} dx \ .$$

Integrating term by term, we then find

(17) $$y_2(x) = y_1(x) \left\{ \ln x - 2x + \frac{5}{4}x^2 - \frac{23}{27}x^3 + \cdots \right\}$$

$$= y_1(x) \ln x + y_1(x) \left(-2x + \frac{5}{4}x^2 - \frac{23}{27}x^3 + \cdots \right) \ .$$

Now we substitute the series for $y_1(x)$ given in (13) and use the Cauchy product to obtain

(18) $$y_2(x) = y_1(x) \ln x + \left(x + x^2 + \frac{1}{4}x^3 + \cdots \right) \left(-2x + \frac{5}{4}x^2 - \frac{23}{27}x^3 + \cdots \right)$$

$$= y_1(x) \ln x + \left(-2x^2 - \frac{3}{4}x^3 - \frac{11}{108}x^4 + \cdots \right) \ .$$

Hence a general solution to equation (12) is

(19) $$y(x) = c_1 y_1(x) + c_2 y_2(x) \ , \qquad x > 0 \ ,$$

where c_1 and c_2 are arbitrary constants and $y_1(x)$ and $y_2(x)$ are given by (13) and (18). See Figure 8.10. ■End

Example 2 illustrates the drawback to using the reduction of order formula to find a second linearly independent solution: We must square a series and then compute its reciprocal. As a result, it is very difficult to determine the general term in the series expansion for $y_2(x)$.

In that example, the roots of the indicial equation were equal and we found a second linearly independent solution (18) that involves the first solution $y_1(x)$ multiplied by $\ln x$. This should not be too surprising if we recall the analogous situation for a Cauchy-Euler equation having an indicial equation with repeated roots (page 480). As we shall soon see, if an indicial equation has roots that differ by an integer, then the expansion of the second solution may also involve the term $y_1(x) \ln x$. In the following theorem, we give the *general form* of two linearly independent solutions for the three cases where the roots of the indicial equation (a) do not differ by an integer, (b) are equal, or (c) differ by a nonzero integer.

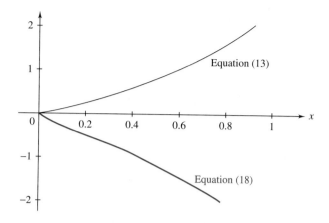

Figure 8.10 Partial sum approximations to solutions of Example 2

FORM OF SECOND LINEARLY INDEPENDENT SOLUTION

Theorem 7.[†] Let x_0 be a regular singular point for $y'' + py' + qy = 0$ and let r_1 and r_2 be the roots of the associated indicial equation, where Re $r_1 \geq$ Re r_2.

(a) If $r_1 - r_2$ is not an integer, then there exist two linearly independent solutions of the form

(20) $\displaystyle y_1(x) = \sum_{n=0}^{\infty} a_n(x - x_0)^{n+r_1}$, $a_0 \neq 0$,

(21) $\displaystyle y_2(x) = \sum_{n=0}^{\infty} b_n(x - x_0)^{n+r_2}$, $b_0 \neq 0$.

(b) If $r_1 = r_2$, then there exist two linearly independent solutions of the form

(22) $\displaystyle y_1(x) = \sum_{n=0}^{\infty} a_n(x - x_0)^{n+r_1}$, $a_0 \neq 0$,

(23) $\displaystyle y_2(x) = y_1(x) \ln(x - x_0) + \sum_{n=1}^{\infty} b_n(x - x_0)^{n+r_1}$.

(c) If $r_1 - r_2$ is a positive integer, then there exist two linearly independent solutions of the form

(24) $\displaystyle y_1(x) = \sum_{n=0}^{\infty} a_n(x - x_0)^{n+r_1}$, $a_0 \neq 0$,

(25) $\displaystyle y_2(x) = C y_1(x) \ln(x - x_0) + \sum_{n=0}^{\infty} b_n(x - x_0)^{n+r_2}$, $b_0 \neq 0$,

where C is a constant that could be zero.

[†]For a proof of Theorem 7, see the text *Ordinary Differential Equations,* 4th edition, by G. Birkhoff and G.-C. Rota (John Wiley & Sons, New York, 1989).

In each case of the theorem, $y_1(x)$ is just the series solution obtained by the method of Frobenius, with $r = r_1$. When $r_1 - r_2$ is not an integer, the method of Frobenius yields a second linearly independent solution by taking $r = r_2$. Let's see how knowing the form of the second solution enables us to obtain it. Again, for simplicity, we consider only indicial equations having real roots.

EXAMPLE 3 Find the first few terms in the series expansion about the regular singular point $x = 0$ for two linearly independent solutions to

(26) $x^2 y''(x) - xy'(x) + (1 - x)y(x) = 0 , \qquad x > 0 .$

SOLUTION This is the same equation we considered in Example 2, where we used the reduction of order formula to find a second linearly independent solution. This time let's try to find a second solution by using the information in Theorem 7. Recall that $r_1 = r_2 = 1$ and so $y_2(x)$ has the form given in (23) (with $x_0 = 0$); that is,

(27) $\displaystyle y_2(x) = y_1(x) \ln x + \sum_{n=1}^{\infty} b_n x^{n+1} ,$

where (cf. (13))

(28) $\displaystyle y_1(x) = \sum_{k=0}^{\infty} \frac{1}{(k!)^2} x^{k+1} .$

Our goal is to determine the coefficients b_n by substituting $y_2(x)$ directly into equation (26). We begin by differentiating $y_2(x)$ in (27) to obtain

$$y_2'(x) = y_1'(x) \ln x + x^{-1} y_1(x) + \sum_{n=1}^{\infty} (n + 1)b_n x^n ,$$

$$y_2''(x) = y_1''(x) \ln x - x^{-2} y_1(x) + 2x^{-1} y_1'(x) + \sum_{n=1}^{\infty} n(n + 1)b_n x^{n-1} .$$

Substituting $y_2(x)$ into (26) yields

(29) $\displaystyle x^2 \left\{ y_1''(x) \ln x - x^{-2} y_1(x) + 2x^{-1} y_1'(x) + \sum_{n=1}^{\infty} n(n + 1)b_n x^{n-1} \right\}$

$$- x \left\{ y_1'(x) \ln x + x^{-1} y_1(x) + \sum_{n=1}^{\infty} (n + 1)b_n x^n \right\}$$

$$+ (1 - x) \left\{ y_1(x) \ln x + \sum_{n=1}^{\infty} b_n x^{n+1} \right\} = 0 ,$$

which simplifies to

(30) $\{x^2 y_1''(x) - xy_1'(x) + (1 - x)y_1(x)\} \ln x - 2y_1(x) + 2xy_1'(x)$

$$+ \sum_{n=1}^{\infty} n(n + 1)b_n x^{n+1} - \sum_{n=1}^{\infty} (n + 1)b_n x^{n+1} + \sum_{n=1}^{\infty} b_n x^{n+1} - \sum_{n=1}^{\infty} b_n x^{n+2} = 0 .$$

Notice that the factor in front of $\ln x$ is just the left-hand side of equation (26) with $y = y_1$. Since y_1 is a solution to (26), this factor is zero. With this observation and a shift in the indices of summation, equation (30) can be rewritten as

(31) $2xy_1'(x) - 2y_1(x) + b_1 x^2 + \sum_{k=2}^{\infty} (k^2 b_k - b_{k-1}) x^{k+1} = 0$.

To identify the coefficients in (31) so that we can set them equal to zero, we must substitute back in the series expansions for $y_1(x)$ and $y_1'(x)$. From (28), we see that $y_1'(x) = \sum_{k=0}^{\infty} (k+1) x^k / (k!)^2$. Inserting this series together with (28) into (31), we find

(32) $\sum_{k=0}^{\infty} \frac{2(k+1) - 2}{(k!)^2} x^{k+1} + b_1 x^2 + \sum_{k=2}^{\infty} (k^2 b_k - b_{k-1}) x^{k+1} = 0$.

Listing separately the $k = 0$ and $k = 1$ terms and combining the remaining terms gives

(33) $(2 + b_1) x^2 + \sum_{k=2}^{\infty} \left[\frac{2k}{(k!)^2} + k^2 b_k - b_{k-1} \right] x^{k+1} = 0$.

Next, we set the coefficients in (33) equal to zero. From the x^2 term we have $2 + b_1 = 0$, so $b_1 = -2$. From the x^{k+1} term, we obtain

$$\frac{2k}{(k!)^2} + k^2 b_k - b_{k-1} = 0$$

or

(34) $b_k = \frac{1}{k^2} \left[b_{k-1} - \frac{2k}{(k!)^2} \right]$, $k \geq 2$.

Taking $k = 2$ and 3, we compute

(35) $b_2 = \frac{1}{2^2} [b_1 - 1] = \frac{-3}{4}$, $b_3 = \frac{1}{9} \left[-\frac{3}{4} - \frac{6}{36} \right] = \frac{-11}{108}$.

Hence a second linearly independent solution to (26) is

(36) $y_2(x) = y_1(x) \ln x - 2x^2 - \frac{3}{4} x^3 - \frac{11}{108} x^4 + \cdots$. ■End

If we compare the different methods used in Examples 2 and 3 for obtaining a second linearly independent solution, we find that the technique of Example 3 has an advantage: It yields a recurrence relation, equation (34), that can be used to solve recursively for the coefficients b_k.

In the next two examples we consider the case when the difference between the roots of the indicial equation is a positive integer. In Example 4, it will turn out that the constant C in formula (25) must be taken to be zero (i.e., no $\ln x$ term is present), while in Example 5, this constant is nonzero (i.e., a $\ln x$ term is present). Since the solutions to these examples require several intermediate computations, we do not display all the details. Rather, we encourage the reader to take an active part by bridging the gaps.

EXAMPLE 4 Find the first few terms in the series expansion about the regular singular point $x = 0$ for a general solution to

(37) $xy''(x) + 4y'(x) - xy(x) = 0$, $x > 0$.

SOLUTION In Example 5 of Section 8.6, we used the method of Frobenius to find a series expansion about $x = 0$ for a solution to equation (37). There we found the indicial equation to be $r^2 + 3r = 0$, which has roots $r_1 = 0$ and $r_2 = -3$. Working with $r_1 = 0$, we obtained the series solution

(38) $y_1(x) = 1 + \dfrac{1}{10}x^2 + \dfrac{1}{280}x^4 + \cdots$

(see equation (53) in Section 8.6, with $a_0 = 1$).

Since $r_1 - r_2 = 3$ is a positive integer, it follows from Theorem 7 that equation (37) has a second linearly independent solution of the form

(39) $y_2(x) = Cy_1(x) \ln x + \displaystyle\sum_{n=0}^{\infty} b_n x^{n-3}$.

When we substitute this expression for y_2 into equation (37), we obtain

(40) $x\left\{ Cy_1''(x) \ln x + 2Cx^{-1}y_1'(x) - Cx^{-2}y_1(x) + \displaystyle\sum_{n=0}^{\infty} (n-3)(n-4)b_n x^{n-5} \right\}$

$+ 4\left\{ Cy_1'(x) \ln x + Cx^{-1}y_1(x) + \displaystyle\sum_{n=0}^{\infty} (n-3)b_n x^{n-4} \right\}$

$- x\left\{ Cy_1(x) \ln x + \displaystyle\sum_{n=0}^{\infty} b_n x^{n-3} \right\} = 0$,

which simplifies to

(41) $\{xy_1''(x) + 4y_1'(x) - xy_1(x)\}C \ln x + 3Cx^{-1}y_1(x) + 2Cy_1'(x)$

$+ \displaystyle\sum_{n=0}^{\infty} (n-3)(n-4)b_n x^{n-4} + \displaystyle\sum_{n=0}^{\infty} 4(n-3)b_n x^{n-4} - \displaystyle\sum_{n=0}^{\infty} b_n x^{n-2} = 0$.

The factor in braces is zero because $y_1(x)$ satisfies equation (37). After we combine the summations and simplify, equation (41) becomes

(42) $3Cx^{-1}y_1(x) + 2Cy_1'(x) - 2b_1x^{-3} + \displaystyle\sum_{k=2}^{\infty} \left[k(k-3)b_k - b_{k-2} \right] x^{k-4} = 0$.

Substituting in the series for $y_1(x)$ and writing out the first few terms of the summation in (42), we obtain

$-2b_1x^{-3} + (-2b_2 - b_0)x^{-2} + (3C - b_1)x^{-1} + (4b_4 - b_2)$

$+ \left(\dfrac{7}{10}C + 10b_5 - b_3 \right)x + (18b_6 - b_4)x^2 + \left(\dfrac{11}{280}C + 28b_7 - b_5 \right)x^3$

$+ \cdots = 0$.

Next, we set the coefficients equal to zero:

$$-2b_1 = 0 \Rightarrow b_1 = 0 \ , \qquad -2b_2 - b_0 = 0 \Rightarrow b_2 = -\frac{1}{2}b_0 \ ,$$

$$3C - b_1 = 0 \Rightarrow C = \frac{1}{3}b_1 = 0 \ , \qquad 4b_4 - b_2 = 0 \Rightarrow b_4 = \frac{1}{4}b_2 = -\frac{1}{8}b_0 \ ,$$

$$\frac{7}{10}C + 10b_5 - b_3 = 0 \Rightarrow b_5 = \frac{b_3 - \frac{7}{10}C}{10} = \frac{1}{10}b_3 \ ,$$

$$18b_6 - b_4 = 0 \Rightarrow b_6 = \frac{1}{18}b_4 = -\frac{1}{144}b_0 \ ,$$

$$\frac{11}{280}C + 28b_7 - b_5 = 0 \Rightarrow b_7 = \frac{b_5 - \frac{11}{280}C}{28} = \frac{1}{280}b_3 \ .$$

Note, in particular, that C must equal zero. Substituting the above values for the b_n's and $C = 0$ back into equation (39) gives

(43) $\qquad y_2(x) = b_0\left\{ x^{-3} - \frac{1}{2}x^{-1} - \frac{1}{8}x - \frac{1}{144}x^3 + \cdots \right\}$

$$+ b_3\left\{ 1 + \frac{1}{10}x^2 + \frac{1}{280}x^4 + \cdots \right\} \ ,$$

where b_0 and b_3 are arbitrary constants. Observe that the expression in braces following b_3 is just the series expansion for $y_1(x)$ given in equation (38). Hence, in order to obtain a second linearly independent solution, we must choose b_0 to be nonzero. Taking $b_0 = 1$ and $b_3 = 0$ gives

(44) $\qquad y_2(x) = x^{-3} - \frac{1}{2}x^{-1} - \frac{1}{8}x - \frac{1}{144}x^3 + \cdots \ .$

Therefore a general solution to equation (37) is

(45) $\qquad y(x) = c_1 y_1(x) + c_2 y_2(x) \ , \qquad x > 0 \ ,$

where $y_1(x)$ and $y_2(x)$ are given in (38) and (44). (Notice that the right-hand side of (43) coincides with (45) if we identify b_0 as c_2 and b_3 as c_1.) See Figure 8.11 on page 504. ■End

EXAMPLE 5 Find the first few terms in the series expansion about the regular singular point $x = 0$ for two linearly independent solutions to

(46) $\qquad xy''(x) + 3y'(x) - xy(x) = 0 \ , \qquad x > 0 \ .$

SOLUTION In Example 6 of Section 8.6, we used the method of Frobenius to find a series expansion about $x = 0$ for a solution to equation (46). The indicial equation turned out to be $r^2 + 2r = 0$,

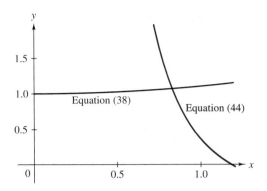

Figure 8.11 Partial sum approximations to solutions of Example 4

which has roots $r_1 = 0$ and $r_2 = -2$. Using $r_1 = 0$, we obtained the series solution

(47) $\qquad y_1(x) = 1 + \dfrac{1}{8}x^2 + \dfrac{1}{192}x^4 + \dfrac{1}{9216}x^6 + \cdots$

(see equation (62) in Section 8.6 and put $a_0 = 1$).

Because $r_1 - r_2 = 2$ is a positive integer, it follows from Theorem 7 that equation (46) has a second linearly independent solution of the form

(48) $\qquad y_2(x) = Cy_1(x)\ln x + \displaystyle\sum_{n=0}^{\infty} b_n x^{n-2} .$

Plugging the expansion for $y_2(x)$ into equation (46) and simplifying yields

(49) $\qquad \{xy_1''(x) + 3y_1'(x) - xy_1(x)\}C\ln x + 2Cx^{-1}y_1(x) + 2Cy_1'(x)$

$\qquad\qquad + \displaystyle\sum_{n=0}^{\infty}(n-2)(n-3)b_n x^{n-3} + \sum_{n=0}^{\infty}3(n-2)b_n x^{n-3} - \sum_{n=0}^{\infty}b_n x^{n-1} = 0 .$

Again the factor in braces is zero because $y_1(x)$ is a solution to equation (46). If we combine the summations and simplify, equation (49) becomes

(50) $\qquad 2Cx^{-1}y_1(x) + 2Cy_1'(x) - b_1 x^{-2} + \displaystyle\sum_{k=2}^{\infty}\big[k(k-2)b_k - b_{k-2}\big]x^{k-3} = 0 .$

Substituting in the series expansions for $y_1(x)$ and $y_1'(x)$ and writing out the first few terms of the summation in (50) leads to

(51) $\qquad -b_1 x^{-2} + (2C - b_0)x^{-1} + (3b_3 - b_1) + \left(\dfrac{3}{4}C + 8b_4 - b_2\right)x$

$\qquad\qquad + (15b_5 - b_3)x^2 + \left(\dfrac{5}{96}C + 24b_6 - b_4\right)x^3 + \cdots = 0 .$

When we set the coefficients in (51) equal to zero, it turns out that we are free to choose C and

b_2 as arbitrary constants:

$$-b_1 = 0 \Rightarrow b_1 = 0 \ ,$$

$$2C - b_0 = 0 \Rightarrow b_0 = 2C \qquad (C \text{ arbitrary}),$$

$$3b_3 - b_1 = 0 \Rightarrow b_3 = \frac{1}{3}b_1 = 0 \ ,$$

$$8b_4 - b_2 + \frac{3}{4}C = 0 \Rightarrow b_4 = \frac{b_2 - \frac{3}{4}C}{8} = \frac{1}{8}b_2 - \frac{3}{32}C \qquad (b_2 \text{ arbitrary}),$$

$$15b_5 - b_3 = 0 \Rightarrow b_5 = \frac{1}{15}b_3 = 0 \ ,$$

$$24b_6 - b_4 + \frac{5}{96}C = 0 \Rightarrow b_6 = \frac{b_4 - \frac{5}{96}C}{24} = \frac{1}{192}b_2 - \frac{7}{1152}C \ .$$

Substituting these values for the b_n's back into (48), we obtain the solution

(52) $$y_2(x) = C\left\{ y_1(x) \ln x + 2x^{-2} - \frac{3}{32}x^2 - \frac{7}{1152}x^4 + \cdots \right\}$$

$$+ b_2\left\{ 1 + \frac{1}{8}x^2 + \frac{1}{192}x^4 + \cdots \right\} \ ,$$

where C and b_2 are arbitrary constants. Since the factor multiplying b_2 is the first solution $y_1(x)$, we can obtain a second linearly independent solution by choosing $C = 1$ and $b_2 = 0$:

(53) $$y_2(x) = y_1(x) \ln x + 2x^{-2} - \frac{3}{32}x^2 - \frac{7}{1152}x^4 + \cdots \ .$$

See Figure 8.12. ∎End

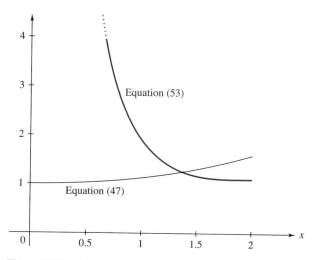

Figure 8.12 Partial sum approximations to the solutions of Example 5

In closing we note that if the roots r_1 and r_2 of the indicial equation associated with a differential equation are complex, then they are complex conjugates. Thus the difference $r_1 - r_2$ is imaginary and, hence, not an integer, and we are in case (a) of Theorem 7. However, rather than employing the display (20)–(21) for the linearly independent solutions, one usually takes the real and imaginary parts of (20); Problem 26 provides an elaboration of this situation.

EXERCISES 8.7

In Problems 1 through 14, find at least the first three nonzero terms in the series expansion about $x = 0$ for a general solution to the given equation for $x > 0$. (These are the same equations as in Problem 19 through 32 of Exercises 8.6.)

1. $9x^2y'' + 9x^2y' + 2y = 0$.
2. $2x(x - 1)y'' + 3(x - 1)y' - y = 0$.
3. $x^2y'' + xy' + x^2y = 0$.
4. $xy'' + y' - 4y = 0$.
5. $x^2z'' + (x^2 + x)z' - z = 0$.
6. $3xy'' + (2 - x)y' - y = 0$.
7. $4x^2y'' + 2x^2y' - (x + 3)y = 0$.
8. $x^2y'' + (x^2 - x)y' + y = 0$.
9. $xw'' - w' - xw = 0$.
10. $3x^2y'' + 8xy' + (x - 2)y = 0$.
11. $xy'' + (x - 1)y' - 2y = 0$.
12. $x(x + 1)y'' + (x + 5)y' - 4y = 0$.
13. $xy'' + (1 - x)y' - y = 0$.
14. $x^2y'' - x(1 + x)y' + y = 0$.

In Problems 15 and 16, determine whether the given equation has a solution that is bounded near the origin, all solutions are bounded near the origin, or none of the solutions are bounded near the origin. (These are the same equations as in Problems 33 and 34 of Exercises 8.6.) Note that you need to analyze only the indicial equation in order to answer the question.

15. $3xy'' + 2(1 - x)y' - 4y = 0$.
16. $xy'' + (x + 2)y' - y = 0$.

In Problems 17 through 20, find at least the first three nonzero terms in the series expansion about $x = 0$ for a general solution to the given linear third order equation

for $x > 0$. (These are the same equations as in Problems 35 through 38 in Exercises 8.6.)

17. $6x^3y''' + 13x^2y'' + (x + x^2)y' + xy = 0$.
18. $6x^3y''' + 11x^2y'' - 2xy' - (x - 2)y = 0$.
19. $6x^3y''' + 13x^2y'' - (x^2 + 3x)y' - xy = 0$.
20. $6x^3y''' + (13x^2 - x^3)y'' + xy' - xy = 0$.

21. **Buckling Columns.** In the study of the buckling of a column whose cross section varies, one encounters the equation

 (54) $\quad x^n y''(x) + \alpha^2 y(x) = 0$, $\quad x > 0$,

 where x is related to the height above the ground and y is the deflection away from the vertical. The positive constant α depends on the rigidity of the column, its moment of inertia at the top, and the load. The positive integer n depends on the type of column. For example, when the column is a truncated cone (see Figure 8.13(a)), we have $n = 4$.

(a) Truncated cone

(b) Truncated pyramid with fixed thickness T and varying width

Figure 8.13 Buckling columns

(a) Use the substitution $x = t^{-1}$ to reduce (54) with $n = 4$ to the form

$$\frac{d^2y}{dt^2} + \frac{2}{t}\frac{dy}{dt} + \alpha^2 y = 0 , \qquad t > 0 .$$

(b) Find at least the first six nonzero terms in the series expansion about $t = 0$ for a general solution to the equation obtained in part (a).

(c) Use the result of part (b) to give an expansion about $x = \infty$ for a general solution to (54).

22. In Problem 21, consider a column with a rectangular cross section with two sides constant and the other two changing linearly (see Figure 8.13(b)). In this case, $n = 1$. Find at least the first four nonzero terms in the series expansion about $x = 0$ for a general solution to equation (54) when $n = 1$.

23. Use the method of Frobenius and the reduction of order formula, equation (11), to find at least the first three nonzero terms in the series expansion about the irregular singular point $x = 0$ for a general solution to the differential equation

$$x^2 y'' + y' - 2y = 0 .$$

24. The equation

$$xy''(x) + (1 - x)y'(x) + ny(x) = 0 ,$$

where n is a nonnegative integer, is called **Laguerre's differential equation.** Show that for each n, this equation has a polynomial solution of degree n. These polynomials are denoted by $L_n(x)$ and are called **Laguerre polynomials.** The first few Laguerre polynomials are

$$L_0(x) = 1 , \qquad L_1(x) = -x + 1 .$$
$$L_2(x) = x^2 - 4x + 2 .$$

25. Use the results of Problem 24 and the reduction of order formula, equation (11), to obtain the first few terms in a series expansion about $x = 0$ for a general solution for $x > 0$ to Laguerre's differential equation for $n = 0$ and 1.

26. To obtain two linearly independent solutions to

$$(55) \qquad x^2 y'' + (x + x^2)y' + y = 0 , \qquad x > 0 ,$$

complete the following steps.

(a) Verify that (55) has a regular singular point at $x = 0$ and that the associated indicial equation has complex roots $\pm i$.

(b) As discussed in Section 8.5, we can express

$$x^{\alpha + i\beta} = x^\alpha x^{i\beta} = x^\alpha \cos(\beta \ln x) + ix^\alpha \sin(\beta \ln x) .$$

Deduce from this formula that $\dfrac{d}{dx}x^{\alpha + i\beta} = (\alpha + i\beta)x^{\alpha - 1 + i\beta}$.

(c) Set $y(x) = \sum_{n=0}^{\infty} a_n x^{n+i}$, where the coefficients now are *complex* constants, and substitute this series into equation (55) using the result of part (b).

(d) Setting the coefficients of like powers equal to zero, derive the recurrence relation

$$a_n = -\frac{n - 1 + i}{(n + i)^2 + 1}a_{n-1} , \qquad \text{for } n \geq 1 .$$

(e) Taking $a_0 = 1$, compute the coefficients a_1 and a_2 and thereby obtain the first few terms of a complex solution to (55).

(f) By computing the real and imaginary parts of the solution obtained in part (e), derive the following linearly independent real solutions to (55):

$$y_1(x) = \left[\cos(\ln x)\right]\left\{1 - \frac{2}{5}x + \frac{1}{10}x^2 + \cdots\right\}$$
$$+ \left[\sin(\ln x)\right]\left\{\frac{1}{5}x - \frac{1}{20}x^2 + \cdots\right\} ,$$

$$y_2(x) = \left[\cos(\ln x)\right]\left\{-\frac{1}{5}x + \frac{1}{20}x^2 + \cdots\right\}$$
$$+ \left[\sin(\ln x)\right]\left\{1 - \frac{2}{5}x + \frac{1}{10}x^2 + \cdots\right\} .$$

8.8 SPECIAL FUNCTIONS

In advanced work in applied mathematics, engineering, and physics, a few special second order equations arise very frequently. These equations have been extensively studied, and volumes have been written on the properties of their solutions, which are often referred to as **special functions.**

For reference purposes we include a brief survey of three of these equations: the hypergeometric equation, Bessel's equation, and Legendre's equation. Bessel's equation governs the radial dependence of the solutions to the classical partial differential equations of physics in spherical coordinate systems (see Section 10.7, page 670); Legendre's equation governs their latitudinal dependence. C. F. Gauss formulated the hypergeometric equation as a generic equation whose solutions include the special functions of Legendre, Chebyshev, Gegenbauer, and Jacobi. For a more detailed study of special functions, we refer the reader to *Basic Hypergeometric Series,* by G. Gasper and M. Rahman (Cambridge University Press, Cambridge, 1990); *Special Functions,* by E. D. Rainville (Macmillan, New York, 1960); and *Higher Transcendental Functions,* by A. Erdelyi (ed.) (McGraw-Hill, New York, 1953), 3 volumes.

Hypergeometric Equation

The linear second order differential equation

$$(1) \qquad x(1 - x)y'' + \left[\gamma - (\alpha + \beta + 1)x\right]y' - \alpha\beta y = 0 \ ,$$

where α, β, and γ are fixed parameters, is called the **hypergeometric equation.** This equation has singular points at $x = 0$ and 1, both of which are regular. Thus a series expansion about $x = 0$ for a solution to (1) obtained by the method of Frobenius will converge at least for $0 < x < 1$ (see Theorem 6, page 490). To find this expansion, observe that the indicial equation associated with $x = 0$ is

$$r(r - 1) + \gamma r = r(r - (1 - \gamma)) = 0 \ ,$$

which has roots 0 and $1 - \gamma$. Let us assume γ is not an integer, and use the root $r = 0$ to obtain a solution to (1) of the form

$$(2) \qquad y_1(x) = \sum_{n=0}^{\infty} a_n x^n \ .$$

Substituting $y_1(x)$ given in (2) into (1), shifting indices, and simplifying ultimately leads to the equation

$$(3) \qquad \sum_{n=1}^{\infty} \left[n(n + \gamma - 1)a_n - (n + \alpha - 1)(n + \beta - 1)a_{n-1}\right]x^{n-1} = 0 \ .$$

Setting the series coefficients equal to zero yields the recurrence relation

$$(4) \qquad n(n + \gamma - 1)a_n - (n + \alpha - 1)(n + \beta - 1)a_{n-1} = 0 \ , \qquad n \geq 1 \ .$$

Since $n \geq 1$ and γ is not an integer, there is no fear of dividing by zero when we rewrite (4) as

$$(5) \qquad a_n = \frac{(n + \alpha - 1)(n + \beta - 1)}{n(n + \gamma - 1)}a_{n-1} \ .$$

Solving recursively for a_n, we obtain

$$(6) \qquad a_n = \frac{\alpha(\alpha + 1)\cdots(\alpha + n - 1)\beta(\beta + 1)\cdots(\beta + n - 1)}{n!\gamma(\gamma + 1)\cdots(\gamma + n - 1)}a_0 \ , \qquad n \geq 1 \ .$$

If we employ the **factorial function** $(t)_n$, which is defined for nonnegative integers n by

(7) $(t)_n := t(t + 1)(t + 2) \cdots (t + n - 1)$, $n \geq 1$,

$(t)_0 := 1$, $t \neq 0$,

then we can express a_n more compactly as

(8) $a_n = \dfrac{(\alpha)_n(\beta)_n}{n!(\gamma)_n} a_0$, $n \geq 1$.

(In fact, we can write $n!$ as $(1)_n$.) If we take $a_0 = 1$ and then substitute the expression for a_n in (8) into (2), we obtain the following solution to the hypergeometric equation:

(9) $y_1(x) = 1 + \displaystyle\sum_{n=1}^{\infty} \frac{(\alpha)_n(\beta)_n}{n!(\gamma)_n} x^n$.

The solution given in (9) is called a **Gaussian hypergeometric function** and is denoted by the symbol $F(\alpha, \beta; \gamma; x)$. [†] That is,

(10) $F(\alpha, \beta; \gamma; x) := 1 + \displaystyle\sum_{n=1}^{\infty} \frac{(\alpha)_n(\beta)_n}{n!(\gamma)_n} x^n$.

Hypergeometric functions are generalizations of the geometric series. To see this, observe that for any constant β that is not zero or a negative integer,

$F(1, \beta; \beta; x) = 1 + x + x^2 + x^3 + \cdots$.

It is interesting to note that many other familiar functions can be expressed in terms of the hypergeometric function. For example,

(11) $F(\alpha, \beta; \beta; x) = (1 - x)^{-\alpha}$,

(12) $F(1, 1; 2; x) = -x^{-1} \ln(1 - x)$,

(13) $F\left(\dfrac{1}{2}, 1; \dfrac{3}{2}; x^2\right) = \dfrac{1}{2} x^{-1} \ln\left(\dfrac{1 + x}{1 - x}\right)$,

(14) $F\left(\dfrac{1}{2}, 1; \dfrac{3}{2}; -x^2\right) = x^{-1} \arctan x$.

We leave the verification of these formulas for the reader.

To obtain a second linearly independent solution to (1) when γ is not an integer, we use the other root, $1 - \gamma$, of the indicial equation and seek a solution of the form

(15) $y_2(x) = \displaystyle\sum_{n=0}^{\infty} b_n x^{n+1-\gamma}$.

[†]*Historical Footnote:* A detailed study of this function was done by Carl Friedrich Gauss in 1813. In *Men of Mathematics* (Dover Publications, Inc., New York, 1937), the mathematical historian E. T. Bell refers to Gauss as the Prince of Mathematicians.

Substituting $y_2(x)$ into equation (1) and solving for b_n, we eventually arrive at

$$(16) \qquad y_2(x) = x^{1-\gamma} + \sum_{n=1}^{\infty} \frac{(\alpha + 1 - \gamma)_n (\beta + 1 - \gamma)_n}{n!(2-\gamma)_n} x^{n+1-\gamma} \ .$$

Factoring out $x^{1-\gamma}$, we see that the second solution $y_2(x)$ can be expressed in terms of a hypergeometric function. That is,

$$(17) \qquad y_2(x) = x^{1-\gamma} F(\alpha + 1 - \gamma, \beta + 1 - \gamma; 2 - \gamma; x) \ .$$

When γ is an integer, one of the formulas given in (9) or (16) (corresponding to the larger root, 0 or $1 - \gamma$) still gives a solution. We then use the techniques of this chapter to obtain a second linearly independent solution, which may or may not involve a logarithmic term. We omit a discussion of these solutions.

In many books, the hypergeometric function is expressed in terms of the gamma function $\Gamma(x)$ instead of the factorial function. Recall that in Section 7.6, we defined

$$(18) \qquad \Gamma(x) := \int_0^{\infty} e^{-u} u^{x-1} \, du \ , \qquad x > 0$$

and showed that

$$(19) \qquad \Gamma(x + 1) = x\Gamma(x) \ , \qquad x > 0 \ .$$

It follows from repeated use of relation (19) that

$$(20) \qquad (t)_n = \frac{\Gamma(t + n)}{\Gamma(t)}$$

for $t > 0$ and n any nonnegative integer. Using relation (20), we can express the hypergeometric function as

$$(21) \qquad F(\alpha, \beta; \gamma; x) = \frac{\Gamma(\gamma)}{\Gamma(\alpha)\Gamma(\beta)} \sum_{n=0}^{\infty} \frac{\Gamma(\alpha + n)\Gamma(\beta + n)}{n!\Gamma(\gamma + n)} x^n \ .$$

Bessel's Equation

The linear second order differential equation

$$(22) \qquad x^2 y'' + xy' + (x^2 - \nu^2)y = 0 \ ,$$

where $\nu \geq 0$ is a fixed parameter, is called **Bessel's equation of order ν**. This equation has a regular singular point at $x = 0$ and no other singular points in the complex plane. Hence a series solution for (22) obtained by the method of Frobenius will converge for $0 < x < \infty$. The indicial equation for (22) is

$$r(r - 1) + r - \nu^2 = (r - \nu)(r + \nu) = 0 \ ,$$

which has roots $r_1 = \nu$ and $r_2 = -\nu$. If 2ν is not an integer, then the method of Frobenius yields two linearly independent solutions, given by

$$(23) \qquad y_1(x) = a_0 \sum_{n=0}^{\infty} \frac{(-1)^n}{2^{2n} n! (1 + \nu)_n} x^{2n+\nu} \ ,$$

(24) $y_2(x) = b_0 \sum\limits_{n=0}^{\infty} \dfrac{(-1)^n}{2^{2n} n! (1 - \nu)_n} x^{2n - \nu}$.

If in (23) we take

$$a_0 = \frac{1}{2^\nu \Gamma(1 + \nu)} \, ,$$

then it follows from relation (20) that the function

(25) $J_\nu(x) := \sum\limits_{n=0}^{\infty} \dfrac{(-1)^n}{n! \Gamma(1 + \nu + n)} \left(\dfrac{x}{2}\right)^{2n + \nu}$

is a solution to (22). We call $J_\nu(x)$ the **Bessel function of the first kind of order** ν.[†] Similarly, taking

$$b_0 = \frac{1}{2^{-\nu} \Gamma(1 - \nu)}$$

in equation (24) gives the solution

(26) $J_{-\nu}(x) := \sum\limits_{n=0}^{\infty} \dfrac{(-1)^n}{n! \Gamma(1 - \nu + n)} \left(\dfrac{x}{2}\right)^{2n - \nu}$,

which is the Bessel function of the first kind of order $-\nu$. When $r_1 - r_2 = 2\nu$ is not an integer, we know by Theorem 7 of Section 8.7 that $J_\nu(x)$ and $J_{-\nu}(x)$ are linearly independent. Moreover, it can be shown that if ν is not an integer, even though 2ν is, then $J_\nu(x)$ and $J_{-\nu}(x)$ are still linearly independent.

What happens in the remaining case when ν is a nonnegative integer, say $\nu = m$? While $J_m(x)$ is still a solution, the function $J_{-m}(x)$ is not even properly defined because formula (26) will involve the gamma function evaluated at a nonpositive integer. From a more in-depth study of the gamma function, it turns out that $1/\Gamma(k) = 0$, for $k = 0, -1, -2, \dots$. Hence (26) becomes

(27) $J_{-m}(x) = \sum\limits_{n=0}^{\infty} \dfrac{(-1)^n}{n! \Gamma(1 - m + n)} \left(\dfrac{x}{2}\right)^{2n - m} = \sum\limits_{n=m}^{\infty} \dfrac{(-1)^n}{n! \Gamma(1 - m + n)} \left(\dfrac{x}{2}\right)^{2n - m}$.

Comparing (27) with the formula (25) for $J_m(x)$, we see (after a shift in index) that

(28) $J_{-m}(x) = (-1)^m J_m(x)$,

which means that $J_{-m}(x)$ and $J_m(x)$ are linearly *dependent*. To resolve the problem of finding a second linearly independent solution in the case when ν is a nonnegative integer, we can use Theorem 7 in Section 8.7. There is, however, another approach to this problem, which we now describe.

For ν not an integer, we can take linear combinations of $J_\nu(x)$ and $J_{-\nu}(x)$ to obtain other

[†]*Historical Footnote:* Frederic Wilhelm Bessel (1784–1846) started his career in commercial navigation and later became an astronomer. In 1817, Bessel introduced the functions $J_\nu(x)$ in his study of planetary orbits.

solutions to (22). In particular, let

(29) $Y_\nu(x) := \dfrac{\cos(\nu\pi)J_\nu(x) - J_{-\nu}(x)}{\sin(\nu\pi)}$, ν not an integer,

for $x > 0$. The function $Y_\nu(x)$ is called the **Bessel function of the second kind of order ν,** and, as can be verified, $J_\nu(x)$ and $Y_\nu(x)$ are linearly independent. Notice that when ν is an integer, the denominator in (29) is zero; but, by formula (28), so is the numerator! Hence it is reasonable to hope that, in a limiting sense, formula (29) is still meaningful. In fact, using L'Hôpital's rule, it is possible to show that for m a nonnegative integer, the function defined by

(30) $Y_m(x) := \lim\limits_{\nu \to m} \dfrac{\cos(\nu\pi)J_\nu(x) - J_{-\nu}(x)}{\sin(\nu\pi)}$

for $x > 0$ is a solution to (22) with $\nu = m$. Furthermore, $J_m(x)$ and $Y_m(x)$ are linearly independent. We again call $Y_m(x)$ the Bessel function of the second kind of order m. In the literature the function $Y_m(x)$ is sometimes denoted by $N_m(x)$; variations of it have been called the **Neumann function** and the **Weber function.**

Figure 8.14 shows the graphs of $J_0(x)$ and $J_1(x)$ and Figure 8.15 the graphs of $Y_0(x)$ and $Y_1(x)$. Notice that the curves for $J_0(x)$ and $J_1(x)$ behave like damped sine waves and have zeros that interlace (see Problem 28). The fact that the Bessel functions J_n have an infinite number of zeros is helpful in certain applications (see Project A in Chapter 10).

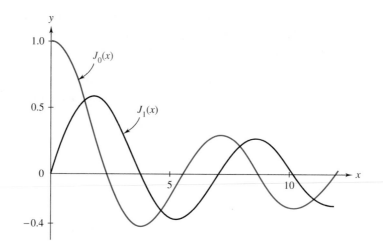

Figure 8.14 Graphs of the Bessel functions $J_0(x)$ and $J_1(x)$

There are several useful recurrence relations involving Bessel functions. For example,

(31) $\dfrac{d}{dx}\big[x^\nu J_\nu(x)\big] = x^\nu J_{\nu-1}(x)$, **(32)** $\dfrac{d}{dx}\big[x^{-\nu}J_\nu(x)\big] = -x^{-\nu}J_{\nu+1}(x)$,

(33) $J_{\nu+1}(x) = \dfrac{2\nu}{x}J_\nu(x) - J_{\nu-1}(x)$, **(34)** $J_{\nu+1}(x) = J_{\nu-1}(x) - 2J_\nu'(x)$.

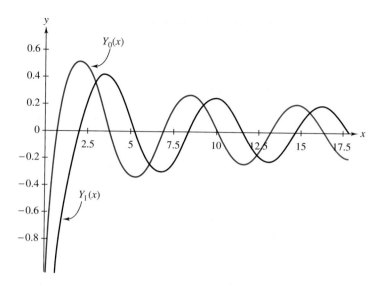

Figure 8.15 Graphs of the Bessel functions $Y_0(x)$ and $Y_1(x)$

Furthermore, analogous equations hold for Bessel functions of the second kind.

To illustrate the techniques involved in proving the recurrence relations, let's verify relation (31). We begin by substituting series (25) for $J_\nu(x)$ into the left-hand side of (31). Differentiating, we get

$$
(35) \quad \frac{d}{dx}\big[x^\nu J_\nu(x)\big] = \frac{d}{dx}\left\{ x^\nu \sum_{n=0}^{\infty} \frac{(-1)^n}{n!\Gamma(1+\nu+n)}\left(\frac{x}{2}\right)^{2n+\nu} \right\}
$$

$$
= \frac{d}{dx}\left\{ \sum_{n=0}^{\infty} \frac{(-1)^n x^{2n+2\nu}}{n!\Gamma(1+\nu+n)2^{2n+\nu}} \right\}
$$

$$
= \sum_{n=0}^{\infty} \frac{(-1)^n (2n+2\nu)x^{2n+2\nu-1}}{n!\Gamma(1+\nu+n)2^{2n+\nu}} \ .
$$

Since $\Gamma(1+\nu+n) = (\nu+n)\Gamma(\nu+n),$ we have from (35)

$$
\frac{d}{dx}\big[x^\nu J_\nu(x)\big] = \sum_{n=0}^{\infty} \frac{(-1)^n 2 x^{2n+2\nu-1}}{n!\Gamma(\nu+n)2^{2n+\nu}} \ .
$$

Factoring out an x^ν gives

$$
\frac{d}{dx}\big[x^\nu J_\nu(x)\big] = x^\nu \sum_{n=0}^{\infty} \frac{(-1)^n}{n!\Gamma(1+(\nu-1)+n)}\left(\frac{x}{2}\right)^{2n+\nu-1} = x^\nu J_{\nu-1}(x) \ ,
$$

as claimed in equation (31). We leave the verifications of the remaining relations as exercises (see Problems 22–24).

The following approximations are useful when analyzing the behavior of Bessel functions for large arguments:

(36) $J_\nu(x) \approx \sqrt{\dfrac{2}{\pi x}} \cos\left(x - \dfrac{\nu\pi}{2} - \dfrac{\pi}{4}\right)$, $Y_\nu(x) \approx \sqrt{\dfrac{2}{\pi x}} \sin\left(x - \dfrac{\nu\pi}{2} - \dfrac{\pi}{4}\right)$, $x \gg 1$.

This oscillatory behavior is demonstrated in Figures 8.14 and 8.15. Problem 41 provides some justification of formula (36). Of course, for small x the leading terms of the power series expansion dominate and we have

(37) $J_\nu(x) \approx x^\nu / [2^\nu \Gamma(1 + \nu)]$, $Y_0(x) \approx (2 \ln x)/\pi$, $Y_{\nu>0}(x) \approx -\Gamma(\nu)2^\nu/(\pi x^\nu)$,
 $0 < x \ll 1$.

Legendre's Equation

The linear second order differential equation

(38) $(1 - x^2)y'' - 2xy' + n(n + 1)y = 0$,

where n is a fixed parameter, is called **Legendre's equation.**[†] This equation has a regular singular point at 1, and hence a series solution for (38) about $x = 1$ may be obtained by the method of Frobenius. By setting $z = x - 1$, equation (38) is transformed into

(39) $z(z + 2)\dfrac{d^2y}{dz^2} + 2(z + 1)\dfrac{dy}{dz} - n(n + 1)y = 0$.

The indicial equation for (39) at $z = 0$ is

$r(r - 1) + r = r^2 = 0$,

which has roots $r_1 = r_2 = 0$. Upon substituting

$$y(z) = \sum_{k=0}^{\infty} a_k z^k$$

into (39) and proceeding as usual, we arrive at the solution

(40) $y_1(x) = 1 + \displaystyle\sum_{k=1}^{\infty} \dfrac{(-n)_k(n + 1)_k}{k!(1)_k}\left(\dfrac{1 - x}{2}\right)^k$,

where we have expressed y_1 in terms of the original variable x and have taken $a_0 = 1$. We have written the solution in the above form because it is now obvious from (40) that

(41) $y_1(x) = F\left(-n, n + 1; 1; \dfrac{1 - x}{2}\right)$,

where F is the Gaussian hypergeometric function defined in (10).

[†]*Historical Footnote:* Solutions to this equation were obtained by Adrien Marie Legendre (1752–1833) in 1785 and are referred to as Legendre functions.

For n a nonnegative integer, the factor

$$(-n)_k = (-n)(-n + 1)(-n + 2) \cdots (-n + k - 1)$$

will be zero for $k \geq n + 1$. Hence the solution given in (40) and (41) is a *polynomial* of degree n. Moreover, $y_1(1) = 1$. These polynomial solutions of equation (38) are called the **Legendre polynomials** or **spherical polynomials** and are traditionally denoted by $P_n(x)$. That is,

(42) $$P_n(x) := 1 + \sum_{k=1}^{\infty} \frac{(-n)_k(n + 1)_k}{k!(1)_k} \left(\frac{1 - x}{2} \right)^k .$$

If we expand about $x = 0$, then $P_n(x)$ takes the form

(43) $$P_n(x) = 2^{-n} \sum_{m=0}^{[n/2]} \frac{(-1)^m(2n - 2m)!}{(n - m)!m!(n - 2m)!} x^{n - 2m} ,$$

where $[n/2]$ is the greatest integer less than or equal to $n/2$ (see Problem 34). The first three Legendre polynomials are

$$P_0(x) = 1 , \qquad P_1(x) = x , \qquad P_2(x) = \frac{3}{2}x^2 - \frac{1}{2} .$$

The Legendre polynomials satisfy the **orthogonality condition**

(44) $$\int_{-1}^{1} P_m(x)P_n(x)dx = 0 \quad \text{for} \quad n \neq m .$$

To see this, we first rewrite equation (38) in what is called the **selfadjoint form:**

(45) $$\left[(1 - x^2)y' \right]' + n(n + 1)y = 0 .$$

Since $P_n(x)$ and $P_m(x)$ satisfy (45) with parameters n and m respectively, we have

(46) $$\left[(1 - x^2)P_n'(x) \right]' + n(n + 1)P_n(x) = 0 ,$$

(47) $$\left[(1 - x^2)P_m'(x) \right]' + m(m + 1)P_m(x) = 0 .$$

Multiplying (46) by $P_m(x)$ and (47) by $P_n(x)$ and then subtracting, we find

$$P_m(x)\left[(1 - x^2)P_n'(x) \right]' - P_n(x)\left[(1 - x^2)P_m'(x) \right]'$$
$$+ \left[n(n + 1) - m(m + 1) \right]P_m(x)P_n(x) = 0 ,$$

which can be rewritten in the form

(48) $$(n^2 - m^2 + n - m)P_m(x)P_n(x) = P_n(x)\left[(1 - x^2)P_m'(x) \right]' - P_m(x)\left[(1 - x^2)P_n'(x) \right]' .$$

It is a straightforward calculation to show that the right-hand side of (48) is just

$$\{ (1 - x^2)\left[P_n(x)P_m'(x) - P_n'(x)P_m(x) \right] \}' .$$

Using this fact and the identity $n^2 - m^2 + n - m = (n - m)(n + m + 1)$, we rewrite equation (48) as

(49) $(n - m)(n + m + 1)P_m(x)P_n(x) = \{(1 - x^2)[P_n(x)P'_m(x) - P'_n(x)P_m(x)]\}'$.

Integrating both sides of (49) from $x = -1$ to $x = 1$ yields

(50) $(n - m)(n + m + 1)\displaystyle\int_{-1}^{1} P_m(x)P_n(x)dx$

$$= \int_{-1}^{1} \{(1 - x^2)[P_n(x)P'_m(x) - P'_n(x)P_m(x)]\}'dx$$
$$= \{(1 - x^2)[P_n(x)P'_m(x) - P'_n(x)P_m(x)]\}\Big|_{-1}^{1}$$
$$= 0$$

because $1 - x^2 = 0$ for $x = \pm 1$. Since n and m are nonnegative integers with $n \neq m$, then $(n - m)(n + m + 1) \neq 0$, and so equation (44) follows from (50).

Legendre polynomials also satisfy the **recurrence formula**

(51) $(n + 1)P_{n+1}(x) = (2n + 1)xP_n(x) - nP_{n-1}(x)$

and **Rodrigues's formula**

(52) $P_n(x) = \dfrac{1}{2^n n!} \dfrac{d^n}{dx^n}\{(x^2 - 1)^n\}$

(see Problems 32 and 33).

The Legendre polynomials are generated by the function $(1 - 2xz + z^2)^{-1/2}$ in the sense that

(53) $(1 - 2xz + z^2)^{-1/2} = \displaystyle\sum_{n=0}^{\infty} P_n(x)z^n$, $|z| < 1$, $|x| < 1$.

That is, if we expand $(1 - 2xz + z^2)^{-1/2}$ in a Taylor series about $z = 0$, treating x as a fixed parameter, then the coefficients of z^n are the Legendre polynomials $P_n(x)$. The function $(1 - 2xz + z^2)^{-1/2}$ is called a **generating function** for $P_n(x)$ and can be derived from the recurrence formula (51) (see Problem 35).

The Legendre polynomials are an example of a class of special functions called **classical orthogonal polynomials.**[†] The latter includes, for example, the **Jacobi polynomials,** $P_n^{(\alpha,\beta)}(x)$; the **Gegenbauer** or **ultraspherical polynomials,** $C_n^\lambda(x)$; the **Chebyshev** (Tchebichef) **polynomials,** $T_n(x)$ and $U_n(x)$; the **Laguerre polynomials,** $L_n^\alpha(x)$; and the **Hermite polynomials,** $H_n(x)$. The properties of the classical orthogonal polynomials can be found in the books mentioned earlier in this section or in the *Handbook of Mathematical Functions with Formulas, Graphs, and Mathematical Tables,* by M. Abramowitz and I. A. Stegun (eds.) (Dover, New York, 1965), Chapter 22.

[†]Here orthogonality is used in the more general sense that $\int_a^b p_n(x)p_m(x)w(x)\,dx = 0$ for $n \neq m$, where $w(x)$ is a weight function on the interval (a, b).

EXERCISES 8.8

In Problems 1 through 4, express a general solution to the given equation using Gaussian hypergeometric functions.

1. $x(1 - x)y'' + \left(\dfrac{1}{2} - 4x\right)y' - 2y = 0$.

2. $3x(1 - x)y'' + (1 - 27x)y' - 45y = 0$.

3. $2x(1 - x)y'' + (1 - 6x)y' - 2y = 0$.

4. $2x(1 - x)y'' + (3 - 10x)y' - 6y = 0$.

In Problems 5 through 8, verify the following formulas by expanding each function in a power series about $x = 0$.

5. $F(1, 1; 2; x) = -x^{-1}\ln(1 - x)$.

6. $F(\alpha, \beta; \beta; x) = (1 - x)^{-\alpha}$.

7. $F\left(\dfrac{1}{2}, 1; \dfrac{3}{2}; x^2\right) = \dfrac{1}{2}x^{-1}\ln\left(\dfrac{1 + x}{1 - x}\right)$.

8. $F\left(\dfrac{1}{2}, 1; \dfrac{3}{2}; -x^2\right) = x^{-1}\arctan x$.

In Problems 9 and 10, use one of the methods discussed in Section 8.7 to obtain two linearly independent solutions to the given hypergeometric equation.

9. $x(1 - x)y'' + (1 - 3x)y' - y = 0$.

10. $x(1 - x)y'' + (2 - 2x)y' - \dfrac{1}{4}y = 0$.

11. Show that the **confluent hypergeometric equation**
$$xy'' + (\gamma - x)y' - \alpha y = 0 ,$$
where α and γ are fixed parameters and γ is not an integer, has two linearly independent solutions
$$y_1(x) = {}_1F_1(\alpha; \gamma; x) := 1 + \sum_{n=1}^{\infty} \frac{(\alpha)_n}{n!(\gamma)_n}x^n$$
and
$$y_2(x) = x^{1-\gamma}{}_1F_1(\alpha + 1 - \gamma; 2 - \gamma; x) .$$

12. Use the property of the gamma function given in (19) to derive relation (20).

In Problems 13 through 18, express a general solution to the given equation using Bessel functions of either the first or second kind.

13. $4x^2y'' + 4xy' + (4x^2 - 1)y = 0$.

14. $9x^2y'' + 9xy' + (9x^2 - 16)y = 0$.

15. $x^2y'' + xy' + (x^2 - 1)y = 0$.

16. $x^2y'' + xy' + x^2y = 0$.

17. $9t^2x'' + 9tx' + (9t^2 - 4)x = 0$.

18. $x^2z'' + xz' + (x^2 - 16)z = 0$.

In Problems 19 and 20, a Bessel equation is given. For the appropriate choice of ν, the Bessel function $J_\nu(x)$ is one solution. Use one of the methods discussed in Section 8.7 to obtain a second linearly independent solution.

19. $x^2y'' + xy' + (x^2 - 1)y = 0$.

20. $x^2y'' + xy' + (x^2 - 4)y = 0$.

21. Show that $x^\nu J_\nu(x)$ satisfies the equation
$$xy'' + (1 - 2\nu)y' + xy = 0 , \qquad x > 0 ,$$
and use this result to find a solution for the equation
$$xy'' - 2y' + xy = 0 , \qquad x > 0 .$$

In Problems 22 through 24, derive the indicated recurrence formulas.

22. Formula (32). **23.** Formula (33).

24. Formula (34).

25. Show that
$$J_{1/2}(x) = (2/\pi x)^{1/2}\sin x \qquad \text{and}$$
$$J_{-1/2}(x) = (2/\pi x)^{1/2}\cos x .$$

26. The Bessel functions of order $\nu = n + 1/2$, n any integer, are related to the **spherical Bessel functions.** Use relation (33) and the results of Problem 25 to show that such Bessel functions can be represented in terms of $\sin x$, $\cos x$, and powers of x. Demonstrate this by determining a closed form for $J_{-3/2}(x)$ and $J_{5/2}(x)$.

27. Use Theorem 7 in Section 8.7 to determine a second linearly independent solution to Bessel's equation of order 0 in terms of the Bessel function $J_0(x)$.

28. Show that between two consecutive positive roots (zeros) of $J_1(x)$, there is a root of $J_0(x)$. This interlacing property of the roots of Bessel functions is illustrated in Figure 8.14. [*Hint:* Use relation (31) and Rolle's theorem from calculus.]

29. Use formula (43) to determine the first five Legendre polynomials.

30. Show that the Legendre polynomials of even degree are even functions of x, while those of odd degree are odd functions.

31. (a) Show that the orthogonality condition (44) for Legendre polynomials implies that

$$\int_{-1}^{1} P_n(x)q(x)dx = 0$$

for *any* polynomial $q(x)$ of degree at most $n - 1$. [*Hint:* The polynomials $P_0, P_1, \ldots, P_{n-1}$ are linearly independent and hence span the space of all polynomials of degree at most $n - 1$. Thus $q(x) = a_0P_0(x) + \cdots + a_{n-1}P_{n-1}(x)$ for suitable constants a_k.]

(b) Prove that if $Q_n(x)$ is a polynomial of degree n such that

$$\int_{-1}^{1} Q_n(x)P_k(x)dx = 0$$

for $k = 0, 1, \ldots, n - 1$, then $Q_n(x) = cP_n(x)$ for some constant c. [*Hint:* Select c so that the coefficient of x^n for $Q_n(x) - cP_n(x)$ is zero. Then, since P_0, \ldots, P_{n-1} is a basis,

$$Q_n(x) - cP_n(x) = a_0P_0(x) + \cdots + a_{n-1}P_{n-1}(x) .$$

Multiply the last equation by $P_k(x)$ $(0 \leq k \leq n - 1)$ and integrate from $x = -1$ to $x = 1$ to show that each a_k is zero.]

32. Deduce the recurrence formula (51) for Legendre polynomials by completing the following steps.

(a) Show that the function

$$Q_{n-1}(x) := (n + 1)P_{n+1}(x) - (2n + 1)xP_n(x)$$

is a polynomial of degree $n - 1$. [*Hint:* Compute the coefficient of the x^{n+1} term using the representation (42). The coefficient of x^n is also zero because $P_{n+1}(x)$ and $xP_n(x)$ are both odd or both even functions, a consequence of Problem 30.]

(b) Using the result of Problem 31(a), show that

$$\int_{-1}^{1} Q_{n-1}(x)P_k(x)dx = 0$$

for $k = 0, 1, \ldots, n - 2$.

(c) From Problem 31(b), conclude that $Q_{n-1}(x) = cP_{n-1}(x)$ and, by taking $x = 1$, show that $c = -n$. [*Hint:* Recall that $P_m(1) = 1$ for all m.] From the definition of $Q_{n-1}(x)$ in part (a), the recurrence formula now follows.

33. To prove Rodrigues's formula (52) for Legendre polynomials, complete the following steps.

(a) Let $v_n := (d^n/dx^n)\{(x^2 - 1)^n\}$ and show that $v_n(x)$ is a polynomial of degree n with the coefficient of x^n equal to $(2n)!/n!$.

(b) Use integration by parts n times to show that, for any polynomial $q(x)$ of degree less than n,

$$\int_{-1}^{1} v_n(x)q(x)dx = 0 .$$

Hint: For example, when $n = 2$,

$$\int_{-1}^{1} \frac{d^2}{dx^2}\{(x^2 - 1)^2\}q(x)dx$$

$$= q(x)\frac{d}{dx}\{(x^2 - 1)^2\}\Big|_{-1}^{1} - \{q'(x)(x^2 - 1)^2\}\Big|_{-1}^{1}$$

$$+ \int_{-1}^{1} q''(x)(x^2 - 1)^2dx .$$

Since $n = 2$, the degree of $q(x)$ is at most 1, and so $q''(x) \equiv 0$. Thus

$$\int_{-1}^{1} \frac{d^2}{dx^2}\{(x^2 - 1)^2\}q(x)dx = 0 .$$

(c) Use the result of Problem 31(b) to conclude that $P_n(x) = cv_n(x)$ and show that $c = 1/2^n n!$ by comparing the coefficients of x^n in $P_n(x)$ and $v_n(x)$.

34. Use Rodrigues's formula (52) to obtain the representation (43) for the Legendre polynomials $P_n(x)$. *Hint:* From the binomial formula,

$$P_n(x) = \frac{1}{2^n n!} \frac{d^n}{dx^n}\{(x^2 - 1)^n\}$$

$$= \frac{1}{2^n n!} \frac{d^n}{dx^n}\left\{\sum_{m=0}^{n} \frac{n!(-1)^m}{(n - m)!m!}x^{2n - 2m}\right\} .$$

35. The generating function in (53) for Legendre polynomials can be derived from the recurrence formula (51) as follows. Let x be fixed and set $f(z) := \sum_{n=0}^{\infty} P_n(x)z^n$. The goal is to determine an explicit formula for $f(z)$.

(a) Show that multiplying each term in the recurrence formula (51) by z^n and summing the terms from $n = 1$ to ∞ leads to the differential equation

$$\frac{df}{dz} = \frac{x - z}{1 - 2xz + z^2} f \ .$$

Hint:

$$\sum_{n=1}^{\infty} (n + 1)P_{n+1}(x)z^n$$

$$= \sum_{n=0}^{\infty} (n + 1)P_{n+1}(x)z^n - P_1(x)$$

$$= \frac{df}{dz} - x \ .$$

(b) Solve the differential equation derived in part (a) and use the initial conditions $f(0) = P_0(x) \equiv 1$ to obtain $f(z) = (1 - 2xz + z^2)^{-1/2}$.

36. Find a general solution about $x = 0$ for the equation

$$(1 - x^2)y'' - 2xy' + 2y = 0$$

by first finding a polynomial solution and then using the reduction of order formula given in equation (11) in Section 8.7, page 497 (or refer to the inside back cover) to find a second (series) solution.

37. The **Hermite polynomials** $H_n(x)$ are polynomial solutions to Hermite's equation

$$y'' - 2xy' + 2ny = 0 \ .$$

The Hermite polynomials are generated by

$$e^{2tx - t^2} = \sum_{n=0}^{\infty} \frac{H_n(x)}{n!}t^n \ .$$

Use this equation to determine the first four Hermite polynomials.

38. The **Chebyshev** (Tchebichef) **polynomials** $T_n(x)$ are polynomial solutions to Chebyshev's equation

$$(1 - x^2)y'' - xy' + n^2y = 0 \ .$$

The Chebyshev polynomials satisfy the recurrence relation

$$T_{n+1}(x) = 2xT_n(x) - T_{n-1}(x) \ ,$$

with $T_0(x) = 1$ and $T_1(x) = x$. Use this recurrence relation to determine the next three Chebyshev polynomials.

39. The **Laguerre polynomials** $L_n(x)$ are polynomial solutions to Laguerre's equation

$$xy'' + (1 - x)y' + ny = 0 \ .$$

The Laguerre polynomials satisfy Rodrigues's formula,

$$L_n(x) = \frac{e^x}{n!}\frac{d^n}{dx^n}(x^n e^{-x}) \ .$$

Use this formula to determine the first four Laguerre polynomials.

40. Reduction to Bessel's Equation. The class of equations of the form

$$(54) \qquad y''(x) + cx^n y(x) = 0 \ , \qquad x > 0 \ ,$$

where c and n are positive constants, can be solved by transforming the equation into Bessel's equation.

(a) First, use the substitution $y = x^{1/2}z$ to transform (54) into an equation involving x and z.

(b) Second, use the substitution

$$s = \frac{2\sqrt{c}}{n + 2}x^{n/2 + 1}$$

to transform the equation obtained in part (a) into the Bessel equation

$$s^2\frac{d^2z}{ds^2} + s\frac{dz}{ds} + \left(s^2 - \frac{1}{(n + 2)^2}\right)z = 0, \ s > 0 \ .$$

(c) A general solution to the equation in part (b) can be given in terms of Bessel functions of the first and second kind. Substituting back in for s and z, obtain a general solution for equation (54).

41. (a) Show that the substitution $z(x) = \sqrt{x}\, y(x)$ renders Bessel's equation (22) in the form

$$(55) \qquad z'' + \left(1 + \frac{1 - 4\nu^2}{4x^2}\right)z = 0 \ .$$

(b) For $x \gg 1$, equation (55) would apparently be approximated by the equation

$$(56) \qquad z'' + z = 0 \ .$$

Write down the general solution to (56), reset $y(x) = z(x)/\sqrt{x}$, and argue the plausibility of formula (36).

(c) For $\nu = \pm 1/2$, equation (55) reduces to equation (56) exactly. Relate this observation to Problem 25.

CHAPTER SUMMARY

Initial value problems that do not fall into the "solvable" categories that have been studied (such as constant-coefficient or equidimensional equations) can often be analyzed by interpreting the differential equation as a prescription for computing the higher derivatives of the unknown function. The **Taylor polynomial method** uses the equation to construct a polynomial approximation matching the initial values of a finite number of derivatives of the unknown. If the equation permits the extrapolation of this procedure to polynomials of arbitrarily high degree, **power series** representations of the solution can be constructed.

Power Series

Every power series $\sum_{n=0}^{\infty} a_n(x - x_0)^n$ has a **radius of convergence** ρ, $0 \leq \rho \leq \infty$, such that the series converges absolutely for $|x - x_0| < \rho$ and diverges when $|x - x_0| > \rho$. By the ratio test,

$$\frac{1}{\rho} = \lim_{n \to \infty} \frac{|a_{n+1}|}{|a_n|} \ ,$$

provided that this limit exists as an extended real number. A function $f(x)$ that is the sum of a power series in some open interval about x_0 is said to be **analytic at x_0**. If f is analytic at x_0, its power series representation about x_0 is the **Taylor series**

$$f(x) = \sum_{n=0}^{\infty} \frac{f^{(n)}(x_0)}{n!} (x - x_0)^n \ .$$

Power Series Method for an Ordinary Point

In the case of a linear equation of the form

(1) $y'' + p(x)y' + q(x)y = 0$,

where p and q are analytic at x_0, the point x_0 is called an **ordinary point,** and the equation has a pair of linearly independent solutions expressible as power series about x_0. The radii of convergence of these series solutions are at least as large as the distance from x_0 to the nearest singularity (real or complex) of the equation. To find power series solutions to (1), we substitute $y(x) = \sum_{n=0}^{\infty} a_n(x - x_0)^n$ into (1), group like terms, and set the coefficients of the resulting power series equal to zero. This leads to a recurrence relation for the coefficients a_n, which, in some cases, may even yield a general formula for the a_n. The same method applies to the nonhomogeneous version of (1), provided the forcing function is also analytic at x_0.

Regular Singular Points

If, in equation (1), either p or q fails to be analytic at x_0, then x_0 is a **singular point** of (1). If x_0 is a singular point for which $(x - x_0)p(x)$ and $(x - x_0)^2 q(x)$ are both analytic at x_0, then x_0 is a **regular singular point.** The Cauchy-Euler equation,

(2) $ax^2 \dfrac{d^2 y}{dx^2} + bx \dfrac{dy}{dx} + cy = 0$, $x > 0$,

has a regular singular point at $x = 0$, and a general solution to (2) can be obtained by substituting $y = x^r$ and examining the roots of the resulting indicial equation $ar^2 + (b - a)r + c = 0$.

Method of Frobenius

For an equation of the form (1) with a regular singular point at x_0, a series solution can be found by the **method of Frobenius.** This is obtained by substituting

$$w(r, x) = (x - x_0)^r \sum_{n=0}^{\infty} a_n(x - x_0)^n$$

into (1), finding a recurrence relation for the coefficients, and choosing $r = r_1$, the larger root of the **indicial equation**

(3) $r(r - 1) + p_0 r + q_0 = 0$,

where $p_0 := \lim_{x \to x_0}(x - x_0)p(x)$, $q_0 := \lim_{x \to x_0}(x - x_0)^2 q(x)$.

Finding a Second Linearly Independent Solution

If the two roots r_1, r_2 of the indicial equation (3) do not differ by an integer, then a second linearly independent solution to (1) can be found by taking $r = r_2$ in the method of Frobenius. However, if $r_1 = r_2$ or $r_1 - r_2$ is a positive integer, then discovering a second solution requires a different approach. This may be a reduction of order procedure or the utilization of Theorem 7, which gives the *forms* of the solutions.

Special Functions

Some special functions in physics and engineering that arise from series solutions to linear second order equations with polynomial coefficients are Gaussian hypergeometric functions, $F(\alpha, \beta; \gamma; x)$; Bessel functions $J_\nu(x)$; and orthogonal polynomials such as those of Legendre, Chebyshev, Laguerre, and Hermite.

REVIEW PROBLEMS

1. Find the first four nonzero terms in the Taylor polynomial approximation for the given initial value problem.

(a) $y' = xy - y^2$; $y(0) = 1$.

(b) $z'' - x^3 z' + xz^2 = 0$; $z(0) = -1$,
 $z'(0) = 1$.

2. Determine all the singular points of the given equation and classify them as regular or irregular.

(a) $(x^2 - 4)^2 y'' + (x - 4)y' + xy = 0$.

(b) $(\sin x)y'' + y = 0$.

3. Find at least the first four nonzero terms in a power series expansion about $x = 0$ for a general solution to the given equation.

(a) $y'' + x^2 y' - 2y = 0$.

(b) $y'' + e^{-x} y' - y = 0$.

4. Find a general formula for the coefficient a_n in a power series expansion about $x = 0$ for a general solution to the given equation.

(a) $(1 - x^2)y'' + xy' + 3y = 0$.

(b) $(x^2 - 2)y'' + 3y = 0$.

5. Find at least the first four nonzero terms in a power series expansion about $x = 2$ for a general solution to
$$w'' + (x - 2)w' - w = 0 .$$

6. Use the substitution $y = x^r$ to find a general solution to the given equation for $x > 0$.

 (a) $2x^2y''(x) + 5xy'(x) - 12y(x) = 0$.

 (b) $x^3y'''(x) + 3x^2y''(x) - 2xy'(x) - 2y(x) = 0$.

7. Use the method of Frobenius to find at least the first four nonzero terms in the series expansion about $x = 0$ for a solution to the given equation for $x > 0$.

 (a) $x^2y'' - 5xy' + (9 - x)y = 0$.

 (b) $x^2y'' + (x^2 + 2x)y' - 2y = 0$.

8. Find the indicial equation and its roots and state (but do not compute) the form of the series expansion about $x = 0$ (as in Theorem 7 on page 499) for two linearly independent solutions of the given equation for $x > 0$.

 (a) $x^2y'' + (\sin x)y' - 4y = 0$.

 (b) $2xy'' + 5y' + xy = 0$.

 (c) $(x \sin x)y'' + xy' + (\tan x)y = 0$.

9. Find at least the first three nonzero terms in the series expansion about $x = 0$ for a general solution to the given equation for $x > 0$.

 (a) $x^2y'' - x(1 + x)y' + y = 0$.

 (b) $xy'' + y' - 2y = 0$.

 (c) $2xy'' + 6y' + y = 0$.

 (d) $x^2y'' + (x - 2)y = 0$.

10. Express a general solution to the given equation using Gaussian hypergeometric functions or Bessel functions.

 (a) $x(1 - x)y'' + \left(\dfrac{1}{2} - 6x\right)y' - 6y = 0$.

 (b) $9\theta^2y'' + 9\theta y' + (9\theta^2 - 1)y = 0$.

TECHNICAL WRITING EXERCISES

1. Knowing that a general solution to a nonhomogeneous linear second order equation can be expressed as a particular solution plus a general solution to the corresponding homogeneous equation, what can you say about the form of a general power series solution to the nonhomogeneous equation about an ordinary point?

2. Discuss advantages and disadvantages of power series solutions over numerical solutions generated by the Euler or Runge-Kutta methods.

3. The factors $x^{0.3}$, $x^{0.3} \ln x$, $x^{-0.3}$, and $x^{0.3 + 12i}$ all can arise in solutions generated using the method of Frobenius to solve a differential equation with 0 as a regular singular point. Discuss the problems encountered in attempting to approximate such solutions using the methods of Euler and of Runge-Kutta.

4. Discuss the issues involved when attempting to use the Taylor polynomials (expanded around $x_0 = 0$) to study the *asymptotic* behavior (as $t \to \pm \infty$) of solutions to the differential equation $y'' + y = 0$.

GROUP PROJECTS FOR CHAPTER 8

A. Spherically Symmetric Solutions to Schrödinger's Equation for the Hydrogen Atom

In quantum mechanics one is interested in determining the wave function and energy states of an atom. These are determined from Schrödinger's equation. In the case of the hydrogen atom, it is possible to find wave functions ψ that are functions only of r, the distance from the proton to the electron. Such functions are called **spherically symmetric** and satisfy the simpler equation

$$(1) \qquad \frac{1}{r}\frac{d^2}{dr^2}(r\psi) = \frac{-8m\pi^2}{h^2}\left(E + \frac{e_0^2}{r}\right)\psi \ ,$$

where e_0^2, m, and h are constants, and E, also a constant, represents the energy of the atom, which we assume here to be negative.

(a) Show that with the substitutions

$$r = \frac{h^2}{4\pi^2 me_0^2}\rho \ , \qquad E = \frac{2\pi^2 me_0^4}{h^2}\varepsilon \ ,$$

where ε is a negative constant, equation (1) reduces to

$$\frac{d^2(\rho\psi)}{d\rho^2} = -\left(\varepsilon + \frac{2}{\rho}\right)\rho\psi \ .$$

(b) If $f := \rho\psi$, then the preceding equation becomes

$$(2) \qquad \frac{d^2 f}{d\rho^2} = -\left(\varepsilon + \frac{2}{\rho}\right)f \ .$$

Show that the substitution $f(\rho) = e^{-\alpha\rho}g(\rho)$, where α is a positive constant, transforms (2) into

$$(3) \qquad \frac{d^2 g}{d\rho^2} - 2\alpha\frac{dg}{d\rho} + \left(\frac{2}{\rho} + \varepsilon + \alpha^2\right)g = 0 \ .$$

(c) If we choose $\alpha^2 = -\varepsilon$ (ε negative), then (3) becomes

$$(4) \qquad \frac{d^2 g}{d\rho^2} - 2\alpha\frac{dg}{d\rho} + \frac{2}{\rho}g = 0 \ .$$

Show that a power series solution $g(\rho) = \sum_{k=1}^{\infty} a_k\rho^k$ (starting with $k = 1$) for (4) must have coefficients a_k that satisfy the recurrence relation

$$(5) \qquad a_{k+1} = \frac{2(\alpha k - 1)}{k(k+1)}a_k \ , \qquad k \geq 1 \ .$$

(d) Now for $a_1 = 1$ and k very large, $a_{k+1} \approx (2\alpha/k)a_k$ and so $a_{k+1} \approx (2\alpha)^k/k!$, which are the coefficients for $\rho e^{2\alpha\rho}$. Hence g acts like $\rho e^{2\alpha\rho}$, so $f(\rho) = e^{-\alpha\rho}g(\rho)$ is like $\rho e^{\alpha\rho}$. Going back further, we then see that $\psi \approx e^{\alpha\rho}$. Therefore, when $r = h^2\rho/4\pi^2 me_0^2$ is large, so is ψ. Roughly speaking, $\psi^2(r)$ is proportional to the probability of finding an

electron a distance r from the proton. Thus the above argument would imply that the electron in a hydrogen atom is more likely to be found at a very large distance from the proton! Since this makes no sense physically, we ask: Do there exist positive values for α for which ψ remains bounded as r becomes large?

Show that when $\alpha = 1/n$, $n = 1, 2, 3, \ldots$, then $g(\rho)$ is a polynomial of degree n and argue that ψ is therefore bounded.

(e) Let E_n and $\psi_n(\rho)$ denote, respectively, the energy state and wave function corresponding to $\alpha = 1/n$. Find E_n (in terms of the constants e_0^2, m, and h) and $\psi_n(\rho)$ for $n = 1, 2$, and 3.

B. *Airy's Equation*

In aerodynamics one encounters the following initial value problem for **Airy's equation:**

$$y'' + xy = 0 , \qquad y(0) = 1 , \qquad y'(0) = 0 .$$

(a) Find the first five nonzero terms in a power series expansion about $x = 0$ for the solution and graph this polynomial for $-10 \le x \le 10$.

(b) Using the Runge-Kutta subroutine (see Section 5.6) with $h = 0.05$, approximate the solution on the interval $[0, 10]$, i.e., at the points $0.05, 0.1, 0.15$, etc.

(c) Using the Runge-Kutta subroutine with $h = 0.05$, approximate the solution on the interval $[-10, 0]$. [*Hint:* With the change of variables $z = -x$, it suffices to approximate the solution to $y'' - zy = 0$; $y(0) = 1$, $y'(0) = 0$, on the interval $[0, 10]$.]

(d) Using your knowledge of constant-coefficient equations as a basis for guessing the behavior of the solutions to Airy's equation, decide whether the power series approximation obtained in part (a) or the numerical approximation obtained in parts (b) and (c) better describes the true behavior of the solution on the interval $[-10, 10]$.

C. *Buckling of a Tower*

A tower is constructed of four angle beams connected by diagonals (see Figure 8.16). The deflection curve $y(x)$ for the tower is governed by the equation

$$\text{(6)} \qquad x^2\frac{d^2y}{dx^2} + \frac{Pa^2}{EI}y = 0 , \qquad a < x < a + L ,$$

where x is the vertical coordinate measured down from the extended top of the tower, y is the deflection from the vertical passing through the center of the unbuckled tower, L is the tower height, a is the length of the truncation, P is the load, E is the modulus of elasticity, and I is the moment of inertia. The appropriate boundary conditions for this design are

$$\text{(7)} \qquad y(a) = 0 ,$$

$$\text{(8)} \qquad y'(a + L) = 0 .$$

Clearly the solution $y(x) \equiv 0$ is always at hand. However, when the load P is heavy enough, the tower can *buckle* and a nontrivial solution $y(x)$ is also possible. We want to predict this phenomenon.

(a) Solve equation (6). [*Hint:* Equation (6) is a Cauchy-Euler equation.]

(a) Unbuckled

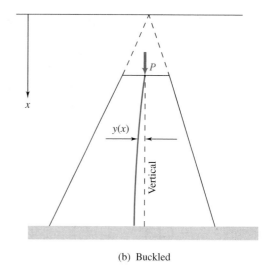

(b) Buckled

Figure 8.16 Buckling tower

(b) Show that the first boundary condition (7) implies
$$y = Ax^{1/2}\sin\left[\beta\ln(x/a)\right] \ ,$$
where A is an arbitrary constant and $\beta := \sqrt{Pa^2/EI - 1/4}$.

(c) Show that the second boundary condition (8) gives
$$A\left\{\tan\left[\beta\ln\left(\frac{a+L}{a}\right)\right] + 2\beta\right\} = 0 \ .$$

(d) Use the result of part (c) to argue that *no* buckling takes place (i.e., the only possibility is $A = 0$) if $0 < \beta < \beta_c$, where β_c is the smallest positive real number that makes the expression in braces zero.

(e) The value of the load corresponding to β_c is called the **critical load** P_c. Solve for P_c in terms of β_c, a, E, and I.

(f) Find the critical load P_c if $a = 10$, $L = 40$, and $EI = 1000$.

D. *Aging Spring and Bessel Functions*

In Problems 30 and 31 in Exercises 8.4, page 479, we discussed a model for a mass-spring system with an aging spring. Without damping, the displacement $x(t)$ at time t is governed by the equation

(9) $mx''(t) + ke^{-\eta t}x(t) = 0 \ ,$

where m, k, and η are positive constants. The general solution to this equation can be expressed using Bessel functions.

(a) The coefficient of x suggests a change of variables of the form $s = \alpha e^{\beta t}$. Show that (9) transforms into

(10) $$\beta^2 s^2 \frac{d^2 x}{ds^2} + \beta^2 s \frac{dx}{ds} + \frac{k}{m}\left(\frac{s}{\alpha}\right)^{-\eta/\beta} x = 0 \ .$$

(b) Show that choosing α and β to satisfy

$$\frac{-\eta}{\beta} = 2 \quad \text{and} \quad \frac{k}{m\beta^2 \alpha^{-\eta/\beta}} = 1$$

transforms (10) into the Bessel equation of order 0 in s and x.

(c) Using the result of part (b), show that a general solution to (9) is given by

$$x(t) = c_1 J_0\left(\frac{2}{\eta}\sqrt{k/m}\, e^{-\eta t/2}\right) + c_2 Y_0\left(\frac{2}{\eta}\sqrt{k/m}\, e^{-\eta t/2}\right) ,$$

where J_0 and Y_0 are the Bessel functions of order 0 of the first and second kind, respectively.

(d) Discuss the behavior of the displacement $x(t)$ for c_2 positive, negative, and zero.

(e) Compare the behavior of the displacement $x(t)$ for η a small positive number and η a large positive number.

Matrix Methods for Linear Systems

INTRODUCTION

In this chapter we return to the analysis of systems of differential equations. When the equations in the system are *linear,* matrix algebra provides a compact notation for expressing the system. In fact, the notation itself suggests new and elegant ways of characterizing the solution properties, as well as novel, efficient techniques for explicitly obtaining solutions.

> In Chapter 5 we analyzed physical situations wherein two fluid tanks containing brine solutions were interconnected and pumped so as ultimately to deplete the salt content in each tank. By accounting for the various influxes and outfluxes of brine, a system of differential equations for the salt contents $\big(x(t)$ and $y(t)\big)$ of each tank was derived; a typical model is
>
> (1) $\quad \begin{aligned} dx/dt &= -4x + 2y \ , \\ dy/dt &= 4x - 4y \ . \end{aligned}$
>
> Express this system in matrix notation as a single equation.

The right-hand side of the first member of (1) possesses a mathematical structure that is familiar from vector calculus; namely, it is the **dot product**[†] of two vectors:

(2) $\qquad -4x + 2y = \begin{bmatrix} -4 & 2 \end{bmatrix} \cdot \begin{bmatrix} x & y \end{bmatrix} \ .$

[†]Recall that the dot product of two vectors **u** and **v** equals the length of **u** times the length of **v** times the cosine of the angle between **u** and **v**. However, it is more conveniently computed from the *components* of **u** and **v** by the "inner product" indicated in equation (2).

Similarly, the second right-hand side in (1) is the dot product

$$4x - 4y = \begin{bmatrix} 4 & -4 \end{bmatrix} \cdot \begin{bmatrix} x & y \end{bmatrix} .$$

The frequent occurrence in mathematics of arrays of dot products, such as evidenced in the system (1), led to the development of **matrix algebra,** a mathematical discipline whose basic operation—the matrix product—is the arrangement of a set of dot products according to the following plan:

$$\begin{bmatrix} -4 & 2 \\ 4 & -4 \end{bmatrix} \begin{bmatrix} x \\ y \end{bmatrix} = \begin{bmatrix} \begin{bmatrix} -4 & 2 \end{bmatrix} \cdot \begin{bmatrix} x & y \end{bmatrix} \\ \begin{bmatrix} 4 & -4 \end{bmatrix} \cdot \begin{bmatrix} x & y \end{bmatrix} \end{bmatrix} = \begin{bmatrix} -4x + 2y \\ 4x - 4y \end{bmatrix} .$$

In general, the product of a **matrix**—i.e., an m by n rectangular array of numbers—and a *column vector* is defined to be the collection of dot products of the *rows* of the matrix with the vector, arranged as a column vector:

$$\begin{bmatrix} \textbf{row \# 1} \\ \textbf{row \# 2} \\ \vdots \\ \textbf{row \# m} \end{bmatrix} \begin{bmatrix} \mathbf{v} \end{bmatrix} = \begin{bmatrix} \begin{bmatrix} \textbf{row \# 1} \end{bmatrix} \cdot \mathbf{v} \\ \begin{bmatrix} \textbf{row \# 2} \end{bmatrix} \cdot \mathbf{v} \\ \vdots \\ \begin{bmatrix} \textbf{row \# m} \end{bmatrix} \cdot \mathbf{v} \end{bmatrix} ,$$

where the vector \mathbf{v} has n components; the dot product of two n-dimensional vectors is computed in the obvious way:

$$\begin{bmatrix} a_1 & a_2 & \cdots & a_n \end{bmatrix} \cdot \begin{bmatrix} x_1 & x_2 & \cdots & x_n \end{bmatrix} = a_1 x_1 + a_2 x_2 + \cdots + a_n x_n .$$

Using the notation for the matrix product, we can write the system (1) for the interconnected tanks as

$$\begin{bmatrix} x' \\ y' \end{bmatrix} = \begin{bmatrix} -4 & 2 \\ 4 & -4 \end{bmatrix} \begin{bmatrix} x \\ y \end{bmatrix} .$$

The following example demonstrates a four-dimensional implementation of this notation. Note that the coefficients in the linear system need not be constants.

EXAMPLE 1 Express the system

$$
\begin{aligned}
x_1' &= 2x_1 + t^2 x_2 + (4t + e^t)x_4 , \\
x_2' &= (\sin t)x_2 + (\cos t)x_3 , \\
x_3' &= x_1 + x_2 + x_3 + x_4 , \\
x_4' &= 0
\end{aligned}
$$

(3)

as a matrix equation.

SOLUTION We express the right-hand side of the first member of (3) as the dot product

$$2x_1 + t^2 x_2 + (4t + e^t)x_4 = \begin{bmatrix} 2 & t^2 & 0 & (4t + e^t) \end{bmatrix} \cdot \begin{bmatrix} x_1 & x_2 & x_3 & x_4 \end{bmatrix} .$$

The other dot products are similarly identified, and the matrix form is given by

$$\begin{bmatrix} x_1' \\ x_2' \\ x_3' \\ x_4' \end{bmatrix} = \begin{bmatrix} 2 & t^2 & 0 & (4t + e^t) \\ 0 & \sin t & \cos t & 0 \\ 1 & 1 & 1 & 1 \\ 0 & 0 & 0 & 0 \end{bmatrix} \begin{bmatrix} x_1 \\ x_2 \\ x_3 \\ x_4 \end{bmatrix} . \quad ∎\,End$$

In general, if a system of differential equations is expressed as

$$x_1' = a_{11}(t)x_1 + a_{12}(t)x_2 + \cdots + a_{1n}(t)x_n$$
$$x_2' = a_{21}(t)x_1 + a_{22}(t)x_2 + \cdots + a_{2n}(t)x_n$$
$$\vdots$$
$$x_n' = a_{n1}(t)x_1 + a_{n2}(t)x_2 + \cdots + a_{nn}(t)x_n ,$$

it is said to be a linear homogeneous system in **normal form**.[†] The matrix formulation of such a system is then

$$\mathbf{x}' = \mathbf{Ax} ,$$

where \mathbf{A} is the **coefficient matrix**

$$\mathbf{A} = \mathbf{A}(t) = \begin{bmatrix} a_{11}(t) & a_{12}(t) & \cdots & a_{1n}(t) \\ a_{21}(t) & a_{22}(t) & \cdots & a_{2n}(t) \\ \vdots & \vdots & & \vdots \\ a_{n1}(t) & a_{n2}(t) & \cdots & a_{nn}(t) \end{bmatrix}$$

and \mathbf{x} is the solution vector

$$\mathbf{x} = \begin{bmatrix} x_1 \\ x_2 \\ \vdots \\ x_n \end{bmatrix} .$$

Note that we have used \mathbf{x}' to denote the vector of derivatives

$$\mathbf{x}' = \begin{bmatrix} x_1 \\ x_2 \\ \vdots \\ x_n \end{bmatrix}' = \begin{bmatrix} x_1' \\ x_2' \\ \vdots \\ x_n' \end{bmatrix} .$$

[†]The normal form was defined for general systems in Section 5.6, page 301.

EXAMPLE 2 Express the differential equation for the undamped, unforced mass-spring oscillator (recall Section 4.1, page 158)

(4) $my'' + ky = 0$

as an equivalent system of first order equations in normal form, expressed in matrix notation.

SOLUTION We have to express the *second* derivative, y'', as a *first* derivative in order to formulate (4) as a first order system. This is easy; the acceleration y'' is the derivative of the *velocity* $v = y'$, so (4) becomes

(5) $mv' + ky = 0$.

The first order system is then assembled by identifying v with y', and appending it to (5):

$$y' = v$$
$$mv' = -ky \ .$$

To put this system in normal form and express it as a matrix equation, we need to divide the second equation by the mass m:

$$\begin{bmatrix} y \\ v \end{bmatrix}' = \begin{bmatrix} 0 & 1 \\ -k/m & 0 \end{bmatrix} \begin{bmatrix} y \\ v \end{bmatrix} \ . \quad \blacksquare \text{End}$$

In general, the customary way to write an nth order linear homogeneous differential equation

$$a_n(t)y^{(n)} + a_{n-1}(t)y^{(n-1)} + \cdots + a_1(t)y' + a_0(t)y = 0$$

as an equivalent system in normal form is to define the first $(n-1)$ derivatives of y (including y, the zeroth derivative, itself) to be new unknowns:

$$x_1(t) = y(t) \ ,$$
$$x_2(t) = y'(t) \ ,$$
$$\vdots$$
$$x_n(t) = y^{(n-1)}(t) \ .$$

Then the system consists of the identification of $x_j(t)$ as the derivative of $x_{j-1}(t)$, together with the original differential equation expressed in these variables (and divided by $a_n(t)$):

$$x_1' \ = x_2 \ ,$$
$$x_2' \ = x_3 \ ,$$
$$\vdots$$
$$x_{n-1}' = x_n \ ,$$
$$x_n' \ = -\frac{a_0(t)}{a_n(t)}x_1 - \frac{a_1(t)}{a_n(t)}x_2 - \cdots - \frac{a_{n-1}(t)}{a_n(t)}x_n \ .$$

For systems of two or more higher-order differential equations, the same procedure is applied to each unknown function in turn; an example will make this clear.

EXAMPLE 3 The coupled mass-spring oscillator depicted in Figure 5.18 on page 287 was shown to be governed by the system

(6)
$$2\frac{d^2x}{dt^2} + 6x - 2y = 0 ,$$
$$\frac{d^2y}{dt^2} + 2y - 2x = 0 .$$

Write (6) in matrix notation.

SOLUTION We introduce notation for the lower-order derivatives:

(7) $x_1 = x , \qquad x_2 = x' , \qquad x_3 = y , \qquad x_4 = y' .$

In these variables, the system (6) states

(8)
$$2x_2' + 6x_1 - 2x_3 = 0 ,$$
$$x_4' + 2x_3 - 2x_1 = 0 .$$

The normal form is then

$$x_1' = x_2 ,$$
$$x_2' = -3x_1 + x_3 ,$$
$$x_3' = x_4 ,$$
$$x_4' = 2x_1 - 2x_3$$

or in matrix notation

$$\begin{bmatrix} x_1 \\ x_2 \\ x_3 \\ x_4 \end{bmatrix}' = \begin{bmatrix} 0 & 1 & 0 & 0 \\ -3 & 0 & 1 & 0 \\ 0 & 0 & 0 & 1 \\ 2 & 0 & -2 & 0 \end{bmatrix} \begin{bmatrix} x_1 \\ x_2 \\ x_3 \\ x_4 \end{bmatrix} . \quad \blacksquare \text{End}$$

EXERCISES 9.1

In Problems 1 through 6, express the given system of differential equations in matrix notation.

1. $x' = 7x + 2y ,$
 $y' = 3x - 2y .$

2. $x' = y ,$
 $y' = -x .$

3. $x' = x + y + z ,$
 $y' = 2z - x ,$
 $z' = 4y .$

4. $x_1' = x_1 - x_2 + x_3 - x_4 ,$
 $x_2' = x_1 + x_4 ,$
 $x_3' = \sqrt{\pi}x_1 - x_3 ,$
 $x_4' = 0 .$

5. $x' = (\sin t)x + e^t y ,$
 $y' = (\cos t)x + (a + bt^3)y .$

6. $x_1' = (\cos 2t)x_1 ,$
 $x_2' = (\sin 2t)x_2 ,$
 $x_3' = x_1 - x_2 .$

In Problems 7 through 10, express the given higher-order differential equation as a matrix system in normal form.

7. The damped mass-spring oscillator equation $my'' + by' + ky = 0 .$

8. Legendre's equation $(1 - t^2)y'' - 2ty' + 2y = 0 .$

9. The Airy equation $y'' - ty = 0 .$

10. Bessel's equation $y'' + \frac{1}{t}y' + \left(1 - \frac{n^2}{t^2}\right)y = 0 .$

In Problems 11 through 13, express the given system of higher-order differential equations as a matrix system in normal form.

11. $x'' + 3x + 2y = 0$,

$\quad y'' - 2x = 0$.

12. $x'' + 3x' - y' + 2y = 0$,

$\quad y'' + x' + 3y' + y = 0$.

13. $x'' - 3x' + t^2 y - (\cos t)x = 0$,

$\quad y''' + y'' - tx' + y' + e^t x = 0$.

9.2 REVIEW 1: LINEAR ALGEBRAIC EQUATIONS

Here and in the next section we review some basic facts concerning linear algebraic systems and matrix algebra that will be useful in solving linear systems of differential equations in normal form. Readers competent in these areas may proceed to Section 9.4.

A set of equations of the form

$$a_{11}x_1 + a_{12}x_2 + \cdots + a_{1n}x_n = b_1 ,$$
$$a_{21}x_1 + a_{22}x_2 + \cdots + a_{2n}x_n = b_2 ,$$
$$\vdots$$
$$a_{n1}x_1 + a_{n2}x_2 + \cdots + a_{nn}x_n = b_n$$

(where the a_{ij}'s and b_i's are given constants) is called a *linear system of n algebraic equations* in the n unknowns x_1, x_2, \ldots, x_n. The procedure for solving the system using elimination methods is well known. Herein we describe a particularly convenient implementation of the method called the *Gauss-Jordan elimination algorithm.*[†] The basic idea of this formulation is to use the first equation to eliminate x_1 in all the other equations; then use the second equation to eliminate x_2 in all the others; and so on. If all goes well, the resulting system will be "uncoupled," and the values of the unknowns x_1, x_2, \ldots, x_n will be apparent. A short example will make this clear.

EXAMPLE 1 Solve the system

$$2x_1 + 6x_2 + 8x_3 = 16 ,$$
$$4x_1 + 15x_2 + 19x_3 = 38 ,$$
$$2x_1 \qquad + 3x_3 = 6 .$$

SOLUTION By subtracting 2 times the first equation from the second, we eliminate x_1 from the latter. Similarly, x_1 is eliminated from the third equation by subtracting 1 times the first equation from it:

$$2x_1 + 6x_2 + 8x_3 = 16 ,$$
$$3x_2 + 3x_3 = 6 ,$$
$$-6x_2 - 5x_3 = -10 .$$

[†]The Gauss-Jordan algorithm is neither the fastest nor the most accurate computer algorithm for solving a linear system of algebraic equations, but for solutions executed by hand it has many pedagogical advantages. Usually it is much faster than *Cramer's rule,* described in Appendix C.

Next we subtract multiples of the second equation from the first and third to eliminate x_2 in them; the appropriate multiples are 2 and -2, respectively:

$$2x_1 \qquad + 2x_3 = 4 \ ,$$
$$3x_2 + 3x_3 = 6 \ ,$$
$$x_3 = 2 \ .$$

Finally we eliminate x_3 from the first two equations by subtracting multiples (2 and 3, respectively) of the third equation:

$$2x_1 \qquad = 0 \ ,$$
$$3x_2 \quad = 0 \ ,$$
$$x_3 = 2 \ .$$

The system is now uncoupled; i.e., we can solve each equation separately:

$$x_1 = 0 \ , \qquad x_2 = 0 \ , \qquad x_3 = 2 \ . \quad \blacksquare \text{End}$$

There are two complications that can disrupt the straightforward execution of the Gauss-Jordan algorithm. The first occurs when the impending variable to be eliminated, say x_j, does not occur in the jth equation. The solution is usually obvious; we employ one of the subsequent equations to eliminate x_j. Example 2 illustrates this maneuver.

EXAMPLE 2 Solve the system

$$x_1 + 2x_2 + 4x_3 + \quad x_4 = 0 \ ,$$
$$-x_1 - 2x_2 - 2x_3 \qquad = 1 \ ,$$
$$-2x_1 - 4x_2 - 8x_3 + 2x_4 = 4 \ ,$$
$$x_1 + 4x_2 + 2x_3 \qquad = -3 \ .$$

SOLUTION The first unknown x_1 is eliminated from the last three equations by subtracting multiples of the first equation:

$$x_1 + 2x_2 + 4x_3 + x_4 = 0 \ ,$$
$$2x_3 + x_4 = 1 \ ,$$
$$4x_4 = 4 \ ,$$
$$2x_2 - 2x_3 - x_4 = -3 \ .$$

Now, we cannot use the second equation to eliminate the second unknown because x_2 is not present. The next equation that *does* contain x_2 is the fourth, so we switch the second and fourth equation:

$$x_1 + 2x_2 + 4x_3 + x_4 = 0 \ ,$$
$$2x_2 - 2x_3 - x_4 = -3 \ ,$$
$$4x_4 = 4 \ ,$$
$$2x_3 + x_4 = 1 \ ,$$

and proceed to eliminate x_2:

$$
\begin{aligned}
x_1 \quad\quad + 6x_3 + 2x_4 &= 3 , \\
2x_2 - 2x_3 - \quad x_4 &= -3 , \\
4x_4 &= 4 , \\
2x_3 + \quad x_4 &= 1 .
\end{aligned}
$$

To eliminate x_3, we have to switch again,

$$
\begin{aligned}
x_1 \quad\quad + 6x_3 + 2x_4 &= 3 , \\
2x_2 - 2x_3 - \quad x_4 &= -3 , \\
2x_3 + \quad x_4 &= 1 , \\
4x_4 &= 4 ,
\end{aligned}
$$

and eliminate, in turn, x_3 and x_4. This gives

$$
\begin{aligned}
x_1 \quad\quad - x_4 &= 0 , & x_1 \quad\quad &= 1 , \\
2x_2 \quad\quad &= -2 , & \text{and} \quad 2x_2 \quad\quad &= -2 , \\
2x_3 + x_4 &= 1 , & 2x_3 \quad &= 0 , \\
4x_4 &= 4 , & 4x_4 &= 4 .
\end{aligned}
$$

The solution to the uncoupled equations is

$$
x_1 = 1 , \quad x_2 = -1 , \quad x_3 = 0 , \quad x_4 = 1 . \quad \blacksquare \text{End}
$$

The other complication that can disrupt the Gauss-Jordan algorithm is much more profound. What if, when we are "scheduled" to eliminate the unknown x_j, it is absent from *all* of the subsequent equations? The first thing to do is to move on to the elimination of the *next* unknown x_{j+1}, as demonstrated in Example 3.

EXAMPLE 3 Apply the Gauss-Jordan algorithm to the system

$$
\begin{aligned}
2x_1 + 4x_2 + x_3 &= 8 , \\
(1) \quad\quad 2x_1 + 4x_2 \quad\quad &= 6 , \\
-4x_1 - 8x_2 + x_3 &= -10 .
\end{aligned}
$$

SOLUTION Elimination of x_1 proceeds as usual:

$$
\begin{aligned}
2x_1 + 4x_2 + x_3 &= 8 , \\
- x_3 &= -2 , \\
3x_3 &= 6 .
\end{aligned}
$$

Now since x_2 is absent from the second *and* third equations, we use the second equation to eliminate x_3 :

$$
\begin{aligned}
2x_1 + 4x_2 \quad\quad &= 6 , \\
(2) \quad\quad\quad -x_3 &= -2 , \\
0 &= 0 .
\end{aligned}
$$

How do we interpret the system (2)? The final equation contains no information, of course, and we ignore it.[†] The second equation implies that $x_3 = 2$.

The first equation implies that $x_1 = 3 - 2x_2$; but there is no equation for x_2. Evidently x_2 is a "free" variable, and we can assign *any* value to it—as long as we take x_1 to be $3 - 2x_2$. Thus (1) has an infinite number of solutions, and a convenient way of characterizing them is

$$x_1 = 3 - 2s , \qquad x_2 = s , \qquad x_3 = 2 ; \quad -\infty < s < \infty .$$

We remark that an equivalent solution can be obtained by treating x_1 as the free variable, say $x_1 = s$, and taking $x_2 = (3 - s)/2, x_3 = 2$. ∎ End

The final example is contrived to demonstrate all the features that we have encountered.

EXAMPLE 4 Find all solutions to the system

$$\begin{aligned}
x_1 - x_2 + 2x_3 + 2x_4 &= 0 , \\
2x_1 - 2x_2 + 4x_3 + 3x_4 &= 1 , \\
3x_1 - 3x_2 + 6x_3 + 9x_4 &= -3 , \\
4x_1 - 4x_2 + 8x_3 + 8x_4 &= 0 .
\end{aligned}$$

SOLUTION We use the first equation to eliminate x_1:

$$\begin{aligned}
x_1 - x_2 + 2x_3 + 2x_4 &= 0 , \\
- x_4 &= 1 , \\
3x_4 &= -3 , \\
0 &= 0 .
\end{aligned}$$

Now, both x_2 *and* x_3 are absent from all subsequent equations, so we use the second equation to eliminate x_4.

$$\begin{aligned}
x_1 - x_2 + 2x_3 \quad\;\; &= 2 , \\
-x_4 &= 1 , \\
0 &= 0 , \\
0 &= 0 .
\end{aligned}$$

There are no constraints on either x_2 or x_3; thus we take them to be free variables and characterize the solutions by

$$x_1 = 2 + s - 2t , \qquad x_2 = s , \qquad x_3 = t , \qquad x_4 = -1, \quad -\infty < s, t < \infty . \quad ∎ \text{End}$$

In closing, we note that if the execution of the Gauss-Jordan algorithm results in a display of the form $0 = 1$ (or $0 = k$, where $k \neq 0$), the original system has no solutions; it is *inconsistent*. This is explored in Problem 12.

[†]The occurrence of the identity $0 = 0$ in the Gauss-Jordan algorithm implies that one of the original equations was *redundant*. In this case, the reader may observe that the final equation in (1) can be derived by subtracting 3 times the second equation from the first.

EXERCISES 9.2

In Problems 1 through 11, find all solutions to the system using the Gauss-Jordan elimination algorithm.

1. $x_1 + 2x_2 + 2x_3 = 6$,
$2x_1 + x_2 + x_3 = 6$,
$x_1 + x_2 + 3x_3 = 6$.

2. $x_1 + x_2 + x_3 + x_4 = 1$,
$x_1 \qquad\qquad + x_4 = 0$,
$2x_1 + 2x_2 - x_3 + x_4 = 0$,
$x_1 + 2x_2 - x_3 + x_4 = 0$.

3. $x_1 + 2x_2 + x_3 = -3$,
$2x_1 + 4x_2 - x_3 = 0$,
$x_1 + 3x_2 - 2x_3 = 3$.

4. $\qquad\qquad x_3 + x_4 = 0$,
$x_1 + x_2 + x_3 + x_4 = 1$,
$2x_1 - x_2 + x_3 + 2x_4 = 0$,
$2x_1 - x_2 + x_3 + x_4 = 0$.

5. $-x_1 + 2x_2 = 0$, **6.** $-2x_1 + 2x_2 - x_3 = 0$,
$2x_1 + 3x_2 = 0$. $x_1 - 3x_2 + x_3 = 0$,
$\qquad\qquad\qquad\qquad$ $4x_1 - 4x_2 + 2x_3 = 0$.

7. $-x_1 + 3x_2 = 0$, **8.** $x_1 + x_2 - x_3 = 0$,
$-3x_1 + 9x_2 = 0$. $-x_1 - x_2 + x_3 = 0$,
$\qquad\qquad\qquad\qquad$ $x_1 + x_2 - x_3 = 0$.

9. $(1-i)x_1 + 2x_2 = 0$, **10.** $x_1 + x_2 + x_3 = i$,
$-x_1 - (1+i)x_2 = 0$. $2x_1 + 3x_2 - ix_3 = 0$,
$\qquad\qquad\qquad\qquad$ $x_1 + 2x_2 + x_3 = i$.

11. $2x_1 \qquad + x_3 = -1$,
$-3x_1 + x_2 + 4x_3 = 1$,
$-x_1 + x_2 + 5x_3 = 0$.

12. Use the Gauss-Jordan elimination algorithm to show that the following systems of equations are inconsistent. That is, demonstrate that the existence of a solution would imply a mathematical contradiction.

(a) $2x_1 - x_2 = 2$,
$-6x_1 + 3x_2 = 4$.

(b) $2x_1 \qquad + x_3 = -1$,
$-3x_1 + x_2 + 4x_3 = 1$,
$-x_1 + x_2 + 5x_3 = 1$.

13. Use the Gauss-Jordan elimination algorithm to show that the following system of equations has a unique solution for $r = 2$, but an infinite number of solutions for $r = 1$.

$2x_1 - 3x_2 = rx_1$,
$x_1 - 2x_2 = rx_2$.

14. Use the Gauss-Jordan elimination algorithm to show that the following system of equations has a unique solution for $r = -1$, but an infinite number of solutions for $r = 2$.

$x_1 + 2x_2 - x_3 = rx_1$,
$x_1 \qquad + x_3 = rx_2$,
$4x_1 - 4x_2 + 5x_3 = rx_3$.

9.3 REVIEW 2: MATRICES AND VECTORS

A **matrix** is a rectangular array of numbers arranged in rows and columns. An $m \times n$ matrix—that is, a matrix with m rows and n columns—is usually denoted by

$$
\mathbf{A} := \begin{bmatrix}
a_{11} & a_{12} & a_{13} & \cdots & a_{1n} \\
a_{21} & a_{22} & a_{23} & \cdots & a_{2n} \\
\vdots & \vdots & \vdots & \cdots & \vdots \\
a_{m1} & a_{m2} & a_{m3} & \cdots & a_{mn}
\end{bmatrix} ,
$$

where the element in the ith row and jth column is a_{ij}. The notation $\left[a_{ij}\right]$ is also used to designate \mathbf{A}. The matrices we will work with usually consist of real numbers, but in certain instances we allow complex-number entries.

Some matrices of special interest are **square matrices,** which have the same number of rows and columns; **diagonal matrices,** which are square matrices with only zero entries off the main diagonal (that is, $a_{ij} = 0$ if $i \neq j$); and (column) **vectors,** which are $n \times 1$ matrices. For example, if

$$\mathbf{A} = \begin{bmatrix} 3 & 4 & -1 \\ 2 & 6 & 5 \\ 0 & 1 & 4 \end{bmatrix}, \qquad \mathbf{B} = \begin{bmatrix} 3 & 0 & 0 \\ 0 & 0 & 0 \\ 0 & 0 & 7 \end{bmatrix}, \qquad \mathbf{x} = \begin{bmatrix} 4 \\ 2 \\ 1 \end{bmatrix},$$

then \mathbf{A} is a square matrix, \mathbf{B} is a diagonal matrix, and \mathbf{x} is a vector. An $m \times n$ matrix whose entries are all zero is called a **zero matrix** and is denoted by $\mathbf{0}$. For consistency, we denote matrices by boldfaced capitals, such as \mathbf{A}, \mathbf{B}, \mathbf{C}, \mathbf{I}, \mathbf{X}, and \mathbf{Y}, and reserve boldfaced lowercase letters, such as \mathbf{c}, \mathbf{x}, \mathbf{y}, and \mathbf{z}, for vectors.

Algebra of Matrices

Matrix Addition and Scalar Multiplication. The operations of matrix addition and scalar multiplication are very straightforward. Addition is performed by adding corresponding elements:

$$\begin{bmatrix} 1 & 2 & 3 \\ 4 & 5 & 6 \end{bmatrix} + \begin{bmatrix} 1 & 1 & 1 \\ 1 & 1 & 1 \end{bmatrix} = \begin{bmatrix} 2 & 3 & 4 \\ 5 & 6 & 7 \end{bmatrix}.$$

Formally, the *sum* of two $m \times n$ matrices is given by

$$\mathbf{A} + \mathbf{B} = \left[a_{ij}\right] + \left[b_{ij}\right] = \left[a_{ij} + b_{ij}\right].$$

(The sole novelty here is that addition is not defined for two matrices whose dimensions m, n differ.)

To multiply a matrix by a scalar (number), we simply multiply each element in the matrix by the number:

$$3\begin{bmatrix} 1 & 2 & 3 \\ 4 & 5 & 6 \end{bmatrix} = \begin{bmatrix} 3 & 6 & 9 \\ 12 & 15 & 18 \end{bmatrix}.$$

In other words, $r\mathbf{A} = r\left[a_{ij}\right] = \left[ra_{ij}\right]$. The notation $-\mathbf{A}$ stands for $(-1)\mathbf{A}$.

Properties of Matrix Addition and Scalar Multiplication. Matrix addition and scalar multiplication are nothing more than mere bookkeeping, and the usual algebraic properties hold. If \mathbf{A}, \mathbf{B}, and \mathbf{C} are $m \times n$ matrices and r, s are scalars, then

$$\mathbf{A} + (\mathbf{B} + \mathbf{C}) = (\mathbf{A} + \mathbf{B}) + \mathbf{C}, \qquad \mathbf{A} + \mathbf{B} = \mathbf{B} + \mathbf{A},$$
$$\mathbf{A} + \mathbf{0} = \mathbf{A}, \qquad \mathbf{A} + (-\mathbf{A}) = \mathbf{0},$$
$$r(\mathbf{A} + \mathbf{B}) = r\mathbf{A} + r\mathbf{B}, \qquad (r + s)\mathbf{A} = r\mathbf{A} + s\mathbf{A},$$
$$r(s\mathbf{A}) = (rs)\mathbf{A} = s(r\mathbf{A}).$$

Matrix Multiplication. The matrix product is what makes matrix algebra interesting and useful. We indicated in Section 9.1 that the product of a matrix \mathbf{A} and a column vector \mathbf{x} is the column vector composed of dot products of the rows of \mathbf{A} with \mathbf{x}:

$$\begin{bmatrix} 1 & 2 & 3 \\ 4 & 5 & 6 \end{bmatrix} \begin{bmatrix} 1 \\ 0 \\ 2 \end{bmatrix} = \begin{bmatrix} 1\cdot 1 + 2\cdot 0 + 3\cdot 2 \\ 4\cdot 1 + 5\cdot 0 + 6\cdot 2 \end{bmatrix} = \begin{bmatrix} 7 \\ 16 \end{bmatrix}.$$

More generally, the product of two matrices \mathbf{A} and \mathbf{B} is formed by taking the array of dot products of the *rows* of the first "factor" \mathbf{A} with the *columns* of the second factor \mathbf{B}; the dot product of the *i*th row of \mathbf{A} with the *j*th column of \mathbf{B} is written as the *ij*th entry of the product \mathbf{AB}:

$$\begin{bmatrix} 1 & 0 & 1 \\ 3 & -1 & 2 \end{bmatrix} \begin{bmatrix} 1 & 2 & x \\ -1 & -1 & y \\ 4 & 1 & z \end{bmatrix} = \begin{bmatrix} 1+0+4 & 2+0+1 & x+0+z \\ 3+1+8 & 6+1+2 & 3x-y+2z \end{bmatrix}$$

$$= \begin{bmatrix} 5 & 3 & x+z \\ 12 & 9 & 3x-y+2z \end{bmatrix}.$$

Note that \mathbf{AB} is only defined when the number of columns of \mathbf{A} matches the number of rows of \mathbf{B}. A useful formula for the product of an $m \times n$ matrix \mathbf{A} and an $n \times p$ matrix \mathbf{B} is

$$\mathbf{AB} := \begin{bmatrix} c_{ij} \end{bmatrix}, \qquad \text{where} \quad c_{ij} := \sum_{k=1}^{n} a_{ik} b_{kj}.$$

The dot product of the *i*th row of \mathbf{A} and the *j*th column of \mathbf{B} is seen in the "sum of products" expression for c_{ij}.

Since \mathbf{AB} is computed in terms of the *rows* of the first factor and the *columns* of the second factor, it should not be surprising that, in general, \mathbf{AB} does not equal \mathbf{BA} (matrix multiplication does not *commute*):

$$\begin{bmatrix} 1 & 2 \\ 3 & 4 \end{bmatrix} \begin{bmatrix} 0 & 1 \\ 1 & 0 \end{bmatrix} = \begin{bmatrix} 2 & 1 \\ 4 & 3 \end{bmatrix}, \qquad \text{but} \qquad \begin{bmatrix} 0 & 1 \\ 1 & 0 \end{bmatrix} \begin{bmatrix} 1 & 2 \\ 3 & 4 \end{bmatrix} = \begin{bmatrix} 3 & 4 \\ 1 & 2 \end{bmatrix}.$$

In fact, the dimensions of \mathbf{A} and \mathbf{B} may render one or the other of these products undefined:

$$\begin{bmatrix} 1 & 2 \\ 3 & 4 \end{bmatrix} \begin{bmatrix} 0 \\ 1 \end{bmatrix} = \begin{bmatrix} 2 \\ 4 \end{bmatrix}; \qquad \begin{bmatrix} 0 \\ 1 \end{bmatrix} \begin{bmatrix} 1 & 2 \\ 3 & 4 \end{bmatrix} \quad \text{not defined}.$$

By the same token, one might not expect $(\mathbf{AB})\mathbf{C}$ to equal $\mathbf{A}(\mathbf{BC})$, since in $(\mathbf{AB})\mathbf{C}$ we take dot products with the *columns* of \mathbf{B}, whereas in $\mathbf{A}(\mathbf{BC})$ we employ the *rows* of \mathbf{B}. So it is a pleasant surprise that *this complication does not arise,* and the "parenthesis grouping" rules are the customary ones:

PROPERTIES OF MATRIX MULTIPLICATION

$(\mathbf{AB})\mathbf{C} = \mathbf{A}(\mathbf{BC})$	(Associativity)
$(\mathbf{A} + \mathbf{B})\mathbf{C} = \mathbf{AC} + \mathbf{BC}$	(Distributivity)
$\mathbf{A}(\mathbf{B} + \mathbf{C}) = \mathbf{AB} + \mathbf{AC}$	(Distributivity)
$(r\mathbf{A})\mathbf{B} = r(\mathbf{AB}) = \mathbf{A}(r\mathbf{B})$	(Associativity)

To summarize, the algebra of matrices proceeds much like the standard algebra of numbers, *except that we should never presume that we can switch the order of matrix factors.*

Matrices as Linear Operators. Let \mathbf{A} be an $m \times n$ matrix and let \mathbf{x} and \mathbf{y} be $n \times 1$ vectors. Then \mathbf{Ax} is an $m \times 1$ vector, and so we can think of multiplication by \mathbf{A} as defining an operator that maps $n \times 1$ vectors into $m \times 1$ vectors. A consequence of the distributivity and associativity properties is that multiplication by \mathbf{A} defines a **linear operator**, since $\mathbf{A}(\mathbf{x} + \mathbf{y}) = \mathbf{Ax} + \mathbf{Ay}$ and $\mathbf{A}(r\mathbf{x}) = r\mathbf{Ax}$. Moreover, if \mathbf{A} is an $m \times n$ matrix and \mathbf{B} is an $n \times p$ matrix, then the $m \times p$ matrix \mathbf{AB} defines a linear operator that is the composition of the linear operator defined by \mathbf{B} with the linear operator defined by \mathbf{A}. That is, $(\mathbf{AB})\mathbf{x} = \mathbf{A}(\mathbf{Bx})$, where \mathbf{x} is a $p \times 1$ vector.

Examples of linear operations are

 (i) stretching or contracting the components of a vector by constant factors;
 (ii) rotating a vector through some angle about a fixed axis;
 (iii) reflecting a vector in a plane mirror.

The Matrix Formulation of Linear Algebraic Systems. Matrix algebra was developed to provide a convenient tool for expressing and analyzing linear algebraic systems. Note that the set of equations

$$
\begin{aligned}
x_1 + 2x_2 + \ x_3 &= 1 \ , \\
x_1 + 3x_2 + 2x_3 &= -1 \ , \\
x_1 \qquad\quad + \ x_3 &= 0
\end{aligned}
$$

can be written using the matrix product

(1)
$$
\begin{bmatrix} 1 & 2 & 1 \\ 1 & 3 & 2 \\ 1 & 0 & 1 \end{bmatrix} \begin{bmatrix} x_1 \\ x_2 \\ x_3 \end{bmatrix} = \begin{bmatrix} 1 \\ -1 \\ 0 \end{bmatrix} .
$$

In general, we express the linear system

$$
\begin{aligned}
a_{11}x_1 + a_{12}x_2 + \cdots + a_{1n}x_n &= b_1 \ , \\
a_{21}x_1 + a_{22}x_2 + \cdots + a_{2n}x_n &= b_2 \ , \\
&\ \ \vdots \\
a_{n1}x_1 + a_{n2}x_2 + \cdots + a_{nn}x_n &= b_n
\end{aligned}
$$

in matrix notation as $\mathbf{Ax = b}$, where \mathbf{A} is the *coefficient matrix,* \mathbf{x} is the vector of unknowns, and \mathbf{b} is the vector of constants occurring on the right-hand side:

$$\mathbf{A} = \begin{bmatrix} a_{11} & a_{12} & \cdots & a_{1n} \\ a_{21} & a_{22} & \cdots & a_{2n} \\ \vdots & \vdots & & \vdots \\ a_{n1} & a_{n2} & \cdots & a_{nn} \end{bmatrix}, \qquad \mathbf{x} = \begin{bmatrix} x_1 \\ x_2 \\ \vdots \\ x_n \end{bmatrix}, \qquad \mathbf{b} = \begin{bmatrix} b_1 \\ b_2 \\ \vdots \\ b_n \end{bmatrix}.$$

If $\mathbf{b = 0}$, the system $\mathbf{Ax = b}$ is said to be *homogeneous* (analogous to the nomenclature of Section 4.2).

Matrix Identity. There is a "multiplicative identity" in matrix algebra, namely, a square diagonal matrix \mathbf{I} with ones down the main diagonal. Multiplying \mathbf{I} on the right or left by any other matrix (with compatible dimensions) reproduces the latter matrix:

$$\begin{bmatrix} 1 & 0 & 0 \\ 0 & 1 & 0 \\ 0 & 0 & 1 \end{bmatrix} \begin{bmatrix} 1 & 2 & 1 \\ 1 & 3 & 2 \\ 1 & 0 & 1 \end{bmatrix} = \begin{bmatrix} 1 & 2 & 1 \\ 1 & 3 & 2 \\ 1 & 0 & 1 \end{bmatrix} = \begin{bmatrix} 1 & 2 & 1 \\ 1 & 3 & 2 \\ 1 & 0 & 1 \end{bmatrix} \begin{bmatrix} 1 & 0 & 0 \\ 0 & 1 & 0 \\ 0 & 0 & 1 \end{bmatrix}.$$

(The notation \mathbf{I}_n is used if it is convenient to specify the dimensions, $n \times n$, of the identity matrix.)

Matrix Inverse. Some *square* matrices \mathbf{A} can be paired with other (square) matrices \mathbf{B} having the property that $\mathbf{BA = I}$:

$$(2) \qquad \begin{bmatrix} \frac{3}{2} & -1 & \frac{1}{2} \\ \frac{1}{2} & 0 & -\frac{1}{2} \\ -\frac{3}{2} & 1 & \frac{1}{2} \end{bmatrix} \begin{bmatrix} 1 & 2 & 1 \\ 1 & 3 & 2 \\ 1 & 0 & 1 \end{bmatrix} = \begin{bmatrix} 1 & 0 & 0 \\ 0 & 1 & 0 \\ 0 & 0 & 1 \end{bmatrix}.$$

When this happens, it can be shown that

(i) \mathbf{B} is the *unique* matrix satisfying $\mathbf{BA = I}$, and
(ii) \mathbf{B} also satisfies $\mathbf{AB = I}$.

In such a case we say that \mathbf{B} is the **inverse** of \mathbf{A} and write $\mathbf{B = A^{-1}}$.

Not every matrix possesses an inverse; the zero matrix $\mathbf{0}$, for example, can never satisfy the equation $\mathbf{0B = I}$. A matrix that has no inverse is said to be **singular.**

If we know an inverse for the coefficient matrix \mathbf{A} in a system of linear equations $\mathbf{Ax = b}$, the solution can be calculated directly by computing $\mathbf{A^{-1}b}$, as the following derivation shows:

$$\mathbf{Ax = b} \text{ implies } \mathbf{A^{-1}Ax = A^{-1}b} \text{ implies } \mathbf{x = A^{-1}b} \ .$$

Using (2), for example, we can solve equation (1) quite efficiently:

$$\begin{bmatrix} x_1 \\ x_2 \\ x_3 \end{bmatrix} = \begin{bmatrix} \frac{3}{2} & -1 & \frac{1}{2} \\ \frac{1}{2} & 0 & -\frac{1}{2} \\ -\frac{3}{2} & 1 & \frac{1}{2} \end{bmatrix} \begin{bmatrix} 1 \\ -1 \\ 0 \end{bmatrix} = \begin{bmatrix} \frac{5}{2} \\ \frac{1}{2} \\ -\frac{5}{2} \end{bmatrix}.$$

When \mathbf{A}^{-1} is known, this calculation is certainly easier than applying the Gauss-Jordan algorithm of the previous section. So it would appear advantageous to be able to find matrix inverses. Some inverses can be obtained directly from the interpretation of the matrix as a linear operator. For example, the inverse of a matrix that rotates a vector is the matrix that rotates it in the opposite direction. A matrix that performs a mirror reflection is its own inverse (what do you get if you reflect twice?). But in general one must employ an algorithm to compute a matrix inverse. The underlying strategy for this algorithm is based on the observation that if \mathbf{X} denotes the inverse of \mathbf{A}, then \mathbf{X} must satisfy the equation $\mathbf{AX} = \mathbf{I}$; finding \mathbf{X} amounts to solving n linear systems of equations for the columns $\{\mathbf{x}_1, \mathbf{x}_2, \ldots, \mathbf{x}_n\}$ of \mathbf{X}:

$$\mathbf{Ax}_1 = \begin{bmatrix} 1 \\ 0 \\ 0 \\ \vdots \\ 0 \\ 0 \end{bmatrix}, \quad \mathbf{Ax}_2 = \begin{bmatrix} 0 \\ 1 \\ 0 \\ \vdots \\ 0 \\ 0 \end{bmatrix}, \quad \cdots, \quad \mathbf{Ax}_n = \begin{bmatrix} 0 \\ 0 \\ 0 \\ \vdots \\ 0 \\ 1 \end{bmatrix}.$$

The implementation of this operation is efficiently executed by the following variation of the Gauss-Jordan algorithm.

Finding the Inverse of a Matrix. By a **row operation,** we mean any one of the following:

(a) Interchanging two rows of the matrix

(b) Multiplying a row of the matrix by a nonzero scalar

(c) Adding a scalar multiple of one row of the matrix to another row.

If the $n \times n$ matrix \mathbf{A} has an inverse, then \mathbf{A}^{-1} can be determined by performing row operations on the $n \times 2n$ matrix $\begin{bmatrix} \mathbf{A} \vdots \mathbf{I} \end{bmatrix}$ obtained by writing \mathbf{A} and \mathbf{I} side by side. In particular, we perform row operations on the matrix $\begin{bmatrix} \mathbf{A} \vdots \mathbf{I} \end{bmatrix}$ until the first n rows and columns form the identity matrix; that is, the new matrix is $\begin{bmatrix} \mathbf{I} \vdots \mathbf{B} \end{bmatrix}$. Then $\mathbf{A}^{-1} = \mathbf{B}$. We remark that if this procedure fails to produce a matrix of the form $\begin{bmatrix} \mathbf{I} \vdots \mathbf{B} \end{bmatrix}$, then \mathbf{A} has no inverse.

EXAMPLE 1 Find the inverse of $\mathbf{A} = \begin{bmatrix} 1 & 2 & 1 \\ 1 & 3 & 2 \\ 1 & 0 & 1 \end{bmatrix}$.

SOLUTION We first form the matrix $\begin{bmatrix} \mathbf{A} \mid \mathbf{I} \end{bmatrix}$ and *row-reduce* the matrix to $\begin{bmatrix} \mathbf{I} \mid \mathbf{A}^{-1} \end{bmatrix}$. Computing, we find the following:

The matrix $\begin{bmatrix} \mathbf{A} \mid \mathbf{I} \end{bmatrix}$

$$\left[\begin{array}{ccc|ccc} 1 & 2 & 1 & 1 & 0 & 0 \\ 1 & 3 & 2 & 0 & 1 & 0 \\ 1 & 0 & 1 & 0 & 0 & 1 \end{array}\right].$$

Subtract the first row from the second and third to obtain

$$\left[\begin{array}{ccc|ccc} 1 & 2 & 1 & 1 & 0 & 0 \\ 0 & 1 & 1 & -1 & 1 & 0 \\ 0 & -2 & 0 & -1 & 0 & 1 \end{array}\right].$$

Add 2 times the second row to the third row to obtain

$$\left[\begin{array}{ccc|ccc} 1 & 2 & 1 & 1 & 0 & 0 \\ 0 & 1 & 1 & -1 & 1 & 0 \\ 0 & 0 & 2 & -3 & 2 & 1 \end{array}\right].$$

Subtract 2 times the second row from the first to obtain

$$\left[\begin{array}{ccc|ccc} 1 & 0 & -1 & 3 & -2 & 0 \\ 0 & 1 & 1 & -1 & 1 & 0 \\ 0 & 0 & 2 & -3 & 2 & 1 \end{array}\right].$$

Multiply the third row by $1/2$ to obtain

$$\left[\begin{array}{ccc|ccc} 1 & 0 & -1 & 3 & -2 & 0 \\ 0 & 1 & 1 & -1 & 1 & 0 \\ 0 & 0 & 1 & -\frac{3}{2} & 1 & \frac{1}{2} \end{array}\right].$$

Add the third row to the first and then subtract the third row from the second to obtain

$$\left[\begin{array}{ccc|ccc} 1 & 0 & 0 & \frac{3}{2} & -1 & \frac{1}{2} \\ 0 & 1 & 0 & \frac{1}{2} & 0 & -\frac{1}{2} \\ 0 & 0 & 1 & -\frac{3}{2} & 1 & \frac{1}{2} \end{array}\right].$$

The matrix shown in color is \mathbf{A}^{-1}. (Compare equation (2).) ■ End

Determinants. For a 2×2 matrix \mathbf{A}, the **determinant of A**, denoted det \mathbf{A} or $|\mathbf{A}|$, is defined by

$$\det \mathbf{A} := \begin{vmatrix} a_{11} & a_{12} \\ a_{21} & a_{22} \end{vmatrix} = a_{11}a_{22} - a_{12}a_{21} \ .$$

We can define the determinant of a 3×3 matrix \mathbf{A} in terms of its cofactor expansion about the first row; that is,

$$\det \mathbf{A} := \begin{vmatrix} a_{11} & a_{12} & a_{13} \\ a_{21} & a_{22} & a_{23} \\ a_{31} & a_{32} & a_{33} \end{vmatrix} = a_{11}\begin{vmatrix} a_{22} & a_{23} \\ a_{32} & a_{33} \end{vmatrix} - a_{12}\begin{vmatrix} a_{21} & a_{23} \\ a_{31} & a_{33} \end{vmatrix} + a_{13}\begin{vmatrix} a_{21} & a_{22} \\ a_{31} & a_{32} \end{vmatrix} \ .$$

For example,

$$\begin{vmatrix} 1 & 2 & 1 \\ 0 & 3 & 5 \\ 2 & 1 & -1 \end{vmatrix} = 1 \begin{vmatrix} 3 & 5 \\ 1 & -1 \end{vmatrix} - 2 \begin{vmatrix} 0 & 5 \\ 2 & -1 \end{vmatrix} + 1 \begin{vmatrix} 0 & 3 \\ 2 & 1 \end{vmatrix}$$

$$= 1(-3 - 5) - 2(0 - 10) + 1(0 - 6) = 6 \ .$$

The determinant of an $n \times n$ matrix can be similarly defined by a cofactor expansion involving $(n - 1)$st order determinants. However, a more practical way to evaluate the determinant when n is large involves the row-reduction of the matrix to upper triangular form. Since here we deal mainly with 2×2 and 3×3 matrices, we refer the reader to an elementary linear algebra text for a further discussion of evaluating determinants.[†]

Determinants—particularly the higher-order ones—are laborious to compute directly. They have a geometric interpretation: det \mathbf{A} is the volume (in n-dimensional space) of the parallelepiped whose edges are given by the column vectors of \mathbf{A}. But their chief value lies in the role they play in the following theorem, which summarizes many of the results from linear algebra that we shall need, and in Cramer's rule, described in Appendix C.

MATRICES AND SYSTEMS OF EQUATIONS

Theorem 1. Let \mathbf{A} be an $n \times n$ matrix. The following statements are equivalent:

(a) \mathbf{A} is singular (does not have an inverse).
(b) The determinant of \mathbf{A} is zero.
(c) $\mathbf{A}\mathbf{x} = \mathbf{0}$ has nontrivial solutions ($\mathbf{x} \neq \mathbf{0}$).
(d) The columns (rows) of \mathbf{A} form a linearly dependent set.

In part (d), the statement that the n columns of \mathbf{A} are linearly dependent means that there exist scalars, c_1, \ldots, c_n *not all zero* such that

$$c_1 \mathbf{a}_1 + c_2 \mathbf{a}_2 + \cdots + c_n \mathbf{a}_n = \mathbf{0} \ ,$$

where \mathbf{a}_j is the vector forming the jth column of \mathbf{A}.

If \mathbf{A} is a singular square matrix (so det $\mathbf{A} = 0$), then $\mathbf{A}\mathbf{x} = \mathbf{0}$ has infinitely many solutions. Indeed, Theorem 1 asserts that there is a vector $\mathbf{x}_0 \neq \mathbf{0}$ such that $\mathbf{A}\mathbf{x}_0 = \mathbf{0}$, and we can get infinitely many other solutions by multiplying \mathbf{x}_0 by any scalar, i.e., taking $\mathbf{x} = c\mathbf{x}_0$. Furthermore, $\mathbf{A}\mathbf{x} = \mathbf{b}$ either has no solutions or it has infinitely many of them of the form

$$\mathbf{x} = \mathbf{x}_p + \mathbf{x}_h \ ,$$

where \mathbf{x}_p is a *particular* solution to $\mathbf{A}\mathbf{x} = \mathbf{b}$ and \mathbf{x}_h is any of the infinity of solutions to $\mathbf{A}\mathbf{x} = \mathbf{0}$ (see Problem 15). The resemblance of this situation to that of solving nonhomogeneous linear differential equations should be quite apparent.

[†]See *Linear Algebra and Its Applications,* 2nd Edition, by David C. Lay (Addison-Wesley, Reading, Mass., 1997).

To illustrate, in Example 3 of Section 9.2 (page 534) we saw that the system

$$\begin{bmatrix} 2 & 4 & 1 \\ 2 & 4 & 0 \\ -4 & -8 & 1 \end{bmatrix} \begin{bmatrix} x_1 \\ x_2 \\ x_3 \end{bmatrix} = \begin{bmatrix} 8 \\ 6 \\ -10 \end{bmatrix}$$

has solutions

$$x_1 = 3 - 2s , \qquad x_2 = s , \qquad x_3 = 2 ; \qquad -\infty < s < \infty .$$

Writing these in matrix notation, we can identify the vectors \mathbf{x}_p and \mathbf{x}_h mentioned above:

$$\mathbf{x} = \begin{bmatrix} 3 - 2s \\ s \\ 2 \end{bmatrix} = \begin{bmatrix} 3 \\ 0 \\ 2 \end{bmatrix} + s \begin{bmatrix} -2 \\ 1 \\ 0 \end{bmatrix} = \mathbf{x}_p + \mathbf{x}_h .$$

Note further that the determinant of \mathbf{A} is indeed zero,

$$\det \mathbf{A} = 2 \begin{vmatrix} 4 & 0 \\ -8 & 1 \end{vmatrix} - 4 \begin{vmatrix} 2 & 0 \\ -4 & 1 \end{vmatrix} + 1 \begin{vmatrix} 2 & 4 \\ -4 & -8 \end{vmatrix} = 2 \cdot 4 - 4 \cdot 2 + 1 \cdot 0 = 0 ,$$

and that the linear dependence of the columns of \mathbf{A} is exhibited by the identity

$$-2 \begin{bmatrix} 2 \\ 2 \\ -4 \end{bmatrix} + 1 \begin{bmatrix} 4 \\ 4 \\ -8 \end{bmatrix} + 0 \begin{bmatrix} 1 \\ 0 \\ 1 \end{bmatrix} = \begin{bmatrix} 0 \\ 0 \\ 0 \end{bmatrix} .$$

If \mathbf{A} is a nonsingular square matrix (i.e., \mathbf{A} has an inverse and $\det \mathbf{A} \neq 0$), then the homogeneous system $\mathbf{Ax} = \mathbf{0}$ has $\mathbf{x} = \mathbf{0}$ as its only solution. More generally, when $\det \mathbf{A} \neq 0$, the system $\mathbf{Ax} = \mathbf{b}$ has a unique solution, (namely, $\mathbf{x} = \mathbf{A}^{-1}\mathbf{b}$).

Calculus of Matrices

If we allow the entries $a_{ij}(t)$ in a matrix $\mathbf{A}(t)$ to be functions of the variable t, then $\mathbf{A}(t)$ is a **matrix function of** t. Similarly, if the entries $x_i(t)$ of a vector $\mathbf{x}(t)$ are functions of t, then $\mathbf{x}(t)$ is a **vector function of** t.

These matrix and vector functions have a calculus much like that of real-valued functions. A matrix $\mathbf{A}(t)$ is said to be **continuous at** t_0 if each entry $a_{ij}(t)$ is continuous at t_0. Moreover, $\mathbf{A}(t)$ is **differentiable at** t_0 if each entry $a_{ij}(t)$ is differentiable at t_0, and we write

$$(3) \qquad \frac{d\mathbf{A}}{dt}(t_0) = \mathbf{A}'(t_0) := \left[a'_{ij}(t_0) \right] .$$

Similarly, we define

$$(4) \qquad \int_a^b \mathbf{A}(t)dt := \left[\int_a^b a_{ij}(t)dt \right] .$$

EXAMPLE 2 Let $\mathbf{A}(t) = \begin{bmatrix} t^2 + 1 & \cos t \\ e^t & 1 \end{bmatrix}$.

Find: **(a)** $\mathbf{A}'(t)$. **(b)** $\displaystyle\int_0^1 \mathbf{A}(t)\,dt$.

SOLUTION Using formulas (3) and (4), we compute

(a) $\mathbf{A}'(t) = \begin{bmatrix} 2t & -\sin t \\ e^t & 0 \end{bmatrix}$. **(b)** $\displaystyle\int_0^1 \mathbf{A}(t)\,dt = \begin{bmatrix} \frac{4}{3} & \sin 1 \\ e - 1 & 1 \end{bmatrix}$. ■ End

EXAMPLE 3 Show that $\mathbf{x}(t) = \begin{bmatrix} \cos \omega t \\ \sin \omega t \end{bmatrix}$ is a solution of the matrix differential equation $\mathbf{x}' = \mathbf{A}\mathbf{x}$, where

$$\mathbf{A} = \begin{bmatrix} 0 & -\omega \\ \omega & 0 \end{bmatrix} .$$

SOLUTION We simply verify that $\mathbf{x}'(t)$ and $\mathbf{A}\mathbf{x}(t)$ are the same vector function:

$$\mathbf{x}'(t) = \begin{bmatrix} -\omega \sin \omega t \\ \omega \cos \omega t \end{bmatrix} ; \qquad \mathbf{A}\mathbf{x} = \begin{bmatrix} 0 & -\omega \\ \omega & 0 \end{bmatrix} \begin{bmatrix} \cos \omega t \\ \sin \omega t \end{bmatrix} = \begin{bmatrix} -\omega \sin \omega t \\ \omega \cos \omega t \end{bmatrix} .$$ ■ End

The basic properties of differentiation are valid for matrix functions.

DIFFERENTIATION FORMULAS FOR MATRIX FUNCTIONS

$$\frac{d}{dt}(\mathbf{C}\mathbf{A}) = \mathbf{C}\frac{d\mathbf{A}}{dt} , \quad \mathbf{C} \text{ a constant matrix.}$$

$$\frac{d}{dt}(\mathbf{A} + \mathbf{B}) = \frac{d\mathbf{A}}{dt} + \frac{d\mathbf{B}}{dt} .$$

$$\frac{d}{dt}(\mathbf{A}\mathbf{B}) = \mathbf{A}\frac{d\mathbf{B}}{dt} + \frac{d\mathbf{A}}{dt}\mathbf{B} .$$

In the last formula, the order in which the matrices are written is very important because, as we have emphasized, matrix multiplication does not always commute.

EXERCISES 9.3

1. Let $\mathbf{A} := \begin{bmatrix} 2 & 1 \\ 3 & 5 \end{bmatrix}$ and $\mathbf{B} := \begin{bmatrix} -1 & 0 \\ 2 & -3 \end{bmatrix}$.

Find: **(a)** $\mathbf{A} + \mathbf{B}$. **(b)** $3\mathbf{A} - \mathbf{B}$.

2. Let $\mathbf{A} := \begin{bmatrix} 2 & 0 & 5 \\ 2 & 1 & 1 \end{bmatrix}$ and $\mathbf{B} := \begin{bmatrix} 1 & -1 & 2 \\ 0 & 3 & -2 \end{bmatrix}$.

Find: **(a)** $\mathbf{A} + \mathbf{B}$. **(b)** $7\mathbf{A} - 4\mathbf{B}$.

3. Let $\mathbf{A} := \begin{bmatrix} 2 & 4 \\ 1 & 1 \end{bmatrix}$ and $\mathbf{B} := \begin{bmatrix} -1 & 3 \\ 5 & 2 \end{bmatrix}$.

Find: **(a)** \mathbf{AB}. **(b)** $\mathbf{A}^2 = \mathbf{AA}$. **(c)** $\mathbf{B}^2 = \mathbf{BB}$.

4. Let $\mathbf{A} := \begin{bmatrix} 2 & 1 \\ 0 & 4 \\ -1 & 3 \end{bmatrix}$ and $\mathbf{B} := \begin{bmatrix} 1 & 1 & -1 \\ 0 & 3 & 1 \end{bmatrix}$.

Find: **(a)** \mathbf{AB}. **(b)** \mathbf{BA}.

5. Let $\mathbf{A} := \begin{bmatrix} 1 & -2 \\ 2 & -3 \end{bmatrix}$, $\mathbf{B} := \begin{bmatrix} 1 & 0 \\ 1 & 1 \end{bmatrix}$, and

$\mathbf{C} := \begin{bmatrix} -1 & 1 \\ 2 & 1 \end{bmatrix}$.

Find: **(a)** \mathbf{AB}. **(b)** \mathbf{AC}. **(c)** $\mathbf{A(B + C)}$.

6. Let $\mathbf{A} := \begin{bmatrix} 1 & 2 \\ 1 & 1 \end{bmatrix}$, $\mathbf{B} := \begin{bmatrix} 0 & 3 \\ 1 & 2 \end{bmatrix}$, and

$\mathbf{C} := \begin{bmatrix} 1 & -4 \\ 1 & 1 \end{bmatrix}$.

Find: **(a)** \mathbf{AB}. **(b)** $\mathbf{(AB)C}$. **(c)** $\mathbf{(A + B)C}$.

7. The matrix obtained from \mathbf{A} by interchanging its rows and columns is called the **transpose** of \mathbf{A} and is denoted by \mathbf{A}^T. For example, if

$$\mathbf{A} = \begin{bmatrix} 1 & 2 & 6 \\ -1 & 2 & -1 \end{bmatrix}, \quad \text{then}$$

$$\mathbf{A}^T = \begin{bmatrix} 1 & -1 \\ 2 & 2 \\ 6 & -1 \end{bmatrix}.$$

In general, we have $\left[a_{ij} \right]^T = \left[b_{ij} \right]$, where $b_{ij} = a_{ji}$.

(a) Show that if \mathbf{u} and \mathbf{v} are each $n \times 1$ column vectors, then the matrix product $\mathbf{u}^T\mathbf{v}$ is the same as the dot product $\mathbf{u} \cdot \mathbf{v}$.

(b) Let \mathbf{v} be a 3×1 column vector with $\mathbf{v}^T = \begin{bmatrix} 2 & 3 & 5 \end{bmatrix}$. Show that, for \mathbf{A} as given above, $(\mathbf{Av})^T = \mathbf{v}^T\mathbf{A}^T$.

(c) Does $(\mathbf{Av})^T = \mathbf{v}^T\mathbf{A}^T$ hold for every $m \times n$ matrix \mathbf{A} and $n \times 1$ vector \mathbf{v}?

(d) Does $(\mathbf{AB})^T = \mathbf{B}^T\mathbf{A}^T$ hold for every pair of matrices \mathbf{A}, \mathbf{B} such that both matrix products are defined?

8. Let $\mathbf{A} := \begin{bmatrix} 2 & -1 \\ -3 & 4 \end{bmatrix}$ and $\mathbf{B} := \begin{bmatrix} 1 & 2 \\ 3 & 2 \end{bmatrix}$.

Verify that $\mathbf{AB} \neq \mathbf{BA}$.

In Problems 9 through 14, compute the inverse of the given matrix, if it exists.

9. $\begin{bmatrix} 2 & 1 \\ -1 & 4 \end{bmatrix}$.

10. $\begin{bmatrix} 4 & 1 \\ 5 & 9 \end{bmatrix}$.

11. $\begin{bmatrix} 1 & 1 & 1 \\ 1 & 2 & 1 \\ 2 & 3 & 2 \end{bmatrix}$.

12. $\begin{bmatrix} 1 & 1 & 1 \\ 1 & 2 & 3 \\ 0 & 1 & 1 \end{bmatrix}$.

13. $\begin{bmatrix} -2 & -1 & 1 \\ 2 & 1 & 0 \\ 3 & 1 & -1 \end{bmatrix}$.

14. $\begin{bmatrix} 1 & 1 & 1 \\ 1 & -1 & 2 \\ 1 & 1 & 4 \end{bmatrix}$.

15. Prove that if \mathbf{x}_p satisfies $\mathbf{Ax}_p = \mathbf{b}$, then every solution to the nonhomogeneous system $\mathbf{Ax} = \mathbf{b}$ is of the form $\mathbf{x} = \mathbf{x}_p + \mathbf{x}_h$, where \mathbf{x}_h is a solution to the corresponding homogeneous system $\mathbf{Ax} = \mathbf{0}$.

16. Let $\mathbf{A} = \begin{bmatrix} 2 & -1 & 1 \\ -1 & 2 & 1 \\ 1 & 1 & 2 \end{bmatrix}$.

(a) Show that \mathbf{A} is singular.

(b) Show that $\mathbf{Ax} = \begin{bmatrix} 3 \\ 1 \\ 3 \end{bmatrix}$ has no solutions.

(c) Show that $\mathbf{Ax} = \begin{bmatrix} 3 \\ 0 \\ 3 \end{bmatrix}$ has infinitely many solutions.

In Problems 17 through 20, find the matrix function $\mathbf{X}^{-1}(t)$ *whose value at t is the inverse of the given matrix* $\mathbf{X}(t)$.

17. $\mathbf{X}(t) = \begin{bmatrix} e^t & e^{4t} \\ e^t & 4e^{4t} \end{bmatrix}$.

18. $\mathbf{X}(t) = \begin{bmatrix} \sin 2t & \cos 2t \\ 2\cos 2t & -2\sin 2t \end{bmatrix}$.

19. $\mathbf{X}(t) = \begin{bmatrix} e^t & e^{-t} & e^{2t} \\ e^t & -e^{-t} & 2e^{2t} \\ e^t & e^{-t} & 4e^{2t} \end{bmatrix}$.

20. $\mathbf{X}(t) = \begin{bmatrix} e^{3t} & 1 & t \\ 3e^{3t} & 0 & 1 \\ 9e^{3t} & 0 & 0 \end{bmatrix}$.

In Problems 21 through 26, evaluate the given determinant.

21. $\begin{vmatrix} 4 & 3 \\ -1 & 2 \end{vmatrix}$.

22. $\begin{vmatrix} 12 & 8 \\ 3 & 2 \end{vmatrix}$.

23. $\begin{vmatrix} 1 & 0 & 0 \\ 3 & 1 & 2 \\ 1 & 5 & -2 \end{vmatrix}$.

24. $\begin{vmatrix} 1 & 0 & 2 \\ 0 & 3 & -1 \\ -1 & 2 & 1 \end{vmatrix}$.

25. $\begin{vmatrix} 1 & 4 & 3 \\ -1 & -1 & 2 \\ 4 & 5 & 2 \end{vmatrix}$.

26. $\begin{vmatrix} 1 & 4 & 4 \\ 3 & 0 & -3 \\ 1 & 6 & 2 \end{vmatrix}$.

In Problems 27 through 29, determine the values of r for which $\det(\mathbf{A} - r\mathbf{I}) = 0$.

27. $\mathbf{A} = \begin{bmatrix} 1 & 1 \\ -2 & 4 \end{bmatrix}$.

28. $\mathbf{A} = \begin{bmatrix} 3 & 3 \\ 2 & 4 \end{bmatrix}$.

29. $\mathbf{A} = \begin{bmatrix} 0 & 0 & 0 \\ 0 & 1 & 0 \\ 1 & 0 & 1 \end{bmatrix}$.

30. Illustrate the equivalence of the assertions (a)–(d) in Theorem 1 (page 543) for the matrix

$$\mathbf{A} = \begin{bmatrix} 4 & -2 & 2 \\ -2 & 4 & 2 \\ 2 & 2 & 4 \end{bmatrix}$$

as follows.
 (a) Show that the row-reduction procedure applied to $[\mathbf{A} \mid \mathbf{I}]$ fails to produce the inverse of \mathbf{A}.
 (b) Calculate $\det \mathbf{A}$.
 (c) Determine a nontrivial solution \mathbf{x} to $\mathbf{Ax} = \mathbf{0}$.
 (d) Find scalars c_1, c_2, and c_3, *not all zero,* so that $c_1\mathbf{a}_1 + c_2\mathbf{a}_2 + c_3\mathbf{a}_3 = \mathbf{0}$, where \mathbf{a}_1, \mathbf{a}_2, and \mathbf{a}_3 are the columns of \mathbf{A}.

In Problems 31 and 32, find $d\mathbf{x}/dt$ for the given vector functions.

31. $\mathbf{x}(t) = \begin{bmatrix} e^{3t} \\ 2e^{3t} \\ -e^{3t} \end{bmatrix}$

32. $\mathbf{x}(t) = \begin{bmatrix} e^{-t}\sin 3t \\ 0 \\ -e^{-t}\sin 3t \end{bmatrix}$.

In Problems 33 and 34, find $d\mathbf{X}/dt$ for the given matrix functions.

33. $\mathbf{X}(t) = \begin{bmatrix} e^{5t} & 3e^{2t} \\ -2e^{5t} & -e^{2t} \end{bmatrix}$.

34. $\mathbf{X}(t) = \begin{bmatrix} \sin 2t & \cos 2t & e^{-2t} \\ -\sin 2t & 2\cos 2t & 3e^{-2t} \\ 3\sin 2t & \cos 2t & e^{-2t} \end{bmatrix}$.

In Problems 35 and 36, verify that the given vector function satisfies the given system.

35. $\mathbf{x}' = \begin{bmatrix} 1 & 1 \\ -2 & 4 \end{bmatrix}\mathbf{x}$, $\mathbf{x}(t) = \begin{bmatrix} e^{3t} \\ 2e^{3t} \end{bmatrix}$.

36. $\mathbf{x}' = \begin{bmatrix} 0 & 0 & 0 \\ 0 & 1 & 0 \\ 1 & 0 & 1 \end{bmatrix}\mathbf{x}$, $\mathbf{x}(t) = \begin{bmatrix} 0 \\ e^t \\ -3e^t \end{bmatrix}$.

In Problems 37 and 38, verify that the given matrix function satisfies the given matrix differential equation.

37. $\mathbf{X}' = \begin{bmatrix} 1 & -1 \\ 2 & 4 \end{bmatrix}\mathbf{X}$, $\mathbf{X}(t) = \begin{bmatrix} e^{2t} & e^{3t} \\ -e^{2t} & -2e^{3t} \end{bmatrix}$.

38. $\mathbf{X}' = \begin{bmatrix} 1 & 0 & 0 \\ 0 & 3 & -2 \\ 0 & -2 & 3 \end{bmatrix}\mathbf{X}$,

$$\mathbf{X}(t) = \begin{bmatrix} e^t & 0 & 0 \\ 0 & e^t & e^{5t} \\ 0 & e^t & -e^{5t} \end{bmatrix}.$$

In Problems 39 and 40, the matrices $\mathbf{A}(t)$ and $\mathbf{B}(t)$ are given. Find

(a) $\displaystyle\int \mathbf{A}(t)\,dt$. **(b)** $\displaystyle\int_0^1 \mathbf{B}(t)\,dt$. **(c)** $\dfrac{d}{dt}[\mathbf{A}(t)\mathbf{B}(t)]$.

39. $\mathbf{A}(t) = \begin{bmatrix} t & e^t \\ 1 & e^t \end{bmatrix}$, $\mathbf{B}(t) = \begin{bmatrix} \cos t & -\sin t \\ \sin t & \cos t \end{bmatrix}$.

40. $\mathbf{A}(t) = \begin{bmatrix} 1 & e^{-2t} \\ 3 & e^{-2t} \end{bmatrix}$, $\mathbf{B}(t) = \begin{bmatrix} e^{-t} & e^{-t} \\ -e^{-t} & 3e^{-t} \end{bmatrix}$.

In Problems 41 through 43, \mathbf{A}^T refers to the transpose matrix discussed in Problem 7.

41. An $n \times n$ matrix \mathbf{A} is called **symmetric** if $\mathbf{A}^T = \mathbf{A}$; that is, if $a_{ij} = a_{ji}$, for all $i, j = 1, \ldots, n$. Show that if \mathbf{A} is an $n \times n$ matrix, then $\mathbf{A} + \mathbf{A}^T$ is a symmetric matrix.

42. Let \mathbf{A} be an $m \times n$ matrix. Show that $\mathbf{A}^T\mathbf{A}$ is a symmetric $n \times n$ matrix and $\mathbf{A}\mathbf{A}^T$ is a symmetric $m \times m$ matrix (see Problem 41).

43. The **inner product** of two vectors is a generalization of the dot product, for vectors with complex entries. It is defined by

$$(\mathbf{x}, \mathbf{y}) := \sum_{i=1}^{n} x_i \bar{y}_i \ , \qquad \text{where}$$

$\mathbf{x} = \text{col}\,(x_1, x_2, \ldots, x_n)$, $\mathbf{y} = \text{col}\,(y_1, y_2, \ldots, y_n)$ are complex vectors and the overbar denotes complex conjugation.

(a) Show that $(\mathbf{x}, \mathbf{y}) = \mathbf{x}^T\bar{\mathbf{y}}$, where $\bar{\mathbf{y}} = \text{col}\,(\bar{y}_1, \bar{y}_2, \ldots, \bar{y}_n)$.

(b) Prove that for any $n \times 1$ vectors $\mathbf{x}, \mathbf{y}, \mathbf{z}$ and any complex number λ, we have

$$(\mathbf{x}, \mathbf{y}) = \overline{(\mathbf{y}, \mathbf{x})} \ ,$$
$$(\mathbf{x}, \mathbf{y} + \mathbf{z}) = (\mathbf{x}, \mathbf{y}) + (\mathbf{x}, \mathbf{z}) \ ,$$
$$(\lambda\mathbf{x}, \mathbf{y}) = \lambda(\mathbf{x}, \mathbf{y}) \ , \quad (\mathbf{x}, \lambda\mathbf{y}) = \bar{\lambda}(\mathbf{x}, \mathbf{y}) \ .$$

9.4 LINEAR SYSTEMS IN NORMAL FORM

In keeping with the introduction presented in Section 9.1, we say that a system of n linear differential equations is in **normal form** if it is expressed as

$$(1) \qquad \mathbf{x}'(t) = \mathbf{A}(t)\mathbf{x}(t) + \mathbf{f}(t) \ ,$$

where $\mathbf{x}(t) = \text{col}\,(x_1(t), \ldots, x_n(t))$, $\mathbf{f}(t) = \text{col}\,(f_1(t), \ldots, f_n(t))$, and $\mathbf{A}(t) = [a_{ij}(t)]$ is an $n \times n$ matrix. As with a scalar linear differential equation, a system is called **homogeneous** when $\mathbf{f}(t) \equiv \mathbf{0}$; otherwise, it is called **nonhomogeneous**. When the elements of \mathbf{A} are all constants, the system is said to have **constant coefficients**. Recall that an nth order linear differential equation

$$(2) \qquad y^{(n)}(t) + p_{n-1}(t)y^{(n-1)}(t) + \cdots + p_0(t)y(t) = g(t)$$

can be rewritten as a first order system in normal form using the substitution $x_1(t) := y(t), x_2(t) := y'(t), \ldots, x_n(t) := y^{(n-1)}(t)$; indeed, equation (2) is equivalent to $\mathbf{x}'(t) = \mathbf{A}(t)\mathbf{x}(t) + \mathbf{f}(t)$, where $\mathbf{x}(t) = \text{col}\,(x_1(t), \ldots, x_n(t))$, $\mathbf{f}(t) := \text{col}\,(0, \ldots, 0, g(t))$, and

$$\mathbf{A}(t) := \begin{bmatrix} 0 & 1 & 0 & \cdots & 0 & 0 \\ 0 & 0 & 1 & & 0 & 0 \\ \vdots & \vdots & \vdots & & \vdots & \vdots \\ 0 & 0 & 0 & \cdots & 0 & 1 \\ -p_0(t) & -p_1(t) & -p_2(t) & \cdots & -p_{n-2}(t) & -p_{n-1}(t) \end{bmatrix} .$$

The theory for systems in normal form parallels very closely the theory of linear differential equations presented in Chapters 4 and 6. In many cases the proofs for scalar linear differential equations carry over to normal systems with appropriate modifications. Conversely, results for normal systems apply to scalar linear equations since, as we showed, any scalar linear equation can be expressed as a normal system. This is the case with the existence and uniqueness theorems for linear differential equations.

The **initial value problem** for the normal system (1) is the problem of finding a differentiable vector function $\mathbf{x}(t)$ that satisfies the system on an interval I and also satisfies the **initial condition** $\mathbf{x}(t_0) = \mathbf{x}_0$, where t_0 is a given point of I and $\mathbf{x}_0 = \mathrm{col}(x_{1,0}, \ldots, x_{n,0})$ is a given vector.

EXISTENCE AND UNIQUENESS

Theorem 2. Suppose $\mathbf{A}(t)$ and $\mathbf{f}(t)$ are continuous on an open interval I that contains the point t_0. Then, for any choice of the initial vector $\mathbf{x}_0 = \mathrm{col}(x_{1,0}, \ldots, x_{n,0})$, there exists a unique solution $\mathbf{x}(t)$ on the whole interval I to the initial value problem

$$\mathbf{x}'(t) = \mathbf{A}(t)\mathbf{x}(t) + \mathbf{f}(t) , \qquad \mathbf{x}(t_0) = \mathbf{x}_0 .$$

We give a proof of this result in Chapter 13[†] and obtain as corollaries the existence and uniqueness theorems for second order equations (Theorem 2, Section 4.2) and higher-order linear equations (Theorem 1, Section 6.1).

If we rewrite system (1) as $\mathbf{x}' - \mathbf{A}\mathbf{x} = \mathbf{f}$ and define the operator $L[\mathbf{x}] := \mathbf{x}' - \mathbf{A}\mathbf{x}$, then we can express system (1) in the operator form $L[\mathbf{x}] = \mathbf{f}$. Here the operator L maps vector functions into vector functions. Moreover, L is a *linear* operator in the sense that for any scalars a, b and differentiable vector functions \mathbf{x}, \mathbf{y}, we have

$$L[a\mathbf{x} + b\mathbf{y}] = aL[\mathbf{x}] + bL[\mathbf{y}] .$$

The proof of this linearity follows from the properties of matrix multiplication (see Problem 25).

As a consequence of the linearity of L, if $\mathbf{x}_1, \ldots, \mathbf{x}_n$ are solutions to the *homogeneous* system $\mathbf{x}' = \mathbf{A}\mathbf{x}$, or $L[\mathbf{x}] = \mathbf{0}$ in operator notation, then any linear combination of these vectors, $c_1\mathbf{x}_1 + \cdots + c_n\mathbf{x}_n$, is also a solution. Moreover, we will see that if the solutions $\mathbf{x}_1, \ldots, \mathbf{x}_n$ are linearly independent, then *every* solution to $L[\mathbf{x}] = \mathbf{0}$ can be expressed as $c_1\mathbf{x}_1 + \cdots + c_n\mathbf{x}_n$ for an appropriate choice of the constants c_1, \ldots, c_n.

LINEAR DEPENDENCE OF VECTOR FUNCTIONS

Definition 1. The m vector functions $\mathbf{x}_1, \ldots, \mathbf{x}_m$ are said to be **linearly dependent on an interval** I if there exist constants c_1, \ldots, c_m, not all zero, such that

$$c_1\mathbf{x}_1(t) + \cdots + c_m\mathbf{x}_m(t) = \mathbf{0}$$

for all t in I. If the vectors are not linearly dependent, they are said to be **linearly independent on** I.

EXAMPLE 1 Show that the vector functions $\mathbf{x}_1(t) = \mathrm{col}(e^t, 0, e^t)$, $\mathbf{x}_2(t) = \mathrm{col}(3e^t, 0, 3e^t)$, and $\mathbf{x}_3(t) = \mathrm{col}(t, 1, 0)$ are linearly dependent on $(-\infty, \infty)$.

[†]All references to Chapters 11–13 refer to the expanded text *Fundamentals of Differential Equations and Boundary Value Problems,* 3rd Edition.

SOLUTION Notice that \mathbf{x}_2 is just 3 times \mathbf{x}_1 and therefore $3\mathbf{x}_1(t) - \mathbf{x}_2(t) + 0 \cdot \mathbf{x}_3(t) = \mathbf{0}$ for all t. Hence, \mathbf{x}_1, \mathbf{x}_2, and \mathbf{x}_3 are linearly dependent on $(-\infty, \infty)$. ∎ End

EXAMPLE 2 Show that the vector functions $\mathbf{x}_1(t) = \text{col}(e^{2t}, 0, e^{2t})$, $\mathbf{x}_2(t) = \text{col}(e^{2t}, e^{2t}, -e^{2t})$, and $\mathbf{x}_3(t) = \text{col}(e^t, 2e^t, e^t)$ are linearly independent on $(-\infty, \infty)$.

SOLUTION To prove independence, we *assume* c_1, c_2, and c_3 are constants for which

(3) $c_1\mathbf{x}_1(t) + c_2\mathbf{x}_2(t) + c_3\mathbf{x}_3(t) = \mathbf{0}$

holds at every t in $(-\infty, \infty)$ and show that this forces $c_1 = c_2 = c_3 = 0$. Let's try $t = 0$ in (3). We have

$$c_1\begin{bmatrix}1\\0\\1\end{bmatrix} + c_2\begin{bmatrix}1\\1\\-1\end{bmatrix} + c_3\begin{bmatrix}1\\2\\1\end{bmatrix} = \mathbf{0} \ ,$$

which is equivalent to the system of linear equations

(4) $\begin{aligned} c_1 + c_2 + c_3 &= 0 \ , \\ c_2 + 2c_3 &= 0 \ , \\ c_1 - c_2 + c_3 &= 0 \ . \end{aligned}$

Either by solving (4) or by checking that the determinant of its coefficients is nonzero (recall Theorem 1 on page 543), we can verify that (4) has only the trivial solution $c_1 = c_2 = c_3 = 0$. Therefore the vector functions \mathbf{x}_1, \mathbf{x}_2, and \mathbf{x}_3 are linearly independent on $(-\infty, \infty)$ (in fact, on any interval containing $t = 0$). ∎ End

As the previous example illustrates, if $\mathbf{x}_1(t), \ldots, \mathbf{x}_n(t)$ are n vector functions each having n components, then these vectors will be linearly independent on an interval I if the determinant

$$\det\big[\mathbf{x}_1(t) \ \ldots \ \mathbf{x}_n(t)\big]$$

is not zero at some point t in I. Because of the analogy with scalar equations, we call this determinant the **Wronskian**.

WRONSKIAN

Definition 2. The **Wronskian** of n vector functions $\mathbf{x}_1(t) = \text{col}(x_{1,1}, \ldots, x_{n,1}), \ldots,$ $\mathbf{x}_n(t) = \text{col}(x_{1,n}, \ldots, x_{n,n})$ is defined to be the real-valued function

$$W[\mathbf{x}_1, \ldots, \mathbf{x}_n](t) := \begin{vmatrix} x_{1,1}(t) & x_{1,2}(t) & \cdots & x_{1,n}(t) \\ x_{2,1}(t) & x_{2,2}(t) & \cdots & x_{2,n}(t) \\ \vdots & \vdots & & \vdots \\ x_{n,1}(t) & x_{n,2}(t) & \cdots & x_{n,n}(t) \end{vmatrix} .$$

We now show that if $\mathbf{x}_1, \mathbf{x}_2, \ldots, \mathbf{x}_n$ are linearly independent *solutions* on I to the homogeneous system $\mathbf{x}' = \mathbf{A}\mathbf{x}$, where \mathbf{A} is an $n \times n$ matrix of continuous functions, then the Wronskian $W(t) := \det[\mathbf{x}_1, \mathbf{x}_2, \ldots, \mathbf{x}_n]$ is never zero on I. For suppose to the contrary that $W(t_0) = 0$ at some point t_0. Then by Theorem 1, page 543, the vanishing of the determinant implies that the column vectors $\mathbf{x}_1(t_0), \mathbf{x}_2(t_0), \ldots, \mathbf{x}_n(t_0)$ are linearly dependent. Thus there exist scalars c_1, \ldots, c_n *not all zero*, such that

$$c_1 \mathbf{x}_1(t_0) + \cdots + c_n \mathbf{x}_n(t_0) = \mathbf{0} \ .$$

However, $c_1 \mathbf{x}_1(t) + \cdots + c_n \mathbf{x}_n(t)$ and the vector function $\mathbf{z}(t) \equiv \mathbf{0}$ are both solutions to $\mathbf{x}' = \mathbf{A}\mathbf{x}$ on I, and they agree at the point t_0. So these solutions must be identical on I according to the existence-uniqueness theorem (Theorem 2). That is,

$$c_1 \mathbf{x}_1(t) + \cdots + c_n \mathbf{x}_n(t) = \mathbf{0}$$

for *all* t in I. But this contradicts the given information that $\mathbf{x}_1, \ldots, \mathbf{x}_n$ are linearly independent on I. We have shown that $W(t_0) \neq 0$, and since t_0 is an arbitrary point, it follows that $W(t) \neq 0$ for all $t \in I$.

The preceding argument has two important implications that parallel the scalar case. First, *the Wronskian of solutions to $\mathbf{x}' = \mathbf{A}\mathbf{x}$ is either identically zero or never zero on I* (see also Problem 31). Second, *a set of n solutions $\mathbf{x}_1, \ldots, \mathbf{x}_n$ to $\mathbf{x}' = \mathbf{A}\mathbf{x}$ on I is linearly independent on I if and only if their Wronskian is never zero on I*. With these facts in hand, we can imitate the proof given for the scalar case in Section 6.1 (Theorem 2) to obtain the following representation theorem for the solutions to $\mathbf{x}' = \mathbf{A}\mathbf{x}$.

REPRESENTATION OF SOLUTIONS (HOMOGENEOUS CASE)

Theorem 3. Let $\mathbf{x}_1, \ldots, \mathbf{x}_n$ be n linearly independent solutions to the homogeneous system

(5) $\mathbf{x}'(t) = \mathbf{A}(t)\mathbf{x}(t)$

on the interval I, where $\mathbf{A}(t)$ is an $n \times n$ matrix function continuous on I. Then every solution to (5) on I can be expressed in the form

(6) $\mathbf{x}(t) = c_1 \mathbf{x}_1(t) + \cdots + c_n \mathbf{x}_n(t) \ ,$

where c_1, \ldots, c_n are constants.

A set of solutions $\{\mathbf{x}_1, \ldots, \mathbf{x}_n\}$ that are linearly independent on I or, equivalently, whose Wronskian does not vanish on I, is called a **fundamental solution set** for (5) on I. The linear combination in (6), written with arbitrary constants, is referred to as a **general solution** to (5).

If we take the vectors in a fundamental solution set and let them form the columns of a matrix $\mathbf{X}(t)$, that is,

$$\mathbf{X}(t) = \begin{bmatrix} \mathbf{x}_1(t) & \mathbf{x}_2(t) & \ldots & \mathbf{x}_n(t) \end{bmatrix} = \begin{bmatrix} x_{1,1}(t) & x_{1,2}(t) & \cdots & x_{1,n}(t) \\ x_{2,1}(t) & x_{2,2}(t) & \cdots & x_{2,n}(t) \\ \vdots & \vdots & & \vdots \\ x_{n,1}(t) & x_{n,2}(t) & \cdots & x_{n,n}(t) \end{bmatrix},$$

then the matrix $\mathbf{X}(t)$ is called a **fundamental matrix** for (5). We can use it to express the general solution (6) as

$$\mathbf{x}(t) = \mathbf{X}(t)\mathbf{c} \ ,$$

where $\mathbf{c} = \mathrm{col}(c_1, \dots, c_n)$ is an arbitrary constant vector. Since $\det \mathbf{X} = W[\mathbf{x}_1, \dots, \mathbf{x}_n]$ is never zero on I, it follows from Theorem 1 in Section 9.3 that $\mathbf{X}(t)$ is invertible for every t in I.

EXAMPLE 3 Verify that the set

$$S = \left\{ \begin{bmatrix} e^{2t} \\ e^{2t} \\ e^{2t} \end{bmatrix}, \begin{bmatrix} -e^{-t} \\ 0 \\ e^{-t} \end{bmatrix}, \begin{bmatrix} 0 \\ e^{-t} \\ -e^{-t} \end{bmatrix} \right\}$$

is a fundamental solution set for the system

$$(7) \qquad \mathbf{x}'(t) = \begin{bmatrix} 0 & 1 & 1 \\ 1 & 0 & 1 \\ 1 & 1 & 0 \end{bmatrix} \mathbf{x}(t)$$

on the interval $(-\infty, \infty)$ and find a fundamental matrix for (7). Also determine a general solution for (7).

SOLUTION Substituting the first vector in the set S into the right-hand side of (7) gives

$$\mathbf{Ax} = \begin{bmatrix} 0 & 1 & 1 \\ 1 & 0 & 1 \\ 1 & 1 & 0 \end{bmatrix} \begin{bmatrix} e^{2t} \\ e^{2t} \\ e^{2t} \end{bmatrix} = \begin{bmatrix} 2e^{2t} \\ 2e^{2t} \\ 2e^{2t} \end{bmatrix} = \mathbf{x}'(t) \ .$$

Hence this vector satisfies system (7) for all t. Similar computations verify that the remaining vectors in S are also solutions to (7) on $(-\infty, \infty)$. For us to show that S is a fundamental solution set, it is enough to observe that the Wronksian

$$W(t) = \begin{vmatrix} e^{2t} & -e^{-t} & 0 \\ e^{2t} & 0 & e^{-t} \\ e^{2t} & e^{-t} & -e^{-t} \end{vmatrix} = e^{2t} \begin{vmatrix} 0 & e^{-t} \\ e^{-t} & -e^{-t} \end{vmatrix} + e^{-t} \begin{vmatrix} e^{2t} & e^{-t} \\ e^{2t} & -e^{-t} \end{vmatrix} = -3$$

is never zero.

A fundamental matrix $\mathbf{X}(t)$ for (7) is just the matrix we used to compute the Wronksian; that is,

$$(8) \qquad \mathbf{X}(t) := \begin{bmatrix} e^{2t} & -e^{-t} & 0 \\ e^{2t} & 0 & e^{-t} \\ e^{2t} & e^{-t} & -e^{-t} \end{bmatrix} \ .$$

A general solution to (7) can now be expressed as

$$\mathbf{x}(t) = \mathbf{X}(t)\mathbf{c} = c_1 \begin{bmatrix} e^{2t} \\ e^{2t} \\ e^{2t} \end{bmatrix} + c_2 \begin{bmatrix} -e^{-t} \\ 0 \\ e^{-t} \end{bmatrix} + c_3 \begin{bmatrix} 0 \\ e^{-t} \\ -e^{-t} \end{bmatrix} . \quad \blacksquare \text{End}$$

It is easy to check that the fundamental matrix in (8) satisfies the equation

$$\mathbf{X}'(t) = \begin{bmatrix} 0 & 1 & 1 \\ 1 & 0 & 1 \\ 1 & 1 & 0 \end{bmatrix} \mathbf{X}(t) \; ;$$

indeed, this is equivalent to showing that $\mathbf{x}' = \mathbf{A}\mathbf{x}$ for each column \mathbf{x} in S. In general, a fundamental matrix for a system $\mathbf{x}' = \mathbf{A}\mathbf{x}$ satisfies the corresponding **matrix differential eqauation** $\mathbf{X}' = \mathbf{A}\mathbf{X}$.

Another consequence of the linearity of the operator L defined by $L[\mathbf{x}] := \mathbf{x}' - \mathbf{A}\mathbf{x}$ is the **superposition principle** for linear systems. It states that if \mathbf{x}_1 and \mathbf{x}_2 are solutions, respectively, to the nonhomogeneous systems

$$L[\mathbf{x}] = \mathbf{g}_1 \quad \text{and} \quad L[\mathbf{x}] = \mathbf{g}_2 \; ,$$

then $c_1\mathbf{x}_1 + c_2\mathbf{x}_2$ is a solution to

$$L[\mathbf{x}] = c_1\mathbf{g}_1 + c_2\mathbf{g}_2 \; .$$

Using the superposition principle and the representation theorem for homogeneous systems, we can prove the following theorem.

REPRESENTATION OF SOLUTIONS (NONHOMOGENEOUS CASE)

Theorem 4. Let \mathbf{x}_p be a particular solution to the nonhomogeneous system

(9) $\mathbf{x}'(t) = \mathbf{A}(t)\mathbf{x}(t) + \mathbf{f}(t)$

on the interval I, and let $\{\mathbf{x}_1, \ldots, \mathbf{x}_n\}$ be a fundamental solution set on I for the corresponding homogeneous system $\mathbf{x}(t) = \mathbf{A}(t)\mathbf{x}(t)$. Then every solution to (9) on I can be expressed in the form

(10) $\mathbf{x}(t) = \mathbf{x}_p(t) + c_1\mathbf{x}_1(t) + \cdots + c_n\mathbf{x}_n(t) \; ,$

where c_1, \ldots, c_n are constants.

The proof of this theorem is almost identical to the proofs of Theorem 6 in Section 4.7 and Theorem 4 in Section 6.1. We leave the proof as an exercise.

The linear combination of $\mathbf{x}_p, \mathbf{x}_1, \ldots, \mathbf{x}_n$ in (10) written with arbitrary constants c_1, \ldots, c_n is called a **general solution** of (9). This general solution can also be expressed as $\mathbf{x} = \mathbf{x}_p + \mathbf{X}\mathbf{c}$, where \mathbf{X} is a fundamental matrix for the homogeneous system and \mathbf{c} is an arbitrary constant vector.

We now summarize the results of this section as they apply to the problem of finding a general solution to a system of n linear first order differential equations in normal form.

APPROACH TO SOLVING NORMAL SYSTEMS

1. To determine a general solution to the $n \times n$ homogeneous system $\mathbf{x}' = \mathbf{Ax}$:

 (a) Find a fundamental solution set $\{\mathbf{x}_1, \ldots, \mathbf{x}_n\}$ that consists of n linearly independent solutions to the homogeneous system.

 (b) Form the linear combination

 $$\mathbf{x} = \mathbf{Xc} = c_1\mathbf{x}_1 + \cdots + c_n\mathbf{x}_n \ ,$$

 where $\mathbf{c} = \mathrm{col}(c_1, \ldots, c_n)$ is any constant vector and $\mathbf{X} = \begin{bmatrix} \mathbf{x}_1 \ldots \mathbf{x}_n \end{bmatrix}$ is the fundamental matrix, to obtain a general solution.

2. To determine a general solution to the nonhomogeneous system $\mathbf{x}' = \mathbf{Ax} + \mathbf{f}$:

 (a) Find a particular solution \mathbf{x}_p to the nonhomogeneous system.

 (b) Form the sum of the particular solution and the general solution $\mathbf{Xc} = c_1\mathbf{x}_1 + \cdots + c_n\mathbf{x}_n$ to the corresponding homogeneous system in part 1,

 $$\mathbf{x} = \mathbf{x}_p + \mathbf{Xc} = \mathbf{x}_p + c_1\mathbf{x}_1 + \cdots + c_n\mathbf{x}_n \ ,$$

 to obtain a general solution to the given system.

We shall devote the rest of this chapter to methods for finding fundamental solution sets for homogeneous systems and particular solutions for nonhomogeneous systems.

EXERCISES 9.4

In Problems 1 through 4, write the given system in the matrix form $\mathbf{x}' = \mathbf{Ax} + \mathbf{f}$.

1. $x'(t) = 3x(t) - y(t) + t^2$,

 $y'(t) = -x(t) + 2y(t) + e^t$.

2. $r'(t) = 2r(t) + \sin t$,

 $\theta'(t) = r(t) - \theta(t) + 1$.

3. $\dfrac{dx}{dt} = t^2 x - y - z + t$,

 $\dfrac{dy}{dt} = e^t z + 5$,

 $\dfrac{dz}{dt} = tx - y + 3z - e^t$.

4. $\dfrac{dx}{dt} = x + y + z$,

 $\dfrac{dy}{dt} = 2x - y + 3z$,

 $\dfrac{dz}{dt} = x + 5z$.

In Problems 5 through 8, rewrite the given scalar equation as a first order system in normal form. Express the system in the matrix form $\mathbf{x}' = \mathbf{Ax} + \mathbf{f}$.

5. $y''(t) - 3y'(t) - 10y(t) = \sin t$.

6. $x''(t) + x(t) = t^2$.

7. $\dfrac{d^4w}{dt^4} + w = t^2$.

8. $\dfrac{d^3y}{dt^3} - \dfrac{dy}{dt} + y = \cos t$.

In Problems 9 through 12, write the given system as a set of scalar equations.

9. $\mathbf{x}' = \begin{bmatrix} 5 & 0 \\ -2 & 4 \end{bmatrix} \mathbf{x} + e^{-2t} \begin{bmatrix} 2 \\ -3 \end{bmatrix}$.

10. $x' = \begin{bmatrix} 2 & 1 \\ -1 & 3 \end{bmatrix} x + e^t \begin{bmatrix} t \\ 1 \end{bmatrix}$.

11. $x' = \begin{bmatrix} 1 & 0 & 1 \\ -1 & 2 & 5 \\ 0 & 5 & 1 \end{bmatrix} x + e^t \begin{bmatrix} 1 \\ 0 \\ 0 \end{bmatrix} + t \begin{bmatrix} 0 \\ 1 \\ 0 \end{bmatrix}$.

12. $x' = \begin{bmatrix} 0 & 1 & 0 \\ 0 & 0 & 1 \\ -1 & 1 & 2 \end{bmatrix} x + t \begin{bmatrix} 1 \\ -1 \\ 2 \end{bmatrix} + \begin{bmatrix} 3 \\ 1 \\ 0 \end{bmatrix}$.

In Problems 13 through 18, determine whether the given vector functions are linearly dependent (LD) or linearly independent (LI) on the interval $(-\infty, \infty)$.

13. $\begin{bmatrix} t \\ 3 \end{bmatrix}, \begin{bmatrix} 4 \\ 1 \end{bmatrix}$.

14. $\begin{bmatrix} te^{-t} \\ e^{-t} \end{bmatrix}, \begin{bmatrix} e^{-t} \\ e^{-t} \end{bmatrix}$.

15. $\begin{bmatrix} \sin t \\ \cos t \end{bmatrix}, \begin{bmatrix} \sin 2t \\ \cos 2t \end{bmatrix}$.

16. $e^t \begin{bmatrix} 1 \\ 5 \end{bmatrix}, e^t \begin{bmatrix} -3 \\ -15 \end{bmatrix}$.

17. $e^{2t} \begin{bmatrix} 1 \\ 0 \\ 5 \end{bmatrix}, e^{2t} \begin{bmatrix} 1 \\ 1 \\ -1 \end{bmatrix}, e^{3t} \begin{bmatrix} 0 \\ 1 \\ 0 \end{bmatrix}$.

18. $\begin{bmatrix} 1 \\ 0 \\ 1 \end{bmatrix}, \begin{bmatrix} t \\ 0 \\ t \end{bmatrix}, \begin{bmatrix} t^2 \\ 0 \\ t^2 \end{bmatrix}$.

In Problems 19 through 22, the given vector functions are solutions to the system $x'(t) = Ax(t)$. Determine whether they form a fundamental solution set. If they do, find a fundamental matrix for the system and give a general solution.

19. $x_1 = e^{2t} \begin{bmatrix} 1 \\ -2 \end{bmatrix}, \quad x_2 = e^{2t} \begin{bmatrix} -2 \\ 4 \end{bmatrix}$.

20. $x_1 = e^{-t} \begin{bmatrix} 3 \\ 2 \end{bmatrix}, \quad x_2 = e^{4t} \begin{bmatrix} 1 \\ -1 \end{bmatrix}$.

21. $x_1 = \begin{bmatrix} e^{-t} \\ 2e^{-t} \\ e^{-t} \end{bmatrix}, \quad x_2 = \begin{bmatrix} e^t \\ 0 \\ e^t \end{bmatrix}$,

$x_3 = \begin{bmatrix} e^{3t} \\ -e^{3t} \\ 2e^{3t} \end{bmatrix}$.

22. $x_1 = \begin{bmatrix} e^t \\ e^t \\ e^t \end{bmatrix}, \quad x_2 = \begin{bmatrix} \sin t \\ \cos t \\ -\sin t \end{bmatrix}$,

$x_3 = \begin{bmatrix} -\cos t \\ \sin t \\ \cos t \end{bmatrix}$.

23. Verify that the vector functions

$$x_1 = \begin{bmatrix} e^t \\ e^t \end{bmatrix} \quad \text{and} \quad x_2 = \begin{bmatrix} e^{-t} \\ 3e^{-t} \end{bmatrix}$$

are solutions to the homogeneous system

$$x' = Ax = \begin{bmatrix} 2 & -1 \\ 3 & -2 \end{bmatrix} x ,$$

on $(-\infty, \infty)$, and that

$$x_p = \frac{3}{2} \begin{bmatrix} te^t \\ te^t \end{bmatrix} - \frac{1}{4} \begin{bmatrix} e^t \\ 3e^t \end{bmatrix} + \begin{bmatrix} t \\ 2t \end{bmatrix} - \begin{bmatrix} 0 \\ 1 \end{bmatrix}$$

is a particular solution to the nonhomogeneous system $x' = Ax + f(t)$, where $f(t) = \text{col}(e^t, t)$. Find a general solution to $x' = Ax + f(t)$.

24. Verify that the vector functions

$$x_1 = \begin{bmatrix} e^{3t} \\ 0 \\ e^{3t} \end{bmatrix}, \quad x_2 = \begin{bmatrix} -e^{3t} \\ e^{3t} \\ 0 \end{bmatrix},$$

$$x_3 = \begin{bmatrix} -e^{-3t} \\ -e^{-3t} \\ e^{-3t} \end{bmatrix}$$

are solutions to the homogeneous system

$$x' = Ax = \begin{bmatrix} 1 & -2 & 2 \\ -2 & 1 & 2 \\ 2 & 2 & 1 \end{bmatrix} x ,$$

on $(-\infty, \infty)$, and that

$$x_p = \begin{bmatrix} 5t + 1 \\ 2t \\ 4t + 2 \end{bmatrix}$$

is a particular solution to $x' = Ax + f(t)$, where $f(t) = \text{col}(-9t, 0, -18t)$. Find a general solution to $x' = Ax + f(t)$.

25. Prove that the operator defined by $L[\mathbf{x}] := \mathbf{x}' - \mathbf{A}\mathbf{x}$, where \mathbf{A} is an $n \times n$ matrix function and \mathbf{x} is an $n \times 1$ differentiable vector function, is a linear operator.

26. Let $\mathbf{X}(t)$ be a fundamental matrix for the system $\mathbf{x}' = \mathbf{A}\mathbf{x}$. Show that $\mathbf{x}(t) = \mathbf{X}(t)\mathbf{X}^{-1}(t_0)\mathbf{x}_0$ is the solution to the initial value problem $\mathbf{x}' = \mathbf{A}\mathbf{x}$, $\mathbf{x}(t_0) = \mathbf{x}_0$.

In Problems 27 and 28, verify that $\mathbf{X}(t)$ is a fundamental matrix for the given system and compute $\mathbf{X}^{-1}(t)$. Use the result of Problem 26 to find the solution to the given initial value problem.

27. $\mathbf{x}' = \begin{bmatrix} 0 & 6 & 0 \\ 1 & 0 & 1 \\ 1 & 1 & 0 \end{bmatrix} \mathbf{x}$, $\quad \mathbf{x}(0) = \begin{bmatrix} -1 \\ 0 \\ 1 \end{bmatrix}$;

$$\mathbf{X}(t) = \begin{bmatrix} 6e^{-t} & -3e^{-2t} & 2e^{3t} \\ -e^{-t} & e^{-2t} & e^{3t} \\ -5e^{-t} & e^{-2t} & e^{3t} \end{bmatrix}.$$

28. $\mathbf{x}' = \begin{bmatrix} 2 & 3 \\ 3 & 2 \end{bmatrix} \mathbf{x}$, $\quad \mathbf{x}(0) = \begin{bmatrix} 3 \\ -1 \end{bmatrix}$;

$$\mathbf{X}(t) = \begin{bmatrix} e^{-t} & e^{5t} \\ -e^{-t} & e^{5t} \end{bmatrix}.$$

29. Show that

$$\begin{vmatrix} t^2 & t|t| \\ 2t & 2|t| \end{vmatrix} \equiv 0$$

on $(-\infty, \infty)$, but that the two vector functions

$$\begin{bmatrix} t^2 \\ 2t \end{bmatrix}, \quad \begin{bmatrix} t|t| \\ 2|t| \end{bmatrix}$$

are linearly independent on $(-\infty, \infty)$.

30. Abel's Formula. If $\mathbf{x}_1, \ldots, \mathbf{x}_n$ are any n solutions to the $n \times n$ system $\mathbf{x}'(t) = \mathbf{A}(t)\mathbf{x}(t)$, then Abel's formula gives a representation for the Wronskian $W(t) := W[\mathbf{x}_1, \ldots, \mathbf{x}_n](t)$. Namely,

$$W(t) = W(t_0)\exp\left(\int_{t_0}^t \{a_{11}(s) + \cdots + a_{nn}(s)\}\,ds\right),$$

where $a_{11}(s), \ldots, a_{nn}(s)$ are the main diagonal elements of $\mathbf{A}(s)$. Prove this formula in the special case

when $n = 3$. [*Hint:* Follow the outline in Problem 36 of Exercises 6.1.]

31. Using Abel's formula, prove that the Wronskian of n solutions to $\mathbf{x}' = \mathbf{A}\mathbf{x}$ on the interval I is either identically zero on I or never zero on I.

32. Prove that a fundamental solution set for the homogeneous system $\mathbf{x}'(t) = \mathbf{A}(t)\mathbf{x}(t)$ always exists on an interval I, provided $\mathbf{A}(t)$ is continuous on I. [*Hint:* Use the existence and uniqueness theorem (Theorem 2) and make judicious choices for \mathbf{x}_0.]

33. Prove Theorem 3 on the representation of solutions of the homogeneous system.

34. Prove Theorem 4 on the representation of solutions of the nonhomogeneous system.

35. To illustrate the connection between a higher-order equation and the equivalent first order system, consider the equation

(11) $\quad y'''(t) - 6y''(t) + 11y'(t) - 6y(t) = 0$.

(a) Show that $\{e^t, e^{2t}, e^{3t}\}$ is a fundamental solution set for (11).

(b) Using the definition of Section 6.1, compute the Wronskian of $\{e^t, e^{2t}, e^{3t}\}$.

(c) Setting $x_1 = y, x_2 = y', x_3 = y''$, show that equation (11) is equivalent to the first order system

(12) $\quad \mathbf{x}' = \mathbf{A}\mathbf{x}$,

where

$$\mathbf{A} := \begin{bmatrix} 0 & 1 & 0 \\ 0 & 0 & 1 \\ 6 & -11 & 6 \end{bmatrix}.$$

(d) The substitution used in part (c) suggests that

$$S := \left\{ \begin{bmatrix} e^t \\ e^t \\ e^t \end{bmatrix}, \begin{bmatrix} e^{2t} \\ 2e^{2t} \\ 4e^{2t} \end{bmatrix}, \begin{bmatrix} e^{3t} \\ 3e^{3t} \\ 9e^{3t} \end{bmatrix} \right\}$$

is a fundamental solution set for system (12). Verify that this is the case.

(e) Compute the Wronskian of S. How does it compare with the Wronskian computed in part (b)?

9.5 HOMOGENEOUS LINEAR SYSTEMS WITH CONSTANT COEFFICIENTS

In this section we discuss a procedure for obtaining a general solution for the homogeneous system

$$(1) \qquad \mathbf{x}'(t) = \mathbf{A}\mathbf{x}(t) \ ,$$

where \mathbf{A} is a (real) *constant* $n \times n$ matrix. The general solution we seek will be defined for all t because the elements of \mathbf{A} are just constant functions, which are continuous on $(-\infty, \infty)$ (recall Theorem 2, page 549). In Section 9.4 we showed that a general solution to (1) can be constructed from a fundamental solution set consisting of n linearly independent solutions to (1). Thus our goal is to find n such vector solutions.

In Chapter 4 we were successful in solving homogeneous linear equations with constant coefficients by guessing that the equation had a solution of the form e^{rt}. Because any scalar linear equation can be expressed as a system, it is reasonable to expect system (1) to have solutions of the form

$$\mathbf{x}(t) = e^{rt}\mathbf{u} \ ,$$

where r is a constant and \mathbf{u} is a constant vector, both of which must be determined. Substituting $e^{rt}\mathbf{u}$ for $\mathbf{x}(t)$ in (1) gives

$$re^{rt}\mathbf{u} = \mathbf{A}e^{rt}\mathbf{u} = e^{rt}\mathbf{A}\mathbf{u} \ .$$

Canceling the factor e^{rt} and rearranging terms, we find that

$$(2) \qquad (\mathbf{A} - r\mathbf{I})\mathbf{u} = \mathbf{0} \ ,$$

where $r\mathbf{I}$ denotes the diagonal matrix with r's along its main diagonal.

The preceding calculation shows that $\mathbf{x}(t) = e^{rt}\mathbf{u}$ is a solution to (1) if and only if r and \mathbf{u} satisfy equation (2). Since the trivial case, $\mathbf{u} = \mathbf{0}$, is of no help in finding linearly independent solutions to (1), we require that $\mathbf{u} \neq \mathbf{0}$. Such vectors are given a special name, as follows.

EIGENVALUES AND EIGENVECTORS

Definition 3. Let $\mathbf{A} = [a_{ij}]$ be an $n \times n$ constant matrix. The **eigenvalues** of \mathbf{A} are those (real or complex) numbers r for which $(\mathbf{A} - r\mathbf{I})\mathbf{u} = \mathbf{0}$ has at least one nontrivial solution[†] \mathbf{u}. The corresponding nontrivial solutions \mathbf{u} are called the **eigenvectors** of \mathbf{A} associated with r.

As stated in Theorem 1 of Section 9.3, a linear homogeneous system of n *algebraic* equations in n unknowns has a nontrivial solution if and only if the determinant of its coefficients is zero. Hence a necessary and sufficient condition for (2) to have a nontrivial solution is that

$$(3) \qquad |\mathbf{A} - r\mathbf{I}| = 0 \ .$$

[†]We will allow \mathbf{u} to have complex-number entries.

Expanding the determinant of $\mathbf{A} - r\mathbf{I}$ in terms of its cofactors, we find that it is an nth degree polynomial in r; that is,

(4) $$|\mathbf{A} - r\mathbf{I}| = p(r) \ .$$

Therefore *finding the eigenvalues of a matrix* \mathbf{A} *is equivalent to finding the zeros of the polynomial* $p(r)$. Equation (3) is called the **characteristic equation** of \mathbf{A}, and $p(r)$ in (4) is the **characteristic polynomial** of \mathbf{A}. The characteristic equation plays a role for systems similar to the role played by the auxiliary equation for scalar equations.

There are many commercially available software packages that can be used to compute the eigenvalues and eigenvectors for a given matrix. Four such packages are DERIVE, available from Soft Warehouse, Inc.; MATLAB, available from The MathWorks, Inc.; MATHEMATICA, available from Wolfram Research; and MAPLE, available from the University of Waterloo. Although the reader is encouraged to make use of such packages, the examples and most exercises in this text can be easily carried out without them. Those exercises for which a computer package is desirable are flagged with the icon ▢ .

EXAMPLE 1 Find the eigenvalues and eigenvectors of the matrix

$$\mathbf{A} := \begin{bmatrix} 2 & -3 \\ 1 & -2 \end{bmatrix} \ .$$

SOLUTION The characteristic equation for \mathbf{A} is

$$|\mathbf{A} - r\mathbf{I}| = \begin{vmatrix} 2 - r & -3 \\ 1 & -2 - r \end{vmatrix} = (2 - r)(-2 - r) + 3 = r^2 - 1 = 0 \ .$$

Hence the eigenvalues of \mathbf{A} are $r_1 = 1$, $r_2 = -1$. To find the eigenvectors corresponding to $r_1 = 1$, we must solve $(\mathbf{A} - r_1\mathbf{I})\mathbf{u} = \mathbf{0}$. Substituting for \mathbf{A} and r_1 gives

(5) $$\begin{bmatrix} 1 & -3 \\ 1 & -3 \end{bmatrix} \begin{bmatrix} u_1 \\ u_2 \end{bmatrix} = \begin{bmatrix} 0 \\ 0 \end{bmatrix} \ .$$

Notice that this matrix equation is equivalent to the single scalar equation $u_1 - 3u_2 = 0$. Therefore the solutions to (5) are obtained by assigning an arbitrary value for u_2, say $u_2 = s$, and setting $u_1 = 3u_2 = 3s$. Consequently, the eigenvectors associated with $r_1 = 1$ can be expressed as

(6) $$\mathbf{u}_1 = s \begin{bmatrix} 3 \\ 1 \end{bmatrix} \ .$$

For $r_2 = -1$, the equation $(\mathbf{A} - r_2\mathbf{I})\mathbf{u} = \mathbf{0}$ becomes

$$\begin{bmatrix} 3 & -3 \\ 1 & -1 \end{bmatrix} \begin{bmatrix} u_1 \\ u_2 \end{bmatrix} = \begin{bmatrix} 0 \\ 0 \end{bmatrix} \ .$$

Solving, we obtain $u_1 = s$ and $u_2 = s$, with s arbitrary. Therefore the eigenvectors associated with the eigenvalue $r_2 = -1$ are

(7) $$\mathbf{u}_2 = s \begin{bmatrix} 1 \\ 1 \end{bmatrix} \ . \quad \blacksquare \text{ End}$$

We remark that in the above example the collection (6) of all eigenvectors associated with $r_1 = 1$ forms a one-dimensional subspace when the zero vector is adjoined. The same is true for $r_2 = -1$. These subspaces are called **eigenspaces.**

EXAMPLE 2 Find the eigenvalues and eigenvectors of the matrix

$$\mathbf{A} := \begin{bmatrix} 1 & 2 & -1 \\ 1 & 0 & 1 \\ 4 & -4 & 5 \end{bmatrix}.$$

SOLUTION The characteristic equation for \mathbf{A} is

$$|\mathbf{A} - r\mathbf{I}| = \begin{vmatrix} 1-r & 2 & -1 \\ 1 & -r & 1 \\ 4 & -4 & 5-r \end{vmatrix} = 0 \ ,$$

which simplifies to $(r-1)(r-2)(r-3) = 0$. Hence the eigenvalues of \mathbf{A} are $r_1 = 1$, $r_2 = 2$, and $r_3 = 3$. To find the eigenvectors corresponding to $r_1 = 1$, we set $r = 1$ in $(\mathbf{A} - r\mathbf{I})\mathbf{u} = \mathbf{0}$. This gives

$$(8) \qquad \begin{bmatrix} 0 & 2 & -1 \\ 1 & -1 & 1 \\ 4 & -4 & 4 \end{bmatrix} \begin{bmatrix} u_1 \\ u_2 \\ u_3 \end{bmatrix} = \begin{bmatrix} 0 \\ 0 \\ 0 \end{bmatrix}.$$

Using elementary row operations (Gaussian elimination), we see that (8) is equivalent to the two equations

$$u_1 - u_2 + u_3 = 0 \ ,$$
$$2u_2 - u_3 = 0 \ .$$

Thus we can obtain the solutions to (8) by assigning an arbitrary value to u_2, say $u_2 = s$, solving $2u_2 - u_3 = 0$ for u_3 to get $u_3 = 2s$, and then solving $u_1 - u_2 + u_3 = 0$ for u_1 to get $u_1 = -s$. Hence the eigenvectors associated with $r_1 = 1$ are

$$(9) \qquad \mathbf{u}_1 = s \begin{bmatrix} -1 \\ 1 \\ 2 \end{bmatrix}.$$

For $r_2 = 2$, we solve

$$\begin{bmatrix} -1 & 2 & -1 \\ 1 & -2 & 1 \\ 4 & -4 & 3 \end{bmatrix} \begin{bmatrix} u_1 \\ u_2 \\ u_3 \end{bmatrix} = \begin{bmatrix} 0 \\ 0 \\ 0 \end{bmatrix}$$

in a similar fashion (cf. Section 9.2) to obtain the eigenvectors

$$(10) \qquad \mathbf{u}_2 = s \begin{bmatrix} -2 \\ 1 \\ 4 \end{bmatrix}.$$

Finally, for $r_3 = 3$, we solve

$$\begin{bmatrix} -2 & 2 & -1 \\ 1 & -3 & 1 \\ 4 & -4 & 2 \end{bmatrix} \begin{bmatrix} u_1 \\ u_2 \\ u_3 \end{bmatrix} = \begin{bmatrix} 0 \\ 0 \\ 0 \end{bmatrix}$$

and get the eigenvectors

(11) $\mathbf{u}_3 = s \begin{bmatrix} -1 \\ 1 \\ 4 \end{bmatrix}$. ∎ End

Let's return to the problem of finding a general solution to a homogeneous system of differential equations. We have already shown that $e^{rt}\mathbf{u}$ is a solution to (1) if r is an eigenvalue and \mathbf{u} a corresponding eigenvector. The question is: Can we obtain n linearly independent solutions to the homogeneous system by finding all the eigenvalues and eigenvectors of \mathbf{A}?

n LINEARLY INDEPENDENT EIGENVECTORS

Theorem 5. Suppose the $n \times n$ constant matrix \mathbf{A} has n linearly independent eigenvectors $\mathbf{u}_1, \mathbf{u}_2, \ldots, \mathbf{u}_n$. Let r_i be the eigenvalue[†] corresponding to \mathbf{u}_i. Then

(12) $\{e^{r_1 t}\mathbf{u}_1, e^{r_2 t}\mathbf{u}_2, \ldots, e^{r_n t}\mathbf{u}_n\}$

is a fundamental solution set on $(-\infty, \infty)$ for the homogeneous system $\mathbf{x}' = \mathbf{A}\mathbf{x}$. Consequently, a general solution of $\mathbf{x}' = \mathbf{A}\mathbf{x}$ is

(13) $\mathbf{x}(t) = c_1 e^{r_1 t}\mathbf{u}_1 + c_2 e^{r_2 t}\mathbf{u}_2 + \cdots + c_n e^{r_n t}\mathbf{u}_n$,

where c_1, \ldots, c_n are arbitrary constants.

Proof. As we have seen, the vector functions listed in (12) are solutions to the homogeneous system. Moreover, their Wronskian is

$$W(t) = \det[e^{r_1 t}\mathbf{u}_1, \ldots, e^{r_n t}\mathbf{u}_n] = e^{(r_1 + \cdots + r_n)t} \det[\mathbf{u}_1, \ldots, \mathbf{u}_n] .$$

Since the eigenvectors are assumed to be linearly independent, it follows from Theorem 1 in Section 9.3 that $\det[\mathbf{u}_1, \ldots, \mathbf{u}_n]$ is not zero. Hence the Wronskian $W(t)$ is never zero. This shows that (12) is a fundamental solution set, and consequently a general solution is given by (13). ∎ End

An application of Theorem 5 is given in the next example.

[†]The eigenvalues r_1, \ldots, r_n may be real or complex and need not be distinct. In this section the cases we discuss have real eigenvalues. We consider complex eigenvalues in Section 9.6.

EXAMPLE 3 Find a general solution of

$$(14) \qquad \mathbf{x}'(t) = \mathbf{A}\mathbf{x}(t) , \quad \text{where} \quad \mathbf{A} = \begin{bmatrix} 2 & -3 \\ 1 & -2 \end{bmatrix}.$$

SOLUTION In Example 1 we showed that the matrix \mathbf{A} has eigenvalues $r_1 = 1$ and $r_2 = -1$. Taking, say, $s = 1$ in equations (6) and (7), we get the corresponding eigenvectors

$$\mathbf{u}_1 = \begin{bmatrix} 3 \\ 1 \end{bmatrix} \quad \text{and} \quad \mathbf{u}_2 = \begin{bmatrix} 1 \\ 1 \end{bmatrix}.$$

Because \mathbf{u}_1 and \mathbf{u}_2 are linearly independent, it follows from Theorem 5 that a general solution to (14) is

$$(15) \qquad \mathbf{x}(t) = c_1 e^t \begin{bmatrix} 3 \\ 1 \end{bmatrix} + c_2 e^{-t} \begin{bmatrix} 1 \\ 1 \end{bmatrix}. \quad \blacksquare \ \text{End}$$

If we sum the vectors on the right-hand side of equation (15) and then write out the expressions for the components of $\mathbf{x}(t) = \text{col}(x_1(t), x_2(t))$, we get

$$x_1(t) = 3c_1 e^t + c_2 e^{-t} ,$$
$$x_2(t) = c_1 e^t + c_2 e^{-t} .$$

This is the familiar form of a general solution for a system, as discussed in Section 5.3.

Example 3 nicely illustrates the geometric role played by the eigenvectors \mathbf{u}_1 and \mathbf{u}_2. If the initial vector $\mathbf{x}(0)$ is a scalar multiple of \mathbf{u}_1 (i.e., $\mathbf{x}(0) = c_1 \mathbf{u}_1$), then the vector solution to the system, $\mathbf{x}(t) = c_1 e^t \mathbf{u}_1$, will always have the same or opposite direction as \mathbf{u}_1. That is, it will lie along the straight line determined by \mathbf{u}_1 (see Figure 9.1). Furthermore, the trajectory of this

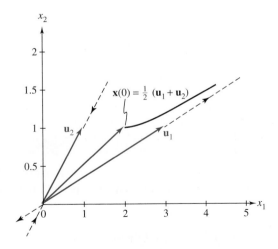

Figure 9.1 Trajectories of solutions for Example 3

solution, as t increases, will tend to infinity, since the corresponding eigenvalue $r_1 = 1$ is positive (observe the e^t term). A similar assertion holds if the initial vector is a scalar multiple of \mathbf{u}_2, except that since $r_2 = -1$ is negative, the trajectory $\mathbf{x}(t) = c_2 e^{-t} \mathbf{u}_2$ will approach the origin as t increases (because of e^{-t}). For an initial vector that involves both \mathbf{u}_1 and \mathbf{u}_2, such as $\mathbf{x}(0) = \frac{1}{2}(\mathbf{u}_1 + \mathbf{u}_2)$, the resulting trajectory is a blend of the above motions, with the contribution due to the larger eigenvalue $r_1 = 1$ dominating as t increases; see Figure 9.1.

The straight-line trajectories in the $x_1 x_2$-plane (the phase plane), then, point along the directions of the eigenvectors of the matrix \mathbf{A}. (See Section 5.2, Figure 5.6, for example.)

A useful property of eigenvectors that concerns their linear independence is stated in the next theorem.

LINEAR INDEPENDENCE OF EIGENVECTORS

Theorem 6. If r_1, \ldots, r_m are *distinct* eigenvalues for the matrix \mathbf{A} and \mathbf{u}_i is an eigenvector associated with r_i, then $\mathbf{u}_1, \ldots, \mathbf{u}_m$ are linearly independent.

Proof. Let's first treat the case $m = 2$. Suppose, to the contrary, that \mathbf{u}_1 and \mathbf{u}_2 are linearly dependent so that

(16) $\mathbf{u}_1 = c\mathbf{u}_2$

for some constant c. Multiplying both sides of (16) by \mathbf{A} and using the fact that \mathbf{u}_1 and \mathbf{u}_2 are eigenvectors with corresponding eigenvalues r_1 and r_2, we obtain

(17) $r_1\mathbf{u}_1 = cr_2\mathbf{u}_2$.

Next we multiply (16) by r_2 and then subtract from (17) to get

$$(r_1 - r_2)\mathbf{u}_1 = \mathbf{0} .$$

Since \mathbf{u}_1 is not the zero vector, we must have $r_1 = r_2$. But this violates the assumption that the eigenvalues are distinct! Hence \mathbf{u}_1 and \mathbf{u}_2 are linearly independent.

The cases $m > 2$ follow by induction. The details of the proof are left as an exercise. ■ End

Combining Theorems 5 and 6, we get the following corollary.

n DISTINCT EIGENVALUES

Corollary 1. If the $n \times n$ constant matrix \mathbf{A} has n distinct eigenvalues r_1, \ldots, r_n and \mathbf{u}_i is an eigenvector associated with r_i, then

$$\left\{ e^{r_1 t}\mathbf{u}_1, \ldots, e^{r_n t}\mathbf{u}_n \right\}$$

is a fundamental solution set for the homogeneous system $\mathbf{x}' = \mathbf{A}\mathbf{x}$.

EXAMPLE 4 Solve the initial value problem

$$\textbf{(18)} \qquad \mathbf{x}'(t) = \begin{bmatrix} 1 & 2 & -1 \\ 1 & 0 & 1 \\ 4 & -4 & 5 \end{bmatrix} \mathbf{x}(t) \,, \qquad \mathbf{x}(0) = \begin{bmatrix} -1 \\ 0 \\ 0 \end{bmatrix}.$$

SOLUTION In Example 2 we showed that the 3×3 coefficient matrix \mathbf{A} has the three distinct eigenvalues $r_1 = 1$, $r_2 = 2$, and $r_3 = 3$. If we set $s = 1$ in equations (9), (10), and (11), we obtain the corresponding eigenvectors

$$\mathbf{u}_1 = \begin{bmatrix} -1 \\ 1 \\ 2 \end{bmatrix}, \qquad \mathbf{u}_2 = \begin{bmatrix} -2 \\ 1 \\ 4 \end{bmatrix}, \qquad \mathbf{u}_3 = \begin{bmatrix} -1 \\ 1 \\ 4 \end{bmatrix},$$

whose linear independence is guaranteed by Theorem 6. Hence a general solution to (18) is

$$\textbf{(19)} \qquad \mathbf{x}(t) = c_1 e^t \begin{bmatrix} -1 \\ 1 \\ 2 \end{bmatrix} + c_2 e^{2t} \begin{bmatrix} -2 \\ 1 \\ 4 \end{bmatrix} + c_3 e^{3t} \begin{bmatrix} -1 \\ 1 \\ 4 \end{bmatrix}$$

$$= \begin{bmatrix} -e^t & -2e^{2t} & -e^{3t} \\ e^t & e^{2t} & e^{3t} \\ 2e^t & 4e^{2t} & 4e^{3t} \end{bmatrix} \begin{bmatrix} c_1 \\ c_2 \\ c_3 \end{bmatrix}.$$

To satisfy the initial condition in (18), we solve

$$\mathbf{x}(0) = \begin{bmatrix} -1 & -2 & -1 \\ 1 & 1 & 1 \\ 2 & 4 & 4 \end{bmatrix} \begin{bmatrix} c_1 \\ c_2 \\ c_3 \end{bmatrix} = \begin{bmatrix} -1 \\ 0 \\ 0 \end{bmatrix}$$

and find that $c_1 = 0$, $c_2 = 1$, and $c_3 = -1$. Inserting these values into (19) gives the desired solution. ∎ End

There is a special class of $n \times n$ matrices that *always* have real eigenvalues and *always* have n linearly independent eigenvectors. These are the real symmetric matrices.

REAL SYMMETRIC MATRICES

Definition 4. A **real symmetric matrix** \mathbf{A} is a matrix with real entries that satisfies $\mathbf{A}^T = \mathbf{A}$.

Taking the transpose of a matrix interchanges its rows and columns. Doing this is equivalent to "flipping" the matrix about its main diagonal. Consequently, $\mathbf{A}^T = \mathbf{A}$ if and only if \mathbf{A} is symmetric about its main diagonal.

If **A** is an $n \times n$ real symmetric matrix, it is known[†] that there always exist n linearly independent eigenvectors. Thus Theorem 5 applies, and a general solution to $\mathbf{x}' = \mathbf{A}\mathbf{x}$ is given by (13).

EXAMPLE 5 Find a general solution of

(20) $\mathbf{x}'(t) = \mathbf{A}\mathbf{x}(t)$, where $\mathbf{A} = \begin{bmatrix} 1 & -2 & 2 \\ -2 & 1 & 2 \\ 2 & 2 & 1 \end{bmatrix}$.

SOLUTION **A** is symmetric, so we are assured that **A** has three linearly independent eigenvectors. To find them, we first compute the characteristic equation for **A**:

$$|\mathbf{A} - r\mathbf{I}| = \begin{vmatrix} 1 - r & -2 & 2 \\ -2 & 1 - r & 2 \\ 2 & 2 & 1 - r \end{vmatrix} = -(r - 3)^2(r + 3) = 0 .$$

Thus the eigenvalues of **A** are $r_1 = r_2 = 3$ and $r_3 = -3$.

Notice that the eigenvalue $r = 3$ has multiplicity 2 when considered as a root of the characteristic equation. Therefore we must find *two* linearly independent eigenvectors associated with $r = 3$. Substituting $r = 3$ in $(\mathbf{A} - r\mathbf{I})\mathbf{u} = \mathbf{0}$ gives

$$\begin{bmatrix} -2 & -2 & 2 \\ -2 & -2 & 2 \\ 2 & 2 & -2 \end{bmatrix} \begin{bmatrix} u_1 \\ u_2 \\ u_3 \end{bmatrix} = \begin{bmatrix} 0 \\ 0 \\ 0 \end{bmatrix} .$$

This system is equivalent to the single equation $-u_1 - u_2 + u_3 = 0$, so we can obtain its solutions by assigning an arbitrary value to u_2, say $u_2 = v$, and an arbitrary value to u_3, say $u_3 = s$. Solving for u_1, we find $u_1 = u_3 - u_2 = s - v$. Therefore the eigenvectors associated with $r_1 = r_2 = 3$ can be expressed as

$$\mathbf{u} = \begin{bmatrix} s - v \\ v \\ s \end{bmatrix} = s \begin{bmatrix} 1 \\ 0 \\ 1 \end{bmatrix} + v \begin{bmatrix} -1 \\ 1 \\ 0 \end{bmatrix} .$$

By first taking $s = 1$, $v = 0$ and then taking $s = 0$, $v = 1$, we get the two linearly independent eigenvectors

(21) $\mathbf{u}_1 = \begin{bmatrix} 1 \\ 0 \\ 1 \end{bmatrix}$, $\mathbf{u}_2 = \begin{bmatrix} -1 \\ 1 \\ 0 \end{bmatrix}$.

For $r_3 = -3$, we solve

$$(\mathbf{A} + 3\mathbf{I})\mathbf{u} = \begin{bmatrix} 4 & -2 & 2 \\ -2 & 4 & 2 \\ 2 & 2 & 4 \end{bmatrix} \begin{bmatrix} u_1 \\ u_2 \\ u_3 \end{bmatrix} = \begin{bmatrix} 0 \\ 0 \\ 0 \end{bmatrix}$$

[†] See *Linear Algebra and Its Applications,* 2nd Edition, by David C. Lay (Addison-Wesley, Reading, Mass., 1997).

to obtain the eigenvectors $\text{col}(-s, -s, s)$. Taking $s = 1$ gives

$$\mathbf{u}_3 = \begin{bmatrix} -1 \\ -1 \\ 1 \end{bmatrix}.$$

Since the eigenvectors \mathbf{u}_1, \mathbf{u}_2, and \mathbf{u}_3 are linearly independent, a general solution to (20) is

$$\mathbf{x}(t) = c_1 e^{3t} \begin{bmatrix} 1 \\ 0 \\ 1 \end{bmatrix} + c_2 e^{3t} \begin{bmatrix} -1 \\ 1 \\ 0 \end{bmatrix} + c_3 e^{-3t} \begin{bmatrix} -1 \\ -1 \\ 1 \end{bmatrix}. \quad \blacksquare \text{ End}$$

If a matrix \mathbf{A} is not symmetric, it is possible for \mathbf{A} to have a repeated eigenvalue but not to have two linearly independent corresponding eigenvectors. In particular, the matrix

(22) $\qquad \mathbf{A} = \begin{bmatrix} 1 & -1 \\ 4 & -3 \end{bmatrix}$

has the repeated eigenvalue $r_1 = r_2 = -1$, but Problem 35 shows that all the eigenvectors associated with $r = -1$ are of the form $\mathbf{u} = s\,\text{col}(1, 2)$. Consequently, no two eigenvectors are linearly independent.

A procedure for finding a general solution in such a case is illustrated in Problems 35–40, but the underlying theory is deferred to Section 9.8, where we discuss the matrix exponential.

EXERCISES 9.5

In Problems 1 through 8, find the eigenvalues and eigenvectors of the given matrix.

1. $\begin{bmatrix} -4 & 2 \\ 2 & -1 \end{bmatrix}$.

2. $\begin{bmatrix} 6 & -3 \\ 2 & 1 \end{bmatrix}$.

3. $\begin{bmatrix} 1 & -1 \\ 2 & 4 \end{bmatrix}$.

4. $\begin{bmatrix} 1 & 5 \\ 1 & -3 \end{bmatrix}$.

5. $\begin{bmatrix} 1 & 0 & 0 \\ 0 & 0 & 2 \\ 0 & 2 & 0 \end{bmatrix}$.

6. $\begin{bmatrix} 0 & 1 & 1 \\ 1 & 0 & 1 \\ 1 & 1 & 0 \end{bmatrix}$.

7. $\begin{bmatrix} 1 & 0 & 0 \\ 2 & 3 & 1 \\ 0 & 2 & 4 \end{bmatrix}$.

8. $\begin{bmatrix} -3 & 1 & 0 \\ 0 & -3 & 1 \\ 4 & -8 & 2 \end{bmatrix}$.

In Problems 9 and 10, some of the eigenvalues of the given matrix are complex. Find all the eigenvalues and eigenvectors.

9. $\begin{bmatrix} 0 & -1 \\ 1 & 0 \end{bmatrix}$.

10. $\begin{bmatrix} 1 & 2 & -1 \\ 0 & 1 & 1 \\ 0 & -1 & 1 \end{bmatrix}$.

In Problems 11 through 16, find a general solution of the system $\mathbf{x}'(t) = \mathbf{A}\mathbf{x}(t)$ for the given matrix \mathbf{A}.

11. $\mathbf{A} = \begin{bmatrix} -1 & \frac{3}{4} \\ -5 & 3 \end{bmatrix}$.

12. $\mathbf{A} = \begin{bmatrix} 1 & 3 \\ 12 & 1 \end{bmatrix}$.

13. $\mathbf{A} = \begin{bmatrix} 1 & 2 & 2 \\ 2 & 0 & 3 \\ 2 & 3 & 0 \end{bmatrix}$.

14. $A = \begin{bmatrix} -1 & 1 & 0 \\ 1 & 2 & 1 \\ 0 & 3 & -1 \end{bmatrix}$.

15. $A = \begin{bmatrix} 1 & 2 & 3 \\ 0 & 1 & 0 \\ 2 & 1 & 2 \end{bmatrix}$. **16.** $A = \begin{bmatrix} -7 & 0 & 6 \\ 0 & 5 & 0 \\ 6 & 0 & 2 \end{bmatrix}$.

17. Consider the system $x'(t) = Ax(t)$, $t \geq 0$, with

$$A = \begin{bmatrix} 1 & \sqrt{3} \\ \sqrt{3} & -1 \end{bmatrix}.$$

(a) Show that the matrix A has eigenvalues $r_1 = 2$ and $r_2 = -2$ with corresponding eigenvectors $u_1 = \text{col}(\sqrt{3}, 1)$ and $u_2 = \text{col}(1, -\sqrt{3})$.
(b) Sketch the trajectory of the solution having initial vector $x(0) = -u_1$.
(c) Sketch the trajectory of the solution having initial vector $x(0) = u_2$.
(d) Sketch the trajectory of the solution having initial vector $x(0) = u_2 - u_1$.

18. Consider the system $x'(t) = Ax(t)$, $t \geq 0$, with

$$A = \begin{bmatrix} -2 & 1 \\ 1 & -2 \end{bmatrix}.$$

(a) Show that the matrix A has eigenvalues $r_1 = -1$ and $r_2 = -3$ with corresponding eigenvectors $u_1 = \text{col}(1, 1)$ and $u_2 = \text{col}(1, -1)$.
(b) Sketch the trajectory of the solution having initial vector $x(0) = u_1$.
(c) Sketch the trajectory of the solution having initial vector $x(0) = -u_2$.
(d) Sketch the trajectory of the solution having initial vector $x(0) = u_1 - u_2$.

In Problems 19 through 24, find a fundamental matrix for the system $x'(t) = Ax(t)$ for the given matrix A.

19. $A = \begin{bmatrix} -1 & 1 \\ 8 & 1 \end{bmatrix}$. **20.** $A = \begin{bmatrix} 5 & 4 \\ -1 & 0 \end{bmatrix}$.

21. $A = \begin{bmatrix} 0 & 1 & 0 \\ 0 & 0 & 1 \\ 8 & -14 & 7 \end{bmatrix}$. **22.** $A = \begin{bmatrix} 3 & 1 & -1 \\ 1 & 3 & -1 \\ 3 & 3 & -1 \end{bmatrix}$.

23. $A = \begin{bmatrix} 2 & 1 & 1 & -1 \\ 0 & -1 & 0 & 1 \\ 0 & 0 & 3 & 1 \\ 0 & 0 & 0 & 7 \end{bmatrix}$.

24. $A = \begin{bmatrix} 4 & -1 & 0 & 0 \\ 0 & 0 & 0 & 0 \\ 0 & 0 & 2 & -3 \\ 0 & 0 & 1 & -2 \end{bmatrix}$.

25. Using matrix algebra techniques, find a general solution of the system

$$x' = x + 2y - z ,$$
$$y' = x + z ,$$
$$z' = 4x - 4y + 5z .$$

26. Using matrix algebra techniques, find a general solution of the system

$$x' = 3x - 4y ,$$
$$y' = 4x - 7y .$$

In Problems 27 through 30, use a linear algebra software package such as DERIVE, MATLAB, MAPLE, or MATHEMATICA to compute the required eigenvalues and eigenvectors and then give a fundamental matrix for the system $x'(t) = Ax(t)$ for the given matrix A.

27. $A = \begin{bmatrix} 0 & 1.1 & 0 \\ 0 & 0 & 1.3 \\ 0.9 & 1.1 & -6.9 \end{bmatrix}$.

28. $A = \begin{bmatrix} 2 & 1 & 1 \\ -1 & 1 & 0 \\ 3 & 3 & 3 \end{bmatrix}$.

29. $A = \begin{bmatrix} 0 & 1 & 0 & 0 \\ 0 & 0 & 1 & 0 \\ 0 & 0 & 0 & 1 \\ 2 & -6 & 3 & 3 \end{bmatrix}$.

30. $A = \begin{bmatrix} 0 & 1 & 0 & 0 \\ 1 & -1 & 0 & 0 \\ 0 & 0 & 0 & 1 \\ 0 & 0 & -2 & 4 \end{bmatrix}$.

In Problems 31 through 34, solve the given initial value problem.

31. $\mathbf{x}'(t) = \begin{bmatrix} 1 & 3 \\ 3 & 1 \end{bmatrix} \mathbf{x}(t), \qquad \mathbf{x}(0) = \begin{bmatrix} 3 \\ 1 \end{bmatrix}.$

32. $\mathbf{x}'(t) = \begin{bmatrix} 6 & -3 \\ 2 & 1 \end{bmatrix} \mathbf{x}(t), \qquad \mathbf{x}(0) = \begin{bmatrix} -10 \\ -6 \end{bmatrix}.$

33. $\mathbf{x}'(t) = \begin{bmatrix} 1 & -2 & 2 \\ -2 & 1 & -2 \\ 2 & -2 & 1 \end{bmatrix} \mathbf{x}(t),$

$\mathbf{x}(0) = \begin{bmatrix} -2 \\ -3 \\ 2 \end{bmatrix}.$

34. $\mathbf{x}'(t) = \begin{bmatrix} 0 & 1 & 1 \\ 1 & 0 & 1 \\ 1 & 1 & 0 \end{bmatrix} \mathbf{x}(t), \qquad \mathbf{x}(0) = \begin{bmatrix} -1 \\ 4 \\ 0 \end{bmatrix}.$

35. (a) Show that the matrix

$$A = \begin{bmatrix} 1 & -1 \\ 4 & -3 \end{bmatrix}$$

has the repeated eigenvalue $r = -1$ and that all the eigenvectors are of the form $\mathbf{u} = s\,\text{col}(1, 2)$.

(b) Use the result of part (a) to obtain a nontrivial solution $\mathbf{x}_1(t)$ to the system $\mathbf{x}' = A\mathbf{x}$.

(c) To obtain a second linearly independent solution to $\mathbf{x}' = A\mathbf{x}$, try $\mathbf{x}_2(t) = te^{-t}\mathbf{u}_1 + e^{-t}\mathbf{u}_2$. [*Hint:* Substitute \mathbf{x}_2 into the system $\mathbf{x}' = A\mathbf{x}$ and derive the relations

$$(A + I)\mathbf{u}_1 = \mathbf{0}, \qquad (A + I)\mathbf{u}_2 = \mathbf{u}_1.$$

Since \mathbf{u}_1 must be an eigenvector, set $\mathbf{u}_1 = \text{col}(1, 2)$ and solve for \mathbf{u}_2.]

(d) What is $(A + I)^2 \mathbf{u}_2$? (In Section 9.8, \mathbf{u}_2 will be identified as a *generalized eigenvector*.)

36. Use the method discussed in Problem 35 to find a general solution to the system

$$\mathbf{x}'(t) = \begin{bmatrix} 5 & -3 \\ 3 & -1 \end{bmatrix} \mathbf{x}(t).$$

37. (a) Show that the matrix

$$A = \begin{bmatrix} 2 & 1 & 6 \\ 0 & 2 & 5 \\ 0 & 0 & 2 \end{bmatrix}$$

has the repeated eigenvalue $r = 2$ with multiplicity 3 and that all the eigenvectors of A are of the form $\mathbf{u} = s\,\text{col}(1, 0, 0)$.

(b) Use the result of part (a) to obtain a solution to the system $\mathbf{x}' = A\mathbf{x}$ of the form $\mathbf{x}_1(t) = e^{2t}\mathbf{u}_1$.

(c) To obtain a second linearly independent solution to $\mathbf{x}' = A\mathbf{x}$, try $\mathbf{x}_2(t) = te^{2t}\mathbf{u}_1 + e^{2t}\mathbf{u}_2$. [*Hint:* Show that \mathbf{u}_1 and \mathbf{u}_2 must satisfy

$$(A - 2I)\mathbf{u}_1 = \mathbf{0}, \qquad (A - 2I)\mathbf{u}_2 = \mathbf{u}_1.]$$

(d) To obtain a third linearly independent solution to $\mathbf{x}' = A\mathbf{x}$, try

$$\mathbf{x}_3(t) = \frac{t^2}{2}e^{2t}\mathbf{u}_1 + te^{2t}\mathbf{u}_2 + e^{2t}\mathbf{u}_3.$$

[*Hint:* Show that \mathbf{u}_1, \mathbf{u}_2, and \mathbf{u}_3 must satisfy

$$(A - 2I)\mathbf{u}_1 = \mathbf{0}, \qquad (A - 2I)\mathbf{u}_2 = \mathbf{u}_1,$$
$$(A - 2I)\mathbf{u}_3 = \mathbf{u}_2.]$$

(e) Show that $(A - 2I)^2 \mathbf{u}_2 = (A - 2I)^3 \mathbf{u}_3 = \mathbf{0}$.

38. Use the method discussed in Problem 37 to find a general solution to the system

$$\mathbf{x}'(t) = \begin{bmatrix} 3 & -2 & 1 \\ 2 & -1 & 1 \\ -4 & 4 & 1 \end{bmatrix} \mathbf{x}(t).$$

39. (a) Show that the matrix

$$A = \begin{bmatrix} 2 & 1 & 1 \\ 1 & 2 & 1 \\ -2 & -2 & -1 \end{bmatrix}$$

has the repeated eigenvalue $r = 1$ of multiplicity 3 and that all the eigenvectors of A are of the form $\mathbf{u} = s\,\text{col}(-1, 1, 0) + v\,\text{col}(-1, 0, 1)$.

(b) Use the result of part (a) to obtain two linearly independent solutions to the system $\mathbf{x}' = A\mathbf{x}$ of the form

$$\mathbf{x}_1(t) = e^t\mathbf{u}_1 \quad \text{and} \quad \mathbf{x}_2(t) = e^t\mathbf{u}_2.$$

(c) To obtain a third linearly independent solution to $\mathbf{x}' = \mathbf{A}\mathbf{x}$, try $\mathbf{x}_3(t) = te^t\mathbf{u}_3 + e^t\mathbf{u}_4$. [*Hint:* Show that \mathbf{u}_3 and \mathbf{u}_4 must satisfy

$$(\mathbf{A} - \mathbf{I})\mathbf{u}_3 = \mathbf{0} , \qquad (\mathbf{A} - \mathbf{I})\mathbf{u}_4 = \mathbf{u}_3 .$$

Choose \mathbf{u}_3, an eigenvector of \mathbf{A}, so that you can solve for \mathbf{u}_4.]

(d) What is $(\mathbf{A} - \mathbf{I})^2\mathbf{u}_4$?

40. Use the method discussed in Problem 39 to find a general solution to the system

$$\mathbf{x}'(t) = \begin{bmatrix} 1 & 3 & -2 \\ 0 & 7 & -4 \\ 0 & 9 & -5 \end{bmatrix} \mathbf{x}(t) .$$

41. Use the substitution $x_1 = y, x_2 = y'$ to convert the linear equation $ay'' + by' + cy = 0$, where a, b, and c are constants, into a normal system. Show that the characteristic equation for this system is the same as the auxiliary equation for the original equation.

42. Show that the **Cauchy-Euler system**

$$t\mathbf{x}'(t) = \mathbf{A}\mathbf{x}(t) ,$$

where \mathbf{A} is a constant matrix, has nontrivial solutions of the form $\mathbf{x}(t) = t^r\mathbf{u}$ if and only if r is an eigenvalue of \mathbf{A} and \mathbf{u} is a corresponding eigenvector.

In Problems 43 and 44, use the result of Problem 42 to find a general solution of the given system.

43. $t\mathbf{x}'(t) = \begin{bmatrix} 1 & 3 \\ -1 & 5 \end{bmatrix} \mathbf{x}(t) , \quad t > 0 .$

44. $t\mathbf{x}'(t) = \begin{bmatrix} -4 & 2 \\ 2 & -1 \end{bmatrix} \mathbf{x}(t) , \quad t > 0 .$

45. Mixing between Interconnected Tanks. Two tanks, each holding 50 L of liquid, are interconnected by pipes with liquid flowing from tank A into tank B at a rate of 4 L/min and from tank B into tank A at 1 L/min (see Figure 9.2). The liquid inside each tank is kept well stirred. Pure water flows into tank A at a rate of 3 L/min, and the solution flows out of tank B at 3 L/min. If, initially, tank A contains 25 kg of salt and tank B contains no salt (only water), determine the mass of salt in each tank at time $t \geq 0$. Graph on the same axes the two quantities $x_1(t)$ and $x_2(t)$, where $x_1(t)$ is the mass of salt in tank A and $x_2(t)$ is the mass in tank B.

Figure 9.2 Mixing problem for interconnected tanks

46. Mixing with a Common Drain. Two tanks, each holding 1 L of liquid, are connected by a pipe through which liquid flows from tank A into tank B at a rate of $3 - \alpha$ L/min $(0 < \alpha < 3)$. The liquid inside each tank is kept well stirred. Pure water flows into tank A at a rate of 3 L/min. Solution flows out of tank A at α L/min and out of tank B at $3 - \alpha$ L/min. If, initially, tank B contains no salt (only water) and tank A contains 1 kg of salt, determine the mass of salt in each tank at time $t \geq 0$. How does the mass of salt in tank A depend on the choice of α? What is the maximum mass of salt in tank B? See Figure 9.3.

Figure 9.3 Mixing problem for a common drain, $0 < \alpha < 3$

47. To find a general solution to the system

$$\mathbf{x}' = \mathbf{A}\mathbf{x} = \begin{bmatrix} 1 & 3 & -1 \\ 3 & 0 & 1 \\ -1 & 1 & 2 \end{bmatrix} \mathbf{x} ,$$

proceed as follows:

(a) Use a numerical root-finding procedure to approximate the eigenvalues.

(b) If r is an eigenvalue, then let $\mathbf{u} = \text{col}\,(u_1, u_2, u_3)$ be an eigenvector associated with r. To solve for

u, assume $u_1 = 1$. (If not u_1, then either u_2 or u_3 may be chosen to be 1. Why?) Now solve the system

$$(\mathbf{A} - r\mathbf{I}) \begin{bmatrix} 1 \\ u_2 \\ u_3 \end{bmatrix} = \begin{bmatrix} 0 \\ 0 \\ 0 \end{bmatrix}$$

for u_2 and u_3. Use this procedure to find approximations for three linearly independent eigenvectors for **A**.

(c) Use these approximations to give a general solution to the system.

48. To complete the proof of Theorem 6, assume the induction hypothesis that $\mathbf{u}_1, \ldots, \mathbf{u}_k$, $2 \leq k$, are linearly independent.

(a) Show that if

$$c_1\mathbf{u}_1 + \cdots + c_k\mathbf{u}_k + c_{k+1}\mathbf{u}_{k+1} = \mathbf{0} \;,$$

then

$$c_1(r_1 - r_{k+1})\mathbf{u}_1 + \cdots + c_k(r_k - r_{k+1})\mathbf{u}_k = \mathbf{0} \;.$$

(b) Use the result of part (a) and the induction hypothesis to conclude that $\mathbf{u}_1, \ldots, \mathbf{u}_{k+1}$ are linearly independent. The theorem follows by induction.

49. Stability. A homogeneous system $\mathbf{x}' = \mathbf{A}\mathbf{x}$ with constant coefficients is **stable** if it has a fundamental matrix whose entries all remain bounded as $t \to +\infty$. (It will follow from Lemma 1 in Section 9.8 that if one fundamental matrix of the system has this property, then all fundamental matrices for the system do.) Otherwise, the system is **unstable**. A stable system is **asymptotically stable** if all solutions approach the zero solution as $t \to +\infty$. Stability is discussed in more detail in Chapter 12.[†]

(a) Show that if **A** has all distinct real eigenvalues, then $\mathbf{x}'(t) = \mathbf{A}\mathbf{x}(t)$ is stable if and only if all eigenvalues are nonpositive.

(b) Show that if **A** has all distinct real eigenvalues, then $\mathbf{x}'(t) = \mathbf{A}\mathbf{x}(t)$ is asymptotically stable if and only if all eigenvalues are negative.

(c) Argue that in parts (a) and (b), we can replace "has distinct real eigenvalues" by "is symmetric" and the statements are still true.

50. In an ice tray, the water level in any particular ice cube cell will change at a rate proportional to the *difference* between that cell's water level and the level in the adjacent cells.

(a) Argue that a reasonable differentiable equation model for the water levels x, y, and z in the simplified three-cell tray depicted in Figure 9.4 is given by

$$x' = y - x \;, \quad y' = x + z - 2y \;, \quad z' = y - z \;.$$

(b) Use eigenvectors to solve this system for the initial conditions $x(0) = 3$, $y(0) = z(0) = 0$.

Figure 9.4 Ice tray

9.6 COMPLEX EIGENVALUES

In the previous section we showed that the homogeneous system

$$(1) \qquad \mathbf{x}'(t) = \mathbf{A}\mathbf{x}(t) \;,$$

where **A** is a constant $n \times n$ matrix, has a solution of the form $\mathbf{x}(t) = e^{rt}\mathbf{u}$ if and only if r is an eigenvalue of **A** and **u** is a corresponding eigenvector. In this section we show how to obtain two real vector solutions to system (1) when **A** is real and has a pair[††] of complex conjugate eigenvalues $\alpha + i\beta$ and $\alpha - i\beta$.

[†]All references to Chapters 11–13 refer to the expanded text *Fundamentals of Differential Equations and Boundary Value Problems,* 3rd Edition.

[††]Recall that the complex roots of a polynomial equation with real coefficients must occur in complex conjugate pairs.

Suppose $r_1 = \alpha + i\beta$ (α and β real numbers) is an eigenvalue of \mathbf{A} with corresponding eigenvector $\mathbf{z} = \mathbf{a} + i\mathbf{b}$, where \mathbf{a} and \mathbf{b} are real constant vectors. We first observe that the complex conjugate of \mathbf{z}, namely, $\overline{\mathbf{z}} := \mathbf{a} - i\mathbf{b}$, is an eigenvector associated with the eigenvalue $r_2 = \alpha - i\beta$. To see this, note that taking the complex conjugate of $(\mathbf{A} - r_1\mathbf{I})\mathbf{z} = \mathbf{0}$ yields $(\mathbf{A} - \overline{r_1}\mathbf{I})\overline{\mathbf{z}} = \mathbf{0}$ because the conjugate of the product is the product of the conjugates and \mathbf{A} and \mathbf{I} have real entries $(\overline{\mathbf{A}} = \mathbf{A}, \overline{\mathbf{I}} = \mathbf{I})$. Since $r_2 = \overline{r_1}$, we see that $\overline{\mathbf{z}}$ is an eigenvector associated with r_2. Therefore two linearly independent complex vector solutions to (1) are

(2) $\mathbf{w}_1(t) = e^{r_1 t}\mathbf{z} = e^{(\alpha + i\beta)t}(\mathbf{a} + i\mathbf{b})$,

(3) $\mathbf{w}_2(t) = e^{r_2 t}\overline{\mathbf{z}} = e^{(\alpha - i\beta)t}(\mathbf{a} - i\mathbf{b})$.

As in Section 4.6, where we handled complex roots to the auxiliary equation, let's use one of these complex solutions and Euler's formula to obtain two real vector solutions. With the aid of Euler's formula, we rewrite $\mathbf{w}_1(t)$ as

$$\mathbf{w}_1(t) = e^{\alpha t}(\cos \beta t + i \sin \beta t)(\mathbf{a} + i\mathbf{b})$$
$$= e^{\alpha t}\{(\cos \beta t\,\mathbf{a} - \sin \beta t\,\mathbf{b}) + i(\sin \beta t\,\mathbf{a} + \cos \beta t\,\mathbf{b})\} .$$

We have thereby expressed $\mathbf{w}_1(t)$ in the form $\mathbf{w}_1(t) = \mathbf{x}_1(t) + i\mathbf{x}_2(t)$, where $\mathbf{x}_1(t)$ and $\mathbf{x}_2(t)$ are the two *real* vector functions

(4) $\mathbf{x}_1(t) := e^{\alpha t}\cos \beta t\,\mathbf{a} - e^{\alpha t}\sin \beta t\,\mathbf{b}$,

(5) $\mathbf{x}_2(t) := e^{\alpha t}\sin \beta t\,\mathbf{a} + e^{\alpha t}\cos \beta t\,\mathbf{b}$.

Since $\mathbf{w}_1(t)$ is a solution to (1), then

$$\mathbf{w}_1'(t) = \mathbf{A}\mathbf{w}_1(t) ,$$
$$\mathbf{x}_1' + i\mathbf{x}_2' = \mathbf{A}\mathbf{x}_1 + i\mathbf{A}\mathbf{x}_2 .$$

Equating the real and imaginary parts yields

$$\mathbf{x}_1'(t) = \mathbf{A}\mathbf{x}_1(t) \quad \text{and} \quad \mathbf{x}_2'(t) = \mathbf{A}\mathbf{x}_2(t) .$$

Hence $\mathbf{x}_1(t)$ and $\mathbf{x}_2(t)$ are real vector solutions to (1) associated with the complex conjugate eigenvalues $\alpha \pm i\beta$. Because \mathbf{a} and \mathbf{b} are not both the zero vector, it can be shown that $\mathbf{x}_1(t)$ and $\mathbf{x}_2(t)$ are linearly independent vector functions on $(-\infty, \infty)$ (see Problem 15).

Let's summarize our findings.

COMPLEX EIGENVALUES

If the real matrix \mathbf{A} has complex conjugate eigenvalues $\alpha \pm i\beta$ with corresponding eigenvectors $\mathbf{a} \pm i\mathbf{b}$, then two linearly independent real vector solutions to $\mathbf{x}'(t) = \mathbf{A}\mathbf{x}(t)$ are

(6) $e^{\alpha t}\cos \beta t\,\mathbf{a} - e^{\alpha t}\sin \beta t\,\mathbf{b}$,

(7) $e^{\alpha t}\sin \beta t\,\mathbf{a} + e^{\alpha t}\cos \beta t\,\mathbf{b}$.

EXAMPLE 1 Find a general solution of

(8) $$\mathbf{x}'(t) = \begin{bmatrix} -1 & 2 \\ -1 & -3 \end{bmatrix} \mathbf{x}(t) \ .$$

SOLUTION The characteristic equation for \mathbf{A} is

$$|\mathbf{A} - r\mathbf{I}| = \begin{vmatrix} -1 - r & 2 \\ -1 & -3 - r \end{vmatrix} = r^2 + 4r + 5 = 0 \ .$$

Hence \mathbf{A} has eigenvalues $r = -2 \pm i$.

To find a general solution, we need only find an eigenvector associated with the eigenvalue $r = -2 + i$. Substituting $r = -2 + i$ into $(\mathbf{A} - r\mathbf{I})\mathbf{z} = \mathbf{0}$ gives

$$\begin{bmatrix} 1 - i & 2 \\ -1 & -1 - i \end{bmatrix} \begin{bmatrix} z_1 \\ z_2 \end{bmatrix} = \begin{bmatrix} 0 \\ 0 \end{bmatrix} \ .$$

The solutions can be expressed as $z_1 = 2s$ and $z_2 = (-1 + i)s$, with s arbitrary. Hence the eigenvectors associated with $r = -2 + i$ are $\mathbf{z} = s \, \mathrm{col}(2, -1 + i)$. Taking $s = 1$ gives the eigenvector

$$\mathbf{z} = \begin{bmatrix} 2 \\ -1 + i \end{bmatrix} = \begin{bmatrix} 2 \\ -1 \end{bmatrix} + i \begin{bmatrix} 0 \\ 1 \end{bmatrix} \ .$$

We have found that $\alpha = -2$, $\beta = 1$, and $\mathbf{z} = \mathbf{a} + i\mathbf{b}$ with $\mathbf{a} = \mathrm{col}(2, -1)$, and $\mathbf{b} = \mathrm{col}(0, 1)$, so a general solution to (8) is

$$\mathbf{x}(t) = c_1 \left\{ e^{-2t} \cos t \begin{bmatrix} 2 \\ -1 \end{bmatrix} - e^{-2t} \sin t \begin{bmatrix} 0 \\ 1 \end{bmatrix} \right\}$$

$$+ c_2 \left\{ e^{-2t} \sin t \begin{bmatrix} 2 \\ -1 \end{bmatrix} + e^{-2t} \cos t \begin{bmatrix} 0 \\ 1 \end{bmatrix} \right\}$$

$$= c_1 \begin{bmatrix} 2e^{-2t} \cos t \\ -e^{-2t}(\cos t + \sin t) \end{bmatrix} + c_2 \begin{bmatrix} 2e^{-2t} \sin t \\ e^{-2t}(\cos t - \sin t) \end{bmatrix} \ . \quad \blacksquare \text{ End}$$

Complex eigenvalues occur in modeling coupled mass-spring systems. For example, the motion of the mass-spring system illustrated in Figure 9.5 on page 572 is governed by the second order system

(9) $$m_1 x_1'' = -k_1 x_1 + k_2 (x_2 - x_1) \ ,$$
$$m_2 x_2'' = -k_2 (x_2 - x_1) - k_3 x_2 \ ,$$

where x_1 and x_2 represent the displacements of the masses m_1 and m_2 to the right of their equilibrium positions and k_1, k_2, k_3 are the spring constants of the three springs (see the discussion

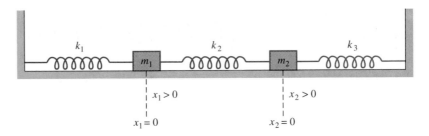

Figure 9.5 Coupled mass-spring system with fixed ends

in Section 5.4). If we introduce the new variables $y_1 := x_1$, $y_2 := x_1'$, $y_3 := x_2$, $y_4 := x_2'$, then we can rewrite the system in the normal form

(10) $\quad \mathbf{y}'(t) = \mathbf{A}\mathbf{y}(t) = \begin{bmatrix} 0 & 1 & 0 & 0 \\ -(k_1 + k_2)/m_1 & 0 & k_2/m_1 & 0 \\ 0 & 0 & 0 & 1 \\ k_2/m_2 & 0 & -(k_2 + k_3)/m_2 & 0 \end{bmatrix} \mathbf{y}(t) \ .$

For such a system, it turns out that \mathbf{A} has only imaginary eigenvalues and they occur in complex conjugate pairs: $\pm i\beta_1$, $\pm i\beta_2$. Hence any solution will consist of sums of sine and cosine functions. The frequencies of these functions

$$f_1 := \frac{\beta_1}{2\pi} \quad \text{and} \quad f_2 := \frac{\beta_2}{2\pi}$$

are called the **normal** or **natural frequencies** of the system (β_1 and β_2 are the **angular frequencies** of the system).

In some engineering applications, the only information that is required about a particular product is a knowledge of its normal frequencies; one must ensure that they are far from the frequencies that occur naturally in the product's operating environment (so that no resonances will be excited).

EXAMPLE 2 Determine the normal frequencies for the coupled mass-spring system governed by system (10) when $m_1 = m_2 = 1$ kg, $k_1 = 1$ N/m, $k_2 = 2$ N/m, and $k_3 = 3$ N/m.

SOLUTION To find the eigenvalues of \mathbf{A}, we must solve the characteristic equation

$$|\mathbf{A} - r\mathbf{I}| = \begin{vmatrix} -r & 1 & 0 & 0 \\ -3 & -r & 2 & 0 \\ 0 & 0 & -r & 1 \\ 2 & 0 & -5 & -r \end{vmatrix} = r^4 + 8r^2 + 11 = 0 \ .$$

From the quadratic formula we find $r^2 = -4 \pm \sqrt{5}$, so the four eigenvalues of \mathbf{A} are $\pm i\sqrt{4 - \sqrt{5}}$ and $\pm i\sqrt{4 + \sqrt{5}}$. Hence the two normal frequencies for this system are

$$\frac{\sqrt{4 - \sqrt{5}}}{2\pi} \approx 0.211 \quad \text{and} \quad \frac{\sqrt{4 + \sqrt{5}}}{2\pi} \approx 0.397 \text{ cycles per second.} \quad \blacksquare \text{ End}$$

EXERCISES 9.6

In Problems 1 through 4, find a general solution of the system $\mathbf{x}'(t) = \mathbf{A}\mathbf{x}(t)$ *for the given matrix* \mathbf{A}.

1. $\mathbf{A} = \begin{bmatrix} 2 & -4 \\ 2 & -2 \end{bmatrix}$.

2. $\mathbf{A} = \begin{bmatrix} -2 & -5 \\ 1 & 2 \end{bmatrix}$.

3. $\mathbf{A} = \begin{bmatrix} 1 & 2 & -1 \\ 0 & 1 & 1 \\ 0 & -1 & 1 \end{bmatrix}$.

4. $\mathbf{A} = \begin{bmatrix} 5 & -5 & -5 \\ -1 & 4 & 2 \\ 3 & -5 & -3 \end{bmatrix}$.

In Problems 5 through 8, find a fundamental matrix for the system $\mathbf{x}'(t) = \mathbf{A}\mathbf{x}(t)$ *for the given matrix* \mathbf{A}.

5. $\mathbf{A} = \begin{bmatrix} -1 & -2 \\ 8 & -1 \end{bmatrix}$.

6. $\mathbf{A} = \begin{bmatrix} -2 & -2 \\ 4 & 2 \end{bmatrix}$.

7. $\mathbf{A} = \begin{bmatrix} 0 & 0 & 1 \\ 0 & 0 & -1 \\ 0 & 1 & 0 \end{bmatrix}$.

8. $\mathbf{A} = \begin{bmatrix} 0 & 1 & 0 & 0 \\ 1 & 0 & 0 & 0 \\ 0 & 0 & 0 & 1 \\ 0 & 0 & -13 & 4 \end{bmatrix}$.

In Problems 9 through 12, use a linear algebra software package to compute the required eigenvalues and eigenvectors for the given matrix \mathbf{A} *and then give a fundamental matrix for the system* $\mathbf{x}'(t) = \mathbf{A}\mathbf{x}(t)$.

9. $\mathbf{A} = \begin{bmatrix} 0 & 1 & 1 \\ -1 & 0 & 1 \\ -1 & -1 & 0 \end{bmatrix}$.

10. $\mathbf{A} = \begin{bmatrix} 0 & 1 & 0 & 0 \\ 0 & 0 & 1 & 0 \\ 0 & 0 & 0 & 1 \\ 13 & -4 & -12 & 4 \end{bmatrix}$.

11. $\mathbf{A} = \begin{bmatrix} 0 & 1 & 0 & 0 \\ 0 & 0 & 1 & 0 \\ 0 & 0 & 0 & 1 \\ -2 & 2 & -3 & 2 \end{bmatrix}$.

12. $\mathbf{A} = \begin{bmatrix} 1 & 0 & 0 & 0 & 0 \\ 0 & 0 & 1 & 0 & 0 \\ 0 & 1 & 0 & 0 & 0 \\ 0 & 0 & 0 & 0 & 1 \\ 0 & 0 & 0 & -29 & -4 \end{bmatrix}$.

In Problems 13 and 14, find the solution to the given system that satisfies the given initial condition.

13. $\mathbf{x}'(t) = \begin{bmatrix} -3 & -1 \\ 2 & -1 \end{bmatrix} \mathbf{x}(t)$,

(a) $\mathbf{x}(0) = \begin{bmatrix} -1 \\ 0 \end{bmatrix}$. **(b)** $\mathbf{x}(\pi) = \begin{bmatrix} 1 \\ -1 \end{bmatrix}$.

(c) $\mathbf{x}(-2\pi) = \begin{bmatrix} 2 \\ 1 \end{bmatrix}$. **(d)** $\mathbf{x}(\pi/2) = \begin{bmatrix} 0 \\ 1 \end{bmatrix}$.

14. $\mathbf{x}'(t) = \begin{bmatrix} 1 & 0 & -1 \\ 0 & 2 & 0 \\ 1 & 0 & 1 \end{bmatrix} \mathbf{x}(t)$,

(a) $\mathbf{x}(0) = \begin{bmatrix} -2 \\ 2 \\ -1 \end{bmatrix}$. **(b)** $\mathbf{x}(-\pi) = \begin{bmatrix} 0 \\ 1 \\ 1 \end{bmatrix}$.

15. Show that $\mathbf{x}_1(t)$ and $\mathbf{x}_2(t)$ given by equations (4) and (5) are linearly independent on $(-\infty, \infty)$, provided $\beta \neq 0$ and \mathbf{a} and \mathbf{b} are not both the zero vector.

16. Show that $\mathbf{x}_1(t)$ and $\mathbf{x}_2(t)$ given by equations (4) and (5) can be obtained as linear combinations of $\mathbf{w}_1(t)$ and $\mathbf{w}_2(t)$ given by equations (2) and (3). [*Hint:* Show that

$$\mathbf{x}_1(t) = \frac{\mathbf{w}_1(t) + \mathbf{w}_2(t)}{2} , \qquad \mathbf{x}_2(t) = \frac{\mathbf{w}_1(t) - \mathbf{w}_2(t)}{2i} .]$$

In Problems 17 and 18, use the results of Problem 42 in Exercises 9.5 to find a general solution to the given Cauchy-Euler system for t > 0.

17. $t\mathbf{x}'(t) = \begin{bmatrix} -1 & -1 & 0 \\ 2 & -1 & 1 \\ 0 & 1 & -1 \end{bmatrix} \mathbf{x}(t)$.

18. $t\mathbf{x}'(t) = \begin{bmatrix} -1 & -1 \\ 9 & -1 \end{bmatrix} \mathbf{x}(t)$.

19. For the coupled mass-spring system governed by system (9), assume $m_1 = m_2 = 1$ kg, $k_1 = k_2 = 2$ N/m, and $k_3 = 3$ N/m. Determine the normal frequencies for this coupled mass-spring system.

20. For the coupled mass-spring system governed by system (9), assume $m_1 = m_2 = 1$ kg, $k_1 = k_2 = k_3 = 1$ N/m, and assume initially that $x_1(0) = 0$ m, $x_1'(0) = 0$ m/sec, $x_2(0) = 2$ m, and $x_2'(0) = 0$ m/sec. Using matrix algebra techniques, solve this initial value problem.

21. *RLC* Network. The currents in the *RLC* network given by the schematic diagram in Figure 9.6 are governed by the following equations:

$$4I_2'(t) + 52q_1(t) = 10 \ ,$$
$$13I_3(t) + 52q_1(t) = 10 \ ,$$
$$I_1(t) = I_2(t) + I_3(t) \ ,$$

where $q_1(t)$ is the charge on the capacitor, $I_1(t) = q_1'(t)$, and initially $q_1(0) = 0$ coulombs and

Figure 9.6 *RLC* network for Problem 21

$I_1(0) = 0$ amps. Solve for the currents I_1, I_2, and I_3. [*Hint:* Differentiate the first two equations, eliminate I_1, and form a normal system with $x_1 = I_2$, $x_2 = I_2'$, and $x_3 = I_3$.]

22. *RLC* Network. The currents in the *RLC* network given by the schematic diagram in Figure 9.7 are governed by the following equations:

$$50I_1'(t) + 80I_2(t) = 160 \ ,$$
$$50I_1'(t) + 800q_3(t) = 160 \ ,$$
$$I_1(t) = I_2(t) + I_3(t) \ ,$$

where $q_3(t)$ is the charge on the capacitor, $I_3(t) = q_3'(t)$, and initially $q_3(0) = 0.5$ coulombs and $I_3(0) = 0$ amps. Solve for the currents I_1, I_2, and I_3. [*Hint:* Differentiate the first two equations, use the third equation to eliminate I_3, and form a normal system with $x_1 = I_1$, $x_2 = I_1'$, and $x_3 = I_2$.]

Figure 9.7 *RLC* network for Problem 22

23. Stability. In Problem 49 of Exercises 9.5, we discussed the notion of stability and asymptotic stability for a linear system of the form $\mathbf{x}'(t) = \mathbf{A}\mathbf{x}(t)$. Assume **A** has all distinct eigenvalues (real or complex).

(a) Show that the system is stable if and only if all the eigenvalues of **A** have nonpositive real part.

(b) Show that the system is asymptotically stable if and only if all the eigenvalues of **A** have negative real part.

9.7 NONHOMOGENEOUS LINEAR SYSTEMS

The techniques discussed in Chapter 4 for finding a particular solution to the nonhomogeneous equation $y'' + p(x)y' + q(x)y = g(x)$ have natural extensions to nonhomogeneous linear systems.

Undetermined Coefficients

The method of undetermined coefficients can be used to find a particular solution to the nonhomogeneous linear system

$$\mathbf{x}'(t) = \mathbf{Ax}(t) + \mathbf{f}(t)$$

when \mathbf{A} is an $n \times n$ *constant* matrix and the entries of $\mathbf{f}(t)$ are polynomials, exponential functions, sines and cosines, or finite sums and products of these functions. We can use Table 4.1 (discussed in Section 4.8 and reproduced on the inside back cover) as a *guide* in choosing the form of a particular solution $\mathbf{x}_p(t)$. Some exceptions are discussed in the exercises (see Problems 25–28).

EXAMPLE 1 Find a general solution of

$$(1) \qquad \mathbf{x}'(t) = \mathbf{Ax}(t) + t\mathbf{g} \; , \quad \text{where} \quad \mathbf{A} = \begin{bmatrix} 1 & -2 & 2 \\ -2 & 1 & 2 \\ 2 & 2 & 1 \end{bmatrix} \quad \text{and} \quad \mathbf{g} = \begin{bmatrix} -9 \\ 0 \\ -18 \end{bmatrix} .$$

SOLUTION In Example 5 in Section 9.5, we found that a general solution to the corresponding homogeneous system $\mathbf{x}' = \mathbf{Ax}$ is

$$(2) \qquad \mathbf{x}_h(t) = c_1 e^{3t} \begin{bmatrix} 1 \\ 0 \\ 1 \end{bmatrix} + c_2 e^{3t} \begin{bmatrix} -1 \\ 1 \\ 0 \end{bmatrix} + c_3 e^{-3t} \begin{bmatrix} -1 \\ -1 \\ 1 \end{bmatrix} .$$

Since the entries in $\mathbf{f}(t) := t\mathbf{g}$ are just linear functions of t, we are inclined to seek a particular solution of the form

$$\mathbf{x}_p(t) = t\mathbf{a} + \mathbf{b} = t \begin{bmatrix} a_1 \\ a_2 \\ a_3 \end{bmatrix} + \begin{bmatrix} b_1 \\ b_2 \\ b_3 \end{bmatrix} ,$$

where the constant vectors \mathbf{a} and \mathbf{b} are to be determined. Substituting this expression for $\mathbf{x}_p(t)$ into system (1) yields

$$\mathbf{a} = \mathbf{A}(t\mathbf{a} + \mathbf{b}) + t\mathbf{g} \; ,$$

which can be written as

$$t(\mathbf{Aa} + \mathbf{g}) + (\mathbf{Ab} - \mathbf{a}) = \mathbf{0} \; .$$

Setting the "coefficients" of this vector polynomial equal to zero yields the two systems

(3) $\mathbf{Aa} = -\mathbf{g}$,

(4) $\mathbf{Ab} = \mathbf{a}$.

By Gaussian elimination or by using a linear algebra software package, we can solve (3) for \mathbf{a} and we find $\mathbf{a} = \text{col}(5, 2, 4)$. Next we substitute for \mathbf{a} in (4) and solve for \mathbf{b} to obtain $\mathbf{b} = \text{col}(1, 0, 2)$. Hence a particular solution for (1) is

(5) $\mathbf{x}_p(t) = t\mathbf{a} + \mathbf{b} = t\begin{bmatrix} 5 \\ 2 \\ 4 \end{bmatrix} + \begin{bmatrix} 1 \\ 0 \\ 2 \end{bmatrix} = \begin{bmatrix} 5t + 1 \\ 2t \\ 4t + 2 \end{bmatrix}$.

 A general solution for (1) is $\mathbf{x}(t) = \mathbf{x}_h(t) + \mathbf{x}_p(t)$, where $\mathbf{x}_h(t)$ is given in (2) and $\mathbf{x}_p(t)$ in (5). ◼ End

 In the preceding example, the nonhomogeneous term $\mathbf{f}(t)$ was a vector polynomial. If, instead, $\mathbf{f}(t)$ has the form

 $$\mathbf{f}(t) = \text{col}(1, t, \sin t) ,$$

then, using the superposition principle, we would seek a particular solution of the form

 $$\mathbf{x}_p(t) = t\mathbf{a} + \mathbf{b} + (\sin t)\mathbf{c} + (\cos t)\mathbf{d} .$$

Similarly, if

 $$\mathbf{f}(t) = \text{col}(t, e^t, t^2) ,$$

we would take

 $$\mathbf{x}_p(t) = t^2\mathbf{a} + t\mathbf{b} + \mathbf{c} + e^t\mathbf{d} .$$

Of course, we must modify our guess should one of the terms be a solution to the corresponding homogeneous system. Problems 25–28 illustrate the strategies.

Variation of Parameters

In Section 4.9, we discussed the method of variation of parameters for a general (not necessarily constant-coefficient) second order linear equation. Simply put, the idea is that if a general solution to the homogeneous equation has the form $x_h(t) = c_1 x_1(t) + c_2 x_2(t)$, where $x_1(t)$ and $x_2(t)$ are linearly independent solutions to the homogeneous equation, then a particular solution to the nonhomogeneous equation would have the form $x_p(t) = v_1(t)x_1(t) + v_2(t)x_2(t)$, where $v_1(t)$ and $v_2(t)$ are certain functions of t. A similar idea can be used for systems.

 Let $\mathbf{X}(t)$ be a fundamental matrix for the homogeneous system

(6) $\mathbf{x}'(t) = \mathbf{A}(t)\mathbf{x}(t)$,

where now *the entries of* \mathbf{A} *may be any continuous functions of t.* Because a general solution to (6) is given by $\mathbf{X}(t)\mathbf{c}$, where \mathbf{c} is a constant $n \times 1$ vector, we seek a particular solution to the nonhomogeneous system

(7) $\mathbf{x}'(t) = \mathbf{A}(t)\mathbf{x}(t) + \mathbf{f}(t)$

of the form

(8) $\mathbf{x}_p(t) = \mathbf{X}(t)\mathbf{v}(t)$,

where $\mathbf{v}(t) = \mathrm{col}\big(v_1(t), \ldots, v_n(t)\big)$ is a vector function of t to be determined.

To derive a formula for $\mathbf{v}(t)$, we first differentiate (8) using the matrix version of the product rule to obtain

$$\mathbf{x}_p'(t) = \mathbf{X}(t)\mathbf{v}'(t) + \mathbf{X}'(t)\mathbf{v}(t) .$$

Substituting the expressions for $\mathbf{x}_p(t)$ and $\mathbf{x}_p'(t)$ into (7) yields

(9) $\mathbf{X}(t)\mathbf{v}'(t) + \mathbf{X}'(t)\mathbf{v}(t) = \mathbf{A}(t)\mathbf{X}(t)\mathbf{v}(t) + \mathbf{f}(t)$.

Since $\mathbf{X}(t)$ satisfies the matrix equation $\mathbf{X}'(t) = \mathbf{A}(t)\mathbf{X}(t)$, equation (9) becomes

$$\mathbf{Xv}' + \mathbf{AXv} = \mathbf{AXv} + \mathbf{f} ,$$
$$\mathbf{Xv}' = \mathbf{f} .$$

Multiplying both sides of the last equation by $\mathbf{X}^{-1}(t)$ (which exists since the columns of $\mathbf{X}(t)$ are linearly independent) gives

$$\mathbf{v}'(t) = \mathbf{X}^{-1}(t)\mathbf{f}(t) .$$

Integrating, we obtain

$$\mathbf{v}(t) = \int \mathbf{X}^{-1}(t)\mathbf{f}(t)\, dt .$$

Hence a particular solution to (7) is

(10) $\mathbf{x}_p(t) = \mathbf{X}(t)\mathbf{v}(t) = \mathbf{X}(t)\displaystyle\int \mathbf{X}^{-1}(t)\mathbf{f}(t)\, dt$.

Combining (10) with the solution $\mathbf{X}(t)\mathbf{c}$ to the homogeneous system yields the following general solution to (7):

(11) $\mathbf{x}(t) = \mathbf{X}(t)\mathbf{c} + \mathbf{X}(t)\displaystyle\int \mathbf{X}^{-1}(t)\mathbf{f}(t)\, dt$.

The elegance of the derivation of the variation of parameters formula (10) for systems becomes evident when one compares it with the more lengthy derivations for the scalar case in Sections 4.9 and 6.4.

Given an initial value problem of the form

(12) $\mathbf{x}'(t) = \mathbf{A}(t)\mathbf{x}(t) + \mathbf{f}(t)$, $\mathbf{x}(t_0) = \mathbf{x}_0$,

we can use the initial condition $\mathbf{x}(t_0) = \mathbf{x}_0$ to solve for \mathbf{c} in (11). Expressing $\mathbf{x}(t)$ using a definite integral, we have

$$\mathbf{x}(t) = \mathbf{X}(t)\mathbf{c} + \mathbf{X}(t)\int_{t_0}^{t} \mathbf{X}^{-1}(s)\mathbf{f}(s)\, ds .$$

Using the initial condition $\mathbf{x}(t_0) = \mathbf{x}_0$, we find

$$\mathbf{x}_0 = \mathbf{x}(t_0) = \mathbf{X}(t_0)\mathbf{c} + \mathbf{X}(t_0) \int_{t_0}^{t_0} \mathbf{X}^{-1}(s)\mathbf{f}(s)\,ds = \mathbf{X}(t_0)\mathbf{c} \ .$$

Solving for \mathbf{c}, we have $\mathbf{c} = \mathbf{X}^{-1}(t_0)\mathbf{x}_0$. Thus the solution to (12) is given by the formula

$$(13) \qquad \mathbf{x}(t) = \mathbf{X}(t)\mathbf{X}^{-1}(t_0)\mathbf{x}_0 + \mathbf{X}(t)\int_{t_0}^{t} \mathbf{X}^{-1}(s)\mathbf{f}(s)\,ds \ .$$

To apply the variation of parameters formulas, we first must determine a fundamental matrix $\mathbf{X}(t)$ for the homogeneous system. In the case when the coefficient matrix \mathbf{A} is constant, we have discussed methods for finding $\mathbf{X}(t)$. However, if the entries of \mathbf{A} depend on t, the determination of $\mathbf{X}(t)$ may be extremely difficult (entailing, perhaps, a matrix power series!).

EXAMPLE 2 Find the solution to the initial value problem

$$(14) \qquad \mathbf{x}'(t) = \begin{bmatrix} 2 & -3 \\ 1 & -2 \end{bmatrix} \mathbf{x}(t) + \begin{bmatrix} e^{2t} \\ 1 \end{bmatrix}, \qquad \mathbf{x}(0) = \begin{bmatrix} -1 \\ 0 \end{bmatrix}.$$

SOLUTION In Example 3 in Section 9.5, we found two linearly independent solutions to the corresponding homogeneous system; namely,

$$\mathbf{x}_1(t) = \begin{bmatrix} 3e^t \\ e^t \end{bmatrix} \quad \text{and} \quad \mathbf{x}_2(t) = \begin{bmatrix} e^{-t} \\ e^{-t} \end{bmatrix}.$$

Hence a fundamental matrix for the homogeneous system is

$$\mathbf{X}(t) = \begin{bmatrix} 3e^t & e^{-t} \\ e^t & e^{-t} \end{bmatrix}.$$

Although the solution to (14) can be found via the method of undetermined coefficients, we shall find it directly from formula (13). For this purpose, we need $\mathbf{X}^{-1}(t)$. One way[†] to obtain $\mathbf{X}^{-1}(t)$ is to form the augmented matrix

$$\begin{bmatrix} 3e^t & e^{-t} & \vdots & 1 & 0 \\ e^t & e^{-t} & \vdots & 0 & 1 \end{bmatrix}$$

and row-reduce this matrix to the matrix $\begin{bmatrix} \mathbf{I} & \vdots & \mathbf{X}^{-1}(t) \end{bmatrix}$. This gives

$$\mathbf{X}^{-1}(t) = \begin{bmatrix} \dfrac{1}{2}e^{-t} & -\dfrac{1}{2}e^{-t} \\[2mm] -\dfrac{1}{2}e^{t} & \dfrac{3}{2}e^{t} \end{bmatrix}.$$

[†]This procedure works for an invertible matrix of any dimension. For an arbitrary 2×2 invertible matrix $\mathbf{U}(t)$, a formula for $\mathbf{U}^{-1}(t)$ is derived in Problem 32.

Substituting into formula (13), we obtain the solution

$$\mathbf{x}(t) = \begin{bmatrix} 3e^t & e^{-t} \\ e^t & e^{-t} \end{bmatrix} \begin{bmatrix} \frac{1}{2} & -\frac{1}{2} \\ -\frac{1}{2} & \frac{3}{2} \end{bmatrix} \begin{bmatrix} -1 \\ 0 \end{bmatrix}$$

$$+ \begin{bmatrix} 3e^t & e^{-t} \\ e^t & e^{-t} \end{bmatrix} \int_0^t \begin{bmatrix} \frac{1}{2}e^{-s} & -\frac{1}{2}e^{-s} \\ -\frac{1}{2}e^s & \frac{3}{2}e^s \end{bmatrix} \begin{bmatrix} e^{2s} \\ 1 \end{bmatrix} ds$$

$$= \begin{bmatrix} -\frac{3}{2}e^t + \frac{1}{2}e^{-t} \\ -\frac{1}{2}e^t + \frac{1}{2}e^{-t} \end{bmatrix} + \begin{bmatrix} 3e^t & e^{-t} \\ e^t & e^{-t} \end{bmatrix} \int_0^t \begin{bmatrix} \frac{1}{2}e^s - \frac{1}{2}e^{-s} \\ -\frac{1}{2}e^{3s} + \frac{3}{2}e^s \end{bmatrix} ds$$

$$= \begin{bmatrix} -\frac{3}{2}e^t + \frac{1}{2}e^{-t} \\ -\frac{1}{2}e^t + \frac{1}{2}e^{-t} \end{bmatrix} + \begin{bmatrix} 3e^t & e^{-t} \\ e^t & e^{-t} \end{bmatrix} \begin{bmatrix} \frac{1}{2}e^t + \frac{1}{2}e^{-t} - 1 \\ \frac{3}{2}e^t - \frac{1}{6}e^{3t} - \frac{4}{3} \end{bmatrix}$$

$$= \begin{bmatrix} -\frac{9}{2}e^t - \frac{5}{6}e^{-t} + \frac{4}{3}e^{2t} + 3 \\ -\frac{3}{2}e^t - \frac{5}{6}e^{-t} + \frac{1}{3}e^{2t} + 2 \end{bmatrix}. \quad \blacksquare \text{ End}$$

EXERCISES 9.7

In Problems 1 through 4, use the method of undetermined coefficients to find a general solution to the system $\mathbf{x}'(t) = \mathbf{A}\mathbf{x}(t) + \mathbf{f}(t)$, *where* \mathbf{A} *and* $\mathbf{f}(t)$ *are given.*

1. $\mathbf{A} = \begin{bmatrix} 6 & 1 \\ 4 & 3 \end{bmatrix}$, $\mathbf{f}(t) = \begin{bmatrix} -11 \\ -5 \end{bmatrix}$.

2. $\mathbf{A} = \begin{bmatrix} 1 & 1 \\ 4 & 1 \end{bmatrix}$, $\mathbf{f}(t) = \begin{bmatrix} -t-1 \\ -4t-2 \end{bmatrix}$.

3. $\mathbf{A} = \begin{bmatrix} 1 & -2 & 2 \\ -2 & 1 & 2 \\ 2 & 2 & 1 \end{bmatrix}$, $\mathbf{f}(t) = \begin{bmatrix} 2e^t \\ 4e^t \\ -2e^t \end{bmatrix}$.

4. $\mathbf{A} = \begin{bmatrix} 2 & 2 \\ 2 & 2 \end{bmatrix}$, $\mathbf{f}(t) = \begin{bmatrix} -4\cos t \\ -\sin t \end{bmatrix}$.

In Problems 5 through 10, use the method of undetermined coefficients (Table 4.1 on inside back cover) to determine only the form of a particular solution for the system $\mathbf{x}'(t) = \mathbf{A}\mathbf{x}(t) + \mathbf{f}(t)$, *where* \mathbf{A} *and* $\mathbf{f}(t)$ *are given.*

5. $\mathbf{A} = \begin{bmatrix} 1 & 1 \\ 0 & 2 \end{bmatrix}$, $\mathbf{f}(t) = e^{-2t}\begin{bmatrix} t \\ 3 \end{bmatrix}$.

6. $\mathbf{A} = \begin{bmatrix} 1 & -1 \\ 0 & -1 \end{bmatrix}$, $\mathbf{f}(t) = \begin{bmatrix} t \\ e^{3t} \end{bmatrix}$.

7. $\mathbf{A} = \begin{bmatrix} 0 & 1 \\ 2 & 0 \end{bmatrix}$, $\mathbf{f}(t) = \begin{bmatrix} \sin 3t \\ t \end{bmatrix}$.

8. $\mathbf{A} = \begin{bmatrix} -1 & 0 \\ 2 & 2 \end{bmatrix}$, $\mathbf{f}(t) = \begin{bmatrix} t^2 \\ t + 1 \end{bmatrix}$.

9. $\mathbf{A} = \begin{bmatrix} 1 & 1 \\ 0 & 1 \end{bmatrix}$, $\mathbf{f}(t) = \begin{bmatrix} e^{2t} \\ e^{3t} \end{bmatrix}$.

10. $\mathbf{A} = \begin{bmatrix} 2 & -1 \\ 1 & 5 \end{bmatrix}$, $\mathbf{f}(t) = \begin{bmatrix} te^{-t} \\ 3e^{-t} \end{bmatrix}$.

In Problems 11 through 16, use the variation of parameters formula (11) to find a general solution of the system $\mathbf{x}'(t) = \mathbf{A}\mathbf{x}(t) + \mathbf{f}(t)$, *where* \mathbf{A} *and* $\mathbf{f}(t)$ *are given.*

11. $\mathbf{A} = \begin{bmatrix} 0 & 1 \\ -1 & 0 \end{bmatrix}$, $\mathbf{f}(t) = \begin{bmatrix} 1 \\ 0 \end{bmatrix}$.

12. $\mathbf{A} = \begin{bmatrix} 1 & 2 \\ 3 & 2 \end{bmatrix}$, $\mathbf{f}(t) = \begin{bmatrix} 1 \\ -1 \end{bmatrix}$.

13. $\mathbf{A} = \begin{bmatrix} 2 & 1 \\ -3 & -2 \end{bmatrix}$, $\mathbf{f}(t) = \begin{bmatrix} 2e^t \\ 4e^t \end{bmatrix}$.

14. $\mathbf{A} = \begin{bmatrix} 0 & -1 \\ 1 & 0 \end{bmatrix}$, $\mathbf{f}(t) = \begin{bmatrix} t^2 \\ 1 \end{bmatrix}$.

15. $\mathbf{A} = \begin{bmatrix} -4 & 2 \\ 2 & -1 \end{bmatrix}$, $\mathbf{f}(t) = \begin{bmatrix} t^{-1} \\ 4 + 2t^{-1} \end{bmatrix}$.

16. $\mathbf{A} = \begin{bmatrix} 0 & 1 \\ -1 & 0 \end{bmatrix}$, $\mathbf{f}(t) = \begin{bmatrix} 8 \sin t \\ 0 \end{bmatrix}$.

In Problems 17 through 20, use the variation of parameters formulas (11) and possibly a linear algebra software package to find a general solution of the system $\mathbf{x}'(t) = \mathbf{A}\mathbf{x}(t) + \mathbf{f}(t)$, *where* \mathbf{A} *and* $\mathbf{f}(t)$ *are given.*

17. $\mathbf{A} = \begin{bmatrix} 0 & 1 & 1 \\ 1 & 0 & 1 \\ 1 & 1 & 0 \end{bmatrix}$, $\mathbf{f}(t) = \begin{bmatrix} 3e^t \\ -e^t \\ -e^t \end{bmatrix}$.

18. $\mathbf{A} = \begin{bmatrix} 1 & -1 & 1 \\ 0 & 0 & 1 \\ 0 & -1 & 2 \end{bmatrix}$, $\mathbf{f}(t) = \begin{bmatrix} 0 \\ e^t \\ e^t \end{bmatrix}$.

19. $\mathbf{A} = \begin{bmatrix} 0 & 1 & 0 & 0 \\ -1 & 0 & 0 & 0 \\ 0 & 0 & 0 & 1 \\ 0 & 0 & 1 & 0 \end{bmatrix}$, $\mathbf{f}(t) = \begin{bmatrix} t \\ 0 \\ e^{-t} \\ t \end{bmatrix}$.

20. $\mathbf{A} = \begin{bmatrix} 0 & 1 & 0 & 0 \\ 0 & 0 & 1 & 0 \\ 0 & 0 & 0 & 1 \\ 8 & -4 & -2 & -1 \end{bmatrix}$, $\mathbf{f}(t) = \begin{bmatrix} e^t \\ 0 \\ 1 \\ 0 \end{bmatrix}$.

In Problems 21 and 22, find the solution to the given system that satisfies the given initial condition.

21. $\mathbf{x}'(t) = \begin{bmatrix} 0 & 2 \\ -1 & 3 \end{bmatrix} \mathbf{x}(t) + \begin{bmatrix} e^t \\ -e^t \end{bmatrix}$,

 (a) $\mathbf{x}(0) = \begin{bmatrix} 5 \\ 4 \end{bmatrix}$. **(b)** $\mathbf{x}(1) = \begin{bmatrix} 0 \\ 1 \end{bmatrix}$.

 (c) $\mathbf{x}(5) = \begin{bmatrix} 1 \\ 0 \end{bmatrix}$. **(d)** $\mathbf{x}(-1) = \begin{bmatrix} -4 \\ 5 \end{bmatrix}$.

22. $\mathbf{x}'(t) = \begin{bmatrix} 0 & 2 \\ 4 & -2 \end{bmatrix} \mathbf{x}(t) + \begin{bmatrix} 4t \\ -4t - 2 \end{bmatrix}$,

 (a) $\mathbf{x}(0) = \begin{bmatrix} 4 \\ -5 \end{bmatrix}$. **(b)** $\mathbf{x}(2) = \begin{bmatrix} 1 \\ 1 \end{bmatrix}$.

23. Using matrix algebra techniques and the method of undetermined coefficients, find a general solution for

$$x''(t) + y'(t) - x(t) + y(t) = -1 ,$$
$$x'(t) + y'(t) - x(t) = t^2 .$$

Compare your solution with the solution in Example 3 in Section 5.3.

24. Using matrix algebra techniques and the method of undetermined coefficients, solve the initial value problem

$$x'(t) - 2y(t) = 4t , \qquad x(0) = 4 ;$$
$$y'(t) + 2y(t) - 4x(t) = -4t - 2 , \qquad y(0) = -5 .$$

Compare your solution with the solution in Example 1 in Section 7.9.

25. To find a general solution to the system

$$\mathbf{x}'(t) = \begin{bmatrix} 0 & 1 \\ -2 & 3 \end{bmatrix} \mathbf{x}(t) + \mathbf{f}(t) \ , \quad \text{where} \quad \mathbf{f}(t) = \begin{bmatrix} e^t \\ 0 \end{bmatrix},$$

proceed as follows:

(a) Find a fundamental solution set for the corresponding homogeneous system.

(b) The obvious choice for a particular solution would be a vector function of the form $\mathbf{x}_p(t) = e^t\mathbf{a}$; however, the homogeneous system has a solution of this form. The next choice would be $\mathbf{x}_p(t) = te^t\mathbf{a}$. Show that this choice does *not* work.

(c) For systems, multiplying by t is not always sufficient. The proper guess is

$$\mathbf{x}_p(t) = te^t\mathbf{a} + e^t\mathbf{b} \ .$$

Use this guess to find a particular solution of the given system.

(d) Use the results of parts (a) and (c) to find a general solution of the given system.

26. For the system of Problem 25, we found that a proper guess for a particular solution is $\mathbf{x}_p(t) = te^t\mathbf{a} + e^t\mathbf{b}$. In some cases **a** or **b** may be zero.

(a) Find a particular solution for the system of Problem 25 if $\mathbf{f}(t) = \text{col}(3e^t, 6e^t)$.

(b) Find a particular solution for the system of Problem 25 if $\mathbf{f}(t) = \text{col}(e^t, e^t)$.

27. Find a general solution of the system

$$\mathbf{x}'(t) = \begin{bmatrix} 0 & 1 & 1 \\ 1 & 0 & 1 \\ 1 & 1 & 0 \end{bmatrix} \mathbf{x}(t) + \begin{bmatrix} -1 \\ -1-e^{-t} \\ -2e^{-t} \end{bmatrix}.$$

[*Hint:* Use superposition to determine the particular solution.]

28. Find a particular solution for the system

$$\mathbf{x}'(t) = \begin{bmatrix} 1 & -1 \\ -1 & 1 \end{bmatrix} \mathbf{x}(t) + \begin{bmatrix} -3 \\ 1 \end{bmatrix}.$$

$\left[Hint: \text{ Try } \mathbf{x}_p(t) = t\mathbf{a} + \mathbf{b}. \right]$

In Problems 29 and 30, find a general solution to the given Cauchy-Euler system for $t > 0$. Remember to
express the system in the form $\mathbf{x}'(t) = \mathbf{A}(t)\mathbf{x}(t) + \mathbf{f}(t)$ before using the variation of parameters formula.

29. $t\mathbf{x}'(t) = \begin{bmatrix} 2 & -1 \\ 3 & -2 \end{bmatrix} \mathbf{x}(t) + \begin{bmatrix} t^{-1} \\ 1 \end{bmatrix}.$

30. $t\mathbf{x}'(t) = \begin{bmatrix} 4 & -3 \\ 8 & -6 \end{bmatrix} \mathbf{x}(t) + \begin{bmatrix} t \\ 2t \end{bmatrix}.$

31. Use the variation of parameters formula (10) to derive a formula for a particular solution y_p to the scalar equation $y'' + p(t)y' + q(t)y = g(t)$ in terms of two linearly independent solutions $y_1(t)$, $y_2(t)$ of the corresponding homogeneous equation. Show that your answer agrees with the formulas derived in Section 4.9. [*Hint:* First write the scalar equation in system form.]

32. Let $\mathbf{U}(t)$ be the invertible 2×2 matrix

$$\mathbf{U}(t) := \begin{bmatrix} a(t) & b(t) \\ c(t) & d(t) \end{bmatrix}.$$

Show that

$$\mathbf{U}^{-1}(t) = \frac{1}{[a(t)d(t) - b(t)c(t)]} \begin{bmatrix} d(t) & -b(t) \\ -c(t) & a(t) \end{bmatrix}.$$

33. *RL* **Network.** The currents in the *RL* network given by the schematic diagram in Figure 9.8 are governed by the following equations:

$$2I_1'(t) + 90I_2(t) = 9 \ ,$$
$$I_3'(t) + 30I_4(t) - 90I_2(t) = 0 \ ,$$
$$60I_5(t) - 30I_4(t) = 0 \ ,$$
$$I_1(t) = I_2(t) + I_3(t) \ ,$$
$$I_3(t) = I_4(t) + I_5(t) \ .$$

Figure 9.8 *RL* network for Problem 33

Assume the currents are initially zero. Solve for the five currents I_1, \ldots, I_5. [*Hint:* Eliminate all unknowns except I_2 and I_5, and form a normal system with $x_1 = I_2$ and $x_2 = I_5$.]

34. Conventional Combat Model. A simplistic model of a pair of conventional forces in combat yields the following system:

$$\mathbf{x}' = \begin{bmatrix} -a & -b \\ -c & -d \end{bmatrix} \mathbf{x} + \begin{bmatrix} p \\ q \end{bmatrix},$$

where $\mathbf{x} = \mathrm{col}(x_1, x_2)$. The variables $x_1(t)$ and $x_2(t)$ represent the strengths of opposing forces at time t. The terms $-ax_1$ and $-dx_2$ represent the *operational loss rates,* and the terms $-bx_2$ and $-cx_1$ represent the *combat loss rates* for the troops x_1 and x_2, respectively. The constants p and q represent the respective rates of reinforcement. Let $a = 1$, $b = 4$, $c = 3$, $d = 2$, and $p = q = 5$. By solving the appropriate initial value problem, determine which forces will win if

(a) $x_1(0) = 20$, $x_2(0) = 20$.
(b) $x_1(0) = 21$, $x_2(0) = 20$.
(c) $x_1(0) = 20$, $x_2(0) = 21$.

35. Mixing Problem. Two tanks A and B, each holding 50 L of liquid, are interconnected by pipes. The

Figure 9.9 Mixing problem for interconnected tanks

liquid flows from tank A into tank B at a rate of 4 L/min and from B into A at a rate of 1 L/min (see Figure 9.9). The liquid inside each tank is kept well stirred. A brine solution that has a concentration of 2 kg/L of salt flows into tank A at a rate of 4 L/min. A brine solution that has a concentraton of 1 kg/L of salt flows into tank B at a rate of 1 L/min. The solutions flow out of the system from both tanks—from tank A at 1 L/min and from tank B at 4 L/min. If, initially, tank A contains pure water and tank B contains 5 kg of salt, determine the mass of salt in each tank at time $t \geq 0$. After several minutes have elapsed, which tank has the higher concentration of salt? What is its limiting concentration?

9.8 THE MATRIX EXPONENTIAL FUNCTION

In this chapter we have developed various ways to extend techniques for scalar differential equations to systems. In this section we take a substantial step further by showing that with the right notation, the formulas for solving normal systems with constant coefficients are identical to the formulas for solving first order equations with constant coefficients. For example, we know that a general solution to the equation $x'(t) = ax(t)$, where a is a constant, is $x(t) = ce^{at}$. Analogously, we shall show that a general solution to the normal system

(1) $\mathbf{x}'(t) = \mathbf{A}\mathbf{x}(t)$,

where \mathbf{A} is a constant $n \times n$ matrix, is $\mathbf{x}(t) = e^{\mathbf{A}t}\mathbf{c}$. Our first task is to define the matrix exponential $e^{\mathbf{A}t}$.

If \mathbf{A} is a constant $n \times n$ matrix, we define $e^{\mathbf{A}t}$ by taking the series expansion for e^{at} and replacing a by \mathbf{A}; that is,

(2) $e^{\mathbf{A}t} := \mathbf{I} + \mathbf{A}t + \mathbf{A}^2 \dfrac{t^2}{2!} + \cdots + \mathbf{A}^n \dfrac{t^n}{n!} + \cdots$.

(Note that we also replace 1 by \mathbf{I}.) By the right-hand side of (2), we mean the $n \times n$ matrix whose elements are power series with coefficients given by the corresponding entries in the matrices $\mathbf{I}, \mathbf{A}, \mathbf{A}^2/2!, \ldots$.

If \mathbf{A} is a diagonal matrix, then the computation of $e^{\mathbf{A}t}$ is straightforward. For example, if

$$\mathbf{A} = \begin{bmatrix} -1 & 0 \\ 0 & 2 \end{bmatrix} ,$$

then

$$\mathbf{A}^2 = \mathbf{A}\mathbf{A} = \begin{bmatrix} 1 & 0 \\ 0 & 4 \end{bmatrix} , \qquad \mathbf{A}^3 = \begin{bmatrix} -1 & 0 \\ 0 & 8 \end{bmatrix} , \qquad \ldots , \qquad \mathbf{A}^n = \begin{bmatrix} (-1)^n & 0 \\ 0 & 2^n \end{bmatrix} ,$$

and so

$$e^{\mathbf{A}t} = \sum_{n=0}^{\infty} \mathbf{A}^n \frac{t^n}{n!} = \begin{bmatrix} \sum_{n=0}^{\infty} (-1)^n \dfrac{t^n}{n!} & 0 \\ 0 & \sum_{n=0}^{\infty} 2^n \dfrac{t^n}{n!} \end{bmatrix} = \begin{bmatrix} e^{-t} & 0 \\ 0 & e^{2t} \end{bmatrix} .$$

More generally, if \mathbf{A} is an $n \times n$ *diagonal* matrix with r_1, r_2, \ldots, r_n down its main diagonal, then $e^{\mathbf{A}t}$ is the diagonal matrix with $e^{r_1 t}, e^{r_2 t}, \ldots, e^{r_n t}$ down its main diagonal (see Problem 26). If \mathbf{A} is not a diagonal matrix, the computation of $e^{\mathbf{A}t}$ is more involved. We deal with this important problem later in this section.

It can be shown that the series (2) converges for all t and has many of the same properties[†] as the scalar exponential e^{at}.

PROPERTIES OF THE MATRIX EXPONENTIAL FUNCTION

Theorem 7. Let \mathbf{A} and \mathbf{B} be $n \times n$ constant matrices and r, s, and t be real (or complex) numbers. Then,

(a) $e^{\mathbf{A}0} = e^{\mathbf{0}} = \mathbf{I}$.

(b) $e^{\mathbf{A}(t+s)} = e^{\mathbf{A}t} e^{\mathbf{A}s}$.

(c) $\left(e^{\mathbf{A}t} \right)^{-1} = e^{-\mathbf{A}t}$.

(d) $e^{(\mathbf{A}+\mathbf{B})t} = e^{\mathbf{A}t} e^{\mathbf{B}t}$, provided that $\mathbf{A}\mathbf{B} = \mathbf{B}\mathbf{A}$.

(e) $e^{r\mathbf{I}t} = e^{rt}\mathbf{I}$.

Property (c) has profound implications. First, it asserts that for any matrix \mathbf{A}, the matrix $e^{\mathbf{A}t}$ *has* an inverse for all t. Moreover, this inverse is obtained by simply replacing t by $-t$. In

[†]For proofs of these and other properties of the matrix exponential function, see *Matrix Computations,* 3rd Edition, by Gene H. Golub and Charles F. van Loan (Johns Hopkins University Press, Baltimore, MD, 1996), Chapter 11. See also the amusing article "Nineteen Dubious Ways to Compute the Exponential of a Matrix," by Cleve Moler and Charles van Loan, *SIAM Review,* Vol. 20, No. 4 (Oct. 1978).

applying property (d) (the law of exponents), one must exercise care because of the stipulation that the matrices \mathbf{A} and \mathbf{B} commute (see Problem 25).

Another important property of the matrix exponential arises from the fact that we can differentiate the series in (2) term by term. This gives

$$\frac{d}{dt}\left(e^{\mathbf{A}t}\right) = \frac{d}{dt}\left(\mathbf{I} + \mathbf{A}t + \mathbf{A}^2\frac{t^2}{2} + \cdots + \mathbf{A}^n\frac{t^n}{n!} + \cdots\right)$$

$$= \mathbf{A} + \mathbf{A}^2 t + \mathbf{A}^3\frac{t^2}{2} + \cdots + \mathbf{A}^n\frac{t^{n-1}}{(n-1)!} + \cdots$$

$$= \mathbf{A}\left[\mathbf{I} + \mathbf{A}t + \mathbf{A}^2\frac{t^2}{2} + \cdots + \mathbf{A}^{n-1}\frac{t^{n-1}}{(n-1)!} + \cdots\right].$$

Hence

$$\frac{d}{dt}\left(e^{\mathbf{A}t}\right) = \mathbf{A}e^{\mathbf{A}t}\,,$$

and so $e^{\mathbf{A}t}$ *is a solution to the matrix differential equation* $\mathbf{X}' = \mathbf{A}\mathbf{X}$. Since $e^{\mathbf{A}t}$ is invertible (property (c)), it follows that the columns of $e^{\mathbf{A}t}$ are linearly independent solutions to system (1). Combining these facts we have the following.

$e^{\mathbf{A}t}$ IS A FUNDAMENTAL MATRIX

Theorem 8. If \mathbf{A} is an $n \times n$ constant matrix, then the columns of the matrix exponential $e^{\mathbf{A}t}$ form a fundamental solution set for the system $\mathbf{x}'(t) = \mathbf{A}\mathbf{x}(t)$. Therefore, $e^{\mathbf{A}t}$ is a fundamental matrix for the system, and a general solution is $\mathbf{x}(t) = e^{\mathbf{A}t}\mathbf{c}$.

Knowing that $e^{\mathbf{A}t}$ is a fundamental matrix is of practical use provided we can calculate $e^{\mathbf{A}t}$. As we observed, if \mathbf{A} is a diagonal matrix, then we simply exponentiate the diagonal elements (times t) to obtain $e^{\mathbf{A}t}$. Also, if \mathbf{B} is a **nilpotent** matrix, that is, $\mathbf{B}^k = \mathbf{0}$ for some positive integer k, then the series for $e^{\mathbf{B}t}$ has only a finite number of terms, since $\mathbf{B}^k = \mathbf{B}^{k+1} = \cdots = \mathbf{0}$. In such cases, $e^{\mathbf{B}t}$ reduces to

$$e^{\mathbf{B}t} = \mathbf{I} + \mathbf{B}t + \cdots + \mathbf{B}^{k-1}\frac{t^{k-1}}{(k-1)!}\,.$$

Using the law of exponents and this fact about nilpotent matrices, we can determine $e^{\mathbf{A}t}$ for a special class of matrices. Let r be a scalar. Since

$$e^{\mathbf{A}t} = e^{r\mathbf{I}t}e^{(\mathbf{A}-r\mathbf{I})t} = e^{rt}e^{(\mathbf{A}-r\mathbf{I})t}\,,$$

we get a finite representation for $e^{\mathbf{A}t}$ if $\mathbf{B} = \mathbf{A} - r\mathbf{I}$ is nilpotent for some r. In fact, when the characteristic polynomial for \mathbf{A} has the form $p(r) = (r_1 - r)^n$, that is, when \mathbf{A} has one eigenvalue r_1

of multiplicity n, it is a consequence of the Cayley-Hamilton theorem[†] that $(r_1\mathbf{I} - \mathbf{A})^n = \mathbf{0}$. Hence $\mathbf{A} - r_1\mathbf{I}$ is nilpotent and

$$e^{\mathbf{A}t} = e^{r_1 t}\left\{ \mathbf{I} + (\mathbf{A} - r_1\mathbf{I})t + \cdots + (\mathbf{A} - r_1\mathbf{I})^{n-1}\frac{t^{n-1}}{(n-1)!} \right\} .$$

EXAMPLE 1 Find the fundamental matrix $e^{\mathbf{A}t}$ for the system

$$(3) \qquad \mathbf{x}' = \mathbf{A}\mathbf{x} , \quad \text{where} \quad \mathbf{A} = \begin{bmatrix} 2 & 1 & 1 \\ 1 & 2 & 1 \\ -2 & -2 & -1 \end{bmatrix} .$$

SOLUTION We begin by computing the characteristic polynomial for \mathbf{A}:

$$p(r) = |\mathbf{A} - r\mathbf{I}| = \begin{vmatrix} 2 - r & 1 & 1 \\ 1 & 2 - r & 1 \\ -2 & -2 & -1 - r \end{vmatrix} = -r^3 + 3r^2 - 3r + 1 = -(r - 1)^3 .$$

Thus $r = 1$ is an eigenvalue of \mathbf{A} with multiplicity 3. By the Cayley-Hamilton theorem, $(\mathbf{A} - \mathbf{I})^3 = \mathbf{0}$, and so

$$(4) \qquad e^{\mathbf{A}t} = e^t e^{(\mathbf{A}-\mathbf{I})t} = e^t\left\{ \mathbf{I} + (\mathbf{A} - \mathbf{I})t + (\mathbf{A} - \mathbf{I})^2\frac{t^2}{2} \right\} .$$

Computing, we find

$$\mathbf{A} - \mathbf{I} = \begin{bmatrix} 1 & 1 & 1 \\ 1 & 1 & 1 \\ -2 & -2 & -2 \end{bmatrix} \quad \text{and} \quad (\mathbf{A} - \mathbf{I})^2 = \begin{bmatrix} 0 & 0 & 0 \\ 0 & 0 & 0 \\ 0 & 0 & 0 \end{bmatrix} .$$

Substituting into (4) yields

$$(5) \qquad e^{\mathbf{A}t} = e^t\begin{bmatrix} 1 & 0 & 0 \\ 0 & 1 & 0 \\ 0 & 0 & 1 \end{bmatrix} + te^t\begin{bmatrix} 1 & 1 & 1 \\ 1 & 1 & 1 \\ -2 & -2 & -2 \end{bmatrix} = \begin{bmatrix} e^t + te^t & te^t & te^t \\ te^t & e^t + te^t & te^t \\ -2te^t & -2te^t & e^t - 2te^t \end{bmatrix} .$$

■ End

In the preceding example we used the nilpotency of the matrix $\mathbf{A} - r\mathbf{I}$ to compute $e^{\mathbf{A}t}$ directly. In general, we cannot expect nilpotency to hold, but we can take advantage of the following relationship between fundamental matrices to help compute $e^{\mathbf{A}t}$.

[†]The Cayley-Hamilton theorem states that a matrix satisfies its own characteristic equation, that is, $p(\mathbf{A}) = \mathbf{0}$. For a discussion of this theorem, see *Matrices and Linear Transformation,* 2nd Edition, by Charles G. Cullen (Dover Publications, New York, 1990), Chapter 5.

RELATIONSHIP BETWEEN FUNDAMENTAL MATRICES

Lemma 1. Let $\mathbf{X}(t)$ and $\mathbf{Y}(t)$ be two fundamental matrices for the same system $\mathbf{x}' = \mathbf{A}\mathbf{x}$. Then, there exists a constant matrix \mathbf{C} such that $\mathbf{X}(t) = \mathbf{Y}(t)\mathbf{C}$.

Proof. Let $\mathbf{x}_1(t), \ldots, \mathbf{x}_n(t)$ be the columns of $\mathbf{X}(t)$ and let $\mathbf{y}_1(t), \ldots, \mathbf{y}_n(t)$ be the columns of $\mathbf{Y}(t)$. Since $\{\mathbf{y}_1(t), \ldots, \mathbf{y}_n(t)\}$ is a fundamental solution set and $\mathbf{x}_j(t), j = 1, \ldots, n$, are solutions to $\mathbf{x}' = \mathbf{A}\mathbf{x}$, there exist constants $c_{1j}, c_{2j}, \ldots, c_{nj}$ such that

$$\mathbf{x}_j(t) = c_{1j}\mathbf{y}_1(t) + c_{2j}\mathbf{y}_2(t) + \cdots + c_{nj}\mathbf{y}_n(t)$$

for each $j = 1, \ldots, n$. But this is equivalent to writing $\mathbf{X}(t) = \mathbf{Y}(t)\mathbf{C}$, where $\mathbf{C} = \begin{bmatrix} c_{ij} \end{bmatrix}$. ∎ End

We now use Lemma 1 to find a formula for $e^{\mathbf{A}t}$ when a fundamental matrix $\mathbf{X}(t)$ for $\mathbf{x}' = \mathbf{A}\mathbf{x}$ is known. Since $e^{\mathbf{A}t}$ is also a fundamental matrix for the system, Lemma 1 asserts that $e^{\mathbf{A}t} = \mathbf{X}(t)\mathbf{C}$ for some constant matrix \mathbf{C}. Setting $t = 0$ yields $\mathbf{I} = \mathbf{X}(0)\mathbf{C}$, and solving for \mathbf{C} we get $\mathbf{C} = \mathbf{X}^{-1}(0)$. Hence

$$(6) \qquad e^{\mathbf{A}t} = \mathbf{X}(t)\mathbf{X}^{-1}(0) \ .$$

Although this formula is useful, it does place the burden of determining $e^{\mathbf{A}t}$ on finding a fundamental matrix $\mathbf{X}(t)$. Fortunately, we can use the *properties* of the matrix exponential $e^{\mathbf{A}t}$ to help simplify this task. Because the columns of a fundamental matrix must have the form $e^{\mathbf{A}t}\mathbf{u}$, let's try to find n vectors \mathbf{u} for which the computation of $e^{\mathbf{A}t}\mathbf{u}$ is tractable.

We begin by using the relation $e^{\mathbf{A}t} = e^{rt}e^{(\mathbf{A}-r\mathbf{I})t}$ to express $e^{\mathbf{A}t}\mathbf{u}$ as

$$(7) \qquad e^{\mathbf{A}t}\mathbf{u} = e^{rt}e^{(\mathbf{A}-r\mathbf{I})t}\mathbf{u}$$

$$= e^{rt}\left\{ \mathbf{u} + t(\mathbf{A}-r\mathbf{I})\mathbf{u} + \cdots + \frac{t^k}{k!}(\mathbf{A}-r\mathbf{I})^k\mathbf{u} + \cdots \right\} \ .$$

Now we know that if r is an eigenvalue of \mathbf{A} and \mathbf{u} is a corresponding eigenvector, then $e^{rt}\mathbf{u}$ is a solution to (1). Indeed, in this situation $(\mathbf{A}-r\mathbf{I})\mathbf{u} = (\mathbf{A}-r\mathbf{I})^2\mathbf{u} = \cdots = \mathbf{0}$. Hence the series in (7) reduces to the first term, $e^{rt}\mathbf{u}$. While it is too much to expect $\mathbf{A} - r\mathbf{I}$ to be nilpotent, it is not too much to ask that $(\mathbf{A}-r\mathbf{I})^k\mathbf{u} = \mathbf{0}$ for some nontrivial vector \mathbf{u} and some positive integer k.

GENERALIZED EIGENVECTORS

Definition 5. Let \mathbf{A} be an $n \times n$ constant matrix and r be an eigenvalue of \mathbf{A}. A nontrivial vector \mathbf{u} that satisfies

$$(\mathbf{A}-r\mathbf{I})^k\mathbf{u} = \mathbf{0}$$

for some positive integer k is called a **generalized eigenvector** associated with r.

A consequence of the primary decomposition theorem in advanced linear algebra[†] is that if the characteristic polynomial for \mathbf{A} is

$$p(r) = (r_1 - r)^{m_1} \cdots (r_k - r)^{m_k} \, ,$$

where the r_i's are distinct eigenvalues of \mathbf{A} and m_i is the multiplicity of the eigenvalue r_i, then for each i there exist m_i linearly independent generalized eigenvectors associated with r_i, and the combined set of $n = m_1 + \cdots + m_k$ generalized eigenvectors is linearly independent. Moreover, if \mathbf{u} is a generalized eigenvector associated with r_i, then $(\mathbf{A} - r_i\mathbf{I})^{m_i}\mathbf{u} = \mathbf{0}$. This leads to the following procedure for finding n linearly independent solutions to system (1).

SOLVING SYSTEMS USING GENERALIZED EIGENVECTORS

To obtain a fundamental solution set for $\mathbf{x}' = \mathbf{Ax}$:

(a) Compute the characteristic polynomial $p(t) = |\mathbf{A} - r\mathbf{I}|$ and find the distinct eigenvalues r_1, \ldots, r_k.

(b) For each eigenvalue r_i, find m_i linearly independent generalized eigenvectors, where m_i is the multiplicity of the eigenvalue r_i.

(c) Use the n linearly independent generalized eigenvectors obtained in (b) to compute the n linearly independent solutions to $\mathbf{x}' = \mathbf{Ax}$ of the form

$$(8) \qquad e^{\mathbf{A}t}\mathbf{u} = e^{rt}\left\{ \mathbf{u} + t(\mathbf{A} - r\mathbf{I})\mathbf{u} + \frac{t^2}{2}(\mathbf{A} - r\mathbf{I})^2\mathbf{u} + \cdots \right\},$$

where r is an eigenvalue and \mathbf{u} is a corresponding generalized eigenvector. If r has multiplicity m_i, then the above series reduces to the first m_i terms.

EXAMPLE 2 Find the fundamental matrix $e^{\mathbf{A}t}$ for the system

$$(9) \qquad \mathbf{x}' = \mathbf{Ax} \, , \quad \text{where} \quad \mathbf{A} = \begin{bmatrix} 1 & 0 & 0 \\ 1 & 3 & 0 \\ 0 & 1 & 1 \end{bmatrix} .$$

SOLUTION We begin by finding the characteristic polynomial for \mathbf{A}:

$$p(r) = |\mathbf{A} - r\mathbf{I}| = \begin{vmatrix} 1 - r & 0 & 0 \\ 1 & 3 - r & 0 \\ 0 & 1 & 1 - r \end{vmatrix} = -(r - 1)^2(r - 3) \, .$$

Hence the eigenvalues of \mathbf{A} are $r = 1$ with multiplicity 2 and $r = 3$ with multiplicity 1.

[†]*Matrices and Linear Transformations,* 2nd Edition, by Charles G. Cullen, Ibid.

Since $r = 1$ has multiplicity 2, we must determine two linearly independent associated generalized eigenvectors. We begin with $(\mathbf{A} - \mathbf{I})\mathbf{u} = \mathbf{0}$; that is,

$$\begin{bmatrix} 0 & 0 & 0 \\ 1 & 2 & 0 \\ 0 & 1 & 0 \end{bmatrix} \begin{bmatrix} u_1 \\ u_2 \\ u_3 \end{bmatrix} = \begin{bmatrix} 0 \\ 0 \\ 0 \end{bmatrix}.$$

Solving, we obtain $u_1 = u_2 = 0$ and $u_3 = s$, where s is arbitrary. Thus there is only one linearly independent eigenvector corresponding to $r = 1$, and with $s = 1$ we choose $\mathbf{u}_1 = \text{col}(0, 0, 1)$. Hence one solution to (9) is

(10) $\mathbf{x}_1(t) = e^t \mathbf{u}_1 = e^t \begin{bmatrix} 0 \\ 0 \\ 1 \end{bmatrix} = \begin{bmatrix} 0 \\ 0 \\ e^t \end{bmatrix}.$

Next we solve $(\mathbf{A} - \mathbf{I})^2 \mathbf{u} = \mathbf{0}$. From

$$(\mathbf{A} - \mathbf{I})^2 \mathbf{u} = \begin{bmatrix} 0 & 0 & 0 \\ 2 & 4 & 0 \\ 1 & 2 & 0 \end{bmatrix} \begin{bmatrix} u_1 \\ u_2 \\ u_3 \end{bmatrix} = \begin{bmatrix} 0 \\ 0 \\ 0 \end{bmatrix},$$

we find $u_2 = s$, $u_1 = -2u_2 = -2s$, and $u_3 = v$, where s and v are arbitrary. Taking $s = 1$ and $v = 0$, we obtain the generalized eigenvector $\mathbf{u}_2 = \text{col}(-2, 1, 0)$, which is linearly independent of \mathbf{u}_1. Now we use \mathbf{u}_2 to obtain a second solution to (9). Since $(\mathbf{A} - \mathbf{I})^2 \mathbf{u}_2 = \mathbf{0}$, formula (8) reduces to

(11) $\mathbf{x}_2(t) = e^{\mathbf{A}t} \mathbf{u}_2 = e^t \{ \mathbf{u}_2 + t(\mathbf{A} - \mathbf{I})\mathbf{u}_2 \}$

$$= e^t \begin{bmatrix} -2 \\ 1 \\ 0 \end{bmatrix} + te^t \begin{bmatrix} 0 & 0 & 0 \\ 1 & 2 & 0 \\ 0 & 1 & 0 \end{bmatrix} \begin{bmatrix} -2 \\ 1 \\ 0 \end{bmatrix}$$

$$= e^t \begin{bmatrix} -2 \\ 1 \\ 0 \end{bmatrix} + te^t \begin{bmatrix} 0 \\ 0 \\ 1 \end{bmatrix} = \begin{bmatrix} -2e^t \\ e^t \\ te^t \end{bmatrix} \dagger$$

For the eigenvalue $r = 3$, we solve $(\mathbf{A} - 3\mathbf{I})\mathbf{u} = \mathbf{0}$, that is,

$$\begin{bmatrix} -2 & 0 & 0 \\ 1 & 0 & 0 \\ 0 & 1 & -2 \end{bmatrix} \begin{bmatrix} u_1 \\ u_2 \\ u_3 \end{bmatrix} = \begin{bmatrix} 0 \\ 0 \\ 0 \end{bmatrix},$$

to obtain the eigenvector $\mathbf{u}_3 = \text{col}(0, 2, 1)$. Hence a third linearly independent solution to (9) is

(12) $\mathbf{x}_3(t) = e^{3t} \mathbf{u}_3 = e^{3t} \begin{bmatrix} 0 \\ 2 \\ 1 \end{bmatrix} = \begin{bmatrix} 0 \\ 2e^{3t} \\ e^{3t} \end{bmatrix}.$

\daggerNote that $\mathbf{x}_2(t)$ has been expressed in the format discussed in Problem 35 of Exercises 9.5, page 567.

The matrix $\mathbf{X}(t)$ whose columns are the vectors $\mathbf{x}_1(t)$, $\mathbf{x}_2(t)$, and $\mathbf{x}_3(t)$ given in equations (10), (11), and (12),

$$\mathbf{X}(t) = \begin{bmatrix} 0 & -2e^t & 0 \\ 0 & e^t & 2e^{3t} \\ e^t & te^t & e^{3t} \end{bmatrix},$$

is a fundamental matrix for (9). Setting $t = 0$ and then computing $\mathbf{X}^{-1}(0)$, we find

$$\mathbf{X}(0) = \begin{bmatrix} 0 & -2 & 0 \\ 0 & 1 & 2 \\ 1 & 0 & 1 \end{bmatrix} \quad \text{and} \quad \mathbf{X}^{-1}(0) = \begin{bmatrix} -\frac{1}{4} & -\frac{1}{2} & 1 \\ -\frac{1}{2} & 0 & 0 \\ \frac{1}{4} & \frac{1}{2} & 0 \end{bmatrix}.$$

It now follows from formula (6) that

$$e^{\mathbf{A}t} = \mathbf{X}(t)\mathbf{X}^{-1}(0) = \begin{bmatrix} 0 & -2e^t & 0 \\ 0 & e^t & 2e^{3t} \\ e^t & te^t & e^{3t} \end{bmatrix} \begin{bmatrix} -\frac{1}{4} & -\frac{1}{2} & 1 \\ -\frac{1}{2} & 0 & 0 \\ \frac{1}{4} & \frac{1}{2} & 0 \end{bmatrix}$$

$$= \begin{bmatrix} e^t & 0 & 0 \\ -\frac{1}{2}e^t + \frac{1}{2}e^{3t} & e^{3t} & 0 \\ -\frac{1}{4}e^t - \frac{1}{2}te^t + \frac{1}{4}e^{3t} & -\frac{1}{2}e^t + \frac{1}{2}e^{3t} & e^t \end{bmatrix}. \quad \blacksquare \text{ End}$$

Use of the fundamental matrix $e^{\mathbf{A}t}$ simplifies many computations. For example, the properties $e^{\mathbf{A}t}e^{-\mathbf{A}s} = e^{\mathbf{A}(t-s)}$ and $(e^{\mathbf{A}t_0})^{-1} = e^{-\mathbf{A}t_0}$ enable us to rewrite the variation of parameters formula (13) in Section 9.7 in a simpler form. Namely, the solution to the initial value problem $\mathbf{x}' = \mathbf{A}\mathbf{x} + \mathbf{f}(t)$, $\mathbf{x}(t_0) = \mathbf{x}_0$ is given by

$$(13) \qquad \mathbf{x}(t) = e^{\mathbf{A}(t-t_0)}\mathbf{x}_0 + \int_{t_0}^{t} e^{\mathbf{A}(t-s)}\mathbf{f}(s)\,ds\ ,$$

which is a system version of the formula for the solution to the scalar initial value problem $x' = ax + f(t)$, $x(t_0) = x_0$.

In closing, we remark that the software packages for eigenvalue computation listed in Section 9.5 (page 558) also contain subroutines for computing the matrix exponential.

EXERCISES 9.8

In Problems 1 through 6, (a) show that the given matrix **A** satisfies $(\mathbf{A} - r\mathbf{I})^k = \mathbf{0}$ for some number r and some positive integer k and (b) use this fact to determine the matrix $e^{\mathbf{A}t}$. [Hint: Compute the characteristic polynomial and use the Cayley-Hamilton theorem.]

1. A = $\begin{bmatrix} 3 & -2 \\ 0 & 3 \end{bmatrix}$.

2. A = $\begin{bmatrix} 1 & -1 \\ 1 & 3 \end{bmatrix}$.

3. A = $\begin{bmatrix} 2 & 1 & -1 \\ -3 & -1 & 1 \\ 9 & 3 & -4 \end{bmatrix}$.

4. A = $\begin{bmatrix} 2 & 1 & 3 \\ 0 & 2 & -1 \\ 0 & 0 & 2 \end{bmatrix}$.

5. A = $\begin{bmatrix} -2 & 0 & 0 \\ 4 & -2 & 0 \\ 1 & 0 & -2 \end{bmatrix}$.

6. A = $\begin{bmatrix} 0 & 1 & 0 \\ 0 & 0 & 1 \\ -1 & -3 & -3 \end{bmatrix}$.

In Problems 7 through 10, determine $e^{\mathbf{A}t}$ by first finding a fundamental matrix $\mathbf{X}(t)$ for $\mathbf{x}' = \mathbf{A}\mathbf{x}$ and then using formula (6).

7. A = $\begin{bmatrix} 0 & 1 \\ -1 & 0 \end{bmatrix}$.

8. A = $\begin{bmatrix} 1 & 1 \\ 4 & 1 \end{bmatrix}$.

9. A = $\begin{bmatrix} 0 & 1 & 0 \\ 0 & 0 & 1 \\ 1 & -1 & 1 \end{bmatrix}$.

10. A = $\begin{bmatrix} 0 & 2 & 2 \\ 2 & 0 & 2 \\ 2 & 2 & 0 \end{bmatrix}$.

In Problems 11 and 12, determine $e^{\mathbf{A}t}$ by using generalized eigenvectors to find a fundamental matrix and then using formula (6).

11. A = $\begin{bmatrix} 5 & -4 & 0 \\ 1 & 0 & 2 \\ 0 & 2 & 5 \end{bmatrix}$.

12. A = $\begin{bmatrix} 1 & 1 & 1 \\ 2 & 1 & -1 \\ 0 & -1 & 1 \end{bmatrix}$.

In Problems 13 through 16, use a linear algebra software package to help in determining $e^{\mathbf{A}t}$.

13. A = $\begin{bmatrix} 0 & 1 & 0 & 0 & 0 \\ 0 & 0 & 1 & 0 & 0 \\ 1 & -3 & 3 & 0 & 0 \\ 0 & 0 & 0 & 0 & 1 \\ 0 & 0 & 0 & -1 & 0 \end{bmatrix}$.

14. A = $\begin{bmatrix} 1 & 0 & 0 & 0 & 0 \\ 0 & 0 & 1 & 0 & 0 \\ 0 & -1 & -2 & 0 & 0 \\ 0 & 0 & 0 & 0 & 1 \\ 0 & 0 & 0 & -1 & 0 \end{bmatrix}$.

15. A = $\begin{bmatrix} 0 & 1 & 0 & 0 & 0 \\ 0 & 0 & 1 & 0 & 0 \\ -1 & -3 & -3 & 0 & 0 \\ 0 & 0 & 0 & 0 & 1 \\ 0 & 0 & 0 & -4 & -4 \end{bmatrix}$.

16. A = $\begin{bmatrix} -1 & 0 & 0 & 0 & 0 \\ 0 & 0 & 1 & 0 & 0 \\ 0 & -1 & -2 & 0 & 0 \\ 0 & 0 & 0 & 0 & 1 \\ 0 & 0 & 0 & -4 & -4 \end{bmatrix}$.

In Problems 17 through 20, use the generalized eigenvectors of **A** to find a general solution to the system $\mathbf{x}'(t) = \mathbf{A}\mathbf{x}(t)$, where **A** is given.

17. A = $\begin{bmatrix} 0 & 1 & 0 \\ 0 & 0 & 1 \\ -2 & -5 & -4 \end{bmatrix}$.

18. A = $\begin{bmatrix} 0 & 0 & 1 \\ 0 & 1 & 2 \\ 0 & 0 & 1 \end{bmatrix}$.

19. A = $\begin{bmatrix} 1 & 0 & 1 & 2 \\ 1 & 1 & 2 & 1 \\ 0 & 0 & 2 & 0 \\ 0 & 0 & 1 & 1 \end{bmatrix}$.

20. $\mathbf{A} = \begin{bmatrix} -1 & -8 & 1 \\ -1 & -3 & 2 \\ -4 & -16 & 7 \end{bmatrix}.$

21. Use the results of Problem 5 to find the solution to the initial value problem

$$\mathbf{x}'(t) = \begin{bmatrix} -2 & 0 & 0 \\ 4 & -2 & 0 \\ 1 & 0 & -2 \end{bmatrix} \mathbf{x}(t), \qquad \mathbf{x}(0) = \begin{bmatrix} 1 \\ 1 \\ -1 \end{bmatrix}.$$

22. Use your answer to Problem 12 to find the solution to the initial value problem.

$$\mathbf{x}'(t) = \begin{bmatrix} 1 & 1 & 1 \\ 2 & 1 & -1 \\ 0 & -1 & 1 \end{bmatrix} \mathbf{x}(t), \qquad \mathbf{x}(0) = \begin{bmatrix} -1 \\ 0 \\ 3 \end{bmatrix}.$$

23. Use the results of Problem 3 and the variation of parameters formula (13) to find the solution to the initial value problem

$$\mathbf{x}'(t) = \begin{bmatrix} 2 & 1 & -1 \\ -3 & -1 & 1 \\ 9 & 3 & -4 \end{bmatrix} \mathbf{x}(t) + \begin{bmatrix} 0 \\ t \\ 0 \end{bmatrix},$$

$$\mathbf{x}(0) = \begin{bmatrix} 0 \\ 3 \\ 0 \end{bmatrix}.$$

24. Use your answer to Problem 9 and the variation of parameters formula (13) to find the solution to the initial value problem

$$\mathbf{x}'(t) = \begin{bmatrix} 0 & 1 & 0 \\ 0 & 0 & 1 \\ 1 & -1 & 1 \end{bmatrix} \mathbf{x}(t) + \begin{bmatrix} 0 \\ 0 \\ t \end{bmatrix},$$

$$\mathbf{x}(0) = \begin{bmatrix} 1 \\ -1 \\ 0 \end{bmatrix}.$$

25. Let

$$\mathbf{A} = \begin{bmatrix} 1 & 2 \\ -1 & 3 \end{bmatrix} \quad \text{and} \quad \mathbf{B} = \begin{bmatrix} 2 & 1 \\ 0 & 1 \end{bmatrix}.$$

(a) Show that $\mathbf{AB} \neq \mathbf{BA}$.

(b) Show that property (d) in Theorem 7 does not hold for these matrices. That is, show that $e^{(\mathbf{A}+\mathbf{B})t} \neq e^{\mathbf{A}t}e^{\mathbf{B}t}$.

26. Let \mathbf{A} be a diagonal $n \times n$ matrix with entries r_1, \ldots, r_n down its main diagonal. To compute $e^{\mathbf{A}t}$, proceed as follows:

(a) Show that \mathbf{A}^k is the diagonal matrix with entries r_1^k, \ldots, r_n^k down its main diagonal.

(b) Use the result of part (a) to show that $e^{\mathbf{A}t}$ is the diagonal matrix with entries $e^{r_1 t}, \ldots, e^{r_n t}$ down its main diagonal.

27. In Problems 35–40 of Exercises 9.5, page 567, some ad hoc formulas were invoked to find general solutions to the system $\mathbf{x}' = \mathbf{A}\mathbf{x}$ when \mathbf{A} had repeated eigenvalues. Using the generalized eigenvector procedure (and especially equation (8)), justify the ad hoc formulas proposed in

(a) Problem 35, Exercises 9.5.

(b) Problem 37, Exercises 9.5.

(c) Problem 39, Exercises 9.5.

CHAPTER SUMMARY

In this chapter we discussed the theory of linear systems in normal form and presented methods for solving such systems. The theory and methods are natural extensions of the development for second order and higher-order linear equations. The important properties and techniques are listed below.

Homogeneous Normal Systems

$$\mathbf{x}'(t) = \mathbf{A}(t)\mathbf{x}(t)$$

The $n \times n$ matrix function $\mathbf{A}(t)$ is assumed to be continuous on an interval I.

Fundamental Solution Set: $\{\mathbf{x}_1, \ldots, \mathbf{x}_n\}$. The n vector solutions $\mathbf{x}_1(t), \ldots, \mathbf{x}_n(t)$ of the homogeneous system on the interval I form a **fundamental solution set,** provided they are linearly independent on I or, equivalently, their **Wronskian**

$$W[\mathbf{x}_1, \ldots, \mathbf{x}_n](t) := \det[\mathbf{x}_1, \ldots, \mathbf{x}_n] = \begin{vmatrix} x_{1,1}(t) & x_{1,2}(t) & \cdots & x_{1,n}(t) \\ x_{2,1}(t) & x_{2,2}(t) & \cdots & x_{2,n}(t) \\ \vdots & \vdots & & \vdots \\ x_{n,1}(t) & x_{n,2}(t) & \cdots & x_{n,n}(t) \end{vmatrix}$$

is never zero on I.

Fundamental Matrix: $\mathbf{X}(t)$. An $n \times n$ matrix function $\mathbf{X}(t)$ whose column vectors form a fundamental solution set for the homogeneous system is called a **fundamental matrix.** The determinant of $\mathbf{X}(t)$ is the Wronskian of the fundamental solution set. Since the Wronskian is never zero on the interval I, then $\mathbf{X}^{-1}(t)$ exists for t in I.

General Solution to Homogeneous System: $\mathbf{X}\mathbf{c} = c_1\mathbf{x}_1 + \cdots + c_n\mathbf{x}_n$. If $\mathbf{X}(t)$ is a fundamental matrix whose column vectors are $\mathbf{x}_1, \ldots, \mathbf{x}_n$, then a general solution to the homogeneous system is

$$\mathbf{x}(t) = \mathbf{X}(t)\mathbf{c} = c_1\mathbf{x}_1(t) + c_2\mathbf{x}_2(t) + \cdots + c_n\mathbf{x}_n(t) \ ,$$

where $\mathbf{c} = \text{col}(c_1, \ldots, c_n)$ is an arbitrary constant vector.

Homogeneous Systems with Constant Coefficients. The form of a general solution for a homogeneous system with constant coefficients depends on the eigenvalues and eigenvectors of the $n \times n$ constant matrix \mathbf{A}. An **eigenvalue** of \mathbf{A} is a number r such that the system $\mathbf{A}\mathbf{u} = r\mathbf{u}$ has a nontrivial solution \mathbf{u}, called an **eigenvector** of \mathbf{A} associated with the eigenvalue r. Finding the eigenvalues of \mathbf{A} is equivalent to finding the roots of the **characteristic equation**

$$|\mathbf{A} - r\mathbf{I}| = 0 \ .$$

The corresponding eigenvectors are found by solving the system $(\mathbf{A} - r\mathbf{I})\mathbf{u} = \mathbf{0}$.

If the matrix \mathbf{A} has n linearly independent eigenvectors $\mathbf{u}_1, \ldots, \mathbf{u}_n$ and r_i is the eigenvalue corresponding to the eigenvector \mathbf{u}_i, then

$$\{e^{r_1 t}\mathbf{u}_1, e^{r_2 t}\mathbf{u}_2, \ldots, e^{r_n t}\mathbf{u}_n\}$$

is a fundamental solution set for the homogeneous system. A class of matrices that always has n linearly independent eigenvectors is the set of **symmetric** matrices—that is, matrices that satisfy $\mathbf{A} = \mathbf{A}^T$.

If \mathbf{A} has complex conjugate eigenvalues $\alpha \pm i\beta$ and associated eigenvectors $\mathbf{z} = \mathbf{a} \pm i\mathbf{b}$, where \mathbf{a} and \mathbf{b} are real vectors, then two linearly independent real vector solutions to the homogeneous system are

$$e^{\alpha t}\cos \beta t\,\mathbf{a} - e^{\alpha t}\sin \beta t\,\mathbf{b} \ , \qquad e^{\alpha t}\sin \beta t\,\mathbf{a} + e^{\alpha t}\cos \beta t\,\mathbf{b}.$$

When \mathbf{A} has a repeated eigenvalue r of multiplicity m, then it is possible that \mathbf{A} does *not* have n linearly independent eigenvectors. However, associated with r are m linearly independent solutions to the homogeneous system of the form

$$\mathbf{x}_1(t) = e^{rt}\mathbf{k}_{1,1} \,, \qquad \mathbf{x}_2(t) = e^{rt}\mathbf{k}_{2,1} + te^{rt}\mathbf{k}_{2,2} \,, \dots \,,$$

$$\mathbf{x}_m(t) = e^{rt}\mathbf{k}_{m,1} + te^{rt}\mathbf{k}_{m,2} + \cdots + \frac{t^{m-1}}{(m-1)!}e^{rt}\mathbf{k}_{m,m} \,.$$

Nonhomogeneous Normal Systems

$$\mathbf{x}'(t) = \mathbf{A}(t)\mathbf{x}(t) + \mathbf{f}(t)$$

The $n \times n$ matrix function $\mathbf{A}(t)$ and the vector function $\mathbf{f}(t)$ are assumed continuous on an interval I.

General Solution to Nonhomogeneous System: $\mathbf{x}_p + \mathbf{Xc}$. If $\mathbf{x}_p(t)$ is any particular solution for the nonhomogeneous system and $\mathbf{X}(t)$ is a fundamental matrix for the associated homogeneous system, then a general solution for the nonhomogeneous system is

$$\mathbf{x}(t) = \mathbf{x}_p(t) + \mathbf{X}(t)\mathbf{c} = \mathbf{x}_p(t) + c_1\mathbf{x}_1(t) + \cdots + c_n\mathbf{x}_n(t) \,,$$

where $\mathbf{x}_1(t), \dots, \mathbf{x}_n(t)$ are the column vectors of $\mathbf{X}(t)$ and $\mathbf{c} = \mathrm{col}(c_1, \dots, c_n)$ is an arbitrary constant vector.

Undetermined Coefficients. If the nonhomogeneous term $\mathbf{f}(t)$ is a vector whose components are polynomials, exponential or sinusoidal functions, and \mathbf{A} is a constant matrix, then one can use an extension of the method of undetermined coefficients to decide the form of a particular solution to the nonhomogeneous system.

Variation of Parameters: $\mathbf{X}(t)\boldsymbol{v}(t)$. Let $\mathbf{X}(t)$ be a fundamental matrix for the homogeneous system. A particular solution to the nonhomogeneous system is given by the **variation of parameters formula**

$$\mathbf{x}_p(t) = \mathbf{X}(t)\boldsymbol{v}(t) = \mathbf{X}(t)\int \mathbf{X}^{-1}(t)\mathbf{f}(t)\, dt.$$

Matrix Exponential Function

If \mathbf{A} is a constant $n \times n$ matrix, then the matrix exponential function

$$e^{\mathbf{A}t} := \mathbf{I} + \mathbf{A}t + \mathbf{A}^2\frac{t^2}{2!} + \cdots + \mathbf{A}^n\frac{t^n}{n!} + \cdots$$

is a fundamental matrix for the homogeneous system $\mathbf{x}'(t) = \mathbf{A}\mathbf{x}(t)$. The matrix exponential has some of the same properties as the scalar exponential e^{at}. In particular,

$$e^{\mathbf{0}} = \mathbf{I} \,, \qquad e^{\mathbf{A}(t+s)} = e^{\mathbf{A}t}e^{\mathbf{A}s} \,, \qquad \left(e^{\mathbf{A}t}\right)^{-1} = e^{-\mathbf{A}t} \,.$$

If $(\mathbf{A} - r\mathbf{I})^k = \mathbf{0}$ for some r and k, then the series for $e^{\mathbf{A}t}$ has only a finite number of terms:

$$e^{\mathbf{A}t} = e^{rt}\left\{\mathbf{I} + (\mathbf{A} - r\mathbf{I})t + \cdots + (\mathbf{A} - r\mathbf{I})^{k-1}\frac{t^{k-1}}{(k-1)!}\right\}.$$

The matrix exponential function $e^{\mathbf{A}t}$ can also be computed from any fundamental matrix $\mathbf{X}(t)$ via the formula

$$e^{\mathbf{A}t} = \mathbf{X}(t)\mathbf{X}^{-1}(0) .$$

Generalized Eigenvectors

If r is an eigenvalue of \mathbf{A}, then a **generalized eigenvector** associated with r is a nonzero vector \mathbf{u} that satisfies $(\mathbf{A} - r\mathbf{I})^k\mathbf{u} = \mathbf{0}$ for some positive integer k. *Every* matrix \mathbf{A} has a set of n linearly independent generalized eigenvectors that can be used to compute a fundamental solution set. In particular, associated with each generalized eigenvector \mathbf{u} and corresponding eigenvalue r is a solution to the homogeneous system $\mathbf{x}'(t) = \mathbf{A}\mathbf{x}(t)$ of the form

$$e^{\mathbf{A}t}\mathbf{u} = e^{rt}\left\{\mathbf{u} + t(\mathbf{A} - r\mathbf{I})\mathbf{u} + \frac{t^2}{2!}(\mathbf{A} - r\mathbf{I})^2\mathbf{u} + \cdots\right\}.$$

If r has multiplicity m, then the above series reduces to the first m terms.

REVIEW PROBLEMS

In Problems 1 through 4, find a general solution for the system $\mathbf{x}'(t) = \mathbf{A}\mathbf{x}(t)$, *where* \mathbf{A} *is given.*

1. $\mathbf{A} = \begin{bmatrix} 6 & -3 \\ 2 & 1 \end{bmatrix}$.

2. $\mathbf{A} = \begin{bmatrix} 3 & 2 \\ -5 & 1 \end{bmatrix}$.

3. $\mathbf{A} = \begin{bmatrix} 1 & 2 & 0 & 0 \\ 2 & 1 & 0 & 0 \\ 0 & 0 & 1 & 2 \\ 0 & 0 & 2 & 1 \end{bmatrix}$.

4. $\mathbf{A} = \begin{bmatrix} 1 & 1 & 0 \\ 0 & 1 & 0 \\ 0 & 0 & 2 \end{bmatrix}$.

In Problems 5 and 6, find a fundamental matrix for the system $\mathbf{x}'(t) = \mathbf{A}\mathbf{x}(t)$, *where* \mathbf{A} *is given.*

5. $\mathbf{A} = \begin{bmatrix} 1 & -1 \\ 2 & 4 \end{bmatrix}$.

6. $\mathbf{A} = \begin{bmatrix} 5 & 0 & 0 \\ 0 & -4 & 3 \\ 0 & 3 & 4 \end{bmatrix}$.

In Problems 7 through 10, find a general solution for the system $\mathbf{x}'(t) = \mathbf{A}\mathbf{x}(t) + \mathbf{f}(t)$, *where* \mathbf{A} *and* $\mathbf{f}(t)$ *are given.*

7. $\mathbf{A} = \begin{bmatrix} 1 & 1 \\ 4 & 1 \end{bmatrix}$, $\mathbf{f}(t) = \begin{bmatrix} 5 \\ 6 \end{bmatrix}$.

8. $\mathbf{A} = \begin{bmatrix} -4 & 2 \\ 2 & -1 \end{bmatrix}$, $\mathbf{f}(t) = \begin{bmatrix} e^{4t} \\ 3e^{4t} \end{bmatrix}$.

9. $\mathbf{A} = \begin{bmatrix} 2 & 1 & -1 \\ -3 & -1 & 1 \\ 9 & 3 & -4 \end{bmatrix}$, $\mathbf{f}(t) = \begin{bmatrix} t \\ 0 \\ 1 \end{bmatrix}$.

10. $\mathbf{A} = \begin{bmatrix} 2 & -2 & 3 \\ 0 & 3 & 2 \\ 0 & -1 & 2 \end{bmatrix}$, $\mathbf{f}(t) = \begin{bmatrix} e^{-t} \\ 2 \\ 1 \end{bmatrix}$.

In Problems 11 and 12, solve the given initial value problem.

11. $\mathbf{x}'(t) = \begin{bmatrix} 0 & 1 \\ -2 & 3 \end{bmatrix} \mathbf{x}(t), \qquad \mathbf{x}(0) = \begin{bmatrix} 1 \\ -1 \end{bmatrix}.$

12. $\mathbf{x}'(t) = \begin{bmatrix} 2 & 1 \\ -4 & 2 \end{bmatrix} \mathbf{x}(t) + \begin{bmatrix} te^{2t} \\ e^{2t} \end{bmatrix}, \qquad \mathbf{x}(0) = \begin{bmatrix} 2 \\ 2 \end{bmatrix}.$

In Problems 13 and 14, find a general solution for the Cauchy-Euler system $t\mathbf{x}'(t) = \mathbf{A}\mathbf{x}(t)$, where \mathbf{A} is given.

13. $\mathbf{A} = \begin{bmatrix} 0 & 3 & 1 \\ 1 & 2 & 1 \\ 1 & 3 & 0 \end{bmatrix}.$

14. $\mathbf{A} = \begin{bmatrix} 1 & 2 & -1 \\ 2 & 1 & 1 \\ -1 & 1 & 0 \end{bmatrix}.$

In Problems 15 and 16, find the fundamental matrix $e^{\mathbf{A}t}$ for the system $\mathbf{x}'(t) = \mathbf{A}\mathbf{x}(t)$, where \mathbf{A} is given.

15. $\mathbf{A} = \begin{bmatrix} 4 & 2 & 3 \\ 2 & 1 & 2 \\ -1 & 2 & 0 \end{bmatrix}.$ **16.** $\mathbf{A} = \begin{bmatrix} 0 & 1 & 4 \\ 0 & 0 & 2 \\ 0 & 0 & 0 \end{bmatrix}.$

TECHNICAL WRITING EXERCISES

1. Explain how the theory of homogeneous linear differential equations (as described in Sections 4.3 and 6.1) follows from the theory of linear systems in normal form (as described in Section 9.4).

2. Discuss the similarities and differences between the method for finding solutions to a linear constant coefficient differential equation (see Sections 4.5, 4.6, and 6.2) and the method for finding solutions to a linear system in normal form that has constant coefficients (see Sections 9.5 and 9.6).

3. Explain how the variation of parameters formulas for linear second order equations derived in Section

4.9 follow from the formulas derived in Section 9.7 for linear systems in normal form.

4. Explain how you would define the matrix functions $\sin \mathbf{A}t$ and $\cos \mathbf{A}t$, where \mathbf{A} is a constant $n \times n$ matrix. How are these functions related to the matrix exponential and how are they connected to the solutions of the system $\mathbf{x}'' + \mathbf{A}^2\mathbf{x} = \mathbf{0}$? The reader will find it informative to consider the cases when \mathbf{A} is

$$\begin{bmatrix} 0 & 0 \\ 0 & 0 \end{bmatrix}, \begin{bmatrix} 0 & 1 \\ 0 & 0 \end{bmatrix}, \begin{bmatrix} 0 & 0 \\ 1 & 0 \end{bmatrix}, \text{ and } \begin{bmatrix} 1 & -1 \\ 1 & -1 \end{bmatrix}.$$

GROUP PROJECTS FOR CHAPTER 9

A. Uncoupling Normal Systems

The easiest normal systems to solve are systems of the form

(1) $\quad \mathbf{x}'(t) = \mathbf{D}\mathbf{x}(t)$,

where \mathbf{D} is an $n \times n$ diagonal matrix. Such a system actually consists of n uncoupled equations

(2) $\quad x_i'(t) = d_{ii}x_i(t)$, $\qquad i = 1, \ldots, n$,

whose solution is

$$x_i(t) = c_i e^{d_{ii}t} ,$$

where the c_i's are arbitrary constants. This raises the following question: When can we *uncouple* a normal system?

To answer this question, we need the following result from linear algebra. An $n \times n$ matrix \mathbf{A} is diagonalizable if and only if \mathbf{A} has n linearly independent eigenvectors $\mathbf{p}_1, \ldots, \mathbf{p}_n$. Moreover, if \mathbf{P} is the matrix whose columns are $\mathbf{p}_1, \ldots, \mathbf{p}_n$, then

(3) $\quad \mathbf{P}^{-1}\mathbf{A}\mathbf{P} = \mathbf{D}$,

where \mathbf{D} is the diagonal matrix whose entry d_{ii} is the eigenvalue associated with the vector \mathbf{p}_i.

(a) Use the above result to show that the system

(4) $\quad \mathbf{x}'(t) = \mathbf{A}\mathbf{x}(t)$,

where \mathbf{A} is an $n \times n$ diagonalizable matrix, is equivalent to an uncoupled system

(5) $\quad \mathbf{y}'(t) = \mathbf{D}\mathbf{y}(t)$,

where $\mathbf{y} = \mathbf{P}^{-1}\mathbf{x}$ and $\mathbf{D} = \mathbf{P}^{-1}\mathbf{A}\mathbf{P}$.

(b) Solve system (5).

(c) Use the results of parts (a) and (b) to show that a general solution to (4) is given by

$$\mathbf{x}(t) = c_1 e^{d_{11}t}\mathbf{p}_1 + c_2 e^{d_{22}t}\mathbf{p}_2 + \cdots + c_n e^{d_{nn}t}\mathbf{p}_n .$$

(d) Use the procedure discussed in parts (a)–(c) to obtain a general solution for the system

$$\mathbf{x}'(t) = \begin{bmatrix} 5 & -4 & 4 \\ 1 & 0 & 1 \\ -1 & 2 & 1 \end{bmatrix} \mathbf{x}(t) .$$

Specify \mathbf{P}, \mathbf{D}, \mathbf{P}^{-1}, and \mathbf{y}.

B. Matrix Laplace Transform Method

The Laplace transform method for solving systems of linear differential equations with constant coefficients was discussed in Chapter 7. To apply the procedure for equations given in matrix

form, we first extend the definition of the Laplace operator \mathcal{L} to a column vector of functions $\mathbf{x} = \text{col}\big(x_1(t), \dots, x_n(t)\big)$ by taking the transform of each component:

$$\mathcal{L}\{\mathbf{x}\}(s) := \text{col}\big(\mathcal{L}\{x_1\}(s), \dots, \mathcal{L}\{x_n\}(s)\big) .$$

With this notation, the vector analogue of the important property relating the Laplace transform of the derivative of a function (see Theorem 4, Chapter 7, page 383) becomes

(6) $\mathcal{L}\{\mathbf{x}'\}(s) = s\mathcal{L}\{\mathbf{x}\}(s) - \mathbf{x}(0) .$

Now suppose we are given the initial value problem

(7) $\mathbf{x}' = \mathbf{A}\mathbf{x} + \mathbf{f}(t) , \qquad \mathbf{x}(0) = \mathbf{x}_0 ,$

where \mathbf{A} is a constant $n \times n$ matrix. Let $\hat{\mathbf{x}}(s)$ denote the Laplace transform of $\mathbf{x}(t)$ and $\hat{\mathbf{f}}(s)$ denote the transform of $\mathbf{f}(t)$. Then, taking the transform of the system and using the relation (6), we get

$$\mathcal{L}\{\mathbf{x}'\} = \mathcal{L}\{\mathbf{A}\mathbf{x} + \mathbf{f}\} ,$$
$$s\hat{\mathbf{x}} - \mathbf{x}_0 = \mathbf{A}\hat{\mathbf{x}} + \hat{\mathbf{f}} .$$

Next we collect the $\hat{\mathbf{x}}$ terms and solve for $\hat{\mathbf{x}}$ by premultiplying by $(s\mathbf{I} - \mathbf{A})^{-1}$:

$$(s\mathbf{I} - \mathbf{A})\hat{\mathbf{x}} = \hat{\mathbf{f}} + \mathbf{x}_0 ,$$
$$\hat{\mathbf{x}} = (s\mathbf{I} - \mathbf{A})^{-1}(\hat{\mathbf{f}} + \mathbf{x}_0) .$$

Finally, we obtain the solution $\mathbf{x}(t)$ by taking the inverse Laplace transform:

(8) $\mathbf{x} = \mathcal{L}^{-1}\{\hat{\mathbf{x}}\} = \mathcal{L}^{-1}\{(s\mathbf{I} - \mathbf{A})^{-1}(\hat{\mathbf{f}} + \mathbf{x}_0)\} .$

In applying the matrix Laplace transform method, it is straightforwaard (but possibly tedious) to compute $(s\mathbf{I} - \mathbf{A})^{-1}$, but the computation of the inverse transform may require some of the special techniques (such as partial fractions) discussed in Chapter 7.

(a) In the above procedure, we used the property that $\mathcal{L}\{\mathbf{A}\mathbf{x}\} = \mathbf{A}\mathcal{L}\{\mathbf{x}\}$ for any constant $n \times n$ matrix \mathbf{A}. Show that this property follows from the linearity of the transform in the scalar case.

(b) Use the matrix Laplace transform method to solve the following initial value problems:

(i) $\mathbf{x}'(t) = \begin{bmatrix} 0 & 2 \\ -1 & 3 \end{bmatrix} \mathbf{x}(t) , \qquad \mathbf{x}(0) = \begin{bmatrix} -1 \\ 3 \end{bmatrix} .$

(ii) $\mathbf{x}'(t) = \begin{bmatrix} 3 & -2 \\ 4 & -1 \end{bmatrix} \mathbf{x}(t) + \begin{bmatrix} \sin t \\ -\cos t \end{bmatrix} , \qquad \mathbf{x}(0) = \begin{bmatrix} 0 \\ 0 \end{bmatrix} .$

(c) By comparing the Laplace transform solution formula (8) with the matrix exponential solution formula given in Section 9.8 (relation (13), page 589) for the homogeneous case $\mathbf{f}(t) \equiv \mathbf{0}$ and $t_0 = 0$, derive the Laplace transform formula for the matrix exponential

(9) $e^{\mathbf{A}t} = \mathcal{L}^{-1}\{(s\mathbf{I} - \mathbf{A})^{-1}\}(t) .$

(d) What is $e^{\mathbf{A}t}$ for the coefficient matrices in part (b) above?

(e) Use (9) to rework Problems 1, 2, 7, and 8 in Exercises 9.8, page 590.

C. *Undamped Second Order Systems*

We have seen that the coupled mass-spring system depicted in Figure 9.5, page 572, is governed by equations (9) of Section 9.6, which we reproduce here:

$$m_1 x_1'' = -k_1 x_1 + k_2(x_2 - x_1) \ ,$$
$$m_2 x_2'' = -k_2(x_2 - x_1) - k_3 x_2 \ .$$

This system was rewritten in normal form as equation (10) in Section 9.6; however, there is some advantage in expressing it as a second order system in the form

$$\begin{bmatrix} m_1 & 0 \\ 0 & m_2 \end{bmatrix} \begin{bmatrix} x_1 \\ x_2 \end{bmatrix}'' = -\begin{bmatrix} k_1 + k_2 & -k_2 \\ -k_2 & k_2 + k_3 \end{bmatrix} \begin{bmatrix} x_1 \\ x_2 \end{bmatrix} \ .$$

This structure,

(10) $\mathbf{Mx}'' = -\mathbf{Kx}$,

with a diagonal *mass matrix* **M** and a symmetric *stiffness matrix* **K**, is typical for most undamped vibrating systems. Our experience with other mass-spring systems (Section 5.4) suggests that we seek solutions to (10) of the form

(11) $\mathbf{x} = \cos \omega t \, \mathbf{v}$ or $\mathbf{x} = \sin \omega t \, \mathbf{v}$,

where **v** is a constant vector and ω is a positive constant.

(a) Show that the system (10) has a nontrivial solution of the form (11) if and only if ω and **v** satisfy the "generealized eigenvalue problem" $\mathbf{Kv} = \omega^2 \, \mathbf{Mv}$.

(b) By employing the inverse of the mass matrix, one can rewrite (10) as

$$\mathbf{x}'' = -\mathbf{M}^{-1}\mathbf{Kx} =: \mathbf{Bx} \ .$$

Show that $-\omega^2$ must be an eigenvalue of **B** if (11) is a nontrivial solution.

(c) If **B** is an $n \times n$ constant matrix, then $\mathbf{x}'' = \mathbf{Bx}$ can be written as a system of $2n$ first order equations in normal form. Thus a general solution can be formed from $2n$ linearly independent solutions. Use the observation in part (b) to find a general solution to the following second order systems:

(i) $\mathbf{x}'' = \begin{bmatrix} 0 & 1 \\ -4 & -5 \end{bmatrix} \mathbf{x}$.

(ii) $\mathbf{x}'' = \begin{bmatrix} -2 & 2 \\ 2 & -5 \end{bmatrix} \mathbf{x}$.

(iii) $\mathbf{x}'' = \begin{bmatrix} -5 & 4 & -4 \\ -1 & 0 & -1 \\ 1 & -2 & -1 \end{bmatrix} \mathbf{x}$.

(iv) $\mathbf{x}'' = \begin{bmatrix} 2 & -1 & -1 \\ -1 & 2 & 0 \\ -1 & 0 & 2 \end{bmatrix} \mathbf{x}$.

D. *Strange Behavior of Competing Species—Part II*

In Project G in Chapter 5, we used numerical experimentation to study the behavior of solutions to the system

$$p_1' = p_1(1 - p_1 - ap_2 - bp_3) \,,$$

(12) $$p_2' = p_2(1 - bp_1 - p_2 - ap_3) \,,$$

$$p_3' = p_3(1 - ap_1 - bp_2 - p_3) \,,$$

where a and b are positive constants and p_i is the population of species S_i, $i = 1, 2, 3$. We found that when $a = b = 0.5$, the populations approach the equilibrium solution $p_1 = p_2 = p_3 = 1/2$ as $t \to +\infty$. However, when $a = 0.5$ and $b = 2$, the populations cycled between the three critical points—$(1, 0, 0)$, $(0, 1, 0)$, $(0, 0, 1)$—getting closer and closer to these points. Moreover, the populations remained near each point for longer and longer periods of time before moving on to the next point. To analyze what is happening near these equilibrium points, we will study the corresponding linearized systems.

(a) Let $f_i(p_1, p_2, p_3)$ denote the function on the right-hand side of the equation for p_i' in (12); in particular, $f_1(p_1, p_2, p_3) = p_1(1 - p_1 - ap_2 - bp_3)$. Now $\mathbf{f} = \text{col}(f_1, f_2, f_3)$ is a mapping from \mathbf{R}^3 into \mathbf{R}^3. Show that the Jacobian matrix for this mapping is

$$\mathbf{A} := \begin{bmatrix} 1 - 2p_1 - ap_2 - bp_3 & -ap_1 & -bp_1 \\ -bp_2 & 1 - 2p_2 - ap_3 - bp_1 & -ap_2 \\ -ap_3 & -bp_3 & 1 - 2p_3 - ap_1 - bp_2 \end{bmatrix}.$$

(b) Show that at the critical point (δ, δ, δ) where $\delta = (1 + a + b)^{-1}$, we have $\mathbf{A} = -\delta\mathbf{B}$, where

$$\mathbf{B} := \begin{bmatrix} 1 & a & b \\ b & 1 & a \\ a & b & 1 \end{bmatrix}.$$

(c) Show that the eigenvalues of \mathbf{B} are

$$1 + a + b \,, \qquad \left(1 - \frac{a + b}{2}\right) \pm i\frac{\sqrt{3}(a - b)}{2} \,.$$

(d) Now, near the critical point $\mathbf{p}^* = (\delta, \delta, \delta)$, the solutions to (12) should behave like the solutions to $\mathbf{p}' = -\delta\mathbf{B}(\mathbf{p} - \mathbf{p}^*)$. Use the results of part (c) to argue that when $a + b < 2$, the solutions to (12) should approach the critical point (δ, δ, δ) and when $a + b > 2$, they should move away.

(e) Using a similar analysis, conclude that when either $a < 1$ or $b < 1$, the solutions to (12) should move away from the critical points $(1, 0, 0)$, $(0, 1, 0)$, and $(0, 0, 1)$ because at least one of the eigenvalues of the linearized system (about each critical point) is positive.

(f) How do the analyses performed in parts (d) and (e) support the conclusions reached from numerical experimentations conducted in Project G in Chapter 5?

10

Partial Differential Equations

10.1 INTRODUCTION: A MODEL FOR HEAT FLOW

Develop a model for the flow of heat through a thin, insulated wire whose ends are kept at a constant temperature of 0°C and whose initial temperature distribution is to be specified.

Suppose the wire is placed along the x-axis with $x = 0$ at the left end of the wire and $x = L$ at the right end (see Figure 10.1). If we let u denote the temperature of the wire, then u depends on the time t and on the position x within the wire. (We will assume the wire is thin and hence u is constant throughout a cross section of the wire corresponding to a fixed value of x.) Because the wire is insulated, we assume no heat enters or leaves through the sides of the wire.

To develop a model for heat flow through the thin wire, let's consider the small volume element V of wire between the two cross-sectional planes A and B that are perpendicular to the x-axis, with plane A located at x and plane B located at $x + \Delta x$ (see Figure 10.1).

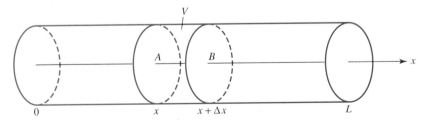

Figure 10.1 Heat flow through a thin piece of wire

The temperature on plane A at time t is $u(x, t)$ and on plane B is $u(x + \Delta x, t)$. We will need the following principles of physics that describe heat flow.[†]

1. **Heat Conduction:** The rate of heat flow (the amount of heat per unit time flowing through a unit of cross-sectional area at A) is proportional to $\partial u/\partial x$, the temperature gradient at A. The proportionality constant k is called the **thermal conductivity** of the material. In general, the thermal conductivity can vary from point to point: $k = k(x)$.

2. **Direction of Heat Flow:** The direction of heat flow is always from points of higher temperature to points of lower temperature.

3. **Specific Heat Capacity:** The amount of heat necessary to raise the temperature of an object of mass m by an amount Δu is $cm\,\Delta u$, where the constant c is the **specific heat capacity** of the material. The specific heat capacity, like the thermal conductivity, can vary with position: $c = c(x)$.

If we let H represent the amount of heat flowing from left to right through the surface A during an interval of time Δt, then the formula for heat conduction becomes

$$H(x) = -k(x)a\,\Delta t \frac{\partial u}{\partial x}(x, t) \ ,$$

where a is the cross-sectional area of the wire. The negative sign follows from the second principle—if $\partial u/\partial x$ is positive, then heat flows from right to left (hotter to colder).

Similarly, the amount of heat flowing from left to right across plane B during an interval of time Δt is

$$H(x + \Delta x) = -k(x + \Delta x)a\,\Delta t \frac{\partial u}{\partial x}(x + \Delta x, t) \ .$$

The net change in the heat ΔE in volume V is the amount entering at end A minus the amount leaving at end B, plus any heat generated by sources (such as electric currents, chemical reactions, heaters, etc.). The latter is modeled by a term $Q(x, t)\Delta x a\,\Delta t$, where Q is the energy rate (power) density. Therefore,

(1) $\quad \Delta E = H(x) - H(x + \Delta x) + Q(x, t)\Delta x a\,\Delta t$

$$= a\,\Delta t \left[k(x + \Delta x)\frac{\partial u}{\partial x}(x + \Delta x, t) - k(x)\frac{\partial u}{\partial x}(x, t) \right] + Q(x, t)\Delta x a\,\Delta t \ .$$

Now by the third principle, the net change is given by $\Delta E = cm\,\Delta u$, where Δu is the change in temperature and c is the specific heat capacity. If we assume the change in temperature in the volume V is essentially equal to the change in temperature at x, that is,

[†]For a discussion of heat transfer, see *University Physics,* 9th edition, by H. D. Young and R. A. Freedman (Addison-Wesley, Reading, Mass., 1996).

$\Delta u = u(x, t + \Delta t) - u(x, t)$, and that the mass of the volume V of wire is $a\rho\,\Delta x$, where $\rho = \rho(x)$ is the density of the wire, then

(2) $\Delta E = c(x)a\rho(x)\,\Delta x\left[u(x, t + \Delta t) - u(x, t)\right]$.

Equating the two expressions for ΔE given in equations (1) and (2) yields

$$a\,\Delta t\left[k(x + \Delta x)\frac{\partial u}{\partial x}(x + \Delta x, t) - k(x)\frac{\partial u}{\partial x}(x, t)\right] + Q(x, t)\,\Delta x a\,\Delta t$$
$$= c(x)a\rho(x)\,\Delta x\left[u(x, t + \Delta t) - u(x, t)\right]$$.

Now dividing both sides by $a\,\Delta x\,\Delta t$ and then taking the limits as Δx and Δt approach zero, we obtain

(3) $\dfrac{\partial}{\partial x}\left[k(x)\dfrac{\partial u}{\partial x}(x, t)\right] + Q(x, t) = c(x)\rho(x)\dfrac{\partial u}{\partial t}(x, t)$.

If the physical parameters k, c, and ρ are uniform along the length of the wire, then (3) reduces to the **one-dimensional heat flow equation**

(4) $\dfrac{\partial u}{\partial t}(x, t) = \beta\dfrac{\partial^2 u}{\partial x^2}(x, t) + P(x, t)$,

where the positive constant $\beta := k/c\rho$ is the **diffusivity** of the material and $P(x, t) := Q(x, t)/c\rho$.

Equation (4) governs the flow of heat in the wire. We have two other constraints in our original problem. First, we are keeping the ends of the wire at $0°\mathrm{C}$. Thus we require that

(5) $u(0, t) = u(L, t) = 0$

for all t. These are called **boundary conditions.** Second, we must be given the initial temperature distribution $f(x)$. That is, we require

(6) $u(x, 0) = f(x)$, $0 < x < L$.

Equation (6) is referred to as the **initial condition** on u.

Combining equations (4), (5), and (6), we have the following mathematical model for the heat flow in a uniform wire without internal sources $(P = 0)$ whose ends are kept at the constant temperature $0°\mathrm{C}$:

(7) $\dfrac{\partial u}{\partial t}(x, t) = \beta\dfrac{\partial^2 u}{\partial x^2}(x, t)$, $0 < x < L$, $t > 0$,

(8) $u(0, t) = u(L, t) = 0$, $t > 0$,

(9) $u(x, 0) = f(x)$, $0 < x < L$.

This model is an example of an **initial-boundary value problem.** Intuitively we expect that equations (7)–(9) completely and unambiguously specify the temperature in the wire. Once we have found a function $u(x, t)$ that meets all three of these conditions, we can be assured that u is the temperature. (Theorem 7 in Section 10.5 will bear this out.)

In higher dimensions, the heat flow equation (or just **heat equation**) has the form

$$\frac{\partial u}{\partial t} = \beta \, \Delta u + P(x, y, z, t)$$

where Δu, in this context, is the **Laplacian** of u.[†] In two and three dimensions the Laplacian is defined, respectively, as

$$\Delta u := \frac{\partial^2 u}{\partial x^2} + \frac{\partial^2 u}{\partial y^2} \quad \text{and} \quad \Delta u := \frac{\partial^2 u}{\partial x^2} + \frac{\partial^2 u}{\partial y^2} + \frac{\partial^2 u}{\partial z^2} \ .$$

When the temperature reaches a steady state, that is when u does not depend on time, and there are no sources, then $\partial u / \partial t = 0$ and the temperature satisfies **Laplace's equation**

$$\Delta u = 0 \ .$$

One classical technique for solving the initial-boundary value problem for the heat equation (7)–(9) is the method of *separation of variables,* which effectively allows us to replace the partial derivatives by ordinary derivatives. This technique is discussed in the next section. In using separation of variables, one is often required to express a given function as a trigonometric series. Such series are called *Fourier series;* their properties are discussed in Sections 10.3 and 10.4. We devote the remaining three sections to the three basic partial differential equations that arise most commonly in applications: the heat equation, the wave equation, and Laplace's equation.

Many computer-based algorithms for solving partial differential equations have been developed. These are based on finite differences, finite elements, variational principles, and projection methods that include the method of moments and, more recently, its wavelet-based implementations. Like the Runge-Kutta or Euler methods for ordinary differential equations, such techniques are more universally applicable than the analytic procedures of separation of variables, but their accuracy is often difficult to assess. Indeed, it is customary practice to assess any newly proposed numerical procedure by comparing its predictions with those of separation of variables.

10.2 METHOD OF SEPARATION OF VARIABLES

The method of separation of variables is a classical technique that is effective in solving several types of partial differential equations. The idea is roughly the following. We think of a solution, say $u(x, t)$, to a partial differential equation as being an infinite linear combination of simple component functions $u_n(x, t)$, $n = 0, 1, 2, \ldots$, which also satisfy the equation and certain boundary conditions. (This is a reasonable assumption provided the partial differential equation and the boundary conditions are linear and homogeneous.) To determine a component solution, $u_n(x, t)$, we assume it can be written with its variables separated; that is, as

$$u_n(x, t) = X_n(x)T_n(t) \ .$$

[†]Regrettably, it is the same notation as we just used for $u(x, t + \Delta t) - u(x, t)$.

Substituting this form for a solution into the partial differential equation and using the boundary conditions leads, in many circumstances, to two *ordinary* differential equations for the unknown functions $X_n(x)$ and $T_n(t)$. In this way we have reduced the problem of solving a partial differential equation to the more familiar problem of solving a differential equation that involves only one variable. In this section we will illustrate this technique for the heat equation and the wave equation.

In the previous section we derived the following initial-boundary value problem as a mathematical model for the sourceless heat flow in a uniform wire whose ends are kept at the constant temperature zero:

(1) $\qquad \dfrac{\partial u}{\partial t}(x, t) = \beta \dfrac{\partial^2 u}{\partial x^2}(x, t) \ , \qquad 0 < x < L \ , \qquad t > 0 \ ,$

(2) $\qquad u(0, t) = u(L, t) = 0 \ , \qquad t > 0 \ ,$

(3) $\qquad u(x, 0) = f(x) \ , \qquad\qquad 0 < x < L \ .$

To solve this problem by the method of separation of variables, we begin by addressing equation (1). We propose that it has solutions of the form

$$u(x, t) = X(x)T(t) \ ,$$

where X is a function of x alone and T is a function of t alone. To determine X and T, we first compute the partial derivatives of u to obtain

$$\frac{\partial u}{\partial t} = X(x)T'(t) \quad \text{and} \quad \frac{\partial^2 u}{\partial x^2} = X''(x)T(t) \ .$$

Substituting these expressions into (1) gives

$$X(x)T'(t) = \beta X''(x)T(t) \ ,$$

and separating variables yields

(4) $\qquad \dfrac{T'(t)}{\beta T(t)} = \dfrac{X''(x)}{X(x)} \ .$

We now observe that the functions on the left-hand side of (4) depend only on t, while those on the right-hand side depend only on x. Since x and t are variables that are independent of one another, the two ratios in (4) must then be *constant*. Thus

$$\frac{X''(x)}{X(x)} = K \quad \text{and} \quad \frac{T'(t)}{\beta T(t)} = K \ ,$$

or

(5) $\qquad X''(x) - KX(x) = 0 \quad \text{and} \quad T'(t) - \beta K T(t) = 0 \ .$

Consequently, for separable solutions, we have reduced the problem of solving the partial differential equation (1) to solving the two *ordinary* differential equations in (5).

Next we consider the boundary conditions in (2). Since $u(x, t) = X(x)T(t)$, these conditions are

$$X(0)T(t) = 0 \quad \text{and} \quad X(L)T(t) = 0 \ , \qquad t > 0 \ .$$

Hence either $T(t) = 0$ for all $t > 0$, which implies that $u(x, t) \equiv 0$, or

(6) $X(0) = X(L) = 0$.

Ignoring the trivial solution $u(x, t) \equiv 0$, we combine the boundary conditions in (6) with the differential equation for X in (5) and obtain the *boundary value problem*

(7) $X''(x) - KX(x) = 0$, $X(0) = X(L) = 0$,

where K can be any constant.

Notice that the function $X(x) \equiv 0$ is a solution of (7) for every K. Depending on the choice of K, this may be the *only* solution to the boundary value problem (7). Thus if we seek a non-trivial solution $u(x, t) = X(x)T(t)$ to (1)–(2), we must first determine those values of K for which the boundary value problem (7) has nontrivial solutions. These special values of K are called **eigenvalues,** and the corresponding nontrivial solutions to (7) are the **eigenfunctions.** This nomenclature arises from the similarity of the operator form of the differential equation, $(D^2 - K)[X] = 0$, with the matrix eigenvector equation $(\mathbf{A} - r\mathbf{I})\mathbf{u} = \mathbf{0}$ (Section 9.5, page 557).

To solve the (constant coefficient) equation in (7), we try $X(x) = e^{rx}$, derive the auxiliary equation $r^2 - K = 0$, and consider three cases.

Case 1. $K > 0$. In this case, the roots of the auxiliary equation are $\pm\sqrt{K}$, so a general solution to the differential equation in (7) is

$$X(x) = C_1 e^{\sqrt{K}x} + C_2 e^{-\sqrt{K}x} .$$

To determine C_1 and C_2, we appeal to the boundary conditions:

$$X(0) = C_1 + C_2 = 0 ,$$
$$X(L) = C_1 e^{\sqrt{K}L} + C_2 e^{-\sqrt{K}L} = 0 .$$

From the first equation, we see that $C_2 = -C_1$. The second equation can then be written as $C_1\left(e^{\sqrt{K}L} - e^{-\sqrt{K}L}\right) = 0$ or $C_1\left(e^{2\sqrt{K}L} - 1\right) = 0$. Since $K > 0$, it follows that $\left(e^{2\sqrt{K}L} - 1\right) > 0$. Therefore C_1, and hence C_2, is zero. Consequently, there is *no* nontrivial solution to (7) for $K > 0$.

Case 2. $K = 0$. Here $r = 0$ is a repeated root to the auxiliary equation, and a general solution to the differential equation is

$$X(x) = C_1 + C_2 x .$$

The boundary conditions in (7) yield $C_1 = 0$ and $C_1 + C_2 L = 0$, which imply that $C_1 = C_2 = 0$. Thus for $K = 0$, there is *no* nontrivial solution to (7).

Case 3. $K < 0$. In this case the roots of the auxiliary equation are $\pm i\sqrt{-K}$. (Note that $-K > 0$ since $K < 0$.) Thus a general solution to $X'' - KX = 0$ is

(8) $X(x) = C_1 \cos \sqrt{-K}x + C_2 \sin \sqrt{-K}x$.

This time the boundary conditions $X(0) = X(L) = 0$ give the system

$$C_1 = 0 ,$$
$$C_1 \cos \sqrt{-K}L + C_2 \sin \sqrt{-K}L = 0 .$$

Because $C_1 = 0$, the system reduces to solving $C_2 \sin \sqrt{-K}L = 0$. Hence, either $\sin \sqrt{-K}L = 0$ or $C_2 = 0$. Now $\sin \sqrt{-K}L = 0$ only when $\sqrt{-K}L = n\pi$, where n is an integer. Therefore (7) has a nontrivial solution $(C_2 \neq 0)$ when $\sqrt{-K}L = n\pi$ or $K = -(n\pi/L)^2$, $n = 1, 2, 3, \ldots$ (we exclude $n = 0$, since it makes $K = 0$). Furthermore, the nontrivial solutions (eigenfunctions) X_n corresponding to the eigenvalue $K = -(n\pi/L)^2$ are given by (cf. (8))

$$(9) \qquad X_n(x) = a_n \sin\left(\frac{n\pi x}{L}\right),$$

where the a_n's are arbitrary nonzero constants.

Having determined that $K = -(n\pi/L)^2$ for any positive integer n, let's consider the second equation in (5):

$$T'(t) + \beta\left(\frac{n\pi}{L}\right)^2 T(t) = 0 .$$

For each $n = 1, 2, 3, \ldots$, a general solution to this linear first order equation is

$$T_n(t) = b_n e^{-\beta(n\pi/L)^2 t} .$$

Combining this with equation (9) we obtain, for each $n = 1, 2, 3, \ldots$, the functions

$$(10) \qquad u_n(x, t) := X_n(x)T_n(t) = a_n \sin(n\pi x/L) b_n e^{-\beta(n\pi/L)^2 t}$$
$$= c_n e^{-\beta(n\pi/L)^2 t} \sin(n\pi x/L) ,$$

where c_n is an arbitrary constant.

We would like to conclude that each $u_n(x, t)$ is a solution to (1)–(2). But, we have shown *only* that *if* (1)–(2) has a solution of the form $u(x, t) = X(x)T(t)$, then u must be one of the functions given in (10). We leave it for the reader to verify, by direct substitution, that the functions in (10) are indeed solutions to (1)–(2). They satisfy the differential equation and the boundary conditions (but not the initial condition).

A simple computation also shows that if u_n and u_m are solutions to (1)–(2), then so is any linear combination $au_n + bu_m$. (This is a consequence of the facts that the operator $\mathcal{L} := \partial/\partial t - \beta\partial^2/\partial x^2$ is a *linear* operator and the boundary conditions in (2) are *homogeneous*.) Furthermore, if we take an *infinite* sum of these functions, that is,

$$(11) \qquad u(x, t) = \sum_{n=1}^{\infty} u_n(x, t) = \sum_{n=1}^{\infty} c_n e^{-\beta(n\pi/L)^2 t} \sin(n\pi x/L) ,$$

then this formal series will again be a solution to (1)–(2), provided the infinite series has the proper convergence behavior.

Now we seek to satisfy the initial condition (3). For a solution u of the form (11) to satisfy (3), we must have

$$u(x, 0) = \sum_{n=1}^{\infty} c_n \sin\left(\frac{n\pi x}{L}\right) = f(x) , \qquad 0 < x < L .$$

The problem (1)–(3) of heat flow in a thin wire has thus been reduced to the problem of determining an expansion for $f(x)$ of the form

(12) $$f(x) = \sum_{n=1}^{\infty} c_n \sin\left(\frac{n\pi x}{L}\right) .$$

Such an expansion is called a **Fourier sine series** and will be discussed in the next two sections. If we can choose the c_n's so that equation (12) holds, then the expansion for $u(x, t)$ in (11) is called a **formal solution** to the heat flow problem (1)–(3). If this expansion converges to a function with continuous second partial derivatives, then the formal solution is an actual (genuine) solution. Moreover, the solution is unique.

EXAMPLE 1 Find the solution to the heat flow problem

(13) $$\frac{\partial u}{\partial t} = 7\frac{\partial^2 u}{\partial x^2} , \qquad\qquad 0 < x < \pi , \qquad t > 0 ,$$

(14) $u(0, t) = u(\pi, t) = 0 , \qquad\qquad t > 0 ,$

(15) $u(x, 0) = 3 \sin 2x - 6 \sin 5x , \qquad 0 < x < \pi .$

SOLUTION Comparing equation (13) with (1), we see that $\beta = 7$ and $L = \pi$. Hence we need only determine the values of c_n in formula (12). That is, we must have

$$u(x, 0) = 3 \sin 2x - 6 \sin 5x = \sum_{n=1}^{\infty} c_n \sin nx .$$

For these data it is easy to solve for c_n. Equating the coefficients of like terms, we find that

$$c_2 = 3 \quad \text{and} \quad c_5 = -6 ,$$

and the remaining c_n's are zero. Hence, from (11), the solution to the heat flow problem (13)–(15) is

$$u(x, t) = c_2 e^{-\beta(2\pi/L)^2 t} \sin(2\pi x/L) + c_5 e^{-\beta(5\pi/L)^2 t} \sin(5\pi x/L)$$
$$= 3e^{-28t} \sin 2x - 6e^{-175t} \sin 5x . \quad \blacksquare \text{End}$$

Another situation in which the separation of variables approach applies occurs in the study of a vibrating string. This concerns the transverse vibrations of a string stretched between two points, such as a guitar string or piano wire. The goal is to find a function $u(x, t)$ that gives the displacement (deflection) of the string at any point x $(0 \le x \le L)$ and any time $t \ge 0$ (see Figure 10.2). In developing the mathematical model, we assume the string is perfectly flexible and

Figure 10.2 Displacement of string at time t

has constant linear density, the tension on the string is constant, gravity is negligible, and no other forces are acting on the string. Under these conditions and the additional assumption that the displacements $u(x, t)$ are small in comparison to the length of the string, it turns out that the motion of the string is governed by the following initial-boundary value problem.[†]

(16) $$\frac{\partial^2 u}{\partial t^2} = \alpha^2 \frac{\partial^2 u}{\partial x^2} \, , \qquad 0 < x < L \, , \qquad t > 0 \, ,$$

(17) $$u(0, t) = u(L, t) = 0 \, , \qquad t \geq 0 \, ,$$

(18) $$u(x, 0) = f(x) \, , \qquad 0 \leq x \leq L \, ,$$

(19) $$\frac{\partial u}{\partial t}(x, 0) = g(x) \, , \qquad 0 \leq x \leq L \, .$$

The constant α^2 appearing in (16) is strictly positive and equals the ratio of the tension to the (linear) density of the string. The physical significance of α, which has units of velocity, will be revealed in Section 10.6. The boundary conditions in (17) reflect the fact that the string is held fixed at the two endpoints $x = 0$ and $x = L$. Equations (18) and (19), respectively, specify the initial displacement of the string and the initial velocity of each point on the string. For the initial and boundary conditions to be consistent, we assume $f(0) = f(L) = 0$ and $g(0) = g(L) = 0$.

Let's apply the method of separation of variables to the initial-boundary value problem for the vibrating string (16)–(19). Thus we begin by assuming equation (16) has a solution of the form

$$u(x, t) = X(x)T(t) \, ,$$

where X is a function of x alone and T is a function of t alone. Differentiating u, we obtain

$$\frac{\partial^2 u}{\partial t^2} = X(x)T''(t) \, , \qquad \frac{\partial^2 u}{\partial x^2} = X''(x)T(t) \, .$$

Substituting these expressions into (16), we have

$$X(x)T''(t) = \alpha^2 X''(x)T(t) \, ,$$

and separating variables gives

$$\frac{T''(t)}{\alpha^2 T(t)} = \frac{X''(x)}{X(x)} \, .$$

Just as before, these ratios must equal some constant K:

(20) $$\frac{X''(x)}{X(x)} = K \quad \text{and} \quad \frac{T''(t)}{\alpha^2 T(t)} = K \, .$$

[†]For a derivation of this mathematical model, see *Partial Differential Equations: Sources and Solutions,* by Arthur D. Snider (Prentice Hall Inc., Upper Saddle River, NJ, 1999).

Furthermore, with $u(x, t) = X(x)T(t)$, the boundary conditions in (17) give

$$X(0)T(t) = 0 , \qquad X(L)T(t) = 0 , \qquad t \geq 0 .$$

In order for these equations to hold for all $t \geq 0$, either $T(t) \equiv 0$, which implies that $u(x, t) \equiv 0$, or

$$X(0) = X(L) = 0 .$$

Ignoring the trivial solution, we combine these boundary conditions with the differential equation for X in (20) and obtain the boundary value problem

(21) $X''(x) - KX(x) = 0 , \qquad X(0) = X(L) = 0 ,$

where K can be any constant.

This is the same boundary value problem that we encountered earlier while solving the heat equation. There we found that the eigenvalues are

$$K = -\left(\frac{n\pi}{L}\right)^2 , \qquad n = 1, 2, 3, \ldots ,$$

with corresponding eigenfunctions (nontrivial solutions)

(22) $X_n(x) = c_n \sin\left(\dfrac{n\pi x}{L}\right) ,$

where the c_n's are arbitrary nonzero constants.

Having determined that $K = -(n\pi/L)^2$ for some positive integer n, let's consider the second equation in (20) for such K:

$$T''(t) + \frac{\alpha^2 n^2 \pi^2}{L^2} T(t) = 0 .$$

For each $n = 1, 2, 3, \ldots$, a general solution is

$$T_n(t) = c_{n,1} \cos\frac{n\pi\alpha}{L}t + c_{n,2} \sin\frac{n\pi\alpha}{L}t .$$

Combining this with equation (22), we obtain, for each $n = 1, 2, 3, \ldots$, the function

$$u_n(x, t) = X_n(x)T_n(t) = \left(c_n \sin\frac{n\pi x}{L}\right)\left(c_{n,1} \cos\frac{n\pi\alpha}{L}t + c_{n,2} \sin\frac{n\pi\alpha}{L}t\right) ,$$

or, reassembling the constants,

(23) $u_n(x, t) = \left(a_n \cos\dfrac{n\pi\alpha}{L}t + b_n \sin\dfrac{n\pi\alpha}{L}t\right) \sin\dfrac{n\pi x}{L} .$

Using the fact that linear combinations of solutions to (16)–(17) are again solutions, we consider the infinite sum (superposition) of the functions in (23):

(24) $u(x, t) = \displaystyle\sum_{n=1}^{\infty} \left[a_n \cos\dfrac{n\pi\alpha}{L}t + b_n \sin\dfrac{n\pi\alpha}{L}t\right] \sin\dfrac{n\pi x}{L} .$

For a solution of the form (24), substitution into the initial conditions (18)–(19) gives

(25) $\qquad u(x, 0) = \sum_{n=1}^{\infty} a_n \sin\dfrac{n\pi x}{L} = f(x) , \qquad\qquad 0 \le x \le L ,$

(26) $\qquad \dfrac{\partial u}{\partial t}(x, 0) = \sum_{n=1}^{\infty} \dfrac{n\pi\alpha}{L} b_n \sin\dfrac{n\pi x}{L} = g(x) , \qquad 0 \le x \le L .$

We have now reduced the vibrating string problem (16)–(19) to the problem of determining the Fourier sine series expansions for $f(x)$ and $g(x)$:

(27) $\qquad f(x) = \sum_{n=1}^{\infty} a_n \sin\dfrac{n\pi x}{L} , \qquad g(x) = \sum_{n=1}^{\infty} B_n \sin\dfrac{n\pi x}{L} ,$

where $B_n = (n\pi\alpha/L)b_n$. If we choose the a_n's and b_n's so that the equations in (25) and (26) hold, then the expansion for $u(x, t)$ in (24) is a **formal solution** to the vibrating string problem (16)–(19). If this expansion converges to a function with continuous second partial derivatives, then the formal solution is an actual (genuine) solution.

EXAMPLE 2 Find the solution to the vibrating string problem

(28) $\qquad \dfrac{\partial^2 u}{\partial t^2} = 4\dfrac{\partial^2 u}{\partial x^2} , \qquad\qquad 0 < x < \pi , \qquad t > 0 ,$

(29) $\qquad u(0, t) = u(\pi, t) = 0 , \qquad\qquad t \ge 0 ,$

(30) $\qquad u(x, 0) = \sin 3x - 4 \sin 10x , \qquad 0 \le x \le \pi ,$

(31) $\qquad \dfrac{\partial u}{\partial t}(x, 0) = 2 \sin 4x + \sin 6x , \qquad 0 \le x \le \pi .$

SOLUTION Comparing equation (28) with equation (16), we see that $\alpha = 2$ and $L = \pi$. Hence we need only determine the values of the coefficients a_n and b_n in formula (24). The a_n's are chosen so that equation (25) holds; that is,

$$u(x, 0) = \sin 3x - 4 \sin 10x = \sum_{n=1}^{\infty} a_n \sin nx .$$

Equating coefficients of like terms, we see that

$$a_3 = 1 , \qquad a_{10} = -4 ,$$

and the remaining a_n's are zero. Similarly, referring to equation (26), we must choose the b_n's so that

$$\dfrac{\partial u}{\partial t}(x, 0) = 2 \sin 4x + \sin 6x = \sum_{n=1}^{\infty} n2b_n \sin nx .$$

Comparing coefficients, we find

$$2 = (4)(2)b_4 \quad\text{or}\quad b_4 = \dfrac{1}{4} ,$$

$$1 = (6)(2)b_6 \quad\text{or}\quad b_6 = \dfrac{1}{12} ,$$

and the remaining b_n's are zero. Hence, from formula (24), the solution to the vibrating string problem (28)–(31) is

$$(32) \qquad u(x, t) = \cos 6t \sin 3x + \frac{1}{4}\sin 8t \sin 4x$$

$$+ \frac{1}{12}\sin 12t \sin 6x - 4 \cos 20t \sin 10x . \quad \blacksquare\text{End}$$

In later sections the method of separation of variables is used to study a wide variety of problems for the heat, wave, and Laplace's equations. However, to use the method effectively, one must be able to compute trigonometric series (or, more generally, eigenfunction expansions) such as the Fourier sine series that we encountered here. These expansions are discussed in the next two sections.

EXERCISES 10.2

In Problems 1 through 8, determine all the solutions, if any, to the given boundary value problem by first finding a general solution to the differential equation.

1. $y'' - y = 0$; $\quad 0 < x < 1$,
$\quad y(0) = 0$, $\quad y(1) = -4$.

2. $y'' - 6y' + 5y = 0$; $\quad 0 < x < 2$,
$\quad y(0) = 1$, $\quad y(2) = 1$.

3. $y'' + 4y = 0$; $\quad 0 < x < \pi$,
$\quad y(0) = 0$, $\quad y'(\pi) = 0$.

4. $y'' + 9y = 0$; $\quad 0 < x < \pi$,
$\quad y(0) = 0$, $\quad y'(\pi) = -6$.

5. $y'' - y = 1 - 2x$; $\quad 0 < x < 1$,
$\quad y(0) = 0$, $\quad y(1) = 1 + e$.

6. $y'' + y = 0$; $\quad 0 < x < 2\pi$,
$\quad y(0) = 0$, $\quad y(2\pi) = 1$.

7. $y'' + y = 0$; $\quad 0 < x < 2\pi$,
$\quad y(0) = 1$, $\quad y(2\pi) = 1$.

8. $y'' - 2y' + y = 0$; $\quad -1 < x < 1$,
$\quad y(-1) = 0$, $\quad y(1) = 2$.

In Problems 9 through 14, find the values of λ (eigenvalues) for which the given problem has a nontrivial solution. Also determine the corresponding nontrivial solutions (eigenfunctions).

9. $y'' + \lambda y = 0$; $\quad 0 < x < \pi$,
$\quad y(0) = 0$, $\quad y'(\pi) = 0$.

10. $y'' + \lambda y = 0$; $\quad 0 < x < \pi$,
$\quad y'(0) = 0$, $\quad y(\pi) = 0$.

11. $y'' + \lambda y = 0$; $\quad 0 < x < 2\pi$,
$\quad y(0) = y(2\pi)$, $\quad y'(0) = y'(2\pi)$.

12. $y'' + \lambda y = 0$; $\quad 0 < x < \pi/2$,
$\quad y'(0) = 0$, $\quad y'(\pi/2) = 0$.

13. $y'' + \lambda y = 0$; $\quad 0 < x < \pi$,
$\quad y(0) - y'(0) = 0$, $\quad y(\pi) = 0$.

14. $y'' - 2y' + \lambda y = 0$; $\quad 0 < x < \pi$,
$\quad y(0) = 0$, $\quad y(\pi) = 0$.

In Problems 15 through 18, solve the heat flow problem (1)–(3) with $\beta = 3, L = \pi$, and the given function $f(x)$.

15. $f(x) = \sin x - 6 \sin 4x$.

16. $f(x) = \sin 3x + 5 \sin 7x - 2 \sin 13x$.

17. $f(x) = \sin x - 7 \sin 3x + \sin 5x$.

18. $f(x) = \sin 4x + 3 \sin 6x - \sin 10x$.

In Problems 19 through 22, solve the vibrating string problem (16)–(19) with $\alpha = 3, L = \pi$, and the given initial functions $f(x)$ and $g(x)$.

19. $f(x) = 3 \sin 2x + 12 \sin 13x$, $\quad g(x) \equiv 0$.

20. $f(x) \equiv 0$,
$\quad g(x) = -2 \sin 3x + 9 \sin 7x - \sin 10x$.

21. $f(x) = 6 \sin 2x + 2 \sin 6x$,
$\quad g(x) = 11 \sin 9x - 14 \sin 15x$.

22. $f(x) = \sin x - \sin 2x + \sin 3x$,
$g(x) = 6 \sin 3x - 7 \sin 5x$.

23. Find the formal solution to the heat flow problem
(1)–(3) with $\beta = 2$ and $L = 1$ if

$$f(x) = \sum_{n=1}^{\infty} \frac{1}{n^2} \sin n\pi x .$$

24. Find the formal solution to the vibrating string problem (16)–(19) with $\alpha = 4$, $L = \pi$, and

$$f(x) = \sum_{n=1}^{\infty} \frac{1}{n^2} \sin nx ,$$

$$g(x) = \sum_{n=1}^{\infty} \frac{(-1)^{n+1}}{n} \sin nx .$$

25. By considering the behavior of the solutions of the equation

$$T'(t) - \beta K T(t) = 0 , \qquad t > 0 ,$$

give an argument that is based on physical grounds to rule out the case where $K > 0$ in equation (5).

26. Verify that $u_n(x, t)$ given in equation (10) satisfies equation (1) and the boundary conditions in (2) by substituting $u_n(x, t)$ directly into the equations involved.

In Problems 27 through 30, a partial differential equation (PDE) is given along with the form of a solution having separated variables. Show that such a solution must satisfy the indicated set of ordinary differential equations.

27. $\dfrac{\partial^2 u}{\partial r^2} + \dfrac{1}{r} \dfrac{\partial u}{\partial r} + \dfrac{1}{r^2} \dfrac{\partial^2 u}{\partial \theta^2} = 0$

with $u(r, \theta) = R(r)T(\theta)$ yields
$$r^2 R''(r) + rR'(r) - \lambda R(r) = 0 ,$$
$$T''(\theta) + \lambda T(\theta) = 0 ,$$
where λ is a constant.

28. $\dfrac{\partial^2 u}{\partial t^2} + \dfrac{\partial u}{\partial t} + u = \alpha^2 \dfrac{\partial^2 u}{\partial x^2}$

with $u(x, t) = X(x)T(t)$ yields
$$X''(x) - \lambda X(x) = 0 ,$$
$$T''(t) + T'(t) + (1 - \lambda\alpha^2)T(t) = 0 ,$$
where λ is a constant.

29. $\dfrac{\partial u}{\partial t} = \beta \left\{ \dfrac{\partial^2 u}{\partial x^2} + \dfrac{\partial^2 u}{\partial y^2} \right\}$

with $u(x, y, t) = X(x)Y(y)T(t)$ yields
$$T'(t) - \beta K T(t) = 0 ,$$
$$X''(x) - JX(x) = 0 ,$$
$$Y''(y) + (J - K)Y(y) = 0 ,$$
where J, K are constants.

30. $\dfrac{\partial^2 u}{\partial r^2} + \dfrac{1}{r} \dfrac{\partial u}{\partial r} + \dfrac{1}{r^2} \dfrac{\partial^2 u}{\partial \theta^2} + \dfrac{\partial^2 u}{\partial z^2} = 0$

with $u(r, \theta, z) = R(r)T(\theta)Z(z)$ yields
$$T''(\theta) + \mu T(\theta) = 0 ,$$
$$Z''(z) + \lambda Z(z) = 0 ,$$
$$r^2 R''(r) + rR'(r) - (r^2 \lambda + \mu)R(r) = 0 ,$$
where μ, λ are constants.

31. For the PDE in Problem 27, assume the following boundary conditions are imposed:

$$u(r, 0) = u(r, \pi) = 0 ,$$
$$u(r, \theta) \text{ remains bounded as } r \to 0^+ .$$

Show that a nontrivial solution of the form $u(r, \theta) = R(r)T(\theta)$ must satisfy the boundary conditions
$$T(0) = T(\pi) = 0 ,$$
$$R(r) \text{ remains bounded as } r \to 0^+ .$$

32. For the PDE in Problem 29, assume the following boundary conditions are imposed:

$$u(0, y, t) = u(a, y, t) = 0 , \quad 0 \le y \le b , \quad t \ge 0 ,$$
$$\dfrac{\partial u}{\partial y}(x, 0, t) = \dfrac{\partial u}{\partial y}(x, b, t) = 0 , \quad 0 \le x \le a , \quad t \ge 0 .$$

Show that a nontrivial solution of the form $u(x, y, t) = X(x)Y(y)T(t)$ must satisfy the boundary conditions
$$X(0) = X(a) = 0 ,$$
$$Y'(0) = Y'(b) = 0 .$$

33. When the temperature in a wire reaches a steady state, that is, when u depends only on x, then $u(x)$ satisfies Laplace's equation $\partial^2 u / \partial x^2 = 0$.

(a) Find the steady-state solution when the ends of the wire are kept at a constant temperature of $50°C$, that is, when $u(0) = u(L) = 50$.

(b) Find the steady-state solution when one end of the wire is kept at $10°C$, while the other is kept at $40°C$, that is, when $u(0) = 10$ and $u(L) = 40$.

10.3 FOURIER SERIES

While analyzing heat flow and vibrating strings in the previous section, we encountered the problem of expressing a function in a trigonometric series (cf. equations (12) and (25) in Section 10.2). In the next two sections, we discuss the theory of Fourier series, which deals with trigonometric series expansions. First, however, we review some function properties that are particularly relevant to this study: piecewise continuity, periodicity, and even and odd symmetry.

In Section 7.2 we defined a **piecewise continuous** function on $[a, b]$ as a function f that is continuous at every point in $[a, b]$, except possibly for a finite number of points at which f has a *jump discontinuity.* Such functions are necessarily integrable over any finite interval on which they are piecewise continuous.

Recall also that a function is **periodic of period** T if $f(x + T) = f(x)$ for all x in the domain of f. The smallest positive value of T is called the **fundamental period.** The trigonometric functions $\sin x$ and $\cos x$ are examples of periodic functions with fundamental period 2π and $\tan x$ is periodic with fundamental period π. A constant function is a periodic function with arbitrary period T.

There are two symmetry properties of functions that will be useful in the study of Fourier series. A function f that satisfies $f(-x) = f(x)$ for all x in the domain of f has a graph that is symmetric with respect to the y-axis (see Figure 10.3(a)). We say that such a function is **even.** A function f that satisfies $f(-x) = -f(x)$ for all x in the domain of f has a graph that is symmetric with respect to the origin (see Figure 10.3(b)). It is said to be an **odd** function. The functions $1, x^2, x^4, \ldots$ are examples of even functions, while the functions x, x^3, x^5, \ldots are odd. The trigonometric functions $\sin x$ and $\tan x$ are odd functions and $\cos x$ is an even function.

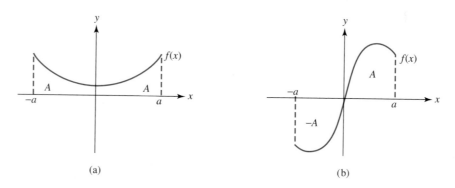

(a) (b)

Figure 10.3 (a) Even function $\int_{-a}^{a} f = A + A = 2\int_{0}^{a} f$ (b) Odd function $\int_{-a}^{a} f = A - A = 0$

EXAMPLE 1 Determine whether the given function is even, odd, or neither.

(a) $f(x) = \sqrt{1 + x^2}$. (b) $g(x) = x^{1/3} - \sin x$. (c) $h(x) = e^x$.

SOLUTION (a) Since $f(-x) = \sqrt{1 + (-x)^2} = \sqrt{1 + x^2} = f(x)$, then $f(x)$ is an even function.

(b) Because $g(-x) = (-x)^{1/3} - \sin(-x) = -x^{1/3} + \sin x = -(x^{1/3} - \sin x) = -g(x)$, it follows that $g(x)$ is an odd function.

(c) Here $h(-x) = e^{-x}$. Since $e^{-x} = e^x$ only when $x = 0$ and e^{-x} is never equal to $-e^x$, then $h(x)$ is neither an even nor an odd function. ■ End

Knowing that a function is even or odd can be useful in evaluating definite integrals.

PROPERTIES OF SYMMETRIC FUNCTIONS

Theorem 1. If f is an even piecewise continuous function on $[-a, a]$, then

(1) $\int_{-a}^{a} f(x)\,dx = 2\int_{0}^{a} f(x)\,dx$.

If f is an odd piecewise continuous function on $[-a, a]$, then

(2) $\int_{-a}^{a} f(x)\,dx = 0$.

Proof. If f is an even function, then $f(-x) = f(x)$. Hence

$$\int_{-a}^{a} f(x)\,dx = \int_{-a}^{0} f(x)\,dx + \int_{0}^{a} f(x)\,dx$$

$$= -\int_{a}^{0} f(u)\,du + \int_{0}^{a} f(x)\,dx = 2\int_{0}^{a} f(x)\,dx ,$$

where we used the change of variables $u = -x$. This is illustrated in Figure 10.3(a). Formula (2) can be proved by a similar argument and is illustrated in Figure 10.3(b). ■ End

The next example deals with certain integrals that are crucial in Fourier series.

EXAMPLE 2 Evaluate the following integrals when m and n are positive integers:

(a) $\int_{-T}^{T} \sin\frac{m\pi x}{T}\cos\frac{n\pi x}{T}\,dx$. (b) $\int_{-T}^{T} \sin\frac{m\pi x}{T}\sin\frac{n\pi x}{T}\,dx$.

(c) $\int_{-T}^{T} \cos\frac{m\pi x}{T}\cos\frac{n\pi x}{T}\,dx$.

SOLUTION **(a)** Since $\sin\dfrac{m\pi(-x)}{T}\cos\dfrac{n\pi(-x)}{T} = -\sin\dfrac{m\pi x}{T}\cos\dfrac{n\pi x}{T}$, the integrand in part (a) is

an odd function. Thus, by Theorem 1,

(3) $$\int_{-T}^{T} \sin\frac{m\pi x}{T}\cos\frac{n\pi x}{T}\,dx = 0 \ .$$

(b) Using the fact that the integrand is even (check this!) and the trigonometric identity

$$2\sin A \sin B = \cos(A - B) - \cos(A + B) \ ,$$

we find for $m \neq n$

$$\int_{-T}^{T} \sin\frac{m\pi x}{T}\sin\frac{n\pi x}{T}\,dx = 2\int_{0}^{T} \sin\frac{m\pi x}{T}\sin\frac{n\pi x}{T}\,dx$$

$$= \int_{0}^{T}\left\{\cos\frac{(m-n)\pi x}{T} - \cos\frac{(m+n)\pi x}{T}\right\}dx$$

$$= \frac{T}{\pi}\left[\frac{\sin\dfrac{(m-n)\pi x}{T}}{m-n} - \frac{\sin\dfrac{(m+n)\pi x}{T}}{m+n}\right]\Bigg|_{0}^{T} = 0 \ .$$

When $m = n$, the integral in (b) becomes

$$\int_{-T}^{T}\left(\sin\frac{n\pi x}{T}\right)^2 dx = 2\int_{0}^{T}\left(\sin\frac{n\pi x}{T}\right)^2 dx = \int_{0}^{T}\left\{1 - \cos\frac{2n\pi x}{T}\right\}dx$$

$$= \left[x - \frac{T}{2n\pi}\sin\frac{2n\pi x}{T}\right]\Bigg|_{0}^{T} = T \ .$$

Hence

(4) $$\int_{-T}^{T} \sin\frac{m\pi x}{T}\sin\frac{n\pi x}{T}\,dx = \begin{cases} 0 \ , & m \neq n \ , \\ T \ , & m = n \ . \end{cases}$$

The integrand in part (c) is also even and can be treated with a similar trigonometric identity. We leave it as an exercise for the reader to verify that (see Problem 8)

(5) $$\int_{-T}^{T} \cos\frac{m\pi x}{T}\cos\frac{n\pi x}{T}\,dx = \begin{cases} 0 \ , & m \neq n \ , \\ T \ , & m = n \ . \end{cases}$$ ■ End

Equations (3)–(5) express an **orthogonality condition**[†] satisfied by the set of trigonometric functions $\{\cos x, \sin x, \cos 2x, \sin 2x, \ldots\}$, where $T = \pi$. We will say more about this later in this section.

It is easy to verify that if each of the functions f_1, \ldots, f_n is periodic of period T, then so is any linear combination

$$c_1 f_1(x) + \cdots + c_n f_n(x) \ .$$

For example, the sum $7 + 3 \cos \pi x - 8 \sin \pi x + 4 \cos 2\pi x - 6 \sin 2\pi x$ has period 2, since each term has period 2. Furthermore, if the infinite series

$$\frac{a_0}{2} + \sum_{n=1}^{\infty} \left(a_n \cos \frac{n\pi x}{T} + b_n \sin \frac{n\pi x}{T} \right)$$

consisting of $2T$-periodic functions converges for all x, then the function to which it converges will be periodic of period $2T$.

Just as we can associate a Taylor series with a function that has derivatives of all orders at a fixed point, we can identify a particular trigonometric series with a piecewise continuous function. To illustrate this, let's assume that $f(x)$ has the series expansion[††]

$$(6) \qquad f(x) = \frac{a_0}{2} + \sum_{n=1}^{\infty} \left\{ a_n \cos \frac{n\pi x}{T} + b_n \sin \frac{n\pi x}{T} \right\} \ ,$$

where the a_n's and b_n's are constants. (Necessarily, f has period $2T$.)

To determine the coefficients $a_0, a_1, b_1, a_2, b_2, \ldots$, we proceed as follows. Let's integrate $f(x)$ from $-T$ to T, assuming we can integrate term by term:

$$\int_{-T}^{T} f(x)\,dx = \int_{-T}^{T} \frac{a_0}{2}\,dx + \sum_{n=1}^{\infty} a_n \int_{-T}^{T} \cos \frac{n\pi x}{T}\,dx + \sum_{n=1}^{\infty} b_n \int_{-T}^{T} \sin \frac{n\pi x}{T}\,dx \ .$$

Since $\sin(n\pi x/T)$ is an odd function, $\int_{-T}^{T} \sin(n\pi x/T)\,dx = 0$. Further, for $n = 1, 2, \ldots$, it is easy to check that $\int_{-T}^{T} \cos(n\pi x/T)\,dx = 0$. Hence

$$\int_{-T}^{T} f(x)\,dx = \int_{-T}^{T} \frac{a_0}{2}\,dx = a_0 T \ ,$$

and so

$$a_0 = \frac{1}{T} \int_{-T}^{T} f(x)\,dx \ .$$

[†]This nomenclature is suggested by the fact that the formulas for the Riemann sums approximating the integrals (3)–(5) look like dot products (Section 9.1, page 528) of higher-dimensional vectors. In most calculus texts, it is shown that the dot product of orthogonal vectors in two dimensions is zero.

[††]The choice of a constant $a_0/2$ instead of just a_0 will make the formulas easier to remember.

(Notice that $a_0/2$ is the average value of f over one period $2T$.) Next, to find the coefficient a_m when $m \geq 1$, we multiply (6) by $\cos(m\pi x/T)$ and integrate:

$$(7) \quad \int_{-T}^{T} f(x) \cos\frac{m\pi x}{T} dx = \frac{a_0}{2} \int_{-T}^{T} \cos\frac{m\pi x}{T} dx + \sum_{n=1}^{\infty} a_n \int_{-T}^{T} \cos\frac{n\pi x}{T} \cos\frac{m\pi x}{T} dx$$

$$+ \sum_{n=1}^{\infty} b_n \int_{-T}^{T} \sin\frac{n\pi x}{T} \cos\frac{m\pi x}{T} dx \ .$$

The orthogonality conditions (3)–(5) render the integrals on the right-hand side quite immediately. We have already observed that

$$\int_{-T}^{T} \cos\frac{m\pi x}{T} dx = 0 \ , \qquad m \geq 1 \ ,$$

and, by formulas (3) and (5), we have

$$\int_{-T}^{T} \sin\frac{n\pi x}{T} \cos\frac{m\pi x}{T} dx = 0 \ ,$$

$$\int_{-T}^{T} \cos\frac{n\pi x}{T} \cos\frac{m\pi x}{T} dx = \begin{cases} 0 \ , & n \neq m \ , \\ T \ , & n = m \ . \end{cases}$$

Hence, in (7) we see that only one term on the right-hand side survives the integration:

$$\int_{-T}^{T} f(x) \cos\frac{m\pi x}{T} dx = a_m T \ .$$

Thus we have a formula for the coefficient a_m:

$$a_m = \frac{1}{T} \int_{-T}^{T} f(x) \cos\frac{m\pi x}{T} dx \ .$$

Similarly, multiplying (6) by $\sin(m\pi x/T)$ and integrating yields

$$\int_{-T}^{T} f(x) \sin\frac{m\pi x}{T} dx = b_m T$$

so that the formula for b_m is

$$b_m = \frac{1}{T} \int_{-T}^{T} f(x) \sin\frac{m\pi x}{T} dx \ .$$

Motivated by the above computations, we now make the following definition.

FOURIER SERIES

Definition 1. Let f be a piecewise continuous function on the interval $[-T, T]$. The **Fourier series**[†] of f is the trigonometric series

(8) $$f(x) \sim \frac{a_0}{2} + \sum_{n=1}^{\infty} \left\{ a_n \cos\frac{n\pi x}{T} + b_n \sin\frac{n\pi x}{T} \right\} ,$$

where the a_n's and b_n's are given by the formulas[††]

(9) $$a_n = \frac{1}{T} \int_{-T}^{T} f(x) \cos\frac{n\pi x}{T} dx , \qquad n = 0, 1, 2, \dots ,$$

(10) $$b_n = \frac{1}{T} \int_{-T}^{T} f(x) \sin\frac{n\pi x}{T} dx , \qquad n = 1, 2, 3, \dots .$$

Formulas (9) and (10) are called the **Euler formulas.** We use the symbol \sim in (8) to remind us that this series is associated with $f(x)$ but may not converge to $f(x)$. We will return to the question of convergence later in this section. First let's consider a few examples of Fourier series.

EXAMPLE 3 Compute the Fourier series for

$$f(x) = \begin{cases} 0 , & -\pi < x < 0 , \\ x , & 0 < x < \pi . \end{cases}$$

SOLUTION Here $T = \pi$. Using formulas (9) and (10), we have

$$a_n = \frac{1}{\pi} \int_{-\pi}^{\pi} f(x) \cos nx\, dx = \frac{1}{\pi} \int_{0}^{\pi} x \cos nx\, dx$$

$$= \frac{1}{\pi n^2} \int_{0}^{\pi n} u \cos u\, du = \frac{1}{\pi n^2} \left[\cos u + u \sin u \right] \Big|_{0}^{\pi n}$$

$$= \frac{1}{\pi n^2} (\cos n\pi - 1) = \frac{1}{\pi n^2} \left[(-1)^n - 1 \right] , \qquad n = 1, 2, 3, \dots ,$$

$$a_0 = \frac{1}{\pi} \int_{-\pi}^{\pi} f(x)\, dx = \frac{1}{\pi} \int_{0}^{\pi} x\, dx = \frac{x^2}{2\pi} \Big|_{0}^{\pi} = \frac{\pi}{2} ,$$

[†]*Historical Footnote:* Joseph B. J. Fourier (1768–1830) developed his series for solving heat flow problems. Lagrange expressed doubts about the validity of the representation, but Dirichlet devised conditions that assured its convergence.
[††]Notice that $f(x)$ need not be defined for every x in $[-T, T]$; we need only that the integrals in (9) and (10) exist.

$$b_n = \frac{1}{\pi} \int_{-\pi}^{\pi} f(x) \sin nx \, dx = \frac{1}{\pi} \int_0^{\pi} x \sin nx \, dx$$

$$= \frac{1}{\pi n^2} \int_0^{\pi n} u \sin u \, du = \frac{1}{\pi n^2} \left[\sin u - u \cos u \right] \Big|_0^{\pi n}$$

$$= \frac{-\cos n\pi}{n} = \frac{(-1)^{n+1}}{n}, \qquad n = 1, 2, 3, \ldots .$$

Therefore,

$$(11) \qquad f(x) \sim \frac{\pi}{4} + \sum_{n=1}^{\infty} \left\{ \frac{1}{\pi n^2} \left[(-1)^n - 1 \right] \cos nx + \frac{(-1)^{n+1}}{n} \sin nx \right\}$$

$$= \frac{\pi}{4} - \frac{2}{\pi} \left\{ \cos x + \frac{1}{9} \cos 3x + \frac{1}{25} \cos 5x + \cdots \right\}$$

$$+ \left\{ \sin x - \frac{1}{2} \sin 2x + \frac{1}{3} \sin 3x + \cdots \right\} .$$

Some partial sums of this series are displayed in Figure 10.4. ∎ End

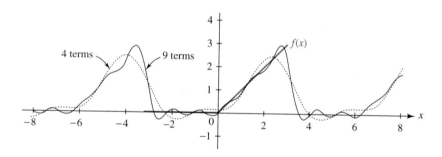

Figure 10.4 Partial sums of Fourier series in Example 3

EXAMPLE 4 Compute the Fourier series for

$$f(x) = \begin{cases} -1 , & -\pi < x < 0 , \\ 1 , & 0 < x < \pi . \end{cases}$$

SOLUTION Again, $T = \pi$. Notice that f is an odd function. Since the product of an odd function and an even function is odd (see Problem 7), $f(x) \cos nx$ is also an odd function. Thus

$$a_n = \frac{1}{\pi} \int_{-\pi}^{\pi} f(x) \cos nx \, dx = 0 , \qquad n = 0, 1, 2, \ldots .$$

Furthermore, $f(x) \sin nx$ is the product of two odd functions and therefore is an even function, so

$$b_n = \frac{1}{\pi} \int_{-\pi}^{\pi} f(x) \sin nx\, dx = \frac{2}{\pi} \int_0^{\pi} \sin nx\, dx$$

$$= \frac{2}{\pi} \left[\frac{-\cos nx}{n} \right] \Bigg|_0^{\pi} = \frac{2}{\pi} \left[\frac{1}{n} - \frac{(-1)^n}{n} \right], \qquad n = 1, 2, 3, \dots ,$$

$$= \begin{cases} 0 , & n \text{ even} , \\ \dfrac{4}{\pi n} , & n \text{ odd} . \end{cases}$$

Thus

$$(12) \qquad f(x) \sim \frac{2}{\pi} \sum_{n=1}^{\infty} \frac{[1 - (-1)^n]}{n} \sin nx = \frac{4}{\pi} \left[\sin x + \frac{1}{3} \sin 3x + \frac{1}{5} \sin 5x + \cdots \right].$$

Some partial sums of (12) are sketched in Figure 10.5. ■ End

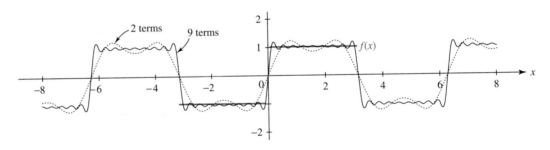

Figure 10.5 Partial sums of Fourier series in Example 4

In Example 4, the odd function f has a Fourier series consisting only of sine functions. It is easy to see that, in general, if f is any odd function, then its Fourier series consists only of sine terms.

EXAMPLE 5 Compute the Fourier series for $f(x) = |x|$, $-1 < x < 1$.

SOLUTION Here $T = 1$. Since f is an even function, $f(x) \sin n\pi x$ is an odd function. Therefore

$$b_n = \int_{-1}^{1} f(x) \sin n\pi x\, dx = 0 , \qquad n = 1, 2, 3, \dots .$$

Since $f(x) \cos n\pi x$ is an even function, we have

$$a_0 = \int_{-1}^{1} f(x)\, dx = 2 \int_{0}^{1} x\, dx = x^2 \Big|_{0}^{1} = 1 \ ,$$

$$a_n = \int_{-1}^{1} f(x) \cos n\pi x\, dx = 2 \int_{0}^{1} x \cos n\pi x\, dx$$

$$= \frac{2}{\pi^2 n^2} \int_{0}^{\pi n} u \cos u\, du = \frac{2}{\pi^2 n^2} \left[\cos u + u \sin u \right] \Big|_{0}^{\pi n} = \frac{2}{\pi^2 n^2} (\cos n\pi - 1)$$

$$= \frac{2}{\pi^2 n^2} \left[(-1)^n - 1 \right] \ , \qquad n = 1, 2, 3, \dots .$$

Therefore

(13) $\qquad f(x) \sim \dfrac{1}{2} + \displaystyle\sum_{n=1}^{\infty} \dfrac{2}{\pi^2 n^2} \left[(-1)^n - 1 \right] \cos n\pi x$

$$= \frac{1}{2} - \frac{4}{\pi^2} \left\{ \cos \pi x + \frac{1}{9} \cos 3\pi x + \frac{1}{25} \cos 5\pi x + \cdots \right\} .$$

Partial sums for (13) are displayed in Figure 10.6. ∎End

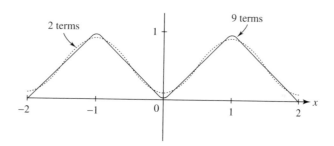

Figure 10.6 Partial sums of Fourier series in Example 5

Notice that the even function f of Example 5 has a Fourier series consisting only of cosine functions and the constant function $1 = \cos 0\pi x$. In general, if f is an even function, then its Fourier series consists only of cosine functions (including $\cos 0\pi x$).

Orthogonal Expansions

Fourier series are examples of orthogonal expansions.[†] A set of functions $\{f_n(x)\}_{n=1}^{\infty}$ is said to be an **orthogonal system** or just **orthogonal** with respect to the nonnegative weight function

[†]Orthogonality is also discussed in Section 8.8 on pages 515–516.

$w(x)$ on the interval $[a, b]$ if

(14) $\displaystyle\int_a^b f_m(x)f_n(x)w(x)\,dx = 0$, whenever $m \neq n$.

As we have seen, the set of trigonometric functions

(15) $\{1, \cos x, \sin x, \cos 2x, \sin 2x, \ldots\}$

is orthogonal on $[-\pi, \pi]$ with respect to the weight function $w(x) \equiv 1$. If we define the **norm** of f as

(16) $\displaystyle \|f\| := \left[\int_a^b f^2(x)w(x)\,dx\right]^{1/2}$,

then we say that a set of functions $\{f_n(x)\}_{n=1}^{\infty}$ $\left(\text{or } \{f_n(x)\}_{n=1}^{N}\right)$ is an **orthonormal system with respect to** $w(x)$ if (14) holds and also $\|f_n\| = 1$ for each n. Equivalently, we say the set is an orthonormal system if

(17) $\displaystyle\int_a^b f_m(x)f_n(x)w(x)\,dx = \begin{cases} 0 , & m \neq n , \\ 1 , & m = n . \end{cases}$

We can always obtain an orthonormal system from an orthogonal system just by dividing each function by its norm. In particular, since

$$\int_{-\pi}^{\pi} \cos^2 nx\,dx = \int_{-\pi}^{\pi} \sin^2 nx\,dx = \pi , \qquad n = 1, 2, 3, \ldots$$

and

$$\int_{-\pi}^{\pi} 1\,dx = 2\pi ,$$

then the orthogonal system (15) gives rise on $[-\pi, \pi]$ to the orthonormal system

$$\{(2\pi)^{-1/2},\ \pi^{-1/2}\cos x,\ \pi^{-1/2}\sin x,\ \pi^{-1/2}\cos 2x,\ \pi^{-1/2}\sin 2x,\ \ldots\} .$$

If $\{f_n(x)\}_{n=1}^{\infty}$ is an orthogonal system with respect to $w(x)$ on $[a, b]$, we might ask if we can expand a function $f(x)$ in terms of these functions; that is, can we express f in the form

(18) $f(x) = c_1 f_1(x) + c_2 f_2(x) + c_3 f_3(x) + \cdots$

for a suitable choice of constants c_1, c_2, \ldots ? Such an expansion is called an **orthogonal expansion**, or a **generalized Fourier series.**

To determine the constants in (18), we can proceed as we did in deriving Euler's formulas for the coefficients of a Fourier series, this time using the orthogonality of the system. Presum-

ing the representation (18) is valid, we multiply by $f_m(x)w(x)$ and integrate to obtain

$$(19) \qquad \int_a^b f(x)f_m(x)w(x)\,dx = c_1 \int_a^b f_1(x)f_m(x)w(x)\,dx + c_2 \int_a^b f_2(x)f_m(x)w(x)\,dx + \cdots$$

$$= \sum_{n=1}^{\infty} c_n \int_a^b f_n(x)f_m(x)w(x)\,dx \ .$$

(Here we have also assumed that we can integrate term by term.) Because the system is orthogonal with respect to $w(x)$, every integral on the right-hand side of (19) is zero except when $n = m$. Solving for c_m gives

$$(20) \qquad c_m = \frac{\displaystyle\int_a^b f(x)f_m(x)w(x)\,dx}{\displaystyle\int_a^b f_m^2(x)w(x)\,dx} = \frac{\displaystyle\int_c^b f(x)f_m(x)w(x)\,dx}{\|f_m\|^2} \ , \qquad n = 1, 2, 3, \dots \ .$$

The derivation of the formula for c_m was only *formal,* since the question of the convergence of the expansion in (18) was not answered. If the series $\sum_{n=1}^{\infty} c_n f_n(x)$ converges *uniformly* to $f(x)$ on $[a, b]$, then each step can be justified, and indeed, the coefficients are given by formula (20). The notion of uniform convergence is discussed in the next subsection and in Section 13.2.[†]

Convergence of Fourier Series

Let's turn to the question of the convergence of a Fourier series. Figures 10.4–10.6 provide some clues. In Example 5 it is possible to use a comparison or limit comparison test to show that the series is absolutely dominated by a p-series of the form $\sum_{n=1}^{\infty} 1/n^2$, which converges. However, this is much harder to do in Example 4, since the terms go to zero like $1/n$. Matters can be even worse, since there exist Fourier series that *diverge.*[††] We state two theorems that deal with the convergence of a Fourier series and two dealing with the properties of termwise differentiation and integration. For proofs of these results, see *Partial Differential Equations of Mathematical Physics,* by Tyn Myint-U (Elsevier North Holland, Inc., New York, 1980), Chapter 5; *Advanced Calculus with Applications,* by N. J. DeLillo (Macmillan, New York, 1982), Chapter 9; or an advanced text on the theory of Fourier series.

Before proceeding, we need the following notation. Let

$$f(x^+) := \lim_{h \to 0^+} f(x + h) \quad \text{and} \quad f(x^-) := \lim_{h \to 0^+} f(x - h) \ .$$

[†]All references to Chapters 11–13 refer to the expanded text *Fundamentals of Differential Equations and Boundary Value Problems,* 3rd edition.

[††]In fact, there are trigonometric series that converge but are *not* Fourier series; an example is $\displaystyle\sum_{n=1}^{\infty} \frac{\sin nx}{\ln(n+1)}$.

POINTWISE CONVERGENCE OF FOURIER SERIES

Theorem 2. If f and f' are piecewise continuous on $[-T, T]$, then for any x in $(-T, T)$

(21) $\dfrac{a_0}{2} + \sum\limits_{n=1}^{\infty} \left\{ a_n \cos\dfrac{n\pi x}{T} + b_n \sin\dfrac{n\pi x}{T} \right\} = \dfrac{1}{2}[f(x^+) + f(x^-)]$,

where the a_n's and b_n's are given by the Euler formulas (9) and (10). For $x = \pm T$, the series converges to $\dfrac{1}{2}[f(-T^+) + f(T^-)]$.[†]

In other words, when f and f' are piecewise continuous on $[-T, T]$, the Fourier series converges to $f(x)$ whenever f is continuous at x and converges to the average of the left- and right-hand limits at points where f is discontinuous.

Observe that the left-hand side of (21) is periodic of period $2T$. This means that if we extend $f(x)$ from the interval $(-T, T)$ to the entire real line using $2T$-periodicity, then equation (21) holds for all x for the $2T$-periodic extension of $f(x)$.

EXAMPLE 6 To which function does the Fourier series for

$$f(x) = \begin{cases} -1 , & -\pi < x < 0 , \\ 1 , & 0 < x < \pi , \end{cases}$$

converge?

SOLUTION In Example 4 we found that the Fourier series for $f(x)$ is given by (12), and in Figure 10.5, we sketched the graphs of two of its partial sums. Now $f(x)$ and $f'(x)$ are piecewise continuous in $[-\pi, \pi]$. Moreover, f is continuous except at $x = 0$. Thus by Theorem 2, the Fourier series of f in (12) converges to the 2π-periodic function $g(x)$, where $g(x) = f(x) = -1$ for $-\pi < x < 0$, $g(x) = f(x) = 1$ for $0 < x < \pi$, $g(0) = [f(0^+) + f(0^-)]/2 = 0$, and at $\pm\pi$ we have $g(\pm\pi) = [f(-\pi^+) + f(\pi^-)]/2 = (-1 + 1)/2 = 0$. The graph of $g(x)$ is given in Figure 10.7. ■ End

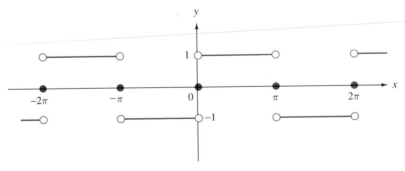

Figure 10.7 The limit function of the Fourier series for $f(x) = \begin{cases} -1 , & -\pi < x < 0 , \\ 1 , & 0 < x < \pi \end{cases}$

[†]From formula (21), we see that it doesn't matter how we define $f(x)$ at its points of discontinuity, since only the left- and right-hand limits are involved. The derivative $f'(x)$, of course, is undefined at such points.

When f is a $2T$-periodic function that is continuous on $(-\infty, \infty)$ and has a piecewise continuous derivative, its Fourier series not only converges at each point—it converges **uniformly on** $(-\infty, \infty)$. This means that for any prescribed tolerance $\varepsilon > 0$, the graph of the partial sum

$$s_N(x) := \frac{a_0}{2} + \sum_{n=1}^{N} \left\{ a_n \cos\frac{n\pi x}{T} + b_n \sin\frac{n\pi x}{T} \right\}$$

will, for all N large, lie in an ε-corridor about the graph of f on $(-\infty, \infty)$ (see Figure 10.8). The property of uniform convergence of Fourier series is particularly helpful when one needs to verify that a formal solution to a partial differential equation is an actual (genuine) solution.

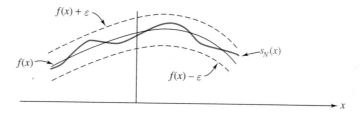

Figure 10.8 An ε-corridor about f

UNIFORM CONVERGENCE OF FOURIER SERIES

Theorem 3. Let f be a continuous function on $(-\infty, \infty)$ and periodic of period $2T$. If f' is piecewise continuous on $[-T, T]$, then the Fourier series for f converges uniformly to f on $[-T, T]$ and hence on any interval. That is, for each $\varepsilon > 0$, there exists an integer N_0 (that depends on ε) such that

$$\left| f(x) - \left\{ \frac{a_0}{2} + \sum_{n=1}^{N} \left\{ a_n \cos\frac{n\pi x}{T} + b_n \sin\frac{n\pi x}{T} \right\} \right\} \right| < \varepsilon \ ,$$

for all $N \geq N_0$, and all $x \in (-\infty, \infty)$.

In Example 5 we obtained the Fourier series expansion given in (13) for $f(x) = |x|$, $-1 < x < 1$. Since $g(x)$, the periodic extension of $f(x)$ (see Figure 10.9 on page 626) is continuous on $(-\infty, \infty)$ and

$$f'(x) = \begin{cases} -1 \ , & -1 < x < 0 \ , \\ 1 \ , & 0 < x < 1 \ , \end{cases}$$

is piecewise continuous on $[-1, 1]$, the Fourier series expansion (13) converges uniformly to $|x|$ on $[-1, 1]$. Compare Figure 10.6.

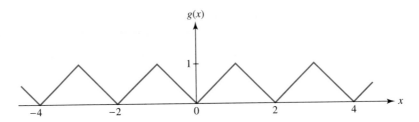

Figure 10.9 Periodic extension of $f(x) = |x|, -1 < x < 1$

The term-by-term differentiation of a Fourier series is not always permissible. For example, the Fourier series for $f(x) = x, -\pi < x < \pi$ (see Problem 9) is

(22) $$f(x) \sim 2 \sum_{n=1}^{\infty} (-1)^{n+1} \frac{\sin nx}{n},$$

which converges for all x, whereas its derived series

$$2 \sum_{n=1}^{\infty} (-1)^{n+1} \cos nx$$

diverges for every x. The following theorem gives sufficient conditions for using termwise differentiation.

DIFFERENTIATION OF FOURIER SERIES

Theorem 4. Let $f(x)$ be continuous on $(-\infty, \infty)$ and $2T$-periodic. Let $f'(x)$ and $f''(x)$ be piecewise continuous on $[-T, T]$. Then, the Fourier series of $f'(x)$ can be obtained from the Fourier series for $f(x)$ by termwise differentiation. In particular, if

$$f(x) = \frac{a_0}{2} + \sum_{n=1}^{\infty} \left\{ a_n \cos \frac{n\pi x}{T} + b_n \sin \frac{n\pi x}{T} \right\},$$

then

$$f'(x) \sim \sum_{n=1}^{\infty} \frac{\pi n}{T} \left\{ -a_n \sin \frac{n\pi x}{T} + b_n \cos \frac{n\pi x}{T} \right\}.$$

Notice that Theorem 4 does not apply to the example function $f(x)$ and its Fourier series expansion shown in (22), since the 2π-periodic extension of this $f(x)$ fails to be continuous on $(-\infty, \infty)$.

Termwise integration of a Fourier series is permissible under much weaker conditions.

INTEGRATION OF FOURIER SERIES

Theorem 5. Let $f(x)$ be piecewise continuous on $\left[-T, T\right]$ with Fourier series

$$f(x) \sim \frac{a_0}{2} + \sum_{n=1}^{\infty} \left\{ a_n \cos \frac{n\pi x}{T} + b_n \sin \frac{n\pi x}{T} \right\} .$$

Then, for any x in $\left[-T, T\right]$, we have

$$\int_{-T}^{x} f(t)\, dt = \int_{-T}^{x} \frac{a_0}{2} dt + \sum_{n=1}^{\infty} \int_{-T}^{x} \left\{ a_n \cos \frac{n\pi t}{T} + b_n \sin \frac{n\pi t}{T} \right\} dt .$$

EXERCISES 10.3

In Problems 1 through 6, determine whether the given function is even, odd, or neither.

1. $f(x) = x^3 + \sin 2x$.
2. $f(x) = \sin^2 x$.
3. $f(x) = (1 - x^2)^{-1/2}$.
4. $f(x) = \sin(x + 1)$.
5. $f(x) = e^{-x} \cos 3x$.
6. $f(x) = x^{1/5} \cos x^2$.

7. Prove the following properties:
 (a) If f and g are even functions, then so is the product fg.
 (b) If f and g are odd functions, then fg is an even function.
 (c) If f is an even function and g is an odd function, then fg is an odd function.

8. Verify formula (5). [*Hint:* Use the identity $2 \cos A \cos B = \cos(A + B) + \cos(A - B)$.]

In Problems 9 through 16, compute the Fourier series for the given function f on the specified interval. Use a computer or graphing calculator to plot a few partial sums of the Fourier series.

9. $f(x) = x$, $-\pi < x < \pi$.

10. $f(x) = |x|$, $-\pi < x < \pi$.

11. $f(x) = \begin{cases} 1 , & -2 < x < 0 , \\ x , & 0 < x < 2 . \end{cases}$

12. $f(x) = \begin{cases} 0 , & -\pi < x < 0 , \\ x^2 , & 0 < x < \pi . \end{cases}$

13. $f(x) = x^2$, $-1 < x < 1$.

14. $f(x) = \begin{cases} x , & 0 < x < \pi , \\ x + \pi , & -\pi < x < 0 . \end{cases}$

15. $f(x) = e^x$, $-\pi < x < \pi$.

16. $f(x) = \begin{cases} 0 , & -\pi < x < -\pi/2 , \\ -1 , & -\pi/2 < x < 0 , \\ 1 , & 0 < x < \pi/2 , \\ 0 , & \pi/2 < x < \pi . \end{cases}$

In Problems 17 through 24, determine the function to which the Fourier series for $f(x)$, given in the indicated problem, converges.

17. Problem 9. **18.** Problem 10.
19. Problem 11. **20.** Problem 12.
21. Problem 13. **22.** Problem 14.
23. Problem 15. **24.** Problem 16.

25. Find the functions represented by the series obtained by the termwise integration of the given series from $-\pi$ to x:

(a) $2 \sum_{n=1}^{\infty} \frac{(-1)^{n+1}}{n} \sin nx \sim x$, $-\pi < x < \pi$.

(b) $\frac{4}{\pi} \sum_{n=0}^{\infty} \frac{\sin(2n + 1)x}{(2n + 1)} \sim f(x)$,

$f(x) = \begin{cases} -1 , & -\pi < x < 0 , \\ 1 , & 0 < x < \pi . \end{cases}$

26. Show that the set of functions

$$\left\{\cos\frac{\pi}{2}x,\ \sin\frac{\pi}{2}x,\ \cos\frac{3\pi}{2}x,\ \sin\frac{3\pi}{2}x,\ \ldots,\right.$$

$$\left.\cos\frac{(2n-1)\pi}{2}x,\ \sin\frac{(2n-1)\pi}{2}x,\ \ldots\right\}$$

is an orthonormal system on $[-1, 1]$ with respect to the weight function $w(x) \equiv 1$.

27. Find the orthogonal expansion (generalized Fourier series) for

$$f(x) = \begin{cases} 0, & -1 < x < 0, \\ 1, & 0 < x < 1, \end{cases}$$

in terms of the orthonormal system of Problem 26.

28. (a) Show that the function $f(x) = x^2$ has the Fourier series, on $-\pi < x < \pi$,

$$f(x) \sim \frac{\pi^2}{3} + 4\sum_{n=1}^{\infty}\frac{(-1)^n}{n^2}\cos nx\ .$$

(b) Use the result of part (a) and Theorem 2 to show that

$$\sum_{n=1}^{\infty}\frac{(-1)^{n+1}}{n^2} = \frac{\pi^2}{12}\ .$$

(c) Use the result of part (a) and Theorem 2 to show that

$$\sum_{n=1}^{\infty}\frac{1}{n^2} = \frac{\pi^2}{6}\ .$$

29. In Section 8.8, it was shown that the Legendre polynomials $P_n(x)$ are orthogonal on the interval $[-1, 1]$ with respect to the weight function $w(x) \equiv 1$. Using the fact that the first three Legendre polynomials are

$$P_0(x) \equiv 1\ ,\quad P_1(x) = x\ ,$$
$$P_2(x) = (3/2)x^2 - (1/2)\ ,$$

find the first three coefficients in the expansion

$$f(x) = a_0 P_0(x) + a_1 P_1(x) + a_2 P_2(x) + \cdots,$$

where $f(x)$ is the function

$$f(x) := \begin{cases} -1, & -1 < x < 0, \\ 1, & 0 < x < 1. \end{cases}$$

30. As in Problem 29, find the first three coefficients in the expansion

$$f(x) = a_0 P_0(x) + a_1 P_1(x) + a_2 P_2(x) + \cdots,$$

when $f(x) = |x|,\ -1 < x < 1$.

31. The Hermite polynomials $H_n(x)$ are orthogonal on the interval $(-\infty, \infty)$ with respect to the weight function $W(x) = e^{-x^2}$. Verify this fact for the first three Hermite polynomials:

$$H_0(x) \equiv 1\ ,\qquad H_1(x) = 2x\ ,$$
$$H_2(x) = 4x^2 - 2\ .$$

32. The Chebyshev (Tchebichef) polynomials $T_n(x)$ are orthogonal on the interval $[-1, 1]$ with respect to the weight function $w(x) = (1 - x^2)^{-1/2}$. Verify this fact for the first three Chebyshev polynomials:

$$T_0(x) \equiv 1\ ,\qquad T_1(x) = x\ ,$$
$$T_2(x) = 2x^2 - 1\ .$$

33. Let $\{f_n(x)\}$ be an orthogonal set of functions on the interval $[a, b]$ with respect to the weight function $w(x)$. Show that they satisfy the **Pythagorean property**

$$\|f_m + f_n\|^2 = \|f_m\|^2 + \|f_n\|^2$$

if $m \neq n$.

34. Norm. The norm of a function $\|f\|$ is like the length of a vector in R^n. In particular, show that the norm defined in (16) satisfies the following properties associated with length $\big($assume f and g are continuous and $w(x) > 0$ on $[a, b]\big)$:

(a) $\|f\| \geq 0$, and $\|f\| = 0$ if and only if $f \equiv 0$.
(b) $\|cf\| = |c|\,\|f\|$, where c is any real number.
(c) $\|f + g\| \leq \|f\| + \|g\|$.

35. Inner Product. The integral in the orthogonality condition (14) is like the dot product of two vectors in R^n. In particular, show that the **inner product** of two functions defined by

(23) $$\langle f, g \rangle := \int_a^b f(x)g(x)w(x)\,dx\ ,$$

where $w(x)$ is a positive weight function, satisfies the following properties associated with the dot product $\big($assume f, g, and h are continuous on $[a, b]\big)$:

(a) $\langle f + g, h \rangle = \langle f, h \rangle + \langle g, h \rangle$.
(b) $\langle cf, h \rangle = c\langle f, h \rangle$ where c is any real number.
(c) $\langle f, g \rangle = \langle g, f \rangle$.

36. Complex Form of the Fourier Series.

(a) Using the Euler formula $e^{i\theta} = \cos\theta + i\sin\theta$, $i = \sqrt{-1}$, prove that

$$\cos nx = \frac{e^{inx} + e^{-inx}}{2} \quad\text{and}\quad \sin nx = \frac{e^{inx} - e^{-inx}}{2i} .$$

(b) Show that the Fourier series

$$f(x) \sim \frac{a_0}{2} + \sum_{n=1}^{\infty} \{a_n \cos nx + b_n \sin nx\}$$

$$= c_0 + \sum_{n=1}^{\infty} \{c_n e^{inx} + c_{-n} e^{-inx}\} ,$$

where

$$c_0 = \frac{a_0}{2} , \quad c_n = \frac{a_n - ib_n}{2} , \quad c_{-n} = \frac{a_n + ib_n}{2} .$$

(c) Finally, use the results of part (b) to show that

$$f(x) \sim \sum_{n=-\infty}^{\infty} c_n e^{inx} , \qquad -\pi < x < \pi ,$$

where

$$c_n = \frac{1}{2\pi} \int_{-\pi}^{\pi} f(x) e^{-inx} dx .$$

37. Least Squares Approximation Property. The Nth partial sum of the Fourier series

$$f(x) \sim \frac{a_0}{2} + \sum_{n=1}^{\infty} \{a_n \cos nx + b_n \sin nx\}$$

gives the best mean-square approximation of f by a trigonometric polynomial. To prove this, let $F_N(x)$ denote an arbitrary trigonometric polynomial of degree N:

$$F_N(x) = \frac{\alpha_0}{2} + \sum_{n=1}^{N} \{\alpha_n \cos nx + \beta_n \sin nx\} ,$$

and define

$$E := \int_{-\pi}^{\pi} [f(x) - F_N(x)]^2 dx ,$$

which is the *total square error.* Expanding the integrand, we get

$$E = \int_{-\pi}^{\pi} f^2(x)\,dx - 2 \int_{-\pi}^{\pi} f(x)F_N(x)\,dx$$

$$+ \int_{-\pi}^{\pi} F_N^2(x)\,dx .$$

(a) Use the orthogonality of the functions $\{1, \cos x, \sin x, \cos 2x, \ldots\}$ to show that

$$\int_{-\pi}^{\pi} F_N^2(x)\,dx = \pi\left(\frac{\alpha_0^2}{2} + \alpha_1^2 + \cdots + \alpha_N^2\right.$$

$$\left. + \beta_1^2 + \cdots + \beta_N^2\right)$$

and

$$\int_{-\pi}^{\pi} f(x)F_N(x)\,dx = \pi\left(\frac{\alpha_0 a_0}{2} + \alpha_1 a_1 + \cdots + \alpha_N a_N\right.$$

$$\left. + \beta_1 b_1 + \cdots + \beta_N b_N\right) .$$

(b) Let E^* be the error when we approximate f by the Nth partial sum of its Fourier series; that is, when we choose $\alpha_n = a_n$ and $\beta_n = b_n$. Show that

$$E^* = \int_{-\pi}^{\pi} f^2(x)\,dx - \pi\left(\frac{a_0^2}{2} + a_1^2 + \cdots + a_N^2\right.$$

$$\left. + b_1^2 + \cdots + b_N^2\right) .$$

(c) Using the results of parts (a) and (b), show that $E - E^* \geq 0$, that is, $E \geq E^*$, by proving that

$$E - E^* = \pi\left\{\frac{(\alpha_0 - a_0)^2}{2} + (\alpha_1 - a_1)^2\right.$$

$$+ \cdots + (\alpha_N - a_N)^2 + (\beta_1 - b_1)^2$$

$$\left. + \cdots + (\beta_N - b_N)^2\right\} .$$

Hence the Nth partial sum of the Fourier series gives the least total square error, since $E \geq E^*$.

38. Bessel's Inequality. Use the fact that E^*, defined in part (b) of Problem 37, is nonnegative to prove **Bessel's inequality**

(24) $$\frac{a_0^2}{2} + \sum_{n=1}^{\infty} \{a_n^2 + b_n^2\} \leq \frac{1}{\pi} \int_{-\pi}^{\pi} f^2(x)\,dx .$$

(If f is piecewise continuous on $[-\pi, \pi]$, then we have equality in (24). This result is called **Parseval's identity.**)

(a) Graph of $f_{11}(x)$

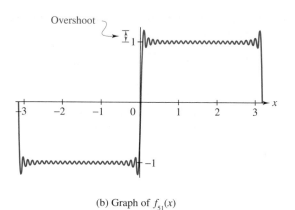

(b) Graph of $f_{51}(x)$

Figure 10.10 Gibbs phenomenon for partial sums of Fourier series

39. Gibbs Phenomenon.[†] The American mathematician Josiah Willard Gibbs (1839–1903) observed that near points of discontinuity of f, the partial sums of the Fourier series for f may overshoot by approximately 9% of the jump, regardless of the number of terms. This is illustrated in Figure 10.10 for the function

$$f(x) = \begin{cases} -1, & -\pi < x < 0, \\ 1, & 0 < x < \pi, \end{cases}$$

whose Fourier series has the partial sums

$$f_{2n-1}(x) = \frac{4}{\pi}\left[\sin x + \frac{1}{3}\sin 3x \right.$$

$$\left. + \cdots + \frac{\sin(2n-1)x}{(2n-1)}\right].$$

To verify this for $f(x)$, proceed as follows:

(a) Show that

$$\pi(\sin x)f'_{2n-1}(x) = 4\sin x \left[\cos x + \cos 3x \right.$$

$$\left. + \cdots + \cos(2n-1)x\right]$$

$$= 2\sin 2nx .$$

(b) Infer from part (a) and the figure that the maximum occurs at $x = \pi/(2n)$ and has the value

$$f_{2n-1}\left(\frac{\pi}{2n}\right) = \frac{4}{\pi}\left[\sin\frac{\pi}{2n} + \frac{1}{3}\sin\frac{3\pi}{2n} \right.$$

$$\left. + \cdots + \frac{1}{2n-1}\sin\frac{(2n-1)\pi}{2n}\right].$$

(c) Show that if one approximates

$$\int_0^\pi \frac{\sin x}{x}\,dx$$

using the partition $x_k := (2k-1)(\pi/2n)$, $k = 1, 2, \ldots, n$, $\Delta x_k = \pi/n$ and choosing the midpoint of each interval as the place to evaluate the integrand, then

$$\int_0^\pi \frac{\sin x}{x}\,dx \approx \frac{\sin(\pi/2n)}{\pi/2n}\frac{\pi}{n} + \cdots$$

$$+ \frac{\sin\left[(2n-1)\pi/2n\right]}{(2n-1)\pi/2n}\frac{\pi}{n}$$

$$= \frac{\pi}{2}f_{2n-1}\left(\frac{\pi}{2n}\right).$$

[†]*Historical Footnote:* Actually, H. Wilbraham discovered this phenomenon some fifty years earlier than Gibbs did. It is more appropriately called the *Gibbs-Wilbraham* phenomenon.

(d) Use the result of part (c) to show that the over-shoot satisfies

$$\lim_{n\to\infty} f_{2n-1}\left(\frac{\pi}{2n}\right) = \frac{2}{\pi}\int_0^\pi \frac{\sin x}{x}dx \ .$$

(e) Using the result of part (d) and a numerical integration algorithm (such as Simpson's rule, Appendix B) for the **sine integral function**

$$Si(z) := \int_0^z \frac{\sin x}{x}dx \ ,$$

show that $\lim_{n\to\infty} f_{2n-1}\big(\pi/(2n)\big) \approx 1.18$. Thus the approximations overshoot the true value of $f(0^+) = 1$ by 0.18 or 9% of the jump from $f(0^-)$ to $f(0^+)$.

10.4 FOURIER COSINE AND SINE SERIES

A typical problem encountered in using separation of variables to solve a partial differential equation is the problem of representing a function defined on some finite interval by a trigonometric series consisting of only sine functions or only cosine functions. For example, in Section 10.2, equation (12), page 607, we needed to express the initial values $u(x, 0) = f(x)$, $0 < x < L$, of the solution to the initial-boundary value problem associated with the problem of heat flow as a trigonometric series of the form

$$(1) \qquad f(x) = \sum_{n=1}^{\infty} c_n \sin\left(\frac{n\pi x}{L}\right) .$$

Recalling that the Fourier series for an odd function defined on $[-L, L]$ consists entirely of sine terms, we might try to achieve (1) by extending the function $f(x), 0 < x < L$, to the interval $(-L, L)$ in such a way that the extended function is odd. This is accomplished by defining the function

$$f_o(x) := \begin{cases} f(x) \ , & 0 < x < L \ , \\ -f(-x) \ , & -L < x < 0 \ . \end{cases}$$

and extending $f_o(x)$ to all x using $2L$-periodicity.[†] Since $f_o(x)$ is an odd function, it has a Fourier series consisting entirely of sine terms. Moreover, $f_o(x)$ is an extension of $f(x)$, since $f_o(x) = f(x)$ on $(0, L)$. This extension is called the **odd $2L$-periodic extension** of $f(x)$. The resulting Fourier series expansion is called a half-range expansion for $f(x)$, since it represents the function $f(x)$ on $(0, L)$, which is half of the interval $(-L, L)$ where it represents $f_o(x)$.

[†]Strictly speaking, we have extended $f_o(x)$ to all x other than the integer multiples of L. Continuity considerations often suggest appropriate values for the extended functions at some of these points, as well. Figure 10.11 on page 633 illustrates this.

In a similar fashion, we can define the **even 2L-periodic extension** of $f(x)$ as the function

$$f_e(x) := \begin{cases} f(x) , & 0 < x < L , \\ f(-x) , & -L < x < 0 , \end{cases}$$

with $f_e(x + 2L) = f_e(x)$.

To illustrate the various extensions, let's consider the function $f(x) = x, 0 < x < \pi$. If we extend $f(x)$ to the interval $(-\pi, \pi)$ using π-periodicity, then the extension \tilde{f} is given by

$$\tilde{f}(x) = \begin{cases} x , & 0 < x < \pi , \\ x + \pi , & -\pi < x < 0 , \end{cases}$$

with $\tilde{f}(x + 2\pi) = \tilde{f}(x)$. In Problem 14 of Exercises 10.3, the reader computed the Fourier series for $\tilde{f}(x)$ and found

$$\tilde{f}(x) \sim \frac{\pi}{2} - \sum_{n=1}^{\infty} \frac{1}{n} \sin 2nx ,$$

which consists of both odd functions (the sine terms) and even functions (the constant term), since the π-periodic extension $\tilde{f}(x)$ is neither an even nor an odd function. The odd 2π-periodic extension of $f(x)$ is just $f_o(x) = x, -\pi < x < \pi$, which has the Fourier series expansion

$$(2) \qquad f_o(x) \sim 2 \sum_{n=1}^{\infty} \frac{(-1)^{n+1}}{n} \sin nx$$

(see Problem 9 in Exercises 10.3). Because $f_o(x) = f(x)$ on the interval $(0, \pi)$, the expansion in (2) is a half-range expansion for $f(x)$. The even 2π-periodic extension of $f(x)$ is the function $f_e(x) = |x|, -\pi < x < \pi$, which has the Fourier series expansion

$$(3) \qquad f_e(x) = \frac{\pi}{2} - \frac{4}{\pi} \sum_{n=0}^{\infty} \frac{1}{(2n + 1)^2} \cos (2n + 1)x$$

(see Problem 10 in Exercises 10.3).

The preceding three extensions, the π-periodic function $\tilde{f}(x)$, the odd 2π-periodic function $f_o(x)$, and the even 2π-periodic function $f_e(x)$, are natural extensions of $f(x)$. There are many other ways of extending $f(x)$. For example, the function

$$g(x) = \begin{cases} x , & 0 < x < \pi , \\ 0 , & -\pi < x < 0 , \end{cases} \qquad g(x + 2\pi) = g(x) ,$$

which we studied in Example 3 of Section 10.3, is also an extension of $f(x)$. However, its Fourier series contains both sine and cosine terms and hence is not as useful as previous extensions. The graphs of these extensions of $f(x)$ are given in Figure 10.11.

(a) π periodic

(b) Odd 2π periodic

(c) Even 2π periodic

(d) Another 2π periodic extension

Figure 10.11 Extensions of $f(x) = x, 0 < x < \pi$

The Fourier series expansions for $f_o(x)$ and $f_e(x)$ given in (2) and (3) represent $f(x)$ on the interval $(0, \pi)$ $\left(\text{actually, they equal } f(x) \text{ on } (0, \pi)\right)$. This motivates the following definitions.

FOURIER COSINE AND SINE SERIES

Definition 2. Let $f(x)$ be piecewise continuous on the interval $[0, T]$. The **Fourier cosine series** of $f(x)$ on $[0, T]$ is

(4) $\dfrac{a_0}{2} + \displaystyle\sum_{n=1}^{\infty} a_n \cos\dfrac{n\pi x}{T}$,

where

(5) $a_n = \dfrac{2}{T}\displaystyle\int_0^T f(x) \cos\dfrac{n\pi x}{T} dx$, $n = 0, 1, \ldots$.

The **Fourier sine series** of $f(x)$ on $[0, T]$ is

(6) $\displaystyle\sum_{n=1}^{\infty} b_n \sin\dfrac{n\pi x}{T}$,

where

(7) $b_n = \dfrac{2}{T}\displaystyle\int_0^T f(x) \sin\dfrac{n\pi x}{T} dx$, $n = 1, 2, \ldots$.

The trigonometric series in (4) is just the Fourier series for $f_e(x)$, the even $2T$-periodic extension of $f(x)$, and that in (6) is the Fourier series for $f_o(x)$, the odd $2T$-periodic extension of $f(x)$. These are called **half-range expansions** for $f(x)$.

EXAMPLE 1 Compute the Fourier series for

$$f(x) = \begin{cases} x , & 0 \le x \le \pi/2 , \\ \pi - x , & \pi/2 \le x \le \pi . \end{cases}$$

SOLUTION Using formula (7) with $T = \pi$, we find

$$b_n = \frac{2}{\pi} \int_0^\pi f(x) \sin nx\, dx = \frac{2}{\pi} \int_0^{\pi/2} x \sin nx\, dx + \frac{2}{\pi} \int_{\pi/2}^\pi (\pi - x) \sin nx\, dx$$

$$= \frac{2}{\pi n^2} \int_0^{\pi n/2} u \sin u\, du + 2 \int_{\pi/2}^\pi \sin nx\, dx - \frac{2}{\pi n^2} \int_{\pi n/2}^{\pi n} u \sin u\, du$$

$$= \frac{2}{\pi n^2} \Big[\sin u - u \cos u \Big] \Big|_0^{\pi n/2} - \frac{2}{n} \Big[\cos \pi n - \cos \frac{n\pi}{2} \Big]$$

$$\qquad - \frac{2}{\pi n^2} \Big[\sin u - u \cos u \Big] \Big|_{\pi n/2}^{\pi n}$$

$$= \frac{4}{\pi n^2} \sin \frac{n\pi}{2} = \begin{cases} 0 , & n \text{ even} , \\ \dfrac{4(-1)^{(n-1)/2}}{\pi n^2} , & n \text{ odd} . \end{cases}$$

So on letting $n = 2k + 1$, we find the Fourier sine series for $f(x)$ to be

$$\textbf{(8)} \qquad \frac{4}{\pi} \sum_{k=0}^\infty \frac{(-1)^k}{(2k + 1)^2} \sin(2k + 1)x = \frac{4}{\pi} \left\{ \sin x - \frac{1}{9} \sin 3x + \frac{1}{25} \sin 5x + \cdots \right\} . \quad \blacksquare \text{End}$$

Since, in Example 1, the function $f(x)$ is continuous and $f'(x)$ is piecewise continuous on $(0, \pi)$, it follows from Theorem 2 on pointwise convergence of Fourier series that

$$f(x) = \frac{4}{\pi} \left\{ \sin x - \frac{1}{9} \sin 3x + \frac{1}{25} \sin 5x - \frac{1}{49} \sin 7x + \cdots \right\}$$

for all x in $[0, \pi]$.

Let's return to the problem of heat flow in one dimension.

EXAMPLE 2 Find the solution to the heat flow problem

$$\textbf{(9)} \qquad \frac{\partial u}{\partial t} = 2 \frac{\partial^2 u}{\partial x^2} , \qquad 0 < x < \pi , \qquad t > 0 ,$$

$$\textbf{(10)} \qquad u(0, t) = u(\pi, t) = 0 , \qquad t > 0 ,$$

$$\textbf{(11)} \qquad u(x, 0) = \begin{cases} x , & 0 < x \le \pi/2 , \\ \pi - x , & \pi/2 \le x < \pi . \end{cases}$$

SOLUTION Comparing equation (9) with equation (1) in Section 10.2 (page 604), we see that $\beta = 2$ and $L = \pi$. Hence we need only represent $u(x, 0) = f(x)$ in a Fourier sine series (cf. equation (12) on page 607) of the form

$$\sum_{n=1}^{\infty} c_n \sin nx \; .$$

In Example 1 we obtained this expansion and showed that

$$c_n = \frac{4}{\pi n^2}\sin\frac{n\pi}{2} = \begin{cases} 0 \; , & n \text{ even} \; , \\ \dfrac{4(-1)^{(n-1)/2}}{\pi n^2} \; , & n \text{ odd} \; . \end{cases}$$

Hence, from equation (11) on page 606, the solution to the heat flow problem (9)–(11) is

$$\begin{aligned} u(x, t) &= \sum_{n=1}^{\infty} c_n e^{-2n^2 t}\sin nx \\ &= \frac{4}{\pi}\sum_{k=0}^{\infty}\frac{(-1)^k}{(2k+1)^2}e^{-2(2k+1)^2 t}\sin(2k+1)x \\ &= \frac{4}{\pi}\left\{ e^{-2t}\sin x - \frac{1}{9}e^{-18t}\sin 3x + \frac{1}{25}e^{-50t}\sin 5x + \cdots \right\} . \end{aligned}$$

A sketch of a partial sum for $u(x, t)$ is displayed in Figure 10.12. ∎ End

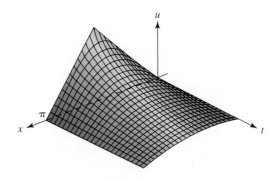

Figure 10.12 Partial sum for $u(x, t)$ in Example 2

EXERCISES 10.4

In Problems 1 through 4, determine (a) the π-periodic extension \tilde{f}, (b) the odd 2π-periodic extension f_o, and (c) the even 2π-periodic extension f_e for the given function f and sketch their graphs.

1. $f(x) = x^2 \; , \qquad 0 < x < \pi \; .$

2. $f(x) = \sin 2x \; , \qquad 0 < x < \pi \; .$

3. $f(x) = \begin{cases} 0 \; , & 0 < x < \pi/2 \; , \\ 1 \; , & \pi/2 < x < \pi \; . \end{cases}$

4. $f(x) = \pi - x \; , \qquad 0 < x < \pi \; .$

In Problems 5 through 10, compute the Fourier sine series for the given function.

5. $f(x) = -1$, $0 < x < 1$.

6. $f(x) = \cos x$, $0 < x < \pi$.

7. $f(x) = x^2$, $0 < x < \pi$.

8. $f(x) = \pi - x$, $0 < x < \pi$.

9. $f(x) = x - x^2$, $0 < x < 1$.

10. $f(x) = e^x$, $0 < x < 1$.

In Problems 11 through 16, compute the Fourier cosine series for the given function.

11. $f(x) = \pi - x$, $0 < x < \pi$.

12. $f(x) = 1 + x$, $0 < x < \pi$.

13. $f(x) = e^x$, $0 < x < 1$.

14. $f(x) = e^{-x}$, $0 < x < 1$.

15. $f(x) = \sin x$, $0 < x < \pi$.

16. $f(x) = x - x^2$, $0 < x < 1$.

In Problems 17 through 19, find the solution to the heat flow problem.

$$\frac{\partial u}{\partial t} = 5\frac{\partial^2 u}{\partial x^2}, \qquad 0 < x < \pi, \qquad t > 0,$$

$$u(0, t) = u(\pi, t) = 0, \qquad t > 0,$$

$$u(x, 0) = f(x), \qquad 0 < x < \pi,$$

where $f(x)$ is given.

17. $f(x) = 1 - \cos 2x$. **18.** $f(x) = x(\pi - x)$.

19. $f(x) = \begin{cases} -x, & 0 < x \le \pi/2, \\ x - \pi, & \pi/2 \le x < \pi. \end{cases}$

10.5 THE HEAT EQUATION

In Section 10.1 we developed a model for heat flow in an insulated uniform wire whose ends are kept at the constant temperature 0°C. In particular, we found that the temperature $u(x, t)$ in the wire is governed by the initial-boundary value problem

$$\text{(1)} \qquad \frac{\partial u}{\partial t} = \beta\frac{\partial^2 u}{\partial x^2}, \qquad 0 < x < L, \qquad t > 0,$$

$$\text{(2)} \qquad u(0, t) = u(L, t) = 0, \qquad t > 0,$$

$$\text{(3)} \qquad u(x, 0) = f(x), \qquad 0 < x < L$$

(see equations (7)–(9) in Section 10.1, page 602). Here equation (2) specifies that the temperature at the ends of the wire is zero, whereas equation (3) specifies the initial temperature distribution.

In Section 10.2 we also derived a formal solution to (1)–(3) using separation of variables. There we found the solution to (1)–(3) to have the form

$$\text{(4)} \qquad u(x, t) = \sum_{n=1}^{\infty} c_n e^{-\beta(n\pi/L)^2 t} \sin\frac{n\pi x}{L},$$

where the c_n's are the coefficients in the Fourier sine series for $f(x)$:

$$\text{(5)} \qquad f(x) = \sum_{n=1}^{\infty} c_n \sin\frac{n\pi x}{L}.$$

In other words, solving (1)–(3) reduces to computing the Fourier sine series for the initial value function $f(x)$.

In this section we discuss heat flow problems where the ends of the wire are insulated or kept at a constant, but nonzero, temperature. (The latter involves nonhomogeneous boundary conditions.) We will also discuss the problem in which a heat source is adding heat to the wire. (This results in a nonhomogeneous partial differential equation.) The problem of heat flow in a rectangular plate is also discussed and leads to the topic of double Fourier series. We conclude this section with a discussion of the existence and uniqueness of solutions to the heat flow problem.

In the model of heat flow in a uniform wire, let's replace the assumption that the ends of the wire are kept at a constant temperature zero and instead assume that the ends of the wire are *insulated,* that is, no heat flows out (or in) at the ends of the wire. It follows from the principle of heat conduction (see Section 10.1) that the temperature gradient must be zero at these endpoints, that is,

$$\frac{\partial u}{\partial x}(0, t) = \frac{\partial u}{\partial x}(L, t) = 0 , \qquad t > 0 .$$

In the next example we obtain the formal solution to the heat flow problem with these boundary conditions.

EXAMPLE 1 Find a formal solution to the heat flow problem governed by the initial-boundary value problem

(6) $$\frac{\partial u}{\partial t} = \beta \frac{\partial^2 u}{\partial x^2} , \qquad 0 < x < L , \qquad t > 0 ,$$

(7) $$\frac{\partial u}{\partial x}(0, t) = \frac{\partial u}{\partial x}(L, t) = 0 , \qquad t > 0 ,$$

(8) $$u(x, 0) = f(x) , \qquad 0 < x < L .$$

SOLUTION Using the method of separation of variables, we first assume

$$u(x, t) = X(x)T(t) .$$

Substituting into equation (6) and separating variables as was done in Section 10.2 (cf. equation (5) on page 604), we get the two equations

(9) $$X''(x) - KX(x) = 0 ,$$
(10) $$T'(t) - \beta KT(t) = 0 ,$$

where K is some constant. The boundary conditions in (7) become

$$X'(0)T(t) = 0 \quad \text{and} \quad X'(L)T(t) = 0 .$$

For these equations to hold for all $t > 0$, either $T(t) \equiv 0$, which implies that $u(x, t) \equiv 0$, or

(11) $$X'(0) = X'(L) = 0 .$$

Combining the boundary conditions in (11) with equation (9) gives the boundary value problem

(12) $$X''(x) - KX(x) = 0 ; \qquad X'(0) = X'(L) = 0 ,$$

where K can be any constant.

To solve for the nontrivial solutions to (12), we begin with the auxiliary equation $r^2 - K = 0$. When $K > 0$, arguments similar to those used in Section 10.2 show that there are no nontrivial solutions to (12).

When $K = 0$, the auxiliary equation has the repeated root 0 and a general solution to the differential equation is

$$X(x) = A + Bx .$$

The boundary conditions in (12) reduce to $B = 0$ with A arbitrary. Thus, for $K = 0$, the nontrivial solutions to (12) are of the form

$$X(x) = c_0 ,$$

where $c_0 = A$ is an arbitrary nonzero constant.

When $K < 0$, the auxiliary equation has the roots $r = \pm i\sqrt{-K}$. Thus a general solution to the differential equation in (12) is

$$X(x) = C_1 \cos \sqrt{-K}x + C_2 \sin \sqrt{-K}x .$$

The boundary conditions in (12) lead to the system

$$\sqrt{-K}C_2 = 0 ,$$
$$-\sqrt{-K}C_1 \sin \sqrt{-K}L + \sqrt{-K}C_2 \cos \sqrt{-K}L = 0 .$$

Hence $C_2 = 0$ and the system reduces to solving $C_1 \sin \sqrt{-K}L = 0$. Since $\sin \sqrt{-K}L = 0$ only when $\sqrt{-K}L = n\pi$, where n is an integer, we obtain a nontrivial solution only when $\sqrt{-K} = n\pi/L$ or $K = -(n\pi/L)^2$, $n = 1, 2, 3, \ldots$. Furthermore, the nontrivial solutions (eigenfunctions) X_n corresponding to the eigenvalue $K = -(n\pi/L)^2$ are given by

(13) $\qquad X_n(x) = c_n \cos \dfrac{n\pi x}{L} ,$

where the c_n's are arbitrary nonzero constants. In fact, formula (13) also holds for $n = 0$, since $K = 0$ has the eigenfunctions $X_0(x) = c_0$.

Having determined that $K = -(n\pi/L)^2$, $n = 0, 1, 2, \ldots$, let's consider equation (10) for such K:

$$T'(t) + \beta(n\pi/L)^2 T(t) = 0 .$$

For $n = 0, 1, 2, \ldots$, the general solution is

$$T_n(t) = b_n e^{-\beta(n\pi/L)^2 t} ,$$

where the b_n's are arbitrary constants. Combining this with equation (13), we obtain the functions

$$u_n(x, t) = X_n(x)T_n(t) = \left[c_n \cos \frac{n\pi x}{L} \right] \left[b_n e^{-\beta(n\pi/L)^2 t} \right] ,$$

$$u_n(x, t) = a_n e^{-\beta(n\pi/L)^2 t} \cos \frac{n\pi x}{L} ,$$

where $a_n = b_n c_n$ is, again, an arbitrary constant.

If we take an infinite series of these functions, we obtain

(14) $$u(x, t) = \frac{a_0}{2} + \sum_{n=1}^{\infty} a_n e^{-\beta(n\pi/L)^2 t} \cos\frac{n\pi x}{L} \, ,$$

which will be a solution to (6)–(7) provided the series has the proper convergence behavior. Notice that in (14) we have altered the constant term and written it as $a_0/2$, thus producing the standard form for cosine expansions.

Assuming a solution to (6)–(7) is given by the series in (14) and substituting into the initial condition (8), we get

(15) $$u(x, 0) = \frac{a_0}{2} + \sum_{n=1}^{\infty} a_n \cos\frac{n\pi x}{L} = f(x) \, , \qquad 0 < x < L \, .$$

This means that if we choose the a_n's as the coefficients in the Fourier cosine series for f,

$$a_n = \frac{2}{L} \int_0^L f(x) \cos\frac{n\pi x}{L} dx \, , \qquad n = 0, 1, 2, \dots \, ,$$

then $u(x, t)$ given in (14) will be a **formal solution** to the heat flow problem (6)–(8). Again, if this expansion converges to a continuous function with continuous second partial derivatives, then the formal solution is an actual solution. ∎ End

Note that the individual terms in the series (4) and (14) are genuine solutions to the heat equation and the associated boundary conditions (but not the initial condition, of course). These solutions are called the **modes** of the system. Observe that the time factor $e^{-\beta(n\pi/L)^2 t}$ decays faster for the more oscillatory (higher n) modes. This makes sense physically; a temperature pattern of many alternating hot and cold strips arranged closely together will equilibrate faster than a more uniform pattern. See Figure 10.13 on page 640.

A good way of expressing how separation of variables "works" is as follows: The initial temperature profile $f(x)$ is decomposed into modes through Fourier analysis—equations (5) or (15). Then each mode is matched with a time decay factor—equations (4) or (14)—and it evolves in time accordingly.

When the ends of the wire are kept at 0°C or when the ends are insulated, the boundary conditions are said to be **homogeneous.** But, when the ends of the wire are kept at constant temperatures different from zero, that is,

(16) $$u(0, t) = U_1 \quad \text{and} \quad u(L, t) = U_2 \, , \qquad t > 0 \, ,$$

then the boundary conditions are called **nonhomogeneous.**

From our experience with vibration problems in Section 4.12 we expect that the solution to the heat flow problem with nonhomogeneous boundary conditions will consist of a **steady-state solution** $v(x)$ that satisfies the nonhomogeneous boundary conditions in (16) plus a **transient solution** $w(x, t)$. That is,

$$u(x, t) = v(x) + w(x, t) \, ,$$

where $w(x, t)$ and its partial derivatives tend to zero as $t \to \infty$. The function $w(x, t)$ will then satisfy homogeneous boundary conditions, as illustrated in the next example.

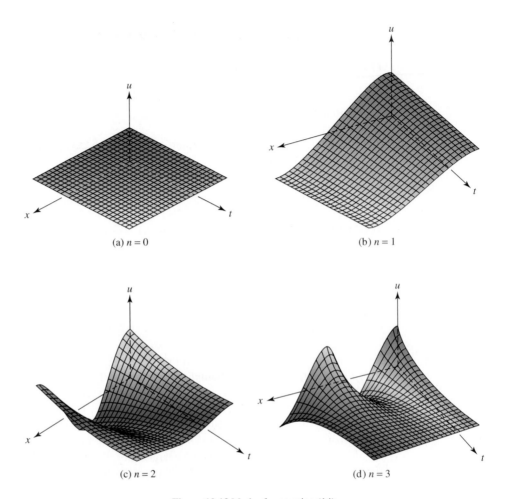

(a) $n = 0$

(b) $n = 1$

(c) $n = 2$

(d) $n = 3$

Figure 10.13 Modes for equation (14)

EXAMPLE 2 Find a formal solution to the heat flow problem governed by the initial-boundary value problem

(17) $\dfrac{\partial u}{\partial t} = \beta \dfrac{\partial^2 u}{\partial x^2}$, $0 < x < L$, $t > 0$,

(18) $u(0, t) = U_1$, $u(L, t) = U_2$, $t > 0$,

(19) $u(x, 0) = f(x)$, $0 < x < L$.

SOLUTION Let's assume the solution $u(x, t)$ consists of a steady-state solution $v(x)$ and a transient solution $w(x, t)$, that is,

(20) $u(x, t) = v(x) + w(x, t)$.

Substituting for $u(x, t)$ in equations (17)–(19) leads to

(21) $\dfrac{\partial u}{\partial t} = \dfrac{\partial w}{\partial t} = \beta v''(x) + \beta \dfrac{\partial^2 w}{\partial x^2}$, $0 < x < L$, $t > 0$,

(22) $v(0) + w(0, t) = U_1$, $v(L) + w(L, t) = U_2$, $t > 0$,

(23) $v(x) + w(x, 0) = f(x)$, $0 < x < L$.

If we allow $t \to \infty$ in (21)–(22), assuming $w(x, t)$ is a transient solution, we obtain the steady-state boundary value problem

$$v''(x) = 0 , \qquad 0 < x < L ,$$
$$v(0) = U_1 , \qquad v(L) = U_2 .$$

Solving for v, we obtain $v(x) = Ax + B$, and choosing A and B so that the boundary conditions are satisfied yields

(24) $v(x) = U_1 + \dfrac{(U_2 - U_1)x}{L}$

as the steady-state solution.

With this choice for $v(x)$, the initial-boundary value problem (21)–(23) reduces to the following initial-boundary value problem for $w(x, t)$:

(25) $\dfrac{\partial w}{\partial t} = \beta \dfrac{\partial^2 w}{\partial x^2}$, $0 < x < L$, $t > 0$,

(26) $w(0, t) = w(L, t) = 0$, $t > 0$,

(27) $w(x, 0) = f(x) - U_1 - \dfrac{(U_2 - U_1)x}{L}$, $0 < x < L$.

Recall that a formal solution to (25)–(27) is given by equation (4). Hence

$$w(x, t) = \sum_{n=1}^{\infty} c_n e^{-\beta(n\pi/L)^2 t} \sin\dfrac{n\pi x}{L} ,$$

where the c_n's are the coefficients of the Fourier sine series expansion

$$f(x) - U_1 - \dfrac{(U_2 - U_1)x}{L} = \sum_{n=1}^{\infty} c_n \sin\dfrac{n\pi x}{L} .$$

Therefore the formal solution to (17)–(19) is

(28) $u(x, t) = U_1 + \dfrac{(U_2 - U_1)x}{L} + \sum_{n=1}^{\infty} c_n e^{-\beta(n\pi/L)^2 t} \sin\dfrac{n\pi x}{L}$,

with

$$c_n = \dfrac{2}{L} \int_0^L \left[f(x) - U_1 - \dfrac{(U_2 - U_1)x}{L} \right] \sin\dfrac{n\pi x}{L} dx . \quad \blacksquare \text{ End}$$

In the next example, we consider the heat flow problem when a heat source is present but is independent of time. (Recall the derivation in Section 10.1.)

EXAMPLE 3 Find a formal solution to the heat flow problem governed by the initial-boundary value problem

$$(29) \qquad \frac{\partial u}{\partial t} = \beta \frac{\partial^2 u}{\partial x^2} + P(x) , \qquad 0 < x < L , \qquad t > 0 ,$$

$$(30) \qquad u(0, t) = U_1 , \qquad u(L, t) = U_2 , \qquad t > 0 ,$$

$$(31) \qquad u(x, 0) = f(x) , \qquad 0 < x < L .$$

SOLUTION We begin by assuming that the solution consists of a steady-state solution $v(x)$ and a transient solution $w(x, t)$, namely,

$$u(x, t) = v(x) + w(x, t) ,$$

where $w(x, t)$ and its partial derivatives tend to zero as $t \to \infty$. Substituting for $u(x, t)$ in (29)–(31) yields

$$(32) \qquad \frac{\partial u}{\partial t} = \frac{\partial w}{\partial t} = \beta v''(x) + \beta \frac{\partial^2 w}{\partial x^2} + P(x) , \qquad 0 < x < L , \qquad t > 0 ,$$

$$(33) \qquad v(0) + w(0, t) = U_1 , \qquad v(L) + w(L, t) = U_2 , \qquad t > 0 ,$$

$$(34) \qquad v(x) + w(x, 0) = f(x) , \qquad 0 < x < L .$$

Letting $t \to \infty$ in (32)–(33), we obtain the steady-state boundary value problem

$$v''(x) = -\frac{1}{\beta} P(x) , \qquad 0 < x < L ,$$

$$v(0) = U_1 , \qquad v(L) = U_2 .$$

The solution to this boundary value problem can be obtained by two integrations using the boundary conditions to determine the constants of integration. The reader can verify that the solution is given by the formula

$$(35) \qquad v(x) = \left[U_2 - U_1 + \int_0^L \left(\int_0^z \frac{1}{\beta} P(s) ds \right) dz \right] \frac{x}{L} + U_1 - \int_0^x \left(\int_0^z \frac{1}{\beta} P(s) ds \right) dz .$$

With this choice for $v(x)$, we find that the initial-boundary value problem (32)–(34) reduces to the following initial-boundary value problem for $w(x, t)$:

$$(36) \qquad \frac{\partial w}{\partial t} = \beta \frac{\partial^2 w}{\partial x^2} , \qquad 0 < x < L , \qquad t > 0 ,$$

$$(37) \qquad w(0, t) = w(L, t) = 0 , \qquad t > 0 ,$$

$$(38) \qquad w(x, 0) = f(x) - v(x) , \qquad 0 < x < L ,$$

where $v(x)$ is given by formula (35). As before, the solution to this initial-boundary value problem is

$$(39) \qquad w(x, t) = \sum_{n=1}^{\infty} c_n e^{-\beta(n\pi/L)^2 t} \sin\frac{n\pi x}{L} \ ,$$

where the c_n's are determined from the Fourier sine series expansion of $f(x) - v(x)$:

$$(40) \qquad f(x) - v(x) = \sum_{n=1}^{\infty} c_n \sin\frac{n\pi x}{L} \ .$$

Thus the formal solution to (29)–(31) is given by

$$u(x, t) = v(x) + w(x, t) \ ,$$

where $v(x)$ is given in (35) and $w(x, t)$ is prescribed by (39)–(40). ∎ End

The method of separation of variables is also applicable to problems in higher dimensions. For example, consider the problem of heat flow in a rectangular plate with sides $x = 0$, $x = L$, $y = 0$, and $y = W$. Assuming the two sides $y = 0$, $y = W$ are kept at a constant temperature of $0°C$ and the two sides $x = 0$, $x = L$ are perfectly insulated, then heat flow is governed by the initial-boundary value problem in the following example (see Figure 10.14).

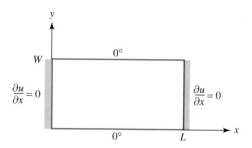

Figure 10.14 Plate with insulated sides

EXAMPLE 4 Find a formal solution $u(x, y, t)$ to the initial-boundary value problem

$$(41) \qquad \frac{\partial u}{\partial t} = \beta\left\{\frac{\partial^2 u}{\partial x^2} + \frac{\partial^2 u}{\partial y^2}\right\} \ , \qquad 0 < x < L \ , \qquad 0 < y < W \ , \qquad t > 0 \ ,$$

$$(42) \qquad \frac{\partial u}{\partial x}(0, y, t) = \frac{\partial u}{\partial x}(L, y, t) = 0 \ , \qquad 0 < y < W \ , \qquad t > 0 \ ,$$

$$(43) \qquad u(x, 0, t) = u(x, W, t) = 0 \ , \qquad 0 < x < L \ , \qquad t > 0 \ ,$$

$$(44) \qquad u(x, y, 0) = f(x, y) \ , \qquad 0 < x < L \ , \qquad 0 < y < W \ .$$

SOLUTION If we assume a solution of the form $u(x, y, t) = V(x, y)T(t)$, then equation (41) separates into the two equations

$$(45) \qquad T'(t) - \beta K T(t) = 0 \ ,$$

$$(46) \qquad \frac{\partial^2 V}{\partial x^2}(x, y) + \frac{\partial^2 V}{\partial y^2}(x, y) - KV(x, y) = 0 \ ,$$

where K can be any constant. To solve equation (46), we again use separation of variables. Here we assume $V(x, y) = X(x)Y(y)$. This allows us to separate equation (46) into the two equations

(47) $\quad X''(x) - JX(x) = 0$,

(48) $\quad Y''(y) + (J - K)Y(y) = 0$,

where J can be any constant (see Problem 29 in Exercises 10.2).

To solve for $X(x)$, we observe that the boundary conditions in (42), in terms of the separated variables, become

$$X'(0)Y(y)T(t) = X'(L)Y(y)T(t) = 0 , \quad 0 < y < W , \quad t > 0 .$$

Hence in order to get a nontrivial solution, we must have

(49) $\quad X'(0) = X'(L) = 0$.

The boundary value problem for X given in equations (47) and (49) was solved in Example 1 (cf. equations (12) and (13)). Here $J = -(m\pi/L)^2$, $m = 0, 1, 2, \ldots$, and

$$X_m(x) = c_m \cos\frac{m\pi x}{L} ,$$

where the c_m's are arbitrary.

To solve for $Y(y)$, we first observe that the boundary conditions in (43) become

(50) $\quad Y(0) = Y(W) = 0$.

Next, substituting $J = -(m\pi/L)^2$ into equation (48) yields

$$Y''(y) - \left(K + (m\pi/L)^2\right)Y(y) = 0 ,$$

which we can rewrite as

(51) $\quad Y''(y) - EY(y) = 0$,

where $E = K + (m\pi/L)^2$. The boundary value problem for Y consisting of (50)–(51) has also been solved before. In Section 10.2 (cf. equations (7) and (9), pages 605 and 606) we showed that $E = -(n\pi/W)^2$, $n = 1, 2, 3, \ldots$, and the nontrivial solutions are given by

$$Y_n(y) = a_n \sin\frac{n\pi y}{W} ,$$

where the a_n's are arbitrary.

Since $K = E - (m\pi/L)^2$, we have

$$K = -(n\pi/W)^2 - (m\pi/L)^2 , \quad m = 0, 1, 2, \ldots , \quad n = 1, 2, 3, \ldots .$$

Substituting K into equation (45), we can solve for $T(t)$ and obtain

$$T_{mn}(t) = b_{mn}e^{-(m^2/L^2 + n^2/W^2)\beta\pi^2 t} .$$

Substituting in for X_m, Y_n, and T_{mn}, we get

$$u_{mn}(x, y, t) = \left(c_m \cos\frac{m\pi x}{L}\right)\left(a_n \sin\frac{n\pi y}{W}\right)\left(b_{mn}e^{-(m^2/L^2+n^2/W^2)\beta\pi^2t}\right) ,$$

$$u_{mn}(x, y, t) = a_{mn}e^{-(m^2/L^2+n^2/W^2)\beta\pi^2t}\cos\frac{m\pi x}{L}\sin\frac{n\pi y}{W} ,$$

where $a_{mn} := a_n b_{mn} c_m$ $(m = 0, 1, 2, \ldots, n = 1, 2, 3, \ldots)$ are arbitrary constants.

If we now take a doubly infinite series of such functions, then we obtain the formal series

$$(52) \qquad u(x, y, t) = \sum_{m=0}^{\infty}\sum_{n=1}^{\infty} a_{mn}e^{-(m^2/L^2+n^2/W^2)\beta\pi^2t}\cos\frac{m\pi x}{L}\sin\frac{n\pi y}{W} .$$

We are now ready to apply the initial conditions (44). Setting $t = 0$, we obtain

$$(53) \qquad u(x, y, 0) = f(x, y) = \sum_{m=0}^{\infty}\sum_{n=1}^{\infty} a_{mn}\cos\frac{m\pi x}{L}\sin\frac{n\pi y}{W} .$$

This is a **double Fourier series.**[†] The formulas for the coefficients a_{mn} are obtained by exploiting the orthogonality conditions twice. Presuming (53) is valid and permits term-by-term integration, we multiply each side by $\cos(p\pi x/L)\sin(q\pi y/W)$ and integrate over x and y:

$$\int_0^L\int_0^W f(x, y)\cos\frac{p\pi x}{L}\sin\frac{q\pi y}{W}dy\,dx$$

$$= \sum_{m=0}^{\infty}\sum_{n=1}^{\infty} a_{mn}\int_0^L\int_0^W \cos\frac{m\pi x}{L}\sin\frac{n\pi y}{W}\cos\frac{p\pi x}{L}\sin\frac{q\pi y}{W}dy\,dx .$$

According to the orthogonality conditions, each integral on the right is zero, except when $m = p$ and $n = q$; thus

$$\int_0^L\int_0^W f(x, y)\cos\frac{p\pi x}{L}\sin\frac{q\pi y}{W}dy\,dx$$

$$= a_{pq}\int_0^L \cos^2\frac{p\pi x}{L}dx\int_0^W \sin^2\frac{q\pi y}{W}dy = \begin{cases} \dfrac{LW}{4}a_{pq} , & p \neq 0 , \\[3mm] \dfrac{LW}{2}a_{pq} , & p = 0 . \end{cases}$$

Hence

$$(54) \qquad a_{0q} = \frac{2}{LW}\int_0^L\int_0^W f(x, y)\sin\frac{q\pi y}{W}dy\,dx , \qquad q = 1, 2, 3, \ldots ,$$

[†]For a discussion of double Fourier series, see *Partial Differential Equations of Mathematical Physics*, by Tyn Myint-U (Elsevier North Holland, Inc., New York, 1980), Section 5.14.

and for $p \geq 1, q \geq 1$,

(55) $a_{pq} = \dfrac{4}{LW} \displaystyle\int_0^L \int_0^W f(x, y) \cos\dfrac{p\pi x}{L} \sin\dfrac{q\pi y}{W} dy\, dx$.

Finally, the solution to the initial-boundary value problem (41)–(44) is given by equation (52), where the coefficients are prescribed by equations (54) and (55). ■ End

Existence and Uniqueness of Solutions

In the examples that we have studied in this section and in Section 10.2, we were able to obtain formal solutions in the sense that we could express the solution in terms of a series expansion consisting of exponentials, sines, and cosines. To prove that these series converge to actual solutions requires results on the convergence of Fourier series and results from real analysis on uniform convergence. We will not go into these details here, but refer the reader to Section 6.5 of the text by Tyn Myint-U (see footnote on page 645) for a proof of the existence of a solution to the heat flow problem discussed in Sections 10.1 and 10.2. (A proof of uniqueness is also given there.)

As might be expected, by using Fourier series and the method of separation of variables one can also obtain "solutions" when the initial data are discontinuous, since the formal solutions only require the existence of a convergent Fourier series. This allows one to study idealized problems in which the initial conditions do not agree with the boundary conditions or the initial conditions involve a jump discontinuity. For example, we may assume that initially one half of the wire is at one temperature, whereas the other half is at a different temperature, that is,

$$f(x) = \begin{cases} U_1, & 0 < x < L/2, \\ U_2, & L/2 < x < L. \end{cases}$$

The formal solution that we obtain will make sense for $0 < x < L, t > 0$, but we must take care in interpreting the results near the points of discontinuity $x = 0, L/2$, and L.

The question of the uniqueness of the solution to the heat flow problem can be answered in various ways. One is tempted to argue that the method of separation of variables yields *formulas* for the solutions and therefore a unique solution. However, this does *not* exclude the possibility of solutions existing that cannot be obtained by the method of separation of variables.

From physical considerations, we know that the *peak* temperature along a wire does not increase spontaneously if there are no heat sources to drive it up. Indeed if hot objects could draw heat from colder objects without external intervention, we would have violations of the second law of thermodynamics. Thus no point on an unheated wire will ever reach a higher temperature than the initial peak temperature. Such statements are called **maximum principles** and can be proved mathematically for the heat equation and Laplace's equation. One such result is the following.[†]

[†]For a discussion of maximum principles and their applications, see *Maximum Principles in Differential Equations,* by M. H. Protter and H. F. Weinberger (Springer-Verlag, New York, 1984).

MAXIMUM PRINCIPLE FOR THE HEAT EQUATION

Theorem 6. Let $u(x, t)$ be a continuously differentiable function that satisfies the heat equation

(56) $\dfrac{\partial u}{\partial t} = \beta \dfrac{\partial^2 u}{\partial x^2}$, $0 < x < L$, $t > 0$,

and the boundary conditions

(57) $u(0, t) = u(L, t) = 0$, $t > 0$.

Then $u(x, t)$ attains its maximum value at $t = 0$, for some x in $[0, L]$, that is,

$$\max_{\substack{t \geq 0 \\ 0 \leq x \leq L}} u(x, t) = \max_{0 \leq x \leq L} u(x, 0) \ .$$

We can use the maximum principle to show that the heat flow problem has a unique solution.

UNIQUENESS OF SOLUTION

Theorem 7. The initial-boundary value problem

(58) $\dfrac{\partial u}{\partial t} = \beta \dfrac{\partial^2 u}{\partial x^2}$, $0 < x < L$, $t > 0$,

(59) $u(0, t) = u(L, t) = 0$, $t > 0$,

(60) $u(x, 0) = f(x)$, $0 < x < L$,

has at most one continuously differentiable solution.

Proof. Assume $u(x, t)$ and $v(x, t)$ are continuously differentiable functions that satisfy the initial-boundary value problem (58)–(60). Let $w = u - v$. Now w is a continuously differentiable solution to the boundary value problem (56)–(57). By the maximum principle, w must attain its maximum at $t = 0$, and since

$$w(x, 0) = u(x, 0) - v(x, 0) = f(x) - f(x) = 0 \ ,$$

we have $w(x, t) \leq 0$. Hence $u(x, t) \leq v(x, t)$ for all $0 \leq x \leq L, t \geq 0$. A similar argument using $\hat{w} = v - u$ yields $v(x, t) \leq u(x, t)$. Therefore we have $u(x, t) = v(x, t)$ for all $0 \leq x \leq L$, $t \geq 0$. Thus there is at most one continuously differentiable solution to the problem (58)–(60). ∎ End

EXERCISES 10.5

In Problems 1 through 10, find a formal solution to the given initial-boundary value problem.

1. $\dfrac{\partial u}{\partial t} = 5\dfrac{\partial^2 u}{\partial x^2}$, $0 < x < 1$, $t > 0$,

 $u(0, t) = u(1, t) = 0$, $t > 0$,
 $u(x, 0) = (1 - x)x^2$, $0 < x < 1$.

2. $\dfrac{\partial u}{\partial t} = \dfrac{\partial^2 u}{\partial x^2}$, $0 < x < \pi$, $t > 0$,

 $u(0, t) = u(\pi, t) = 0$, $t > 0$,
 $u(x, 0) = x^2$, $0 < x < \pi$.

3. $\dfrac{\partial u}{\partial t} = 3\dfrac{\partial^2 u}{\partial x^2}$, $0 < x < \pi$, $t > 0$,

 $\dfrac{\partial u}{\partial x}(0, t) = \dfrac{\partial u}{\partial x}(\pi, t) = 0$, $t > 0$,
 $u(x, 0) = x$, $0 < x < \pi$.

4. $\dfrac{\partial u}{\partial t} = 2\dfrac{\partial^2 u}{\partial x^2}$, $0 < x < 1$, $t > 0$,

 $\dfrac{\partial u}{\partial x}(0, t) = \dfrac{\partial u}{\partial x}(1, t) = 0$, $t > 0$,
 $u(x, 0) = x(1 - x)$, $0 < x < 1$.

5. $\dfrac{\partial u}{\partial t} = \dfrac{\partial^2 u}{\partial x^2}$, $0 < x < \pi$, $t > 0$,

 $\dfrac{\partial u}{\partial x}(0, t) = \dfrac{\partial u}{\partial x}(\pi, t) = 0$, $t > 0$,
 $u(x, 0) = e^x$, $0 < x < \pi$.

6. $\dfrac{\partial u}{\partial t} = 7\dfrac{\partial^2 u}{\partial x^2}$, $0 < x < \pi$, $t > 0$,

 $\dfrac{\partial u}{\partial x}(0, t) = \dfrac{\partial u}{\partial x}(\pi, t) = 0$, $t > 0$,
 $u(x, 0) = 1 - \sin x$, $0 < x < \pi$.

7. $\dfrac{\partial u}{\partial t} = 2\dfrac{\partial^2 u}{\partial x^2}$, $0 < x < \pi$, $t > 0$,

 $u(0, t) = 5$, $u(\pi, t) = 10$, $t > 0$,
 $u(x, 0) = \sin 3x - \sin 5x$, $0 < x < \pi$.

8. $\dfrac{\partial u}{\partial t} = \dfrac{\partial^2 u}{\partial x^2}$, $0 < x < \pi$, $t > 0$,

 $u(0, t) = 0$, $u(\pi, t) = 3\pi$, $t > 0$,
 $u(x, 0) = 0$, $0 < x < \pi$.

9. $\dfrac{\partial u}{\partial t} = \dfrac{\partial^2 u}{\partial x^2} + e^{-x}$, $0 < x < \pi$, $t > 0$,

 $u(0, t) = u(\pi, t) = 0$, $t > 0$,
 $u(x, 0) = \sin 2x$, $0 < x < \pi$.

10. $\dfrac{\partial u}{\partial t} = 3\dfrac{\partial^2 u}{\partial x^2} + x$, $0 < x < \pi$, $t > 0$,

 $u(0, t) = u(\pi, t) = 0$, $t > 0$,
 $u(x, 0) = \sin x$, $0 < x < \pi$.

11. Find a formal solution to the initial-boundary value problem

$$\dfrac{\partial u}{\partial t} = 4\dfrac{\partial^2 u}{\partial x^2}$$, $0 < x < \pi$, $t > 0$,

$$\dfrac{\partial u}{\partial x}(0, t) = 0$$, $u(\pi, t) = 0$, $t > 0$,

$$u(x, 0) = f(x)$$, $0 < x < \pi$.

12. Find a formal solution to the initial-boundary value problem

$$\dfrac{\partial u}{\partial t} = \dfrac{\partial^2 u}{\partial x^2}$$, $0 < x < \pi$, $t > 0$,

$$u(0, t) = 0$$, $u(\pi, t) + \dfrac{\partial u}{\partial x}(\pi, t) = 0$, $t > 0$,

$$u(x, 0) = f(x)$$, $0 < x < \pi$.

13. Find a formal solution to the initial-boundary value problem

$$\dfrac{\partial u}{\partial t} = 2\dfrac{\partial^2 u}{\partial x^2} + 4x$$, $0 < x < \pi$, $t > 0$,

$$u(0, t) = u(\pi, t) = 0$$, $t > 0$,

$$u(x, 0) = \sin x$$, $0 < x < \pi$.

14. Find a formal solution to the initial-boundary value problem

$$\dfrac{\partial u}{\partial t} = 3\dfrac{\partial^2 u}{\partial x^2} + 5$$, $0 < x < \pi$, $t > 0$,

$$u(0, t) = u(\pi, t) = 1$$, $t > 0$,

$$u(x, 0) = 1$$, $0 < x < \pi$.

In Problems 15 through 18, find a formal solution to the initial-boundary value problem.

$$\frac{\partial u}{\partial t} = \frac{\partial^2 u}{\partial x^2} + \frac{\partial^2 u}{\partial y^2} \ , \quad 0 < x < \pi \ , \ 0 < y < \pi \ , \ t > 0 \ ,$$

$$\frac{\partial u}{\partial x}(0, y, t) = \frac{\partial u}{\partial x}(\pi, y, t) = 0 \ , \quad 0 < y < \pi \ , \ t > 0 \ ,$$

$$u(x, 0, t) = u(x, \pi, t) = 0 \ , \quad 0 < x < \pi \ , \quad t > 0 \ ,$$

$$u(x, y, 0) = f(x, y) \ , \quad 0 < x < \pi \ , \quad 0 < y < \pi \ ,$$

for the given function $f(x, y)$.

15. $f(x, y) = \cos 6x \sin 4y - 3 \cos x \sin 11y$.

16. $f(x, y) = \cos x \sin y + 4 \cos 2x \sin y$
$\qquad - 3 \cos 3x \sin 4y$.

17. $f(x, y) = y$.

18. $f(x, y) = x \sin y$.

19. Chemical Diffusion. Chemical diffusion through a thin layer is governed by the equation

$$\frac{\partial C}{\partial t} = k \frac{\partial^2 C}{\partial x^2} - LC \ ,$$

where $C(x, t)$ is the concentration in moles/cm^3, the diffusivity k is a positive constant with units cm^2/sec, and $L > 0$ is a consumption rate with units sec^{-1}. Assume the boundary conditions are

$$C(0, t) = C(a, t) = 0 \ , \qquad t > 0 \ ,$$

and the initial concentration is given by

$$C(x, 0) = f(x) \ , \qquad 0 < x < a \ .$$

Use the method of separation of variables to solve formally for the concentration $C(x, t)$. What happens to the concentration as $t \to +\infty$?

10.6 THE WAVE EQUATION

In Section 10.2 we presented a model for the motion of a vibrating string. If $u(x, t)$ represents the displacement (deflection) of the string and the ends of the string are held fixed, then the motion of the string is governed by the initial-boundary value problem

(1) $$\frac{\partial^2 u}{\partial t^2} = \alpha^2 \frac{\partial^2 u}{\partial x^2} \ , \qquad 0 < x < L \ , \qquad t > 0 \ ,$$

(2) $$u(0, t) = u(L, t) = 0 \ , \qquad t > 0 \ ,$$

(3) $$u(x, 0) = f(x) \ , \qquad 0 < x < L \ ,$$

(4) $$\frac{\partial u}{\partial t}(x, 0) = g(x) \ , \qquad 0 < x < L \ .$$

Equation (1) is called the **wave equation**.

The constant α^2 appearing in (1) is strictly positive and depends on the linear density and tension of the string. The boundary conditions in (2) reflect the fact that the string is held fixed at the two endpoints $x = 0$ and $x = L$.

Equations (3) and (4) specify, respectively, the initial displacement and the initial velocity of each point on the string. For the initial and boundary conditions to be consistent, we assume $f(0) = f(L) = 0$ and $g(0) = g(L) = 0$.

Using the method of separation of variables, we found in Section 10.2 that a formal solution to (1)–(4) is given by (cf. equations (24)–(26) on pages 609–610)

(5) $$u(x, t) = \sum_{n=1}^{\infty} \left[a_n \cos \frac{n\pi\alpha}{L} t + b_n \sin \frac{n\pi\alpha}{L} t \right] \sin \frac{n\pi x}{L} \ ,$$

where the a_n's and b_n's are determined from the Fourier sine series

(6) $$f(x) = \sum_{n=1}^{\infty} a_n \sin\frac{n\pi x}{L} \,,$$

(7) $$g(x) = \sum_{n=1}^{\infty} b_n \left(\frac{n\pi\alpha}{L}\right) \sin\frac{n\pi x}{L} \,.$$

See Figure 10.15 for a sketch of partial sums for (5).

Figure 10.15 Partial sums for equation (5) with $f(x) = \begin{cases} x, & 0 < x \le \pi/2 \\ \pi - x, & \pi/2 \le x < \pi \end{cases}$ and $g(x) = 0$

Each term (or mode) in expansion (5) can be viewed as a **standing wave** (a wave that vibrates in place without lateral motion along the string). For example, the first term,

$$\left(a_1 \cos\frac{\pi\alpha}{L}t + b_1 \sin\frac{\pi\alpha}{L}t\right) \sin\frac{\pi x}{L} \,,$$

consists of a sinusoidal shape function $\sin(\pi x/L)$ multiplied by a time-varying amplitude. The second term is also a sinusoid $\sin(2\pi x/L)$ with a time-varying amplitude. In the latter case, there is a *node* in the middle at $x = L/2$ that never moves. For the nth term, we have a sinusoid $\sin(n\pi x/L)$ with a time-varying amplitude and $(n-1)$ nodes. This is illustrated in Figure 10.16. Thus separation of variables decomposes the initial data into sinusoids or modes (equations (6)–(7)), assigns each mode a frequency at which to vibrate (equation (5)), and represents the solution as the superposition of infinitely many standing waves. Modes with more nodes vibrate at higher frequencies. By choosing the initial configuration of the string to have the same shape as one of the individual terms in the solution, we can activate only that mode.

The different modes of vibration of a guitar string are distinguishable to the human ear by the frequency of the sound they generate; different frequencies are discerned as different "pitches." The *fundamental* is the frequency of the lowest mode, and integer multiples of the fundamental frequency are called "harmonics." A cello and a trombone sound different even if they are playing the same fundamental, say $F^{\#}$, because of the difference in the intensities of the harmonics they generate.

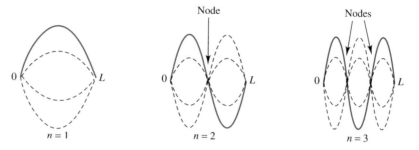

Figure 10.16 Standing waves. Time-varying amplitudes are shown by dashed curves.

Many engineering devices operate better in some modes than in others. For example, certain modes propagate down a waveguide or an optical fiber with less attenuation than the others. In such cases, engineers design the systems to suppress the undesirable modes, either by shaping the initial excitation or by introducing devices (like resistor cards in a waveguide) which damp out certain modes preferentially.

The fundamental mode on a guitar string can be suppressed by holding the finger lightly in contact with the midpoint, thereby creating a stationary point or node there. According to equation (5), the pitch of the note then doubles, thus producing an "octave." This style of fingering, called "playing harmonics," is used frequently in musical performance.

An important characteristic of an engineering device is the set of angular frequencies supported by its eigenmodes—its "eigenfrequencies." According to equation (5), the eigenfrequencies of the vibrating string are the harmonics $\omega_n = n\pi\alpha/L$, $n = 1, 2, 3, \ldots$. These blend together pleasantly to the human ear, and we enjoy melodies played on string instruments. The eigenfrequencies of a drum are not harmonics (see Problem 21). Therefore drums are used for rhythm, not for melody.

As we have seen in the preceding section, the method of separation of variables can be used to solve problems with nonhomogeneous boundary conditions and nonhomogeneous equations where the forcing term is time independent. In the next example, we will consider a problem with a time-dependent forcing term $h(x, t)$.

EXAMPLE 1 For given functions f, g, and h, find a formal solution to the initial-boundary value problem

(8) $$\frac{\partial^2 u}{\partial t^2} = \alpha^2 \frac{\partial^2 u}{\partial x^2} + h(x, t) , \qquad 0 < x < L , \qquad t > 0 ,$$

(9) $$u(0, t) = u(L, t) = 0 , \qquad t > 0 ,$$

(10) $$u(x, 0) = f(x) , \qquad 0 < x < L ,$$

(11) $$\frac{\partial u}{\partial t}(x, 0) = g(x) , \qquad 0 < x < L .$$

SOLUTION The boundary conditions in (9) certainly require that the solution be zero for $x = 0$ and $x = L$. Motivated by the fact that the solution to the corresponding homogeneous system (1)–(4) consists of a superposition of standing waves, let's try to find a solution to (8)–(11) of the form

(12) $$u(x, t) = \sum_{n=1}^{\infty} u_n(t) \sin\frac{n\pi x}{L} ,$$

where the $u_n(t)$'s are functions of t to be determined.

For each fixed t, we can compute a Fourier sine series for $h(x, t)$. If we assume the series is convergent to $h(x, t)$, then

(13) $\qquad h(x, t) = \sum_{n=1}^{\infty} h_n(t) \sin\frac{n\pi x}{L} \; ,$

where the coefficient $h_n(t)$ is given by (recall equation (6) on page 633)

$$h_n(t) = \frac{2}{L} \int_0^L h(x, t) \sin\frac{n\pi x}{L} dx \; , \qquad n = 1, 2, \dots \; .$$

If the series in (13) has the proper convergence properties, then we can substitute (12) and (13) into equation (8) and obtain

$$\sum_{n=1}^{\infty} \left[u_n''(t) + \left(\frac{n\pi\alpha}{L}\right)^2 u_n(t) \right] \sin\frac{n\pi x}{L} = \sum_{n=1}^{\infty} h_n(t) \sin\frac{n\pi x}{L} \; .$$

Equating the coefficients in each series (why?), we have

$$u_n''(t) + \left(\frac{n\pi\alpha}{L}\right)^2 u_n(t) = h_n(t) \; .$$

This is a nonhomogeneous, constant coefficient equation that can be solved using variation of parameters. The reader should verify that

$$u_n(t) = a_n \cos\frac{n\pi\alpha}{L}t + b_n \sin\frac{n\pi\alpha}{L}t + \frac{L}{n\pi\alpha} \int_0^t h_n(s) \sin\left[\frac{n\pi\alpha}{L}(t - s)\right] ds$$

(cf. Problem 28 of Exercises 4.9.) Hence, with this choice of $u_n(t)$, the series in (12) is a formal solution to the partial differential equation (8).

Since[†]

$$u_n(0) = a_n \quad \text{and} \quad u_n'(0) = b_n\left(\frac{n\pi\alpha}{L}\right) \; ,$$

substituting (12) into the initial conditions (10)–(11) yields

(14) $\qquad u(x, 0) = f(x) = \sum_{n=1}^{\infty} a_n \sin\frac{n\pi x}{L} \; ,$

(15) $\qquad \frac{\partial u}{\partial t}(x, 0) = g(x) = \sum_{n=1}^{\infty} b_n\left(\frac{n\pi\alpha}{L}\right) \sin\frac{n\pi x}{L} \; .$

Thus, if we choose the a_n's and b_n's so that equations (14) and (15) are satisfied, a formal solution to (8)–(11) is given by

(16) $\qquad u(x, t) = \sum_{n=1}^{\infty} \Bigg\{ a_n \cos\frac{n\pi\alpha}{L}t + b_n \sin\frac{n\pi\alpha}{L}t$

$$+ \frac{L}{n\pi\alpha} \int_0^t h_n(s) \sin\left[\frac{n\pi\alpha}{L}(t - s)\right] ds \Bigg\} \sin\frac{n\pi x}{L} \; . \quad \blacksquare \text{ End}$$

[†]To compute $u_n'(0)$, we use the fact that $\dfrac{d}{dt} \displaystyle\int_0^t G(s, t)\, ds = G(t, t) + \displaystyle\int_0^t \dfrac{\partial G}{\partial t}(s, t)\, ds.$

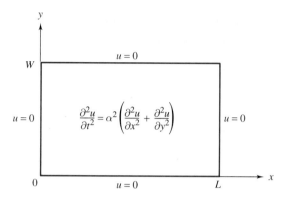

Figure 10.17 Vibrating membrane

The method of separation of variables can also be used to solve initial-boundary value problems for the wave equation in higher dimensions. For example, a vibrating rectangular membrane of length L and width W (see Figure 10.17) is governed by the following initial-boundary value problem for $u(x, y, t)$:

(17) $$\frac{\partial^2 u}{\partial t^2} = \alpha^2 \left(\frac{\partial^2 u}{\partial x^2} + \frac{\partial^2 u}{\partial y^2} \right), \qquad 0 < x < L, \qquad 0 < y < W, \qquad t > 0,$$

(18) $\quad u(0, y, t) = u(L, y, t) = 0, \qquad 0 < y < W, \qquad t > 0,$

(19) $\quad u(x, 0, t) = u(x, W, t) = 0, \qquad 0 < x < L, \qquad t > 0,$

(20) $\quad u(x, y, 0) = f(x, y), \qquad 0 < x < L, \qquad 0 < y < W,$

(21) $\quad \dfrac{\partial u}{\partial t}(x, y, 0) = g(x, y), \qquad 0 < x < L, \qquad 0 < y < W.$

Using an argument similar to the one given for the problem of heat flow in a rectangular plate (Example 4 in Section 10.5), we find that the initial-boundary value problem (17)–(21) has a formal solution

(22) $$u(x, y, t) = \sum_{m=1}^{\infty} \sum_{n=1}^{\infty} \left\{ a_{mn} \cos \left(\sqrt{\frac{m^2}{L^2} + \frac{n^2}{W^2}}\, \alpha \pi t \right) \right.$$

$$\left. + b_{mn} \sin \left(\sqrt{\frac{m^2}{L^2} + \frac{n^2}{W^2}}\, \alpha \pi t \right) \right\} \sin \frac{m \pi x}{L} \sin \frac{n \pi y}{W},$$

where the constants a_{mn} and b_{mn} are determined from the double Fourier series

$$f(x, y) = \sum_{m=1}^{\infty} \sum_{n=1}^{\infty} a_{mn} \sin \frac{m \pi x}{L} \sin \frac{n \pi y}{W},$$

$$g(x, y) = \sum_{m=1}^{\infty} \sum_{n=1}^{\infty} \alpha \pi \sqrt{\frac{m^2}{L^2} + \frac{n^2}{W^2}}\, b_{mn} \sin \frac{m \pi x}{L} \sin \frac{n \pi y}{W}.$$

In particular,

(23) $a_{mn} = \dfrac{4}{LW} \displaystyle\int_0^L \int_0^W f(x, y) \sin\dfrac{m\pi x}{L} \sin\dfrac{n\pi y}{W} \, dy\, dx$,

(24) $b_{mn} = \dfrac{4}{LW\pi\alpha\sqrt{\dfrac{m^2}{L^2} + \dfrac{n^2}{W^2}}} \displaystyle\int_0^L \int_0^W g(x, y) \sin\dfrac{m\pi x}{L} \sin\dfrac{n\pi y}{W} \, dy\, dx$.

We leave the derivation of this solution as an exercise (see Problem 19).

We mentioned earlier that the solution to the vibrating string problem (1)–(4) consisted of a superposition of standing waves. There are also "traveling waves" associated with the wave equation. Traveling waves arise naturally out of d'Alembert's solution to the wave equation for an "infinite" string.

To obtain d'Alembert's solution to the wave equation

$$\frac{\partial^2 u}{\partial t^2} = \alpha^2 \frac{\partial^2 u}{\partial x^2} \, ,$$

we use the change of variables

$$\psi = x + \alpha t \, , \qquad \eta = x - \alpha t \, .$$

If u has continuous second partial derivatives, then $\partial u/\partial x = \partial u/\partial\psi + \partial u/\partial\eta$ and $\partial u/\partial t = \alpha(\partial u/\partial\psi - \partial u/\partial\eta)$, from which we obtain

$$\frac{\partial^2 u}{\partial x^2} = \frac{\partial^2 u}{\partial\psi^2} + 2\frac{\partial^2 u}{\partial\psi\partial\eta} + \frac{\partial^2 u}{\partial\eta^2} \, ,$$

$$\frac{\partial^2 u}{\partial t^2} = \alpha^2\left\{ \frac{\partial^2 u}{\partial\psi^2} - 2\frac{\partial^2 u}{\partial\psi\partial\eta} + \frac{\partial^2 u}{\partial\eta^2} \right\} \, .$$

Substituting these expressions into the wave equation and simplifying yields

$$\frac{\partial^2 u}{\partial\psi\partial\eta} = 0 \, .$$

We can solve this equation directly by first integrating with respect to ψ to obtain

$$\frac{\partial u}{\partial\eta} = b(\eta) \, ,$$

where $b(\eta)$ is an arbitrary function of η, and then integrating with respect to η to find

$$u(\psi, \eta) = A(\psi) + B(\eta) \, ,$$

where $A(\psi)$ and $B(\eta) = \int b(\eta)\, d\eta$ are arbitrary functions. Substituting the original variables x and t gives **d'Alembert's solution**

(25) $u(x, t) = A(x + \alpha t) + B(x - \alpha t)$.

It is easy to check by direct substitution that $u(x, t)$, defined by formula (25), is indeed a solution to the wave equation, provided A and B are twice-differentiable functions.

EXAMPLE 2 Using d'Alembert's formula (25), find a solution to the initial value problem

$$
(26) \qquad \frac{\partial^2 u}{\partial t^2} = \alpha^2 \frac{\partial^2 u}{\partial x^2} \, , \qquad -\infty < x < \infty \, , \qquad t > 0 \, ,
$$

$$
(27) \qquad u(x, 0) = f(x) \, , \qquad -\infty < x < \infty \, ,
$$

$$
(28) \qquad \frac{\partial u}{\partial t}(x, 0) = g(x) \, , \qquad -\infty < x < \infty \, .
$$

SOLUTION A solution to (26) is given by formula (25), so we need only choose the functions A and B so that the initial conditions (27)–(28) are satisfied. For this we need

$$
(29) \qquad u(x, 0) = A(x) + B(x) = f(x) \, ,
$$

$$
(30) \qquad \frac{\partial u}{\partial t}(x, 0) = \alpha A'(x) - \alpha B'(x) = g(x) \, .
$$

Integrating equation (30) from x_0 to x $\left(x_0 \text{ arbitrary}\right)$ and dividing by α gives

$$
(31) \qquad A(x) - B(x) = \frac{1}{\alpha} \int_{x_0}^{x} g(s)ds + C \, ,
$$

where C is also arbitrary. Solving the system (29) and (31), we obtain

$$
A(x) = \frac{1}{2}f(x) + \frac{1}{2\alpha} \int_{x_0}^{x} g(s)ds + \frac{C}{2} \, ,
$$

$$
B(x) = \frac{1}{2}f(x) - \frac{1}{2\alpha} \int_{x_0}^{x} g(s)ds - \frac{C}{2} \, .
$$

Using these functions in formula (25) gives

$$
u(x, t) = \frac{1}{2}\left[f(x + \alpha t) + f(x - \alpha t)\right] + \frac{1}{2\alpha}\left[\int_{x_0}^{x+\alpha t} g(s)ds - \int_{x_0}^{x-\alpha t} g(s)ds\right] \, ,
$$

which simplifies to

$$
(32) \qquad u(x, t) = \frac{1}{2}\left[f(x + \alpha t) + f(x - \alpha t)\right] + \frac{1}{2\alpha} \int_{x-\alpha t}^{x+\alpha t} g(s)ds \, . \quad \blacksquare \text{End}
$$

EXAMPLE 3 Find the solution to the initial value problem

$$
(33) \qquad \frac{\partial^2 u}{\partial t^2} = 4\frac{\partial^2 u}{\partial x^2} \, , \qquad -\infty < x < \infty \, , \qquad t > 0 \, ,
$$

$$
(34) \qquad u(x, 0) = \sin x \, , \qquad -\infty < x < \infty \, ,
$$

$$
(35) \qquad \frac{\partial u}{\partial t}(x, 0) = 1 \, , \qquad -\infty < x < \infty \, .
$$

SOLUTION This is just a special case of the preceding example where $\alpha = 2, f(x) = \sin x$, and $g(x) = 1$. Substituting into (32), we obtain the solution

(36) $$u(x, t) = \frac{1}{2}\left[\sin(x + 2t) + \sin(x - 2t)\right] + \frac{1}{4}\int_{x-2t}^{x+2t} ds$$

$$= \sin x \cos 2t + t . \quad \blacksquare \text{End}$$

We now use d'Alembert's formula to show that the solution to the "infinite" string problem consists of traveling waves.

Let $h(x)$ be a function defined on $(-\infty, \infty)$. The function $h(x + a)$, where $a > 0$, is a translation of the function $h(x)$ in the sense that its "shape" is the same as $h(x)$, but its position has been shifted to the left by an amount a. This is illustrated in Figure 10.18 for a function $h(x)$ whose graph consists of a triangular "bump." If we let $t \geq 0$ be a parameter (say, time), then the functions $h(x + \alpha t)$ represent a family of functions with the same shape but shifted farther and farther to the left as $t \to \infty$ (Figure 10.19). We say that $h(x + \alpha t)$ is a **traveling wave** moving to the left with speed α. In a similar fashion, $h(x - \alpha t)$ is a traveling wave moving to the right with speed α.

If we refer to formula (25), we find that the solution to $\partial^2 u/\partial t^2 = \alpha^2 \partial^2 u/\partial x^2$ consists of traveling waves $A(x + \alpha t)$ moving to the left with speed α and $B(x - \alpha t)$ moving to the right at the same speed.

In the special case when the initial velocity $g(x) \equiv 0$, we have

$$u(x, t) = \frac{1}{2}\left[f(x + \alpha t) + f(x - \alpha t)\right] .$$

Hence $u(x, t)$ is the sum of the traveling waves

$$\frac{1}{2}f(x + \alpha t) \quad \text{and} \quad \frac{1}{2}f(x - \alpha t) .$$

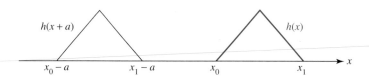

Figure 10.18 Graphs of $h(x)$ and $h(x + a)$

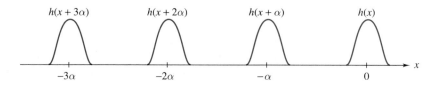

Figure 10.19 Traveling wave

These waves are initially superimposed, since

$$u(x, 0) = \frac{1}{2}f(x) + \frac{1}{2}f(x) = f(x) .$$

As t increases, the two waves move away from each other with speed 2α. This is illustrated in Figure 10.20 for a triangular wave.

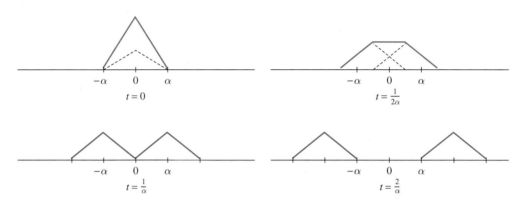

Figure 10.20 Decomposition of initial displacement into traveling waves

EXAMPLE 4 Express the standing wave $\cos\dfrac{\pi\alpha}{L}t \sin\dfrac{\pi x}{L}$ as a superposition of traveling waves.

SOLUTION By a familiar trigonometric identity,

$$\cos\frac{\pi\alpha}{L}t \sin\frac{\pi x}{L} = \frac{1}{2}\sin\frac{\pi}{L}(x - \alpha t) + \frac{1}{2}\sin\frac{\pi}{L}(x + \alpha t)$$
$$= h(x - \alpha t) + h(x + \alpha t) ,$$

where

$$h(x) = \frac{1}{2}\sin\frac{\pi x}{L} .$$

This standing wave results from adding a wave traveling to the right to the same waveform traveling to the left. ■ End

Existence and Uniqueness of Solutions

In Example 1, the method of separation of variables was used to derive a formal solution to the given initial-boundary value problem. To show that these series converge to an actual solution requires results from real analysis just as was the case for the formal solutions to the heat equation in Section 10.5. In Examples 2 and 3, we can establish the validity of d'Alembert's solution by direct substitution into the initial value problem, assuming sufficient differentiability of the initial functions. We leave it as an exercise for the reader to show that if f has a continuous

second derivative and g has a continuous first derivative, then d'Alembert's solution is a true solution (see Problem 12).

The question of the uniqueness of the solution to the initial-boundary value problem (1)–(4) can be answered using an **energy argument.**

UNIQUENESS OF THE SOLUTION TO THE VIBRATING STRING PROBLEM

Theorem 8. The initial-boundary value problem

(37) $$\frac{\partial^2 u}{\partial t^2} = \alpha^2 \frac{\partial^2 u}{\partial x^2} \ , \qquad 0 < x < L \ , \qquad t > 0 \ ,$$

(38) $\quad u(0, t) = u(L, t) = 0 \ , \qquad t > 0 \ ,$

(39) $\quad u(x, 0) = f(x) \ , \qquad 0 < x < L \ ,$

(40) $\quad \dfrac{\partial u}{\partial t}(x, 0) = g(x) \ , \qquad 0 < x < L \ ,$

has at most one twice continuously differentiable solution.

Proof. Assume both $u(x, t)$ and $v(x, t)$ are twice continuously differentiable solutions to (37)–(40) and let $w(x, t) := u(x, t) - v(x, t)$. It is easy to check that $w(x, t)$ satisfies the initial-boundary value problem (37)–(40) with zero initial data; that is, for $0 \le x \le L$,

(41) $w(x, 0) = 0 \quad \text{and} \quad \dfrac{\partial w}{\partial t}(x, 0) = 0 \ .$

We now show that $w(x, t) \equiv 0$ for $0 \le x \le L, t \ge 0$.

If $w(x, t)$ is the displacement of the vibrating string at location x for time t, then with the appropriate units, the total energy $E(t)$ of the vibrating string at time t is defined by the integral

(42) $E(t) := \dfrac{1}{2} \displaystyle\int_0^L \left[\alpha^2 \left(\dfrac{\partial w}{\partial x} \right)^2 + \left(\dfrac{\partial w}{\partial t} \right)^2 \right] dx \ .$

(The first term in the integrand relates to the stretching of the string at location x and represents the potential energy. The second term is the square of the velocity of the vibrating string at x and represents the kinetic energy.)

We now consider the derivative of $E(t)$:

$$\frac{dE}{dt} = \frac{d}{dt} \left\{ \frac{1}{2} \int_0^L \left[\alpha^2 \left(\frac{\partial w}{\partial x} \right)^2 + \left(\frac{\partial w}{\partial t} \right)^2 \right] dx \right\} \ .$$

Since w has continuous second partial derivatives (because u and v do), we can interchange the order of integration and differentiation. This gives

(43) $\dfrac{dE}{dt} = \displaystyle\int_0^L \left[\alpha^2 \dfrac{\partial w}{\partial x} \dfrac{\partial^2 w}{\partial t \partial x} + \dfrac{\partial w}{\partial t} \dfrac{\partial^2 w}{\partial t^2} \right] dx \ .$

Again the continuity of the second partials of w guarantees that the mixed partials are equal; that is,

$$\frac{\partial^2 w}{\partial t \partial x} = \frac{\partial^2 w}{\partial x \partial t} .$$

Combining this fact with integration by parts, we obtain

(44) $$\int_0^L \alpha^2 \frac{\partial w}{\partial x} \frac{\partial^2 w}{\partial t \partial x} dx = \int_0^L \alpha^2 \frac{\partial w}{\partial x} \frac{\partial^2 w}{\partial x \partial t} dx$$

$$= \alpha^2 \frac{\partial w}{\partial x}(L, t) \frac{\partial w}{\partial t}(L, t) - \alpha^2 \frac{\partial w}{\partial x}(0, t) \frac{\partial w}{\partial t}(0, t)$$

$$- \int_0^L \alpha^2 \frac{\partial^2 w}{\partial x^2} \frac{\partial w}{\partial t} dx .$$

The boundary conditions $w(0, t) = w(L, t) = 0$, $t \geq 0$, imply that $(\partial w/\partial t)(0, t) = (\partial w/\partial t)(L, t) = 0$, $t \geq 0$. This reduces equation (44) to

$$\int_0^L \alpha^2 \frac{\partial w}{\partial x} \frac{\partial^2 w}{\partial t \partial x} dx = -\int_0^L \alpha^2 \frac{\partial w}{\partial t} \frac{\partial^2 w}{\partial x^2} dx .$$

Substituting this in for the first integrand in (43), we find

$$\frac{dE}{dt} = \int_0^L \frac{\partial w}{\partial t} \left[\frac{\partial^2 w}{\partial t^2} - \alpha^2 \frac{\partial^2 w}{\partial x^2} \right] dx .$$

Since w satisfies equation (37), the integrand is zero for all x. Thus $dE/dt = 0$, and so $E(t) \equiv C$, where C is a constant. This means that the total energy is conserved within the vibrating string.

The first boundary condition in (41) states that $w(x, 0) = 0$ for $0 \leq x \leq L$. Hence $(\partial w/\partial x)(x, 0) = 0$ for $0 < x < L$. Combining this with the second boundary condition in (41), we find that when $t = 0$, the integrand in (42) is zero for $0 < x < L$. Therefore, $E(0) = 0$. Since $E(t) \equiv C$, we must have $C = 0$. Hence

(45) $$E(t) = \frac{1}{2} \int_0^L \left[\alpha^2 \left(\frac{\partial w}{\partial x} \right)^2 + \left(\frac{\partial w}{\partial t} \right)^2 \right] dx \equiv 0 .$$

That is, the total energy of the system is zero.

Because the integrand in (45) is nonnegative and continuous and the integral is zero, the integrand must be zero for $0 \leq x \leq L$. Moreover, the integrand is the sum of two squares and so each term must be zero. Hence

$$\frac{\partial w}{\partial x}(x, t) = 0 \quad \text{and} \quad \frac{\partial w}{\partial t}(x, t) = 0$$

for all $0 \leq x \leq L$, $t \geq 0$. Thus $w(x, t) = K$, where K is a constant. Physically, this says that there is no motion in the string.

Finally, since w is constant and w is zero when $t = 0$, then $w(x, t) \equiv 0$. Consequently, $u(x, t) = v(x, t)$ and the initial-boundary value problem has at most one solution. ■ End

EXERCISES 10.6

In Problems 1 through 4, find a formal solution to the vibrating string problem governed by the given initial-boundary value problem.

1. $\dfrac{\partial^2 u}{\partial t^2} = \dfrac{\partial^2 u}{\partial x^2}$, $0 < x < 1$, $t > 0$,

$u(0, t) = u(1, t) = 0$, $t > 0$,

$u(x, 0) = x(1 - x)$, $0 < x < 1$,

$\dfrac{\partial u}{\partial t}(x, 0) = \sin 7\pi x$, $0 < x < 1$.

2. $\dfrac{\partial^2 u}{\partial t^2} = 16\dfrac{\partial^2 u}{\partial x^2}$, $0 < x < \pi$, $t > 0$,

$u(0, t) = u(\pi, t) = 0$, $t > 0$,

$u(x, 0) = \sin^2 x$, $0 < x < \pi$,

$\dfrac{\partial u}{\partial t}(x, 0) = 1 - \cos x$, $0 < x < \pi$.

3. $\dfrac{\partial^2 u}{\partial t^2} = 4\dfrac{\partial^2 u}{\partial x^2}$, $0 < x < \pi$, $t > 0$,

$u(0, t) = u(\pi, t) = 0$, $t > 0$,

$u(x, 0) = x^2(\pi - x)$, $0 < x < \pi$,

$\dfrac{\partial u}{\partial t}(x, 0) = 0$, $0 < x < \pi$.

4. $\dfrac{\partial^2 u}{\partial t^2} = 9\dfrac{\partial^2 u}{\partial x^2}$, $0 < x < \pi$, $t > 0$,

$u(0, t) = u(\pi, t) = 0$, $t > 0$,

$u(x, 0) = \sin 4x + 7 \sin 5x$, $0 < x < \pi$,

$\dfrac{\partial u}{\partial t}(x, 0) = \begin{cases} x , & 0 < x < \pi/2 , \\ \pi - x , & \pi/2 < x < \pi . \end{cases}$

5. The Plucked String. A vibrating string is governed by the initial-boundary value problem (1)–(4). If the string is lifted to a height h_0 at $x = a$ and released, then the initial conditions are

$f(x) = \begin{cases} h_0 x/a , & 0 < x \leq a , \\ h_0(L - x)/(L - a) , & a < x < L , \end{cases}$

and $g(x) \equiv 0$. Find a formal solution.

6. The Struck String. A vibrating string is governed by the initial-boundary value problem (1)–(4). If the string is struck at $x = a$, then the initial conditions may be approximated by $f(x) \equiv 0$ and

$g(x) = \begin{cases} v_0 x/a , & 0 < x \leq a , \\ v_0(L - x)/(L - a) , & a < x < L , \end{cases}$

where v_0 is a constant. Find a formal solution.

In Problems 7 and 8, find a formal solution to the vibrating string problem governed by the given nonhomogeneous initial-boundary value problem.

7. $\dfrac{\partial^2 u}{\partial t^2} = \dfrac{\partial^2 u}{\partial x^2} + tx$, $0 < x < \pi$, $t > 0$,

$u(0, t) = u(\pi, t) = 0$, $t > 0$,

$u(x, 0) = \sin x$, $0 < x < \pi$,

$\dfrac{\partial u}{\partial t}(x, 0) = 5 \sin 2x - 3 \sin 5x$, $0 < x < \pi$.

8. $\dfrac{\partial^2 u}{\partial t^2} = \dfrac{\partial^2 u}{\partial x^2} + x \sin t$, $0 < x < \pi$, $t > 0$,

$u(0, t) = u(\pi, t) = 0$, $t > 0$,

$u(x, 0) = 0$, $0 < x < \pi$,

$\dfrac{\partial u}{\partial t}(x, 0) = 0$, $0 < x < \pi$.

9. If one end of a string is held fixed while the other is free, then the motion of the string is governed by the initial-boundary value problem

$$\dfrac{\partial^2 u}{\partial t^2} = \alpha^2 \dfrac{\partial^2 u}{\partial x^2} , \qquad 0 < x < L , \qquad t > 0 ,$$

$u(0, t) = 0$ and $\dfrac{\partial u}{\partial x}(L, t) = 0$, $t > 0$,

$u(x, 0) = f(x)$, $0 < x < L$,

$\dfrac{\partial u}{\partial t}(x, 0) = g(x)$, $0 < x < L$.

Derive a formula for a formal solution.

10. Derive a formula for the solution to the following initial-boundary value problem involving nonhomogeneous boundary conditions

$$\frac{\partial^2 u}{\partial t^2} = \alpha^2 \frac{\partial^2 u}{\partial x^2} , \qquad 0 < x < L , \qquad t > 0 ,$$

$$u(0, t) = U_1 , \qquad u(L, t) = U_2 , \qquad t > 0 ,$$
$$u(x, 0) = f(x) , \qquad 0 < x < L ,$$

$$\frac{\partial u}{\partial t}(x, 0) = g(x) , \qquad 0 < x < L ,$$

where U_1 and U_2 are constants.

11. **The Telegraph Problem.**[†] Use the method of separation of variables to derive a formal solution to the telegraph problem

$$\frac{\partial^2 u}{\partial t^2} + \frac{\partial u}{\partial t} + u = \alpha^2 \frac{\partial^2 u}{\partial x^2} , \qquad 0 < x < L , \qquad t > 0 ,$$

$$u(0, t) = u(L, t) = 0 , \qquad t > 0 ,$$
$$u(x, 0) = f(x) , \qquad 0 < x < L ,$$

$$\frac{\partial u}{\partial t}(x, 0) = 0 , \qquad 0 < x < L .$$

12. Verify d'Alembert's solution (32) to the initial value problem (26)–(28) when $f(x)$ has a continuous second derivative and $g(x)$ has a continuous first derivative by substituting it directly into the equations.

In Problems 13 through 18, find the solution to the initial value problem.

$$\frac{\partial^2 u}{\partial t^2} = \alpha^2 \frac{\partial^2 u}{\partial x^2} , \qquad -\infty < x < \infty , \qquad t > 0 ,$$

$$u(x, 0) = f(x) , \qquad -\infty < x < \infty ,$$

$$\frac{\partial u}{\partial t}(x, 0) = g(x) , \qquad -\infty < x < \infty ,$$

for the given functions $f(x)$ and $g(x)$.

13. $f(x) \equiv 0 , \qquad g(x) = \cos x .$
14. $f(x) = x^2 , \qquad g(x) \equiv 0 .$
15. $f(x) = x , \qquad g(x) = x .$
16. $f(x) = \sin 3x , \qquad g(x) \equiv 1 .$
17. $f(x) = e^{-x^2} , \qquad g(x) = \sin x .$
18. $f(x) = \cos 2x , \qquad g(x) = 1 - x .$

19. Derive the formal solution given in equations (22)–(24) to the vibrating membrane problem governed by the initial-boundary value problem (17)–(21).

20. **Long Water Waves.** The motion of long water waves in a channel of constant depth is governed by the **linearized Korteweg and de Vries (KdV) equation**

$$(46) \qquad u_t + \alpha u_x + \beta u_{xxx} = 0 ,$$

where $u(x, t)$ is the displacement of the water from its equilibrium depth at location x and at time t, and α and β are positive constants.

(a) Show that equation (46) has a solution of the form

$$(47) \qquad u(x, t) = V(z) ,$$
$$z = kx - w(k)t ,$$

where k is a fixed constant and $w(k)$ is a function of k, provided V satisfies

$$(48) \qquad -w\frac{dV}{dz} + \alpha k\frac{dV}{dz} + \beta k^3 \frac{d^3 V}{dz^3} = 0 .$$

These solutions, defined by (47), are called **uniform waves.**

(b) Physically, we are interested only in solutions $V(z)$ that are bounded and nonconstant on the infinite interval $(-\infty, \infty)$. Show that such solutions exist only if $\alpha k - w > 0$.

(c) Let $\lambda^2 = (\alpha k - w)/(\beta k^3)$. Show that the solutions from part (b) can be expressed in the form

$$V(x, t) = A \sin\left[\lambda kx - (\alpha \lambda k - \beta \lambda^3 k^3)t + B\right] ,$$

where A and B are arbitrary constants. [*Hint:* Solve for w in terms of λ and k and use (47).]

(d) Since both λ and k can be chosen arbitrarily and they always appear together as the product λk, we can set $\lambda = 1$ without loss of generality. Hence we have

$$V(x, t) = A \sin\left[kx - (\alpha k - \beta k^3)t + B\right] ,$$

as a uniform wave solution to (46). The defining relation $w(k) = \alpha k - \beta k^3$ is called the

[†]For a discussion of the telegraph problem, see *Methods of Mathematical Physics,* by R. Courant and D. Hibert, Vol. II (Wiley-Interscience, New York, 1989).

dispersion relation, the ratio $w(k)/k = \alpha - \beta k^2$ is called the **phase velocity,** and the derivative $dw/dk = \alpha - 3\beta k^2$ is called the **group velocity.** When the group velocity is not constant, the waves are called **dispersive.** Show that the standard wave equation $u_{tt} = \alpha^2 u_{xx}$ has only nondispersive waves.

21. Vibrating Drum. A vibrating circular membrane of unit radius whose edges are held fixed in a plane and whose displacement $u(r, t)$ depends only on the radial distance r from the center and on the time t is governed by the initial-boundary value problem.

$$\frac{\partial^2 u}{\partial t^2} = \alpha^2 \left(\frac{\partial^2 u}{\partial r^2} + \frac{1}{r}\frac{\partial u}{\partial r} \right) ,$$

$$0 < r < 1 , \quad t > 0 ,$$

$$u(1, t) = 0 , \qquad t > 0 ,$$

$$u(r, t) \quad \text{remains finite as } r \to 0^+ ,$$

$$u(r, 0) = f(r) , \qquad 0 < r < 1 ,$$

$$\frac{\partial u}{\partial t}(r, 0) = g(r) , \qquad 0 < r < 1 ,$$

where f and g are the initial displacements and velocities, respectively. Use the method of separation of variables to derive a formal solution to the vibrating drum problem. [*Hint:* Show that there is a family of solutions of the form

$$u_n(r, t) = \left[a_n \cos(k_n \alpha t) + b_n \sin(k_n \alpha t) \right] J_0(k_n r) ,$$

where J_0 is the Bessel function of the first kind of order zero and $0 < k_1 < k_2 < \cdots < k_n < \cdots$ are the positive zeros of J_0. Now use superposition.] See Figure 10.21.

Figure 10.21 Mode shapes for vibrating drum. (a) $J_0(2.405r)$, (b) $J_0(5.520r)$, (c) $J_0(8.654r)$

10.7 LAPLACE'S EQUATION

In Section 10.1 we showed how Laplace's equation,

$$\frac{\partial^2 u}{\partial x^2} + \frac{\partial^2 u}{\partial y^2} = 0 ,$$

arises in the study of steady-state or time-independent solutions to the heat equation. Because these solutions do not depend on time, initial conditions are irrelevant and only boundary conditions are specified. Other applications include the static displacement $u(x, y)$ of a stretched membrane fastened in space along the boundary of a region (here u must satisfy Laplace's equation inside the region); the electrostatic and gravitational potentials in certain force fields (here u must satisfy Laplace's equation in any region that is free of electrical charges or mass);

and, in fluid mechanics for an idealized fluid, the stream function $u(x, y)$ whose level curves (stream lines), $u(x, y) = $ constant, represent the path of particles in the fluid (again u satisfies Laplace's equation in the flow region).

There are two basic types of boundary conditions that are usually associated with Laplace's equation: **Dirichlet boundary conditions,** where the solution $u(x, y)$ to Laplace's equation in a domain D is required to satisfy

$$u(x, y) = f(x, y) \qquad \text{on } \partial D \ ,$$

with $f(x, y)$ a specified function defined on the boundary ∂D of D; and **Neumann boundary conditions,** where the directional derivative $\partial u / \partial n$ along the outward normal to the boundary is required to satisfy

$$\frac{\partial u}{\partial n}(x, y) = g(x, y) \qquad \text{on } \partial D \ ,$$

with $g(x, y)$ a specified function defined on ∂D. We say that the boundary conditions are **mixed** if the solution is required to satisfy $u(x, y) = f(x, y)$ on part of the boundary and $(\partial u / \partial n)(x, y) = g(x, y)$ on the remaining portion of the boundary.

In this section we use the method of separation of variables to find solutions to Laplace's equation with various boundary conditions for rectangular, circular, and cylindrical domains. We also discuss the existence and uniqueness of such solutions.

EXAMPLE 1 Find a solution to the following mixed boundary value problem for a rectangle (see Figure 10.22):

(1) $$\frac{\partial^2 u}{\partial x^2} + \frac{\partial^2 u}{\partial y^2} = 0 \ , \qquad 0 < x < a \ , \qquad 0 < y < b \ ,$$

(2) $$\frac{\partial u}{\partial x}(0, y) = \frac{\partial u}{\partial x}(a, y) = 0 \ , \qquad 0 \le y \le b \ ,$$

(3) $$u(x, b) = 0 \ , \qquad 0 \le x \le a \ ,$$

(4) $$u(x, 0) = f(x) \ , \qquad 0 \le x \le a \ .$$

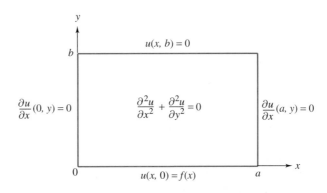

Figure 10.22 Mixed boundary value problem

SOLUTION Separating variables, we first let $u(x, y) = X(x)Y(y)$. Substituting into equation (1), we have

$$X''(x)Y(y) + X(x)Y''(y) = 0 ,$$

which separates into

$$\frac{X''(x)}{X(x)} = -\frac{Y''(y)}{Y(y)} = K ,$$

where K is some constant. This leads to the two ordinary differential equations

(5) $X''(x) - KX(x) = 0 ,$

(6) $Y''(y) + KY(y) = 0 .$

From the boundary condition (2), we observe that

(7) $X'(0) = X'(a) = 0 .$

We have encountered the eigenvalue problem in (5) and (7) before (see Example 1 in Section 10.5). The eigenvalues are $K = K_n = -(n\pi/a)^2$, $n = 0, 1, 2, \ldots$, with corresponding solutions

(8) $X_n(x) = a_n \cos\left(\dfrac{n\pi x}{a}\right) ,$

where the a_n's are arbitrary constants.

Setting $K = K_n = -(n\pi/a)^2$ in equation (6) and solving for Y gives[†]

$$Y_0(y) = A_0 + B_0 y ,$$

(9)
$$Y_n(y) = A_n \cosh\left(\frac{n\pi y}{a}\right) + B_n \sinh\left(\frac{n\pi y}{a}\right) , \qquad n = 1, 2, \ldots .$$

Now the boundary condition $u(x, b) = 0$ in (3) will be satisfied if $Y(b) = 0$. Setting $y = b$ in (9), we see that we want $A_0 = -bB_0$ and

$$A_n = -B_n \tanh\left(\frac{n\pi b}{a}\right), \text{ or equivalently, } Y_n(y) = C_n \sinh\left[\frac{n\pi}{a}(y - b)\right], n = 1, 2, \ldots .$$

(see Problem 18). Thus we find that there are solutions to (1)–(3) of the form

$$u_0(x, y) = X_0(x)Y_0(y) = a_0 B_0(y - b) = E_0(y - b) ,$$

$$u_n(x, y) = X_n(x)Y_n(y) = a_n \cos\left(\frac{n\pi x}{a}\right) C_n \sinh\left[\frac{n\pi}{a}(y - b)\right]$$

$$= E_n \cos\left(\frac{n\pi x}{a}\right) \sinh\left[\frac{n\pi}{a}(y - b)\right] , \qquad n = 1, 2, \ldots ,$$

[†]We usually express $Y_n(y) = a_n e^{n\pi y/a} + b_n e^{-n\pi y/a}$. However, computation is simplified in this case by using the hyperbolic functions $\cosh z = (e^z + e^{-z})/2$ and $\sinh z = (e^z - e^{-z})/2$.

where the E_n's are constants. In fact, by the superposition principle,

(10) $$u(x, y) = E_0(y - b) + \sum_{n=1}^{\infty} E_n \cos\left(\frac{n\pi x}{a}\right) \sinh\left[\frac{n\pi}{a}(y - b)\right]$$

is a formal solution to (1)–(3).

Applying the remaining nonhomogeneous boundary condition in (4), we have

$$u(x, 0) = f(x) = -E_0 b + \sum_{n=1}^{\infty} E_n \sinh\left(-\frac{n\pi b}{a}\right) \cos\left(\frac{n\pi x}{a}\right) .$$

This is a Fourier cosine series for $f(x)$ and hence the coefficients are given by the formulas

(11)

$$E_0 = \frac{1}{(-ba)} \int_0^a f(x)\, dx \ ,$$

$$E_n = \frac{2}{a \sinh\left(-\dfrac{n\pi b}{a}\right)} \int_0^a f(x) \cos\left(\frac{n\pi x}{a}\right) dx \ , \qquad n = 1, 2, \dots .$$

Thus a formal solution is given by (10) with the constants E_n given by (11). See Figure 10.23 for a sketch of partial sums for (10). ■ End

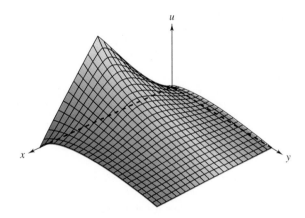

Figure 10.23 Partial sums for Example 1 with $f(x) = \begin{cases} x, & 0 < x \leq \pi/2 \\ \pi - x, & \pi/2 \leq x < \pi \end{cases}$

In Example 1 the boundary conditions were homogeneous on three sides of the rectangle and nonhomogeneous on the fourth side, $\{(x, y): y = 0, 0 \leq x \leq a\}$. It is important to note that the method used in Example 1 can also be used to solve problems for which the boundary conditions are nonhomogeneous on all sides. This is accomplished by solving four separate boundary value problems in which three sides have homogeneous boundary conditions and only one side is nonhomogeneous. The solution is then obtained by summing these four solutions (see Problem 5).

For problems involving circular domains, it is usually more convenient to use polar coordinates. In rectangular coordinates the Laplacian has the form

$$\Delta u = \frac{\partial^2 u}{\partial x^2} + \frac{\partial^2 u}{\partial y^2} \ .$$

In polar coordinates (r, θ), we let

$$x = r \cos \theta \ , \qquad y = r \sin \theta$$

so that

$$r = \sqrt{x^2 + y^2} \ , \qquad \tan \theta = y/x \ .$$

With patience and a little care in applying the chain rule, one can show that the Laplacian in polar coordinates is

$$(12) \qquad \Delta u = \frac{\partial^2 u}{\partial r^2} + \frac{1}{r} \frac{\partial u}{\partial r} + \frac{1}{r^2} \frac{\partial^2 u}{\partial \theta^2}$$

(see Problem 6). In the next example we obtain a solution to the **Dirichlet problem** in a disk of radius a.

EXAMPLE 2 A circular metal disk of radius a has its top and bottom insulated. The edge $(r = a)$ of the disk is kept at a specified temperature that depends on its location (varies with θ). The steady-state temperature inside the disk satisfies Laplace's equation. Determine the temperature distribution $u(r, \theta)$ inside the disk by finding the solution to the following Dirichlet boundary value problem, depicted in Figure 10.24:

$$(13) \qquad \frac{\partial^2 u}{\partial r^2} + \frac{1}{r} \frac{\partial u}{\partial r} + \frac{1}{r^2} \frac{\partial^2 u}{\partial \theta^2} = 0 \ , \qquad 0 \le r < a \ , \qquad -\pi \le \theta \le \pi \ ,$$

$$(14) \qquad u(a, \theta) = f(\theta) \ , \qquad -\pi \le \theta \le \pi \ .$$

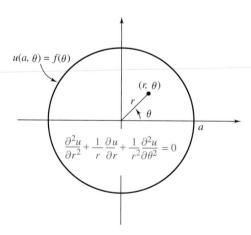

Figure 10.24 Steady-state temperature distribution in a disk

SOLUTION To use the method of separation of variables, we first set

$$u(r, \theta) = R(r)T(\theta) \ ,$$

where $0 \leq r < a$ and $-\pi \leq \theta \leq \pi$. Substituting into (13) and separating variables give

$$\frac{r^2 R''(r) + rR'(r)}{R(r)} = -\frac{T''(\theta)}{T(\theta)} = \lambda \ ,$$

where λ is any constant. This leads to the two ordinary differential equations

(15) $r^2 R''(r) + rR'(r) - \lambda R(r) = 0 \ ,$

(16) $T''(\theta) + \lambda T(\theta) = 0 \ .$

For $u(r, \theta)$ to be continuous in the disk $0 \leq r < a$, we need $T(\theta)$ to be 2π-periodic; in particular, we require

(17) $T(-\pi) = T(\pi) \quad \text{and} \quad T'(-\pi) = T'(\pi) \ .$

Therefore we seek nontrivial solutions to the eigenvalue problem (16)–(17).

When $\lambda < 0$, the general solution to (16) is the sum of two exponentials. Hence we have only trivial 2π-periodic solutions.

When $\lambda = 0$, we find $T(\theta) = A\theta + B$ to be the solution to (16). This linear function is periodic only when $A = 0$, that is, $T_0(\theta) = B$ is the only 2π-periodic solution corresponding to $\lambda = 0$.

When $\lambda > 0$, the general solution to (16) is

$$T(\theta) = A \cos \sqrt{\lambda}\theta + B \sin \sqrt{\lambda}\theta \ .$$

Here we get a nontrivial 2π-periodic solution only when $\sqrt{\lambda} = n, n = 1, 2, \ldots$. (The reader can check this using (17).) Hence we obtain the nontrivial 2π-periodic solutions

(18) $T_n(\theta) = A_n \cos n\theta + B_n \sin n\theta$

corresponding to $\sqrt{\lambda} = n, n = 1, 2, \ldots$.

Now for $\lambda = n^2, n = 0, 1, 2, \ldots$, equation (15) is the Cauchy-Euler equation

(19) $r^2 R''(r) + rR'(r) - n^2 R(r) = 0$

(see Section 4.5, page 187). When $n = 0$, the general solution is

$$R_0(r) = C + D \ln r \ .$$

Since $\ln r \to -\infty$ as $r \to 0^+$, this solution is unbounded near $r = 0$ when $D \neq 0$. Therefore we must choose $D = 0$ if $u(r, \theta)$ is to be continuous at $r = 0$. We now have $R_0(r) = C$ and so $u_0(r, \theta) = R_0(r)T(\theta) = CB$, which for convenience we write in the form

(20) $u_0(r, \theta) = \dfrac{A_0}{2} \ ,$

where A_0 is an arbitrary constant.

When $\lambda = n^2, n = 1, 2, \ldots$, the reader should verify that equation (19) has the general solution

$$R_n(r) = C_n r^n + D_n r^{-n} .$$

Since $r^{-n} \to \infty$ as $r \to 0^+$, we must set $D_n = 0$ in order for $u(r, \theta)$ to be bounded at $r = 0$. Thus

$$R_n(r) = C_n r^n .$$

Now for each $n = 1, 2, \ldots$, we have the solutions

(21) $\qquad u_n(r, \theta) = R_n(r) T_n(\theta) = C_n r^n (A_n \cos n\theta + B_n \sin n\theta) ,$

and by forming an infinite series from the solutions in (20) and (21), we get the following formal solution to (13):

$$u(r, \theta) = \frac{A_0}{2} + \sum_{n=1}^{\infty} C_n r^n (A_n \cos n\theta + B_n \sin n\theta) .$$

It is more convenient to write this series in the equivalent form

(22) $\qquad u(r, \theta) = \frac{a_0}{2} + \sum_{n=1}^{\infty} \left(\frac{r}{a}\right)^n (a_n \cos n\theta + b_n \sin n\theta) ,$

where the a_n's and b_n's are constants. These constants can be determined from the boundary condition; indeed with $r = a$ in (22), condition (14) becomes

$$f(\theta) = \frac{a_0}{2} + \sum_{n=1}^{\infty} (a_n \cos n\theta + b_n \sin n\theta) .$$

Hence if $f(\theta)$ is 2π-periodic, we recognize that a_n, b_n are Fourier coefficients. Thus

(23) $\qquad a_n = \frac{1}{\pi} \int_{-\pi}^{\pi} f(\theta) \cos n\theta \, d\theta , \qquad n = 0, 1, \ldots ,$

(24) $\qquad b_n = \frac{1}{\pi} \int_{-\pi}^{\pi} f(\theta) \sin n\theta \, d\theta , \qquad n = 1, 2, \ldots .$

To summarize, if a_n and b_n are defined by formulas (23) and (24), then $u(r, \theta)$ given in (22) is a formal solution to the Dirichlet problem (13)–(14). ∎ End

The procedure in Example 2 can also be used to study the **Neumann problem** in a disk:

(25) $\qquad \Delta u = 0 , \qquad 0 \le r < a , \qquad -\pi \le \theta \le \pi ,$

(26) $\qquad \dfrac{\partial u}{\partial r}(a, \theta) = f(\theta) , \qquad -\pi \le \theta \le \pi .$

For this problem there is no unique solution, since if u is a solution, then the function u plus a constant is also a solution. Moreover, f must satisfy the **consistency condition**

(27) $\qquad \displaystyle\int_{-\pi}^{\pi} f(\theta) d\theta = 0 .$

If we interpret the solution $u(r, \theta)$ of equation (25) as the steady-state temperature distribution inside a circular disk that does not contain either a heat source or heat sink, then equation (26) specifies the flow of heat across the boundary of the disk. Here the consistency condition (27) is simply the requirement that the net flow of heat across the boundary is zero. (If we keep pumping heat in, the temperature won't reach equilibrium!) We leave the solution of the Neumann problem and the derivation of the consistency condition for the exercises.

The technique used in Example 2 also applies to annular domains, $\{(r, \theta): 0 < a < r < b\}$, and to exterior domains, $\{(r, \theta): a < r\}$. We also leave these applications as exercises.

Laplace's equation in cylindrical coordinates arises in the study of steady-state temperature distributions in a solid cylinder and in determining the electric potential inside a cylinder. In cylindrical coordinates,

$$x = r\cos\theta \, , \qquad y = r\sin\theta \, , \qquad z = z \, ,$$

Laplace's equation becomes

$$(28) \qquad \Delta u = \frac{\partial^2 u}{\partial r^2} + \frac{1}{r}\frac{\partial u}{\partial r} + \frac{1}{r^2}\frac{\partial^2 u}{\partial\theta^2} + \frac{\partial^2 u}{\partial z^2} = 0 \, .$$

The Dirichlet problem for the cylinder $\{(r, \theta, z): 0 \le r \le a, 0 \le z \le b\}$ has the boundary conditions

$$(29) \qquad u(a, \theta, z) = f(\theta, z) \, , \qquad -\pi \le \theta \le \pi \, , \qquad 0 \le z \le b \, ,$$

$$(30) \qquad u(r, \theta, 0) = g(r, \theta) \, , \qquad 0 \le r \le a \, , \qquad -\pi \le \theta \le \pi \, ,$$

$$(31) \qquad u(r, \theta, b) = h(r, \theta) \, , \qquad 0 \le r \le a \, , \qquad -\pi \le \theta \le \pi \, .$$

To solve the Dirichlet boundary value problem (28)–(31), we first solve the three boundary value problems corresponding to: (i) $g \equiv 0$ and $h \equiv 0$; (ii) $f \equiv 0$ and $h \equiv 0$; and (iii) $f \equiv 0$ and $g \equiv 0$. Then by the superposition principle, the solution to (28)–(31) will be the sum of these three solutions. This is the same method that was discussed in dealing with Dirichlet problems on rectangular domains. (See the remarks following Example 1.) In the next example we solve the Dirichlet problem when $g \equiv 0$ and $h \equiv 0$.

EXAMPLE 3 The base $(z = 0)$ and the top $(z = b)$ of a charge-free cylinder are grounded and therefore are at zero potential. The potential on the lateral surface $(r = a)$ of the cylinder is given by $u(a, \theta, z) = f(\theta, z)$ where $f(\theta, 0) = f(\theta, b) = 0$. Inside the cylinder, the potential $u(r, \theta, z)$ satisfies Laplace's equation. Determine the potential u inside the cylinder by finding a solution to the Dirichlet boundary value problem

$$(32) \qquad \frac{\partial^2 u}{\partial r^2} + \frac{1}{r}\frac{\partial u}{\partial r} + \frac{1}{r^2}\frac{\partial^2 u}{\partial\theta^2} + \frac{\partial^2 u}{\partial z^2} = 0 \, , \quad 0 \le r < a \, , \quad -\pi \le \theta \le \pi \, , \quad 0 < z < b \, ,$$

$$(33) \qquad u(a, \theta, z) = f(\theta, z) \, , \qquad -\pi \le \theta \le \pi \, , \qquad 0 \le z \le b \, ,$$

$$(34) \qquad u(r, \theta, 0) = u(r, \theta, b) = 0 \, , \qquad 0 \le r < a \, , \qquad -\pi \le \theta \le \pi \, .$$

SOLUTION Using the method of separation of variables, we first assume

$$u(r, \theta, z) = R(r)T(\theta)Z(z) \, .$$

Substituting into equation (32) and separating out the Z's, we find

$$\frac{R''(r) + (1/r)R'(r)}{R(r)} + \frac{1}{r^2}\frac{T''(\theta)}{T(\theta)} = -\frac{Z''(z)}{Z(z)} = \lambda ,$$

where λ can be any constant. Separating further the R's and T's gives

$$\frac{r^2 R''(r) + rR'(r)}{R(r)} - r^2\lambda = -\frac{T''(\theta)}{T(\theta)} = \mu ,$$

where μ can also be any constant. We now have the three ordinary differential equations:

(35) $r^2 R''(r) + rR'(r) - (r^2\lambda + \mu)R(r) = 0 ,$

(36) $T''(\theta) + \mu T(\theta) = 0 ,$

(37) $Z''(z) + \lambda Z(z) = 0 .$

For u to be continuous in the cylinder, $T(\theta)$ must be 2π-periodic. Thus let's begin with the eigenvalue problem

$$T''(\theta) + \mu T(\theta) = 0 , \qquad -\pi < \theta < \pi ,$$
$$T(-\pi) = T(\pi) \quad \text{and} \quad T'(-\pi) = T'(\pi) .$$

In Example 2 we showed that this problem has nontrivial solutions for $\mu = n^2$, $n = 0, 1, 2, \ldots$, that are given by

(38) $T_n(\theta) = A_n \cos n\theta + B_n \sin n\theta ,$

where the A_n's and B_n's are arbitrary constants.

The boundary conditions in (34) imply that $Z(0) = Z(b) = 0$. Therefore Z must satisfy the eigenvalue problem

$$Z''(z) + \lambda Z(z) = 0 , \qquad 0 < z < b , \qquad Z(0) = Z(b) = 0 .$$

We have seen this eigenvalue problem several times before. Nontrivial solutions exist for $\lambda = (m\pi/b)^2$, $m = 1, 2, 3, \ldots$ and are given by

(39) $Z_m(z) = C_m \sin\left(\dfrac{m\pi z}{b}\right) ,$

where the C_m's are arbitrary constants.

Substituting for μ and λ in equation (35) gives

$$r^2 R''(r) + rR'(r) - \left(r^2\left(\frac{m\pi}{b}\right)^2 + n^2\right)R(r) = 0 , \qquad 0 \leq r < a .$$

The reader should verify that the change of variables $s = (m\pi r/b)$ transforms this equation into the **modified Bessel's equation of order n**[†]

(40) $s^2 R''(s) + sR'(s) - (s^2 + n^2)R(s) = 0 , \qquad 0 \leq s < \dfrac{m\pi a}{b} .$

[†]The modified Bessel's equation of order n arises in many applications and has been studied extensively. We refer the reader to the text *Special Functions* by E. D. Raineville (Chelsea Publishing Co., Inc., New York, 1972), for details about its solution.

The modified Bessel's equation of order n has two linearly independent solutions—the **modified Bessel function of the first kind:**

$$I_n(s) = \sum_{k=0}^{\infty} \frac{(s/2)^{2k+n}}{k!\,\Gamma(k+n+1)}\,,$$

which remains bounded near zero, and the **modified Bessel function of the second kind:**

$$K_n(s) = \lim_{v \to n} \frac{\pi}{2} \frac{I_{-v}(s) - I_v(s)}{\sin v\pi}\,,$$

which becomes unbounded as $s \to 0$. (Recall that Γ is the gamma function discussed in Section 7.6.) A general solution to (40) has the form $CK_n + DI_n$, where C and D are constants. Since u must remain bounded near $s = 0$, we must take $C = 0$. Thus the desired solutions to (40) have the form

(41) $$R_{mn}(r) = D_{mn} I_n\left(\frac{m\pi r}{b}\right)\,, \qquad n = 0, 1, \ldots, \qquad m = 1, 2, \ldots,$$

where the D_{mn}'s are arbitrary constants.

If we multiply the functions in (38), (39), and (41) and then sum over m and n, we obtain the following series solution to (32) and (34):

(42) $$u(r, \theta, z) = \sum_{m=1}^{\infty} a_{m0} I_0\left(\frac{m\pi r}{b}\right) \sin\left(\frac{m\pi z}{b}\right)$$

$$+ \sum_{n=1}^{\infty} \sum_{m=1}^{\infty} (a_{mn} \cos n\theta + b_{mn} \sin n\theta) I_n\left(\frac{m\pi r}{b}\right) \sin\left(\frac{m\pi z}{b}\right)\,,$$

where the a_{mn}'s and b_{mn}'s are arbitrary constants.

The constants in (42) can be obtained by imposing the boundary condition (33). Setting $r = a$ and rearranging terms, we have

(43) $$f(\theta, z) = \sum_{m=1}^{\infty} a_{m0} I_0\left(\frac{m\pi a}{b}\right) \sin\left(\frac{m\pi z}{b}\right)$$

$$+ \sum_{n=1}^{\infty} \left[\sum_{m=1}^{\infty} a_{mn} I_n\left(\frac{m\pi a}{b}\right) \sin\left(\frac{m\pi z}{b}\right) \right] \cos n\theta$$

$$+ \sum_{n=1}^{\infty} \left[\sum_{m=1}^{\infty} b_{mn} I_n\left(\frac{m\pi a}{b}\right) \sin\left(\frac{m\pi z}{b}\right) \right] \sin n\theta\,.$$

Treating this double Fourier series like the one in Section 10.5 (page 645), we find

(44) $$a_{m0} = \int_0^b \int_{-\pi}^{\pi} f(\theta, z) \sin\left(\frac{m\pi z}{b}\right) d\theta\, dz \Big/ \pi b I_0\left(\frac{m\pi a}{b}\right)\,,$$

(45) $$a_{mn} = 2 \int_0^b \int_{-\pi}^{\pi} f(\theta, z) \sin\left(\frac{m\pi z}{b}\right) \cos n\theta\, d\theta\, dz \Big/ \pi b I_n\left(\frac{m\pi a}{b}\right)\,, \qquad n \geq 1\,,$$

(46) $$b_{mn} = 2 \int_0^b \int_{-\pi}^{\pi} f(\theta, z) \sin\left(\frac{m\pi z}{b}\right) \sin n\theta\, d\theta\, dz \Big/ \pi b I_n\left(\frac{m\pi a}{b}\right)\,, \qquad n \geq 1\,.$$

Consequently, a formal solution to (32)–(34) is given by equation (42) with the constants a_{mn} and b_{mn} determined by equations (44)–(46). ■End

Existence and Uniqueness of Solutions

The existence of solutions to the boundary value problems for Laplace's equation can be established by studying the convergence of the formal solutions that we obtained using the method of separation of variables.

To answer the question of the uniqueness of the solution to a Dirichlet boundary value problem for Laplace's equation, recall that Laplace's equation arises in the search for steady-state solutions to the heat equation. Just as there are maximum principles for the heat equation, there are also maximum principles for Laplace's equation. We state one such result here.[†]

MAXIMUM PRINCIPLE FOR LAPLACE'S EQUATION

Theorem 9. Let $u(x, y)$ be a solution to Laplace's equation in a bounded domain D with $u(x, y)$ continuous in \overline{D}, the closure of D. $(\overline{D} = D \cup \partial D$, where ∂D denotes the boundary of D.) Then $u(x, y)$ attains its maximum value on ∂D.

The uniqueness of the solution to the Dirichlet boundary value problem follows from the maximum principle. We state this result in the next theorem but leave its proof as an exercise (see Problem 19).

UNIQUENESS OF SOLUTION

Theorem 10. Let D be a bounded domain. If there is a continuous solution to the Dirichlet boundary value problem

$$\Delta u(x, y) = 0 \quad \text{in } D ,$$
$$u(x, y) = f(x, y) \quad \text{on } \partial D ,$$

then the solution is unique.

Solutions to Laplace's equation in two variables are called **harmonic** functions. These functions arise naturally in the study of analytic functions of a single complex variable. Moreover, complex analysis provides many useful results about harmonic functions. For a discussion of this interaction, we refer the reader to an introductory text on complex analysis such as *Fundamentals of Complex Analysis,* Second Edition, by E. B. Saff and A. D. Snider (Prentice-Hall, Inc., Englewood Cliffs, New Jersey, 1993).

[†]A proof can be found in Section 8.2 of the text *Partial Differential Equations of Mathematical Physics,* 2nd edition, by Tyn Myint-U (Elsevier North Holland, Inc., New York, 1980).

EXERCISES 10.7

In Problems 1 through 5, find a formal solution to the given boundary value problem.

1. $\dfrac{\partial^2 u}{\partial x^2} + \dfrac{\partial^2 u}{\partial y^2} = 0$, $0 < x < \pi$, $0 < y < 1$,

$\dfrac{\partial u}{\partial x}(0, y) = \dfrac{\partial u}{\partial x}(\pi, y) = 0$, $0 \le y \le 1$,

$u(x, 0) = 4 \cos 6x + \cos 7x$, $0 \le x \le \pi$,

$u(x, 1) = 0$, $0 \le x \le \pi$.

2. $\dfrac{\partial^2 u}{\partial x^2} + \dfrac{\partial^2 u}{\partial y^2} = 0$, $0 < x < \pi$, $0 < y < \pi$,

$\dfrac{\partial u}{\partial x}(0, y) = \dfrac{\partial u}{\partial x}(\pi, y) = 0$, $0 \le y \le \pi$,

$u(x, 0) = \cos x - 2 \cos 4x$, $0 \le x \le \pi$,

$u(x, \pi) = 0$, $0 \le x \le \pi$.

3. $\dfrac{\partial^2 u}{\partial x^2} + \dfrac{\partial^2 u}{\partial y^2} = 0$, $0 < x < \pi$, $0 < y < \pi$,

$u(0, y) = u(\pi, y) = 0$, $0 \le y \le \pi$,

$u(x, 0) = f(x)$, $0 \le x \le \pi$,

$u(x, \pi) = 0$, $0 \le x \le \pi$.

4. $\dfrac{\partial^2 u}{\partial x^2} + \dfrac{\partial^2 u}{\partial y^2} = 0$, $0 < x < \pi$, $0 < y < \pi$,

$u(0, y) = u(\pi, y) = 0$, $0 \le y \le \pi$,

$u(x, 0) = \sin x + \sin 4x$, $0 \le x \le \pi$,

$u(x, \pi) = 0$, $0 \le x \le \pi$.

5. $\dfrac{\partial^2 u}{\partial x^2} + \dfrac{\partial^2 u}{\partial y^2} = 0$, $0 < x < \pi$, $0 < y < 1$,

$\dfrac{\partial u}{\partial x}(0, y) = \dfrac{\partial u}{\partial x}(\pi, y) = 0$, $0 \le y \le 1$,

$u(x, 0) = \cos x - \cos 3x$, $0 \le x \le \pi$,

$u(x, 1) = \cos 2x$, $0 \le x \le \pi$.

6. Derive the polar coordinate form of the Laplacian given in equation (12).

In Problems 7 and 8, find a solution to the Dirichlet boundary value problem for a disk:

$$\dfrac{\partial^2 u}{\partial r^2} + \dfrac{1}{r}\dfrac{\partial u}{\partial r} + \dfrac{1}{r^2}\dfrac{\partial^2 u}{\partial \theta^2} = 0 \ ,$$

$$0 \le r < 2 \ , -\pi \le \theta \le \pi \ ,$$

$$u(2, \theta) = f(\theta) \ , -\pi \le \theta \le \pi$$

for the given function $f(\theta)$.

7. $f(\theta) = |\theta|$, $-\pi \le \theta \le \pi$.

8. $f(\theta) = \cos^2 \theta$, $-\pi \le \theta \le \pi$.

9. Find a solution to the Neumann boundary value problem for a disk:

$$\dfrac{\partial^2 u}{\partial r^2} + \dfrac{1}{r}\dfrac{\partial u}{\partial r} + \dfrac{1}{r^2}\dfrac{\partial^2 u}{\partial \theta^2} = 0 \ ,$$

$$0 \le r < a \ , -\pi \le \theta \le \pi \ ,$$

$$\dfrac{\partial u}{\partial r}(a, \theta) = f(\theta) \ , -\pi \le \theta \le \pi \ .$$

10. A solution to the Newmann problem (25)–(26) must also satisfy the consistency condition in (27). To show this, use **Green's second formula**

$$\iint_D (v \Delta u - u \Delta v)\,dx\,dy = \int_{\partial D}\left(v\dfrac{\partial u}{\partial n} - u\dfrac{\partial v}{\partial n}\right)ds \ ,$$

where $\partial/\partial n$ is the outward normal derivative and ds is the differential of arc length. $\left[\textit{Hint:}\ \text{Take}\ v \equiv 1 \right.$ and observe that $\partial u/\partial n = \partial u/\partial r.\big]$

11. Find a solution to the following Dirichlet problem for an annulus:

$$\dfrac{\partial^2 u}{\partial r^2} + \dfrac{1}{r}\dfrac{\partial u}{\partial r} + \dfrac{1}{r^2}\dfrac{\partial^2 u}{\partial \theta^2} = 0 \ ,$$

$$1 < r < 2 \ , -\pi \le \theta \le \pi \ ,$$

$$u(1, \theta) = \sin 4\theta - \cos \theta \ , -\pi \le \theta \le \pi \ ,$$

$$u(2, \theta) = \sin \theta \ , -\pi \le \theta \le \pi \ .$$

12. Find a solution to the following Dirichlet problem for an annulus:

$$\dfrac{\partial^2 u}{\partial r^2} + \dfrac{1}{r}\dfrac{\partial u}{\partial r} + \dfrac{1}{r^2}\dfrac{\partial^2 u}{\partial \theta^2} = 0 \ ,$$

$$1 < r < 3 \ , -\pi \le \theta \le \pi \ ,$$

$$u(1, \theta) = 0 \ , -\pi \le \theta \le \pi \ ,$$

$$u(3, \theta) = \cos 3\theta + \sin 5\theta \ , -\pi \le \theta \le \pi \ .$$

13. Find a solution to the following Dirichlet problem for an exterior domain:

$$\frac{\partial^2 u}{\partial r^2} + \frac{1}{r}\frac{\partial u}{\partial r} + \frac{1}{r^2}\frac{\partial^2 u}{\partial \theta^2} = 0 ,$$

$$1 < r , \qquad -\pi \le \theta \le \pi ,$$

$$u(1, \theta) = f(\theta) , \qquad -\pi \le \theta \le \pi ,$$

$$u(r, \theta) \quad \text{remains bounded as } r \to \infty.$$

14. Find a solution to the following Newmann problem for an exterior domain:

$$\frac{\partial^2 u}{\partial r^2} + \frac{1}{r}\frac{\partial u}{\partial r} + \frac{1}{r^2}\frac{\partial^2 u}{\partial \theta^2} = 0 ,$$

$$1 < r , \qquad -\pi \le \theta \le \pi ,$$

$$\frac{\partial u}{\partial r}(1, \theta) = f(\theta) , \qquad -\pi \le \theta \le \pi ,$$

$$u(r, \theta) \quad \text{remains bounded as } r \to \infty.$$

15. Find a solution to the following Dirichlet problem for a half disk:

$$\frac{\partial^2 u}{\partial r^2} + \frac{1}{r}\frac{\partial u}{\partial r} + \frac{1}{r^2}\frac{\partial^2 u}{\partial \theta^2} = 0 ,$$

$$0 < r < 1 , \qquad 0 < \theta < \pi ,$$

$$u(r, 0) = 0 , \qquad 0 \le r \le 1 ,$$

$$u(r, \pi) = 0 , \qquad 0 \le r \le 1 ,$$

$$u(1, \theta) = \sin 3\theta , \qquad 0 \le \theta \le \pi ,$$

$$u(0, \theta) \quad \text{bounded.}$$

16. Find a solution to the following Dirichlet problem for a half annulus:

$$\frac{\partial^2 u}{\partial r^2} + \frac{1}{r}\frac{\partial u}{\partial r} + \frac{1}{r^2}\frac{\partial^2 u}{\partial \theta^2} = 0 ,$$

$$\pi < r < 2\pi , \qquad 0 < \theta < \pi .$$

$$u(r, 0) = \sin r , \qquad \pi \le r \le 2\pi .$$

$$u(r, \pi) = 0 , \qquad \pi \le r \le 2\pi ,$$

$$u(\pi, \theta) = u(2\pi, \theta) = 0 , \qquad 0 \le \theta \le \pi .$$

17. Find a solution to the mixed boundary value problem

$$\frac{\partial^2 u}{\partial r^2} + \frac{1}{r}\frac{\partial u}{\partial r} + \frac{1}{r^2}\frac{\partial^2 u}{\partial \theta^2} = 0 ,$$

$$1 < r < 3 , \qquad -\pi \le \theta \le \pi ,$$

$$u(1, \theta) = f(\theta) , \qquad -\pi \le \theta \le \pi ,$$

$$\frac{\partial u}{\partial r}(3, \theta) = g(\theta) , \qquad -\pi \le \theta \le \pi .$$

18. Show that

$$-B_n \tanh\left(\frac{n\pi b}{a}\right)\cosh\left(\frac{n\pi y}{a}\right) + B_n \sinh\left(\frac{n\pi y}{a}\right)$$

$$= C_n \sinh\left[\frac{n\pi}{a}(y - b)\right] ,$$

where

$$C_n = B_n/\cosh\left(\frac{n\pi b}{a}\right) .$$

19. Prove Theorem 10 on the uniqueness of the solution to the Dirichlet problem.

20. Stability. Use the maximum principle to prove the following theorem on the continuous dependence of the solution on the boundary conditions:

Theorem. Let f_1 and f_2 be continuous functions on ∂D, where D is a bounded domain. For $i = 1$ and 2, let u_i be the solution to the Dirichlet problem

$$\Delta u = 0 , \qquad \text{in } D ,$$

$$u = f_i , \qquad \text{on } \partial D .$$

If the boundary values satisfy

$$|f_1(x, y) - f_2(x, y)| \le \varepsilon \qquad \text{for all } (x, y) \text{ on } \partial D ,$$

where $\varepsilon > 0$ is some constant, then

$$|u_1(x, y) - u_2(x, y)| \le \varepsilon \qquad \text{for all } (x, y) \text{ in } D .$$

21. For the Dirichlet problem described in Example 3, let $a = b = \pi$ and assume the potential on the lateral side $(r = \pi)$ of the cylinder is $f(\theta, z) = \sin z$. Use equations (44)–(46) to compute the solution given by equation (42).

22. Invariance of Laplace's Equation. A complex-valued function $f(z)$ of the complex variable $z = x + iy$ can be written in the form $f(z) = u(x, y) + iv(x, y)$, where u and v are real-valued functions. If $f(z)$ is analytic in a planar region D, then its real and imaginary parts satisfy the Cauchy-Riemann equations in D; that is, in D

$$\frac{\partial u}{\partial x} = \frac{\partial v}{\partial y} , \qquad \frac{\partial u}{\partial y} = -\frac{\partial v}{\partial x} .$$

Let f be analytic and one-to-one in D and assume its inverse f^{-1} is analytic in D', where D' is the image of D under f. Then

$$\frac{\partial x}{\partial u} = \frac{\partial y}{\partial v} , \qquad \frac{\partial x}{\partial v} = -\frac{\partial y}{\partial u}$$

are just the Cauchy-Riemann equations for f^{-1}. Show that if $\phi(x, y)$ satisfies Laplace's equation for (x, y) in D, then $\psi(u, v) := \phi(x(u, v), y(u, v))$ satisfies Laplace's equation for (u, v) in D'.

23. **Fluid Flow around a Corner.** The stream lines that describe the fluid flow around a corner (see Figure 10.25) are given by $\phi(x, y) = k$, where k is a constant and ϕ, the stream function, satisfies the boundary value problem

$$\frac{\partial^2 \phi}{\partial x^2} + \frac{\partial^2 \phi}{\partial y^2} = 0 , \qquad x > 0 , \qquad y > 0 ,$$

$$\phi(x, 0) = 0 , \qquad 0 \le x ,$$

$$\phi(0, y) = 0 , \qquad 0 \le y .$$

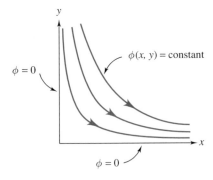

Figure 10.25 Flow around a corner

(a) Using the results of Problem 22, show that this problem can be reduced to finding the flow above a flat plate (see Figure 10.26). That is,

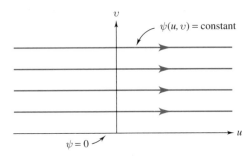

Figure 10.26 Flow above a flat surface

show that the problem reduces to finding the solution to

$$\frac{\partial^2 \psi}{\partial u^2} + \frac{\partial^2 \psi}{\partial v^2} = 0 ,$$

$$v > 0 , \qquad -\infty < u < \infty ,$$

$$\psi(u, 0) = 0 , \qquad -\infty < u < \infty ,$$

where ψ and ϕ are related as follows: $\psi(u, v) = \phi(x(u, v), y(u, v))$ with the mapping between (u, v) and (x, y) given by the analytic function $f(z) = z^2$.

(b) Verify that a nonconstant solution to the problem in part (a) is given by $\psi(u, v) = v$.

(c) Using the result of part (b), find a stream function $\phi(x, y)$ for the original problem.

24. **Unbounded Domain.** Using separation of variables, find a solution of Laplace's equation in the infinite rectangle $0 < x < \pi, 0 < y < \infty$ that is zero on the sides $x = 0$ and $x = \pi$, approaches zero as $y \to \infty$, and equals $f(x)$ for $y = 0$.

CHAPTER SUMMARY

Separation of Variables: A classical technique that is effective in solving boundary value problems for partial differential equations is the **method of separation of variables.** Briefly, the idea is to assume first that there exists a solution that can be written with the variables separated: e.g., $u(x, t) = X(x)T(t)$. Substituting $u = XT$ into the partial differential equation and then imposing the boundary conditions leads to the problem of finding the eigenvalues and eigenfunctions for a boundary value problem for an ordinary differential equation. Solving for the eigenvalues and eigenfunctions, one eventually obtains solutions $u_n(x, t) = X_n(x)T_n(t)$, $n = 1, 2, 3, \ldots$ that solve the partial differential equation and the boundary conditions. Taking

infinite linear combinations of the $u_n(x, t)$ yields solutions to the partial differential equation and boundary conditions of the form

$$u(x, t) = \sum_{n=1}^{\infty} c_n u_n(x, t) \ .$$

The coefficients c_n are then obtained using the initial conditions or some other boundary conditions.

Fourier Series: Let $f(x)$ be a piecewise continuous function on the interval $[-T, T]$. The **Fourier series** of f is the trigonometric series

$$f(x) \sim \frac{a_0}{2} + \sum_{n=1}^{\infty} \left\{ a_n \cos \frac{n\pi x}{T} + b_n \sin \frac{n\pi x}{T} \right\} \ ,$$

where the a_n's and b_n's are given by **Euler's formulas:**

$$a_n = \frac{1}{T} \int_{-T}^{T} f(x) \cos \frac{n\pi x}{T} dx \ , \qquad n = 0, 1, 2, \dots \ ,$$

$$b_n = \frac{1}{T} \int_{-T}^{T} f(x) \sin \frac{n\pi x}{T} dx \ , \qquad n = 1, 2, 3, \dots \ .$$

Let f be piecewise continuous on the half-interval $[0, T]$. The **Fourier cosine series** of $f(x)$ on $[0, T]$ is

$$f(x) \sim \frac{a_0}{2} + \sum_{n=1}^{\infty} a_n \cos \frac{n\pi x}{T} \ ,$$

where

$$a_n = \frac{2}{T} \int_{0}^{T} f(x) \cos \frac{n\pi x}{T} dx \ .$$

The **Fourier sine series** of $f(x)$ on $[0, T]$ is

$$f(x) \sim \sum_{n=1}^{\infty} b_n \sin \frac{n\pi x}{T} \ ,$$

where

$$b_n = \frac{2}{T} \int_{0}^{T} f(x) \sin \frac{n\pi x}{T} dx \ .$$

Fourier series and the method of separation of variables are used to solve boundary value problems and initial-boundary value probems for the three classical equations:

Heat equation	$\dfrac{\partial u}{\partial t} = \beta \dfrac{\partial^2 u}{\partial x^2} \ .$
Wave equation	$\dfrac{\partial^2 u}{\partial t^2} = \alpha^2 \dfrac{\partial^2 u}{\partial x^2} \ .$
Laplace's equation	$\dfrac{\partial^2 u}{\partial x^2} + \dfrac{\partial^2 u}{\partial y^2} = 0 \ .$

For the heat equation, there is a unique solution $u(x, t)$ when the boundary values at the ends of a conducting wire are specified along with the initial temperature distribution. The wave equation yields the displacement $u(x, t)$ of a vibrating string when the ends are held fixed and the initial displacement and initial velocity at each point of the string are given. In such a case, separation of variables yields a solution that is the sum of standing waves. For an infinite string with specified initial displacement and velocity, d'Alembert's solution yields traveling waves. Laplace's equation arises in the study of steady-state solutions to the heat equation where either Dirichlet or Neumann boundary conditions are specified.

TECHNICAL WRITING EXERCISES

1. The method of separation of variables is an important technique in solving initial-boundary value problems and boundary value problems for *linear* partial differential equations. Explain where the linearity of the differential equation plays a crucial role in the method of separation of variables.

2. In applying the method of separation of variables, we have encountered a variety of *special functions,* such as sines, cosines, Bessel functions, and modified Bessel functions. Describe three or four examples of partial differential equations that involve other special functions, such as Legendre polynomials, Hermite polynomials, and Laguerre polynomials. (Some exploring in the library may be needed.)

3. A constant coefficient second order partial differential equation of the form
$$a\frac{\partial^2 u}{\partial x^2} + b\frac{\partial^2 u}{\partial x \partial y} + c\frac{\partial^2 u}{\partial y^2} = 0$$

can be classified using the discriminant $D := b^2 - 4ac$. In particular, the equation is called

hyperbolic if $D > 0$, and
elliptic if $D < 0$.

Verify that the wave equation is hyperbolic and Laplace's equation is elliptic. It can be shown that such hyperbolic (elliptic) equations can be transformed by a linear change of variables into the wave (Laplace's) equation. Based on your knowledge of the latter equations, describe which types of problems (initial value, boundary value, etc.) are appropriate for hyperbolic equations and elliptic equations.

GROUP PROJECTS FOR CHAPTER 10

A. Steady-State Temperature Distribution in a Circular Cylinder

When the temperature u inside a circular cylinder reaches a steady state, it satisfies Laplace's equation $\Delta u = 0$. If the temperature on the lateral surface $(r = a)$ is kept at zero, the temperature on the top $(z = b)$ is kept at zero, and the temperature on the bottom $(z = 0)$ is given by $u(r, \theta, 0) = f(r, \theta)$, then the steady-state temperature satisfies the boundary value problem

$$\frac{\partial^2 u}{\partial r^2} + \frac{1}{r}\frac{\partial u}{\partial r} + \frac{1}{r^2}\frac{\partial^2 u}{\partial \theta^2} + \frac{\partial^2 u}{\partial z^2} = 0 , \qquad 0 \le r < a , \qquad -\pi \le \theta \le \pi , \qquad 0 < z < b ,$$

$$u(a, \theta, z) = 0 , \qquad -\pi \le \theta \le \pi , \qquad 0 \le z \le b ,$$

$$u(r, \theta, b) = 0 , \qquad 0 \le r < a , \qquad -\pi \le \theta \le \pi ,$$

$$u(r, \theta, 0) = f(r, \theta) , \qquad 0 \le r < a , \qquad -\pi \le \theta \le \pi ,$$

where $f(a, \theta) = 0$ for $-\pi \le \theta \le \pi$ (see Figure 10.27). To find a solution to this boundary value problem, proceed as follows:

(a) Let $u(r, \theta, z) = R(r)T(\theta)Z(z)$. Show that R, T, and Z must satisfy the three ordinary differential equations

$$r^2 R'' + rR' - (r^2\lambda + \mu)R = 0 ,$$

$$T'' + \mu T = 0 ,$$

$$Z'' + \lambda Z = 0 .$$

(b) Show that $T(\theta)$ has the form

$$T(\theta) = A \cos n\theta + B \sin n\theta$$

for $\mu = n^2$, $n = 0, 1, 2, \ldots$.

(c) Show that $Z(z)$ has the form

$$Z(z) = C \sinh\left[\beta(b - z)\right]$$

for $\lambda = -\beta^2$, where $\beta > 0$.

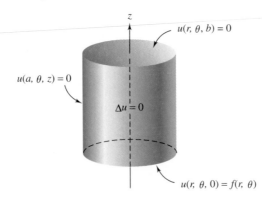

Figure 10.27 Dirichlet problem for cylinder

(d) Show that $R(r)$ has the form, for each n,

$$R_n(r) = DJ_n(\beta r) \ ,$$

where J_n is the Bessel function of the first kind.

(e) Show that the boundary conditions require $R(a) = 0$, and so

$$J_n(\beta a) = 0 \ .$$

Hence, for each n, if $0 < \alpha_{n1} < \alpha_{n2} < \cdots < \alpha_{nm} < \cdots$ are the zeros of J_n, then

$$\beta_{nm} = \alpha_{nm}/a \ .$$

Moreover,

$$R_n(r) = DJ_n(\alpha_{nm}r/a) \ .$$

(f) Use the preceding results to show that $u(r, \theta, z)$ has the form

$$u(r, \theta, z) = \sum_{n=0}^{\infty} \sum_{m=1}^{\infty} J_n\left(\frac{\alpha_{nm}r}{a}\right)(a_{nm} \cos n\theta + b_{nm} \sin n\theta) \sinh\left(\frac{\alpha_{nm}(b-z)}{a}\right) \ ,$$

where a_{nm} and b_{nm} are constants.

(g) Use the final boundary condition and a double orthogonal expansion involving Bessel functions and trigonometric functions to derive the formulas

$$a_{0m} = \frac{1}{\pi a^2 \sinh(\alpha_{0m}b/a)[J_1(\alpha_{0m})]^2} \int_0^a \int_0^{2\pi} f(r, \theta) J_0\left(\frac{\alpha_{0m}r}{a}\right) r \, dr \, d\theta \ ,$$

for $m = 1, 2, 3, \ldots$, and for $n, m = 1, 2, 3, \ldots$,

$$a_{nm} = \frac{2}{\pi a^2 \sinh(\alpha_{nm}b/a)[J_{n+1}(\alpha_{nm})]^2} \int_0^a \int_0^{2\pi} f(r, \theta) J_n\left(\frac{\alpha_{nm}r}{a}\right) \cos(n\theta) r \, dr \, d\theta \ ,$$

$$b_{nm} = \frac{2}{\pi a^2 \sinh(\alpha_{nm}b/a)[J_{n+1}(\alpha_{nm})]^2} \int_0^a \int_0^{2\pi} f(r, \theta) J_n\left(\frac{\alpha_{nm}r}{a}\right) \sin(n\theta) r \, dr \, d\theta \ .$$

B. A Laplace Transform Solution of the Wave Equation

Laplace transforms can be used to solve certain partial differential equations. To illustrate this technique, consider the initial-boundary value problem

(1) $$\frac{\partial^2 u}{\partial t^2} = \alpha^2 \frac{\partial^2 u}{\partial x^2} \ ; \qquad 0 < x < \infty , \qquad t > 0 ,$$

(2) $$u(0, t) = h(t) \ , \qquad t > 0 \ ,$$

(3) $$u(x, 0) = 0 \ , \qquad 0 < x < \infty \ ,$$

(4) $$\frac{\partial u}{\partial t}(x, 0) = 0 \ , \qquad 0 < x < \infty \ ,$$

(5) $$\lim_{x \to \infty} u(x, t) = 0 \ , \qquad t \geq 0 \ .$$

This problem arises in studying a semi-infinite string that is initially horizontal and at rest and where one end is being moved vertically. Let $u(x, t)$ be the solution to (1)–(5). For each x, let

$$U(x, s) := \mathcal{L}\{u(x, t)\}(x, s) = \int_0^\infty e^{-st} u(x, t)\, dt \ .$$

(a) Using the fact that

$$\mathcal{L}\left\{\frac{\partial^2 u}{\partial x^2}\right\} = \frac{\partial^2}{\partial x^2}\mathcal{L}\{u\} \ ,$$

show that $U(x, s)$ satisfies the equation

(6) $$s^2 U(x, s) = \alpha^2 \frac{\partial^2 U}{\partial x^2} \ , \qquad 0 < x < \infty \ .$$

(b) Show that the general solution to (6) is

$$U(x, s) = A(s)e^{-sx/\alpha} + B(s)e^{sx/\alpha} \ ,$$

where $A(s)$ and $B(s)$ are arbitrary functions of s.

(c) Since $u(x, t) \to 0$ as $x \to \infty$ for all $0 \le t < \infty$, we have $U(x, s) \to 0$ as $x \to \infty$. Use this fact to show that the $B(s)$ in part (b) must be zero.

(d) Using equation (2), show that

$$A(s) = H(s) = \mathcal{L}\{h\}(s) \ ,$$

where $A(s)$ is given in part (b).

(e) Use the results of parts (b), (c), and (d) to obtain a formal solution to (1)–(5).

C. *Green's Function*

Let Ω be a region in the xy-plane having a smooth boundary $\partial\Omega$. Associated with Ω is a **Green's function** $G(x, y; \xi, \eta)$ defined for pairs of distinct points (x, y), (ξ, η) in Ω. The function $G(x, y; \xi, \eta)$ has the following property.

Let $\Delta u := \partial^2 u/\partial x^2 + \partial^2 u/\partial y^2$ denote the Laplacian operator on Ω; let $h(x, y)$ be a given continuous function on Ω; and let $f(x, y)$ be a given continuous function on $\partial\Omega$. Then, a continuous solution to the Dirichlet boundary value problem

$$\Delta u(x, y) = h(x, y) \ , \qquad \text{in } \Omega \ ,$$
$$u(x, y) = f(x, y) \ , \qquad \text{on } \partial\Omega \ ,$$

is given by

(7) $$u(x, y) = \iint_\Omega G(x, y; \xi, \eta)h(\xi, \eta)\, d\xi\, d\eta$$

$$+ \int_{\partial\Omega} f(\xi, \eta)\frac{\partial G}{\partial n}(x, y; \xi, \eta)\, d\sigma(\xi, \eta) \ ,$$

where n is the outward normal to the boundary $\partial\Omega$ of Ω and the second integral is the line integral around the boundary of Ω with the interior of Ω on the left as the boundary is traversed. In (7) we assume that $\partial\Omega$ is sufficiently smooth so that the integrands and integrals exist.

When Ω is the upper half-plane, the Green's function is[†]

$$G(x, y; \xi, \eta) = \frac{1}{4\pi}\ln\left[\frac{(x-\xi)^2 + (y-\eta)^2}{(x-\xi)^2 + (y+\eta)^2}\right] .$$

(a) Using (7), show that a solution to

(8) $\Delta u(x, y) = h(x, y) ,$ $-\infty < x < \infty ,$ $0 < y ,$

(9) $u(x, 0) = f(x) ,$ $-\infty < x < \infty ,$

is given by

$$u(x, y) = \frac{y}{\pi}\int_{-\infty}^{\infty}\frac{f(\xi)}{(x-\xi)^2 + y^2}d\xi$$

$$+ \frac{1}{4\pi}\int_0^{\infty}\int_{-\infty}^{\infty}\ln\left[\frac{(x-\xi)^2 + (y-\eta)^2}{(x-\xi)^2 + (y+\eta)^2}\right]h(\xi, \eta)d\xi\, d\eta .$$

(b) Use the result of part (a) to determine a solution to (8)–(9) when $h \equiv 0$ and $f \equiv 1$.

(c) When Ω is the interior of the unit circle $x^2 + y^2 = 1$, the Green's function, in polar coordinates (r, θ) and (ρ, ϕ), is given by

$$G(r, \theta; \rho, \phi) = \frac{1}{4\pi}\ln\left[r^2 + \rho^2 - 2r\rho\cos(\phi - \theta)\right]$$

$$- \frac{1}{4\pi}\ln\left[r^2 + \rho^{-2} - 2r\rho^{-1}\cos(\phi - \theta)\right] - \frac{1}{4\pi}\ln\rho .$$

Using (7), show that the solution to

$$\frac{\partial^2 u}{\partial r^2} + \frac{1}{r}\frac{\partial u}{\partial r} + \frac{1}{r^2}\frac{\partial^2 u}{\partial\theta^2} = 0 , \qquad 0 \leq r < 1 , \qquad 0 \leq \theta \leq 2\pi ,$$

$$u(1, \theta) = f(\theta) , \qquad 0 \leq \theta \leq 2\pi ,$$

is given by

$$u(r, \theta) = \frac{1}{2\pi}\int_0^{2\pi}\frac{1 - r^2}{1 + r^2 - 2r\cos(\phi - \theta)}f(\phi)d\phi .$$

This is known as **Poisson's integral formula.**

(d) Use Poisson's integral formula to derive the following theorem for solutions to Laplace's equation.

[†]Techniques for determining Green's functions can be found in texts on partial differential equations such as *Partial Differential Equations of Mathematical Physics,* 2nd edition, by Tyn Myint-U (Elsevier North Holland, Inc., New York, 1980), Chapter 10.

MEAN VALUE PROPERTY

Theorem 11. Let $u(x, y)$ satisfy $\Delta u = 0$ in a bounded domain Ω in \mathbf{R}^2 and let (x_0, y_0) lie in Ω. Then

(10) $$u(x_0, y_0) = \frac{1}{2\pi} \int_0^{2\pi} u(x_0 + r\cos\theta, y_0 + r\sin\theta)\,d\theta$$

for all $r > 0$ for which the disk $\overline{B}(x_0, y_0; r) = \{(x, y): (x - x_0)^2 + (y - y_0)^2 \leq r^2\}$ lies entirely in Ω.

[*Hint:* Use a change of variables that maps the disk $\overline{B}(x_0, y_0; r)$ to the unit disk $\overline{B}(0, 0; 1)$.]

D. Numerical Method for $\Delta u = f$ on a Rectangle

Let R denote the open rectangle

$$R = \{(x, y): a < x < b, \quad c < y < d\}$$

and ∂R be its boundary. Here we describe a numerical technique for solving the generalized Dirichlet problem

(11) $$\frac{\partial^2 u}{\partial x^2} + \frac{\partial^2 u}{\partial y^2} = f(x, y), \qquad \text{for } (x, y) \text{ in } R,$$

$$u(x, y) = g(x, y), \qquad \text{for } (x, y) \text{ on } \partial R.$$

The method is similar to the finite-difference technique discussed in Chapter 4. We begin by selecting positive integers m and n and step sizes h and k so that $b - a = hm$ and $d - c = kn$. The interval $[a, b]$ is now partitioned into m equal subintervals and $[c, d]$ into n equal subintervals. The partition points are

$$x_i = a + ih, \qquad 0 \leq i \leq m,$$
$$y_j = c + jk, \qquad 0 \leq j \leq n$$

(see Figure 10.28). The (dashed) lines $x = x_i$ and $y = y_j$ are called **grid lines** and their intersections (x_i, y_j) are the **mesh points** of the partition. Our goal is to obtain approximations to the solu-

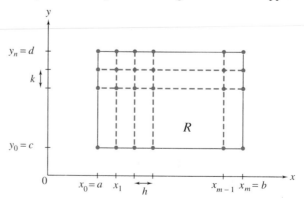

Figure 10.28 Rectangular grid

tion $u(x, y)$ of problem (11) at each *interior* mesh point, i.e., at (x_i, y_j) where $1 \leq i \leq m - 1$, $1 \leq j \leq n - 1$. (Of course, from (11), we are given the values of $u(x, y)$ at the boundary mesh points, e.g., $u(x_0, y_j) = g(x_0, y_j)$, $0 \leq j \leq n$.)

The next step is to approximate the partial derivatives $\partial^2 u/\partial x^2$ and $\partial^2 u/\partial y^2$ using the *centered-difference approximations*

(12) $\dfrac{\partial^2 u}{\partial x^2}(x_i, y_j) \approx \dfrac{1}{h^2}[u(x_{i+1}, y_j) - 2u(x_i, y_j) + u(x_{i-1}, y_j)]$ for $1 \leq i \leq m - 1$,

and

(13) $\dfrac{\partial^2 u}{\partial y^2}(x_i, y_j) \approx \dfrac{1}{k^2}[u(x_i, y_{j+1}) - 2u(x_i, y_j) + u(x_i, y_{j-1})]$ for $1 \leq j \leq n - 1$.

These approximations are based on the generic formula

$$y''(x) = [y(x + h) - 2y(x) + y(x - h)]/h^2 + [\text{terms involving } h^2, h^3, \dots] ,$$

which follows from the Taylor series expansion for $y(x)$.

(a) Show that substituting the approximations (12) and (13) into the Dirichlet problem (11) yields the following system:

(14) $2\left[\left(\dfrac{h}{k}\right)^2 + 1\right]u_{i,j} - (u_{i+1,j} + u_{i-1,j}) - \left(\dfrac{h}{k}\right)^2(u_{i,j+1} + u_{i,j-1}) = -h^2 f(x_i, y_j)$

for $i = 1, 2, \dots, m - 1$, and $j = 1, 2, \dots, n - 1$;

$u_{0,j} = g(a, y_j)$, $u_{m,j} = g(b, y_j)$, $j = 0, 1, \dots, n$,
$u_{i,0} = g(x_i, c)$, $u_{i,n} = g(x_i, d)$, $i = 1, 2, \dots, m - 1$,

where $u_{i,j}$ approximates $u(x_i, y_j)$.

Notice that each equation in (14) involves approximations to the solution that appear in a cross centered at a mesh point (see Figure 10.29).

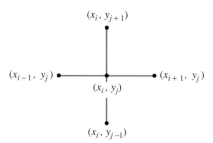

(x_i, y_{j+1})

(x_{i-1}, y_j) (x_{i+1}, y_j)
(x_i, y_j)

(x_i, y_{j-1})

Figure 10.29 Approximate solution $u_{i,j}$ at (x_i, y_j) is obtained from values at the ends of the cross

(b) Show that the system in part (a) is a linear system of $(m - 1)(n - 1)$ unknowns in $(m - 1)(n - 1)$ equations.

(c) For Laplace's equation where $f(x, y) \equiv 0$, show that when $h = k$, equation (14) yields

$$u_{i,j} = \frac{1}{4}(u_{i-1,j} + u_{i+1,j} + u_{i,j-1} + u_{i,j+1}) .$$

Compare this averaging formula with the mean value property of Theorem 11 (Project C).

(d) For the square plate

$$R = \{(x, y): 0 \le x \le 0.4, 0 \le y \le 0.4\} ,$$

the boundary is maintained at the following temperatures:

$$u(0, y) = 0 , \qquad u(0.4, y) = 10y , \qquad 0 \le y \le 0.4 .$$
$$u(x, 0) = 0 , \qquad u(x, 0.4) = 10x , \qquad 0 \le x \le 0.4 .$$

Using the system in part (a) with $f(x, y) \equiv 0$, $m = n = 4$, and $h = k = 0.1$, find approximations to the steady-state temperatures at the mesh points of the plate. [*Hint:* It is helpful to label these grid points with a single index, say p_1, p_2, \ldots, p_q, choosing the ordering in a book-reading sequence.]

Appendices

A NEWTON'S METHOD

To solve an equation $g(x) = 0$, we must find the point or points where the graph of $y = g(x)$ meets the x-axis. One procedure for approximating a solution is **Newton's method.**

To motivate Newton's method geometrically, we let \tilde{x} be a root of $g(x) = 0$ and let x_1 be our guess at the value of \tilde{x}. If $g(x_1) = 0$, we are done. If $g(x_1) \neq 0$, then we are off by some amount that we call dy (see Figure A.1). Then

$$\frac{dy}{dx} = g'(x_1) \; ,$$

and so

(1) $\qquad dx = \dfrac{dy}{g'(x_1)} \; .$

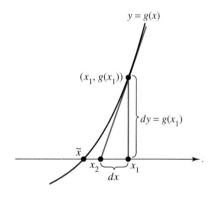

Figure A.1 Tangent line approximation of root

Now $dx = x_1 - x_2$ or $x_2 = x_1 - dx$, where x_2 is at the point where the tangent line through $(x_1, g(x_1))$ intersects the x-axis (Figure A.1). Using equation (1) and the fact that $dy = g(x_1)$, we obtain

$$x_2 = x_1 - \frac{dy}{g'(x_1)} = x_1 - \frac{g(x_1)}{g'(x_1)},$$

which we use as the next approximation to the root \tilde{x}.

Repeating this process with x_2 in place of x_1, we obtain the next approximation x_3 to the root \tilde{x}. In general, we find the next approximation x_{n+1} by the formula

(2) $$x_{n+1} = x_n - \frac{g(x_n)}{g'(x_n)}, \qquad n = 1, 2, \dots .$$

The process is illustrated in Figure A.2.

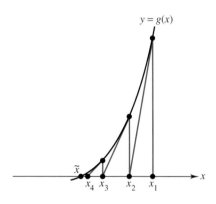

Figure A.2 Sequence of iterations converging to root

If the initial guess x_1 is sufficiently close to a root \tilde{x}, then the sequence of iterations $\{x_n\}_{n=1}^{\infty}$ usually converges to the root \tilde{x}. However, if we make a bad guess for x_1, then the process may lead away from \tilde{x}.

EXAMPLE 1 Find a root to four decimal places of the equation

(3) $$x^3 + 2x - 4 = 0 .$$

SOLUTION Setting $g(x) = x^3 + 2x - 4$, we find that $g'(x) = 3x^2 + 2$ is positive for all x. Hence g is increasing and has at most one zero. Furthermore, since $g(1) = -1$ and $g(2) = 8$, this zero must lie between 1 and 2. Thus we begin the procedure with the initial guess $x_1 = 1.5$. For $g(x) = x^3 + 2x - 4$, equation (2) becomes

(4) $$x_{n+1} = x_n - \frac{x_n^3 + 2x_n - 4}{3x_n^2 + 2}, \qquad n = 1, 2, \dots .$$

With $x_1 = 1.5$, equation (4) gives

$$x_2 = 1.5 - \frac{(1.5)^3 + 2(1.5) - 4}{3(1.5)^2 + 2} = 1.5 - \frac{2.375}{8.75} \approx 1.22857 \ .$$

Using x_2 to compute x_3 and so on, we find

$$x_3 = 1.18085 \ ,$$
$$x_4 = 1.17951 \ ,$$
$$x_5 = 1.17951 \ ,$$

where we have rounded off the computations to five decimal places. Since x_4 and x_5 agree to four decimal places and we are uncertain of the fifth decimal place because of roundoff, we surmise that the root \tilde{x} of (3) agrees with 1.1795 to four decimal places. Indeed,

$$g(1.1795) = -0.00005 \ldots \quad \text{and} \quad g(1.1796) = 0.00056 \ldots \ ,$$

so $1.1795 < \tilde{x} < 1.1796$. Consequently, $\tilde{x} = 1.1795 \ldots$. ■ End

Observe that Newton's method transforms the problem of finding a root to the equation $g(x) = 0$ into the problem of finding a fixed point for the function $h(x) = x - g(x)/g'(x)$; that is, finding a number x such that $x = h(x)$. (See equation (2).)

Several theorems give conditions that guarantee that the sequence of iterations $\{x_n\}_{n=1}^{\infty}$ defined by (2) will converge to a zero of $g(x)$. We mention one such result.

CONVERGENCE OF NEWTON'S METHOD

Theorem 1. Suppose a zero \tilde{x} of $g(x)$ lies in the interval (a, b) and that in this interval

$$g'(x) > 0 \quad \text{and} \quad g''(x) > 0 \ .$$

If we select x_1 so that $\tilde{x} < x_1 < b$, then the sequence of iterations defined by (2) will decrease to \tilde{x}.

We do not give a proof of this theorem, but refer the reader to an introductory numerical analysis text such as *Numerical Analysis,* 6th edition, by R. Burden and J. Faires (Brooks/Cole Publishing Co., Pacific Grove, CA, 1997).

B SIMPSON'S RULE

A useful procedure for approximating the value of a definite integral is **Simpson's rule.**

Let the interval $[a, b]$ be divided into $2n$ equal parts and let x_0, x_1, \ldots, x_{2n} be the points of the partition, that is,

$$x_k := a + kh \ , \qquad k = 0, 1, \ldots, 2n \ ,$$

where $h := (b - a)/(2n)$. If

$$y_k := f(x_k) , \qquad k = 0, 1, \ldots, 2n ,$$

then the Simpson's rule approximation I_S for the value of the definite integral

$$\int_a^b f(x)dx$$

is given by

(1) $\qquad I_S = \dfrac{h}{3}\left[y_0 + 4y_1 + 2y_2 + 4y_3 + \cdots + 2y_{2n-2} + 4y_{2n-1} + y_{2n}\right]$

$$= \dfrac{h}{3}\sum_{k=1}^{n} (y_{2k-2} + 4y_{2k-1} + y_{2k}) .$$

If

$$E := \int_a^b f(x)dx - I_S$$

is the error that results from using Simpson's rule to approximate the value of the definite integral, then

(2) $\qquad |E| \leq \dfrac{(b-a)}{180}h^4 M ,$

where $M := \max|f^{(4)}(x)|$ for all x in $[a, b]$.

EXAMPLE 1 Use Simpson's rule with $n = 4$ to approximate the value of the definite integral

(3) $\qquad \displaystyle\int_0^1 \dfrac{1}{1 + x^2}dx .$

SOLUTION Here $h = 1/8$, $x_k = k/8$, $k = 0, 1, \ldots, 8$, and

$$y_k = (1 + x_k^2)^{-1} = \dfrac{1}{1 + \dfrac{k^2}{64}} = \dfrac{64}{64 + k^2} .$$

By Simpson's rule (1), we find

$$I_S = \dfrac{\left(\frac{1}{8}\right)}{3}\left[1 + 4\left(\dfrac{64}{64 + 1}\right) + 2\left(\dfrac{64}{64 + 4}\right) + 4\left(\dfrac{64}{64 + 9}\right) \right.$$

$$+ 2\left(\dfrac{64}{64 + 16}\right) + 4\left(\dfrac{64}{64 + 25}\right) + 2\left(\dfrac{64}{64 + 36}\right)$$

$$\left. + 4\left(\dfrac{64}{64 + 49}\right) + \left(\dfrac{64}{64 + 64}\right) \right] = 0.7854 .$$

Hence the value of the definite integral in (3) is approximately $I_S = 0.7854$. ∎ End

For a more detailed discussion of Simpson's rule, we refer the reader to a numerical analysis book such as *Numerical Analysis,* 6th edition, by R. Burden and J. Faires (Brooks/Cole Publishing, Pacific Grove, CA, 1997).

C CRAMER'S RULE

When a system of n linear equations in n unknowns has a unique solution, determinants can be used to obtain a formula for the unknowns. This procedure is called **Cramer's rule.** When n is small, these formulas provide a simple procedure for solving the system.

Suppose that for a system of n linear equations in n unknowns,

$$
\begin{aligned}
a_{11}x_1 + a_{12}x_2 + \cdots + a_{1n}x_n &= b_1 , \\
a_{21}x_1 + a_{22}x_2 + \cdots + a_{2n}x_n &= b_2 , \\
\vdots \qquad\qquad \vdots \qquad\qquad \vdots \\
a_{n1}x_1 + a_{n2}x_2 + \cdots + a_{nn}x_n &= b_n ,
\end{aligned}
$$

(1)

the coefficient matrix

(2)
$$
\mathbf{A} := \begin{bmatrix} a_{11} & a_{12} & \cdots & a_{1n} \\ a_{21} & a_{22} & \cdots & a_{2n} \\ \vdots & \vdots & & \vdots \\ a_{n1} & a_{n2} & \cdots & a_{nn} \end{bmatrix}
$$

has a nonzero determinant. Then Cramer's rule gives the solutions

(3)
$$
x_i = \frac{\det \mathbf{A}_i}{\det \mathbf{A}} , \qquad i = 1, 2, \ldots, n ,
$$

where \mathbf{A}_i is the matrix obtained from \mathbf{A} by replacing the ith column of \mathbf{A} by the column vector

$$
\begin{bmatrix} b_1 \\ b_2 \\ \vdots \\ b_n \end{bmatrix}
$$

consisting of the constants on the right-hand side of system (1).

EXAMPLE 1 Use Cramer's rule to solve the system

$$
\begin{aligned}
x_1 + 2x_2 - x_3 &= 0 , \\
2x_1 + x_2 + x_3 &= 9 , \\
x_1 - x_2 - 2x_3 &= 1 .
\end{aligned}
$$

SOLUTION We first compute the determinant of the coefficient matrix:

$$\det \begin{bmatrix} 1 & 2 & -1 \\ 2 & 1 & 1 \\ 1 & -1 & -2 \end{bmatrix} = 12 \ .$$

Using formula (3), we find

$$x_1 = \frac{1}{12} \det \begin{bmatrix} 0 & 2 & -1 \\ 9 & 1 & 1 \\ 1 & -1 & -2 \end{bmatrix} = \frac{48}{12} = 4 \ ,$$

$$x_2 = \frac{1}{12} \det \begin{bmatrix} 1 & 0 & -1 \\ 2 & 9 & 1 \\ 1 & 1 & -2 \end{bmatrix} = \frac{-12}{12} = -1 \ ,$$

$$x_3 = \frac{1}{12} \det \begin{bmatrix} 1 & 2 & 0 \\ 2 & 1 & 9 \\ 1 & -1 & 1 \end{bmatrix} = \frac{24}{12} = 2 \ . \quad \blacksquare \text{End}$$

For a more detailed discussion of Cramer's rule, we refer the reader to an introductory linear algebra text such as *Linear Algebra and Its Applications,* 2nd Edition, by David C. Lay (Addison-Wesley, Reading, Mass., 1997).

D METHOD OF LEAST SQUARES

The **method of least squares** is a procedure for fitting a straight line to a set of measured data. Consider the *scatter diagram* in Figure D.1, consisting of an x, y-plot of some data points $\{(x_i, y_i); i = 1, 2, \ldots, N\}$. It is desired to construct the straight line $y = \alpha + \beta x$ that best fits the data points, in the sense that the sum of the squares of the vertical deviations from the points to the line is minimized.

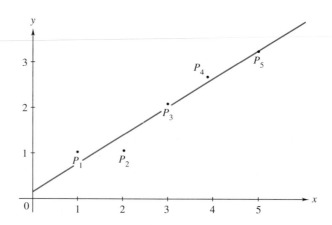

Figure D.1 Scatter diagram and least-squares linear fit

If the y-axis intercept α and the slope β of the line are known, then the y value on the line corresponding to the measured x value x_i is given by $\alpha + \beta x_i$. The corresponding measured y value, y_i, thus deviates from the line by $[y_i - (\alpha + \beta x_i)]$, and the total sum of squares of deviations is

$$S := \sum_{i=1}^{N} [y_i - (\alpha + \beta x_i)]^2 .$$

The interesting feature of the function S is that the symbols x_i and y_i are constants, while α and β are the (unknown) variables. The values of α and β that minimize S will force its partial derivatives to be zero:

$$0 = \frac{\partial S}{\partial \alpha} = \sum_{i=1}^{N} 2[y_i - (\alpha + \beta x_i)](-1) ,$$

$$0 = \frac{\partial S}{\partial \beta} = \sum_{i=1}^{N} 2[y_i - (\alpha + \beta x_i)](-x_i) .$$

Displaying these conditions in a form that emphasizes the roles of α and β, we rewrite them after a little algebra as

$$\alpha N + \beta \sum_{i=1}^{N} x_i = \sum_{i=1}^{N} y_i ,$$

$$\alpha \sum_{i=1}^{N} x_i + \beta \sum_{i=1}^{N} x_i^2 = \sum_{i=1}^{N} x_i y_i ,$$

and obtain the formulas for the optimal values of intercept and slope:

$$\alpha = \frac{\left(\sum_{i=1}^{N} x_i^2\right)\left(\sum_{i=1}^{N} y_i\right) - \left(\sum_{i=1}^{N} x_i\right)\left(\sum_{i=1}^{N} x_i y_i\right)}{N \sum_{i=1}^{N} x_i^2 - \left(\sum_{i=1}^{N} x_i\right)^2} ,$$

$$\beta = \frac{N \sum_{i=1}^{N} x_i y_i - \left(\sum_{i=1}^{N} x_i\right)\left(\sum_{i=1}^{N} y_i\right)}{N \sum_{i=1}^{N} x_i^2 - \left(\sum_{i=1}^{N} x_i\right)^2} .$$

EXAMPLE 1 Find the least-squares linear fit to the data points $P_1(1, 1)$, $P_2(2, 1)$, $P_3(3, 2)$, $P_4(4, 2.5)$, and $P_5(5, 3.1)$, which are plotted in Figure D.1.

SOLUTION Arranging the data as in Table D.1 on page A-8 yields

$$\alpha = \frac{55(9.6) - 15(34.5)}{5(55) - (15)^2} = \frac{10.5}{50} = 0.21 ,$$

$$\beta = \frac{5(34.5) - 15(9.6)}{5(55) - (15)^2} = \frac{28.5}{50} = 0.57 .$$

Thus $y = 0.21 + 0.57x$ is the equation for the best-fitting line, which is graphed in Figure D.1.

■ End

TABLE D.1				
i	x_i	y_i	$x_i y_i$	x_i^2
1	1	1	1	1
2	2	1	2	4
3	3	2	6	9
4	4	2.5	10	16
5	5	3.1	15.5	25
Sums	15	9.6	34.5	55

E RUNGE-KUTTA PROCEDURE FOR *n* EQUATIONS

The following subroutine and algorithm generalize the programs given in Section 5.6 (pages 306–307) to systems of *n* first order ordinary differential equations.

CLASSICAL 4TH ORDER RUNGE-KUTTA SUBROUTINE (*n* EQUATIONS)

Purpose To approximate the solution to the initial value problem

$$x_i' = f_i(t, x_1, \ldots, x_n) \; ;$$
$$x_i(t_0) = a_i \;, \quad i = 1, 2, \ldots, n$$

for $t_0 \le t \le c$.

INPUT $n, t_0, a_1, \ldots, a_n, c, N$ (number of steps), PRNTR ($=1$ to print table).

Step 1 Set step size

$$h = (c - t_0)/N \;, \quad t = t_0 \;, \quad x_1 = a_1 \;, \quad \ldots \;, \quad x_n = a_n$$

Step 2 For $j = 1$ to N do Steps 3–5

Step 3 Set

$$k_{1,i} = hf_i(t, x_1, \ldots, x_n) \;,$$
$$i = 1, \ldots, n \;;$$
$$k_{2,i} = hf_i\left(t + \frac{h}{2}, x_1 + \frac{1}{2}k_{1,1}, \ldots, x_n + \frac{1}{2}k_{1,n}\right) \;,$$
$$i = 1, \ldots, n \;;$$
$$k_{3,i} = hf_i\left(t + \frac{h}{2}, x_1 + \frac{1}{2}k_{2,1}, \ldots, x_n + \frac{1}{2}k_{2,n}\right) \;,$$
$$i = 1, \ldots, n \;;$$
$$k_{4,i} = hf_i(t + h, x_1 + k_{3,1}, \ldots, x_n + k_{3,n}) \;,$$
$$i = 1, \ldots, n$$

Step 4 Set

$$t = t + h \;;$$
$$x_i = x_i + \frac{1}{6}(k_{1,i} + 2k_{2,i} + 2k_{3,i} + k_{4,i}) \;, \quad i = 1, \ldots, n$$

Step 5 If PRNTR $= 1$, print t, x_1, x_2, \ldots, x_n

CLASSICAL 4TH ORDER RUNGE-KUTTA ALGORITHM WITH TOLERANCE (n EQUATIONS)

Purpose To approximate the solution to the initial value problem

$$x_i' = f_i(t, x_1, \ldots, x_n) \; ;$$
$$x_i(t_0) = a_i \; , \quad i = 1, 2, \ldots, n$$

at $t = c$, with tolerance ε.

INPUT $n, t_0, a_1, \ldots, a_n, c$
ε (tolerance)
M (maximum number of iterations)

Step 1 Set $z_i = a_i, \quad i = 1, 2, \ldots, n$; set PRNTR $= 0$
Step 2 For $m = 0$ to M do Steps 3–7 (or, to save time, start with $m > 0$)
Step 3 Set $N = 2^m$
Step 4 Call FOURTH ORDER RUNGE-KUTTA SUBROUTINE (n EQUATIONS)
Step 5 Print h, x_1, x_2, \ldots, x_n
Step 6 If $|z_i - x_i| < \varepsilon$ for $i = 1, \ldots, n$, go to Step 10
Step 7 Set $z_i = x_i, \quad i = 1, \ldots, n$
Step 8 Print "$x_i(c)$ is approximately"; x_i (for $i = 1, \ldots, n$); "but may not be within the tolerance"; ε
Step 9 Go to Step 11
Step 10 Print "$x_i(c)$ is approximately"; x_i (for $i = 1, \ldots, n$); "with tolerance"; ε
Step 11 STOP
OUTPUT Approximations of the solution to the initial value problem at $t = c$, using 2^m steps.

Answers to Odd-Numbered Problems

CHAPTER 1

Exercises 1.1, page 5

1. ODE, 2nd order, ind. var. t, dep. var. x, linear.
3. ODE, 1st order, ind. var. x, dep. var. y, nonlinear.
5. ODE, 1st order, ind. var. t, dep. var. p, nonlinear.
7. ODE, 1st order, ind. var. x, dep. var. y, nonlinear.
9. ODE, 2nd order, ind. var. x, dep. var. y, linear.
11. PDE, 2nd order, ind. var. t, r, dep. var. N.
13. $dp/dt = kp$, where k is the proportionality constant.
15. $dT/dt = k(M - T)$, where k is the proportionality constant.
17. Alison wins by $6\sqrt{3} - 4\sqrt{6} \approx 0.594$ seconds.

Exercises 1.2, page 14

3. Yes. 5. No. 7. Yes. 9. No. 11. Yes.
13. Yes. 19. The left-hand side is always ≥ 3.
21. (a) $1/3, -3$. (b) $1 \pm \sqrt{6}$. 23. Yes. 25. Yes.
27. No. 31. (a) No.

Exercises 1.3, page 22

1.

Figure B.1
Solution to Problem 1(b)

Figure B.2
Solution to Problem 1(c)

1. (d) The solutions in parts (b) and (c) both become infinite and have the line $y = 2x$ as an asymptote as $x \to \infty$. As $x \to -\infty$, the solution in part (b) becomes infinite and has $y = -2x$ as an asymptote, while the solution in part (c) does not even exist for x negative.
3. All solutions have limiting value 8 as $t \to +\infty$.

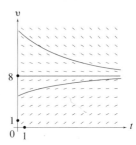

Figure B.3 Solutions to Problem 3 satisfying
$v(0) = 5$, $v(0) = 8$, $v(0) = 15$

5. (a)

Figure B.4 Direction field for Problem 5(a)

(b) $\lim_{t \to \infty} p(t) = 3/2$. (c) $\lim_{t \to \infty} p(t) = 3/2$.
(d) No.

7. (a)

Figure B.5 Direction field for Problem 7(a)

(b) 2. **(c)** 2. **(d)** 0. **(e)** No.

9. (d) It increases and asymptotically approaches the line
$y = x - 1$.
(f), (g) See Figure B.6.

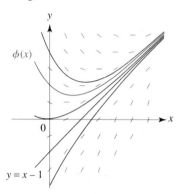

Figure B.6 Direction field and sketch of $\phi(x)$ for Problem 9(f), (g)

11.

Figure B.7
Solution to Problem 11

13.

Figure B.8
Solution to Problem 13

15.

Figure B.9 Solution to Problem 15

17. It approaches 3.

Exercises 1.4, page 31

1.

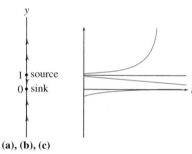

(a), (b), (c)
Figure B.10 Solution to Problem 1

3.

(a), (b), (c)
Figure B.11 Solution to Problem 3

5.

(a), (b), (c)
Figure B.12 Solution to Problem 5

7. (b), (c) 0 is source, $\pm\left(\dfrac{2k + 1}{2}\right)\pi$ is sink if $k \geq 0$ is even integer, source if k is odd integer.

(a)

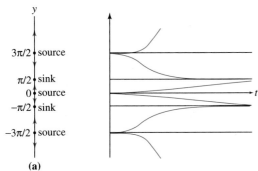

(a)
Figure B.13 Solution to Problem 7(a)

9.

(a), (b), (c)
Figure B.14 Solution to Problem 9

11. $y(t) \to 2$.
13. $y(t) \to \pi/2$.
15. $y(t) \to \pi$.
17. $y(t) \to 1$.
19. More than 50.
21.

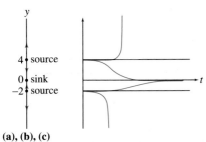

Figure B.15 Solution to Problem 21

23.

Figure B.16 Solution to Problem 23

25.

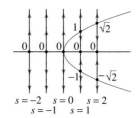

Figure B.17 Solution to Problem 25

Exercises 1.5, page 35

1.

x_n	0.1	0.2	0.3	0.4	0.5
y_n	-1.000	-1.010	-1.030	-1.059	-1.097

(rounded to three decimal places).

3.

x_n	0.1	0.2	0.3	0.4	0.5
y_n	2.700	2.511	2.383	2.292	2.225

(rounded to three decimal places).

5.

x_n	1.2	1.4	1.6	1.8
y_n	1.400	1.960	2.789	4.110

7.

N	$\approx x(1)$
1	1
2	1.06185
4	1.13920
8	1.19157

9.

x_n	y_n
1.1	-0.9
1.2	-0.81654
1.3	-0.74572
1.4	-0.68480
1.5	-0.63176
1.6	-0.58511
1.7	-0.54371
1.8	-0.50669
1.9	-0.47335
2.0	-0.44314

11. 1.46635.

15. $T(1) \approx 80.460°$, $T(2) \approx 73.647°$.

CHAPTER 2

Exercises 2.2, page 51

1. Yes. **3.** Yes. **5.** No.
7. $y = (3x - x^3 + C)^{1/3}$.

9. $y = Ce^{2x - \cos x}$. **11.** $2y + \sin(2y) = 4 \arctan x + C$.

13. $x = \dfrac{Ce^t}{Ce^t - 1}$, $x \equiv 1$. **15.** $y = 1/(C - e^{\cos x})$.

17. $y = 2e^{-x^4/4} + 1$. **19.** $y = -3e^{-1 - \cos\theta}$.

21. $y = \sin^2 x + 2\sin x$. **23.** $y = \arctan(1 + x^2)$.

25. $y = 4e^{x^3/3} - 1$.

27. (a) $y(x) = \displaystyle\int_0^x e^{t^2} dt$.

(b) $y(x) = \left(1 + 3\displaystyle\int_0^x e^{t^2} dt\right)^{1/3}$.

(c) $y(x) = \tan\left(\displaystyle\int_0^x \sqrt{1 + \sin t}\, dt + \pi/4\right)$.

(d) $y(0.5) \approx 1.381$.

29. (d) $\partial f/\partial y$ is not continuous at $(0,0)$. **33.** 281 kg.

35. (a) 82.2 min. **(b)** 31.8 min.
(c) Never attains desired temperature.

37. (a) $1105.17 . **(b)** 27.73 years.
(c) $4427.59 .

39. *A* wins.

Exercises 2.3, page 59

1. Linear. **3.** Neither. **5.** Both.

7. $y = (1/2)e^{3x} + Ce^x$. **9.** $r = \sin\theta + C\cos\theta$.

11. $y = -t - 2 + Ce^t$. **13.** $x = y^3 + Cy^{-2}$.

15. $y = 1 + C(x^2 + 1)^{-1/2}$. **17.** $y = xe^x - x$.

19. $x = t^{-1}/2 - 2t^{-3}$. **21.** $y = x^2\cos x - \pi^2\cos x$.

23. $y(t) = (38/3)e^{-5t} - (8/3)e^{-20t}$.

25. (b) $y(3) \approx 0.183$.

27. (b) 0.9960 . **(c)** 0.9486 , 0.9729 .

29. $x = e^{4y}/2 + Ce^{2y}$.

31. (a) $y = x - 1 + Ce^{-x}$. **(b)** $y = x - 1 + 2e^{-x}$.
(c) $y = x/3 - 1/9 + Ce^{-3x}$.

(d) $y = \begin{cases} x - 1 + 2e^{-x}, & 0 \le x \le 2, \\ x/3 - 1/9 + (4e^6/9 + 2e^4)e^{-3x}, & 2 < x. \end{cases}$

(e)

Figure B.18 Solution to Pròblem 31(e)

33. (a) $y = x$ is only solution in a neighborhood of $x = 0$.
(b) $y = -3x + Cx^2$ satisfies $y(0) = 0$ for any C .

35. (a) 0.281 kg/L. **(b)** 0.598 kg/L.

37. $x(t) = \dfrac{1}{2} - \dfrac{2\cos(\pi t/12)}{4 + (\pi/12)^2} - \dfrac{(\pi/12)\sin(\pi t/12)}{4 + (\pi/12)^2}$

$+ \left(\dfrac{19}{2} + \dfrac{2}{4 + (\pi/12)^2}\right)e^{-2t}$.

39. 71.8°F at noon; 26.9°F at 5 P.M.

Exercises 2.4, page 70

1. Linear with y as dep. var. **3.** Exact.

5. Exact, also linear with x as dep. var.

7. Linear with r as dep. var.

9. $y = (C - 3x)/(x^2 - 1)$.

11. $\sin x \cos y + x^2 - y^2 = C$. **13.** $t \ln y + t = C$.

15. $r = (C - e^\theta)\sec\theta$. **17.** Not exact.

19. $x^2 - y^2 + \arctan(xy) = C$.

21. $\ln x + x^2 y^2 - \sin y = \pi^2$. **23.** $y = -2/(te^t + 2)$.

25. $\sin x - x \cos x = \ln y + 1/y + \pi - 1$ (equation is separable, not exact).

27. (a) $-\ln|y| + f(x)$.
(b) $\cos x \sin y - y^2/2 + f(x)$, where f is a function of x only.

29. (c) $y = x^2/(C - x)$. **(d)** Yes, $y \equiv 0$.

33. (a) $x = cy^2$; $x \equiv 0$; $y \equiv 0$.
(b) $x^2 + 4y^2 = c$. **(c)** $2y^2\ln y - y^2 + 2x^2 = c$.
(d) $2x^2 + y^2 = c$.

Exercises 2.5, page 76

1. Separable, linear with x as dep. var., has integrating factor depending on x alone.

3. Integrating factor depending on y alone.

5. Integrating factor depending on y alone.

7. $\mu = x^{-2}$; $y^2/2 - y/x + 3x = C$ and $x \equiv 0$.

9. $\mu = x$; $x^2 y^2 + x^2 y + x^4 = C$.

11. $\mu = y^{-2}$; $x^2 y^{-1} + x = C$ and $y \equiv 0$.

13. $\mu = xy$; $x^2 y^3 - 2x^3 y^2 = C$.

15. $\mu(z) = \exp\left(\int H(z)\, dz\right)$; $z = xy$.

17. (b) $\mu = e^y$; $x = y - 1 + Ce^{-y}$.

Exercise s 2.6, page 83

1. $y' = G(ax + by)$. **3.** Linear coefficients.

5. Bernoulli. **7.** Homogeneous.

9. $\ln(y^2/x^6) - y^2/x^2 = C$.

11. $y = x/(\ln|x| + C)$ and $y \equiv 0$.

13. $\sqrt{1 + x^2/t^2} = \ln|t| + C$.

15. $(x^2 - 4y^2)^3 x^2 = C$.

17. $y = (x + C)^2/4 - x$ and $y = -x$.

19. $y = x + (6 + 4Ce^{2x})/(1 + Ce^{2x})$ and $y = x + 4$.

21. $y = 2/(Cx - x^3)$ and $y \equiv 0$.

23. $y = 5x^2/(x^5 + C)$ and $y \equiv 0$.

25. $x^{-2} = 2t^2\ln|t| + Ct^2$ and $x \equiv 0$.

27. $r = \theta^2/(C - \theta)$ and $r \equiv 0$.

29. $(y + 2)^2 + 2(x + 1)(y + 2) - 3(x + 1)^2 = C$.

31. $(2x + 2y - 3)^3 = C(2x + y - 2)^2$.

33. $y = 4x + (3 + Ce^{4x})/(1 - Ce^{4x})$.

35. $\ln\left[(x + 3)^2 + 3(t - 1)^2\right]$
$\quad + (2/\sqrt{3})\arctan\left[(x + 3)/\sqrt{3}(t - 1)\right] = C$.

37. $y^{-2} = -e^{2x}/2 + Ce^{-2x}$ and $y \equiv 0$.

39. $\theta y^2 = C(\theta + y)^2$ and $y = -\theta$.

41. $(x - y + 2)^2 = Ce^{2x} + 1$.

45. $(y - 4x)^2(y + x)^3 = C$.

Review Problems, page 87

1. $e^x + ye^{-y} = C$. **3.** $x^2y - x^3 + y^{-2} = C$.

5. $y + x\sin(xy) = C$. **7.** $y = (7\ln|t| + C)^{-1/7}$.

9. $(x^2 + 4y^2)^3x^2 = C$. **11.** $\tan(t - x) + t = C$.

13. $y = -(x^2/2)\cos(2x) + (x/4)\sin(2x) + Cx$.

15. $y = 2x + 3 - (x + C)^2/4$.

17. $y = 2/(1 + Ce^{2\theta})$ and $y \equiv 0$.

19. $y^2 = x^2 + Cx^3$ and $x \equiv 0$.

21. $xy - x^2 - x + y^2/2 - 4y = C$.

23. $y^2 + 2xy - x^2 = C$.

25. $x^2y^{-2} - 2xy^{-1} - 4xy^{-2} = C$ and $y \equiv 0$.

27. $\left[(y - 4)^2 - 3(x - 3)^2\right]\left\{\left[\sqrt{3}(x - 3) + (y - 4)\right]\right.$
$\quad \left./\left[\sqrt{3}(x - 3) - (y - 4)\right]\right\}^{1/\sqrt{3}} = C$.

29. $x^4y^3 - 3x^3y^2 + x^4y^2 = C$.

31. $y = -x^3/2 + 7x/2$.

33. $x = -t - 2 + 3e^{-t}$. **35.** $y = -2x\sqrt{2x^2 - 1}$.

37. $\ln\left[(y - 2)^2 + 2(x - 1)^2\right]$
$\quad + \sqrt{2}\arctan\left[\dfrac{y - 2}{\sqrt{2}(x - 1)}\right] = \ln 2$.

39. $y = \sqrt{(19x^4 - 1)/2}$.

CHAPTER 3

Exercises 3.2, page 104

1. $50 - 45e^{-2t/25}$ kg , 5.07 min.

3. $(0.4)(100 - t) - (39 \times 10^{-8})(100 - t)^4$ L ; 19.96 min.

5. 0.0097%; 73.24 hr.

7. 20 min later; $1/e$ times as salty.

9. 43,236. **13.** 5882; 6000.

15. 1527; 1527.

17. (a)

Year	$\dfrac{1}{p}\dfrac{dp}{dt}$	Logistic (Least Squares)
1790	0.0351	3.06
1800	0.0363	4.23
1810	0.0331	5.86
1820	0.0335	8.08
1830	0.0326	11.12
1840	0.0359	15.22
1850	0.0356	20.72
1860	0.0267	27.99
1870	0.0260	37.42
1880	0.0255	49.37
1890	0.0207	64.06
1900	0.0210	81.49
1910	0.0149	101.29
1920	0.0161	122.72
1930	0.0072	144.71
1940	0.0149	166.09
1950	0.0185	185.82
1960	0.0133	203.16
1970	0.0115	217.76
1980	0.0102	229.61
1990		238.96
2000		246.15

(b) $p_1 = 266.657$, $A = 0.000123985$.

(c) $p_0 = 3.05548$ (using all data, including 1990).

(d) See table in part (a).

19. $(1/2)\ln 15 \approx 1.354$ years; 14 million tons per year.

21. 1 hr; 2 hr. **23.** 11.7% **25.** 31,606 yr.

27. e^{-2t} kg of Hh, $2e^{-t} - 2e^{-2t}$ kg of It, and
$\quad 1 - 2e^{-t} + e^{-2t}$ kg of Bu.

Exercises 3.3, page 113

1. 20.7 min. **3.** 22.6 min. **5.** 9:08 A.M.

7. 28.3°C; 32.5°C; 1:16 P.M.

9. 16.3°C; 19.1°C; 31.7°C; 28.9°C. **11.** 30.4 min.

13. 148.6°F.

15. $T - M = C(T + M)\exp\left[2\arctan(T/M) - 4M^3kt\right]$;
for T near M, $M^4 - T^4 \approx 4M^3(M - T)$, and so
$dT/dt \approx k_1(M - T)$, where $k_1 = 4M^3k$.

Exercises 3.4, page 121

1. $(0.981)t + (0.0981)e^{-10t} - 0.0981$ m; 1019 sec.

3. 18.6 sec. **5.** $4.91t + 22.55 - 22.55e^{-2t}$ m; 97.3 sec.

7. 242 sec.

9. $95.65t + 956.5e^{-t/10} - 956.5$ m; 13.2 sec.

11. $e^{bv}(bv - mg)^{mg} = e^{v_0 b}(bv_0 - mg)^{mg}e^{-b^2 x/m}$.

13. 2.69 sec.; 101.19 m. **15.** $(\omega_0 - T/k)e^{-kt/I} + T/k$.

17. 300 sec.

19. $2636e^{-t/20} + 131.8t - 2636$ m; 1.768 sec.

21. $5e^{-2t}/2 + 6t - 5/2$; 6 m/sec. **23.** Sailboat B.

25. (e) 11.18 km/sec. **(f)** 2.38 km/sec.

Exercises 3.5, page 133

3.

h	"e"
1	3
0.1	2.72055
0.01	2.71830
0.001	2.71828
0.0001	2.71828

7.

x_n	y_n
1.1	0.10450
1.2	0.21668
1.3	0.33382
1.4	0.45300
1.5	0.57135

9.

x_n	y_n
0.2	0.61784
0.4	1.23864
0.6	1.73653
0.8	1.98111
1.0	1.99705
1.2	1.88461
1.4	1.72447
1.6	1.56184
1.8	1.41732
2.0	1.29779

11. $\phi(1) \approx x(1; 2^{-3}) = 1.25494$.

13. $\phi(1) \approx y(1; 2^{-3}) = 0.71698$.

15. $x = 1.27$.

17.

x_n	$y_n(h = 0.2)$	$y_n(h = 0.1)$	$y_n(h = 0.025)$
0.1		-1	0.06250
0.2	-3	1	0.00391
0.3		-1	0.00024
0.4	9	1	0.00002
0.5		-1	0.00000
0.6	-27	1	0.00000
0.7		-1	0.00000
0.8	81	1	0.00000
0.9		-1	0.00000
1.0	-243	1	0.00000

We conclude that step size can dramatically affect convergence.

19.

	y_n		
x_n	$r = 1.5$	$r = 2$	$r = 3$
0.25	1.58286	1.53125	1.39063
0.5	2.35144	2.04960	1.55347
0.75	3.26750	2.44003	1.62885
1.0	4.25316	2.68675	1.66999
1.25	5.21675	2.82920	1.69406
1.5	6.08340	2.90804	1.70858
1.75	6.81163	2.95080	1.71748
2.0	7.39215	2.97377	1.72298
2.25	7.83709	2.98604	1.72640
2.5	8.16851	2.99257	1.72852
2.75	8.41036	2.99605	1.72985
3.0	8.58432	2.99790	1.73067
3.25	8.70817	2.99889	1.73119
3.5	8.79571	2.99941	1.73151
3.75	8.85729	2.99969	1.73171
4.0	8.90044	2.99983	1.73184
4.25	8.93062	2.99991	1.73192
4.5	8.95168	2.99995	1.73197
4.75	8.96637	2.99997	1.73200
5.0	8.97660	2.99999	1.73202

Limiting population is $3^{1/(r-1)}$.

21.

	T_n		
Time	$K = 0.2$	$K = 0.4$	$K = 0.6$
Midnight	65.0000	65.0000	65.0000
4 A.M.	69.1639	68.5644	68.1299
8 A.M.	71.4836	72.6669	73.6678
Noon	72.9089	75.1605	76.9783
4 P.M.	72.0714	73.5977	74.7854
8 P.M.	69.8095	69.5425	69.2832
Midnight	68.3852	67.0500	65.9740

Exercises 3.6, page 143

1. $y_{n+1} = y_n + h\cos(x_n + y_n)$
$$- \frac{h^2}{2}\sin(x_n + y_n)\left[1 + \cos(x_n + y_n)\right]$$.

3. $y_{n+1} = y_n + h(x_n - y_n) + \frac{h^2}{2}(1 - x_n + y_n)$
$$- \frac{h^3}{6}(1 - x_n + y_n) + \frac{h^4}{24}(1 - x_n + y_n)$$.

5. Order 2, $\phi(1) \approx 1.3725$; order 4, $\phi(1) \approx 1.3679$.

7. -11.7679. **9.** 1.36789. **11.** $x = 1.41$.

13. $x = 0.50$.

15.

x_n	y_n
0.5	0.21462
1.0	0.13890
1.5	−0.02668
2.0	−0.81879
2.5	−1.69491
3.0	−2.99510

19. $v(3) \approx 0.24193$ with $h = 0.0625$.

21. $z(1) \approx 2.87083$ with $h = 0.03125$.

CHAPTER 4

Exercises 4.1, page 160

1. $y(t) = \dfrac{-63}{541} \sin 5t - \dfrac{30}{541} \cos 5t$.

3. $y(t) = -\cos 2t + \dfrac{3}{4} \sin 2t$.

Exercises 4.2, page 166

1. Linear, homogeneous, variable coefficients.

3. Nonlinear.

5. Linear, nonhomogeneous, variable coefficients.

7. Linear, nonhomogeneous, constant coefficients.

9. (a) $3x \sin x - (x^2 + 5) \cos x$. **(b)** 0.

 (c) $(r^2 - 4r - 5)x^r$.

13. (a) $2e^{2x}\cos x - 3e^{2x}\sin x$.

 (b) $-4e^{2x}\cos x + 3e^{2x}\sin x$.

15. Unique solution on $(-\pi/2, \pi/2)$.

17. Unique solution on $(0, 3)$.

19. Does not apply; equation is nonlinear.

21. Unique solution on $(-\infty, 1)$. **23.** $D^2 + 7D + 12$.

25. $D^2 + x$.

27. (a) 2. **(b)** $x \sin x + \cos x$. **31.** $-2; 1; 5$.

33. (a) $3x^2 - x^3$. **(b)** $6x + 3x^2 - 2x^3$.

 (c) $3x^2 + 2x^3$. **(d)** $6x + 3x^2 - 2x^3$.

 (e) $D^2 + D - 2$. **(f)** $6x + 3x^2 - 2x^3$.

35. (a) $(D + 4)(D - 1)$. **(b)** $(D + 3)(D - 2)$.

 (c) $(2D - 1)(D + 5)$.

 (d) $\left(D + \sqrt{2}\right)\left(D - \sqrt{2}\right)$.

37. (a) $\phi'(x_0) = \lim\limits_{n \to \infty} \dfrac{\phi(x_n) - \phi(x_0)}{x_n - x_0} = \lim\limits_{n \to \infty} \dfrac{0 - 0}{x_n - x_0} = 0$.

Exercises 4.3, page 176

1. Lin. indep.; $2e^{-2x}$. **3.** Lin. indep.; $-e^{4x}$.

5. Lin. dep.; 0 .

7. (b) $y = c_1 x^2 + c_2 x^{-1}$. **(c)** $y = -3x^2 + x^{-1}$.

9. (b) $y = c_1 e^x \cos 2x + c_2 e^x \sin 2x$.

 (c) $y = 2e^x \cos 2x - e^x \sin 2x$.

11. (b) $y = c_1 e^t + c_2(t^2 + 2t + 2)$.

 (c) $y = 5e^{t-1} - t^2 - 2t - 2$.

13. (c) $\phi(x) = (1)e^x + (-1)(e^x - e^{-6x})$,

 $\phi(x) = (-3)e^x + (1)(3e^x + e^{-6x})$.

15. The function in (b) cannot be a Wronskian of solutions.

17. (a) True. **(b)** False. **19.** $W = Cxe^{-x}$.

25. (a) Lin. indep. **(b)** Lin. dep. **(c)** Lin. indep.

 (d) Lin. dep.

27. (a) 2. **(b)** 0. **(c)** $2e^{3x}$.

31. (e) No, because coefficient of y'' vanishes at $x = 0$ and equation cannot be written in standard form.

Exercises 4.4, page 181

1. $y = e^{-5x}$. **3.** $y = x^4$. **5.** $x = t + 1$.

7. $xw'' + 2xw' + (x + 1)w = 0$.

9. (a) $y = x \displaystyle\int x^{-2} e^{x^2/2} dx$.

 (b) $y = \displaystyle\sum_{n=0}^{\infty} \dfrac{x^{2n}}{n! 2^n (2n - 1)} = -1 + \dfrac{x^2}{2} + \dfrac{x^4}{24} + \cdots$.

13. (a) $y = (1 - 2x^2) \displaystyle\int (1 - 2x^2)^{-2} e^{x^2} dx$.

 (b) $y = (3x - 2x^3) \displaystyle\int (3x - 2x^3)^{-2} e^{x^2} dx$.

15. (a) $y = (x - 1) \displaystyle\int e^x x^{-1} (x - 1)^{-2} dx$.

 (b) $y = (x^2 - 4x + 2) \displaystyle\int e^x x^{-1} (x^2 - 4x + 2)^{-2} dx$.

17. (b) $g(2) \approx 10.7983$.

Exercises 4.5, page 189

1. $c_1 e^{-3x} + c_2 e^{-2x}$. **3.** $c_1 e^{-4x} + c_2 x e^{-4x}$.

5. $c_1 e^{(-1-\sqrt{5})x/2} + c_2 e^{(-1+\sqrt{5})x/2}$.

7. $c_1 e^{x/2} + c_2 e^{-4x}$.

9. $c_1 e^{(1+3\sqrt{5})x/2} + c_2 e^{(1-3\sqrt{5})x/2}$.

11. $c_1 e^{-5x/2} + c_2 x e^{-5x/2}$. **13.** $3e^{-4x}$.

15. $e^{-x} - 2xe^{-x}$. **17.** $\left(\sqrt{3}/2\right)\left[e^{(1+\sqrt{3})x} - e^{(1-\sqrt{3})x}\right]$.

19. $2e^{5(x+1)} + e^{-(x+1)}$.

21. (a) $ar + b = 0$. **(b)** $ce^{-bx/a}$. **23.** $ce^{-4x/5}$.

25. $ce^{13x/6}$. **27.** $c_1 e^{-x} + c_2 e^x + c_3 e^{6x}$.

29. $c_1 e^{-x} + c_2 e^{3x} + c_3 e^{5x}$.

31. $c_1 e^x + c_2 e^{(-1+\sqrt{3})x} + c_3 e^{(-1-\sqrt{3})x}$.

33. $c_1 x + c_2 x^{-7}$. **35.** $c_1 x^{-1} + c_2 x^{-4}$.

37. $c_1 x^{-1/3} + c_2 x^{-1/3} \ln x$. **39.** $x - 3x^4$.
41. $c_1(x-2) + c_2(x-2)^7$. **43.** No.
45. $c_1 x^{-2} + c_2(-x)^{-1/2}$. **47.** $c_1 x + c_2 x \ln x + c_3 x^3$.
49. (a) $c_1 e^{r_1 x} + c_2 e^{r_2 x} + c_3 e^{r_3 x}$, where $r_1 = -4.832$,
 $r_2 = -1.869$, and $r_3 = 0.701$.
(b) $c_1 e^{r_1 x} + c_2 e^{-r_1 x} + c_3 e^{r_2 x} + c_4 e^{-r_2 x}$,
 where $r_1 = 1.176$, $r_2 = 1.902$.
(c) $c_1 e^{-x} + c_2 e^{-2x} + c_3 e^x + c_4 e^{2x} + c_5 e^{3x}$.
53. $b = 2000\sqrt{3}$, multiply by the factor $\sqrt{2}$.

Exercises 4.6, page 198

1. $c_1 \cos 3x + c_2 \sin 3x$. **3.** $c_1 e^{3x}\cos x + c_2 e^{3x}\sin x$.
5. $c_1 e^{-2x}\cos\sqrt{2}x + c_2 e^{-2x}\sin\sqrt{2}x$.
7. $c_1 e^{x/2}\cos(5x/2) + c_2 e^{x/2}\sin(5x/2)$. **9.** $c_1 e^x + c_2 e^{7x}$.
11. $c_1 e^{-5x} + c_2 x e^{-5x}$.
13. $c_1 e^{-x}\cos 2x + c_2 e^{-x}\sin 2x$.
15. $c_1 e^{-5x}\cos 4x + c_2 e^{-5x}\sin 4x$.
17. $c_1 e^{x/2}\cos(3\sqrt{3}x/2) + c_2 e^{x/2}\sin(3\sqrt{3}x/2)$.
19. $c_1 e^x + c_2 e^{-x}\cos 2x + c_3 e^{-x}\sin 2x$.
21. $2e^{-x}\cos x + 3e^{-x}\sin x$.
23. $\left(\sqrt{2}/4\right)\left[e^{(2+\sqrt{2})x} - e^{(2-\sqrt{2})x}\right]$.
25. $e^x\sin x - e^x\cos x$. **27.** $e^{2x} - \sqrt{2}e^x\sin\sqrt{2}x$.
29. (a) $c_1 e^{-x} + c_2 e^x\cos\sqrt{2}x + c_3 e^x\sin\sqrt{2}x$.
(b) $c_1 e^{2x} + c_2 e^{-2x}\cos 3x + c_3 e^{-2x}\sin 3x$.
(c) $c_1 \cos 2x + c_2 \sin 2x + c_3 \cos 3x + c_4 \sin 3x$.
31. (a) Oscillatory. **(b)** Tends to zero.
(c) Tends to $-\infty$. **(d)** Tends to $-\infty$.
(e) Tends to $+\infty$.
33. $c_1 x^2\cos\left(\sqrt{2}\ln x\right) + c_2 x^2\sin\left(\sqrt{2}\ln x\right)$.
35. $c_1 x^{-1}\cos(2\ln x) + c_2 x^{-1}\sin(2\ln x)$.
37. (a) $y(t) = 0.3e^{-3t}\cos 4t + 0.2e^{-3t}\sin 4t$ m.
(b) $2/\pi$.
(c) Decreases the frequency of oscillation, introduces the factor e^{-3t} , causing the solution to decay to zero.
39. $b \geq 2\sqrt{Ik}$.
41. (a) $c_1 \cos x + c_2 \sin x + c_3 x \cos x + c_4 x \sin x$.
(b) $(c_1 + c_2 x)e^{-x}\cos\left(\sqrt{3}x\right)$
 $+ (c_3 + c_4 x)e^{-x}\sin\left(\sqrt{3}x\right)$.

Exercises 4.7, page 202

1. $c_1 e^x + c_2 e^{-x} - x$. **3.** $c_1 e^{2x} + c_2 e^{-x} + x - 1$.
5. $e^{-x}\left(c_1 \cos\sqrt{3}x + c_2 \sin\sqrt{3}x\right) + \sin 2x$.
7. $e^{t/2}\left[c_1 \cos\left(\sqrt{3}t/2\right) + c_2 \sin\left(\sqrt{3}t/2\right)\right] + \cos t$.
9. $c_1 e^x + c_2 x e^x + x^2 e^x$.

11. (a) $5\cos x$. **(b)** $\cos x - e^{2x}$.
(c) $4\cos x + 6e^{2x}$.
13. (a) $x^2 - x$, $x^3 - x$, (so are $x - x^2, x^2 - x^3$).
(b) $y(x) = 5x - 6x^2 + 2x^3$.
(c) $\dfrac{2 - 4x}{x(x-1)}$.
17. (a) $c_1 e^x + c_2 e^{2x} + c_3 e^{-5x} - x^3 + 2x^2$.
(b) $c_1 e^x + c_2 e^{2x} + c_3 e^{-5x} + 2x^3 + x^2$.
19. (a) $v = e^x/2 + c_1 e^{-x}$.
(b) $y = -e^x/2 + c_1 e^{-x} + c_2 e^{2x}$.

Exercises 4.8, page 211

1. $y_p \equiv -3$. **3.** e^{2x} . **5.** $3t^2 - 2t + 4$.
7. $\cos 3x$. **9.** $re^r/2 + 3e^r/4$.
11. $2^x\left[(\ln 2)^2 + 1\right]^{-1}$.
13. $4x^2 e^x$. **15.** $c_1 e^x + c_2 e^{-x} + 11x - 1$.
17. $c_1 e^t + c_2 e^{2t} + e^t(\cos t - \sin t)/2$.
19. $e^{-\theta}(c_1 \cos\theta + c_2 \sin\theta) + (\theta e^{-\theta}\sin\theta)/2$.
21. $c_1 e^{2t} + c_2 t e^{2t} + t^3 e^{2t}/6$.
23. $c_1 e^{\sqrt{3}x} + c_2 e^{-\sqrt{3}x} - x^2/3 - 2/9 + e^x/2$.
25. $e^x - 1$. **27.** $e^{-x} + \sin x - \cos x$.
29. $(3/20)\sin 2x - (1/20)\cos 2x - (3/10)\cos x$
 $- (1/10)\sin x$.
31. $-(1/2)\sin\theta - e^{2\theta}/3 + 3e^{\theta}/4 + 7e^{-\theta}/12$.
33. $(Ax^2 + Bx)\sin x + (Cx^2 + Dx)\cos x + E10^x$.
35. $e^t(A\cos t + B\sin t) + Ct^2 + Dt + E$.
37. $Ae^{5x} + (Bx + C)\cos 3x + (Dx + E)\sin 3x$.
39. No. **41.** Yes. **43.** No.
45. $(1/5)\cos x + (2/5)\sin x$.
47. $-(1/2) - (4/25)xe^x + (1/10)x^2 e^x$.
49. $c_1 x + c_2 x^{-3} + x^3/12$.
51. $c_1 x^3 + c_2 x^{-2} - (1/5)x^{-2}\ln x + 1$.
53. $c_1 \cos(\ln x) + c_2 \sin(\ln x) - (1/4)(\ln x)^2\cos(\ln x)$
 $+ (1/4)(\ln x)\sin(\ln x)$.
55. (a) $-e^{-x}\sin 2x - 2e^{-x}\cos 2x + 2$.
(b) $e^{-x}(c_1 \sin 2x + c_2 \cos 2x)$.
(c) $y =$
$$\begin{cases} -e^{-x}\sin 2x - 2e^{-x}\cos 2x + 2 , & 0 \leq x \leq 3\pi/2 ; \\ (-1 - e^{3\pi/2})e^{-x}\sin 2x + (-2 - 2e^{3\pi/2})e^{-x}\cos 2x , & x \geq 3\pi/2 . \end{cases}$$
57. (a) $y(t) = \dfrac{2V\cos(\pi/2V)}{V^2 - 1}\sin t$.
(b) $V \approx 0.73$.

Exercises 4.9, page 217

1. $c_1 \cos 2x + c_2 \sin 2x$
 $- (1/4)(\cos 2x)\ln|\sec 2x + \tan 2x|$.

3. $c_1 e^{2t} + c_2 e^{-t} + e^{3t}/4$. 5. $c_1 e^x + c_2 x e^x + x e^x \ln|x|$.

7. $c_1 \cos 4\theta + c_2 \sin 4\theta + (\theta/4)\sin 4\theta$
 $+ (1/16)(\cos 4\theta)\ln|\cos 4\theta|$.

9. $c_1 \cos 2x + c_2 \sin 2x$
 $+ [(\cos 2x)\ln|\csc 2x + \cot 2x| - 1]/4$.

11. $c_1 \cos(3 \ln x) + c_2 \sin(3 \ln x)$
 $+ (1/9)\cos(3 \ln x)\ln|\sec(3 \ln x) + \tan(3 \ln x)|$.

13. $c_1 \cos x + c_2 \sin x + (\sin x)\ln|\sec x + \tan x| - 2$.

15. $c_1 \cos 2x + c_2 \sin 2x + (1/24)\sec^2 2x - 1/8$
 $+ (1/8)(\sin 2x)\ln|\sec 2x + \tan 2x|$.

17. $c_1 \cos x + c_2 \sin x - x^2 + 3 + 3x \sin x$
 $+ 3(\cos x)\ln|\cos x|$.

19. $c_1 \cos 2x + c_2 \sin 2x - e^x/5$
 $- (1/2)(\cos 2x)\ln|\sec 2x + \tan 2x|$.

21. $c_1 x + c_2 x \ln|x| + (1/2)x(\ln|x|)^2$
 $+ 3x(\ln|x|)[\ln(\ln|x|)]$.

23. $c_1 e^x + c_2(x + 1) - x^2$.

25. $c_1(5x - 1) + c_2 e^{-5x} - x^2 e^{-5x}/10$.

27. $y = e^{1-x} - e^{x-1} + \dfrac{e^x}{2}\displaystyle\int_1^x \dfrac{e^{-t}}{t}dt - \dfrac{e^{-x}}{2}\displaystyle\int_1^x \dfrac{e^t}{t}dt$,

 $y(2) \approx -1.93$.

Exercises 4.10, page 228

1. Let $Y(t) = y(-t)$. Then $Y'(t) = -y'(-t)$, $Y''(t) = y''(-t)$.
 But $y''(s) - sy(s) = 0$, so $y''(-t) + ty(-t) = 0$ or
 $Y''(t) + tY(t) = 0$.

3. The spring stiffness is $(-6y)$, so it opposes negative dis-
 placements $(y < 0)$ and reinforces positive displacements
 $(y > 0)$. Initially $y < 0$ and $y' < 0$, so the (positive)
 stiffness reverses the negative velocity and restores y to 0.
 Thereafter, $y > 0$ and the negative stiffness drives y to
 $+\infty$.

5. (a) $y'' = 2y^3 = \dfrac{d}{dy}(y^4/2)$. Thus by setting $K = 0$ and

 choosing the $(-)$ sign in equation (11), we get

 $t = -\displaystyle\int \dfrac{dy}{\sqrt{2y^4/2}} + c = \dfrac{1}{y} + c$, or $y = 1/(t - c)$.

 (b) Linear dependence would imply
 $\dfrac{y_1(t)}{y_2(t)} = \dfrac{1/(t - c_1)}{1/(t - c_2)} = \dfrac{t - c_2}{t - c_1} \equiv$ constant
 in a neighborhood of 0, which is false if $c_1 \neq c_2$.

 (c) If $y(t) = 1/(t - c)$, then $y(0) = -1/c$,
 $y'(0) = -1/c^2 = -y(0)^2$, which is false for the given
 data.

7. (a) The velocity, which is always perpendicular to the
 lever arm, is $\ell \, d\theta/dt$. Thus (lever arm) times (perpen-
 dicular momentum) $= \ell \, m\ell \, d\theta/dt = m\ell^2 d\theta/dt$.

(b) The component of the gravitational force perpendicu-
 lar to lever arm is $mg \sin \theta$, and is directed towards
 decreasing θ. Thus torque $= -\ell mg \sin \theta$.

(c) Torque $= \dfrac{d}{dt}$ (angular momentum) or
 $-\ell mg \sin \theta = (m\ell^2 \theta')' = m\ell^2 \theta''$.

9. 2 or -2 .

11. The sign of the damping coefficient $(y')^2 - 1$ indicates
 that low velocities are boosted by negative damping but
 that high velocities are slowed. Hence one expects a
 limit cycle.

13. (a) Airy. (b) Duffing. (c) van der Pol.

15. (a) Yes ($t^2 = $ positive stiffness).
 (b) No ($-t^2 = $ negative stiffness).
 (c) Yes ($y^4 = $ positive stiffness).
 (d) No ($y^5 = $ negative stiffness for $y < 0$).
 (e) Yes ($4 + 2 \cos t = $ positive stiffness).
 (f) Yes (positive stiffness and damping).
 (g) No (negative stiffness and damping).

17. $1/4\sqrt{2}$.

Exercises 4.11, page 238

1. $y(t) = -(1/2)\cos 4t + (1/2)\sin 4t$
 $= (\sqrt{2}/2)\sin(4t - \pi/4)$;
 amplitude $= \sqrt{2}/2$; period $= \pi/2$;
 frequency $= 2/\pi$; $\pi/16$ sec.

3. $b = 0$: $y(t) = \cos 4t$.

Figure B.19 $b = 0$

$b = 6$: $y(t) = e^{-3t}\cos\sqrt{7}t + (3/\sqrt{7})e^{-3t}\sin\sqrt{7}t$
 $= (4/\sqrt{7})e^{-3t}\sin(\sqrt{7}t + \phi)$, where
 $\phi = \arctan\sqrt{7}/3 \approx 0.723$.

Figure B.20 $b = 6$

$b = 8$: $y(t) = (1 + 4t)e^{-4t}$.

Figure B.21 $b = 8$

$b = 10$: $y(t) = (4/3)e^{-2t} - (1/3)e^{-8t}$.

Figure B.22 $b = 10$

5. $k = 20$: $y(t) = [(1 + \sqrt{5})/2]e^{(-5+\sqrt{5})t}$
$+ [(1 - \sqrt{5})/2]e^{(-5-\sqrt{5})t}$.

Figure B.23 $k = 20$

$k = 25$: $y(t) = (1 + 5t)e^{-5t}$.

Figure B.24 $k = 25$

$k = 30$: $y(t) = e^{-5t}\cos\sqrt{5}t + \sqrt{5}e^{-5t}\sin\sqrt{5}t$
$= \sqrt{6}e^{-5t}\sin(\sqrt{5}t + \phi)$, where
$\phi = \arctan(1/\sqrt{5}) \approx 0.421$.

Figure B.25 $k = 30$

7. $y(t) = (-3/4)e^{-8t}\cos 8t - e^{-8t}\sin 8t$
$= (5/4)e^{-8t}\sin(8t + \phi)$, where
$\phi = \pi + \arctan(3/4) \approx 3.785$;
damp. factor $= (5/4)e^{-8t}$;
quasiperiod $= \pi/4$; quasifreq. $= 4/\pi$.
9. 0.242 m.
11. $(10/\sqrt{9999})\arctan(\sqrt{9999}) \approx 0.156$ sec.
13. Relative extrema at
$t = [\pi/3 + n\pi - \arctan(\sqrt{3}/2)]/(2\sqrt{3})$
for $n = 0, 1, 2, \ldots$; but touches curves $\pm\sqrt{7/12}\,e^{-2t}$
at $t = [\pi/2 + m\pi - \arctan(\sqrt{3}/2)]/(2\sqrt{3})$
for $m = 0, 1, 2, \ldots$.
15. First measure half the quasiperiod P as the time between
two successive zero crossings. Then compute the ratio
$y(t + P)/y(t) = e^{-(b/2m)P}$.

Exercises 4.12, page 247
1. $M(\gamma) = 1/\sqrt{(1 - 4\gamma^2)^2 + 4\gamma^2}$.

Figure B.26

3. $y(t) = \cos 3t + (1/3)t \sin 3t$.

Figure B.27

5. (a) $y(t) = -[F_0/(k - m\gamma^2)]\cos(\sqrt{k/m}\,t)$
$+ [F_0/(k - m\gamma^2)]\cos \gamma t$
$= (F_0/[m(\omega^2 - \gamma^2)])(\cos \gamma t - \cos \omega t)$.
(c) $y(t) = \sin 8t \sin t$.

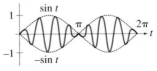

Figure B.28

7. $y(t) = c_1 e^{r_1 t} + c_2 e^{r_2 t} + \dfrac{F_0 \sin(\gamma t + \theta)}{\sqrt{(k - m\gamma^2)^2 + b^2 \gamma^2}}$,

where $r_1, r_2 = -(b/2m) \pm (1/2m)\sqrt{b^2 - 4mk}$
and $\tan \theta = (k - m\gamma^2)/(b\gamma)$ as in equation (7).

9. $y_p(t) = (0.08)\cos 2t + (0.06)\sin 2t$
$\qquad = (0.1)\sin(2t + \theta)$, where
$\qquad \theta = \arctan(4/3) \approx 0.927$.

11. $y(t) = -(18/85)e^{-2t}\cos 6t - (22/255)e^{-2t}\sin 6t$
$\qquad + \left(2/\sqrt{85}\right)\sin(2t + \theta)$, where
$\qquad \theta = \arctan(9/2) \approx 1.352$;
\qquad res. freq. $= 2\sqrt{2}/\pi$ cycles/sec.

Review Problems, page 250

1. $c_1 e^{-9x} + c_2 e^x$.

3. $c_1 e^{x/2}\cos(3x/2) + c_2 e^{x/2}\sin(3x/2)$.

5. $c_1 e^{3x/2} + c_2 e^{x/3}$.

7. $c_1 e^{-x/3}\cos(x/6) + c_2 e^{-x/3}\sin(x/6)$.

9. $c_1 e^{7x/4} + c_2 x e^{7x/4}$.

11. $t^{1/2}\left\{ c_1 \cos\left[\left(\sqrt{19}/2\right)\ln t\right] + c_2 \sin\left[\left(\sqrt{19}/2\right)\ln t\right]\right\}$.

13. $c_1 \cos 4x + c_2 \sin 4x + (1/17)x e^x - (2/289)e^x$.

15. $c_1 e^{-2x} + c_2 e^{-x} + c_3 e^{-x/3}$.

17. $c_1 e^x + c_2 e^{-x/2}\cos\left(\sqrt{43}x/2\right) + c_3 e^{-x/2}\sin\left(\sqrt{43}x/2\right)$.

19. $c_1 e^{-3x} + c_2 e^{x/2} + c_3 x e^{x/2}$.

21. $c_1 e^{3x/2}\cos\left(\sqrt{19}x/2\right) + c_2 e^{3x/2}\sin\left(\sqrt{19}x/2\right) - e^x/5$
$\qquad + x^2 + 6x/7 + 4/49$.

23. $c_1 \cos 4\theta + c_2 \sin 4\theta$
$\qquad - (1/16)(\cos 4\theta)\ln|\sec 4\theta + \tan 4\theta|$.

25. $c_1 e^{3x/2} + c_2 x e^{3x/2} + e^{3x}/9 + e^{5x}/49$.

27. $c_1 x + c_2 x^{-2} - 2x^{-2}\ln x + x \ln x$.

29. $e^{-2x}\cos\left(\sqrt{3}x\right)$.

31. $2e^x \cos 3x - (7/3)e^x \sin 3x - \sin 3x$.

33. $-e^{-x} - 3e^{5x} + e^{8x}$.

35. $\cos \theta + 2\sin \theta + \theta \sin \theta + (\cos \theta)\ln|\cos \theta|$.

37. $c_1 e^{-x} + c_2 (x - 1)$.

39. $c_1 x + c_2\left[\dfrac{x}{2}\ln\left(\dfrac{1 + x}{1 - x}\right) - 1\right]$.

41. (a), (c), (e), and (f) have all solutions bounded as
$\qquad t \to +\infty$.

43. $y_p(t) = (1/4)\sin 8t$; $\sqrt{62}/2\pi$.

CHAPTER 5

Exercises 5.2, page 274

1. $x = y^3$, $y > 0$.

Figure B.29

3. $\left(1/\sqrt{2}, 1/\sqrt{2}\right)$, $\left(-1/\sqrt{2}, -1/\sqrt{2}\right)$. **5.** $(0, 0)$.

7. $e^x + y e^{-y} = c$. **9.** $e^x + xy - y^2 = c$.

11. $y^2 - x^2 = c$. **13.** $(x - 1)^2 + (y - 1)^2 = c$.

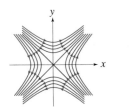

Figure B.30 **Figure B.31**

15. $(-2, 1)$ is a saddle point (unstable).

Figure B.32

17. $(0, 0)$ is a center (stable).

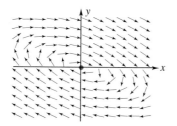

Figure B.33

19. $(0, 0)$ is a saddle point (unstable).

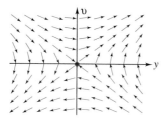

Figure B.34

21. $(0, 0)$ is a center (stable).

Figure B.35

23. $(0, 0)$ is a center (stable); $(1, 0)$ is a saddle point (unstable).

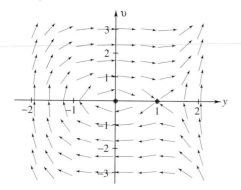

Figure B.36

25. (a) Periodic. **(b)** Not periodic.
 (c) Critical point (periodic).

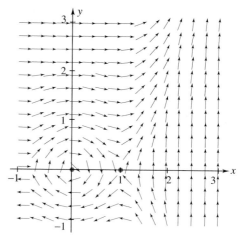

Figure B.37

27. $(x(t), y(t))$ approaches $(0, 0)$.

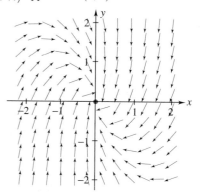

Figure B.38

29. (a) Unstable node. **(b)** Center (stable).
 (c) Stable node. **(d)** Unstable spiral.
 (e) Saddle (unstable).
 (f) Asymptotically stable spiral.

31. (a) $y' = v$, $v' = f(y)$.

 (b) $\dfrac{dv}{dy} = \dfrac{f(y)}{v}$ implies $\displaystyle\int v\,dv = \int f(y)\,dy$ + constant, or

 $v^2/2 = F(y) + K$.

33. (a)

Figure B.39

(b) 295 persons.

35. (a) $x' = v, v' = -x + 1/(\lambda - x)$.

(b) $\dfrac{dv}{dx} = \dfrac{-x + 1/(\lambda - x)}{v}$ implies

$$\int v\,dv = \int \left(-x + \frac{1}{\lambda - x}\right)dx + \text{constant, or}$$

$v^2/2 = -x^2/2 - \ln(\lambda - x) + \text{constant}$. The integral curve equation follows with the choice $C/2$ for the constant.

(c) At a critical point $v = 0$ and $-x + 1/(\lambda - x) = 0$.

Solutions for the latter are $x = \dfrac{\lambda \pm \sqrt{\lambda^2 - 4}}{2}$, and

are real only for $\lambda \geq 2$.

(d)

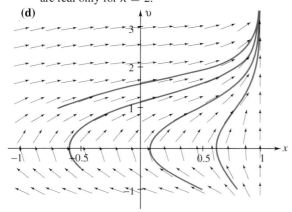

Figure B.40 $\lambda = 1$

(e) See Figures B.40 and B.41. When $\lambda = 3$, one critical point is a center and the other is a saddle. For $\lambda = 1$, the bar is attracted to the magnet. For $\lambda = 3$, the bar may oscillate periodically, or (rarely) come to rest at the saddle critical point.

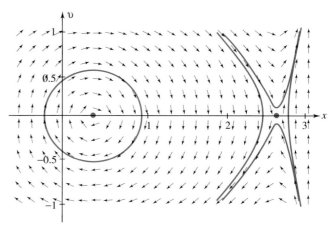

Figure B.41 $\lambda = 3$

37. (c) For upper half plane, center is at $v = 0$, $y = -1$; for lower half plane, center is at $v = 0$, $y = 1$.

(d) All points on the segment $v = 0$, $-1 \leq y \leq 1$.

(e) $y = -0.5$; see Figure B.42.

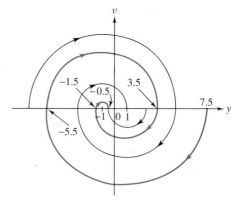

Figure B.42

Exercises 5.3, page 284

1. $x = c_2 e^{-t}$; $y = c_1 e^{-2t} + c_2 e^{-t}$.

3. $x = c_1 + c_2 e^{-2t}$; $y = c_2 e^{-2t}$.

5. $x \equiv -5$; $y \equiv 1$.

7. $u = c_1 + c_2 e^{-t} + (1/2)e^t + (5/3)t$;
$v = c_1 - 2c_2 e^{-t} + (5/3)t$.

9. $x = c_1 e^t + (1/4)\cos t - (1/4)\sin t$;
$y = -3c_1 e^t - (3/4)\cos t - (1/4)\sin t$.

11. $u = c_1 \cos 2t + c_2 \sin 2t + c_3 e^{\sqrt{3}t} + c_4 e^{-\sqrt{3}t}$
 $- (3/10)e^t$;

 $v = c_1 \cos 2t + c_2 \sin 2t - (2/5)c_3 e^{\sqrt{3}t} - (2/5)e^{-\sqrt{3}t}$
 $+ (1/5)e^t$.

13. $x = 2c_2 e^t \cos 2t - 2c_1 e^t \sin 2t$;
 $y = c_1 e^t \cos 2t + c_2 e^t \sin 2t$.

15. $w = -(2/3)c_1 e^{2t} + c_2 e^{7t} + t + 1$;
 $z = c_1 e^{2t} + c_2 e^{7t} - 5t - 2$.

17. $x = c_1 \cos t + c_2 \sin t - 4c_3 \cos\left(\sqrt{6}t\right)$
 $- 4c_4 \sin\left(\sqrt{6}t\right)$;

 $y = c_1 \cos t + c_2 \sin t + c_3 \cos\left(\sqrt{6}t\right) + c_4 \sin\left(\sqrt{6}t\right)$.

19. $x = 2e^{3t} - e^{2t}$;
 $y = -2e^{3t} + 2e^{2t}$.

21. $x = e^t + e^{-t} + \cos t + \sin t$;
 $y = e^t + e^{-t} - \cos t - \sin t$.

23. Infinitely many solutions satisfying
 $x + y = e^t + e^{-2t}$.

25. $x(t) = -c_1 e^t - 2c_2 e^{2t} - c_3 e^{3t}$,
 $y(t) = c_1 e^t + c_2 e^{2t} + c_3 e^{3t}$,
 $z(t) = 2c_1 e^t + 4c_2 e^{2t} + 4c_3 e^{3t}$.

27. $x(t) = c_1 e^{8t} + c_2 e^{4t} + c_3$,

 $y(t) = \frac{1}{2}\left(c_1 e^{8t} - c_2 e^{4t} + c_3\right)$,

 $z(t) = -c_1 e^{8t} + c_3$.

29. $-3 \le \lambda \le -1$.

31. $x(t) = -\left(100 + \frac{200}{\sqrt{7}}\right)e^{r_1 t} - \left(100 - \frac{200}{\sqrt{7}}\right)e^{r_2 t}$

 $+ 200$ kg ,

 $y(t) = \frac{300}{\sqrt{7}}e^{r_1 t} - \frac{300}{\sqrt{7}}e^{r_2 t} + 200$ kg, where

 $r_1 = \frac{-5 - \sqrt{7}}{100} \approx -0.0765$,

 $r_2 = \frac{-5 + \sqrt{7}}{100} \approx -0.0235$.

33. $x = \left[\frac{200 - 100\sqrt{5}}{\sqrt{5}}\right]e^{(-3 + \sqrt{5})t/100}$

 $- \left[\frac{200 + 100\sqrt{5}}{\sqrt{5}}\right]e^{(-3 - \sqrt{5})t/100} + 200$;

 $y = -\left(\frac{100}{\sqrt{5}}\right)e^{(-3 + \sqrt{5})t/100} + \left(\frac{100}{\sqrt{5}}\right)e^{(-3 - \sqrt{5})t/100}$

 $+ 200$.

35. $90.4°F$. **37.** $460/11 \approx 41.8°F$.

Exercises 5.4, page 291

1. $m_1 x'' = k_1(y - x)$,
 $m_2 y'' = -k_1(y - x) - k_2 y$;
 $x(0) = -1$, $x'(0) = 0$, $y(0) = 0$, $y'(0) = 0$.
 $x(t) = -(8/17)\cos t - (9/17)\cos\left(\sqrt{20/3}\,t\right)$,
 $y(t) = -(6/17)\cos t + (6/17)\cos\left(\sqrt{20/3}\,t\right)$.

3. $mx'' = -kx + k(y - x)$,
 $my'' = -k(y - x) + k(z - y)$,
 $mz'' = -k(z - y) - kz$;
 The normal frequency, $(1/2\pi)\sqrt{(2 + \sqrt{2})(k/m)}$, has the
 mode $x(t) = z(t) = -(1/\sqrt{2})\,y(t)$; the normal frequency
 $(1/2\pi)\sqrt{(2 - \sqrt{2})(k/m)}$ has the mode
 $x(t) = z(t) = (1/\sqrt{2})\,y(t)$; and the normal frequency
 $(1/2\pi)\sqrt{2k/m}$ has the mode $x(t) = -z(t)$, $y(t) \equiv 0$.

5. $x(t) = -e^{-t} - te^{-t} - \cos t$; $y(t) = e^{-t} + te^{-t} - \cos t$.

7. $x(t) = (2/5)\cos t + (4/5)\sin t - (2/5)\cos \sqrt{6}t$
 $+ \left(\sqrt{6}/5\right)\sin \sqrt{6}t - \sin 2t$;

 $y(t) = (4/5)\cos t + (8/5)\sin t + (1/5)\cos \sqrt{6}t$
 $- \left(\sqrt{6}/10\right)\sin \sqrt{6}t - (1/2)\sin 2t$.

9. $(1/2\pi)\sqrt{g/l}$; $(1/2\pi)\sqrt{(g/l) + (2k/m)}$.

Exercises 5.5, page 300

1. $I(t) = \left(19/\sqrt{21}\right)\left[e^{(-25 - 5\sqrt{21})t/2} - e^{(-25 + 5\sqrt{21})t/2}\right]$.

3. $I_p(t) = (4/51)\cos 20t - (1/51)\sin 20t$; resonance
 frequency is $5/\pi$.

5. $M(\gamma) = 1/\sqrt{(100 - 4\gamma^2)^2 + 100\gamma^2}$.

Figure B.43

7. $L = 35$ henrys, $R = 10$ ohms, $C = 1/15$ farads, and
 $E(t) = 50 \cos 10t$ volts.

11. $I_1 = (3/5)e^{-3t/2} - (8/5)e^{-2t/3} + 1$,
 $I_2 = (1/5)e^{-3t/2} - (6/5)e^{-2t/3} + 1$,
 $I_3 = (2/5)e^{-3t/2} - (2/5)e^{-2t/3}$.

13. $(1/2)I_1' + 2q_3 = \cos 3t$ (where $I_3 = q_3'$) ,

$(1/2)I_1' + I_2 = 0, I_1 = I_2 + I_3$:

$I_1(0) = I_2(0) = I_3(0) = 0$;

$I_1 = -(36/61)e^{-t}\cos\sqrt{3}t$

$\qquad - \left(42\sqrt{3}/61\right)e^{-t}\sin\sqrt{3}t + (36/61)\cos 3t$

$\qquad + (30/61)\sin 3t$,

$I_2 = (45/61)e^{-t}\cos\sqrt{3}t - \left(39\sqrt{3}/61\right)e^{-t}\sin\sqrt{3}t$

$\qquad - (45/61)\cos 3t + (54/61)\sin 3t$,

$I_3 = -(81/61)e^{-t}\cos\sqrt{3}t$

$\qquad - \left(3\sqrt{3}/61\right)e^{-t}\sin\sqrt{3}t + (81/61)\cos 3t$

$\qquad - (24/61)\sin 3t$.

Exercises 5.6, page 311

1. $x_1' = x_2$, $x_2' = 3x_1 - tx_2 + t^2$; $x_1(0) = 3$,
$x_2(0) = -6$.

3. $x_1' = x_2$, $x_2' = x_3$, $x_3' = x_4$,
$x_4' = x_4 - 7x_1 + \cos t$; $x_1(0) = x_2(0) = 1$,
$x_3(0) = 0$, $x_4(0) = 2$.

5. $x_1' = x_2$, $x_2' = x_2 - x_3 + 2t$,
$x_3' = x_4$, $x_4' = x_1 - x_3 - 1$;
$x_1(3) = 5$, $x_2(3) = 2$, $x_3(3) = 1$, $x_4(3) = -1$.

7. $x_1' = x_2$, $x_2' = x_3$, $x_3' = x_4 + t$, $x_4' = x_5$,
$x_5' = (2x_4 - 2x_3 + 1)/5$; $x_1(0) = x_2(0) = x_3(0) = 4$,
$x_4(0) = x_5(0) = 1$.

9. $t_{n+1} = t_n + h$, $n = 0, 1, \ldots$;
$x_{n+1,1} = x_{n,1} + hf_1(t_n, x_{n,1}, x_{n,2})$,
$x_{n+1,2} = x_{n,2} + hf_2(t_n, x_{n,1}, x_{n,2})$.

11.

i	t_i	$y(t_i)$
1	0.250	0.96924
2	0.500	0.88251
3	0.750	0.75486
4	1.000	0.60656

13.

i	t_i	$y(t_i)$
1	1.250	0.80761
2	1.500	0.71351
3	1.750	0.69724
4	2.000	0.74357

15. $y(8) \approx 24.01531$.

17. $y(1) \approx x_1\left(1, 2^{-3}\right) = 2.77493$.

19. $u\left(1; 2^{-2}\right) = v\left(1; 2^{-2}\right) = 0.36789$.

21.

i	t_i	$x(t_i)$	$y(t_i)$
1	0.250	0.98661	0.75492
2	0.500	0.91749	1.04259
3	0.750	0.79823	1.38147
4	1.000	0.63303	1.77948

23.

		Part (a)		Part (b)		Part (c)	
i	t_i	$x(t_i)$	$y(t_i)$	$x(t_i)$	$y(t_i)$	$x(t_i)$	$y(t_i)$
1	0.5	1.95247	2.25065	1.48118	2.42311	0.91390	2.79704
2	1.0	3.34588	1.83601	2.66294	1.45358	1.63657	1.13415
3	1.5	4.53662	3.36527	5.19629	2.40348	4.49334	1.07811
4	2.0	2.47788	4.32906	3.10706	4.64923	5.96115	5.47788
5	2.5	1.96093	2.71900	1.92574	3.32426	1.51830	5.93110
6	3.0	2.86412	1.96166	2.34143	2.05910	0.95601	2.18079
7	3.5	4.28449	2.77457	3.90106	2.18977	2.06006	0.98131
8	4.0	3.00965	4.11886	3.83241	3.89043	5.62642	1.38072
9	4.5	2.18643	3.14344	2.32171	3.79362	5.10594	5.10462
10	5.0	2.63187	2.25824	2.21926	2.49307	1.74187	5.02491

25.

i	t_i	$x_1(t_i) \approx H(t_i)$
1	0.5	0.09573
2	1.0	0.37389
3	1.5	0.81045
4	2.0	1.37361
5	2.5	2.03111
6	3.0	2.75497
7	3.5	3.52322
8	4.0	4.31970
9	4.5	5.13307
10	5.0	5.95554

27. Yes, yes.

29. $y(1) \approx x_1(1; 2^{-3}) = 1.25958$.

31. $y(0.1) \approx 0.00647, \ldots, y(2.0) \approx 1.60009$.

33. **(a)** $P_1(10) \approx 0.567, P_2(10) \approx 0.463, P_3(10) \approx 0.463$.
(b) $P_1(10) \approx 0.463, P_2(10) \approx 0.567, P_3(10) \approx 0.463$.
(c) $P_1(10) \approx 0.463, P_2(10) \approx 0.463, P_3(10) \approx 0.567$.
All populations approach 0.5.

Exercises 5.7, page 323

1. For $\omega = 3/2$: The Poincaré map alternates between the
points $(0.8, 1.5)$ and $(0.8, -1.5)$. There is a subharmonic
solution of period 4π. For $\omega = 3/5$: The Poincaré map
cycles through the points
$(-1.5625, 0.6)$, $(-2.1503, -0.4854)$,
$(-0.6114, 0.1854)$, $(-2.5136, 0.1854)$, and
$(-0.9747, -0.4854)$. There is a subharmonic solution of
period 10π.

3. The points become unbounded.

5. The attractor is the point $(-1.0601, 0.2624)$.

9. (a) $\{1/7, 2/7, 4/7, 1/7, \ldots\}$,
 $\{3/7, 6/7, 5/7, 3/7, \ldots\}$.
(b) $\{1/15, 2/15, 4/15, 8/15, 1/15, \ldots\}$,
 $\{1/5, 2/5, 4/5, 3/5, 1/5, \ldots\}$,
 $\{1/3, 2/3, 1/3, \ldots\}$,
 $\{7/15, 14/15, 13/15, 11/15, 7/15, \ldots\}$.
(c) $x_n = 0$ for $n \geq j$.

13.

Figure B.44

Review Problems, page 327

1. A trajectory is a path traced by an actual solution pair $(x(t), y(t))$ as t increases; thus it is a directed (oriented) curve. An integral curve is the graph of a solution to the phase plane equation; it has no direction. All trajectories lie along (parts of) integral curves. A given integral curve can be the underlying point set for several different trajectories.

3. $x^2 - (y - 1)^2 = c$; critical point $(0, 1)$ is saddle (unstable).

5. Asymptotically stable spiral point.

7. $x(t) = -(c_1/3)t^3 - (c_2/2)t^2 - (c_3 + 2c_1)t + c_4$,
 $y(t) = c_1 t^2 + c_2 t + c_3$.

9. $x(t) = c_1 \cos 3t + c_2 \sin 3t + e^t/10$,
 $y(t) = (3/2)(c_1 + c_2)\cos 3t - (3/2)(c_1 - c_2)\sin 3t$
 $\qquad - (11/20)e^t - (1/4)e^{-t}$.

11. $x(t) = 2 \sin t$, $y(t) = e^t - \cos t + \sin t$,
 $z(t) = e^t + \cos t + \sin t$.

13. $x(t) = -(139/4)e^{-t/6} - (49/4)e^{-t/2} + 48$,
 $y(t) = -(139/2)e^{-t/6} + (49/2)e^{-t/2} + 48$.

15. $I_1 + I_2 + I_3 = 0$, $q/C = R_2 I_2$,
 $R_2 I_2 = R_1 I_3 + L \, dI_3/dt$, where q is the charge on the capacitor $(I_1 = dq/dt)$;
 $I_3 = e^{-t}(A \cos t + B \sin t)$,
 $I_2 = e^{-t}(B \cos t - A \sin t)$,
 $I_1 = e^{-t}[(A - B)\sin t - (A + B)\cos t]$.

17. $x_1' = x_2$, $x_2' = x_3$, $x_3' = (1/3)(5 + e^t x_1 - 2x_2)$.

19. $x_1' = x_2$, $x_2' = x_3$, $x_3' = t - x_5 - x_6$,
 $x_4' = x_5$, $x_5' = x_6$, $x_6' = x_2 - x_3$.

CHAPTER 6

Exercises 6.1, page 345

1. $(-\infty, 0)$. **3.** $(3\pi/2, 5\pi/2)$. **5.** $(0, \infty)$.
7. Lin. indep.; $48e^{7x}$. **9.** Lin. dep.; 0.
11. Lin. indep.; $(3/2)x^{-5/2}$. **13.** Lin. indep.; $12x^4$.
15. $c_1 e^{3x} + c_2 e^{-x} + c_3 e^{-4x}$. **17.** $c_1 x + c_2 x^2 + c_3 x^3$.
19. (a) $c_1 e^x + c_2 e^{-x}\cos 2x + c_3 e^{-x}\sin 2x + x^2$.
 (b) $-e^x + e^{-x}\sin 2x + x^2$.
21. (a) $c_1 x + c_2 x \ln x + c_3 x(\ln x)^2 + \ln x$.
 (b) $3x - x \ln x + x(\ln x)^2 + \ln x$.
23. (a) $2 \sin x - x$. **(b)** $4x - 6 \sin x$.
29. (b) Let $f_1(x) = |x - 1|$ and $f_2(x) = x - 1$.
33. e^{2x} , $(\sin x - 2 \cos x)/5$, $-(2 \sin x + \cos x)/5$.
35. $xy''' - y'' + xy' - y = 0$.

Exercises 6.2, page 352

1. $c_1 + c_2 e^{2x} + c_3 e^{-4x}$. **3.** $c_1 e^{-x} + c_2 e^{-2x/3} + c_3 e^{x/2}$.
5. $c_1 e^{-x} + c_2 e^{-x}\cos 5x + c_3 e^{-x}\sin 5x$.
7. $c_1 e^{-x} + c_2 e^{(3+\sqrt{65})x/4} + c_3 e^{(3-\sqrt{65})x/4}$.
9. $c_1 e^{3x} + c_2 x e^{3x} + c_3 x^2 e^{3x}$.
11. $c_1 e^{-x} + c_2 x e^{-x} + c_3 x^2 e^{-x} + c_4 x^3 e^{-x}$.
13. $c_1 \cos \sqrt{2}x + c_2 x \cos \sqrt{2}x + c_3 \sin \sqrt{2}x + c_4 x \sin \sqrt{2}x$.
15. $c_1 e^x + c_2 x e^x + c_3 e^{-3x} + (c_4 + c_5 x)e^{-x}\cos 2x$
 $\qquad + (c_6 + c_7 x)e^{-x}\sin 2x$.
17. $c_1 e^{-4x} + c_2 e^{3x} + (c_3 + c_4 x + c_5 x^2)e^{-2x}$
 $\qquad + (c_6 + c_7 x)e^{-2x}\cos x + (c_8 + c_9 x)e^{-2x}\sin x + c_{10}$
 $\qquad + c_{11}x + c_{12}x^2 + c_{13}x^3 + c_{14}x^4$.
19. $e^x - 2e^{-2x} - 3e^{2x}$. **21.** $e^{2x} - \sqrt{2}e^x \sin \sqrt{2}x$.
23. $x(t) = c_1 + c_2 t + c_3 e^t$, $y(t) = c_1 - c_2 + c_2 t$.
27. $c_1 e^{1.120x} + c_2 e^{0.296x} + c_3 e^{-0.520x} + c_4 e^{-2.896x}$.
29. $c_1 e^{-0.5x}\cos(0.866x) + c_2 e^{-0.5x}\sin(0.866x)$
 $\qquad + c_3 e^{-0.5x}\cos(1.323x) + c_4 e^{-0.5x}\sin(1.323x)$.
31. (a) $\{x, x^{-1}, x^2\}$. **(b)** $\{x, x^2, x^{-1}, x^{-2}\}$.
 (c) $\{x, x^2\cos(3 \ln x), x^2\sin(3 \ln x)\}$.
33. (b) $x(t) = c_1 \cos t + c_2 \sin t + c_3 \cos \sqrt{6}t$
 $\qquad + c_4 \sin \sqrt{6}t$.
 (c) $y(t) = 2c_1 \cos t + 2c_2 \sin t - (c_3/2) \cos \sqrt{6}t$
 $\qquad - (c_4/2)\sin \sqrt{6}t$.
 (d) $x(t) = (3/5)\cos t + (2/5)\cos \sqrt{6}t$,
 $\qquad y(t) = (6/5)\cos t - (1/5)\cos \sqrt{6}t$.
35. $c_1 \cosh rx + c_2 \sinh rx + c_3 \cos rx + c_4 \sin rx$, where
 $r^4 = k/(EI)$.

Exercises 6.3, page 359

1. $c_1 x e^x + c_2 + c_3 x + c_4 x^2$.
3. $c_1 x^2 e^{-2x}$.

5. $c_1e^x + c_2e^{3x} + c_3e^{-2x} - (1/6)xe^x + (1/6)x^2$
$\quad + (5/18)x + 37/108$.

7. $c_1e^x + c_2e^{-2x} + c_3xe^{-2x} - (1/6)x^2e^{-2x}$.

9. $c_1e^x + c_2xe^x + c_3x^2e^x + (1/6)x^3e^x$.

11. D^5 . 13. $D + 7$. 15. $(D-2)(D-1)$.

17. $[(D+1)^2 + 4]^3$. 19. $(D+2)^2[(D+5)^2 + 9]^2$.

21. $c_3\cos 2x + c_4\sin 2x + c_5$.

23. $c_3xe^{3x} + c_4x^2 + c_5x + c_6$.

25. $c_3 + c_4x + c_5\cos 2x + c_6\sin 2x$.

27. $c_3xe^{-x}\cos x + c_4xe^{-x}\sin x + c_5x^2 + c_6x + c_7$.

29. $c_2x + c_3x^2 + c_6x^2e^x$.

31. $-2e^{3x} + e^{-2x} + x^2 - 1$. 33. $x^2e^{-2x} - x^2 + 3$.

39. $x(t) = -(1/63)e^{3t} + c_1 + c_2t - c_3e^{\sqrt{2}t} - c_4e^{-\sqrt{2}t}$,
$\quad y(t) = (8/63)e^{3t} + c_1 + c_2t + c_3e^{\sqrt{2}t} + c_4e^{-\sqrt{2}t}$.

Exercises 6.4, page 363

1. $(1/6)x^2e^{2x}$. 3. $e^{2x}/16$.

5. $\ln(\sec x) - (\sin x)\ln(\sec x + \tan x)$.

7. $c_1x + c_2x^2 + c_3x^3 - (1/24)x^{-1}$.

9. $-(1/2)e^x\int e^{-x}g(x)\,dx + (1/6)e^{-x}\int e^xg(x)\,dx$
$\quad + (1/3)e^{2x}\int e^{-2x}g(x)\,dx$.

11. $c_1x + c_2x^{-1} + c_3x^3 - x\sin x - 3\cos x + 3x^{-1}\sin x$.

Review Problems, page 366

1. **(a)** $(0, \infty)$. **(b)** $(-4, -1)$, $(-1, 1)$, $(1, \infty)$.

5. **(a)** $e^{-5x}(c_1 + c_2x) + e^{2x}(c_3 + c_4x + c_5x^2)$
$\quad\quad + (\cos x)(c_6 + c_7x) + (\sin x)(c_8 + c_9x)$.

 (b) $c_1 + c_2x + c_3x^2 + c_4x^3 + e^x(c_5 + c_6x)$
$\quad\quad + (e^{-x}\cos\sqrt{3}x)(c_7 + c_8x)$
$\quad\quad + (e^{-x}\sin\sqrt{3}x)(c_9 + c_{10}x)$.

7. **(a)** D^3 . **(b)** $D^2(D-3)$. **(c)** $[D^2 + 4]^2$.

 (d) $[(D+2)^2 + 9]^3$.

 (e) $D^3(D+1)^2(D^2 + 4)(D^2 + 9)$.

9. $c_1x + c_2x^5 + c_3x^{-1} - (1/21)x^{-2}$.

CHAPTER 7

Exercises 7.2, page 380

1. $\dfrac{1}{s^2}$, $s > 0$. 3. $\dfrac{1}{s-6}$, $s > 6$.

5. $\dfrac{s}{s^2 + 4}$, $s > 0$.

7. $\dfrac{s-2}{(s-2)^2 + 9}$, $s > 2$.

9. $e^{-2s}\left(\dfrac{2s+1}{s^2}\right)$, $s > 0$.

11. $\dfrac{e^{-\pi s} + 1}{s^2 + 1}$, all s .

13. $\dfrac{6}{s+3} - \dfrac{2}{s^3} + \dfrac{2}{s^2} - \dfrac{8}{s}$, $s > 0$.

15. $\dfrac{6}{s^4} - \dfrac{1}{(s-1)^2} + \dfrac{s-4}{(s-4)^2 + 1}$, $s > 4$.

17. $\dfrac{6}{(s-3)^2 + 36} - \dfrac{6}{s^4} + \dfrac{1}{s-1}$, $s > 3$.

19. $\dfrac{24}{(s-5)^5} - \dfrac{s-1}{(s-1)^2 + 7}$, $s > 5$.

21. Continuous (hence piecewise continuous).

23. Piecewise continuous.

25. Continuous (hence piecewise continuous).

27. Neither.

29. All but functions (c), (e), and (h).

Exercises 7.3, page 386

1. $\dfrac{2}{s^3} + \dfrac{2}{(s-1)^2 + 4}$.

3. $\dfrac{s+1}{(s+1)^2 + 9} + \dfrac{1}{s-6} - \dfrac{1}{s}$.

5. $\dfrac{4}{(s+1)^3} - \dfrac{1}{s^2} + \dfrac{s}{s^2 + 16}$.

7. $\dfrac{24}{s^5} - \dfrac{24}{s^4} + \dfrac{12}{s^3} - \dfrac{4}{s^2} + \dfrac{1}{s}$.

9. $\dfrac{4(s+1)}{[(s+1)^2 + 4]^2}$. 11. $\dfrac{s}{s^2 - b^2}$.

13. $\dfrac{1}{2s} - \dfrac{s}{2(s^2 + 4)}$. 15. $\dfrac{3s}{4(s^2 + 1)} + \dfrac{s}{4(s^2 + 9)}$.

17. $\dfrac{s}{2(s^2 + 9)} - \dfrac{s}{2(s^2 + 49)}$.

19. $\dfrac{n+m}{2[s^2 + (n+m)^2]} + \dfrac{m-n}{2[s^2 + (m-n)^2]}$.

21. $\dfrac{s-a}{(s-a)^2 + b^2}$.

25. **(a)** $\dfrac{s^2 - b^2}{(s^2 + b^2)^2}$. **(b)** $\dfrac{2s^3 - 6sb^2}{(s^2 + b^2)^3}$.

29. $\dfrac{1}{s^2 + 6s + 10}$. 33. e^{-s}/s^2 .

35. $e^{-(\pi/2)s}/(s^2 + 1)$.

Exercises 7.4, page 396

1. $e^t t^3$. **3.** $e^{-t}\cos 3t$. **5.** $(1/2)e^{-2t}\sin 2t$.

7. $2e^{-2t}\cos 3t + 4e^{-2t}\sin 3t$.

9. $(3/2)e^t\cos 2t - 3e^t\sin 2t$.

11. $\dfrac{6}{s+5} - \dfrac{1}{s+2} - \dfrac{4}{s-1}$. **13.** $\dfrac{1}{(s+1)^2} - \dfrac{2}{s}$.

15. $-\dfrac{3}{s+1} + \dfrac{(s-1)+2}{(s-1)^2+4}$.

17. $-\dfrac{5}{6s} + \dfrac{11}{10(s-2)} - \dfrac{4}{15(s+3)}$.

19. $\dfrac{1}{17}\left[\dfrac{1}{(s-3)} - \dfrac{s+1}{(s+1)^2+1} - \dfrac{4}{(s+1)^2+1}\right]$.

21. $(1/3) + e^t + (14/3)e^{6t}$.

23. $-e^{-3t} + 2te^{-3t} + 6e^{-t}$.

25. $8e^{2t} - e^{-t}\cos 2t + 3e^{-t}\sin 2t$.

27. $-(5/3)e^{-t} + (5/12)e^{2t} + (5/4)e^{-2t}$.

29. $3e^{-2t} + 7e^t\cos t + 11e^t\sin t$.

31. $F_1(s) = F_2(s) = F_3(s) = 1/s^2$,
 $\mathcal{L}^{-1}\{1/s^2\}(t) = f_3(t) = t$.

33. $e^{5t}/t - e^{-2t}/t$. **35.** $2(\cos t - \cos 3t)/t$.

39. $\dfrac{A}{s} + \dfrac{B}{s-1} + \dfrac{C}{s+2}$, where

$$A = \left.\dfrac{2s+1}{(s-1)(s+2)}\right|_{s=0} = \dfrac{-1}{2} ,$$

$$B = \left.\dfrac{2s+1}{s(s+2)}\right|_{s=1} = 1 ,$$

$$C = \left.\dfrac{2s+1}{s(s-1)}\right|_{s=-2} = \dfrac{-1}{2} .$$

41. $2e^{-t} - 4e^{3t} + 5e^{2t}$.

43. $\dfrac{4}{s+2} + \dfrac{2(s-1)+2(3)}{(s-1)^2+2^2}$.

Exercises 7.5, page 405

1. $2e^t\cos 2t + e^t\sin 2t$. **3.** $-e^{-3t} + 3te^{-3t}$.

5. $t^2 + \cos t - \sin t$. **7.** $\cos t - 4e^{5t} + 8e^{2t}$.

9. $3te^t - e^{-6t}$. **11.** $2 - t + e^{2-t} + 2e^{t-2}$.

13. $(7/5)\sin t + (11/5)\cos t + (3/5)e^{2t-\pi} - e^{-(t-\pi/2)}$.

15. $\dfrac{-s^2 + s - 1}{(s^2+1)(s-1)(s-2)}$.

17. $\dfrac{s^5 + s^4 + 6}{s^4(s^2+s-1)}$.

19. $\dfrac{s^3 + 5s^2 - 6s + 1}{s(s-1)(s^2+5s-1)}$.

21. $\dfrac{s^3 + s^2 + 2s}{(s^2+1)(s-1)^2}$.

23. $\dfrac{-s^3 + 1 + 3se^{-2s} - e^{-2s}}{s^2(s^2+4)}$.

25. $2e^t - \cos t - \sin t$.

27. $(t^2-4)e^{-t}$. **29.** $(3a-b)e^t/2 + (b-a)e^{3t}/2$.

31. $5/2 + (a-5/2)e^{-t}\cos t + (a+b-5/2)e^{-t}\sin t$.

35. $t^2/2$.

37. $\cos t + t\sin t + c(\sin t - t\cos t)$, (c arbitrary) .

39. $e(t) = -a\cos\sqrt{k/I}\,t$.

41. $e(t) = \left(-2aI/\sqrt{4Ik-\mu^2}\right)e^{-\mu t/2I}\sin\left(\sqrt{4Ik-\mu^2}\,t/2I\right)$.

Exercises 7.6, page 417

1. $2e^{-s}/s^3$. **3.** $e^{-2s}(4s^2+4s+2)/s^3$.

5. $(2e^{-s} - e^{-2s} + 2e^{-3s})/s$.

7. $\left[(e^{-s} - e^{-2s})(s+1)\right]/s^2$.

9. $(e^{-s} - 2e^{-2s} + e^{-3s})/s^2$. **11.** $e^{t-2}u(t-2)$.

13. $e^{-2(t-2)}u(t-2) - 3e^{-2(t-4)}u(t-4)$.

15. $e^{-2(t-3)}\left[\cos(t-3) - 2\sin(t-3)\right]u(t-3)$.

17. $(7e^{6-2t} - 6e^{3-t})u(t-3)$.

19. $10 - 10u(t-3\pi)\left[1 + e^{-(t-3\pi)}(\cos t + \sin t)\right]$
 $+ 10u(t-4\pi)\left[1 - e^{-(t-4\pi)}(\cos t + \sin t)\right]$.

21. $\dfrac{1 - 2se^{-2s} - e^{-2s}}{s^2(1 - e^{-2s})}$.

Figure B.45

23. $\dfrac{1}{1 - e^{-2s}}\left[\dfrac{1 - e^{-s-1}}{s+1} + \dfrac{e^{-s} - e^{-2s}}{s}\right]$.

Figure B.46

25. $\dfrac{1}{s(1 + e^{-as})}$. **27.** $\dfrac{1 - e^{-as}}{as^2(1 + e^{-as})}$.

29. $\sin t + [1 - \cos(t - 3)]u(t - 3)$.

Figure B.47

31. $t + [4 - t + \sin(t - 2) - 2\cos(t - 2)]u(t - 2)$.

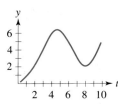

Figure B.48

33. $e^{-t}\cos t + 2e^{-t}\sin t$
$\qquad + (1/2)[1 - e^{2\pi-t}(\cos t + \sin t)]u(t - 2\pi)$
$\qquad - (1/2)[1 - e^{4\pi-t}(\cos t + \sin t)]u(t - 4\pi)$.

35. $e^{-t} + e^{-2t} + (1/2)[e^{-3t} - 2e^{-2(t+1)} + e^{-(t+4)}]u(t - 2)$.

37. $\cos 2t + (1/3)[1 - u(t - 2\pi)]\sin t$
$\qquad + (1/6)[8 + u(t - 2\pi)]\sin 2t$.

39. $2e^{-2t} - 2e^{-3t}$
$\qquad + [1/36 + (1/6)(t - 1) - (1/4)e^{-2(t-1)}$
$\qquad\qquad + (2/9)e^{-3(t-1)}]u(t - 1)$
$\qquad - [19/36 + (1/6)(t - 5)$
$\qquad\qquad - (7/4)e^{-2(t-5)} + (11/9)e^{-3(t-5)}]u(t - 5)$.

45. $\dfrac{e^{t-n}}{6} - \dfrac{e^{-t}}{2}\left[\dfrac{e^{n+1} - 1 - e^{n+2} + e^2}{e - 1}\right]$
$\qquad + \dfrac{e^{-2t}}{3}\left[\dfrac{e^{2n+2} - 1 - e^{2n+3} + e^3}{e^2 - 1}\right]$,
\qquad for $n < t < n + 1$.

47. $\displaystyle\sum_{n=1}^{\infty}\dfrac{1}{s^n} = \dfrac{1}{s - 1}$.

49. $\displaystyle\sum_{n=1}^{\infty}\dfrac{(-1)^{n+1}}{2ns^{2n}} = \dfrac{1}{2}\ln\left(1 + \dfrac{1}{s^2}\right)$.

51. (a) $\sqrt{\pi/s}$. **(b)** $105\sqrt{\pi}/(16s^{9/2})$.

59. $e^t - e^{-t} + (1/2)[e^{t-1} + e^{1-t} - 2]u(t - 1)$
$\qquad - (1/2)[e^{t-4} + e^{4-t} - 2]u(t - 4)$.

61. $4 - 2e^{-3t/125} + 2u(t - 10)[1 - e^{-3(t-10)/125}]$.

63. The resulting differential equation has variable coefficients, so the Laplace transform method will result in a differential equation for the transform.

Exercises 7.7, page 427

1. $2te^t - e^t + \displaystyle\int_0^t e^{t-v}(t - v)g(v)dv$.

3. $\displaystyle\int_0^t g(v)e^{2v-2t}\sin(t - v)dv + e^{-2t}\cos t + 3e^{-2t}\sin t$.

5. $1 - \cos t$. **7.** $2e^{5t} - 2e^{-2t}$. **9.** $(t/2)\sin t$.

11. $(2/3)e^{-2t} + (1/3)e^t$. **13.** $s^{-2}(s - 3)^{-1}$.

15. $t/4 + (3/8)\sin 2t$. **17.** $\cos t$. **19.** 3 .

21. $e^{-t/2}\cos(\sqrt{3}t/2) - (1/\sqrt{3})e^{-t/2}\sin(\sqrt{3}t/2)$.

23. $H(s) = (s^2 + 9)^{-1}$; $h(t) = (1/3)\sin 3t$;
$\qquad y_k(t) = 2\cos 3t - \sin 3t$;
$\qquad y(t) = (1/3)\displaystyle\int_0^t (\sin 3(t - v))g(v)dv + 2\cos 3t - \sin 3t$.

25. $H(s) = (s^2 - s - 6)^{-1}$; $h(t) = (e^{3t} - e^{-2t})/5$;
$\qquad y_k(t) = 2e^{3t} - e^{-2t}$;
$\qquad y(t) = (1/5)\displaystyle\int_0^t [e^{3(t-v)} - e^{-2(t-v)}]g(v)dv + 2e^{3t} - e^{-2t}$.

27. $H(s) = (s^2 - 2s + 5)^{-1}$; $h(t) = (1/2)e^t\sin 2t$;
$\qquad y_k(t) = e^t\sin 2t$;
$\qquad y(t) = (1/2)\displaystyle\int_0^t e^{(t-v)}(\sin 2(t - v))g(v)dv + e^t\sin 2t$.

29. $(1/30)\displaystyle\int_0^t e^{-2(t-v)}(\sin 6(t - v))e(v)dv - e^{-2t}\cos 6t$
$\qquad + e^{-2t}\sin 6t$.

31. $t^2/2$.

Exercises 7.8, page 434

1. -1. **3.** -1. **5.** e^{-2}. **7.** $e^{-s} - e^{-3s}$.

9. e^{-s}. **11.** 0. **13.** $-(\sin t)u(t - \pi)$.

15. $e^t + e^{-3t} + (1/4)(e^{t-1} - e^{3-3t})u(t - 1)$
$\qquad - (1/4)(e^{t-2} - e^{6-3t})u(t - 2)$.

17. $2(e^{t-2} - e^{-(t-2)})u(t - 2) + 2e^t - t^2 - 2$.

19. $e^{-t} - e^{-5t} + (e/4)(e^{1-t} - e^{5-5t})u(t - 1)$.

21. $\sin t + (\sin t)u(t - 2\pi)$.

Figure B.49

23. $\sin t + (\sin t)u(t - \pi) + (\sin t)u(t - 2\pi)$.

Figure B.50

25. $(1/2)e^{-2t}\sin 2t$. **27.** $(1/2)e^{t}\sin 2t$.

29. The mass remains stopped at $x(t) \equiv 0$, $t > \pi/2$.

35. $\dfrac{L}{6EI}\left[3\lambda x^2 - x^3 + (x - \lambda)^3 u(x - \lambda)\right]$.

Exercises 7.9, page 438

1. $x = e^{t}$; $y = e^{t}$. **3.** $z = e^{t}$; $w \equiv 0$.

5. $x = (7/4)e^{t} + (7/4)e^{-t} - (3/2)\cos t$;
$y = (7/4)e^{t} - (7/4)e^{-t} + (1/2)\sin t$.

7. $x = -(150/17)e^{5t/2}\cos\left(\sqrt{15}t/2\right)$
$\quad - \left(334\sqrt{15}/85\right)e^{5t/2}\sin\left(\sqrt{15}t/2\right) - (3/17)e^{-3t}$;
$y = (46/17)e^{5t/2}\cos\left(\sqrt{15}t/2\right)$
$\quad - \left(146\sqrt{15}/85\right)e^{5t/2}\sin\left(\sqrt{15}t/2\right) + (22/17)e^{-3t}$.

9. $x = 4e^{-2t} - e^{-t} - \cos t$; $y = 5e^{-2t} - e^{-t}$.

11. $x = (e^{t} - e^{-t})/2 - (1/2)\left[e^{t-2} - e^{-(t-2)}\right]u(t - 2)$;
$y = 1 - (e^{t} + e^{-t})/2$
$\quad - \left[1 - (e^{t-2} + e^{-(t-2)})/2\right]u(t - 2)$.

13. $x = e^{-t} - (1/2)\left[e^{-(t-\pi)} + \cos t - \sin t\right]u(t - \pi)$;
$y = e^{-t} + \left[1 - (1/2)e^{-(t-\pi)} + (1/2)\cos t\right.$
$\qquad\qquad\left. + (1/2)\sin t\right]u(t - \pi)$.

15. $x = t^2$; $y = t - 1$.

17. $x = (t - 2)e^{t-2}$; $y = e^{t-2}$.

19. $x = -7e^{-t} + e^{t}$; $y = 2e^{-t}$; $z = -13e^{-t} + e^{t}$.

21. $x = -e^{-t/2} + e^{-t/6}$
$\quad + \left[48 - 36e^{-(t-5)/6} - 12e^{-(t-5)/2}\right]u(t - 5)$;
$y = 2e^{-t/2} + 2e^{-t/6}$
$\quad + \left[48 - 72e^{-(t-5)/6} + 24e^{-(t-5)/2}\right]u(t - 5)$.

23. $2I_1 + (0.1)I_3' + (0.2)I_1' = 6$, $(0.1)I_3' - I_2 = 0$,
$I_1 = I_2 + I_3$; $I_1(0) = I_2(0) = I_3(0) = 0$;
$I_1 = -e^{-20t} - 2e^{-5t} + 3$,
$I_2 = -2e^{-20t} + 2e^{-5t}$,
$I_3 = e^{-20t} - 4e^{-5t} + 3$.

Review Problems, page 440

1. $\dfrac{3}{s} + e^{-2s}\left[\dfrac{1}{s} - \dfrac{1}{s^2}\right]$. **3.** $\dfrac{2}{(s + 9)^3}$.

5. $\dfrac{1}{s - 2} - \dfrac{6}{s^4} + \dfrac{2}{s^3} - \dfrac{5}{s^2 + 25}$. **7.** $\dfrac{s^2 - 36}{(s^2 + 36)^2}$.

9. $2e^{-4s}\left[\dfrac{1}{s^3} + \dfrac{4}{s^2} + \dfrac{8}{s}\right]$. **11.** $(7/2)t^2 e^{-3t}$.

13. $2e^{t} + 2e^{-2t}\cos 3t + e^{-2t}\sin 3t$.

15. $e^{-2t} + e^{-t} - 2te^{-t}$.

17. $\left[2e^{t-2} + 2e^{4-2t}\right]u(t - 2)$.

19. $e^{2t} - e^{5t}$.

21. $-(3/2) + t + t^2/2 + (3/2)e^{-t}\cos t - (1/2)e^{-t}\sin t$.

23. $\left(2/\sqrt{7}\right)e^{-3t/2}\sin\left(\sqrt{7}t/2\right)$
$\quad + \left\{(1/4) - \left(3/(4\sqrt{7})\right)e^{-3(t-1)/2}\sin\left(\sqrt{7}(t - 1)/2\right)\right.$
$\qquad\left. - (1/4)e^{-3(t-1)/2}\cos\left(\sqrt{7}(t - 1)/2\right)\right\}u(t - 1)$.

25. $c\left[t + te^{-2t} + e^{-2t} - 1\right]$.

27. $(9/10)e^{-3t} + (1/10)\cos t - (3/10)\sin t$.

29. $(s^2 - 5s + 6)^{-1}$; $e^{3t} - e^{2t}$.

31. $x = 1 - (e^{t} + e^{-t})/2$
$\quad - \left[1 - (e^{t-2} + e^{-(t-2)})/2\right]u(t - 2)$,
$y = (e^{t} - e^{-t})/2 - (1/2)\left[e^{t-2} - e^{-(t-2)}\right]u(t - 2)$.

CHAPTER 8

Exercises 8.1, page 452

1. $1 + x + x^2 + \cdots$. **3.** $x + x^2 + (1/2)x^3 + \cdots$.

5. $1 - (1/6)t^3 + (1/180)t^6 + \cdots$.

7. $(1/6)\theta^3 - (1/120)\theta^5 + (1/5040)\theta^7 + \cdots$.

9. **(a)** $p_3(x) = (x - 1) - \dfrac{1}{2}(x - 1)^2 + \dfrac{1}{3}(x - 1)^3$.

(b) $|\varepsilon_3(1.5)| = |f^{(4)}(\xi)(1.5 - 1)^4/4!| \le 6(0.5)^4/4!$.

(c) $|\ln(1.5) - p_3(1.5)| = 0.011202\ldots$.

(d)

Figure B.51

13. $t + (1/2)t^2 - (1/6)t^3 + \cdots$.

15. $1 - x^2/6$.

Exercises 8.2, page 460

1. $[-1, 3)$. **3.** $(-4, 0)$. **5.** $[1, 3]$.

9. $\displaystyle\sum_{n=0}^{\infty}\left[\dfrac{1}{n + 1} + 2^{-n-1}\right]x^n$.

11. $x + x^2 + (1/3)x^3 + \cdots$.

13. $1 - 2x + (5/2)x^2 + \cdots$.

15. (c) $1 - (1/2)x + (1/4)x^2 - (1/24)x^3 + \cdots$.

17. $\displaystyle\sum_{n=1}^{\infty} (-1)^n n x^{n-1}$. **19.** $\displaystyle\sum_{n=1}^{\infty} a_n n x^{n-1}$.

21. $\ln(x + 1) = \displaystyle\sum_{n=0}^{\infty} \frac{(-1)^n}{n+1} x^{n+1}$.

23. $\displaystyle\sum_{k=0}^{\infty} (k + 1)a_{k+1}x^k$.

25. $\displaystyle\sum_{k=1}^{\infty} a_{k-1}x^k$. **29.** $\displaystyle\sum_{n=0}^{\infty} \frac{(-1)^{n+1}(x - \pi)^{2n}}{(2n)!}$.

31. $1 + \displaystyle\sum_{n=1}^{\infty} 2x^n$.

33. $6(x - 1) + 3(x - 1)^2 + (x - 1)^3$:

35. (a) $\displaystyle\sum_{n=0}^{\infty} (-1)^n (x - 1)^n$.

(b) $\displaystyle\sum_{n=1}^{\infty} \frac{(-1)^{n-1}}{n}(x - 1)^n$.

Exercises 8.3, page 471

1. -1 . **3.** $\pm\sqrt{2}$. **5.** $-1, 2$.

7. $x = n\pi$, n an integer.

9. $\theta \le 0$ and $\theta = n\pi$, $n = 1, 2, 3, \ldots$.

11. $y = a_0(1 - 2x + (3/2)x^2 - x^3/3 + \cdots)$.

13. $a_0(1 + x^4/12 + \cdots) + a_1(x + x^5/20 + \cdots)$.

15. $a_0(1 - x^2/2 - x^3/6 + \cdots)$
$\qquad + a_1(x + x^2/2 - x^3/6 + \cdots)$.

17. $a_0(1 - x^2/2 + \cdots) + a_1(x - x^3/6 + \cdots)$.

19. $a_0 \displaystyle\sum_{n=0}^{\infty} \frac{1}{n!}x^{2n}$.

21. $a_0(1 - 2x^2 + x^4/3)$
$\qquad + a_1\left(x + \displaystyle\sum_{k=1}^{\infty} \frac{(-3)(-1)\cdots(2k - 5)}{(2k + 1)!}x^{2k+1}\right)$.

23. $a_{3k+2} = 0$, $k = 0, 1, \ldots$.
$$a_0\left(1 + \sum_{k=1}^{\infty} \frac{[1\cdot4\cdot7\cdots(3k - 2)]^2}{(3k)!}x^{3k}\right)$$
$$+ a_1\left(x + \sum_{k=1}^{\infty} \frac{[2\cdot5\cdot8\cdots(3k - 1)]^2}{(3k + 1)!}x^{3k+1}\right)$$.

25. $2 + x^2 - (5/12)x^4 + (11/72)x^6 + \cdots$.

27. $x + (1/6)x^3 - (1/12)x^4 + (7/120)x^5 + \cdots$.

29. $1 - 2x + x^2 - (1/6)x^3$.

31. $1 + 2x - (3/4)x^2 - (5/6)x^3$.

35. $10 - 25t^2 + (250/3)t^3 - (775/4)t^4 + \cdots$.

Exercises 8.4, page 478

1. 2. **3.** $\sqrt{3}$. **5.** $\pi/2$.

7. $a_0\big[1 - (x - 1)^2 + (1/2)(x - 1)^4$
$\qquad - (1/6)(x - 1)^6 + \cdots\big]$.

9. $a_0\big[1 + (x - 1)^2 + \cdots\big]$
$\qquad + a_1\big[(x - 1) + (1/3)(x - 1)^3 + \cdots\big]$.

11. $a_0\big[1 - (1/8)(x - 2)^2 + (1/32)(x - 2)^3 + \cdots\big]$
$\qquad + a_1\big[(x - 2) + (1/8)(x - 2)^2$
$\qquad\qquad - (7/96)(x - 2)^3 + \cdots\big]$.

13. $1 - (1/2)t^2 + (1/6)t^4 - (31/720)t^6 + \cdots$.

15. $1 + x + (1/24)x^4 + (1/60)x^5 + \cdots$.

17. $1 - (1/6)(x - \pi)^3 + (1/120)(x - \pi)^5$
$\qquad + (1/180)(x - \pi)^6 + \cdots$.

19. $-1 + x + x^2 + (1/2)x^3 + \cdots$.

21. $a_0\big[1 + (1/2)x^2 + (1/8)x^4 + (1/48)x^6 + \cdots\big]$
$\qquad + \big[(1/2)x^2 + (1/12)x^4 + (11/720)x^6 + \cdots\big]$.

23. $a_0\big[1 - (1/2)x^2 + \cdots\big] + a_1\big[x - (1/3)x^3 + \cdots\big]$
$\qquad + \big[(1/2)x^2 + (1/3)x^3 + \cdots\big]$.

25. $a_0\big[1 - (1/2)x^2 + \cdots\big] + a_1\big[x + \cdots\big]$
$\qquad + \big[(1/2)x^2 - (1/6)x^3 + \cdots\big]$.

27. $a_0\big[1 - (1/2)x^2 - (1/6)x^3 + \cdots\big]$
$\qquad + a_1\big[x + (1/2)x^2 + \cdots\big] + \big[(1/6)x^3 + \cdots\big]$.

29. (a) $a_0\left[1 + \displaystyle\sum_{k=1}^{\infty} (-1)^k\frac{n(n - 2)(n - 4)\cdots(n - 2k + 2)(n + 1)(n + 3)\cdots(n + 2k - 1)}{(2k)!}x^{2k}\right]$
$\qquad + a_1\left[x + \displaystyle\sum_{k=1}^{\infty} (-1)^k\frac{(n - 1)(n - 3)\cdots(n - 2k + 1)(n + 2)(n + 4)\cdots(n + 2k)}{(2k + 1)!}x^{2k+1}\right]$.

(c) $P_0(x) = 1$, $P_1(x) = x$, $P_2(x) = (1/2)(3x^2 - 1)$.

31. (a) $1 - (1/2)t^2 + (\eta/6)t^3 + \big[(1 - \eta^2)/24\big]t^4 + \cdots$. **(b)** Yes.

Exercises 8.5, page 482

1. $c_1 x^{-2} + c_2 x^{-3}$. **3.** $c_1 x \cos(4 \ln x) + c_2 x \sin(4 \ln x)$.

5. $c_1 x^3 \cos(2 \ln x) + c_2 x^3 \sin(2 \ln x)$.

7. $c_1 x + c_2 x^{-1} \cos(3 \ln x) + c_3 x^{-1} \sin(3 \ln x)$.

9. $c_1 x + c_2 x^{-1/2} \cos\left[\left(\sqrt{19}/2 \right) \ln x \right]$
$\qquad + c_3 x^{-1/2} \sin\left[\left(\sqrt{19}/2 \right) \ln x \right]$.

11. $c_1 (x - 3)^{-2} + c_2 (x - 3)^{1/2}$.

13. $c_1 x + c_2 x^2 + (4/15) x^{-1/2}$. **15.** $2t^4 + t^{-3}$.

17. $(31/17) x + (3/17) x^{-2} \cos(5 \ln x)$
$\qquad - (76/85) x^{-2} \sin(5 \ln x)$.

Exercises 8.6, page 494

1. ± 1 are regular. **3.** $\pm i$ are regular.

5. 1 is regular, -1 is irregular.

7. 2 is regular, -1 is irregular.

9. -4 is regular and 2 is irregular.

11. $r^2 - 3r - 10 = 0$; $r_1 = 5$, $r_2 = -2$.

13. $r^2 - 5r/9 - 4/3 = 0$; $r_1 = \left(5 + \sqrt{457} \right)/18$,
$\quad r_2 = \left(5 - \sqrt{457} \right)/18$.

15. $r^2 - 1 = 0$; $r_1 = 1$, $r_2 = -1$.

17. $r^2 + r - 12 = 0$; $r_1 = 3$, $r_2 = -4$.

19. $a_0 \left[x^{2/3} - (1/2) x^{5/3} + (5/28) x^{8/3} - (1/21) x^{11/3} + \cdots \right]$.

21. $a_0 \left[1 - (1/4) x^2 + (1/64) x^4 - (1/2304) x^6 + \cdots \right]$.

23. $a_0 \left[x - (1/3) x^2 + (1/12) x^3 - (1/60) x^4 + \cdots \right]$.

25. $a_0 \displaystyle\sum_{n=0}^{\infty} \frac{(-1)^n x^{n+(3/2)}}{2^{n-1}(n+2)!}$. **27.** $a_0 \displaystyle\sum_{n=0}^{\infty} \frac{x^{2n+2}}{2^{2n}(n+1)! n!}$.

29. $a_0 x^2$. **31.** $a_0 \displaystyle\sum_{n=0}^{\infty} \frac{x^n}{n!} = a_0 e^x$; yes, $a_0 < 0$.

33. $a_0 \left[x^{1/3} + \displaystyle\sum_{n=1}^{\infty} \frac{2^{n-2}(3n+4) x^{n+(1/3)}}{3^n n!} \right]$; yes, $a_0 < 0$.

35. $a_0 \left[x^{5/6} - (1/11) x^{11/6} + (1/374) x^{17/6} \right.$
$\qquad \left. - (1/25,806) x^{23/6} + \cdots \right]$.

37. $a_0 \left[x^{4/3} + (1/17) x^{7/3} + (1/782) x^{10/3} \right.$
$\qquad \left. + (1/68,034) x^{13/3} + \cdots \right]$.

39. The expansion $\displaystyle\sum_{n=0}^{\infty} n! x^n$ diverges for $x \neq 0$.

41. The transformed equation is
$z \, d^2 y/dz^2 + 3 \, dy/dz - y = 0$. Also $zp(z) = 3$ and
$z^2 q(z) = -z$ are analytic at $z = 0$; hence $z = 0$ is a
regular singular point.
$y_1(x) = a_0 \left[1 + (1/3) x^{-1} + (1/24) x^{-2} \right.$
$\qquad \left. + (1/360) x^{-3} + \cdots \right]$.

45. $a_0 \left[x^{-3} - (1/2) x^{-1} - (1/8) x - (1/144) x^3 + \cdots \right]$
$\qquad + a_3 \left[1 + (1/10) x^2 + (1/280) x^4 \right.$
$\qquad\qquad \left. + (1/15,120) x^6 + \cdots \right]$.

Exercises 8.7, page 506

1. $c_1 y_1(x) + c_2 y_2(x)$, where
$\quad y_1(x) = x^{2/3} - (1/2) x^{5/3} + (5/28) x^{8/3} + \cdots$ and
$\quad y_2(x) = x^{1/3} - (1/2) x^{4/3} + (1/5) x^{7/3} + \cdots$.

3. $c_1 y_1(x) + c_2 y_2(x)$, where
$\quad y_1(x) = 1 - (1/4) x^2 + (1/64) x^4 + \cdots$ and
$\quad y_2(x) = y_1(x) \ln x + (1/4) x^2 - (3/128) x^4$
$\qquad\qquad + (11/13,824) x^6 + \cdots$.

5. $c_1 \left[x - (1/3) x^2 + (1/12) x^3 + \cdots \right] + c_2 \left[x^{-1} - 1 \right]$.

7. $c_1 y_1(x) + c_2 y_2(x)$, where
$\quad y_1(x) = x^{3/2} - (1/6) x^{5/2} + (1/48) x^{7/2} + \cdots$ and
$\quad y_2(x) = x^{-1/2} - (1/2) x^{1/2}$.

9. $c_1 w_1 + c_2 w_2$, where
$\quad w_1(x) = x^2 + (1/8) x^4 + (1/192) x^6 + \cdots$ and
$\quad w_2(x) = w_1(x) \ln x + 2 - (3/32) x^4 - (7/1152) x^6 + \cdots$.

11. $c_1 y_1(x) + c_2 y_2(x)$, where $y_1(x) = x^2$ and
$\quad y_2(x) = x^2 \ln x - 1 + 2x - (1/3) x^3 + (1/24) x^4 + \cdots$.

13. $c_1 y_1(x) + c_2 y_2(x)$, where
$\quad y_1(x) = 1 + x + (1/2) x^2 + \cdots$ and
$\quad y_2(x) = y_1(x) \ln x - \left[x + (3/4) x^2 + (11/36) x^3 + \cdots \right]$.

15. $c_1 y_1(x) + c_2 y_2(x)$, where
$\quad y_1(x) = x^{1/3} + (7/6) x^{4/3} + (5/9) x^{7/3} + \cdots$ and
$\quad y_2(x) = 1 + 2x + (6/5) x^2 + \cdots$; all solutions are
\quad bounded near the origin.

17. $c_1 y_1(x) + c_2 y_2(x) + c_3 y_3(x)$, where
$\quad y_1(x) = x^{5/6} - (1/11) x^{11/6} + (1/374) x^{17/6} + \cdots$,
$\quad y_2(x) = 1 - x + (1/14) x^2 + \cdots$, and
$\quad y_3(x) = y_2(x) \ln x + 7x - (117/196) x^2$
$\qquad\qquad + (4997/298116) x^3 + \cdots$.

19. $c_1 y_1(x) + c_2 y_2(x) + c_3 y_3(x)$, where
$\quad y_1(x) = x^{4/3} + (1/17) x^{7/3} + (1/782) x^{10/3} + \cdots$,
$\quad y_2(x) = 1 - (1/3) x - (1/30) x^2 + \cdots$, and
$\quad y_3(x) = x^{-1/2} - (1/5) x^{1/2} - (1/10) x^{3/2} + \cdots$.

21. (b) $c_1 y_1(t) + c_2 y_2(t)$, where
$\quad y_1(t) = 1 - (1/6)(\alpha t)^2 + (1/120)(\alpha t)^4 + \cdots$,
\quad and
$\quad y_2(t) = t^{-1} \left[1 - (1/2)(\alpha t)^2 + (1/24)(\alpha t)^4 + \cdots \right]$.

(c) $c_1 y_1(x) + c_2 y_2(x)$, where
$\quad y_1(x) = 1 - (1/6)(\alpha/x)^2 + (1/120)(\alpha/x)^4 + \cdots$,
\quad and
$\quad y_2(x) = x \left[1 - (1/2)(\alpha/x)^2 + (1/24)(\alpha/x)^4 + \cdots \right]$.

23. $c_1 y_1(x) + c_2 y_2(x)$, where
$y_1(x) = 1 + 2x + 2x^2$ and
$y_2(x) = -(1/6)x^{-1} - (1/24)x^{-2} - (1/120)x^{-3} + \cdots$.

25. For $n = 0$,
$c_1 + c_2[\ln x + x + (1/4)x^2 + (1/18)x^3 + \cdots]$
and for $n = 1$,
$c_1(1 - x) + c_2[(1 - x)\ln x + 3x - (1/4)x^2$
$\qquad\qquad - (1/36)x^3 + \cdots]$.

Exercises 8.8, page 517

1. $c_1 F\left(1, 2; \dfrac{1}{2}; x\right) + c_2 x^{1/2} F\left(\dfrac{3}{2}, \dfrac{5}{2}; \dfrac{3}{2}; x\right)$.

3. $c_1 F\left(1, 1; \dfrac{1}{2}; x\right) + c_2 x^{1/2} F\left(\dfrac{3}{2}, \dfrac{3}{2}; \dfrac{3}{2}; x\right)$.

5. $F(1, 1; 2; x) = \displaystyle\sum_{n=0}^{\infty} \dfrac{1}{(n+1)} x^n = -x^{-1} \ln(1 - x)$.

7. $F\left(\dfrac{1}{2}, 1; \dfrac{3}{2}; x^2\right) = \displaystyle\sum_{n=0}^{\infty} (1/2)_n x^{2n}/(3/2)_n$

$\qquad\qquad = \displaystyle\sum_{n=0}^{\infty} x^{2n}/(2n + 1)$

$\qquad\qquad = \dfrac{1}{2} x^{-1} \ln\left(\dfrac{1+x}{1-x}\right)$.

9. $(1 - x)^{-1}$, $(1 - x)^{-1} \ln x$.

13. $c_1 J_{1/2}(x) + c_2 J_{-1/2}(x)$. **15.** $c_1 J_1(x) + c_2 Y_1(x)$.

17. $c_1 J_{2/3}(x) + c_2 J_{-2/3}(x)$.

19. $J_1(x)\ln x - x^{-1} + (3/64)x^3 - (7/2304)x^5 + \cdots$.

21. $x^{3/2} J_{3/2}(x)$.

27. $J_0(x)\ln x + \displaystyle\sum_{n=1}^{\infty} \dfrac{(-1)^{n+1}}{2^{2n}(n!)^2}\left[1 + \dfrac{1}{2} + \cdots + \dfrac{1}{n}\right] x^{2n}$.

29. 1 , x , $(3x^2 - 1)/2$,
$(5x^3 - 3x)/2$, $(35x^4 - 30x^2 + 3)/8$.

37. 1 , $2x$, $4x^2 - 2$, $8x^3 - 12x$.

39. 1 , $1 - x$, $(2 - 4x + x^2)/2$,
$(6 - 18x + 9x^2 - x^3)/6$.

41. (b) $z = c_1 \cos x + c_2 \sin x$, so for x large,
$y(x) \approx c_1 x^{-1/2} \cos x + c_2 x^{-1/2} \sin x$.

Review Problems, page 521

1. (a) $1 - x + (3/2)x^2 - (5/3)x^3 + \cdots$.
(b) $-1 + x - (1/6)x^3 + (1/6)x^4 + \cdots$.

3. (a) $a_0\left[1 + x^2 + \cdots\right] + a_1\left[x + (1/3)x^3 + \cdots\right]$.
(b) $a_0\left[1 + (1/2)x^2 - (1/6)x^3 + \cdots\right]$
$\qquad + a_1\left[x - (1/2)x^2 + (1/2)x^3 + \cdots\right]$.

5. $a_0\left[1 + (1/2)(x - 2)^2 - (1/24)(x - 2)^4 + \cdots\right]$
$\qquad + a_1(x - 2)$.

7. (a) $a_0\left[x^3 + x^4 + (1/4)x^5 + (1/36)x^6 + \cdots\right]$.
(b) $a_0\left[x - (1/4)x^2 + (1/20)x^3 - (1/120)x^4 + \cdots\right]$.

9. (a) $c_1 y_1(x) + c_2 y_2(x)$, where
$y_1(x) = x + x^2 + (1/2)x^3 + \cdots = xe^x$ and
$y_2(x) = y_1(x)\ln x - x^2 - (3/4)x^3 - (11/36)x^4 + \cdots$.
(b) $c_1 y_1(x) + c_2 y_2(x)$, where
$y_1(x) = 1 + 2x + x^2 + \cdots$ and
$y_2(x) = y_1(x)\ln x - \left[4x + 3x^2 + (22/27)x^3 + \cdots\right]$.
(c) $c_1 y_1(x) + c_2 y_2(x)$, where
$y_1(x) = 1 - (1/6)x + (1/96)x^2 + \cdots$ and
$y_2(x) = y_1(x)\ln x - 8x^{-2} - 4x^{-1} + (29/36) + \cdots$.
(d) $c_1 y_1(x) + c_2 y_2(x)$, where
$y_1(x) = x^2 - (1/4)x^3 + (1/40)x^4 + \cdots$ and
$y_2(x) = x^{-1} + (1/2) + (1/4)x + \cdots$.

CHAPTER 9

Exercises 9.1, page 531

1. $\begin{bmatrix} x \\ y \end{bmatrix}' = \begin{bmatrix} 7 & 2 \\ 3 & -2 \end{bmatrix} \begin{bmatrix} x \\ y \end{bmatrix}$.

3. $\begin{bmatrix} x \\ y \\ z \end{bmatrix}' = \begin{bmatrix} 1 & 1 & 1 \\ -1 & 0 & 2 \\ 0 & 4 & 0 \end{bmatrix} \begin{bmatrix} x \\ y \\ z \end{bmatrix}$.

5. $\begin{bmatrix} x \\ y \end{bmatrix}' = \begin{bmatrix} \sin t & e^t \\ \cos t & a + bt^3 \end{bmatrix} \begin{bmatrix} x \\ y \end{bmatrix}$.

7. $\begin{bmatrix} y \\ v \end{bmatrix}' = \begin{bmatrix} 0 & 1 \\ -k/m & -b/m \end{bmatrix} \begin{bmatrix} y \\ v \end{bmatrix}$.

9. $\begin{bmatrix} x_1 \\ x_2 \end{bmatrix}' = \begin{bmatrix} 0 & 1 \\ t & 0 \end{bmatrix} \begin{bmatrix} x_1 \\ x_2 \end{bmatrix}$; $x_1 = y$, $x_2 = y'$.

11. $\begin{bmatrix} x_1 \\ x_2 \\ x_3 \\ x_4 \end{bmatrix}' = \begin{bmatrix} 0 & 1 & 0 & 0 \\ -3 & 0 & -2 & 0 \\ 0 & 0 & 0 & 1 \\ 2 & 0 & 0 & 0 \end{bmatrix} \begin{bmatrix} x_1 \\ x_2 \\ x_3 \\ x_4 \end{bmatrix}$;
$x_1 = x$, $x_2 = x'$, $x_3 = y$, $x_4 = y'$.

13. $\begin{bmatrix} x_1 \\ x_2 \\ x_3 \\ x_4 \\ x_5 \end{bmatrix} = \begin{bmatrix} 0 & 1 & 0 & 0 & 0 \\ \cos t & 3 & -t^2 & 0 & 0 \\ 0 & 0 & 0 & 1 & 0 \\ 0 & 0 & 0 & 0 & 1 \\ -e^t & t & 0 & -1 & -1 \end{bmatrix} \begin{bmatrix} x_1 \\ x_2 \\ x_3 \\ x_4 \\ x_5 \end{bmatrix}$;
$x_1 = x$, $x_2 = x'$, $x_3 = y$, $x_4 = y'$, $x_5 = y''$.

Exercises 9.2, page 536

1. $x_1 = 2$, $x_2 = 1$, $x_3 = 1$.
3. $x_1 = -1$, $x_2 = 0$, $x_3 = -2$.
5. $x_1 = 0$, $x_2 = 0$.
7. $x_1 = 3s$, $x_2 = s$, $-\infty < s < \infty$.
9. $x_1 = 2s$, $x_2 = (-1 + i)s$, with s any complex number.
11. $x_1 = -(s + 1)/2$, $x_2 = -(11s + 1)/2$, $x_3 = s$, $-\infty < s < \infty$.
13. For $r = 2$, unique solution is $x_1 = x_2 = 0$.
 For $r = 1$, solutions are $x_1 = 3s, x_2 = s$,
 $-\infty < s < \infty$.

Exercises 9.3, page 545

1. (a) $\begin{bmatrix} 1 & 1 \\ 5 & 2 \end{bmatrix}$. **(b)** $\begin{bmatrix} 7 & 3 \\ 7 & 18 \end{bmatrix}$.

3. (a) $\begin{bmatrix} 18 & 14 \\ 4 & 5 \end{bmatrix}$. **(b)** $\begin{bmatrix} 8 & 12 \\ 3 & 5 \end{bmatrix}$. **(c)** $\begin{bmatrix} 16 & 3 \\ 5 & 19 \end{bmatrix}$.

5. (a) $\begin{bmatrix} -1 & -2 \\ -1 & -3 \end{bmatrix}$. **(b)** $\begin{bmatrix} -5 & -1 \\ -8 & -1 \end{bmatrix}$. **(c)** $\begin{bmatrix} -6 & -3 \\ -9 & -4 \end{bmatrix}$.

7. (c) Yes. **(d)** Yes.

9. $\begin{bmatrix} 4/9 & -1/9 \\ 1/9 & 2/9 \end{bmatrix}$. **11.** Doesn't exist.

13. $\begin{bmatrix} 1 & 0 & 1 \\ -2 & 1 & -2 \\ 1 & 1 & 0 \end{bmatrix}$.

17. $\begin{bmatrix} (4/3)e^{-t} & -(1/3)e^{-t} \\ -(1/3)e^{-4t} & (1/3)e^{-4t} \end{bmatrix}$.

19. $\begin{bmatrix} e^{-t} & (1/2)e^{-t} & -(1/2)e^{-t} \\ (1/3)e^{t} & -(1/2)e^{t} & (1/6)e^{t} \\ -(1/3)e^{-2t} & 0 & (1/3)e^{-2t} \end{bmatrix}$.

21. 11. **23.** -12. **25.** 25. **27.** 2, 3.

29. 0, 1, 1. **31.** $\begin{bmatrix} 3e^{3t} \\ 6e^{3t} \\ -3e^{3t} \end{bmatrix}$. **33.** $\begin{bmatrix} 5e^{5t} & 6e^{2t} \\ -10e^{5t} & -2e^{2t} \end{bmatrix}$.

39. (a) $\begin{bmatrix} (1/2)t^2 + c_1 & e^t + c_2 \\ t + c_3 & e^t + c_4 \end{bmatrix}$.

(b) $\begin{bmatrix} \sin 1 & -1 + \cos 1 \\ 1 - \cos 1 & \sin 1 \end{bmatrix}$.

(c) $\begin{bmatrix} (1 + e^t)\cos t + (e^t - t)\sin t & (e^t - t)\cos t - (e^t + 1)\sin t \\ (e^t - 1)\sin t + e^t\cos t & (e^t - 1)\cos t - e^t\sin t \end{bmatrix}$.

Exercises 9.4, page 554

1. $\begin{bmatrix} x'(t) \\ y'(t) \end{bmatrix} = \begin{bmatrix} 3 & -1 \\ -1 & 2 \end{bmatrix}\begin{bmatrix} x(t) \\ y(t) \end{bmatrix} + \begin{bmatrix} t^2 \\ e^t \end{bmatrix}$.

3. $\begin{bmatrix} dx/dt \\ dy/dt \\ dz/dt \end{bmatrix} = \begin{bmatrix} t^2 & -1 & -1 \\ 0 & 0 & e^t \\ t & -1 & 3 \end{bmatrix}\begin{bmatrix} x \\ y \\ z \end{bmatrix} + \begin{bmatrix} t \\ 5 \\ -e^t \end{bmatrix}$.

5. $\begin{bmatrix} x_1'(t) \\ x_2'(t) \end{bmatrix} = \begin{bmatrix} 0 & 1 \\ 10 & 3 \end{bmatrix}\begin{bmatrix} x_1(t) \\ x_2(t) \end{bmatrix} + \begin{bmatrix} 0 \\ \sin t \end{bmatrix}$;

 $x_1 = y$, $x_2 = y'$.

7. $\begin{bmatrix} x_1'(t) \\ x_2'(t) \\ x_3'(t) \\ x_4'(t) \end{bmatrix} = \begin{bmatrix} 0 & 1 & 0 & 0 \\ 0 & 0 & 1 & 0 \\ 0 & 0 & 0 & 1 \\ -1 & 0 & 0 & 0 \end{bmatrix}\begin{bmatrix} x_1(t) \\ x_2(t) \\ x_3(t) \\ x_4(t) \end{bmatrix} + \begin{bmatrix} 0 \\ 0 \\ 0 \\ t^2 \end{bmatrix}$;

 $x_1 = w$, $x_2 = w'$, $x_3 = w''$, $x_4 = w'''$.

9. $x_1'(t) = 5x_1(t) + 2e^{-2t}$;
 $x_2'(t) = -2x_1(t) + 4x_2(t) - 3e^{-2t}$.

11. $x_1'(t) = x_1(t) + x_3(t) + e^t$;
 $x_2'(t) = -x_1(t) + 2x_2(t) + 5x_3(t) + t$;
 $x_3'(t) = 5x_2(t) + x_3(t)$.

13. LI. **15.** LI. **17.** LI.

19. Not a fundamental solution set.

21. Yes; $\begin{bmatrix} e^{-t} & e^t & e^{3t} \\ 2e^{-t} & 0 & -e^{3t} \\ e^{-t} & e^t & 2e^{3t} \end{bmatrix}$;

 $c_1\begin{bmatrix} e^{-t} \\ 2e^{-t} \\ e^{-t} \end{bmatrix} + c_2\begin{bmatrix} e^t \\ 0 \\ e^t \end{bmatrix} + c_3\begin{bmatrix} e^{3t} \\ -e^{3t} \\ 2e^{3t} \end{bmatrix}$.

23. $\frac{3}{2}\begin{bmatrix} te^t \\ te^t \end{bmatrix} - \frac{1}{4}\begin{bmatrix} e^t \\ 3e^t \end{bmatrix} + \begin{bmatrix} t \\ 2t \end{bmatrix} - \begin{bmatrix} 0 \\ 1 \end{bmatrix} + c_1\begin{bmatrix} e^t \\ e^t \end{bmatrix}$

 $+ c_2\begin{bmatrix} e^{-t} \\ 3e^{-t} \end{bmatrix}$.

27. $\mathbf{X}^{-1}(t) = \begin{bmatrix} 0 & (1/4)e^t & -(1/4)e^t \\ -(1/5)e^{2t} & (4/5)e^{2t} & -(2/5)e^{2t} \\ (1/5)e^{-3t} & (9/20)e^{-3t} & (3/20)e^{-3t} \end{bmatrix}$;

 $\mathbf{x}(t) = \begin{bmatrix} -(3/2)e^{-t} + (3/5)e^{-2t} - (1/10)e^{3t} \\ (1/4)e^{-t} - (1/5)e^{-2t} - (1/20)e^{3t} \\ (5/4)e^{-t} - (1/5)e^{-2t} - (1/20)e^{3t} \end{bmatrix}$.

35. (b) $2e^{6t}$. **(e)** $2e^{6t}$; same.

Exercises 9.5, page 565

1. Eigenvalues are $r_1 = 0$ and $r_2 = -5$ with associated eigenvectors $\mathbf{u}_1 = \begin{bmatrix} s \\ 2s \end{bmatrix}$ and $\mathbf{u}_2 = \begin{bmatrix} 2s \\ -s \end{bmatrix}$.

3. Eigenvalues are $r_1 = 2$ and $r_2 = 3$ with associated eigenvectors $\mathbf{u}_1 = s\begin{bmatrix} 1 \\ -1 \end{bmatrix}$ and $\mathbf{u}_2 = s\begin{bmatrix} 1 \\ -2 \end{bmatrix}$.

5. Eigenvalues are $r_1 = 1$, $r_2 = 2$, and $r_3 = -2$ with associated eigenvectors $\mathbf{u}_1 = s\begin{bmatrix} 1 \\ 0 \\ 0 \end{bmatrix}$, $\mathbf{u}_2 = s\begin{bmatrix} 0 \\ 1 \\ 1 \end{bmatrix}$,

and $\mathbf{u}_3 = s\begin{bmatrix} 0 \\ -1 \\ 1 \end{bmatrix}$.

7. Eigenvalues are $r_1 = 1$, $r_2 = 2$, and $r_3 = 5$ with associated eigenvectors $\mathbf{u}_1 = s\begin{bmatrix} 2 \\ -3 \\ 2 \end{bmatrix}$, $\mathbf{u}_2 = s\begin{bmatrix} 0 \\ -1 \\ 1 \end{bmatrix}$,

and $\mathbf{u}_3 = s\begin{bmatrix} 0 \\ 1 \\ 2 \end{bmatrix}$.

9. Eigenvalues are $r_1 = i$ and $r_2 = -i$, with associated eigenvectors $\mathbf{u}_1 = s\begin{bmatrix} i \\ 1 \end{bmatrix}$ and $\mathbf{u}_2 = s\begin{bmatrix} -i \\ 1 \end{bmatrix}$.

11. $c_1 e^{3t/2}\begin{bmatrix} 3 \\ 10 \end{bmatrix} + c_2 e^{t/2}\begin{bmatrix} 1 \\ 2 \end{bmatrix}$.

13. $c_1 e^{-t}\begin{bmatrix} -2 \\ 1 \\ 1 \end{bmatrix} + c_2 e^{-3t}\begin{bmatrix} 0 \\ -1 \\ 1 \end{bmatrix} + c_3 e^{5t}\begin{bmatrix} 1 \\ 1 \\ 1 \end{bmatrix}$.

15. $c_1 e^{-t}\begin{bmatrix} 3 \\ 0 \\ -2 \end{bmatrix} + c_2 e^{t}\begin{bmatrix} 1 \\ -6 \\ 4 \end{bmatrix} + c_3 e^{4t}\begin{bmatrix} 1 \\ 0 \\ 1 \end{bmatrix}$.

17. (b)

$(-\sqrt{3}, -1)$

Figure B.52

(c)

$(1, -\sqrt{3})$

Figure B.53

(d)

$(1 - \sqrt{3}, -\sqrt{3} - 1)$

Figure B.54

19. $\begin{bmatrix} e^{3t} & e^{-3t} \\ 4e^{3t} & -2e^{-3t} \end{bmatrix}$.

21. $\begin{bmatrix} e^{t} & e^{2t} & e^{4t} \\ e^{t} & 2e^{2t} & 4e^{4t} \\ e^{t} & 4e^{2t} & 16e^{4t} \end{bmatrix}$.

23. $\begin{bmatrix} e^{2t} & e^{-t} & e^{3t} & -e^{7t} \\ 0 & -3e^{-t} & 0 & e^{7t} \\ 0 & 0 & e^{3t} & 2e^{7t} \\ 0 & 0 & 0 & 8e^{7t} \end{bmatrix}$.

25. $x = -c_1 e^{t} - 2c_2 e^{2t} - c_3 e^{3t}$,
$y = c_1 e^{t} + c_2 e^{2t} + c_3 e^{3t}$,
$z = 2c_1 e^{t} + 4c_2 e^{2t} + 4c_3 e^{3t}$.

27. $\begin{bmatrix} e^{-0.3473t} & e^{0.5237t} & 0.0286e^{-7.0764t} \\ -0.3157e^{-0.3473t} & 0.4761e^{0.5237t} & -0.1837e^{-7.0764t} \\ 0.0844e^{-0.3473t} & 0.1918e^{0.5237t} & e^{-7.0764t} \end{bmatrix}$.

29. $\begin{bmatrix} 0.0251e^{3.4142t} & -0.2361e^{-1.6180t} & e^{0.6180t} & e^{0.5858t} \\ 0.0858e^{3.4142t} & 0.3820e^{-1.6180t} & 0.6180e^{0.6180t} & 0.5858e^{0.5858t} \\ 0.2929e^{3.4142t} & -0.6180e^{-1.6180t} & 0.3820e^{0.6180t} & 0.3431e^{0.5858t} \\ e^{3.4142t} & e^{-1.6180t} & 0.2361e^{0.6180t} & 0.2010e^{0.5858t} \end{bmatrix}$.

31. $\begin{bmatrix} 2e^{4t} + e^{-2t} \\ 2e^{4t} - e^{-2t} \end{bmatrix}$. **33.** $\begin{bmatrix} -3e^{-t} + e^{5t} \\ -2e^{-t} - e^{5t} \\ e^{-t} + e^{5t} \end{bmatrix}$.

35. (b) $\mathbf{x}_1(t) = e^{-t} \begin{bmatrix} 1 \\ 2 \end{bmatrix}$.

(c) $\mathbf{x}_2(t) = te^{-t} \begin{bmatrix} 1 \\ 2 \end{bmatrix} + e^{-t} \begin{bmatrix} 1 \\ 1 \end{bmatrix}$.

37. (b) $\mathbf{x}_1(t) = e^{2t} \begin{bmatrix} 1 \\ 0 \\ 0 \end{bmatrix}$.

(c) $\mathbf{x}_2(t) = te^{2t} \begin{bmatrix} 1 \\ 0 \\ 0 \end{bmatrix} + e^{2t} \begin{bmatrix} 0 \\ 1 \\ 0 \end{bmatrix}$.

(d) $\mathbf{x}_3(t) = \dfrac{t^2}{2} e^{2t} \begin{bmatrix} 1 \\ 0 \\ 0 \end{bmatrix} + te^{2t} \begin{bmatrix} 0 \\ 1 \\ 0 \end{bmatrix} + e^{2t} \begin{bmatrix} 0 \\ -6/5 \\ 1/5 \end{bmatrix}$.

39. (b) $\mathbf{x}_1(t) = e^t \begin{bmatrix} -1 \\ 1 \\ 0 \end{bmatrix}$; $\mathbf{x}_2(t) = e^t \begin{bmatrix} -1 \\ 0 \\ 1 \end{bmatrix}$.

(c) $\mathbf{x}_3(t) = te^t \begin{bmatrix} 1 \\ 1 \\ -2 \end{bmatrix} + e^t \begin{bmatrix} 1 \\ 0 \\ 0 \end{bmatrix}$. **(d) 0.**

43. $c_1 \begin{bmatrix} 3t^2 \\ t^2 \end{bmatrix} + c_2 \begin{bmatrix} t^4 \\ t^4 \end{bmatrix}$.

45. $x_1(t) = (25/2)(e^{-3t/25} + e^{-t/25})$,
$x_2(t) = (25)(e^{-t/25} - e^{-3t/25})$.

47. (a) $r_1 = 2.39091$, $r_2 = -2.94338$, $r_3 = 3.55247$.

(b) $\mathbf{u}_1 = \begin{bmatrix} 1 \\ -2.64178 \\ -9.31625 \end{bmatrix}$; $\mathbf{u}_2 = \begin{bmatrix} 1 \\ -1.16825 \\ 0.43862 \end{bmatrix}$;

$\mathbf{u}_3 = \begin{bmatrix} 1 \\ 0.81004 \\ -0.12236 \end{bmatrix}$.

(c) $c_1 e^{r_1 t} \mathbf{u}_1 + c_2 e^{r_2 t} \mathbf{u}_2 + c_3 e^{r_3 t} \mathbf{u}_3$, where the r_i's and the \mathbf{u}_i's are given in parts (a) and (b).

Exercises 9.6, page 573

1. $c_1 \begin{bmatrix} 2\cos 2t \\ \cos 2t + \sin 2t \end{bmatrix} + c_2 \begin{bmatrix} 2\sin 2t \\ \sin 2t - \cos 2t \end{bmatrix}$.

3. $c_1 e^t \cos t \begin{bmatrix} -1 \\ 1 \\ 0 \end{bmatrix} - c_1 e^t \sin t \begin{bmatrix} -2 \\ 0 \\ 1 \end{bmatrix} + c_2 e^t \sin t \begin{bmatrix} -1 \\ 1 \\ 0 \end{bmatrix} + c_2 e^t \cos t \begin{bmatrix} -2 \\ 0 \\ 1 \end{bmatrix} + c_3 e^t \begin{bmatrix} 1 \\ 0 \\ 0 \end{bmatrix}$.

Note: Equivalent answer is obtained if col $(-1, 1, 0)$ and col $(-2, 0, 1)$ are replaced by col $(-5, 1, 2)$ and col $(0, -2, 1)$, respectively.

5. $\begin{bmatrix} e^{-t}\cos 4t & e^{-t}\sin 4t \\ 2e^{-t}\sin 4t & -2e^{-t}\cos 4t \end{bmatrix}$. **7.** $\begin{bmatrix} 1 & \cos t & \sin t \\ 0 & -\cos t & -\sin t \\ 0 & -\sin t & \cos t \end{bmatrix}$.

9. $\begin{bmatrix} 1 & \sqrt{3}\sin(\sqrt{3}t) - \cos(\sqrt{3}t) & -\sin(\sqrt{3}t) - \sqrt{3}\cos(\sqrt{3}t) \\ -1 & \sqrt{3}\sin(\sqrt{3}t) + \cos(\sqrt{3}t) & \sin(\sqrt{3}t) - \sqrt{3}\cos(\sqrt{3}t) \\ 1 & 2\cos(\sqrt{3}t) & 2\sin(\sqrt{3}t) \end{bmatrix}$.

11. $\begin{bmatrix} e^t\sin t - e^t\cos t & -e^t\sin t - e^t\cos t & -\cos t & -\sin t \\ 2e^t\sin t & -2e^t\cos t & \sin t & -\cos t \\ 2e^t\sin t + 2e^t\cos t & 2e^t\sin t - 2e^t\cos t & \cos t & \sin t \\ 4e^t\cos t & 4e^t\sin t & -\sin t & \cos t \end{bmatrix}$.

13. (a) $\begin{bmatrix} e^{-2t}(\sin t - \cos t) \\ -2e^{-2t}\sin t \end{bmatrix}$. **(b)** $\begin{bmatrix} -e^{-2(t-\pi)}\cos t \\ e^{-2(t-\pi)}(\cos t - \sin t) \end{bmatrix}$. **(c)** $\begin{bmatrix} e^{-2(t+2\pi)}(2\cos t - 3\sin t) \\ e^{-2(t+2\pi)}(\cos t + 5\sin t) \end{bmatrix}$.

(d) $\begin{bmatrix} e^{\pi-2t}\cos t \\ e^{\pi-2t}(\sin t - \cos t) \end{bmatrix}$.

17. $c_1 \begin{bmatrix} t^{-1} \\ 0 \\ -2t^{-1} \end{bmatrix} + c_2 \begin{bmatrix} t^{-1}\cos(\ln t) \\ t^{-1}\sin(\ln t) \\ -t^{-1}\cos(\ln t) \end{bmatrix} + c_3 \begin{bmatrix} t^{-1}\sin(\ln t) \\ -t^{-1}\cos(\ln t) \\ -t^{-1}\sin(\ln t) \end{bmatrix}$.

19. $\dfrac{\sqrt{9 - \sqrt{17}}}{2\sqrt{2}\,\pi} \approx 0.249$; $\dfrac{\sqrt{9 + \sqrt{17}}}{2\sqrt{2}\,\pi} \approx 0.408$.

21. $I_1 = (5/6)e^{-2t}\sin 3t$,
$I_2 = -(10/13)e^{-2t}\cos 3t + (25/78)e^{-2t}\sin 3t$,
$I_3 = (10/13)e^{-2t}\cos 3t + (20/39)e^{-2t}\sin 3t$.

Exercises 9.7, page 579

1. $c_1 e^{7t}\begin{bmatrix} 1 \\ 1 \end{bmatrix} + c_2 e^{2t}\begin{bmatrix} 1 \\ -4 \end{bmatrix} + \begin{bmatrix} 2 \\ -1 \end{bmatrix}$.

3. $c_1 e^{-3t}\begin{bmatrix} 1 \\ 1 \\ -1 \end{bmatrix} + c_2 e^{3t}\begin{bmatrix} 1 \\ 0 \\ 1 \end{bmatrix} + c_3 e^{3t}\begin{bmatrix} -1 \\ 1 \\ 0 \end{bmatrix} + e^{t}\begin{bmatrix} 1 \\ 0 \\ -1 \end{bmatrix}$.

5. $\mathbf{x}_p = e^{-2t}\big[\mathbf{a} + t\mathbf{b}\big]$.

7. $\mathbf{x}_p = t\mathbf{a} + \mathbf{b} + \sin 3t\,\mathbf{c} + \cos 3t\,\mathbf{d}$.

9. $\mathbf{x}_p = e^{2t}\mathbf{a} + e^{3t}\mathbf{b}$.

11. $c_1\begin{bmatrix} \sin t \\ \cos t \end{bmatrix} + c_2\begin{bmatrix} \cos t \\ -\sin t \end{bmatrix} + \begin{bmatrix} 0 \\ -1 \end{bmatrix}$.

13. $c_1 e^{t}\begin{bmatrix} 1 \\ -1 \end{bmatrix} + c_2 e^{-t}\begin{bmatrix} 1 \\ -3 \end{bmatrix} + \begin{bmatrix} 5te^{t} + (3/4)e^{t} \\ -5te^{t} + (9/4)e^{t} \end{bmatrix}$.

15. $c_1\begin{bmatrix} 1 \\ 2 \end{bmatrix} + c_2 e^{-5t}\begin{bmatrix} -2 \\ 1 \end{bmatrix} + \begin{bmatrix} \ln|t| + (8/5)t - 8/25 \\ 2\ln|t| + (16/5)t + 4/25 \end{bmatrix}$.

17. $c_1 e^{2t}\begin{bmatrix} 1 \\ 1 \\ 1 \end{bmatrix} + c_2 e^{-t}\begin{bmatrix} -1 \\ 0 \\ 1 \end{bmatrix} + c_3 e^{-t}\begin{bmatrix} -1 \\ 1 \\ 0 \end{bmatrix} + e^{t}\begin{bmatrix} 1 \\ -1 \\ -1 \end{bmatrix}$.

19. $\begin{bmatrix} c_1\cos t + c_2\sin t + 1 \\ -c_1\sin t + c_2\cos t - t \\ c_3 e^{t} + c_4 e^{-t} - (1/4)e^{-t} - t + (1/2)te^{-t} \\ c_3 e^{t} - c_4 e^{-t} - (1/4)e^{-t} - 1 - (1/2)te^{-t} \end{bmatrix}$.

21. (a) $\begin{bmatrix} 4te^{t} + 5e^{t} \\ 2te^{t} + 4e^{t} \end{bmatrix}$.

(b) $\begin{bmatrix} -2e^{t-1} + 2e^{2(t-1)} - 3e^{2t-1} + (4t - 1)e^{t} \\ -e^{t-1} + 2e^{2(t-1)} - 3e^{2t-1} + (2t + 1)e^{t} \end{bmatrix}$.

(c) $\begin{bmatrix} (-20 + 2e^{-5})e^{t} + (-3e^{-5} - e^{-10})e^{2t} + (4t + 3)e^{t} \\ (-10 + e^{-5})e^{t} + (-3e^{-5} - e^{-10})e^{2t} + (2t + 3)e^{t} \end{bmatrix}$.

(d) $\begin{bmatrix} -18e^{t+1} + 14e^{2t+2} - 3e^{2t+1} + (4t + 7)e^{t} \\ -9e^{t+1} + 14e^{2t+2} - 3e^{2t+1} + (2t + 5)e^{t} \end{bmatrix}$.

23. $x = (c_1 + c_2 t)e^{t} + c_3 e^{-t} - t^2 - 4t - 6$;
$y = -c_2 e^{t} - 2c_3 e^{-t} - t^2 - 2t - 3$.

25. (a) $\left\{ \begin{bmatrix} e^{t} \\ e^{t} \end{bmatrix}, \begin{bmatrix} e^{2t} \\ 2e^{2t} \end{bmatrix} \right\}$. **(c)** $te^{t}\begin{bmatrix} 2 \\ 2 \end{bmatrix} + e^{t}\begin{bmatrix} 0 \\ 1 \end{bmatrix}$.

(d) $c_1 e^{t}\begin{bmatrix} 1 \\ 1 \end{bmatrix} + c_2 e^{2t}\begin{bmatrix} 1 \\ 2 \end{bmatrix} + te^{t}\begin{bmatrix} 2 \\ 2 \end{bmatrix} + e^{t}\begin{bmatrix} 0 \\ 1 \end{bmatrix}$.

27. $c_1 e^{2t}\begin{bmatrix} 1 \\ 1 \\ 1 \end{bmatrix} + c_2 e^{-t}\begin{bmatrix} -1 \\ 0 \\ 1 \end{bmatrix} + c_3 e^{-t}\begin{bmatrix} -1 \\ 1 \\ 0 \end{bmatrix}$

$\quad + te^{-t}\begin{bmatrix} 1 \\ 0 \\ -1 \end{bmatrix} + e^{-t}\begin{bmatrix} 1 \\ 0 \\ 0 \end{bmatrix} + \begin{bmatrix} 0 \\ 0 \\ 1 \end{bmatrix}$.

29. $c_1 t\begin{bmatrix} 1 \\ 1 \end{bmatrix} + c_2 t^{-1}\begin{bmatrix} 1 \\ 3 \end{bmatrix}$

$\quad + \begin{bmatrix} -(3/4)t^{-1} - (1/2)t^{-1}\ln t + 1 \\ -(3/4)t^{-1} - (3/2)t^{-1}\ln t + 2 \end{bmatrix}$.

33. $I_1 = I_2 + 3I_5, I_3 = 3I_5, I_4 = 2I_5$, where
$I_2 = \big(1/\big(20\sqrt{817}\big)\big)\big[\big(13 - \sqrt{817}\big)e^{-(31+\sqrt{817})5t/2}$
$\quad - \big(13 + \sqrt{817}\big)e^{(-31+\sqrt{817})5t/2}\big] + 1/10$,
$I_5 = \big(3/\big(40\sqrt{817}\big)\big)\big[\big(31 - \sqrt{817}\big)e^{-(31+\sqrt{817})5t/2}$
$\quad - \big(31 + \sqrt{817}\big)e^{(-31+\sqrt{817})5t/2}\big] + 3/20$.

35. $x_1 = \big(2050 - (5845/4)e^{-3t/50} - (2355/4)e^{-7t/50}\big)/21$,
$x_2 = \big(1850 - (5845/2)e^{-3t/50} + (2355/2)e^{-7t/50}\big)/21$.
Tank A will have the higher concentration. The limiting concentration in tank A is 41/21 kg/L and in tank B is 37/21 kg/L.

Exercises 9.8, page 590

1. (a) $r = 3$; $k = 2$. **(b)** $e^{3t}\begin{bmatrix} 1 & -2t \\ 0 & 1 \end{bmatrix}$.

3. (a) $r = -1$; $k = 3$. **(b)** $e^{-t}\begin{bmatrix} 1 + 3t - (3/2)t^2 & t & -t + (1/2)t^2 \\ -3t & 1 & t \\ 9t - (9/2)t^2 & 3t & 1 - 3t + (3/2)t^2 \end{bmatrix}$.

5. (a) $r = -2$; $k = 2$. **(b)** $e^{-2t}\begin{bmatrix} 1 & 0 & 0 \\ 4t & 1 & 0 \\ t & 0 & 1 \end{bmatrix}$. **7.** $\begin{bmatrix} \cos t & \sin t \\ -\sin t & \cos t \end{bmatrix}$.

9. $\dfrac{1}{2}\begin{bmatrix} e^t + \cos t - \sin t & 2\sin t & e^t - \cos t - \sin t \\ e^t - \sin t - \cos t & 2\cos t & e^t + \sin t - \cos t \\ e^t - \cos t + \sin t & -2\sin t & e^t + \cos t + \sin t \end{bmatrix}$.

11. $\dfrac{1}{25}\begin{bmatrix} -4 + 29e^{5t} - 20te^{5t} & 20 - 20e^{5t} & -8 + 8e^{5t} - 40te^{5t} \\ -5 + 5e^{5t} & 25 & -10 + 10e^{5t} \\ 2 - 2e^{5t} + 10te^{5t} & -10 + 10e^{5t} & 4 + 21e^{5t} + 20te^{5t} \end{bmatrix}$.

13. $\begin{bmatrix} (1 - t + t^2/2)e^t & (t - t^2)e^t & (t^2/2)e^t & 0 & 0 \\ (t^2/2)e^t & (1 - t + t^2)e^t & (t + t^2/2)e^t & 0 & 0 \\ (t + t^2/2)e^t & (-3t - t^2)e^t & (1 + 2t + t^2/2)e^t & 0 & 0 \\ 0 & 0 & 0 & \cos t & \sin t \\ 0 & 0 & 0 & -\sin t & \cos t \end{bmatrix}$.

15. $\begin{bmatrix} (1 + t + t^2/2)e^{-t} & (t + t^2)e^{-t} & (t^2/2)e^{-t} & 0 & 0 \\ (-t^2/2)e^{-t} & (1 + t - t^2)e^{-t} & (t - t^2/2)e^{-t} & 0 & 0 \\ (-t + t^2/2)e^{-t} & (-3t + t^2)e^{-t} & (1 - 2t + t^2/2)e^{-t} & 0 & 0 \\ 0 & 0 & 0 & (1 + 2t)e^{-2t} & te^{-2t} \\ 0 & 0 & 0 & -4te^{-2t} & (1 - 2t)e^{-2t} \end{bmatrix}$.

17. $c_1e^{-t}\begin{bmatrix} 1 \\ -1 \\ 1 \end{bmatrix} + c_2e^{-t}\begin{bmatrix} -t \\ -1 + t \\ 2 - t \end{bmatrix} + c_3e^{-2t}\begin{bmatrix} 1 \\ -2 \\ 4 \end{bmatrix}$. **19.** $c_1e^{2t}\begin{bmatrix} 3 \\ 6 \\ 1 \\ 1 \end{bmatrix} + c_2e^t\begin{bmatrix} 0 \\ 1 \\ 0 \\ 0 \end{bmatrix} + c_3e^t\begin{bmatrix} 1 \\ t \\ 0 \\ 0 \end{bmatrix} + c_4e^t\begin{bmatrix} 2t \\ t + t^2 \\ 0 \\ 1 \end{bmatrix}$.

21. $\begin{bmatrix} e^{-2t} \\ 4te^{-2t} + e^{-2t} \\ te^{-2t} - e^{-2t} \end{bmatrix}$. **23.** $e^{-t}\begin{bmatrix} 3t \\ 3 \\ 9t \end{bmatrix} + e^{At}\begin{bmatrix} 2 - e^t(t^2 - 2t + 2) \\ 1 + e^t(t - 1) \\ 6 - 3e^t(t^2 - 2t + 2) \end{bmatrix}$, where e^{At} is the matrix in the answer to Problem 3.

Review Problems, page 594

1. $c_1e^{3t}\begin{bmatrix} 1 \\ 1 \end{bmatrix} + c_2e^{4t}\begin{bmatrix} 3 \\ 2 \end{bmatrix}$. **3.** $c_1e^{-t}\begin{bmatrix} 1 \\ -1 \\ 0 \\ 0 \end{bmatrix} + c_2e^{-t}\begin{bmatrix} 0 \\ 0 \\ 1 \\ -1 \end{bmatrix} + c_3e^{3t}\begin{bmatrix} 1 \\ 1 \\ 0 \\ 0 \end{bmatrix} + c_4e^{3t}\begin{bmatrix} 0 \\ 0 \\ 1 \\ 1 \end{bmatrix}$. **5.** $\begin{bmatrix} e^{2t} & e^{3t} \\ -e^{2t} & -2e^{3t} \end{bmatrix}$.

7. $c_1e^{-t}\begin{bmatrix} 1 \\ -2 \end{bmatrix} + c_2e^{3t}\begin{bmatrix} 1 \\ 2 \end{bmatrix} + \begin{bmatrix} -1/3 \\ -14/3 \end{bmatrix}$. **9.** $c_1e^{-t}\begin{bmatrix} 1 \\ 0 \\ 3 \end{bmatrix} + c_2e^{-t}\begin{bmatrix} t \\ 1 \\ 3t \end{bmatrix} + c_3e^{-t}\begin{bmatrix} -t + (1/2)t^2 \\ t \\ 1 - 3t + (3/2)t^2 \end{bmatrix} + \begin{bmatrix} 2 + t \\ 7 - 3t \\ 10 \end{bmatrix}$.

11. $\begin{bmatrix} 3e^t - 2e^{2t} \\ 3e^t - 4e^{2t} \end{bmatrix}$. **13.** $c_1 t^{-1} \begin{bmatrix} -3 \\ 1 \\ 0 \end{bmatrix} + c_2 t^{-1} \begin{bmatrix} -1 \\ 0 \\ 1 \end{bmatrix} + c_3 t^4 \begin{bmatrix} 1 \\ 1 \\ 1 \end{bmatrix}$.

15. $\begin{bmatrix} (1/2)e^t + (2/3)e^{5t} - (1/6)e^{-t} & -e^t + (2/3)e^{5t} + (1/3)e^{-t} & -(1/2)e^t + (2/3)e^{5t} - (1/6)e^{-t} \\ (1/3)e^{5t} - (1/3)e^{-t} & (1/3)e^{5t} + (2/3)e^{-t} & (1/3)e^{5t} - (1/3)e^{-t} \\ -(1/2)e^t + (1/2)e^{-t} & e^t - e^{-t} & (1/2)e^t + (1/2)^{-t} \end{bmatrix}$.

CHAPTER 10

Exercises 10.2, page 611

1. $y = \left[4/(e - e^{-1}) \right](e^{-x} - e^x)$. **3.** $y \equiv 0$.

5. $y = e^x + 2x - 1$. **7.** $y = \cos x + c \sin x$; c arbitrary.

9. $\lambda_n = (2n - 1)^2/4$ and $y_n = c_n \sin\left[(2n - 1)x/2\right]$, where $n = 1, 2, 3, \ldots$ and the c_n's are arbitrary.

11. $\lambda_n = n^2$, $n = 0, 1, 2, \ldots$; $y_0 = a_0$ and
$y_n = a_n \cos nx + b_n \sin nx$, $n = 1, 2, 3, \ldots$, where a_0, a_n, and b_n are arbitrary.

13. The eigenvalues are the roots of
$$\tan\left(\sqrt{\lambda_n}\, \pi\right) + \sqrt{\lambda_n} = 0, \text{ where } \lambda_n > 0.$$
For n large, $\lambda_n \approx (2n - 1)^2/4$, n is a positive integer. The eigenfunctions are
$$y_n = c_n\left[\sin(\sqrt{\lambda_n}x) + \sqrt{\lambda_n}\cos(\sqrt{\lambda_n}x)\right],$$
where the c_n's are arbitrary.

15. $u(x, t) = e^{-3t}\sin x - 6e^{-48t}\sin 4x$.

17. $u(x, t) = e^{-3t}\sin x - 7e^{-27t}\sin 3x + e^{-75t}\sin 5x$.

19. $u(x, t) = 3 \cos 6t \sin 2x + 12 \cos 39t \sin 13x$.

21. $u(x, t) = 6 \cos 6t \sin 2x + 2 \cos 18t \sin 6x$
$\qquad + (11/27) \sin 27t \sin 9x$
$\qquad - (14/45) \sin 45t \sin 15x$.

23. $u(x, t) = \displaystyle\sum_{n=1}^{\infty} a\, n^{-2}e^{-2\pi^2 n^2 t} \sin n\pi x$.

25. If $K > 0$, then $T(t)$ becomes unbounded as $t \to \infty$, and so the temperature $u(x, t) = X(x)T(t)$ becomes unbounded at each position x. Since the temperature must remain bounded for all time, $K \not> 0$.

33. (a) $u(x) \equiv 50$. **(b)** $u(x) = 30x/L + 10$.

Exercises 10.3, page 627

1. Odd. **3.** Even. **5.** Neither.

9. $f(x) \sim \displaystyle\sum_{n=1}^{\infty} a\, \frac{2(-1)^{n+1}}{n} \sin nx$.

11. $f(x) \sim 1 + \displaystyle\sum_{n=1}^{\infty} a \left[\frac{2}{\pi^2 n^2}\left(-1 + (-1)^n\right) \cos\frac{n\pi x}{2} \right.$
$\qquad\qquad\qquad \left. + \frac{1}{\pi n}\left((-1)^{n+1} - 1\right) \sin\frac{n\pi x}{2} \right]$.

13. $f(x) \sim \dfrac{1}{3} + \displaystyle\sum_{n=1}^{\infty} a\, \frac{4(-1)^n}{n^2 \pi^2} \cos n\pi x$.

15. $f(x) \sim \left[(\sinh \pi)/\pi\right]\left(1 + 2 \displaystyle\sum_{n=1}^{\infty} a \left[\frac{(-1)^n}{1 + n^2} \cos nx \right.\right.$
$\qquad\qquad\qquad\qquad \left.\left. + \frac{(-1)^{n+1}n}{1 + n^2} \sin nx\right]\right)$.

17. The 2π-periodic function $g(x)$, where
$$g(x) = \begin{cases} x , & -\pi < x < \pi , \\ 0 , & x = \pm\pi . \end{cases}$$

19. The 4-periodic function $g(x)$, where
$$g(x) = \begin{cases} 1 , & -2 < x < 0 , \\ x , & 0 < x < 2 , \\ 1/2 , & x = 0 , \\ 3/2 , & x = \pm 2 . \end{cases}$$

21. The 2-periodic function $g(x)$, where
$g(x) = x^2$, $-1 \leq x \leq 1$.

23. The 2π-periodic function $g(x)$, where
$$g(x) = \begin{cases} e^x , & -\pi < x < \pi , \\ (e^\pi + e^{-\pi})/2 , & x = \pm\pi . \end{cases}$$

25. (a) $F(x) = (x^2 - \pi^2)/2$. **(b)** $F(x) = |x| - \pi$.

27. $f(x) \sim \displaystyle\sum_{n=1}^{\infty} a\, \frac{2}{(2n - 1)\pi}\left[(-1)^{n+1}\cos\left(\frac{(2n - 1)\pi x}{2}\right) \right.$
$\qquad\qquad\qquad\qquad \left. + \sin\left(\frac{(2n - 1)\pi x}{2}\right)\right]$.

29. $a_0 = 0$; $a_1 = 3/2$; $a_2 = 0$.

Exercises 10.4, page 635

1. (a) The π-periodic function $\tilde{f}(x)$, where
$$\tilde{f}(x) = x^2 , \quad 0 < x < \pi .$$

(b) The 2π-periodic function $f_o(x)$, where
$$f_o(x) = \begin{cases} x^2 , & 0 < x < \pi , \\ -x^2 , & -\pi < x < 0 . \end{cases}$$

(c) The 2π-periodic function $f_e(x)$, where
$$f_e(x) = \begin{cases} x^2 , & 0 < x < \pi , \\ x^2 , & -\pi < x < 0 . \end{cases}$$

3. (a) The π-periodic function $\tilde{f}(x)$, where
$$\tilde{f}(x) = \begin{cases} 0 , & 0 < x < \pi/2 , \\ 1 , & \pi/2 < x < \pi . \end{cases}$$

(b) The 2π-periodic function $f_o(x)$, where
$$f_o(x) = \begin{cases} -1 , & -\pi < x < -\pi/2 , \\ 0 , & -\pi/2 < x < 0 , \\ 0 , & 0 < x < \pi/2 , \\ 1 , & \pi/2 < x < \pi . \end{cases}$$

(c) The 2π-periodic function $f_e(x)$, where
$$f_e(x) = \begin{cases} 1 , & -\pi < x < -\pi/2 , \\ 0 , & -\pi/2 < x < 0 , \\ 0 , & 0 < x < \pi/2 , \\ 1 , & \pi/2 < x < \pi . \end{cases}$$

5. $f(x) \sim -\dfrac{4}{\pi} \displaystyle\sum_{k=1}^{\infty} \dfrac{1}{2k-1} \sin(2k-1)\pi x$.

7. $f(x) \sim \displaystyle\sum_{n=1}^{\infty} \left[\dfrac{2\pi(-1)^{n+1}}{n} + \dfrac{4}{\pi n^3}\left((-1)^n - 1\right) \right] \sin nx$.

9. $f(x) \sim \displaystyle\sum_{k=0}^{\infty} \dfrac{8}{(2k+1)^3 \pi^3} \sin(2k+1)\pi x$.

11. $f(x) \sim \dfrac{\pi}{2} + \dfrac{4}{\pi} \displaystyle\sum_{k=1}^{\infty} \dfrac{1}{(2k-1)^2} \cos(2k-1)x$.

13. $f(x) \sim e - 1 + 2 \displaystyle\sum_{n=1}^{\infty} \dfrac{(-1)^n e - 1}{1 + \pi^2 n^2} \cos n\pi x$.

15. $f(x) \sim \dfrac{2}{\pi} + \dfrac{2}{\pi} \displaystyle\sum_{k=1}^{\infty} \left(\dfrac{1}{2k+1} - \dfrac{1}{2k-1} \right) \cos 2kx$.

17. $u(x, t) =$
$$\dfrac{2}{\pi} \sum_{k=1}^{\infty} \left[\dfrac{2}{2k-1} - \dfrac{1}{2k+1} - \dfrac{1}{2k-3} \right] e^{-5(2k-1)^2 t} \sin(2k-1)x .$$

19. $u(x, t) = \dfrac{4}{\pi} \displaystyle\sum_{k=0}^{\infty} \dfrac{(-1)^{k+1}}{(2k+1)^2} e^{-5(2k+1)^2 t} \sin(2k+1)x$.

Exercises 10.5, page 648

1. $u(x, t) = \displaystyle\sum_{n=1}^{\infty} \dfrac{8(-1)^{n+1} - 4}{\pi^3 n^3} e^{-5\pi^2 n^2 t} \sin n\pi x$.

3. $u(x, t) = \dfrac{\pi}{2} - \displaystyle\sum_{k=0}^{\infty} \dfrac{4}{\pi(2k+1)^2} e^{-3(2k+1)^2 t} \cos(2k+1)x$.

5. $u(x, t) = \dfrac{2(e^\pi - 1)}{\pi} + \displaystyle\sum_{n=1}^{\infty} \dfrac{2e^\pi(-1)^n - 2}{\pi(1 + n^2)} e^{-n^2 t} \cos nx$.

7. $u(x, t) = 5 + \dfrac{5}{\pi}x - \dfrac{30}{\pi} e^{-2t} \sin x + \dfrac{5}{\pi} e^{-8t} \sin 2x$

$\qquad + \left(1 - \dfrac{10}{\pi}\right) e^{-18t} \sin 3x$

$\qquad + \dfrac{5}{2\pi} e^{-32t} \sin 4x - \left(1 + \dfrac{6}{\pi}\right) e^{-50t} \sin 5x$

$\qquad + \displaystyle\sum_{n=6}^{\infty} \dfrac{10}{\pi n} \left[2(-1)^n - 1\right] e^{-2n^2 t} \sin nx$.

9. $u(x, t) = \dfrac{e^{-\pi} - 1}{\pi}x - e^{-x} + 1 + \displaystyle\sum_{n=1}^{\infty} c_n e^{-n^2 t} \sin nx$,

where

$$c_n = \begin{cases} \dfrac{2e^{-\pi} - 2}{\pi n}(-1)^n + \dfrac{2n}{\pi(1 + n^2)}\left((-1)^{n+1} e^{-\pi} + 1\right) \\ \qquad + \dfrac{2}{\pi n}\left[(-1)^n - 1\right] , \quad n \neq 2 , \\ \dfrac{e^{-\pi} - 1}{\pi} + \dfrac{4}{5\pi}(1 - e^{-\pi}) + 1 , \quad n = 2 . \end{cases}$$

11. $u(x, t) = \displaystyle\sum_{n=0}^{\infty} a_n e^{-4(n+1/2)^2 t} \cos(n + 1/2)x$, where

$f(x) = \displaystyle\sum_{n=0}^{\infty} a_n \cos(n + 1/2)x$.

13. $u(x, t) = \dfrac{\pi^2}{3}x - \dfrac{1}{3}x^3 - 3e^{-2t} \sin x$

$\qquad + \displaystyle\sum_{n=2}^{\infty} \dfrac{4(-1)^n}{n^3} e^{-2n^2 t} \sin nx$.

15. $u(x, y, t) = e^{-52t} \cos 6x \sin 4y - 3e^{-122t} \cos x \sin 11y$.

17. $u(x, y, t) = 2 \displaystyle\sum_{n=1}^{\infty} \dfrac{(-1)^{n+1}}{n} e^{-n^2 t} \sin ny$.

19. $C(x, t) = \displaystyle\sum_{n=1}^{\infty} c_n e^{-[L + kn^2 \pi^2 / a^2] t} \sin\left(\dfrac{n\pi x}{a}\right)$,

where

$c_n = \dfrac{2}{a} \displaystyle\int_0^a f(x) \sin\left(\dfrac{n\pi x}{a}\right) dx$.

Concentration goes to zero as $t \to +\infty$.

Exercises 10.6, page 660

1. $u(x, t) = \dfrac{1}{7\pi}\sin 7\pi t \sin 7\pi x$

$\qquad + \sum\limits_{k=0}^{\infty} \dfrac{8}{\big((2k+1)\pi\big)^3}\cos(2k+1)\pi t \sin(2k+1)\pi x$.

3. $u(x, t) = \sum\limits_{n=1}^{\infty} \dfrac{4}{n^3}\big[2(-1)^{n+1} - 1\big]\cos 2nt \sin nx$.

5. $u(x, t) = \dfrac{2h_0 L^2}{\pi^2 a(L - a)} \sum\limits_{n=1}^{\infty} \dfrac{1}{n^2} \sin\dfrac{n\pi a}{L} \sin\dfrac{n\pi x}{L} \cos\dfrac{n\pi at}{L}$.

7. $u(x, t) = \cos t \sin x + \dfrac{5}{2}\sin 2t \sin 2x - \dfrac{3}{5}\sin 5t \sin 5x$

$\qquad + 2\sum\limits_{n=1}^{\infty} \dfrac{(-1)^{n+1}}{n^3}\left[t - \dfrac{\sin nt}{n}\right]\sin nx$.

9. $u(x, t) = \sum\limits_{n=0}^{\infty}\left(a_n \cos\dfrac{(2n+1)\pi\alpha t}{2L}\right.$

$\qquad \left. + b_n \sin\dfrac{(2n+1)\pi\alpha t}{2L}\right)\sin\dfrac{(2n+1)\pi x}{2L}$, where

$\quad f(x) = \sum\limits_{n=0}^{\infty} a_n \sin\dfrac{(2n+1)\pi x}{2L}$ and

$\quad g(x) = \sum\limits_{n=0}^{\infty} b_n\dfrac{(2n+1)\pi\alpha}{2L}\sin\dfrac{(2n+1)\pi x}{2L}$.

11. $u(x, t) = \sum\limits_{n=1}^{\infty} a_n T_n(t)\sin\dfrac{n\pi x}{L}$, where

$\quad a_n = \dfrac{2}{L}\int_0^L f(x)\sin\dfrac{n\pi x}{L}dx$,

\quad and

$\quad T_n(t) = e^{-t/2}\left(\cos\beta_n t + \dfrac{1}{2\beta_n}\sin\beta_n t\right)$,

\quad where

$\quad \beta_n = \dfrac{1}{2L}\sqrt{3L^2 + 4\alpha^2\pi^2 n^2}$.

13. $u(x, t) = \dfrac{1}{2\alpha}\big[\sin(x + \alpha t) - \sin(x - \alpha t)\big]$

$\qquad = \dfrac{1}{\alpha}\sin \alpha t \cos x$.

15. $u(x, t) = x + tx$.

17. $u(x, t) = \dfrac{1}{2}\left[e^{-(x+\alpha t)^2} + e^{-(x-\alpha t)^2}\right.$

$\qquad \left. + \dfrac{\cos(x - \alpha t) - \cos(x + \alpha t)}{\alpha}\right]$.

21. $u(r, t) = \sum\limits_{n=1}^{\infty}\big[a_n \cos(k_n\alpha t) + b_n \sin(k_n\alpha t)\big]J_0(k_n r)$,

\quad where

$\quad a_n = \dfrac{1}{c_n}\int_0^1 f(r)J_0(k_n r)r\,dr$,

\quad and

$\quad b_n = \dfrac{1}{\alpha k_n c_n}\int_0^1 g(r)J_0(k_n r)r\,dr$,

\quad with

$\quad c_n = \int_0^1 J_0^2(k_n r)r\,dr$.

Exercises 10.7, page 673

1. $u(x, y) = \dfrac{4\cos 6x \sinh\big(6(y - 1)\big)}{\sinh(-6)}$

$\qquad + \dfrac{\cos 7x \sinh\big(7(y - 1)\big)}{\sinh(-7)}$.

3. $u(x, y) = \sum\limits_{n=1}^{\infty} A_n \sin nx \sinh(ny - n\pi)$, where

$\quad A_n = \dfrac{2}{\pi \sinh(-n\pi)}\int_0^{\pi} f(x)\sin nx\,dx$.

5. $u(x, y) = \dfrac{\cos x \sinh(y - 1)}{\sinh(-1)} - \dfrac{\cos 3x \sinh(3y - 3)}{\sinh(-3)}$

$\qquad + \dfrac{\cos 2x \sinh 2y}{\sinh(2)}$.

7. $u(r, \theta) = \dfrac{\pi}{2} - \sum\limits_{k=0}^{\infty} \dfrac{r^{2k+1}}{(2k+1)^2\pi 2^{2k-1}}\cos(2k+1)\theta$.

9. $u(r, \theta) = \dfrac{a_0}{2} + \sum\limits_{n=1}^{\infty}\left(\dfrac{r}{a}\right)^n(a_n \cos n\theta + b_n \sin n\theta)$, where

$\quad a_0$ is arbitrary, and for $n = 1, 2, 3, \ldots$

$\quad a_n = \dfrac{a}{\pi n}\int_{-\pi}^{\pi} f(\theta)\cos n\theta\,d\theta$,

$\quad b_n = \dfrac{a}{n\pi}\int_{-\pi}^{\pi} f(\theta)\sin n\theta\,d\theta$.

11. $u(r, \theta) = \left(\dfrac{1}{3}r - \dfrac{4}{3}r^{-1}\right)\cos\theta + \left(\dfrac{2}{3}r - \dfrac{2}{3}r^{-1}\right)\sin\theta$

$\qquad + \left(-\dfrac{1}{255}r^4 + \dfrac{256}{255}r^{-4}\right)\sin 4\theta$.

13. $u(r, \theta) = \dfrac{a_0}{2} + \displaystyle\sum_{n=1}^{\infty} r^{-n}(a_n \cos n\theta + b_n \sin n\theta)$, where

$$a_n = \frac{1}{\pi} \int_{-\pi}^{\pi} f(\theta) \cos n\theta \, d\theta \, , \qquad n = 0, 1, 2, \ldots \, ,$$

and

$$b_n = \frac{1}{\pi} \int_{-\pi}^{\pi} f(\theta) \sin n\theta \, d\theta \, , \qquad n = 1, 2, 3, \ldots \, .$$

15. $u(r, \theta) = r^3 \sin 3\theta$.

17. $u(r, \theta) = a_0 + b_0 \ln r + \displaystyle\sum_{n=1}^{\infty} \Big[(a_n r^n + b_n r^{-n}) \cos n\theta$

$$+ (c_n r^n + d_n r^{-n}) \sin n\theta \Big] \, ,$$

where

$$a_0 = \frac{1}{2\pi} \int_{-\pi}^{\pi} f(\theta) \, d\theta \, ,$$

$$b_0 = \frac{3}{2\pi} \int_{-\pi}^{\pi} g(\theta) \, d\theta \, ,$$

and for $n = 1, 2, 3, \ldots$

$$a_n + b_n = \frac{1}{\pi} \int_{-\pi}^{\pi} f(\theta) \cos n\theta \, d\theta \, ,$$

$$n3^{n-1} a_n - n3^{-n-1} b_n = \frac{1}{\pi} \int_{-\pi}^{\pi} g(\theta) \cos n\theta \, d\theta \, ,$$

and

$$c_n + d_n = \frac{1}{\pi} \int_{-\pi}^{\pi} f(\theta) \sin n\theta \, d\theta \, ,$$

$$n3^{n-1} c_n - n3^{-n-1} d_n = \frac{1}{\pi} \int_{-\pi}^{\pi} g(\theta) \sin n\theta \, d\theta \, .$$

21. $u(r, \theta, z) = \big[I_0(r)/I_0(\pi) \big] \sin z$.

23. **(c)** $\phi(x, y) = 2xy$.

Index